S0-CBI-349

Methods in Enzymology

Volume 360

BIOPHOTONICS

Part A

METHODS IN ENZYMOLOGY

Methods in Enzymology

Volume 360

Biophotonics

Part A

EDITED BY

Gerard Marriott

DEPARTMENT OF PHYSIOLOGY
UNIVERSITY OF WISCONSIN
BIOPHOTONICS CLUSTER PROGRAM
MADISON, WISCONSIN

Ian Parker

DEPARTMENT OF NEUROBIOLOGY AND BEHAVIOR
UNIVERSITY OF CALIFORNIA
IRVINE, CALIFORNIA

ACADEMIC PRESS

An imprint of Elsevier Science

Amsterdam Boston London New York Oxford Paris
San Diego San Francisco Singapore Sydney Tokyo

This book is printed on acid-free paper. ∞

Academic Press
An imprint of Elsevier Science.
525 B Street, Suite 1900, San Diego, California 92101-4495, USA
http://www.academicpress.com

Academic Press
84 Theobalds Road, London WC1X 8RR, UK
http://www.academicpress.com

International Standard Book Number: 0-12-182263-X

PRINTED IN THE UNITED STATES OF AMERICA
03 04 05 06 07 SB 9 8 7 6 5 4 3 2 1

Table of Contents

Contributors to Volume 360

Article numbers are in parentheses following the names of contributors.
Affiliations listed are current.

LUIS A. BAGATOLLI (20), *Department of Physics, MEMPHYS–Center for Biomembrane Physics, University of Southern Denmark, DK-5230 Odense M, Denmark*

JIHONG BAI (9), *Department of Physiology, University of Wisconsin, Madison, Wisconsin 53706*

GABRIEL S. BRANDT (10), *Division of Chemistry and Chemical Engineering, California Institute of Technology, Pasadena, California 91125*

JULIE C. CANMAN (26), *Department of Biology, University of North Carolina, Chapel Hill, North Carolina 27599*

VICTORIA E. CENTONZE (23), *Department of Cellular and Structural Biology, University of Texas Health Science Center, San Antonio, Texas 78229*

EDWIN R. CHAPMAN (9), *Department of Physiology, University of Wisconsin, Madison, Wisconsin 53706*

ANDREW H. A. CLAYTON (6), *Ludwig Institute for Cancer Research, Royal Melbourne Hospital, Parkville, Victoria 3050, Australia*

ROBERT M. CLEGG (22), *Department of Physics, Laboratory for Fluorescence Dynamics, University of Illinois Urbana–Champaign, Urbana, Illinois 61801*

PINA COLARUSSO (16), *Department of Physiology, Health Sciences Center, University of Calgary, Calgary, Alberta, Canada T2N 4N1*

JACQUES COPPEY (25), *Institut Jacques Monod, UMR 7592, CNRS, Universités P6/P7, 75251 Paris Cedex 05, France*

MAÏTÉ COPPEY-MOISAN (25), *Institut Jacques Monod, UMR 7592, CNRS, Universités P6/P7, 75251 Paris Cedex 05, France*

CHRISTINE R. CREMO (5), *Department of Biochemistry, University of Nevada, Reno, Nevada 89557*

JOHN C. CRONEY (1), *Department of Cell and Molecular Biology, John A. Burns School of Medicine, University of Hawaii, Honolulu, Hawaii 96822*

DENNIS A. DOUGHERTY (10), *Division of Chemistry and Chemical Engineering, California Institute of Technology, Pasadena, California 91125*

CHRISTIANE DURIEUX (25), *Institut Jacques Monod, UMR 7592, CNRS, Universités P6/P7, 75251 Paris Cedex 05, France*

GRAHAM C. R. ELLIS-DAVIES (8), *Department of Pharmacology and Physiology, Drexel University College of Medicine, Philadelphia, Pennsylvania 19102*

BRENDA K. EUSTACE (29), *Department of Physiology, Tufts University School of Medicine, Boston, Massachusetts 02111*

ISABELLE GAUTIER (25), *Institut Jacques Monod, UMR 7592, CNRS, Universités P6/P7, 75251 Paris Cedex 05, France*

HANS GERRITSEN (23), *Department of Molecular Biophysics, Debye Institute, Utrecht University, Utrecht, The Netherlands*

CHRISTOPHER GOHLKE (22), *Department of Physics, Laboratory for Fluorescence Dynamics, University of Illinois Urbana–Champaign, Urbana, Illinois 61801*

ENRICO GRATTON (20), *Department of Physics, Laboratory for Fluorescence Dynamics, University of Illinois Urbana–Champaign, Urbana, Illinois 61801*

IGNACY GRYCZYNSKI (2), *Department of Biochemistry and Molecular Biology, Center for Fluorescence Spectroscopy, University of Maryland School of Medicine, Baltimore, Maryland 21201*

ZYGMUNT GRYCZYNSKI (2), *Department of Biochemistry and Molecular Biology, Center for Fluorescence Spectroscopy, University of Maryland School of Medicine, Baltimore, Maryland 21201*

QUENTIN S. HANLEY (6), *Department of Chemical and Biological Sciences, Cave Hill Campus, University of the West Indies, St. Michael, Barbados*

J. WOODLAND HASTINGS (3), *Department of Molecular and Cellular Biology, Harvard University, Cambridge, Massachusetts 02138*

PHILIP G. HAYDON (21), *Department of Neuroscience, University of Pennsylvania School of Medicine, Philadelphia, Pennsylvania 19104*

THEODORE HAZLETT (20), *Department of Physics, Laboratory for Fluorescence Dynamics, University of Illinois Urbana–Champaign, Urbana, Illinois 61801*

BRIAN HERMAN (23), *Department of Cellular and Structural Biology, University of Texas Health Science Center, San Antonio, Texas 78229*

OLIVER HOLUB (22), *Department of Physics, Laboratory of Fluorescence Dynamics, University of Urbana–Champaign, Urbana, Illinois 61801*

KENNETH JACOBSON (11), *Department of Cell and Developmental Biology, University of North Carolina, Chapel Hill, North Carolina 27599*

DAVID M. JAMESON (1), *Department of Cell and Molecular Biology, John A. Burns School of Medicine, University of Hawaii, Honolulu, Hawaii 96822*

DANIEL G. JAY (29), *Department of Physiology, Tufts University School of Medicine, Boston, Massachusetts 02111*

CARL HIRSCHIE JOHNSON (3, 12), *Department of Biological Sciences, Vanderbilt University, Nashville, Tennessee 37235*

JAMES E. N. JONKMAN (18), *Advanced Optical Microscopy Facility, Ontario Cancer Institute, Princess Margaret Hospital, Toronto, Ontario, Canada M5G 2M9*

THOMAS M. JOVIN (6), *Department of Molecular Biology, Max Planck Institute for Biophysical Chemistry, 37077 Goettingen, Germany*

AKIHITO KANAUCHI (12), *Department of Biological Sciences, Vanderbilt University, Nashville, Tennessee 37235*

KLAUS KEMNITZ (25), *EuroPhoton GmbH, D-12247 Berlin, Germany*

HOLGER KRESS (18), *Cell Biology and Biophysics Program, European Molecular Biology Laboratory, D-69117 Heidelberg, Germany*

AKIHIRO KUSUMI (27), *Department of Biological Sciences, Nagoya University, Nagoya 464-8602, Japan*

JOSEPH R. LAKOWICZ (2), *Department of Biochemistry and Molecular Biology, Center for Fluorescence Spectroscopy, University of Maryland School of Medicine, Baltimore, Maryland 21201*

RANDY W. LARSEN (13), *Department of Chemistry, University of South Florida, Tampa, Florida 33620*

HENRY A. LESTER (10), *Division of Biology, California Institute of Technology, Pasadena, California 91125*

MOSHE LEVI (14), *Department of Medicine, Division of Nephrology, University of Colorado Health Sciences Center, Denver, Colorado 80262*

PAUL S. MADDOX (26), *Department of Biology, University of North Carolina, Chapel Hill, North Carolina 27599*

GERARD MARRIOTT (11, 24), *Department of Physiology, University of Wisconsin, Madison, Wisconsin 53706*

ATSUSHI MASUDA (23), *Department of Cellular and Structural Biology, University of Texas Health Science Center, San Antonio, Texas 78229*

VINCENT MIGNOTTE (25), *Department of Hematology, Institut Cochin de Genetique Moleculaire, 75014 Paris, France*

JAROSLAVA MIKŠOVSKÁ (13), *Department of Chemistry, University of South Florida, Tampa, Florida 33620*

ATSUSHI MIYAWAKI (7), *Laboratory for Cell Function and Dynamics, Advanced Technology Development Group, Brain Science Institute, Institute of Physical and Chemical Research, Wako City, Saitama 351-0198, Japan*

HIDEAKI MIZUNO (7), *Laboratory for Cell Function and Dynamics, Advanced Technology Development Group, Brain Science Institute, Institute of Physical and Chemical Research, Wako City, Saitama 351-0198, Japan*

PIERRE D. J. MOENS (1), *Department of Cell and Molecular Biology, John A. Burns School of Medicine, University of Hawaii, Honolulu, Hawaii 96822*

BEN MOREE (26), *Department of Biology, University of North Carolina, Chapel Hill, North Carolina 27599*

TAKEHARU NAGAI (7), *Laboratory for Cell Function and Dynamics, Advanced Technology Development Group, Brain Science Institute, Institute of Physical and Chemical Research, Wako City, Saitama 351-0198, Japan*

TIZIANA PARASASSI (14), *Institute of Neurobiology and Molecular Medicine, National Research Council, 00137 Rome, Italy*

IAN PARKER (15, 19), *Department of Neurobiology and Behavior, University of California, Irvine, California 92697*

E. JAMES PETERSSON (10), *Division of Chemistry and Chemical Engineering, California Institute of Technology, Pasadena, California 91125*

TRISTAN PIOLOT (25), *EuroPhoton GmbH, D-12247 Berlin, Germany*

DAVID W. PISTON (12), *Department of Molecular Physiology and Biophysics, Vanderbilt University, Nashville, Tennessee 37235*

KEN RITCHIE (27), *Department of Biological Sciences, Nagoya University, Nagoya 464-8602, Japan*

ALEXANDER ROHRBACH (18), *Cell Biology and Biophysics Program, European Molecular Biology Laboratory, D-69117 Heidelberg, Germany*

PARTHA ROY (11), *Department of Cell and Developmental Biology, University of North Carolina, Chapel Hill, North Carolina 27599*

E. D. SALMON (26), *Department of Biology, University of North Carolina, Chapel Hill, North Carolina 27599*

SUSANA A. SANCHEZ (20), *Department of Physics, Laboratory for Fluorescence Dynamics, University of Illinois Urbana–Champaign, Urbana, Illinois 61801*

MICHAEL J. SANDERSON (19), *Department of Physiology, University of Massachusetts Medical School, Worcester, Massachusetts 01655*

ASAKO SAWANO (7), *Laboratory for Cell Function and Dynamics, Advanced Technology Development Group, Brain Science Institute, Institute of Physical and Chemical Research, Wako City, Saitama 351-0198, Japan*

KENNETH R. SPRING (16), *Laboratory of Kidney and Electrolyte Metabolism, National Heart, Lung, and Blood Institute, National Institutes of Health, Bethesda, Maryland 20892*

C. MICHAEL STANLEY (17), *Chroma Technology Corp., Brattleboro, Vermont 05301*

ERNST H. K. STELZER (18), *Cell Biology and Biophysics Program, European Molecular Biology Laboratory, D-69117 Heidelberg, Germany*

VINOD SUBRAMANIAM (6), *Advanced Science and Technology Laboratory, AstraZeneca Research and Development Charnwood, Loughborough LE11 5RH, United Kingdom*

MAO SUN (23), *Texas Tech University Medical School, Lubbock, Texas 79409*

JIM SWOGER (18), *Cell Biology and Biophysics Program, European Molecular Biology Laboratory, D-69117 Heidelberg, Germany*

ARTHUR G. SZABO (4), *Department of Chemistry, Wilfrid Laurier University, Waterloo, Ontario, Canada N2L 4H1*

MARC TRAMIER (25), *Institut Jacques Monod, UMR 7592, CNRS, Universités P6/P7, 75251 Paris Cedex 05, France*

SUSAN M. TWINE (4), *Department of Chemistry, Wilfrid Laurier University, Waterloo, Ontario, Canada N2L 4H1*

ALAN S. VERKMAN (28), *University of California, San Francisco, California 94143*

ALBRECHT G. VON ARNIM (12), *Department of Botany, University of Tennessee, Knoxville, Tennessee 37996*

YAO XU (12), *Department of Biological Sciences, Vanderbilt University, Nashville, Tennessee 37235*

YULING YAN (24), *Department of Mechanical Engineering, University of Hawaii at Manoa, Honolulu, Hawaii 96822*

NIKI M. ZACHARIAS (10), *Division of Chemistry and Chemical Engineering, California Institute of Technology, Pasadena, California 91125*

HUBERT ZAJICEK (14), *Department of Internal Medicine, University of Texas Southwestern Medical Center, Dallas, Texas 75235*

Preface

The use of optical techniques in experimental biology dates back over 300 years, and has advanced through a series of technological breakthroughs exemplified by the development of diffraction-limited microscopy at the end of the nineteenth century. We are now, once again, at a time of explosive progress, and the newly christened field of biophotonics is emerging from innovative research at the interface of the physical, biological, medical, and engineering sciences. Biophotonics encompasses a broad range of techniques and methodologies developed in areas as diverse as photophysics, photochemistry, optical spectroscopy, and microscopy. It is thus highly interdisciplinary in nature. For example, an imaging technique based on concepts and principles borrowed from physics, chemistry, and engineering may be used to investigate a biological system at the level of a single molecule on an engineered surface within a living cell or even within an animal.

Distinct from previous publications devoted to specific topics, such as imaging or spectroscopy, this 2-volume work of *Methods in Enzymology* is the first to attempt a comprehensive coverage of the broad field of biophotonics research. We believe it will prove a valuable resource for researchers interested in developing and applying photonic technologies to solve biological problems at all levels, from single biomolecules to the living organism. The chapters are written by internationally renowned researchers, and provide technically detailed coverage of the basic principles of optical spectroscopy and microscopy with numerous applications of specialized techniques and probes in biology, medicine, and biotechnology.

Given the interdisciplinary nature of biophotonics, we failed in our initial attempt to classify the chapters within defined categories of technology, applications, and biological systems. Nevertheless, chapters in the two volumes are roughly equally divided among three main areas. Biophotonics, Part A (Volume 360) covers the basic principles and practice of biological spectroscopy and imaging microscopy, with supporting chapters on photophysics, optical design, and biological and synthetic probes. Biophotonics, Part B (Volume 361) focuses on the development and application of imaging technologies to understand the mechanisms underlying biomolecular structure, function, and dynamics in diverse molecular and cellular systems. It also illustrates how biophotonic technologies are being used to power new breakthroughs in medical imaging and diagnosis and in biotechnology.

We would like to thank Mary Ellen Perry for her professional administrative skills and Shirley Light for her excellent editorial assistance.

GERARD MARRIOTT
IAN PARKER

METHODS IN ENZYMOLOGY

VOLUME I. Preparation and Assay of Enzymes
Edited by SIDNEY P. COLOWICK AND NATHAN O. KAPLAN

VOLUME II. Preparation and Assay of Enzymes
Edited by SIDNEY P. COLOWICK AND NATHAN O. KAPLAN

VOLUME III. Preparation and Assay of Substrates
Edited by SIDNEY P. COLOWICK AND NATHAN O. KAPLAN

VOLUME IV. Special Techniques for the Enzymologist
Edited by SIDNEY P. COLOWICK AND NATHAN O. KAPLAN

VOLUME V. Preparation and Assay of Enzymes
Edited by SIDNEY P. COLOWICK AND NATHAN O. KAPLAN

VOLUME VI. Preparation and Assay of Enzymes (*Continued*)
Preparation and Assay of Substrates
Special Techniques
Edited by SIDNEY P. COLOWICK AND NATHAN O. KAPLAN

VOLUME VII. Cumulative Subject Index
Edited by SIDNEY P. COLOWICK AND NATHAN O. KAPLAN

VOLUME VIII. Complex Carbohydrates
Edited by ELIZABETH F. NEUFELD AND VICTOR GINSBURG

VOLUME IX. Carbohydrate Metabolism
Edited by WILLIS A. WOOD

VOLUME X. Oxidation and Phosphorylation
Edited by RONALD W. ESTABROOK AND MAYNARD E. PULLMAN

VOLUME XI. Enzyme Structure
Edited by C. H. W. HIRS

VOLUME XII. Nucleic Acids (Parts A and B)
Edited by LAWRENCE GROSSMAN AND KIVIE MOLDAVE

VOLUME XIII. Citric Acid Cycle
Edited by J. M. LOWENSTEIN

VOLUME XIV. Lipids
Edited by J. M. LOWENSTEIN

VOLUME XV. Steroids and Terpenoids
Edited by RAYMOND B. CLAYTON

VOLUME XVI. Fast Reactions
Edited by KENNETH KUSTIN

VOLUME XVII. Metabolism of Amino Acids and Amines (Parts A and B)
Edited by HERBERT TABOR AND CELIA WHITE TABOR

VOLUME XVIII. Vitamins and Coenzymes (Parts A, B, and C)
Edited by DONALD B. MCCORMICK AND LEMUEL D. WRIGHT

VOLUME XIX. Proteolytic Enzymes
Edited by GERTRUDE E. PERLMANN AND LASZLO LORAND

VOLUME XX. Nucleic Acids and Protein Synthesis (Part C)
Edited by KIVIE MOLDAVE AND LAWRENCE GROSSMAN

VOLUME XXI. Nucleic Acids (Part D)
Edited by LAWRENCE GROSSMAN AND KIVIE MOLDAVE

VOLUME XXII. Enzyme Purification and Related Techniques
Edited by WILLIAM B. JAKOBY

VOLUME XXIII. Photosynthesis (Part A)
Edited by ANTHONY SAN PIETRO

VOLUME XXIV. Photosynthesis and Nitrogen Fixation (Part B)
Edited by ANTHONY SAN PIETRO

VOLUME XXV. Enzyme Structure (Part B)
Edited by C. H. W. HIRS AND SERGE N. TIMASHEFF

VOLUME XXVI. Enzyme Structure (Part C)
Edited by C. H. W. HIRS AND SERGE N. TIMASHEFF

VOLUME XXVII. Enzyme Structure (Part D)
Edited by C. H. W. HIRS AND SERGE N. TIMASHEFF

VOLUME XXVIII. Complex Carbohydrates (Part B)
Edited by VICTOR GINSBURG

VOLUME XXIX. Nucleic Acids and Protein Synthesis (Part E)
Edited by LAWRENCE GROSSMAN AND KIVIE MOLDAVE

VOLUME XXX. Nucleic Acids and Protein Synthesis (Part F)
Edited by KIVIE MOLDAVE AND LAWRENCE GROSSMAN

VOLUME XXXI. Biomembranes (Part A)
Edited by SIDNEY FLEISCHER AND LESTER PACKER

VOLUME XXXII. Biomembranes (Part B)
Edited by SIDNEY FLEISCHER AND LESTER PACKER

VOLUME XXXIII. Cumulative Subject Index Volumes I-XXX
Edited by MARTHA G. DENNIS AND EDWARD A. DENNIS

VOLUME XXXIV. Affinity Techniques (Enzyme Purification: Part B)
Edited by WILLIAM B. JAKOBY AND MEIR WILCHEK

VOLUME XXXV. Lipids (Part B)
Edited by JOHN M. LOWENSTEIN

VOLUME XXXVI. Hormone Action (Part A: Steroid Hormones)
Edited by BERT W. O'MALLEY AND JOEL G. HARDMAN

VOLUME XXXVII. Hormone Action (Part B: Peptide Hormones)
Edited by BERT W. O'MALLEY AND JOEL G. HARDMAN

VOLUME XXXVIII. Hormone Action (Part C: Cyclic Nucleotides)
Edited by JOEL G. HARDMAN AND BERT W. O'MALLEY

VOLUME XXXIX. Hormone Action (Part D: Isolated Cells, Tissues, and Organ Systems)
Edited by JOEL G. HARDMAN AND BERT W. O'MALLEY

VOLUME XL. Hormone Action (Part E: Nuclear Structure and Function)
Edited by BERT W. O'MALLEY AND JOEL G. HARDMAN

VOLUME XLI. Carbohydrate Metabolism (Part B)
Edited by W. A. WOOD

VOLUME XLII. Carbohydrate Metabolism (Part C)
Edited by W. A. WOOD

VOLUME XLIII. Antibiotics
Edited by JOHN H. HASH

VOLUME XLIV. Immobilized Enzymes
Edited by KLAUS MOSBACH

VOLUME XLV. Proteolytic Enzymes (Part B)
Edited by LASZLO LORAND

VOLUME XLVI. Affinity Labeling
Edited by WILLIAM B. JAKOBY AND MEIR WILCHEK

VOLUME XLVII. Enzyme Structure (Part E)
Edited by C. H. W. HIRS AND SERGE N. TIMASHEFF

VOLUME XLVIII. Enzyme Structure (Part F)
Edited by C. H. W. HIRS AND SERGE N. TIMASHEFF

VOLUME XLIX. Enzyme Structure (Part G)
Edited by C. H. W. HIRS AND SERGE N. TIMASHEFF

VOLUME L. Complex Carbohydrates (Part C)
Edited by VICTOR GINSBURG

VOLUME LI. Purine and Pyrimidine Nucleotide Metabolism
Edited by PATRICIA A. HOFFEE AND MARY ELLEN JONES

VOLUME LII. Biomembranes (Part C: Biological Oxidations)
Edited by SIDNEY FLEISCHER AND LESTER PACKER

VOLUME LIII. Biomembranes (Part D: Biological Oxidations)
Edited by SIDNEY FLEISCHER AND LESTER PACKER

VOLUME LIV. Biomembranes (Part E: Biological Oxidations)
Edited by SIDNEY FLEISCHER AND LESTER PACKER

VOLUME LV. Biomembranes (Part F: Bioenergetics)
Edited by SIDNEY FLEISCHER AND LESTER PACKER

VOLUME LVI. Biomembranes (Part G: Bioenergetics)
Edited by SIDNEY FLEISCHER AND LESTER PACKER

VOLUME LVII. Bioluminescence and Chemiluminescence
Edited by MARLENE A. DELUCA

VOLUME LVIII. Cell Culture
Edited by WILLIAM B. JAKOBY AND IRA PASTAN

VOLUME LIX. Nucleic Acids and Protein Synthesis (Part G)
Edited by KIVIE MOLDAVE AND LAWRENCE GROSSMAN

VOLUME LX. Nucleic Acids and Protein Synthesis (Part H)
Edited by KIVIE MOLDAVE AND LAWRENCE GROSSMAN

VOLUME 61. Enzyme Structure (Part H)
Edited by C. H. W. HIRS AND SERGE N. TIMASHEFF

VOLUME 62. Vitamins and Coenzymes (Part D)
Edited by DONALD B. MCCORMICK AND LEMUEL D. WRIGHT

VOLUME 63. Enzyme Kinetics and Mechanism (Part A: Initial Rate and Inhibitor Methods)
Edited by DANIEL L. PURICH

VOLUME 64. Enzyme Kinetics and Mechanism (Part B: Isotopic Probes and Complex Enzyme Systems)
Edited by DANIEL L. PURICH

VOLUME 65. Nucleic Acids (Part I)
Edited by LAWRENCE GROSSMAN AND KIVIE MOLDAVE

VOLUME 66. Vitamins and Coenzymes (Part E)
Edited by DONALD B. MCCORMICK AND LEMUEL D. WRIGHT

VOLUME 67. Vitamins and Coenzymes (Part F)
Edited by DONALD B. MCCORMICK AND LEMUEL D. WRIGHT

VOLUME 68. Recombinant DNA
Edited by RAY WU

VOLUME 69. Photosynthesis and Nitrogen Fixation (Part C)
Edited by ANTHONY SAN PIETRO

VOLUME 70. Immunochemical Techniques (Part A)
Edited by HELEN VAN VUNAKIS AND JOHN J. LANGONE

VOLUME 71. Lipids (Part C)
Edited by JOHN M. LOWENSTEIN

VOLUME 72. Lipids (Part D)
Edited by JOHN M. LOWENSTEIN

VOLUME 73. Immunochemical Techniques (Part B)
Edited by JOHN J. LANGONE AND HELEN VAN VUNAKIS

VOLUME 74. Immunochemical Techniques (Part C)
Edited by JOHN J. LANGONE AND HELEN VAN VUNAKIS

VOLUME 75. Cumulative Subject Index Volumes XXXI, XXXII, XXXIV–LX
Edited by EDWARD A. DENNIS AND MARTHA G. DENNIS

VOLUME 76. Hemoglobins
Edited by ERALDO ANTONINI, LUIGI ROSSI-BERNARDI, AND EMILIA CHIANCONE

VOLUME 77. Detoxication and Drug Metabolism
Edited by WILLIAM B. JAKOBY

VOLUME 78. Interferons (Part A)
Edited by SIDNEY PESTKA

VOLUME 79. Interferons (Part B)
Edited by SIDNEY PESTKA

VOLUME 80. Proteolytic Enzymes (Part C)
Edited by LASZLO LORAND

VOLUME 81. Biomembranes (Part H: Visual Pigments and Purple Membranes, I)
Edited by LESTER PACKER

VOLUME 82. Structural and Contractile Proteins (Part A: Extracellular Matrix)
Edited by LEON W. CUNNINGHAM AND DIXIE W. FREDERIKSEN

VOLUME 83. Complex Carbohydrates (Part D)
Edited by VICTOR GINSBURG

VOLUME 84. Immunochemical Techniques (Part D: Selected Immunoassays)
Edited by JOHN J. LANGONE AND HELEN VAN VUNAKIS

VOLUME 85. Structural and Contractile Proteins (Part B: The Contractile Apparatus and the Cytoskeleton)
Edited by DIXIE W. FREDERIKSEN AND LEON W. CUNNINGHAM

VOLUME 86. Prostaglandins and Arachidonate Metabolites
Edited by WILLIAM E. M. LANDS AND WILLIAM L. SMITH

VOLUME 87. Enzyme Kinetics and Mechanism (Part C: Intermediates, Stereochemistry, and Rate Studies)
Edited by DANIEL L. PURICH

VOLUME 88. Biomembranes (Part I: Visual Pigments and Purple Membranes, II)
Edited by LESTER PACKER

VOLUME 89. Carbohydrate Metabolism (Part D)
Edited by WILLIS A. WOOD

VOLUME 90. Carbohydrate Metabolism (Part E)
Edited by WILLIS A. WOOD

VOLUME 91. Enzyme Structure (Part I)
Edited by C. H. W. HIRS AND SERGE N. TIMASHEFF

VOLUME 92. Immunochemical Techniques (Part E: Monoclonal Antibodies and General Immunoassay Methods)
Edited by JOHN J. LANGONE AND HELEN VAN VUNAKIS

VOLUME 93. Immunochemical Techniques (Part F: Conventional Antibodies, Fc Receptors, and Cytotoxicity)
Edited by JOHN J. LANGONE AND HELEN VAN VUNAKIS

VOLUME 94. Polyamines
Edited by HERBERT TABOR AND CELIA WHITE TABOR

VOLUME 95. Cumulative Subject Index Volumes 61–74, 76–80
Edited by EDWARD A. DENNIS AND MARTHA G. DENNIS

VOLUME 96. Biomembranes [Part J: Membrane Biogenesis: Assembly and Targeting (General Methods; Eukaryotes)]
Edited by SIDNEY FLEISCHER AND BECCA FLEISCHER

VOLUME 97. Biomembranes [Part K: Membrane Biogenesis: Assembly and Targeting (Prokaryotes, Mitochondria, and Chloroplasts)]
Edited by SIDNEY FLEISCHER AND BECCA FLEISCHER

VOLUME 98. Biomembranes (Part L: Membrane Biogenesis: Processing and Recycling)
Edited by SIDNEY FLEISCHER AND BECCA FLEISCHER

VOLUME 99. Hormone Action (Part F: Protein Kinases)
Edited by JACKIE D. CORBIN AND JOEL G. HARDMAN

VOLUME 100. Recombinant DNA (Part B)
Edited by RAY WU, LAWRENCE GROSSMAN, AND KIVIE MOLDAVE

VOLUME 101. Recombinant DNA (Part C)
Edited by RAY WU, LAWRENCE GROSSMAN, AND KIVIE MOLDAVE

VOLUME 102. Hormone Action (Part G: Calmodulin and Calcium-Binding Proteins)
Edited by ANTHONY R. MEANS AND BERT W. O'MALLEY

VOLUME 103. Hormone Action (Part H: Neuroendocrine Peptides)
Edited by P. MICHAEL CONN

VOLUME 104. Enzyme Purification and Related Techniques (Part C)
Edited by WILLIAM B. JAKOBY

VOLUME 105. Oxygen Radicals in Biological Systems
Edited by LESTER PACKER

VOLUME 106. Posttranslational Modifications (Part A)
Edited by FINN WOLD AND KIVIE MOLDAVE

VOLUME 107. Posttranslational Modifications (Part B)
Edited by FINN WOLD AND KIVIE MOLDAVE

VOLUME 126. Biomembranes (Part N: Transport in Bacteria, Mitochondria, and Chloroplasts: Protonmotive Force)
Edited by SIDNEY FLEISCHER AND BECCA FLEISCHER

VOLUME 127. Biomembranes (Part O: Protons and Water: Structure and Translocation)
Edited by LESTER PACKER

VOLUME 128. Plasma Lipoproteins (Part A: Preparation, Structure, and Molecular Biology)
Edited by JERE P. SEGREST AND JOHN J. ALBERS

VOLUME 129. Plasma Lipoproteins (Part B: Characterization, Cell Biology, and Metabolism)
Edited by JOHN J. ALBERS AND JERE P. SEGREST

VOLUME 130. Enzyme Structure (Part K)
Edited by C. H. W. HIRS AND SERGE N. TIMASHEFF

VOLUME 131. Enzyme Structure (Part L)
Edited by C. H. W. HIRS AND SERGE N. TIMASHEFF

VOLUME 132. Immunochemical Techniques (Part J: Phagocytosis and Cell-Mediated Cytotoxicity)
Edited by GIOVANNI DI SABATO AND JOHANNES EVERSE

VOLUME 133. Bioluminescence and Chemiluminescence (Part B)
Edited by MARLENE DELUCA AND WILLIAM D. MCELROY

VOLUME 134. Structural and Contractile Proteins (Part C: The Contractile Apparatus and the Cytoskeleton)
Edited by RICHARD B. VALLEE

VOLUME 135. Immobilized Enzymes and Cells (Part B)
Edited by KLAUS MOSBACH

VOLUME 136. Immobilized Enzymes and Cells (Part C)
Edited by KLAUS MOSBACH

VOLUME 137. Immobilized Enzymes and Cells (Part D)
Edited by KLAUS MOSBACH

VOLUME 138. Complex Carbohydrates (Part E)
Edited by VICTOR GINSBURG

VOLUME 139. Cellular Regulators (Part A: Calcium- and Calmodulin-Binding Proteins)
Edited by ANTHONY R. MEANS AND P. MICHAEL CONN

VOLUME 140. Cumulative Subject Index Volumes 102–119, 121–134

VOLUME 141. Cellular Regulators (Part B: Calcium and Lipids)
Edited by P. MICHAEL CONN AND ANTHONY R. MEANS

VOLUME 142. Metabolism of Aromatic Amino Acids and Amines
Edited by SEYMOUR KAUFMAN

Volume 213. Carotenoids (Part A: Chemistry, Separation, Quantitation, and Antioxidation)
Edited by Lester Packer

Volume 214. Carotenoids (Part B: Metabolism, Genetics, and Biosynthesis)
Edited by Lester Packer

Volume 215. Platelets: Receptors, Adhesion, Secretion (Part B)
Edited by Jacek J. Hawiger

Volume 216. Recombinant DNA (Part G)
Edited by Ray Wu

Volume 217. Recombinant DNA (Part H)
Edited by Ray Wu

Volume 218. Recombinant DNA (Part I)
Edited by Ray Wu

Volume 219. Reconstitution of Intracellular Transport
Edited by James E. Rothman

Volume 220. Membrane Fusion Techniques (Part A)
Edited by Nejat Düzguünes

Volume 221. Membrane Fusion Techniques (Part B)
Edited by Nejat Düzgünes

Volume 222. Proteolytic Enzymes in Coagulation, Fibrinolysis, and Complement Activation (Part A: Mammalian Blood Coagulation Factors and Inhibitors)
Edited by Laszlo Lorand and Kenneth G. Mann

Volume 223. Proteolytic Enzymes in Coagulation, Fibrinolysis, and Complement Activation (Part B: Complement Activation, Fibrinolysis, and Nonmammalian Blood Coagulation Factors)
Edited by Laszlo Lorand and Kenneth G. Mann

Volume 224. Molecular Evolution: Producing the Biochemical Data
Edited by Elizabeth Anne Zimmer, Thomas J. White, Rebecca L. Cann, and Allan C. Wilson

Volume 225. Guide to Techniques in Mouse Development
Edited by Paul M. Wassarman and Melvin L. DePamphilis

Volume 226. Metallobiochemistry (Part C: Spectroscopic and Physical Methods for Probing Metal Ion Environments in Metalloenzymes and Metalloproteins)
Edited by James F. Riordan and Bert L. Vallee

Volume 227. Metallobiochemistry (Part D: Physical and Spectroscopic Methods for Probing Metal Ion Environments in Metalloproteins)
Edited by James F. Riordan and Bert L. Vallee

Volume 228. Aqueous Two-Phase Systems
Edited by Harry Walter and Göte Johansson

VOLUME 229. Cumulative Subject Index Volumes 195–198, 200–227

VOLUME 230. Guide to Techniques in Glycobiology
Edited by WILLIAM J. LENNARZ AND GERALD W. HART

VOLUME 231. Hemoglobins (Part B: Biochemical and Analytical Methods)
Edited by JOHANNES EVERSE, KIM D. VANDEGRIFF, AND ROBERT M. WINSLOW

VOLUME 232. Hemoglobins (Part C: Biophysical Methods)
Edited by JOHANNES EVERSE, KIM D. VANDEGRIFF, AND ROBERT M. WINSLOW

VOLUME 233. Oxygen Radicals in Biological Systems (Part C)
Edited by LESTER PACKER

VOLUME 234. Oxygen Radicals in Biological Systems (Part D)
Edited by LESTER PACKER

VOLUME 235. Bacterial Pathogenesis (Part A: Identification and Regulation of Virulence Factors)
Edited by VIRGINIA L. CLARK AND PATRIK M. BAVOIL

VOLUME 236. Bacterial Pathogenesis (Part B: Integration of Pathogenic Bacteria with Host Cells)
Edited by VIRGINIA L. CLARK AND PATRIK M. BAVOIL

VOLUME 237. Heterotrimeric G Proteins
Edited by RAVI IYENGAR

VOLUME 238. Heterotrimeric G-Protein Effectors
Edited by RAVI IYENGAR

VOLUME 239. Nuclear Magnetic Resonance (Part C)
Edited by THOMAS L. JAMES AND NORMAN J. OPPENHEIMER

VOLUME 240. Numerical Computer Methods (Part B)
Edited by MICHAEL L. JOHNSON AND LUDWIG BRAND

VOLUME 241. Retroviral Proteases
Edited by LAWRENCE C. KUO AND JULES A. SHAFER

VOLUME 242. Neoglycoconjugates (Part A)
Edited by Y. C. LEE AND REIKO T. LEE

VOLUME 243. Inorganic Microbial Sulfur Metabolism
Edited by HARRY D. PECK, JR., AND JEAN LEGALL

VOLUME 244. Proteolytic Enzymes: Serine and Cysteine Peptidases
Edited by ALAN J. BARRETT

VOLUME 245. Extracellular Matrix Components
Edited by E. RUOSLAHTI AND E. ENGVALL

VOLUME 246. Biochemical Spectroscopy
Edited by KENNETH SAUER

VOLUME 247. Neoglycoconjugates (Part B: Biomedical Applications)
Edited by Y. C. LEE AND REIKO T. LEE

VOLUME 305. Bioluminescence and Chemiluminescence (Part C)
Edited by THOMAS O. BALDWIN AND MIRIAM M. ZIEGLER

VOLUME 306. Expression of Recombinant Genes in Eukaryotic Systems
Edited by JOSEPH C. GLORIOSO AND MARTIN C. SCHMIDT

VOLUME 307. Confocal Microscopy
Edited by P. MICHAEL CONN

VOLUME 308. Enzyme Kinetics and Mechanism (Part E: Energetics of Enzyme Catalysis)
Edited by DANIEL L. PURICH AND VERN L. SCHRAMM

VOLUME 309. Amyloid, Prions, and Other Protein Aggregates
Edited by RONALD WETZEL

VOLUME 310. Biofilms
Edited by RON J. DOYLE

VOLUME 311. Sphingolipid Metabolism and Cell Signaling (Part A)
Edited by ALFRED H. MERRILL, JR., AND YUSUF A. HANNUN

VOLUME 312. Sphingolipid Metabolism and Cell Signaling (Part B)
Edited by ALFRED H. MERRILL, JR., AND YUSUF A. HANNUN

VOLUME 313. Antisense Technology (Part A: General Methods, Methods of Delivery, and RNA Studies)
Edited by M. IAN PHILLIPS

VOLUME 314. Antisense Technology (Part B: Applications)
Edited by M. IAN PHILLIPS

VOLUME 315. Vertebrate Phototransduction and the Visual Cycle (Part A)
Edited by KRZYSZTOF PALCZEWSKI

VOLUME 316. Vertebrate Phototransduction and the Visual Cycle (Part B)
Edited by KRZYSZTOF PALCZEWSKI

VOLUME 317. RNA–Ligand Interactions (Part A: Structural Biology Methods)
Edited by DANIEL W. CELANDER AND JOHN N. ABELSON

VOLUME 318. RNA–Ligand Interactions (Part B: Molecular Biology Methods)
Edited by DANIEL W. CELANDER AND JOHN N. ABELSON

VOLUME 319. Singlet Oxygen, UV-A, and Ozone
Edited by LESTER PACKER AND HELMUT SIES

VOLUME 320. Cumulative Subject Index Volumes 290–319

VOLUME 321. Numerical Computer Methods (Part C)
Edited by MICHAEL L. JOHNSON AND LUDWIG BRAND

VOLUME 322. Apoptosis
Edited by JOHN C. REED

VOLUME 323. Energetics of Biological Macromolecules (Part C)
Edited by MICHAEL L. JOHNSON AND GARY K. ACKERS

VOLUME 358. Bacterial Pathogenesis (Part C: Identification, Regulation, and Function of Virulence Factors)
Edited by VIRGINIA L. CLARK AND PATRIK M. BAVOIL

VOLUME 359. Nitric Oxide (Part D: Nitric Oxide Detection, Mitochondria and Cell Functions, and Peroxynitrite Reactions)
Edited by ENRIQUE CADENAS AND LESTER PACKER

VOLUME 360. Biophotonics (Part A)
Edited by GERARD MARRIOTT AND IAN PARKER

VOLUME 361. Biophotonics (Part B) (in preparation)
Edited by GERARD MARRIOTT AND IAN PARKER

VOLUME 362. Recognition of Carbohydrates in Biological Systems (Part A: General Procedures) (in preparation)
Edited by YUAN C. LEE AND REIKO T. LEE

VOLUME 363. Recognition of Carbohydrates in Biological Systems (Part B: Specific Applications) (in preparation)
Edited by YUAN C. LEE AND REIKO T. LEE

[1] Fluorescence: Basic Concepts, Practical Aspects, and Some Anecdotes

By DAVID M. JAMESON, JOHN C. CRONEY, and PIERRE D. J. MOENS

Introduction

The theoretical foundations of fluorescence spectroscopy were established in the first half of the twentieth century by pioneers including Enrique Gaviola, Jean and Francis Perrin (father and son), Peter Pringsheim, Sergei Vavilov, F. Weigert, F. Dushinsky, Alexander Jabłoński, Theodor Förster, and, more recently, Gregorio Weber.[1–7] In the last quarter of the twentieth century, advances in electronics, lasers, computers, and molecular biology have allowed fluorescence methodologies to assume an important role in diverse disciplines including chemistry, cell biology, and the biomedical sciences. This volume of *Methods in Enzymology* covers many of the most exciting new developments in fluorescence spectroscopy—developments and techniques that presently define the state of the art. In this article, however, we wish to remind readers of the origins of several important aspects of fluorescence spectroscopy. We also wish to discuss some practical aspects of fluorescence determinations which are sometimes forgotten as those new to these methods often focus on learning the software associated with commercial instrumentation. Much of the modern "point-and-click" software approach allows the novice to immediately apply fluorescence methods to their particular research problems, taking advantage of the highly sophisticated instrumentation and probe chemistries that are now readily available. However, kits and user-friendly software should not dissuade beginners from learning the fundamentals of fluorescence methodologies and instrumentation. Such knowledge not only allows avoidance of potential pitfalls, recognition of artifacts, and fuller appreciation of the applicability of fluorescence techniques, it also makes the research more interesting and fun! We should also point out that although fluorescence measurements are usually carried out with sophisticated and expensive instrumentation,

[1] G. Weber, *Methods Enzymol.* **278,** 1 (1997).
[2] B. Nickel, *EPA Newslett.* **58,** 9 (1996).
[3] B. Nickel, *EPA Newslett.* **61,** 27 (1997).
[4] B. Nickel, *EPA Newslett.* **64,** 19 (1998).
[5] M. N. Berberan-Santos, *in* "New Trends in Fluorescence Spectroscopy" (B. Valeur and J.-C. Brochon, eds.), p. 7. Springer-Verlag, Heidelberg, Germany, 2001.
[6] D. M. Jameson, *in* "New Trends in Fluorescence Spectroscopy" (B. Valeur and J.-C. Brochon, eds.), p. 35. Springer-Verlag, Heidelberg, Germany, 2001.
[7] B. Valeur, "Molecular Fluorescence: Principles and Applications." Wiley-VCH, Weinheim, Germany, 2002.

0076-6879/03 $35.00

many of the basic principles of fluorescence can be appreciated and demonstrated with a simple ultraviolet (UV) hand lamp.[8]

Virtually all fluorescence data required for any research project will fall into one of the following categories: (1) the fluorescence emission spectrum, (2) the excitation spectrum of the fluorescence, (3) the quantum yield, (4) the fluorescence lifetime, or (5) the polarization (anisotropy) of the emission.

In this article, we examine each of these categories and briefly discuss historical developments, underlying concepts, and practical considerations. We preface these discussions with an overview of the most salient aspects of three essential pieces of fluorescence hardware, namely, light sources, monochromators, and detectors (specifically photomultipliers). Finally, we discuss the phenomenon of fluorescence resonance energy transfer (FRET)—a topic in vogue presently. Our discussion of FRET, however, does not follow the path already trodden by numerous reviews but instead focuses sharply, and in some detail, on energy transfer between like molecules, that is, homo-FRET.

Before delving into our discussion of the theory and practice of fluorescence measurements, we should briefly mention the realm of potential fluorophores. Fluorescence probes can be broadly placed into two categories: intrinsic probes and extrinsic probes. By intrinsic we mean any naturally occurring molecule which exhibits sufficient fluorescence to be of practical utility. Examples include the aromatic amino acids, especially tryptophan and tyrosine, NADH, FAD, FMN, some porphyrins, some fatty acids and lipids, some modified nucleic acids (such as the Wye base in some tRNAs), pyridoxal phosphate, chlorophylls, pteridines, and some others. By extrinsic we mean probes that can be introduced into the target system to form a complex, either noncovalent or covalent. Up until the late 1970s, the choice of commercially available probes was limited; typically, researchers [such as one of the authors (D.M.J.) of this article] had to design and synthesize fluorophores for particular types of studies. The choice of fluorophore was dictated by the properties of the instruments [light sources and photomultiplier tube (PMT) sensitivities] as well as by the lifetimes and spectral properties required. The first probe made for physicochemical studies of proteins was dimethylaminonaphthalene sulfonyl chloride (dansyl choride), synthesized by G. Weber and designed to be conjugated to proteins and used for fluorescence polarization measurements.[9,10] In the last two decades, however, the availability of fluorescence probes, principally by the company Molecular Probes (Eugene, OR; www.probes.com), has expanded incredibly. Literally thousands of probes are now available. Examples of common noncovalent probes are 8-anilino-1-naphthalene sulfonate (ANS), 1,6-diphenyl-1,3,5-hexatriene (DPH), 2-dimethylamino-6-lauroylnaphthalene (Laurdan), ethidium bromide, Hoechst

[8] J. C. Croney, D. M. Jameson, and R. P. Learmonth, *Biochem. Mol. Biol. Ed.* **29,** 60 (2001).

[9] G. Weber, *Biochem. J.* **51,** 145 (1952).

[10] G. Weber, *Biochem. J.* **51,** 155 (1952).

dyes, 4′,6-diamidino-2-phenylindole (DAPI), to name but a few, while covalent probes functionalized with reactive groups such as sulfonyl choride, isothiocyanate, succinimidyl ester, iodoacetamide, maleimide, hydrazine derivatives, and others are available to target a wide variety of biomolecules. In addition to probes specifically designed to target biomolecules for solution studies, a wide range of probes exist which are specialized for fluorescence microscopy studies on living cells.

Basic Instrumentation

A typical modern spectrofluorimeter is shown in Fig. 1 (adapted with permission from ISS, Inc.), which illustrates the principal components and the general layout (we have adopted the nomenclature of "fluorimeter" for steady-state instrumentation and "fluorometer" for lifetime instrumentation, a custom that originated

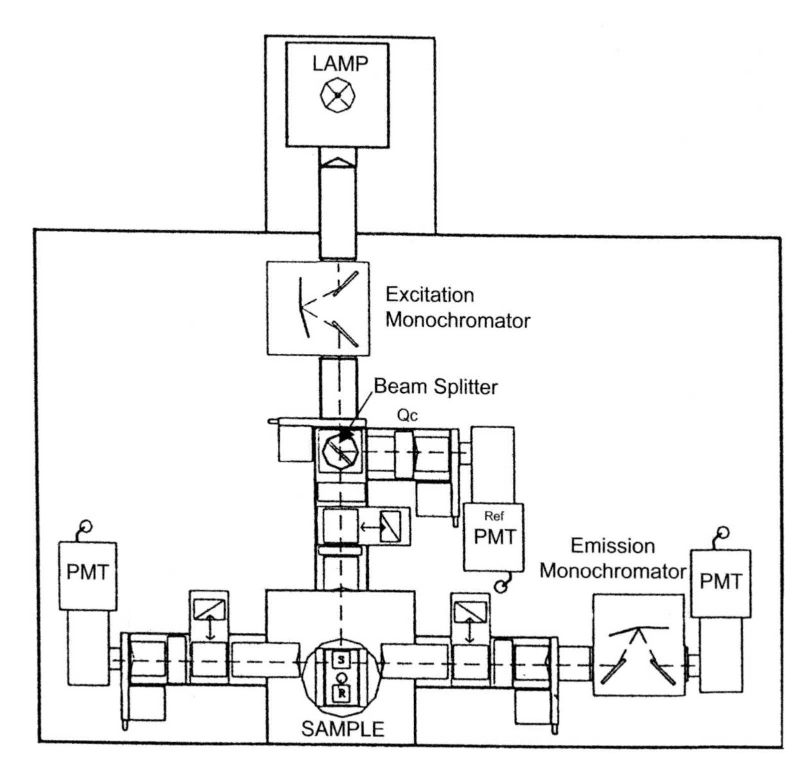

FIG. 1. Schematic diagram of a spectrofluorimeter (revised with permission from commercial literature from ISS, Champaign, IL). The principal components include the xenon arc lamp, the excitation and emission monochromators, a quartz beam splitter, a quantum counting solution (Qc), three photomultiplier tubes (PMT), and excitation (P_{ex}) and emission (P_{em}) calcite prism polarizers.

with Gaviola,[11] who termed his lifetime instrument a fluorometer). Clearly the starting point for any fluorescence observation is the light source. Fortunately, we no longer need to rely on sunlight as the principal excitation source for fluorescence. A comprehensive discussion of the development of light sources is beyond the scope of this article, but a good overview of a wide range of light sources (and many other relevant topics such as filters, monochromators, and photodetectors) can be found in the book by Moore *et al.*[12] We shall restrict ourselves to consideration of the light sources most commonly employed in commercial fluorescence instrumentation, namely the xenon arc lamp, the xenon–mercury arc lamp, lasers, light-emitting diodes (LEDs), and laser diodes. We should note that the most common light sources used in absorption spectrophotometry, deuterium and tungsten lamps, are rarely used in fluorescence since they are relatively weak photon sources.

For our present discussion the most relevant aspect of light sources is the useful wavelength range. From this point of view the xenon arc lamp is by far the most common light source in commercial instruments since it produces usable light from the ultraviolet to the infrared. This range is adequate for most fluorescence studies of biological samples since such studies are usually limited by the absorption characteristics of water at either end of this spectral range, and by photodamage in the deep ultraviolet. An example of the light distribution from a xenon arc lamp, from 220 to 590 nm, is shown in Fig. 2 (this spectrum is convoluted by the characteristics of the excitation monochromator and quartz beam splitter—topics which will be discussed later). Although the spectral range shown is relatively narrow, we should note that xenon arc lamps provide significant illumination out to around 2000 nm. Clearly the intensity of this light source depends dramatically on the wavelength—a fact which has a significant impact on the excitation spectrum, as discussed later. The xenon–mercury arc has the characteristics of the xenon arc source but is dominated by very prominent lines due to the mercury transitions. The more prominent mercury lines are near 254, 297, 302, 313, 365, 405, 436, 546, and 578 nm.[13] Much of the early work on proteins carried out in the former Soviet Union utilized 296.7 nm as the excitation wavelength since mercury lamps were a common light source at that time; the choice of this wavelength was appropriate since tyrosine absorption is negligible above 295 nm and hence tryptophan residues could be preferentially excited.

The use of lasers in fluorescence has primarily been restricted to time-resolved instrumentation, due to the intensity and spatial characteristics of laser sources [with the notable exception of fluorescence-activated cell sorters (FACS)]. Of course, laser sources have distinct emission lines characteristic of the atomic

[11] E. Gaviola, *Z. Physik.* **35,** 748 (1926).

[12] J. H. Moore, C. C. Davis, and M. A. Coplan, "Building Scientific Apparatus: A Practial Guide to Design and Construction." Addison-Wesley, Reading, MA, 1983.

[13] C. J. Sansonetti, M. L. Salit, and J. Reader, *Appl. Optics* **35,** 74 (1996).

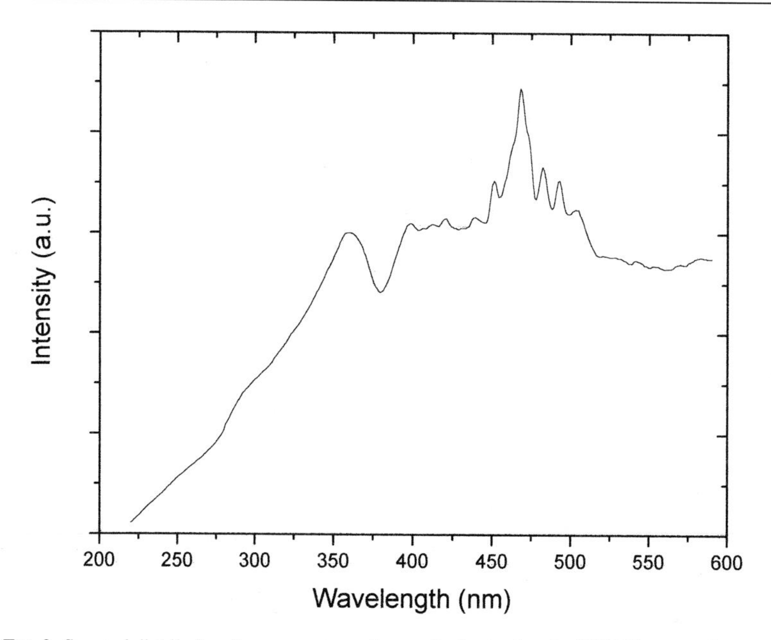

FIG. 2. Spectral distribution from a xenon arc lamp: obtained using the ISS PC1 spectrofluorimeter shown in Fig. 1.

processes involved. The most commonly used lasers in modern biological fluorescence spectroscopy are the argon ion laser, the helium–cadmium laser, the neodymium:yttrium–aluminum–garnet (Nd:YAG) laser and, more recently, the titanium–sapphire laser. The most commonly used argon ion laser lines are near 488 and 514 nm. Large-frame, high-power argon ion lasers [e.g., the Spectra-Physics (Mountain View, CA) model 2045 laser in the authors' laboratory] produce lines in the deep UV at 275 nm and between 300 and 305 nm as well as the mid-UV near 334, 351, and 364 nm and in the visible near 457, 476, 488, 497, 501, 514, and 528 nm. Helium–cadmium lasers produce lines near 325 and 442 nm. Nd:YAG lasers emit at 1064 nm and are typically doubled or quadrupled to 532 and 266 nm, respectively (frequency-doubled Nd:YAG lasers are now readily available as green-emitting laser pointers). Often, the doubled Nd:YAG output at 532 nm is used to pump a dye laser (typically a rhodamine-based dye) whose output in a range around 600 nm can then be doubled to produce UV light over a range around 300 nm. The titanium–sapphire laser, which emits over a range of about 700–1000 nm, is presently the light source of choice for multiphoton excitation. A relevant characteristic of laser sources is the time profile of their output; for example, typical argon ion or helium–cadmium lasers are operated as continuous sources (termed CW; intensity essentially time invariant) whereas Nd:YAG and titanium–sapphire lasers are usually operated as pulsed sources.

LEDs are becoming more popular since the list of available wavelengths is growing and, more importantly, extending deeper into the UV. At the time of this writing, LED sources down to 370 nm are available commercially. LEDs are much less intense than lasers (and not collimated) but have the advantages that they are relatively inexpensive, low-power, solid-state devices that generate little heat and which provide usable intensities over a narrow (but not discrete) spectral range. Their energy output can also be directly modulated, which suggests time-resolved applications (discussed below). Another solid-state device, the laser diode, provides monochromatic radiation; near-UV laser diodes have become available which are much more intense than LEDs but which—at the time of this article—only extend down to about 400 nm.

If the excitation light source is not at a discrete wavelength, that is, a laser or laser diode, then a device capable of wavelength selectivity is required, typically either an optical filter or a monochromator. For excitation purposes, the most useful optical filter is usually an interference-type filter, which typically can isolate wavelengths with a resolution of a few nanometers. The figure of merit for these types of filters is their full width at half-maximum (FWHM). For example, an interference filter centered at 400 nm with an FWHM of 5 nm transmits 50% of the intensity at 395 and 405 nm that it transmits at 400 nm. Such filters are not always very efficient, especially as the wavelength decreases. For example, the peak light transmission efficiency of a typical UV interference filter may be ~20% compared to ~70 to 80% for an interference filter centered in the visible wavelength region. Other types of filters used in fluorescence studies will be discussed when the emission side of the fluorimeter is considered.

Monochromators are the most common and versatile devices used to isolate specific wavelengths of light from broad-band sources such as xenon arc lamps. Monochromators operate by dispersing the incident light—most people are very aware of the light-dispersing qualities of a prism (we note that most people have observed the excellent light dispersion qualities of raindrops, which result in rainbows, but far fewer notice that the weaker of the double rainbows, which can occur under particularly sunny conditions, has the order of the colors reversed due to the additional raindrop-interior reflection). The spectral region selected by a monochromator depends on the design of the monochromator and, ultimately, on the physical size of the monochromator slits; the key consideration here is the dispersion of the monochromator, which allows conversion of the physical width of a slit (e.g., in millimeters) to the FWHM of the spectral region passed. For example, the monochromators in the instrument shown in Fig. 1 utilize fixed slits (as opposed to infinitely variable slits) and have dispersions of 8 nm per millimeter. Slits ranging from 0.025 to 4 mm are commonly available, which thus supply spectral resolutions ranging from 0.2 to 32 nm. Different monochromators, of course, have different dispersion factors but the common feature is that the smaller the slit, the higher will be the spectral resolution. This resolution comes with a cost, namely, a reduction in the light intensity: a 2-fold reduction in each slit width (entrance or exit) results in an approximate 4-fold decrease in light intensity. In commercial spectrofluorimeters, prism-based monochromators are not commonly used, one reason being that a linear scan of the prism assembly will not result in a linear dispersion of wavelengths. For many years, commercially available monochromators [such as those from Bausch & Lomb (Rochester, NY) or Thermo Jarrell Ash (Franklin, MA)] used planar ruled diffraction gratings as the dispersive element. These devices worked well but exhibited parasitic light levels due to imperfections in the ruling process, which could seriously hamper measurements of turbid samples. This source of stray light was dramatically reduced when concave holographic gratings became available. These types of diffraction gratings are made using interference patterns generated onto photoresist substrates using laser sources. One of the first widespread commercial uses of holographic gratings in spectrofluorimeters was by SLM in the early 1970s [the gratings themselves were from Jobin Yvon (Edison, NJ)]. Regardless of the type of grating utilized, the efficiency with which a monochromator transmits light will show both wavelength and polarization dependence (discussed below).

After the excitation wavelength is isolated it is sometimes passed through a polarizing device to select one plane of polarization (polarization is discussed in more detail below). A variety of devices, both naturally occurring and man-made, have been used to polarize light. The Vikings, in fact, used a "sunstone" to observe the location of the sun on foggy or overcast days (magnetic compasses were unreliable at higher latitudes)—days that lasted a long time at the high northern latitudes! This stone is now thought to have been composed of the mineral cordierite, a natural polarizing material. By taking advantage of the fact (even if they did not know the reasons behind it) that scattered sunlight was highly polarized whereas light coming along the direction of the sun was not, Vikings could observe the distribution of the sky's brightness through the sunstone and localize the position of the sun and, if the time of day were known, the compass directions. Nowadays, the most common polarizers are either dichroic devices, which operate by effectively absorbing one plane of polarization (e.g., Polaroid type-H sheets based on stretched polyvinyl alcohol impregnated with iodine, invented in 1926 by E. H. Land while he was a freshman at Harvard), or double-refracting calcite ($CaCO_3$) crystal polarizers (discovered independently by Erasmus Bartholin and Christiaan Huygens in the seventeenth century), which differentially disperse the two planes of polarization (examples of this class of polarizers are Nicol polarizers, Wollaston prisms, and Glan-type polarizers such as the Glan–Foucault, Glan–Thompson, and Glan–Taylor polarizers). [Typically film polarizers are inexpensive but do not transmit efficiently in the ultraviolet. (Quiz: The interested reader should work out which direction of polarized light polarizing sunglasses pass; Hint—glare or light reflected from surfaces is largely polarized in a horizontal direction).] Calcite prism polarizers, on the other hand, transmit well into the ultraviolet (<240 nm) but are expensive (in the range of U.S. $1000) and have small apertures and angular acceptance tolerances, which means that incident light cannot be tightly focused (a comparison of the wavelength-dependent polarizing efficiencies of several types of polarizing devices is given in Jameson *et al.*[14]

In our typical fluorimeter, the exciting light impinges on the sample and the emission is viewed at right angles to the direction of excitation (Fig. 1). This 90° observation angle is primarily intended to reduce the extent of exciting light that passes to the detection side. Only when samples are relatively turbid, for example in the case of lipid suspensions or membrane samples, will significant levels of exciting light be scattered and potentially reach the photodetector (since optical filters and monochromators are not perfect). We should also note that the focal lengths of the lenses just before and after the sample have a small influence on the measured polarization. Specifically, the larger the numerical aperture of the lenses focusing the excitation and collecting the emitted light (i.e., the shorter the focal length and hence the larger the cone of collected light), the lower the measured polarization compared to the true polarization.[15] This effect is most serious in microscope optics, having only a small influence on measurements taken with normal spectrofluorimeters. A typical instrument such as that shown in Fig. 1 may yield a polarization lower by only a few percent from the true value. After excitation by the polarized light, the emitted light can (if desired) also be passed through a polarizing device and the spectral region of interest can be isolated by an optical filter or a second monochromator. The usual filters used in the emission side are either bandpass filters (which transmit a broader spectral region than interference filters; FWHMs can be several tens of nanometers or greater) or cut-on filters. This latter filter is often referred to as a cut-off filter, depending on the viewpoint of whether the transmission commences sharply at a given wavelength (cuts on) or, equivalently, if the optical density decreases sharply at that wavelength (cuts off). Regardless, the operational principle of these types of filters, which are also known as longpass filters, is that they can be used to block any excitation light scattered toward the emission direction and then collect a large percentage of the total emission. Web sites that contain transmission data for many types of filters include www.mellesgriot.com and www.corion.com. A useful handbook containing information about filters for fluorescence microscopy can be found at www.chroma.com.

Most modern instruments use photomultiplier tubes (PMTs) for detection and quantification of the emitted light. These devices are, of course, based on the photoelectric effect, that is, the ejection of electrons from metallic surfaces as a consequence of incident light. The theory for this effect was developed by Albert Einstein, who received the Nobel Prize for this work and not for his more famous theory of relativity. The original phototubes were based on a simple arrangement to collect the emitted photoelectrons and produce an electric current, which could then be quantified. These early devices were not much of an

[14] D. M. Jameson, G. Weber, R. D. Spencer, and G. Mitchell, *Rev. Sci. Instrum.* **49,** 510 (1978).
[15] G. Weber, *J. Opt. Soc. Am.* **46,** 962 (1956).

improvement over the human eye, although they did offer the considerable advantage of protecting the observer from the deleterious effect of UV or infrared (IR) radiation on the visual system. PMT devices were soon developed, however, that had multiple plates after the photosensitive cathode, called dynodes, held at progressively more positive voltages, which acted as secondary electron-emitting surfaces and which would eject several electrons for each incident electron and hence multiply the effect manyfold (practical gains above 10^9 anode electrons per photoelectron can be achieved for short light pulses although continuous gains of about 10^7 are typical, due to thermal loading in the final dynodes). In the last few decades, significant progress had been made in the commercialization of PMTs with "extended red response," which essentially means PMTs that can efficiently detect light out to beyond 800 nm. One of the first widely used PMTs in this category was the Hamamatsu (Bridgewater, NJ) R928 (a side-on, nine-stage PMT with a multialkali photocathode rated from 185 to 900 nm). One of the authors (D.M.J.) vividly remembers being handed one long ago by Gregorio Weber, with instructions to "try it out!" At that time it seemed miraculous because it so greatly minimized the correction factors (discussed below) needed to acquire true molecular spectra extending above 600 nm. For detailed information about a variety of PMTs and other types of detectors, the reader can refer to http://usa.hamamatsu.com.

Emission Spectra

Using complementary filters, George Gabriel Stokes realized in 1852[16] that the fluorescence phenomenon, which he originally termed "dispersive reflexion" but, in a subsequent article renamed "fluorescence," resulted in the emission of light at longer wavelengths than the absorbed light, and his name is still used to describe this wavelength shift (for an illuminating account of the origins of the terms fluorescence, phosphorescence, and luminescence see Valeur[7]). Early observations on fluorescence relied on various types of filters (usually chemical solutions, sometimes even wine!) to block the exciting light (interestingly, C. V. Raman, in his pioneering work on the effect that now bears his name, had to use filters to block out fluorescence). Modern filters are usually made of sophisticated glasses or plastics and include cut-on, cut-off, bandpass, and interference type filters (discussed above). It is interesting to note that most of the important original observations on fluorescence were made almost exclusively by visual methods since reliable and sensitive photomultipliers came into popular use only after World War II.[17,18]

[16] G. G. Stokes, *Philos. Trans.* **142,** 463 (1852).

[17] G. Weber, *in* "Time-Resolved Fluorescence Spectroscopy in Biochemistry and Biology" (R. B. Cundall and R. E. Dale, eds.), Vol. 69, p. 1. Plenum Press, New York, 1983.

[18] G. Weber, *in* "Fluorescent Biomolecules" (D. M. Jameson and G. D. Reinhart, eds.), p. 343. Plenum Press, New York, 1989.

Consequently, many of the pioneers in this field, such as Gregorio Weber, suffered from acute eye ailments in later life.[6]

In the middle of the twentieth century commercial fluorescence instrumentation began to appear. The earliest commercial instruments were essentially attachments for spectrophotometers, such as the Beckman (Fullerton, CA) DU spectrophotometer[19]; this attachment allowed the emitted light (excited by the mercury vapor source through a filter) to be reflected into the monochromator of the spectrophotometer. The first commercial spectrofluorimeters with monochromators for both excitation and emission were inspired by the work of Bowman *et al.*,[20] and were produced by AMINCO-Bowman and Farrand. These early instruments allowed biologists to use fluorescence to develop clinically relevant assays for a wide variety of biological molecules.[19] Among the first emission spectra with direct significance for protein chemists were those of the aromatic amino acids, published in 1957 by Teale and Weber.[21]

Early examination of a large number of emission spectra resulted in the formulation of certain general rules.

1. *In a pure substance existing in solution in a unique form, the fluorescence spectrum is invariant, remaining the same independent of the excitation wavelength.* In fact, if the fluorescence spectrum changes as the excitation wavelength varies, one may usually conclude that there are more than one emitting species present (exceptions may be due to dipolar relaxation or other excited state effects such as deprotonation, which will not be discussed here).

2. *The fluorescence spectrum lies at longer wavelengths than the absorption.* If the absorption maximum of the band of least frequency is at wavelength λ_{abs} and the maximum of the fluorescence emission is at λ_{fl} it is always found that $\lambda_{abs} < \lambda_{fl}$ or $\nu_{abs} > \nu_{fl}$, where ν corresponds to c/λ.

3. *The fluorescence spectrum is, to a good approximation, a mirror image of the absorption band of least frequency.* The wavelength of reflection is found midway between ν_{abs} and ν_{fl} and corresponds to the energy of the pure electronic transition $(0 \rightarrow 0')$.

These general observations follow from consideration of the Perrin–Jabłoński diagram shown in Fig. 3 (we use this term to recognize the fact, as pointed out by Nickel,[2] that F. Perrin actually utilized energy level diagrams to describe fluorescence phenomena before Jabłoński; the most recent text on fluorescence by B. Valeur[7] also adopts this nomenclature). Specifically, although the fluorophore may be excited into different singlet state energy levels (e.g., S_1, S_2, etc.) rapid

[19] S. Udenfriend, "Fluorescence Assay in Biology and Medicine." Academic Press, New York, 1962.

[20] R. L. Bowman, P. A. Caufield, and S. Udenfriend, *Science* **122,** 32 (1955).

[21] F. W. J. Teale and G. Weber, *Biochem. J.* **53,** 476 (1957).

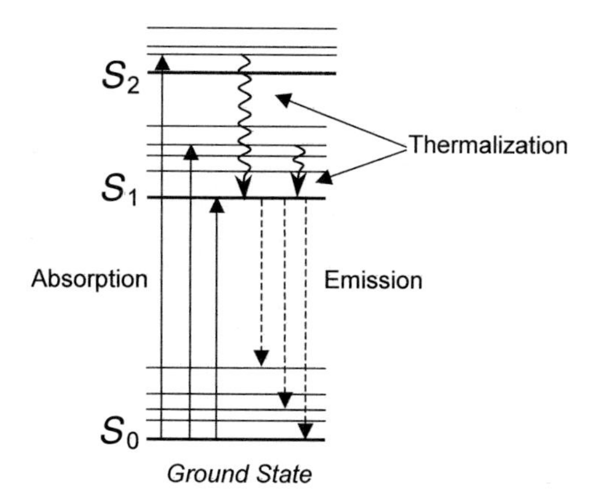

FIG. 3. Perrin–Jabłoński diagram. S_0 is the ground state, whereas S_1 and S_2 are electronically excited states. The time scales of the absorption, thermalization, and emission processes are typically in the range of $\sim 10^{-15}$, $\sim 10^{-12}$, and $\sim 10^{-9}$ sec, respectively.

thermalization invariably occurs and emission takes place from the lowest vibrational level of the first excited electronic state (S_1). Only in a few rare cases, such as azulene, does emission occur from the S_2 level (see, e.g., Liu[22]). This fact accounts for the independence of the emission spectrum from the excitation wavelength. The fact that ground state fluorophores, at room temperature, are predominantly in the lowest vibrational level of the ground electronic state (as required from the Boltzmann distribution law) accounts for the Stokes shift. Finally, the fact that the spacings of the energy levels in the vibrational manifolds of the ground state and first excited electronic states are usually similar accounts for the fact that the emission and absorption spectra (plotted in energy units such as reciprocal wavenumbers) are approximately mirror images (although emission spectra are usually plotted in terms of wavelengths, most modern software allows for facile conversion and display in terms of wavenumbers).

Virtually all modern spectrofluorimeters use right-angle geometry, that is, the direction of observation of the emitted light is 90° to that of the excitation. As mentioned above, this geometry was chosen to allow for a more complete elimination of exciting light from the observed signal. In other words, the unabsorbed excitation light will continue through the cuvette and be absorbed by the blackened sample compartment. This right-angle observation geometry is the reason, of course, that fluorescence cuvettes, as opposed to absorption cuvettes, have all four sides polished (occasionally novices are observed using absorption cuvettes in spectrofluorimeters, which usually leads to disappointing results!). While on the topic

[22] R. S. H. Liu, *J. Chem. Ed.* **79,** 183 (2002).

FIG. 4. Rhodamine B in ethanol excited by Nd : YAG laser pointer device (in each case, the light is passing through the cuvette from the right to the left). Optical densities are 0.04, 1.0, 3.0, and >30, from left to right, respectively.

of cuvettes, we should also note that quartz (or fused silica) cuvettes are normally used because they transmit down to the deep UV, and if glass or plastic cuvettes are being used one must be aware of their wavelength cutoff points. This right-angle geometry also immediately demands consideration of the effect of sample optical density on the observed signal. This effect is illustrated in Fig. 4, which depicts excitation light impinging on four samples of rhodamine B in ethanol, with optical densities (at 532 nm, the wavelength from the green laser pointer utilized to excite the fluorescence) ranging from 0.04 to greater than 30. It is readily apparent that the excitation light traverses, virtually undiminished in intensity, the sample with the low (0.04) optical density, whereas it is significantly absorbed by sample with an optical density equal to 1, largely absorbed by sample with optical density equal to 3, and completely absorbed by the most concentrated sample. Since the emission optics are typically focused on the center of the cuvette, one can easily see that the observed signal will suffer as a consequence of absorption of the exciting light by the initial layers of the solution. In such a case, increasing the concentration of the sample in an effort to increase the signal will lead to even less detected fluorescence. One simple way to mitigate this problem—if a high concentration must be utilized, for example, to study weak binding equilibria, or if the quantum yield is low—is to use a cuvette with a reduced pathlength. A T-shaped cuvette, for example, with one long and one short dimension, can be used with the short direction oriented to receive the excitation. Front-face observation, either using a specially designed sample compartment or a triangular cuvette, may also be used to circumvent inner filter effects (see, e.g., Hirsch[23]). Web sites describing a wide variety of cuvettes include www.optiglass.com and www.nsgpci.com. If the Stokes shift of the fluorophore is small, it may also be necessary to worry about absorption of the emitted light by the solution along the emission path. Various mathematical methods to correct the observed fluorescence intensity for these so-called "inner filter" effects have been described.[7,24] Such inner filter effects are less important for either polarization or lifetime measurements because, in these cases, the property being measured does not depend (up to a point) on the absorption by the solution.

In a typical emission spectrum, the excitation wavelength is fixed and the fluorescence intensity versus wavelength is obtained (we note that some analytical techniques actually use synchronous scanning, wherein both excitation and emission wavelength are scanned simultaneously; however, this specialized method is not discussed further). Typically, the intensity axis of spectra is reported in "arbitrary units." This nomenclature recognizes the fact that the observed intensity depends not only on the concentration of the fluorophore and on molecular factors such as the extinction coefficient and quantum yield of the fluorophore, but also on instrumental factors such as slit widths, PMT voltage, and amplifier gains (photon-counting methods are discussed below), considerations that arise because the normal spectrofluorimeter is not a double-beam instrument like a typical spectrophotometer. Fluorescence spectra recorded in this manner are "technical" or "uncorrected" emission spectra. Uncorrected spectra do not take into account the wavelength-dependent response of the emission optics (principally the monochromator) and the photodetector (most commonly a photomultiplier tube). Up until the mid-1970s or so, most of the PMTs (such as the popular EMI 6256S) used in commercial and even homebuilt instruments had responses that fell off dramatically at wavelengths above 600 nm. Consequently, instrument correction factors (discussed in more detail below) for compounds such as porphyrins and rhodamines (with emission maxima above 600 nm) or even fluorescein (with an emission maximum near 520 nm) were not insignificant. In the late-1970s, however, PMTs with extended red responses (such as the Hamamatsu 928 mentioned above) became almost standard and made emission correction factors much less dramatic. At about the same time holographic diffraction gratings began to appear in commercial instruments—such as the SLM 8000—which featured significantly less parasitic or stray light than the conventionally ruled gratings. Even these modern monochromators, however, exhibit nonideal behavior, which can seriously confuse the novice. One of the more dramatic effects is the so-called Wood's anomaly. In 1902, R. W. Wood observed that diffraction gratings exhibited some anomalous transmission characteristics in narrow spectral regions. These anomalies, known

[23] R. E. Hirsch, *Methods Enzymol.* **232,** 231 (1994).

[24] J. R. Lakowicz, "Principles of Fluorescence Spectroscopy." Kluwer Academic, New York, 1999.

as "Wood's anomalies," are highly dependent on the polarization of the incident light, being almost entirely polarized perpendicular to the rulings of the grating. They occur in diffraction gratings when a diffracted order becomes tangent to the plane of the grating and are believed to result from a resonant interaction of the incident light with surface plasmons, which are electronic excitations confined to the metal surface.[25] The Wood's anomalies in the monochromators used in the SLM 8000 were discussed in a Ph.D. thesis[26] and an earlier review.[27] (The Wood's anomaly near 380 nm in these monochromators gave a characteristic shoulder in the uncorrected spectrum of intrinsic protein fluorescence, which made it easy to identify the instrument used to obtain the spectrum; in fact, the true "connoisseur" of fluorescence should be able to identify the fluorimeter used from observation of the Wood's anomalies in spectra.) Figure 5 shows scans of a tungsten lamp (in a quartz bulb), which illustrates the Wood's anomalies in an SLM 8000 spectrofluorimeter. The four scans shown correspond to the case of no emission polarizer (Fig. 5A), an emission polarizer oriented parallel to the laboratory axis (Fig. 5B), an emission polarizer oriented perpendicular to the laboratory axis (Fig. 5C), and scan B divided by scan C (Fig. 5D). These scans demonstrate that the Wood's anomalies are located near 380 and 620 nm and, moreover, that they occur only in the perpendicular component. [*Note:* On some instruments, such as the Cary Eclipse (Varian, Palo Alto, CA), the monochromators are mounted with a 90° rotation to render the slit image horizontal, and the resulting Wood's anomalies appear in the parallel component.] Scan D in Fig. 5 furthermore illustrates the fact that the response of the monochromator has a general dependence on the polarization of the observed light.

These considerations bring us to a discussion of emission correction factors. The wavelength and polarization response of the components of any given spectrofluorimeter can be taken into account by generating correction factors specific for that instrument. The correction factors for the emission monochromator, for instance, can be obtained by using a standard lamp. For example, the standard lamp used in D.M.J's thesis[26] was a 45-W halogen–tungsten coiled-coil filament lamp, in a quartz envelope; the spectral irradiance of this lamp, at a distance of 50 cm when the lamp was operated at 6.50 amperes current, was determined by the National Bureau of Standards (the precise current was required because it determined the filament temperature and hence the blackbody radiation spectrum). Thus this lamp was mounted such that light would reach the sample compartment and be reflected from a freshly coated MgO surface (the reflectivity of which was known as a function of wavelength), and one then simply recorded the spectrum.

[25] P. Murdin, *ING La Palma Technical Note* **76,** 1990.

[26] D. M. Jameson, *in* "Biochemistry." University of Illinois, Champaign-Urbana, IL, 1978.

[27] D. M. Jameson, *in* "Fluorescein Hapten: An Immunological Probe" (E. Voss, ed.), p. 23. CRC Press, Boca Raton, FL, 1984.

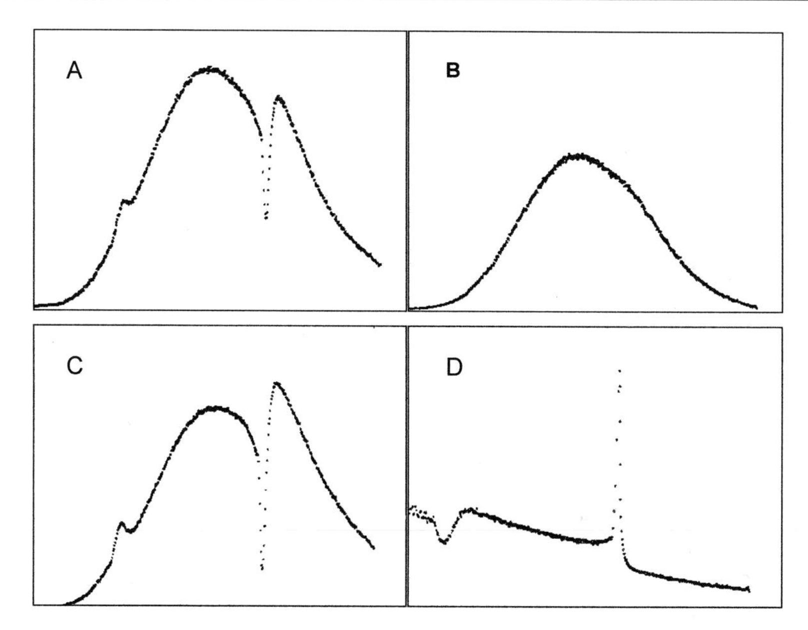

FIG. 5. Scans of a standard tungsten lamp showing Wood's anomalies in an SLM monochromator. Scans A–C are from 251 to 800 nm; scan D is from 351 to 800 nm. Scan A, No polarizer; scan B, parallel emission polarizer; scan C, perpendicular emission polarizer; scan D, ratio of scan B to scan A. Adapted from D. M. Jameson, *in* "Biochemistry." University of Illinois, Champaign-Urbana, IL, 1978.

Comparison of this spectrum with the known spectral irradiance yielded the correction factors. This approach is essentially used by most instrument manufacturers, who now typically supply customers with the appropriate correction factors, usually in a software file. Given our previous discussion of monochromator effects, one can appreciate that the correction factors for parallel polarized light will differ from that for perpendicularly polarized light, as shown in Fig. 6A. To obtain a corrected spectrum for a sample, then one can obtain the technical or uncorrected spectrum through a parallel polarizer and then apply the parallel (vertical) correction factors, as shown in Fig. 6C (technical emission spectra obtained without any polarizer and through a parallel polarizer are shown in Fig. 6B and C, respectively).

Before one applies instrument correction factors, however, one may first have to correct the spectrum for solvent background. Usually this step is not required if the fluorophore has a decent quantum yield and is present at reasonable concentrations, and if a minimum of care was used in preparation of the buffer or solvent. However, if the sample is dilute or the fluorophore is highly quenched then care must be taken, especially regarding the infamous Raman peak. Even though the molar absorptivity due to the O–H stretching mode—which occurs near 3400 cm^{-1}—is weak, the fact that water is 55 *M* means that a measurable quasi-elastic scatter peak can be seen in most fluorimeters (in fact, the ability to resolve Raman

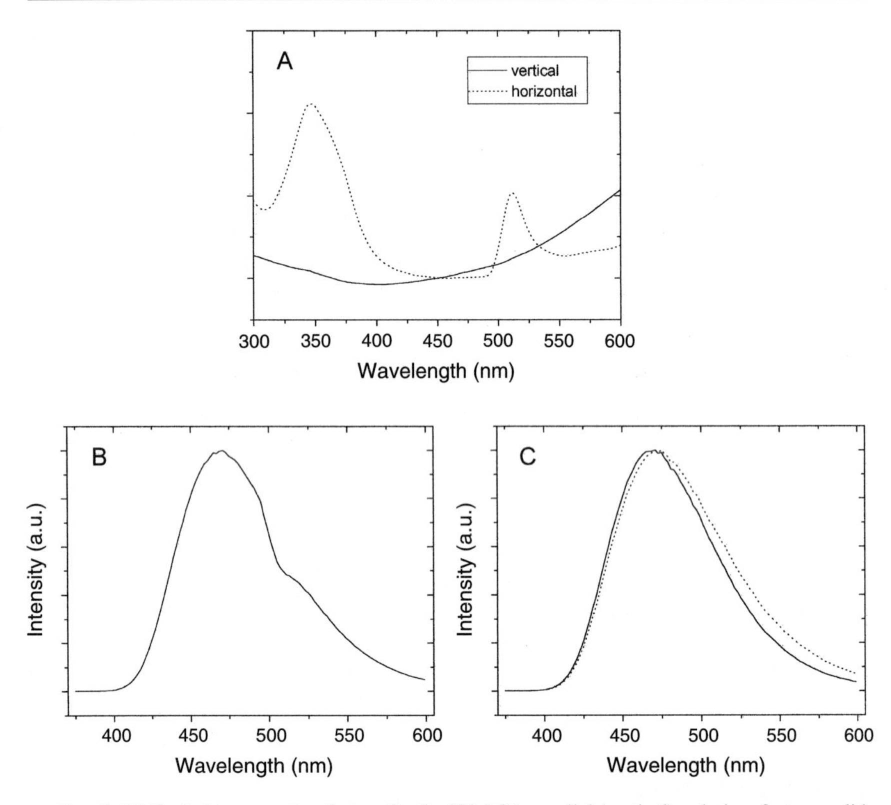

FIG. 6. (A) Emission correction factors for the ISS PC1; parallel (vertical) polarizer factors, solid line; perpendicular (horizontal) polarizer, dotted line. (B) Uncorrected (technical) spectrum for ANS in ethanol, no polarizer. (C) Uncorrected emission spectra (with parallel polarizer), solid line; corrected, dotted line.

peaks is a necessary indication of an instrument's sensitivity). The position of the Raman peak depends, of course, on the excitation wavelength as shown in Fig. 7B (Fig. 7A shows the Rayleigh scatter, the Raman scatter, and scatter due to the second order of the monochromator). Moreover, because the wavelength scale is not linear with energy, the wavelength difference between the Rayleigh (elastic scatter) and Raman peaks is not fixed. To a useful approximation, the position of the Raman peak (in aqueous solutions) can be calculated by the expression

$$\frac{1}{\lambda_{R}} = \frac{1}{\lambda_{EX}} - 0.00034 \quad (1)$$

where λ_R and λ_{EX} are the wavelengths of the Raman and excitation peaks, respectively. Figures 7B and 7C make the point that the Raman peaks for aqueous buffers stay fairly constant in intensity and whether they present a concern will depend on

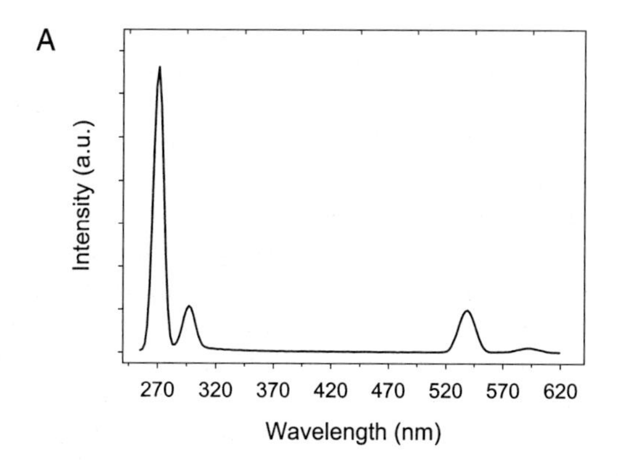

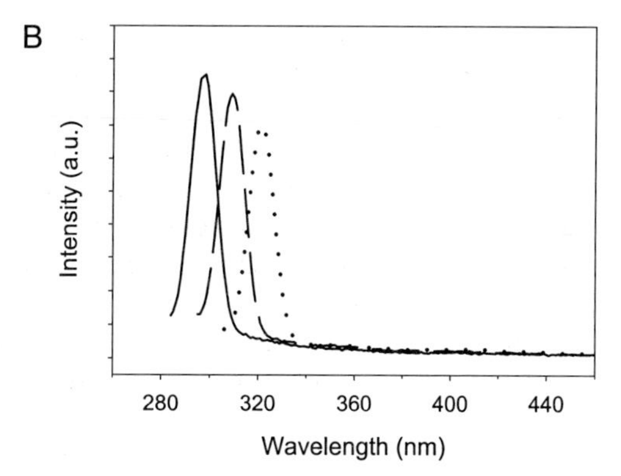

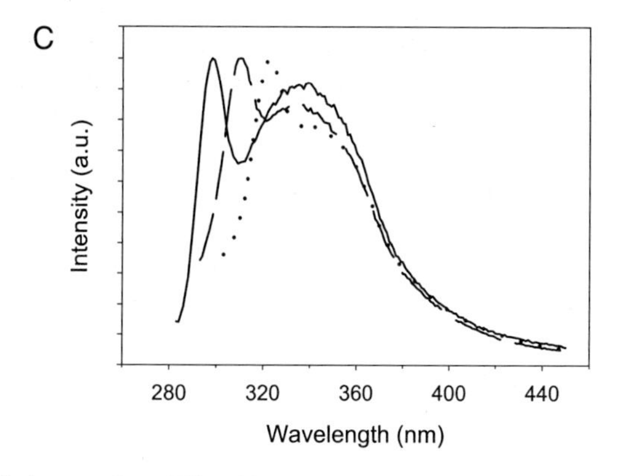

FIG. 7. (A) Emission scan from 250 to 650 nm with excitation at 270 nm; peaks correspond (from left to right) to Rayleigh scatter (270 nm), Raman scatter (297 nm), second-order Rayleigh (540 nm), and second-order Raman (594 nm). (B) Scans of 50 m*M* phosphate buffer, excited at 270 nm (solid line), 280 nm (dashed line), and 290 nm (dotted line). (C) Scans of $\sim 3 \times 10^{-8}$ *M* bovine serum albumin excited at 270 nm (solid line), 280 nm (dashed line), 290 nm (dotted line).

the fluorophore concentration (Fig. 7B also indicates how a "clean" buffer should look relative to the Raman peak for UV excitation). Also shown in Fig. 7C are emission spectra of $\sim 3 \times 10^{-8}$ *M* bovine serum albumin (BSA) excited at three different wavelengths (270, 280, and 290 nm), which show (1) that fluorescence does not change position while the Raman peaks shift and (2) the relative magnitude of the Raman peaks relative to the fluorescence. In the case of tryptophan, one usually can go to about 10^{-7} *M* before worrying about interference from Raman peaks; in the case of fluorophores such as fluorescein, which enjoy robust extinction coefficients and quantum yields, it is possible to work down to about 10^{-9} *M* probe before the Raman peak becomes a concern. (We note that because the Raman peak is highly polarized, care must be taken in some types of polarization studies, involving sample dilution, that increasing polarizations upon decreasing sample signals is not due to the Raman.[14])

It is also worth noting that the most sensitive fluorescence instruments use "photon-counting" methods. Photon counting was originally developed in the 1950s for astronomical photometric observations. It was first applied in fluorescence in the area of lifetime determinations and then in the mid-1970s to steady-state fluorescence measurements.[14,28] The underlying concept of the photon-counting technique is illustrated in Fig. 8 (adapted from Jameson[26]). The advantages of photon counting over direct current measurements include (1) improvement in the signal-to-noise ratio at low light levels, (2) long-term stability, that is, less sensitivity to power supply and amplifier drifts, (3) direct acquisition of data in the digital domain, which facilitates data processing, (4) and effective dynamic range of many orders of magnitude (as determined by the count rate and collection time), and (5) the absolute character of the unit of measurement, that is, photons per unit time. Although photon counting thus offers significant advantages over analog electronics, there is one common pitfall. Namely, it is necessary to be cognizant of the count rate. As shown in Fig. 8, the pulse from the discriminator output has a finite width, which results in an effective instrument dead time. Specifically, if more photons are incident on the system during the time corresponding to a photon pulse they will not be registered, a situation known as "pulse pileup." This fact means that the response of the detector system to incident photons will deviate from linearity as the count rate increases. With modern, fast discriminators the linear count range can extend up to hundreds of thousands of counts per second, but one should be aware of the linear response range for the particular instrument being utilized. Polarization determinations are particularly sensitive to pulse pileup because the parallel intensity component will usually be larger than the perpendicular intensity component and hence will "pile up" first, resulting in an apparent decrease in polarization as the count rate increases.

[28] D. M. Jameson, R. D. Spencer, and G. Weber, *Rev. Sci. Instrum.* **47,** 1034 (1976).

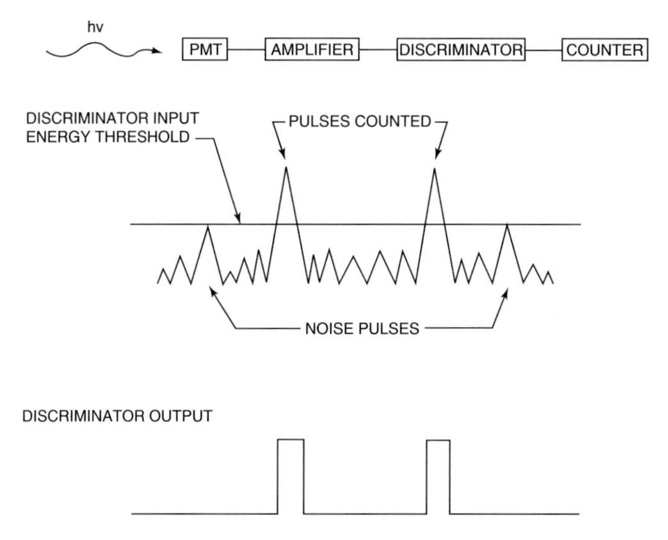

FIG. 8. Schematic illustration of the photon-counting method. Adapted from D. M. Jameson, *in* "Biochemistry." University of Illinois, Champaign-Urbana, IL, 1978.

Excitation Spectra

The relative efficiencies of different wavelengths of incident light to excite fluorophores is determined as the excitation spectrum. In this case, the excitation monochromator is varied while the emission wavelength is kept constant if a monochromator is utilized, or the emitted light can be observed through a filter. If the system is "well-behaved," that is, if the three general rules outlined above hold, the excitation spectrum would be expected to match the absorption spectrum. In this case, however, as in the case of the emission spectrum, corrections for instrumentation factors are required. In fact, excitation spectra not corrected for instrument parameters will deviate seriously from "corrected" excitation spectra, much more so than in the case of emission spectra. This situation is caused by the wavelength-dependent variation in light intensity due to the energy profile of the xenon arc source (Fig. 2). The magnitude of this effect can be seen in Fig. 9A, which depicts the absorption spectrum and the uncorrected excitation spectrum for ANS in ethanol (acquired with an ISS PC1 spectrofluorimeter). Since the xenon arc source produces less light at 280 nm than, for example, 360 nm, the relative heights of these peaks will be distorted from the actual absorption peaks. This effect does not appear in absorption spectra since virtually all absorption spectrophotometers are, in effect, double-beam instruments, which correct for the wavelength dependence of the light source. In the case of the fluorescence excitation spectrum, an approximate correction can be accomplished by using a ratio mode wherein the signal plotted is the sample fluorescence divided by

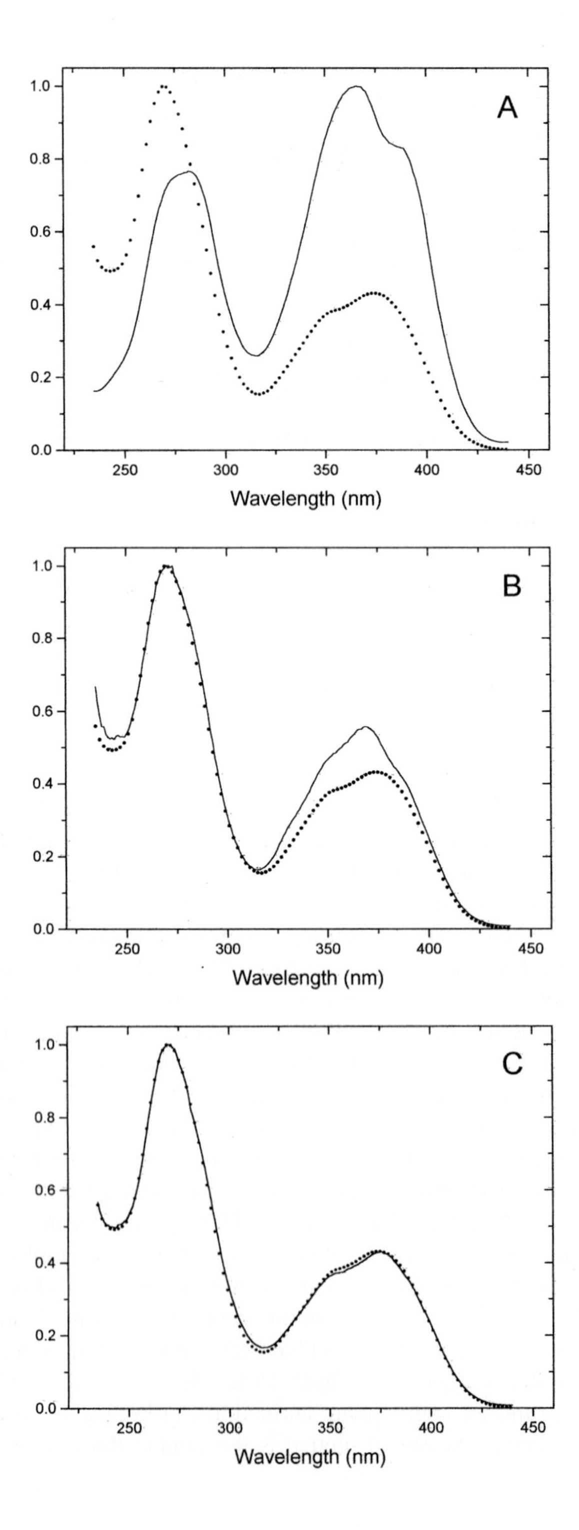

A
1.0
0.8
0.6
0.4
0.2
0.0
250
300
350
400
450
Wavelength (nm)
B
1.0
0.8
0.6
0.4
0.2
0.0
250
300
350
400
450
Wavelength (nm)
C
1.0
0.8
0.6
0.4
0.2
0.0
250
300
350
400
450
Wavelength (nm)

a reference signal. The reference signal is derived from a small percentage of the excitation beam, which is directed by a quartz beam splitter to a quantum counter (Fig. 1). This quantum counter is typically a concentrated solution of rhodamine B in ethanol placed in a triangular cuvette oriented such that the exciting light strikes the center of the front face of the cuvette while the fluorescence is viewed through a red cut-on filter blocking all light below about 590 nm. In this way, all of the exciting light impinging on the quantum-counting solution is absorbed and the fluorescence intensity in the reference channel will reflect the incident light intensity at all excitation wavelengths. The ratio-corrected excitation spectrum for ANS in ethanol is shown in Fig. 9B along with the absorption spectrum. Note that the correspondence between these two spectra is much better than in Fig. 9A but still is not perfect. This small deviation occurs because the quartz beam splitter, used to deflect some light to the reference channel, is not ideal. In fact, the reflectivity of the beam splitter depends on the polarization of the incident light and, as already discussed, the light coming from monochromators will show a wavelength-dependent polarization. Consequently, at wavelengths corresponding to the Wood's anomalies in the excitation monochromator there will be a different ratio of reflected to transmitted light compared with other wavelengths and the "corrected" excitation spectrum will not be properly corrected. To obtain a true corrected excitation spectrum it is necessary to determine the actual lamp profile in the sample compartment, using a quantum counter/front-face arrangement, for example, as was done to obtain the lamp profile shown in Fig. 1. One can then divide the uncorrected sample excitation spectrum by this lamp profile and generate the corrected excitation spectrum as shown in Fig. 9C.

Quantum Yield

The quantum yield (q) of a fluorophore is essentially a measure of how efficiently absorbed light is emitted. The definition of quantum yield and its first direct measurements were due to Vavilov,[29–31] who discovered that fluorescein was an almost perfect emitter with a quantum yield close to unity. One way to view this quantity is that q equals the number of quanta absorbed divided by the number of quanta emitted.

[29] S. I. Vavilov, *Philos. Mag.* **43,** 307 (1922).
[30] S. I. Vavilov, *Z. Physik* **22,** 266 (1924).
[31] S. I. Vavilov, *Z. Physik* **42,** 311 (1927).

FIG. 9. (A) Absorption (dotted line) and uncorrected excitation spectra for ANS in ethanol. (B) Absorption (dotted line) and "ratio"-corrected excitation spectra. (C) Absorption (dotted line) and excitation spectra corrected by a lamp curved as described in text.

A more insightful expression relates the quantum yield to the relative rates of the radiative and nonradiative pathways, which deactivate the excited state. Hence,

$$q = \frac{k_r}{k_r + \sum k_{nr}} \tag{2}$$

where k_r and k_{nr} correspond to radiative and nonradiative processes, respectively. In Eq. (2), $\sum k_{nr}$ designates the sum of the rate constants for the various processes that compete with the emission process. These processes include photochemical and dissociative processes in which the products are well-characterized chemical species (electron, proton, radical, molecular isomer) as well as less well-characterized changes that result in a return to the ground state with simultaneous dissipation of the energy of the excited state into heat. These latter processes are collectively called nonradiative transitions and two types have been clearly recognized: intersystem crossing and internal conversion. Intersystem crossing refers to the radiationless spin inversion of S_1 in the excited state that results in the isoenergetic, or almost isoenergetic, conversion into a triplet state. When spin inversion occurs in S_1 the resulting triplet reached is T_1, which is characterized by an energy intermediate between the ground state and first excited singlet state. Another process that is nonradiative from the point of view of the original excited molecule is resonance energy transfer.

In the past, measurements of quantum yields usually relied on comparison of the intensity of the fluorescence with that of the exciting radiation. The first method of this kind, due to S. I. Vavilov,[30] also had historical interest as it led to the demonstration that fluorescent compounds in solution (fluorescein in the original case) could reradiate most of the absorbed energy rather than a small fraction of it, as many at that time believed. Vavilov, and later Melhuish,[32] used integrating spheres to determine quantum yields. This device is a closed chamber, not necessarily spherical, coated with a light-diffusing material of negligible absorption for the fluorescence and exciting radiation. In practice, a coating of MgO provides good diffusivity and low absorption both in the ultraviolet and visible regions of the spectrum. A detector placed somewhere in the spherical wall samples the radiation density, and it is assumed that conditions are such that homogeneous radiation distribution inside the sphere is reached regardless of the directional emission of the source (diffusing plate or fluorescent sample). A second method of the same type, proposed by Weber and Teale,[33] eliminated the integrating sphere and compared the fluorescent signal with the signal from a nonfluorescent scattering solution, both of which were adjusted to give the same apparent absorbance at the exciting wavelength. These methods of comparison of excitation and fluorescence signals require correction for the difference in spectral response of the photomultiplier

[32] W. H. Melhuish, *J. Phys. Chem.* **65,** 229 (1961).

[33] G. Weber and F. W. J. Teale, *Trans. Faraday Soc.* **53,** 646 (1957).

over the wavelength of excitation and emission, which in turn calls for an exact determination of the molecular fluorescence spectrum. To obviate these sources of error, often underestimated, Weber and Teale interposed between source and photomultiplier a solution of rhodamine B in ethylene glycol intended to act as a proportional quantum counter for the scattering and fluorescent emission.[33] As rhodamine B solutions do not absorb strongly beyond 600 nm this arrangement was limited to fluorophores that did not emit at longer wavelengths.

Interestingly, it is not always necessary to measure emitted light to determine quantum yields because nonradiative processes will lead to dissipation of heat to the solvent, which can be determined calorimetrically. Seybold *et al.*,[34] for example, used an ingenious calorimetric method to measure the thermal expansion of the illuminated fluorescent solution, which was then compared with the larger expansion owing to the nonfluorescent absorber (such as India ink); this volume change could then be directly related to the quantum yield of the fluorophore.

Because the absolute quantum yields of a number of common fluorophores have now been determined, it is usually possible to ascertain the quantum yield of a fluorophore, with adequate accuracy, by comparing its yield relative to a known standard. For example, the quantum yield of tryptophan at 20° at neutral pH is generally taken as 0.14, that of quinine sulfate in 0.1 *N* H_2SO_4 is generally taken as 0.55, and that of fluorescein in 0.01 *N* NaOH is generally taken as 0.94. To ascertain the quantum yield of the fluorophore in question, ideally a solution is made of a standard (such as those listed above) with an optical density matching that of the target fluorophore (note that this step should usually be done via dilutions because accurate optical density measurements are best realized in a range, e.g., near 1, which would cause severe inner filter effects in fluorescence measurements). One then acquires the corrected spectra of standard and target fluorophores, integrate the spectral areas and thus obtain the quantum yield of the target fluorophore is obtained from the ratio of these areas. Tables listing quantum yields with appropriate references for a number of compounds are given on page 160 of Valeur[7] and page 53 of Lakowicz.[24] Vavilov noted[31] that the quantum yield was independent of the excitation wavelength, a conclusion now known as Vavilov's Law. We now appreciate that exceptions to Vavilov's Law can occur. For example, electron ejection from the upper (S_2) level of indole leads to a decrease in the quantum yield at wavelengths below 240 nm.[35,36]

Excited State Lifetimes

In most cases of interest, it is virtually impossible to predict *a priori* the excited state lifetime of a fluorescent molecule. The true molecular lifetime, that is,

[34] P. G. Seybold, M. Gouterman, and J. Callis, *Photochem. Photobiol.* **9,** 229 (1969).
[35] H. B. Steen, *J. Chem. Phys.* **61,** 3997 (1974).
[36] I. Tatischeff and R. Klein, *Photochem. Photobiol.* **22,** 221 (1975).

the lifetime expected in the absence of any excited state deactivation processes, can be approximated by the Strickler–Berg equation.[37]

$$\tau_m^{-1} = 2.88 \times 10^{-9} n^2 \langle \overline{\nu_f^{-3}} \rangle \int_{\Delta\nu_a} \varepsilon(\bar{\nu}) d \ln \bar{\nu} \tag{3}$$

where

$$\langle \overline{\nu_f^{-3}} \rangle = \frac{\int_{\Delta\nu_e} F(\bar{\nu})\, d\nu}{\int_{\Delta\nu_a} F(\bar{\nu}) \nu^{-3}\, d\nu} \tag{4}$$

In these equations, τ_m is the molecular lifetime, n is the refractive index of the solvent, $\Delta\nu_a$ and $\Delta\nu_e$ correspond to the experimental limits of the absorption and emission bands (S_0 to S_1 transitions), respectively, ε is the molar absorption, and $F(\nu)$ describes the spectral distribution of the emission in photons per wavelength interval.

However, the effect of competing radiationless processes that deactivate the excited state (such as quenching, energy transfer, and intersystem crossing) must be taken into account. The relationship between the molecular lifetime (τ_m) and the experimentally observed lifetime (τ_e) is then

$$\tau_e = q\tau_m \tag{5}$$

where q is the quantum yield.

How well do these equations actually work? For NADH in water the equation of Strickler and Berg [Eq. (3)] gives $\tau_m = 76$ ns. Direct measurements yield $\tau = 0.4$ ns and $q = 0.025$, which combined give $\tau_m = 16$ ns. In fact, agreement is rarely better than ~20%.

Fluorescence lifetime measurements are traditionally realized using either the impulse response method (in which excitation is by a brief pulse of light, after which the direct decay of the fluorescence is observed) or the harmonic response method (in which the intensity of the excitation light is modulated sinusoidally and the phase shift of the fluorescence, relative to the excitation, is determined). Direct measurements of fluorescence lifetimes were first successfully realized by the Argentinean E. Gaviola, in Berlin in 1926,[11] following inconclusive efforts by R. W. Wood[38] and P. F. Gottling.[39] Gaviola used a phase fluorometer, which operated at 10 MHz; Gaviola used visual compensation methods to measure the lifetime of aqueous rhodamine (obtaining a value of 2 ns). The mathematical theory behind phase fluorometry did not actually appear until 1933 and was due to Duschinsky.[40] The first use of pulse methods to determine fluorescence lifetimes

[37] S. J. Strickler and R. A. Berg, *J. Chem. Phys.* **37,** 814 (1962).
[38] R. W. Wood, *Proc. R. Soc. A.* **99,** 362 (1921).
[39] P. F. Gottling, *Phys. Rev.* **22,** 566 (1923).
[40] F. Duschinsky, *Z. Physik* **81,** 23 (1933).

is generally credited to Brody.[41] Since these early beginnings, both methods, phase and pulse, have advanced tremendously. Perhaps the most important developments in phase fluorometry were the cross-correlation instrument of Spencer and Weber[42] and the first true variable multifrequency phase and modulation fluorometer built by E. Gratton (originally while a postdoctoral fellow with G. Weber) in the late 1970s.[43] Indeed, the approach taken by Gratton, which revolutionized frequency domain methodologies, was soon adopted by other laboratories and also led to the commercial development of the method by ISS (Champaign, IL). Important developments in the pulse method included the use of time-correlated single photon counting and the appearance of ultrafast pulsed laser sources, which rendered moot much of the effort that had been expended on deconvolution techniques. Despite these impressive advances in the pulse method, however, it must be said that W. R. Ware erred when he predicted at the 1980 NATO *Advanced Studies Meeting on Time-Resolved Fluorescence in Biochemistry and Biology*[44] that "the use of mode-locked, synchronously pumped dye laser excitation sources, with or without frequency doubling, in conjunction with single-photon detection, would seem to be the final nail in the phase-shift coffin. . . ." In fact, multifrequency phase and modulation fluorometry is widely utilized both for normal solution studies as well as for fluorescence lifetime microscopy. These frequency domain methods have also found increasing applications in diagnostic imaging (see, e.g., Stankovic *et al.*[45] or the web site www.iss.com).

Interestingly, one form of phase and modulation fluorometry utilizes pulsed sources for excitation. The possibility of using the harmonic content of high-repetition rate-pulsed light sources for phase/modulation fluorometry was discussed by Gratton and Lopez-Delgado[46] and one of the first implementations of the method used the Frascati synchrotron radiation source ADONE.[47] The concept underlying this approach is that repetitive pulse sources have a harmonic content, as dictated by Fourier and Laplace transforms, which can be used to generate multiple frequencies for phase and modulation measurements. The main difference between the use of external modulating devices, such as Pockels cells, and the harmonic content of a repetitive light source, such as synchrotron radiation or a mode-locked laser, is that in the latter case the researcher is limited to frequencies that are multiples of the fundamental repetition rate (of course, these frequencies

[41] S. S. Brody, *Rev. Sci. Instrum.* **28,** 1021 (1957).

[42] R. D. Spencer and G. Weber, *Ann. N.Y. Acad. Sci.* **158,** 361 (1969).

[43] E. Gratton and M. Limkeman, *Biophys. J.* **44,** 315 (1983).

[44] W. R. Ware, *in* "Time-Resolved Spectroscopy in Biochemistry and Biology" (R. B. Cundall and R. E. Dale, eds.), Vol. 69, p. 23. Plenum Press, New York, 1983.

[45] M. R. Stankovic, D. Maulik, W. Rosenfeld, P. G. Stubblefield, A. D. Kofinas, E. Gratton, M. A. Franceschini, S. Fantini, and D. Hueber, *J. Matern. Fetal Med.* **9,** 142 (2000).

[46] E. Gratton and R. Lopez-Delgado, *Il Nuovo Cimento* **56B,** 110 (1980).

[47] E. Gratton, D. M. Jameson, N. Rosato, and G. Weber, *Rev. Sci. Instrum.* **55,** 486 (1984).

can also be modulated externally by using electro-optical devices). For example, the fundamental frequency of the original source at the Frascati Electron Storage Ring, ADONE (operating in the single bunch mode), was 2.886167 MHz, which then gave rise to a set of harmonic frequencies (a comb function) set 2.886167 MHz apart. The intensities characteristic of this frequency set fit a Gaussian envelope with a half-width of 500 MHz (the inverse of 2 ns, the FWHM of the Gaussian pulse width). After the original proof of concept at Frascati, the harmonic content approach to phase and modulation measurements was used in conjunction with a mode-locked laser-based instrument.[48]

The basic principles of both frequency domain and time domain approaches to fluorescence lifetime determinations have been reviewed many times and we refer readers with a sustaining interest to these accounts (e.g., see Refs. 7, 24, and 49–51). We shall, instead, remind the reader of a fundamental aspect of lifetime determinations that is relevant to either method. Specifically, we refer to the use of the so-called magic angles. This effect was discussed by Spencer and Weber,[52] who drew attention to the fact that observation of the fluorescence at 90° to the excitation will weigh the vertically polarized component relative to horizontally polarized components (because when viewed in this manner one of the horizontal components cannot be observed), so that the measured lifetime will deviate slightly from the true molecular lifetime. This effect is usually small but can, in extreme cases, be as much as 10–15%.[52] This artifact, however, can be avoided through the use of specific conditions for excitation and emission. Namely, four "magic angle" conditions that eliminate polarization bias are as follows: (1) 35° excitation, no emission polarizer; (2) natural excitation, 35° emission polarizer; (3) 55° excitation polarizer, parallel (0°) emission polarizer; and (4) parallel (0°) excitation polarizer, 55° emission polarizer. In practice, at least with the frequency domain approach, researchers typically use method 4 because the modulated excitation is usually generated with a Pockels cell and a polarizer, which will produce a parallel polarization. The advantage of this method is also that the sample and reference signals (used to generate the phase shift and demodulation ratio) both will be viewed through a polarizer oriented at 55°, which can eliminate any polarization bias in the PMT.

As time-resolved instrumentation improved over the years, so did the mathematical approaches to analysis. In the 1970s, a tremendous effort was expended on development of methods to deconvolute and fit pulse decay data; many of these

[48] R. J. Alcala, E. Gratton, and D. M. Jameson, *Anal. Instrum.* **14,** 225 (1985).

[49] J. N. Demas, "Excited State Lifetime Measurements." Academic Press, New York, 1983.

[50] D. M. Jameson, E. Gratton, and R. D. Hall, *Appl. Spectrosc. Rev.* **20,** 55 (1984).

[51] D. M. Jameson and T. L. Hazlett, *in* "Biophysical and Biochemical Aspects of Fluorescence" (G. Dewey, ed.), p. 105. Plenum Press, New York, 1991.

[52] R. D. Spencer and G. Weber, *J. Chem. Phys.* **52,** 1654 (1970).

methods are discussed in the book that resulted from the 1978 NATO conference,[53] which was one of the first large international meetings devoted exclusively to time-resolved fluorescence. As flash lamps began to give way to shorter and shorter laser pulses, however, many of these deconvolution methods became moot because the majority of the systems being studied had lifetimes that were long, compared with the duration of the exciting pulse. The present availability of femtosecond pulses means that special deconvolution techniques are required only in rare cases.

Analysis of lifetime data has, in fact, been developing in other directions than originally envisaged at that NATO conference. For example, software has been developed (largely from the Gratton laboratory[54–56]) that allows the fitting of lifetime data to distribution models, as opposed only to sums of discrete exponentials. The original motivation for this development was to try to rationalize the lifetime heterogeneity observed in single-tryptophan proteins in terms of dynamic aspects of the protein matrix. Also, both frequency domain and time domain approaches have taken advantage of global analysis approaches.[57,58] These approaches allow the linking of various lifetime properties among sets of data, which often reveals trends and associations that were not originally apparent and that often significantly improve the precision of the resolved components. Another relatively recent approach is the maximum entropy method (MEM), which uses "model-less" fits of the lifetime data (usually time domain data) to reveal the underlying decay kinetics.[59–61] The choice of whether to use either the MEM approach or direct model-fitting approaches[50,54–56,58] is actually more of a philosophical choice because some people prefer to evaluate their data against a particular model whereas others prefer to avoid the constraints of models altogether.

Finally, we should note that time-resolved methods, using essentially the same instrumentation used to acquire intensity decay data, have also been developed to acquire direct hydrodynamic information about fluorescent systems. In other words, rotation properties of fluorophores and the system they monitor can be obtained by so-called time-decay anisotropy or dynamic polarization measurements. These methods have been reviewed extensively[7,24,51,62] and we shall not discuss them further.

[53] R. B. Cundall and R. E. Dale, eds., "NATO ASI Series, Series A: Life Sciences," Vol. 69. Plenum Press, New York, 1983.

[54] J. R. Alcala, E. Gratton, and F. G. Prendergast, *Biophys. J.* **51,** 925 (1987).

[55] J. R. Alcala, E. Gratton, and F. G. Prendergast, *Biophys. J.* **51,** 597 (1987).

[56] J. R. Alcala, E. Gratton, and F. G. Prendergast, *Biophys. J.* **51,** 587 (1987).

[57] J. M. Beechem, J. R. Knutson, and L. Brand, *Biochem. Soc. Trans.* **14,** 832 (1986).

[58] J. M. Beechem, M. Gratton, J. R. Knutson, and W. W. Mantulin, *in* "Topics in Fluorescence Spectroscopy II" (J. R. Lakowicz, ed.), p. 241. Plenum Press, New York, 1991.

[59] A. K. Livesey and J.-C. Brochon, *Biophys. J.* **52,** 693 (1987).

[60] J.-C. Brochon, A. K. Livesey, J. Pouget, and B. Valeur, *Chem. Phys. Lett.* **174,** 517 (1990).

[61] J.-C. Brochon, *Chem. Phys. Lett.* **174,** 517 (1990).

[62] E. Gratton, D. M. Jameson, and R. D. Hall, *Annu. Rev. Biophys. Bioeng.* **13,** 105 (1984).

Polarization/Anisotropy

The linear polarization (p) of a beam of light may be defined as the ratio of the intensity of the polarized light (P) to the total light, that is, natural (N) plus polarized, N being characterized by $P = 0$.

$$p = \frac{P}{P + N} \tag{6}$$

This definition presupposes the absence of elliptically polarized light in the light beam, which can then be considered for our purpose as a mixture of natural and linearly polarized light. This case is entirely correct for the radiation emitted by optically inactive molecules and all but correct for emission by an optically active molecule, as in this case the fraction of circularly polarized light in the emission does not generally exceed one-thousandth of the total.

If $I_{\parallel}$ is defined as the intensity seen through a polarizer set to maximally transmit P, and $I_{\perp}$ is at 90° to this direction, thus excluding P, then

$$I_{\parallel} - I_{\perp} = P \tag{7}$$

$$I_{\parallel} + I_{\perp} = P + N \tag{8}$$

and hence

$$p = \frac{I_{\parallel} - I_{\perp}}{I_{\parallel} + I_{\perp}} \tag{9}$$

Thus, the linear polarization, or fraction of linearly polarized light in a beam, is measured by setting a polarizer in two positions at 90° from each other, one of these two positions being set so as to obtain maximum intensity. The sign of the polarization is a matter of convention. It is possible to speak of positive and negative polarizations of the fluorescence because $I_{\parallel}$ is chosen to coincide with the electric vector of the exciting light and the definition of p of Eq. (9) is used in every case. The polarization of the fluorescence from a source is regarded as positive if the greater fluorescence intensity is polarized parallel to the polarization of the excitation, that is, $I_{\parallel} > I_{\perp}$ and negative if the greater fluorescence intensity is polarized normal to the polarization of the excitation that is, $I_{\parallel} < I_{\perp}$. As a matter of practical significance, to properly determine polarization values the researcher must correct for any bias in the detector arm (either due to monochomator or PMT response) and this operation is accomplished by measuring the parallel and perpendicular polarized intensities of the emission while exciting with perpendicularly polarized light, that is, in a direction that is orthogonal to both emission components (due to the right-angle observation). This symmetrical excitation arrangement means that the two emission polarization components should be equal, and hence any deviation from equality can be ascribed to a bias and corrected. Although this principle was recognized early on in polarization research it is now usually referred to as the

"G factor" after the nomenclature of Azumi and McGlynn,[63] who used this term to refer to the bias of the grating in the emission monochromator.

The polarization of dilute solutions of fluorescein in water was first reported by F. Weigert in 1920,[64] who pointed out that the polarization value depended on the molecular size of the fluorophore and the viscosity of the medium. The correct quantitative relation between the observed polarization and parameters such as the excited state lifetime, the size of the fluorophore, and the viscosity of the solution was enunciated by F. Perrin in 1926[65] (for an excellent discussion of the outstanding contributions of J. and F. Perrin to fluorescence the reader is directed to the article by Berberan-Santos[5]). In this classic 1926 article, the equation, now known as the Perrin equation, first appears as

$$p = p_0 \frac{1}{1 + \left(1 - \frac{1}{3} p_0\right) \frac{RT}{V\eta} \tau} \tag{10}$$

where p is the polarization, p_0 is the limiting polarization (in the absence of rotation), R is the universal gas constant, T is the absolute temperature, V is the molar volume of the rotating unit, η is the solvent viscosity, and τ is the excited state lifetime. Equation (10) is now usually written as

$$\frac{1}{p} - \frac{1}{3} = \left(\frac{1}{p_0} - \frac{1}{3}\right)\left(1 + \frac{RT}{V\eta}\tau\right) \tag{11}$$

This expression is often further simplified to

$$\frac{1}{p} - \frac{1}{3} = \left(\frac{1}{p_0} - \frac{1}{3}\right)\left(1 + \frac{3\tau}{\rho}\right) \tag{12}$$

where ρ is the Debye rotational relaxation time, which for a sphere is given as

$$\rho_0 = \frac{3\eta V}{RT} \tag{13}$$

In the case of a protein wherein the partial specific volume (ν) and hydration (h) are known, Eq. (14) can then be written

$$\rho_0 = \frac{3\eta M(\nu + h)}{RT} \tag{14}$$

where M is the molecular weight, ν is the partial specific volume, and h is the degree of hydration. For a spherical protein of molecular mass 44 kDa, with a

[63] T. Azumi and S. P. McGlynn, *J. Phys. Chem.* **37,** 2413 (1962).

[64] F. Weigert, *Verh. Dtsch. Chem. Ger.* **23,** 100 (1920).

[65] F. Perrin, *J. Phys.* **7,** 390 (1926).

partial specific volume of 0.74 and hydration of 0.3 ml/mg at room temperature, $\rho = (3)(0.01)(44000)(0.74 + 0.3)/(8.31 \times 10^7)(293) = {\sim}56$ ns. So to a rough approximation, the Debye rotational relaxation time (in nanoseconds) for a spherical protein is close to its molecular mass. Of course, for nonspherical particles, which may be approximated as hydrodynamically equivalent prolate or oblate ellipsoids, the equation is more complicated and considers the harmonic mean of the rotational relaxation times about the three principle axes of rotation.[9,66] We should comment here on the term "rotational correlation time," often denoted as τ_c. In fact, $\rho = 3\tau_c$, a fact that stems from the original definitions of these terms. As pointed out previously,[67] the rotational relaxation time was originally defined by Debye in relation to dielectric dispersion, in which the relevant orientational distribution is a function of cos θ, where θ is the angle between the direction of the electric field and the molecular dipole axis. Perrin used this approach of Debye when deriving the characteristic time for molecular rotation of spheres. Much later, Bloch derived equations for computing the decay of nuclear polarizations and used a slightly different approach, which yielded a characteristic time that was one-third the Debye rotational relaxation time, that is, the correlation time. It is, in fact, possible to use the function known as anisotropy (r), which is defined as

$$r = \frac{I_{\parallel} - I_{\perp}}{I_{\parallel} + 2I_{\perp}} \tag{15}$$

and recast the Perrin equation as

$$\frac{r}{r_0} = 1 + \frac{\tau}{\tau_c} \tag{16}$$

where r_0 is the limiting anisotropy and τ_c is the rotational correlation time as discussed above. The simplified form of this equation, compared with the original Perrin equation, has attracted users. We simply want to stress that the information content of the various terms, polarization/anisotropy or relaxation time/correlation time, is identical and the most important consideration is to clearly specify which terms are being used. We also leave to the reader to prove, given the definition of anisotropy, that the following two equations hold:

$$r = \frac{2}{3}\left(\frac{1}{P} - \frac{1}{3}\right)^{-1} \tag{17}$$

$$r = \frac{2P}{3 - P} \tag{18}$$

[66] D. M. Jameson and S. E. Seifried, *Methods* **19,** 222 (1999).
[67] D. M. Jameson and W. H. Sawyer, *Methods Enzymol.* **246,** 283 (1995).

Francis Perrin had actually also introduced the quantity $2p/(3-p)$,[5,68] and this function was later named "anisotropy" by Jabłoński.[69] Curiously, Jabłoński never acknowledged the use by Perrin of this function. Similarly, when Jabłoński published in 1960 on the additivity of anisotropy,[70] he neglected to cite the 1952 article by G. Weber,[9] which explicitly gave the formulation for the additivity of polarization, and from which the Jabłoński equation followed in a trivial fashion.

Polarization measurements are usually used to acquire information about the rotational motions of probes, either to gain knowledge about the viscosity of the medium or to understand hydrodynamic aspects of the system. This latter consideration has been widely applied to study macromolecular associations (such as protein–protein or protein–nucleic acid) as well as ligand binding.[66,67] In fact, one of the most widely used techniques in clinical assays for drugs or metabolites is the fluorescence polarization immunoassay, using the Abbott (Abbott Park, IL) TDx instrument (www.abbottdiagnostics.com). Interestingly, in this approach Abbott adopted the nomenclature of the "millipolarization" unit, a term that is also now widely used in fluorescence plate readers and that is, simply, the polarization times 1000. When the TDx system first came out in the early 1980s one of us (D.M.J.) asked some of the principal developers of the system why the term millipolarization was used, as, of course, polarization is a unitless number. The response was that many of the potential end-users of the TDx system were concerned about the traditional nomenclature because changes in polarization of, say, 0.20 to 0.30 did not seem very large. But when these numbers were given as millipolarization units, that is, 200 to 300 mps, then these same end-users were impressed!

As discussed below, polarization data can also be used to study energy transfer processes. One other use of polarization data seems worthy of comment, however, because it lies at the heart of the method. Namely, the so-called limiting polarization (P_0 in the Perrin equation), that is, the polarization in the absence of rotation or energy transfer, gives a measure of the angle between the absorption and emission dipoles. As derived elsewhere,[7,71] the limits for polarization from a solution of randomly oriented fluorophores are $+1/2$ and $-1/3$, depending on whether the absorption and emission dipoles are collinear ($+1/2$) or orthogonal ($-1/3$). The expression relating the limiting polarization to this angle is

$$\frac{1}{P_0} - \frac{1}{3} = \frac{5}{3}\left(\frac{2}{3\cos^2\phi - 1}\right) \tag{19}$$

This method is, in fact, one of the few that can be used to gain knowledge of the relative orientation of these dipoles (see, e.g., Valeur[7] for some examples). The

[68] F. Perrin, *Acta Phys. Polon.* **5,** 335 (1936).

[69] A. Jabłoński, *Acta Phys. Polon.* **17,** 471 (1957).

[70] A. Jabłoński, *Bull. Acad. Polon. Sci. Serie Sci. Math. Astron. Phys.* **6,** 259 (1960).

[71] G. Weber, *in* "Fluorescence and Phosphorescence" (D. Hercules, ed.), p. 217. John Wiley & Sons, New York, 1966.

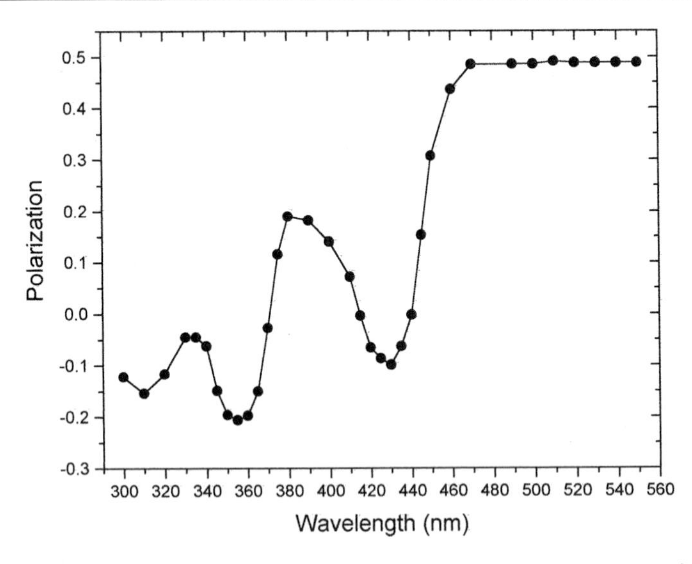

FIG. 10. Excitation polarization spectra of rhodamine B embedded in a Lucite matrix at room temperature. Emission was viewed through a cut-on filter passing wavelengths longer than 560 nm; slits were ~4 nm. The instrumentation utilized is described in D. M. Jameson, G. Weber, R. D. Spencer, and G. Mitchell, *Rev. Sci. Instrum.* **49,** 510 (1978).

basic idea is that as the excitation wavelength is varied and different electronic states are reached (Fig. 3), the direction of the corresponding absorption transition dipoles varies and, because emission is always from the same dipole regardless of which level is initially excited, the observed polarization will vary depending on the original angle between the absorption and emission dipoles. This fact means, of course, that care must be exercised concerning which excitation wavelength is used for polarization studies—a point often overlooked because most practitioners simply excite into the final absorption band, which will usually have the highest P_0. As shown in Fig. 10, for the case of immobilized rhodamine B, the variation of polarization with wavelength (known as the excitation polarization spectrum) can be profound. Yet, if the limiting polarization at the wavelength utilized is known, the same type of measurements can be carried out even if the limiting polarization is negative (see, e.g., VanderMeulen *et al.*[72]). Finally, we should mention that the polarization across the emission spectrum is almost always constant (which is why cut-on filters can be used to collect the entire emission and hence gain in sensitivity), but a few cases occur (e.g., pyrene) wherein the polarization will vary with emission wavelength and hence the choice of observation wavelength or filter is critical.

[72] D. L. VanderMeulen, D. G. Nealon, E. Gratton, and D. M. Jameson, *Biophys. Chem.* **36,** 177 (1990).

Fluorescence Resonance Energy Transfer, Specifically Fluorescence Resonance Energy Transfer between Identical Molecules

As mentioned earlier, a wealth of articles dealing with heterotransfer have appeared, and readers interested in this area may consult various sources.[7,24,73–77] Energy transfer was noted as early as 1922 by Cario and Franck,[78] who observed that irradiation of mixtures of mercury and thallium vapors, with a wavelength absorbed by mercury but not by thallium, resulted in emission from both atoms. Hence, a nonradiative transfer mechanism was required. One topic in energy transfer that has not been extensively discussed, however, is homo-FRET (fluorescence resonance energy transfer between identical molecules). In the case of homo-FRET, as opposed to hetero-FRET, the transfer efficiency cannot be quantified by observation of the decrease in the donor intensity (or lifetime) and/or increase in the acceptor intensity, because donor and acceptor are the same. Nonetheless, transfer efficiency, and hence distance, information can be obtained by monitoring the polarization of the emission. One of the few reviews in this area was provided in 1983 by Kawski,[79] who dealt mainly with homotransfer between fluorophores in isotropic media. Valeur also considers homo-FRET.[7] We should also mention that the use of the word "fluorescence" in the phrase "fluorescence resonance energy transfer" is somewhat of a misnomer because the "fluorescence" itself is not transferring.

The first report on homo-FRET was by Gaviola and Pringsheim,[80] who noted that an increase in the concentration of dyes in viscous solvent was accompanied by a progressive depolarization, even at concentrations where self-quenching was still negligible. Francis Perrin recognized this phenomenon as a special case of fluorescence energy transfer: *"L'existence de transferts d'activation est expérimentalement prouvée pour de telles molécules par la décroissance de la polarisation de la lumière de fluorescence quand la concentration croit (aucune dépolarisation ne peut résulter dans une solution très visqueuse d'une rotation des molécules, qui exigerait, pour se produire en un temps de l'ordre de la durée moyenne d'émission, une énergie plus grande que l'énergie d'activation toute entière)"*. ["The existence

[73] H. C. Cheung, *in* "Topics in Fluorescence Spectroscopy" (J. R. Lakowicz, ed.), Vol. 3, p. 127. Plenum Press, New York, 1991.

[74] R. M. Clegg, *in* "Fluorescence Imaging Spectroscopy and Microscopy" (X. F. Wang and B. Herman, eds.), Vol. 137, p. 179. John Wiley & Sons, New York, 1996.

[75] C. G. dos Remedios and P. D. J. Moens, *in* "Resonance Energy Transfer" (D. L. Andrews and A. A. Demidov, eds.), p. 1. John Wiley & Sons, New York, 1999.

[76] B. W. Van Der Meer, G. Coker, and S.-Y. S. Chen, "Resonance Energy Transfer: Theory and Data." Wiley-VCH, New York, 1991.

[77] P. Wu and L. Brand, *Anal. Biochem.* **218,** 1 (1994).

[78] G. Cario and J. Franck, *Z. Physik* **11,** 161 (1922).

[79] A. Kawski, *Photochem. Photobiol.* **38,** 487 (1983).

[80] E. Gaviola and P. Pringsheim, *Z. Physik* **24,** 24 (1924).

of transfer of activation is proven experimentally for such molecules by the decrease in polarization of the fluorescent light when the concentration is increased (in a very viscous solution, no depolarization can result from the rotation of the molecules, which would require, to occur in a time of the order of the mean lifetime of the emission, a greater energy than the whole activation energy).”] [81] In this article, F. Perrin gave a quantum mechanical theory of energy transfer between like molecules in solution. He also presented a qualitative discussion of the effect of the spectral overlap between the emission spectrum of the donor and the absorption spectrum of the acceptor (in the case of like molecules), which could explain the difference between the calculated distance at which the transfer of activation occurs and the distance inferred from the polarization observations. *“Il est vraisemblable que cet écart est dû à l'étalement des fréquences d'absorption et d'émission, sur des bandes spectrales assez larges n'empiétant qu'assez peu l'une sur l'autre. Cet étalement, dû aux possibilités de vibrations mécaniques internes des molécules, et au couplage très fort de ces vibrations avec l'agitation thermique du solvant, diminue considérablement la grandeur des moments associés aux transitions à la fois pour les deux molécules en présence.”* [“It is reasonable to think that this gap is due to the spreading of the absorption and emission frequency, over spectral bands sufficiently large and overlapping only slightly. This spreading, resulting from possible internal mechanical vibrations, and from the very strong coupling of these vibrations to the thermal agitation of the solvent, considerably reduces the size of the moment associated with the possible simultaneous transitions for the two molecules considered.”][81] Several years later, Förster[82,83] published the first quantitative theory of molecular resonance energy transfer.

Depolarization Due to Energy Transfer

At the Tenth Spiers Memorial Lecture of the Faraday Society, Förster reported that “excitation transfer between alike molecules can occur in repeated steps. So the excitation may *migrate* from the absorbing molecule over a considerable number of other ones before deactivation occurs by fluorescence or some other process. Though this kind of transfer cannot be recognized from fluorescence spectra, it may be observed by the decrease of fluorescence polarization. . . . ”[84]

Depolarization of fluorescence can occur either because of Brownian rotations (discussed earlier) or because of energy transfer (Fig. 11); the excited fluorophore transfers energy, in a radiationless process, to an acceptor (with absorption and emission dipole moments oriented differently from the donor molecule), which

[81] F. Perrin, *Ann. Phys.* **XVII,** 283 (1932).

[82] T. Förster, *Naturwissenschaften* **6,** 166 (1946).

[83] T. Förster, *Ann. Phys. (Leipzig)* **2,** 55 (1948).

[84] T. Förster, “Discussions of the Faraday Society,” Vol. 27, p. 7. Aberdeen University Press, Aberdeen, Scotland, 1960.

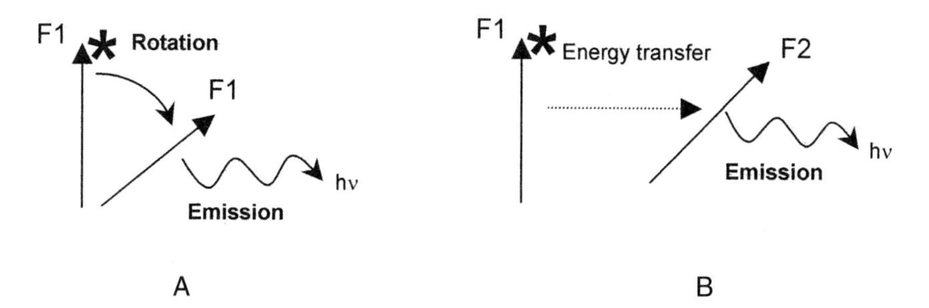

FIG. 11. (A) Depolarization resulting from rotational diffusion of a fluorophore. (B) Depolarization resulting from homo-FRET.

in turn emits a photon. Of course, these two processes can occur simultaneously and hence may both contribute to depolarization of the emitted light. Unlike heterotransfer, and apart from the fact that multiple energy transfer events can occur before an acceptor molecule finally emits the photon, back transfer can also occur from the excited acceptor molecule to the originally excited donor, a process that limits the depolarization of the emitted light. If the rate of transfer from the donor molecule to the acceptor is equal to the rate of back transfer from the acceptor molecule to the donor molecule, the minimum value of anisotropy reached, in the case of 100% transfer efficiency, is roughly equal to the anisotropy in the absence of energy transfer divided by the number of chromophores (see below). However, the rate of back transfer is not necessarily equal to the rate of forward transfer. For instance, energy transfer from fluorophores with higher quantum yields, and therefore longer lifetimes, to those with smaller quantum yields and shorter lifetimes will occur more often than the converse process.

Homo-FRET does not, to a first approximation, result in changes in the overall fluorescence intensities and lifetimes of the fluorophores. Also, unlike depolarization due to Brownian motions, depolarization from homo-FRET is temperature independent. For some fluorophores (see below), there is a failure of energy transfer when the fluorophores are excited on the red edge of their excitation band[85,86] (also reviewed in Valeur[7]).

Homo-FRET for Fluorophores in Isotropic Media

When homo-FRET occurs in viscous solution, it can be detected by the concentration depolarization of fluorescence.[80] In 1954, Weber showed that the reciprocal of the polarization is a linear function of the concentration of the fluorescent

[85] G. Weber, *Biochem. J.* **75,** 335 (1960).
[86] G. Weber and M. Shinitzky, *Proc. Natl. Acad. Sci. U.S.A.* **65,** 823 (1970).

molecules.[87] Weber also showed[71,87] that it was possible to calculate the distance R_0 (i.e., the critical Förster distance; the distance at which the probability of emission equals the probability of transfer) by the following equations:

$$\frac{1}{P} - \frac{1}{3} \approx \left(\frac{1}{P_\infty} - \frac{1}{3}\right)\left[1 + 1.68\left(\frac{R_0}{2a}\right)^6 C\right] \tag{20}$$

where P is the polarization at concentration C, P_∞ is the polarization at infinite dilution, and $2a$ is the distance of closest approach of any molecule to the excited molecule (a is the molecular radius). R_0 can be obtained from the slope of the straight line obtained (in a thin layer in which radiative transfer is negligible) when $(1/P) - (1/3)$ is plotted against the concentration C. Depolarization in thick layers gives rise to an upward curvature of the line (instead of a straight line) when $(1/P) - (1/3)$ is plotted against the concentration because of the radiative transfer in more concentrated solution, which increases the degree of depolarization. This approach was used to calculate the R_0 values for various fluorophores in solution.[71,85,87]

Another approach was described by Kawski and Nowaczyk.[88] They generalized the simple active sphere model proposed by Jabłoński[89] to take into account the mutual orientation of the donor and acceptor transition dipole moment. In this treatment, a luminescent center is assumed to consist of donor surrounded by a shell of volume V within which a number of identical unexcited molecules (acceptors) can be found [$V = (4/3)\pi(R_s^3 - a^3)$, where a is the minimum separation at which the contribution of the Förster interaction to the excitation energy transfer is maximum and R_s is the effective range of the Förster interaction[79]]. Assuming that the rate of transfer is equal to τ^{-1}, the normalized anisotropy is given by

$$\frac{r}{r_0} = \frac{2(\nu - 1 + e^{-\nu})}{\nu^2} \tag{21}$$

where the mean number of acceptors (ν) within the shell is expressed by

$$\nu = \frac{4}{3}\pi R_s^3\left[1 - \langle\kappa^2\rangle\left(\frac{R_0}{R_s}\right)^6\right]n \tag{22}$$

where n is the fluorophore concentration per milliliter and κ^2 is the well-known orientation factor (see, e.g., Van Der Meer *et al.*[76]).

Kawski[79] calculated R_0 for several fluorophores, using this modified simple sphere model, and showed that there was good agreement between the calculated values and those obtained from the absorption and emission spectral overlap of these dyes.

[87] G. Weber, *Trans. Faraday Soc.* **50,** 552 (1954).
[88] A. Kawski and K. Nowaczyk, *Acta Phys. Polon.* **A54,** 777 (1978).
[89] A. Jabłoński, *Acta Phys. Polon.* **14,** 295 (1955).

For two-dimensional systems, the mean number of acceptor molecules is given by Eq. (23)[88]:

$$\nu = \pi R_s^2 \left\{ 1 - \frac{R_0^3 \langle \kappa^2 \rangle}{6R_s^6} \left[\frac{R_0^3}{2} + \left(\frac{R_0^6}{4} + \frac{2R_s^6}{\langle \kappa^2 \rangle} \right)^{1/2} \right] \right\} n \tag{23}$$

Homotransfer in Proteins and Peptides

Weber[90] was the first to study the polarization spectra of proteins containing tyrosine and/or tryptophan. He showed that homo-FRET between tyrosine residues was responsible for the low polarization values of the proteins compared with the polarization values of tyrosine solutions. Similarly, in proteins containing tryptophan, Weber noted that the polarization values at 270 nm were lower than that of *N*-glycyltryptophan. Homotransfer was also found to be the sole factor contributing to the decrease in polarization with the increase in the number of 1-anilino-8-naphthalene sulfonate molecules adsorbed on the same bovine serum albumin molecule.[91–93]

The homotransfer method was applied to protein oligomers to study dissociation and subunit exchange under pressure.[94–96] Both heterotransfer and homotransfer permit qualitative detection of the dissociation and subunit exchange of the oligomers, but only homotransfer allows a direct quantitative estimate of both processes.[94] When labeled proteins are in the presence of a large excess of unlabeled proteins and the pressure is raised from atmospheric to that required for half-dissociation, the polarization increases because of two different processes: (1) dissociation (which places the dissociated fraction at distances at which transfer is negligible) and (2) exchange of labeled subunits with unlabeled subunits. Assuming the independence of these two causes of change in polarization,[94]

$$r_x(t) = r_1 + (r_s - r_1)\left(1 - \exp^{-t/\tau_x}\right) \tag{24}$$

$$r_\alpha(t) = r_1 + (r_s - r_1)\left(1 - \exp^{-t/t_1}\right) \tag{25}$$

where $r_\alpha(t)$ is the anisotropy change due to dissociation with a time constant t_1, $r_x(t)$ is the anisotropy change following the replacement of fully labeled oligomers by others carrying a single labeled subunit with a characteristic time (τ_x) for reduction of the unscrambled subunit fraction to e^{-1} of its original value, r_1 is the observed decreased emission anisotropy, and r_s characterizes the emission

[90] G. Weber, *Biochem. J.* **75,** 345 (1960).
[91] G. Weber and L. Young, *J. Biol. Chem.* **239,** 1415 (1964).
[92] G. Weber and E. Daniel, *Biochemistry* **5,** 1900 (1966).
[93] G. Weber and S. R. Anderson, *Biochemistry* **8,** 361 (1969).
[94] L. Erijman and G. Weber, *Biochemistry* **30,** 1595 (1991).
[95] L. Erijman and G. Weber, *Photochem. Photobiol.* **57,** 411 (1993).
[96] K. Ruan and G. Weber, *Biochemistry* **32,** 6295 (1993).

anisotropy of both dissociation and complete exchange. The observed emission anisotropy $r(t)$ is a weighted average of $r_\alpha(t)$ and $r_x(t)$:

$$r(t) = \alpha r_\alpha(t) + (1 - \alpha) r_x(t) \tag{26}$$

Self-Association of Peptides within Lipid Bilayer

MacPhee *et al.*[97] showed evidence of peptide self-association, resulting from an increase in the peptide-to-surface density, using homotransfer between amphipathic α-helices derived from apolipoprotein C-II labeled with 7-nitrobenz-2-oxa-1,3-diazole (NBD). They followed the work of Wobler and Hudson,[98] which predicted the energy transfer efficiencies for heterotransfer between donors and acceptors randomly distributed on a two-dimensional surface as a function of fluorophore surface density. Several assumptions were made to accommodate for homotransfer.

1. *The fluorescence anisotropy is primarily determined by the initially excited donor molecule and a single transfer event is sufficient to cause complete depolarization of the light emitted from the acceptor.* This assumption followed from the work of Agranovich and Galanin,[99] who calculated that when the electronic transition dipoles of two molecules are randomly oriented the emission anisotropy after a single energy transfer event is close to 0, when the absorption and emission dipoles are parallel and the probes do not rotate. Also, in the case of randomly oriented chromophores undergoing efficient energy transfer, the acceptor steady-state anisotropy is expected to be close to 0.[100]
2. *The orientational distribution of the probes is random.*
3. *The distance of closest approach (R_e) between donor and acceptor is less than the Förster critical distance (R_0).*

The efficiency of transfer can be calculated from the expression for the decay of the intensity parallel and perpendicular to the excitation direction[101]:

$$I(t)_{\text{par}} = \left(\frac{I_0}{3}\right)[1 + r_{01}(1 + e^{-Kt}) + r_{02}(1 - e^{-Kt})]e^{-\Gamma t} \tag{27}$$

$$I(t)_{\text{perp}} = \left(\frac{I_0}{3}\right)\left[1 + \frac{r_{01}(1 + e^{-Kt})}{2} - \frac{r_{02}(1 - e^{-Kt})}{2}\right]e^{-\Gamma t} \tag{28}$$

[97] C. E. MacPhee, G. J. Howlett, W. H. Sawyer, and A. H. Clayton, *Biochemistry* **38,** 10878 (1999).
[98] P. K. Wobler and B. S. Hudson, *Biophys. J.* **28,** 197 (1979).
[99] V. M. Agranovich and M. D. Galanin, "Electronic Excitation Energy Transfer in Condensed Matter." North-Holland Publishing, New York, 1982.
[100] M. N. Berberan-Santos and B. Valeur, *J. Chem. Phys.* **95,** 8048 (1991).
[101] B. D. Hamman, A. V. Oleinikov, G. G. Jokhadze, R. R. Traut, and D. M. Jameson, *Biochemistry* **35,** 16680 (1996).

where I_0 is the initial intensity, r_{01} and r_{02} are, respectively, the anisotropy decays of the donors and acceptors only, Γ is the fluorescence decay rate, and K is the rate of transfer between donor and acceptor. On integration, and assuming that the rate of back transfer is equal to the rate of forward transfer, a steady-state expression for the efficiency of transfer E is obtained by

$$E = \frac{2(r_{01} - \langle r \rangle)}{r_{01}} \tag{29}$$

where $\langle r \rangle$ is the observed anisotropy (assuming $\langle \kappa^2 \rangle = 2/3$).

Using the Wobler and Hudson equation[98]

$$E = 1 - \left(A_1 \exp^{-k_1 c} + A_2 \exp^{-k_2 c} \right) \tag{30}$$

where A_1, A_2, k_1, and k_2 are constants that depend on the ratio Re/R_0.[98] MacPhee *et al.*[97] showed that the data obtained for their peptide could not be fitted to values of energy transfer expected from fluorophores randomly distributed and oriented on a planar surface. Therefore, they concluded that the high transfer efficiencies they observed were the result of self-association of the peptides on the lipid surface.

Evaluation of Number of Subunits in Oligomer

Runnels and Scarlata[102] described general equations to calculate the expected anisotropy for complexes composed of varying numbers of labeled subunits and applied them to study the oligomerization of the Gag and matrix domains of human immunodeficiency virus type 1 (HIV-1).[103] In the simplest case in which the distances between the fluorophores in the cluster are equal, there is no change in the lifetime of the individual fluorophores, and the anisotropy following a single transfer event is equal to 0, the anisotropy for N molecules (r_ν) can be calculated as

$$r_N = r_1 \left(\frac{1 + F\tau}{1 + NF\tau} \right) \tag{31}$$

with the Förster transfer rate, F, assuming that the rates of forward and back transfer are equal ($F = F_{jk} = F_{kj}$):

$$F = \frac{1}{\tau} \left(\frac{R_0}{R} \right)^6 \tag{32}$$

The authors demonstrated that it is possible to determine the number of subunits in a cluster for values of $R/R_0 < 0.8$ (R being the distance separating the centers of the donor and acceptor fluorophores). Blackman *et al.*,[104] on the other hand,

[102] L. W. Runnels and S. F. Scarlata, *Biophys. J.* **69,** 1569 (1995).
[103] S. Scarlata, L. S. Ehrlich, and C. A. Carter, *J. Mol. Biol.* **277,** 161 (1998).
[104] S. M. Blackman, D. W. Piston, and A. H. Beth, *Biophys. J.* **75,** 1117 (1998).

used a Monte Carlo numerical approach to study the oligomeric state of erythrocyte anion-exchange protein (band 3) and showed that the anisotropy decay was consistent with a dimeric and/or tetrameric protein.

Distance Estimation by Homotransfer

Using Eq. (29) and Förster's equation,

$$E = \frac{R_0^6}{\left(R_0^6 + R^6\right)} \tag{33}$$

relating the efficiency of transfer E to the critical distance R_0 and R, the distance between the centers of the donor and acceptor fluorophores,[83,101] Hamman *et al.* used homo-FRET to study the separation between subunit domains of the dimeric ribosomal stalk protein L7/L12. Using cysteine mutations, placed by site-directed mutagenesis and labeled with fluorescein, they demonstrated that the C-terminal domains of the protein are, on average, significantly separated, contrary to the model proposed from the crystal structure of the C-terminal domain of the protein.[105] These homo-FRET studies also demonstrated rapid and facile subunit exchange between populations of L7/L12 dimers at concentrations significantly above the dimer/monomer dissociation constants.

It is interesting to note that distances between fluorophores can also be calculated from Eqs. (31) and (32) when the number N of fluorophores is known and for R/R_0 values <0.8 and <1.8. When the distance between two fluorophores is determined as a function of the anisotropy, using both methods [i.e., using Eqs. (29) and (33) or Eqs. (31) and (32)] and the results are compared (Fig. 12), the equations used by Hamman *et al.*[101] will give a slightly smaller distance between the fluorophores than the equations of Runnels and Scarlata.[102] Also, for short distances, when the energy transfer is maximum, and the rates of forward and back transfer are equal, the lowest value for the anisotropy between two fluorophores using both methods will be one-half that of the anisotropy without transfer.

Homotransfer in Vivo

Using a fusion protein between the herpes simplex virus thymidine kinase and the green fluorescent protein (GFP), Gautier *et al.*[106] demonstrated the presence of homotransfer within aggregates that formed in cells transfected with the

[105] M. Leijonmarck and A. Liljas, *J. Mol. Biol.* **195,** 555 (1987).

[106] I. Gautier, M. Tramier, C. Durieux, J. Coppey, R. B. Pansu, J. C. Nicolas, K. Kemnitz, and M. Coppey-Moisan, *Biophys. J.* **80,** 3000 (2001).

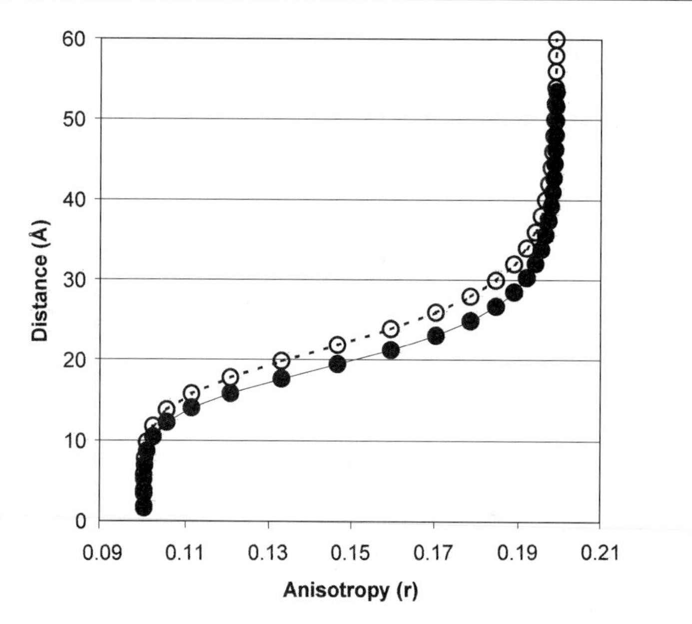

FIG. 12. Distance between two fluorophores calculated from anisotropy after FRET. Open circles represent the results obtained with Eqs. (31) and (32). Solid circles represent the results obtained with Eqs. (29) and (33). The limiting anisotropy was set to 0.2 and the Förster critical distance R_0 was 20 Å.

fusion protein. These researchers uses the equation described by Tanaka and Mataga[107]:

$$r(t) = \frac{3}{20}(2\cos^2\delta - \cos^2\theta_{ij} - \cos^2\theta_{ji})e^{-2\omega t} + \frac{1}{20}(6\cos^2\delta + 3\cos^2\theta_{ij} + 3\cos^2\theta_{ji} - 4) \quad (34)$$

with

$$\omega = \frac{3}{2}\langle\kappa^2\rangle\left(\frac{R_0}{R}\right)^6\tau^{-1} \quad (35)$$

where δ is the intramolecular angle between the absorption and emission transition moments, and θ_{ij} and θ_{ij} are the static mutual orientations of the transition moments of absorption to state i and of emission from state j, and between those of absorption to state j and emission from state i, respectively. Taking into account that (1) in GFP the emission and absorption dipole moments are parallel,[108]

[107] F. Tanaka and N. Mataga, *Photochem. Photobiol.* **29,** 1091 (1979).

[108] A. Volkmer, V. Subramaniam, D. J. Birch, and T. M. Jovin, *Biophys. J.* **78,** 1589 (2000).

(2) rotation of the protein dimer during the fluorescence lifetime is negligible, (3) there is no reorientation of the transition moment during the fluorescence lifetime because the chromophore of GFP is rigidly fixed inside the barrel, and (4) it was likely that the orientation of GFP chromophores was symmetrical within the dimer, Gautier *et al.*[106] calculated, for a dimer, the mutual orientation between the two chromophores and the upper limit of the distance separating the two chromophores.

Failure of Homotransfer on Excitation at Long-Wave Edge of Absorption Spectrum: Red Edge Effect

In 1960, in a study of homotransfer in solutions of tyrosine, tryptophan, and other compounds, Weber showed that concentrated solutions of indole in propylene glycol at $-70°$ did not show significant depolarization when excited at wavelengths longer than 305 nm, whereas depolarization was evident on excitation between 240 and 300 nm.[85] The ratio of the polarization on excitation at 305 and 270 nm (p_{305}/p_{270}), which varies from 1.4 to 1.7 in the simple indole derivatives, was found to be greater than 2 in 9 of the 10 tryptophan-containing proteins studied.[90] A similar observation showing the dependence of energy transfer on the wavelength of the exciting light was reported for 1-anilino-8-naphthalene sulfonate in propylene glycol at $-70°$ or when adsorbed to bovine serum albumin in aqueous solution.[109] In 1970, Weber and Shinitzky[86] studied this effect in more detail and found that in the many aromatic fluorophores they examined (including fluorescein, indole, etc.), unlike a few nonaromatic fluorophores (such as the antibiotic filipin, biacetyl, 4-hydro-*N*-methylnicotinamide, and *N*-butyltriazoline dione), transfer was much decreased or undetectable on excitation at the red edge of the absorption spectrum. More recently, Helms *et al.*[110] made use of homo-FRET and the red edge effect to demonstrate proximity between specific tryptophan residues in a multitryptophan protein. The red-edge drop in energy transfer efficiency (named Weber's red edge effect by Eisinger *et al.*[111]) is characteristic of homo-FRET and is not observed in the case of heterotransfer.[86,111] Berberan-Santos *et al.*[112] have proposed that the lack of energy transfer at the red edge is due to inhomogeneous spectral broadening and the resulting directed energy transfer. The absorption and emission spectrum can be represented as composed of a distribution of spectra corresponding to different fluorophore–solvent configurations. By exciting at the red edge, a subpopulation of fluorophores having a "red" spectrum can be preferentially excited over a subpopulation having a "blue" spectrum. Because the efficiency

[109] S. R. Anderson and G. Weber, *Biochemistry* **8,** 371 (1969).

[110] M. K. Helms, T. L. Hazlett, H. Mizuguchi, C. A. Hasemann, K. Uyeda, and D. M. Jameson, *Biochemistry* **37,** 14057 (1998).

[111] J. Eisinger, A. A. Lamola, J. W. Longworth, and W. B. Gratzer, *Nature (London)* **226,** 113 (1970).

[112] M. N. Berberan-Santos, J. Pouget, and B. Valeur, *J. Phys. Chem.* **97,** 11376 (1993).

of transfer is related to the spectral overlap between the emission spectrum of the donor and the absorption spectrum of the acceptor, the probability of transfer from a "blue" subpopulation to a "red" one is higher than the probability of transfer from a red subpopulation to a blue one. Therefore, on red edge excitation, as the excitation wavelength increases beyond the absorption maximum, the probability of energy transfer from the directly excited chomophore (red subpopulation) decreases because the proportion of the blue subpopulation to which transfer is weak or impossible drastically increases, resulting in the failure of energy transfer. An excellent discussion of red edge effects is given by Valeur.[7]

Summary

We hope that we have conveyed information of interest and value to present and future fluorescence practitioners. Those readers with a sustaining interest in this topic may wish to consult more comprehensive sources such as *Molecular Fluorescence: Principles and Applications,* an excellent text by Valeur,[7] or *Principles of Fluorescence Spectroscopy* by Lakowicz.[24] Many specialized fluorescence topics are covered in the series *Topics in Fluorescence Spectroscopy* (Volumes 1–6), and several volumes of *Methods in Enzymology* (e.g., Volumes 246 and 278) have dealt with issues in fluorescence spectroscopy. Proceedings from the International Conference on Methods and Applications of Fluorescence Spectroscopy, 1997 (MAFS 97)[113] and MAFS 98 (in press) also present fluorescence work on many different topics in biological and chemical fields. The *Molecular Probes Handbook* and web site (www.probes.com) are also rich sources of useful information. Finally, any reader with a question or seeking advice on some topic related to fluorescence is welcome to e-mail D.M.J. at djameson@hawaii.edu.

Acknowledgments

The authors wish to acknowledge support from the National Science Foundation (MCB9808427) and the American Heart Association (9950020N and 0151578Z). They also wish to sincerely thank Gerard Marriott, the long-suffering editor of this volume, for his vast patience.

[113] B. Valeur and J.-C. Brochon, eds., "New Trends in Fluorescence Spectroscopy." Springer-Verlag, Heidelberg, Germany, 2001.

[2] Fluorescence-Sensing Methods

By ZYGMUNT GRYCZYNSKI, IGNACY GRYCZYNSKI, and JOSEPH R. LAKOWICZ

Introduction

There have been significant advances in the methodology for fluorescence sensing.[1–11] Examples of some remarkable improvements in chemical and biomedical fluorescence-sensing technology are the introduction of time-resolved fluorescence sensing,[1,2] long-lifetime metal–ligand complexes with microsecond decay times,[12–15] and protein-based sensors.[16–19] Also, numerous fluorescence probes for detecting metal ions, chloride, calcium, phosphate, citrate, oxygen, and glucose[20–40] have been developed, widely expanding the applications of

[1] J. R. Lakowicz, "Principles of Fluorescence Spectroscopy," 2nd Ed. Kluwer Academic, New York, 1999.

[2] H. Szmacinski and J. R. Lakowicz, *in* "Topics in Fluorescence Spectroscopy" (J. R. Lakowicz, ed.), Vol. 4, p. 295. Plenum Press, New York, 1994.

[3] R. E. Kunz, ed., *Sensors Actuators B Chem.* **1** (1997).

[4] R. B. Thompson, ed., *SPIE Proc.* **2980** (1997).

[5] J. R. Lakowicz and R. B. Thompson, eds., *SPIE Proc.* **3602** (1999).

[6] J. N. Miller and D. J. S. Birch, eds., *J. Fluoresc.* **7,** 1S (1997).

[7] I. Klimant and O. S. Wolfbeis, "Europt(r)ode IV," p. 125. German Chemical Society.

[8] L. Tolosa, I. Gryczynski, L. R. Eichhorn, J. D. Dattelbaum, F. N. Castellano, G. Rao, and J. R. Lakowicz, *Anal. Biochem.* **267,** 114 (1999).

[9] J. R. Lakowicz, F. N. Castellano, J. D. Dattelbaum, L. Tolosa, and I. Gryczynski, *Anal. Chem.* **70,** 5115 (1998).

[10] J. Lakowicz, I. Gryczynski, Z. Gryczynski, L. Tolosa, J. Dattelbaum, and G. Rao, *Appl. Spectrosc.* **53,** 1149 (1999).

[11] I. Gryczynski, Z. Gryczynski, and J. Lakowicz, *Anal. Chem.* **71,** 1241 (1999).

[12] J. N. Demas and B. A. DeGraff, *in* "Topics in Fluorescence Spectroscopy" (J. R. Lakowicz, ed.), Vol. 4, p. 71. Plenum Press, New York, 1994.

[13] E. Terpetschnig, H. Szmacinski, H. Malak, and J. R. Lakowicz, *Biophys. J.* **68,** 342 (1995).

[14] H. Szmacinski, F. N. Castellano, E. Terpetschnig, J. D. Dattelbaum, J. R. Lakowicz, and G. J. Meyer, *Biochim. Biophys. Acta* **1383,** 151 (1998).

[15] H. Szmacinski, E. Terpetschnig, and J. R. Lakowicz, *Biophys. Chem.* **62,** 109 (1996).

[16] K. R. Rogers, *J. Mol. Biotechnol.* **14,** 109 (2000).

[17] S. DeMarcos, J. Galban, C. Alonso, and J. R. Castillo, *Analyst* **122,** 355 (1997).

[18] S. D'Auria, N. Di Cesare, Z. Gryczynski, I. Gryczynski, M. Rossi, and J. Lakowicz, *Biochem. Biophys. Res. Commun.* **274,** 727 (2000).

[19] S. D'Auria, Z. Gryczynski, I. Gryczynski, M. Rossi, and J. Lakowicz, *Anal. Biochem.* **283,** 83 (2000).

[20] U. E. Spichiger-Keller, "Chemical Sensors and Biosensors for Medical and Biological Applications," p. 413. Wiley-VCH, New York, 1998.

[21] J. R. Lakowicz, ed., "Topics in Fluorescence Spectroscopy," Vol. 4, p. 501. Plenum Press, New York, 1994.

0076-6879/03 $35.00

fluorescence sensing. There is growing interest in constructing simple, inexpensive electronic devices capable of accurate detection of various analytes in blood, tissue, or other real-world environments. Use of reference fluorophores in addition to the sensing fluorophores[7–11] is a technological advance that significantly expands the scope of possible applications enabling sensing in real-world conditions. These new methods are based on the use of a fluorescence reference either within the sample, or external to the sample, but within the optical path. Two types of sensing methods that significantly simplify measurements have been developed: modulation sensing with the use of microsecond lifetime probes such as metal–ligand complexes (MLCs)[12–15] and polarization sensing[10,11,41,42] with the use of highly polarized references such as oriented thin polymer films.[43,44] With modern electronics it is easy to imagine portable battery-powered instruments for measuring blood gasses, electrolytes, glucose, lactate, or other analytes (Scheme 1).

Sensing methods with external or internal reference standards have a number of advantages over simple fluorescence intensity sensing. The measurement becomes a ratiometric parameter that is independent of variation in the excitation source and in many cases makes possible correction for ambient light contribution and sample

[22] B. Valeur, J. Bourson, and J. Pouget, *in* "Fluorescent Chemosensors for Ion and Molecule Recognition" (A. W. Czarnik, ed.). *ACS Symp. Ser.* **538,** 25 (1993).

[23] R. Y. Tsien, T. J. Rink, and M. Poenie, *Cell Calcium* **6,** 145 (1985).

[24] H. Iatridou, E. Foukaraki, M. A. Kuhn, E. M. Marcus, R. P. Haugland, and H. E. Kateriinopoulos, *Cell Calcium* **15,** 190 (1994).

[25] E. U. Akkaya and J. R. Lakowicz, *Anal. Biochem.* **213,** 285 (1993).

[26] A. Minta, J. P. Y. Kao, and R. Y. Tsien, *J. Biol. Chem.* **264,** 8171 (1989).

[27] J. P. Y. Kao, *Methods Cell Biol.* **40,** 155 (1994).

[28] L. Fabbrizzi, M. Licchelli, L. Parodi, A. Poggi, and A. Taglietti, *J. Fluoresc.* **8,** 263 (1998).

[29] L. Prodi, F. Bolletta, N. Zaccheroni, C. I. F. Watt, and N. J. Mooney, *Chem. Eur. J.* **4,** 1090 (1998).

[30] C. R. Chenthamarakshari and A. Ajayaghosh, *Tetrahedron Lett.* **39,** 1795 (1998).

[31] A. Metzger and E. V. Anslyn, *Angew. Chem. Int. Ed.* **37,** 649 (1998).

[32] J. E. Whitaker, R. P. Haugland, and G. Prendergast, *Anal. Biochem.* **194,** 330 (1991).

[33] O. S. Wolfbeis, *in* "Fiber Optic Chemical Sensors and Biosensors" (O. S. Wolfbeis, ed.), Vol. II, p. 19. CRC Press, Boca Raton, FL, 1991.

[34] M. E. Lippitsch, J. Pusterhofer, M. J. P. Leiner, and O. S. Wolfbeis, *Anal. Chim. Acta* **205,** 1 (1988).

[35] S. Draxler, M. E. Lippitsch, I. Klimant, H. Kraus, and O. S. Wolfbeis, *J. Phys. Chem.* **99,** 3162 (1995).

[36] H. M. Heise, R. Marbach, T. Koschinsky, and F. A. Gries, *Artif. Organs* **18,** 439 (1994).

[37] J. J. Burmeister, H. Chung, and M. A. Arnold, *Photochem. Photobiol.* **67,** 50 (1998).

[38] N. Di Cesare and J. R. Lakowicz, *J. Photochem. Photobiol. A* **143,** 39 (2001).

[39] H. Shinmori, M. Takeuchi, and S. Shinkai, *Tetrahedron* **51,** 1893 (1995).

[40] C. R. Cooper and T. D. James, *J. Chem. Soc. Perkin Trans I.*

[41] J. R. Lakowicz, I. Gryczynski, Z. Gryczynski, L. Tolosa, L. Renders-Eichhorn, and G. Rao, *J. Biomed. Opt.* **4,** 445 (1999).

[42] I. Gryczynski, Z. Gryczynski, G. Rao, and J. Lakowicz, *Analyst* **124,** 1041 (1999).

[43] Z. Gryczynski and A. Kawski, *Z. Naturforsch.* **42a,** 1396 (1987).

[44] A. Kawski, Z. Gryczynski, I. Gryczynski, J. R. Lakowicz, and G. Piszczek, *Z. Naturforsch.* **51a,** 1037 (1996).

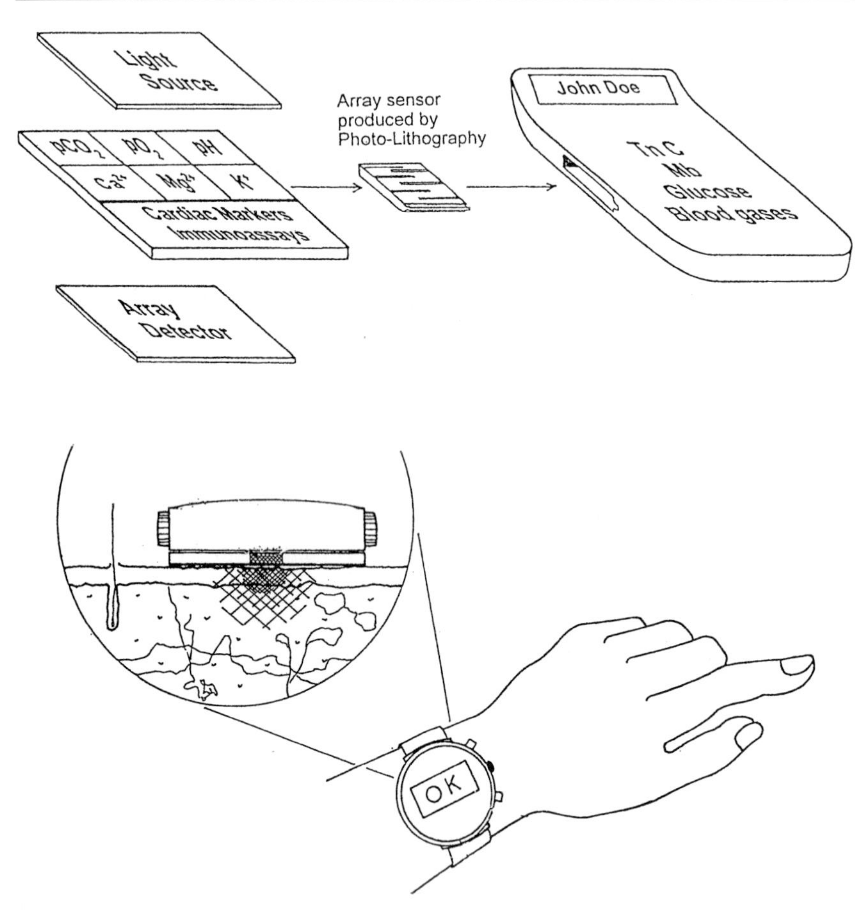

SCHEME 1. Novel sensing for blood chemistry and transdermal measurements.

imperfections such as scattering or reabsorption. Such methods may considerably simplify the measurement, allowing sensing with relatively inexpensive devices.

The method of polarization sensing represents an innovative and simple approach to fluorescence sensing. The use of a stretched polymer film as a highly polarized reference allows a significant enlargement of the dynamic range for polarization changes, increasing the resolution of a practical device. Introduction of a dual-polarizer system together with a polarizer analyzer enabled easy visual determination of the analyte concentration.[11] Elimination of advanced electronic detectors allows enormous simplification of testing devices. The use of inexpensive photodiode detectors with that technology strongly increases the sensitivity and practicability of electronic devices.[45]

[45] Z. Gryczynski, I. Gryczynski, and J. Lakowicz, *Opt. Eng.* **39,** 2351 (2000).

Fluorescence-Sensing Methods

It is widely recognized that ratiometric or lifetime-based methods possess intrinsic advantages for chemical and biomedical sensing.[1,2,20,21] Fluorescence intensity measurements are typically unreliable outside the laboratory and require frequent recalibration because of a variety of chemical, optical, or other instrumental factors. Unfortunately, most sensing fluorophores display changes in intensity in response to analytes, and relatively few wavelength ratiometric and lifetime-based probes are available. Some useful wavelength ratiometric probes are available for pH, Ca^{2+}, and Mg^{2+},[21–29] but the probes for Na^+ and K^+ display small spectral shifts and negligible fluorescence lifetime changes that are often inadequate for quantitative measurements. In general, collisional quenching by species such as O_2 and Cl^- occurs with a change in intensity and lifetime but without an emission spectral shift.

One obstacle for fluorescence sensing is the difficulty of performing the measurements in real-world situations. The difficulties of intensity-based measurements such as sample scattering or absorption can be circumvented by the use of lifetime-based probes or ratiometric probes. The success of lifetime-based sensing for calcium and other intracellular analytes has stimulated efforts to develop new approaches to fluorescence sensing. We have developed polarization- and modulation-sensing methods that use fluorescent internal/external standards to convert the probe intensity changes into ratiometric observables. Modulation sensing converts the fluorescence change of the sensing probe to a modulation change. Polarization sensing converts a fluorescence intensity change of the probe into a change in polarization.

Lifetime-Based Sensing

Lifetime-based sensing was introduced more then 10 years ago and is an active area of research.[1,2] The main advantage of lifetime-based sensing is the fact that intensity decay (fluorescence lifetime) is independent of its absolute signal, which may vary because of external factors like scattering and absorption.

The fluorescence lifetime is a fundamental characteristic of the fluorophore. The lifetime of the excited state is the average time the fluorophore molecule spends in the excited state after excitation. The fluorescence lifetime is governed by the radiative (Γ) and nonradiative (k_{nr}) rate constants[1]

$$\tau = \frac{1}{\Gamma + k_{nr}} \tag{1}$$

Lifetime-based sensing utilizes molecular interactions that result in a change in the fluorescence lifetime. The most common mechanisms that affect fluorescence lifetimes are collisional quenching and resonance energy transfer (RET). Both mechanisms exclusively modify nonradiative decay pathways without changing the radiative rate constant Γ. In contrast to fluorescence intensity, which strongly depends on the geometry and optical quality of the sample, fluorescence lifetimes depend only on intrinsic molecular constants.

Fluorescence intensity measurements are precise and reliable only when transparent optical containers are used. Fluorescence lifetimes are practically independent of optical sample quality. The important advantage of lifetime-based sensing is the fact that as long as the fluorescence intensity is great enough to measure the signal, the fluorescence lifetime provides useful information about the sensing probe.

When fluorescence lifetime-based sensing was first introduced,[1,2] we predicted that it would be easier to produce probes that display changes in lifetime in response to the analyte than to design probes that display spectral shifts. This assumption was based on the fact that a wide variety of quenchers and molecular interactions result in changes in the fluorescence lifetime of the fluorophores. This prediction has been proved correct for a number of analytes. A number of fluorescent probes that display a change in lifetime in response to the analyte have been developed. Examples include the Calcium Green series[2] and its analogous Mg^{2+} probes, which display significant lifetime changes in response to the binding of specific cations,[2] or pH probes such as the seminaphthofluorescein (SNAFL) and seminaphthorhodafluor (SNARF) series, which display changes in lifetime on pH-induced ionization.[2] Collisional quenchers such as O_2, Br^-, and Cl^- also cause changes in the fluorescence lifetime.

Lifetime-based sensing has been well established for experimental conditions in which quantitative fluorescence intensity measurements would be difficult to make, for example, blood, tissue, cells, or oceanographic studies.[1] However, lifetime-based sensing is still regarded as a moderately complex method that requires a moderate level of instrumentation.

Modulation Sensing

A variety of novel approaches to fluorescence-based sensing have appeared. One of the most innovative approaches was the construction of fluorescence sensor from two emitting species: one fluorophore does not change its emission and the other changes fluorescence intensity in response to the analyte. If the intrinsic spectral properties of both fluorophores are different it is convenient to measure the signal at two wavelengths and to convert the intensity data to ratiometric measurements. If the intrinsic fluorescence lifetimes are different for both fluorophores and one exclusively changes the fluorescence intensity in response to analyte, fluorescence sensors constructed from them will display a change in the apparent fluorescence lifetime induced only by the change in relative intensities.[7–9]

There are two factors that result in apparent fluorescence lifetime changes (modulation and phase changes) in the emitting system: (1) an intrinsic lifetime change in the sensing probe due to a physical process such as resonance energy

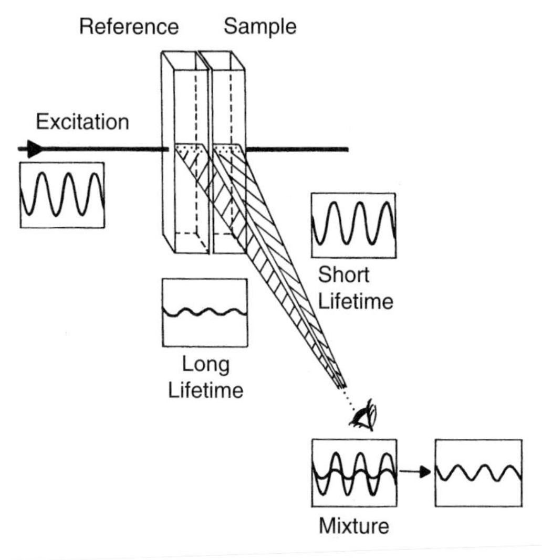

FIG. 1. Concept of modulation sensing.

transfer or diffusional quenching and (2) a change in apparent lifetime resulting only from modification of the relative intensity fractions of the two emitting species (of different fluorescence lifetimes) that comprise the sensor. There are two variables, phase and modulation, in frequency domain lifetime measurements. We have developed a method based only on the modulation data, a method that introduces a calibrated amount of background fluorescence of significantly different fluorescence lifetime into the probe signal.

The concept of modulation-based sensing is schematically depicted in Fig. 1. Amplitude-modulated light excites both the sample and reference (background) fluorophores. The sensing fluorophore (probe) changes its fluorescence intensity in response to the analyte. Its fluorescence lifetime is in the nanosecond range, as is typical of organic fluorophores. The fluorescence lifetime of the reference fluorophore is much longer (more then 10 times longer), on the order of hundreds or thousands of nanoseconds. For a wide range of modulation frequencies (1–10 MHz) the signal from the probe is highly modulated ($m_s \approx 1$) and the signal from the reference is practically totally demodulated ($m_r \approx 0$). If both signals are mixed on the detector, the total signal coming out will have a modulation proportional to the intensity fractions of the probe, f_p. As shown in Abugo *et al.*,[46] the measured modulation, m_{obs}, at the properly chosen frequency can be approximated by Eq. (2):

$$m_{obs} = f_p \tag{2}$$

[46] O. Abugo, Z. Gryczynski, and J. Lakowicz, *J. Biomed. Opt.* **4,** 429 (1999).

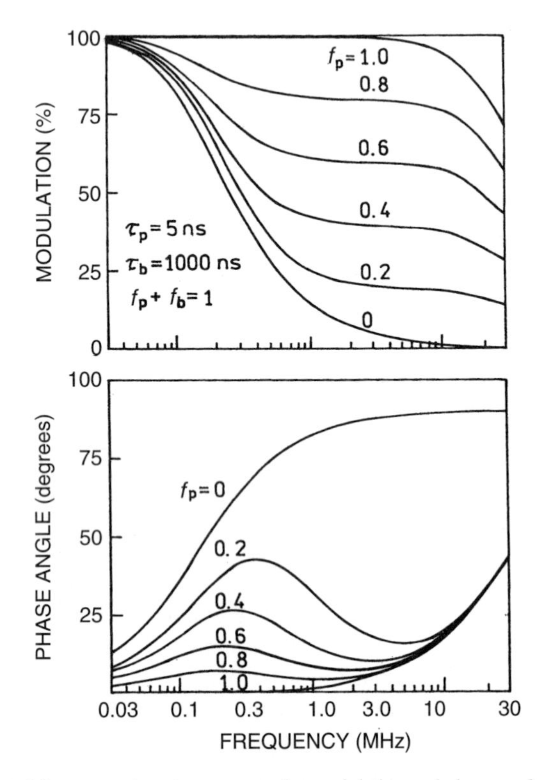

FIG. 2. Simulated frequency domain responses for modulation and phase angle for the mixture of two fluorescence lifetimes (5 and 1000 ns) as a function of increasing steady state intensity fraction of a long-lifetime component.

To understand the dependence of the modulation on the intensity fraction of the probe signal, let us consider a fluorescence probe with a lifetime of 5 ns mixed with a fluorescence probe with a long lifetime of 1000 ns. Figure 2 shows simulated frequency domain responses of the mixed signal for different fractions of probe fluorescence. In the wide frequency range between 1 and 10 MHz the modulation response is flat. The modulation signal (proportional to the sample intensity fraction) covers almost the full dynamic range between 0 and 1, as the probe intensity fraction changes from 0 to 1. These allow easy measurements and practical construction of sensing devices. We note that the intensity fraction is not directly proportional to the intensity of the probe.

There are several ways to introduce a fluorescence reference into the probe signal. One is to mix a long-lifetime fluorophore (insensitive to analyte) with the probe that responds to analyte with an intensity change.[7,8] Another, more practical way is to use a long-lifetime reference dye embedded in thin polymer film.[9,46] Films made out of polyvinyl alcohol (PVA) or polyethylene are easy to prepare

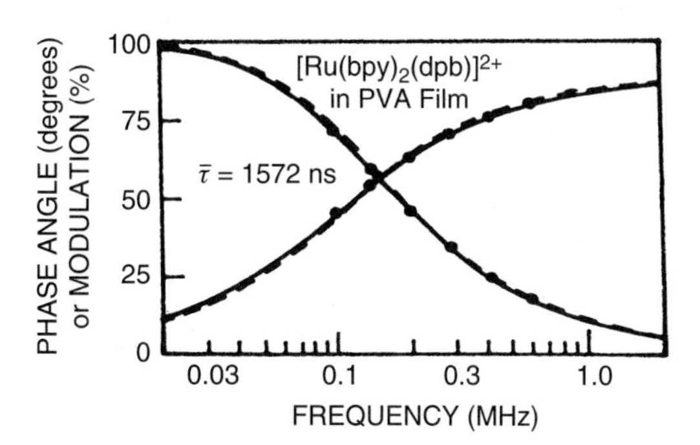

FIG. 3. Frequency domain response of $[Ru(bpy)_2(dpb)]^{2+}$ embedded in polyvinyl alcohol (PVA) film.

and are stable over long periods of time. These reference films are convenient to use and easy to apply in front of samples. An example of the time-resolved fluorescence response of $[Ru(bpy)_2(dpb)]^{2+}$ [bpy is 2,2′-bipyridine, dpb is 2,3-bis(2-pyridyl)benzoquinoxaline] embedded in PVA is shown in Fig. 3. This film has an almost homogeneous, long fluorescence lifetime that is stable for years.

A schematic configuration for sensing with a polymer film in front of the cuvette containing the sample is shown in Fig. 4. The combined signal from reference and probe will track changes in the probe intensity relative to the constant background. The change in modulation for the sensor depends on the relative initial intensity of the probe in reference to the background. By regulating the concentration of the reference fluorophore, or by choosing a different observation filter (observation wavelength) and/or excitation wavelength, it is possible to regulate the initial ratio between the probe and the reference signal. Because the intensity response of the probe to the analyte is known, it is easy to optimize the initial conditions in

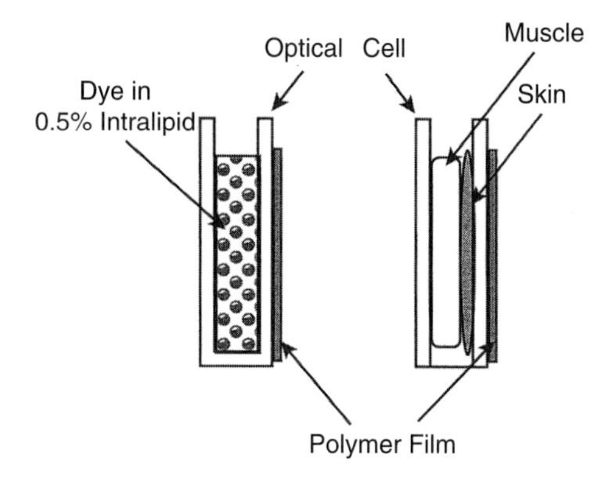

FIG. 4. Sample configuration for fluorescence sensing, with the polymer film in front of the cuvette. Cuvette with Intralipid (*left*) and tissue (*right*).

such a way that the dynamic range of the sensor will be maximal over the analyte concentration range of interest.

Consider a probe that varies its fluorescence intensity n-fold in response to a full range of analyte concentrations. To find optimal initial conditions in which the dynamic response of the sensor will be maximal we consider the change in observed modulation as a function of two parameters, the probe intensity change (n) and the initial ratio of signals of the probe to the reference fluorophore (k). The ratio k is given by

$$k = I_{\mathrm{p}}^{0}/I_{\mathrm{r}}^{0} \tag{3}$$

where I_{p}^{0} and I_{r}^{0} are initial fluorescence intensities of probe and reference, respectively. The initial observed modulation will be

$$m_{\mathrm{obs}}^{0} = f_{\mathrm{p}}^{0}$$

where

$$f_{\mathrm{p}}^{0} = \frac{I_{\mathrm{p}}^{0}}{I_{\mathrm{p}}^{0} + I_{\mathrm{r}}^{0}} \tag{4}$$

and f_{p}^{0} is the initial intensity fraction of the probe fluorescence. If the probe intensity is changed n-fold in response to analyte ($I_{\mathrm{p}} = nI_{\mathrm{p}}^{0}$), the new fraction of the probe fluorescence will be

$$f_{\mathrm{p}} = \frac{I_{\mathrm{p}}}{I_{\mathrm{p}} + I_{\mathrm{r}}^{0}} = \frac{nI_{\mathrm{p}}^{0}}{nI_{\mathrm{p}}^{0} + I_{\mathrm{r}}^{0}} \tag{5}$$

Dividing the right-hand side of Eqs. (4) and (5) by I_{r}^{0} yields

$$f_{\mathrm{p}}^{0} = \frac{k}{k+1} \quad \text{and} \quad f_{\mathrm{p}} = \frac{nk}{nk+1} \tag{6}$$

The observed modulation will follow the change in probe intensity. The total change in observed modulation, Δm, when the probe changes its intensity n-fold, is

$$\Delta m = f_{\mathrm{p}} - f_{\mathrm{p}}^{0} = \frac{nk}{nk+1} - \frac{k}{k+1} \tag{7}$$

Figure 5 shows the change in modulation, Δm, as a function of the ratio of the initial intensities, k, for different assumed values of probe intensity change, n. It is interesting to observe that for each value of total probe intensity change, n, an optimal initial ratio of probe-to-reference intensity can be selected. It is possible to differentiate Eq. (7) and find the relation between n and k yielding maximal

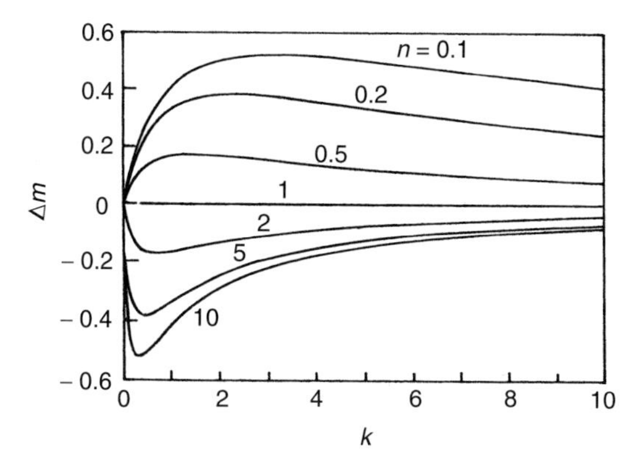

FIG. 5. Expected modulation change, Δm, as a function of initial intensities ratio, k, for different values of expected probe intensity change, n.

modulation change:

$$k = \left(\frac{n-1}{n^2-n}\right)^{1/2} \tag{8}$$

Equation (8) may be used to adjust the reference concentration, excitation wavelength, and/or observation wavelength so that the observed modulation change will be maximal. For example, for a probe in which the intensity increases fivefold due to the analyte, the optimal starting probe/background intensity ratio will be 0.45.

In practical applications, modulation sensing typically includes a fluorescent reference within the sample or placed immediately adjacent to the sample. In the first case the fluorescence probe may simply be mixed with the inert fluorophore. Absorption properties of the probe and the long-lifetime background fluorophore should be such that they can be excited with one excitation wavelength and their emission spectra should overlap. Examples of convenient, long-lifetime background fluorophores are metal–ligand complexes. Their fluorescence lifetimes are usually in the hundreds of nanoseconds, much longer than those of most fluorescence probes. Their absorption and emission spectra are broad and easy to overlap with most visible or red fluorescent probes.

Figure 6 shows an experimental calibration curve and emission spectra for fluorescein and $[Ru(bpy)_2(dpb)]^{2+}$ mixed in one cuvette. In response to pH, fluorescein changes its fluorescence intensity, modifying its relative contribution to the combined signal. By choosing different excitation and/or observation filters it is possible to vary relative initial fluorescence intensities between the probe and background. This is also a simple and convenient way to change and adjust the sensitivity range of the sensor.

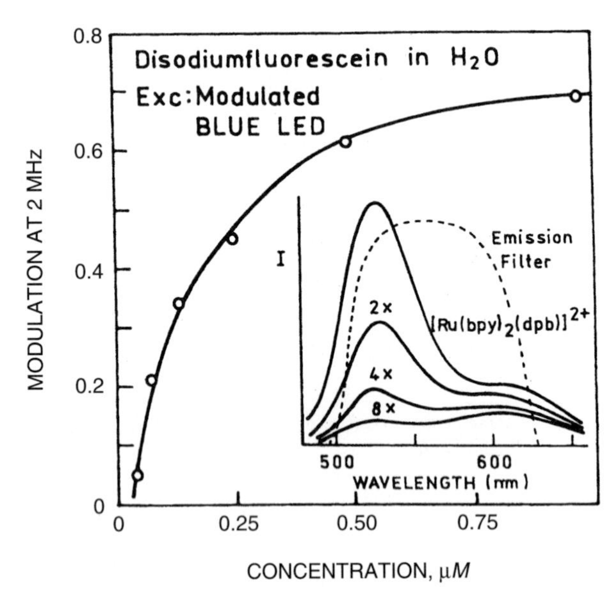

FIG. 6. Experimental calibration curve and emission spectra of fluorescein and $[Ru(bpy)_2(dpb)]^{2+}$ mixed in one cuvette.

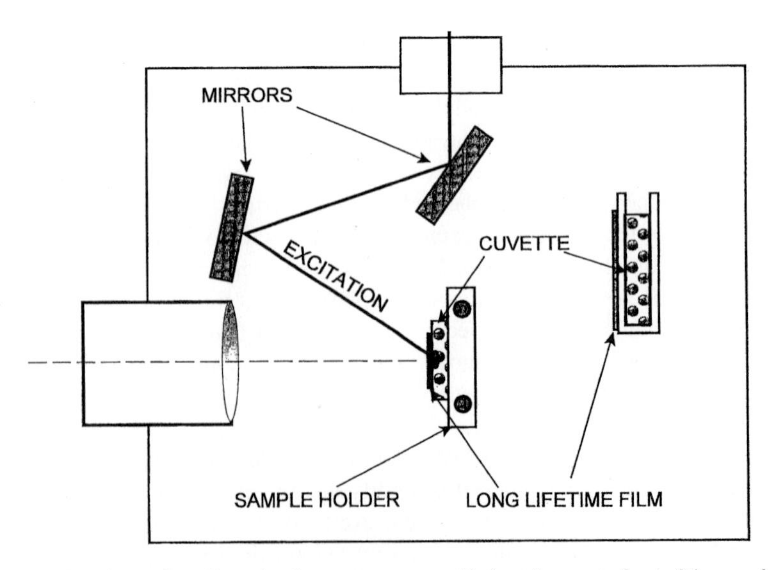

FIG. 7. Experimental configuration for measurements with the reference in front of the sample.

Figure 7 shows an experimental configuration for measurements with the reference in front of the sample. In this approach the reference film contains a luminescent metal–ligand complex (MLC) that displays red fluorescence and a decay time in the microsecond range. The reference is adequately diluted so as to transmit both the excitation light and emission from the probe. Such a configuration is convenient for measurements in scattered media, blood, or tissue.

One possible application of modulation sensing is measurements made through the skin. Because tissue and skin are translucent for red and near-infrared light, red fluorescent probes may be easily used for various diagnostics purposes. An example of such measurements for the purpose of drug compliance monitoring was discussed in Abugo *et al.*[46] To control patient compliance, red inactive dye mixed with or coated on the medication is used. Preliminary experiments to prove the principles of the method were performed with the configuration shown in Fig. 7. Figure 8 shows the experimental data (modulation and phase as a function

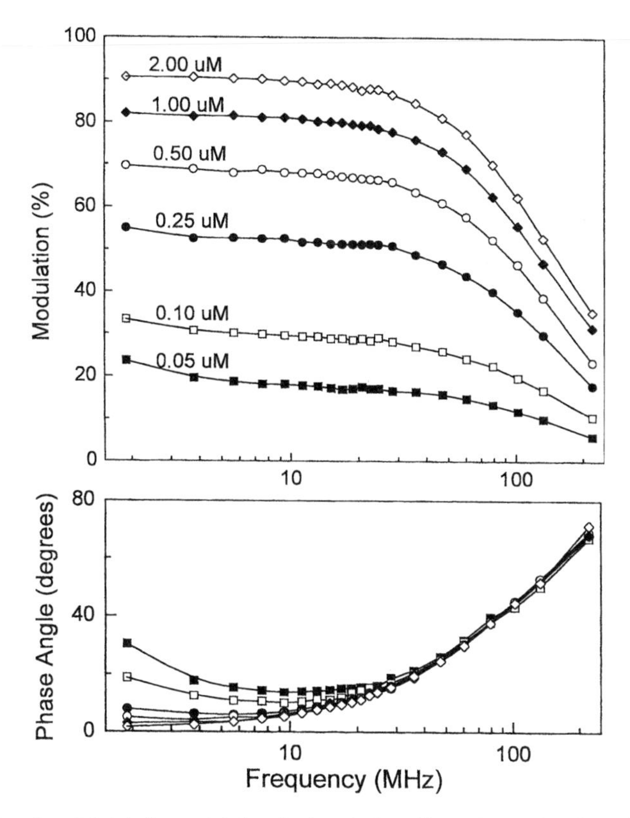

FIG. 8. Experimental data for modulation (*top*) and phase (*bottom*) as a function of modulation frequency, collected for various dilutions of rhodamine 800 in Intralipid.

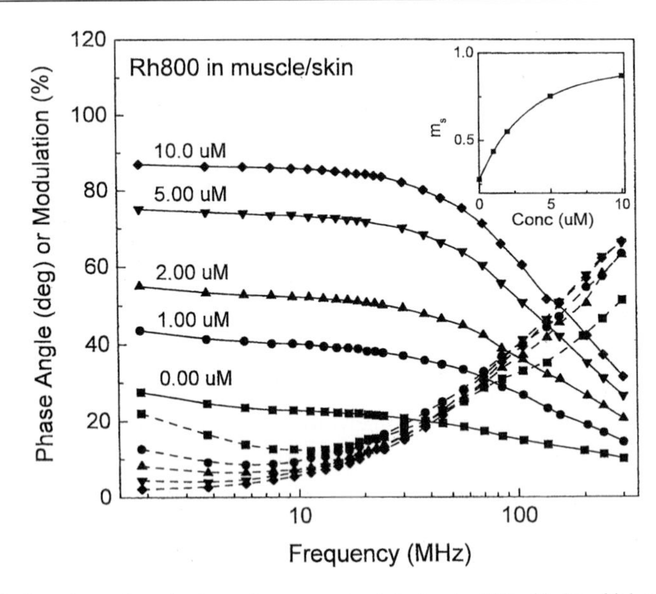

FIG. 9. Experimental results for various amounts of rhodamine 800 added to chicken tissue.

of modulation frequency) collected for different dilutions of rhodamine 800 in Intralipid. Experimental results for different amounts of rhodamine 800 added to chicken tissue and measured through the skin are presented in Fig. 9.[46]

The use of front-face optics makes it possible to construct simple devices easy to use for sensing through the skin. An example of a prototypical device for drug compliance monitoring using modulation sensing is shown in Fig. 10. As an excitation source a number of frequency-modulated light-emitting diodes (LEDs) or laser diodes may be used. The modulation of the combined emission from the sample and reference can be used to calculate the sample signal. An advantage of this approach is that the measurement is ratiometric relative to a highly stable reference signal from the film.

Polarization Sensing

Another innovative approach to sensing is the method of polarization sensing.[10,11,41,42] Polarization sensing is based on the additivity property of anisotropies.[1] In the polarization sensing method a fluorescence reference with high polarization is introduced. The polarization of the combined emission depends on the relative intensities of the polarized reference and emission from the probe, and can thus be used to quantify the analyte concentration.

An advantage of this method is the fact that measurements can be performed with unmodulated light sources such as an LED, laser diode, electroluminescent light (ELL), or even sunlight.

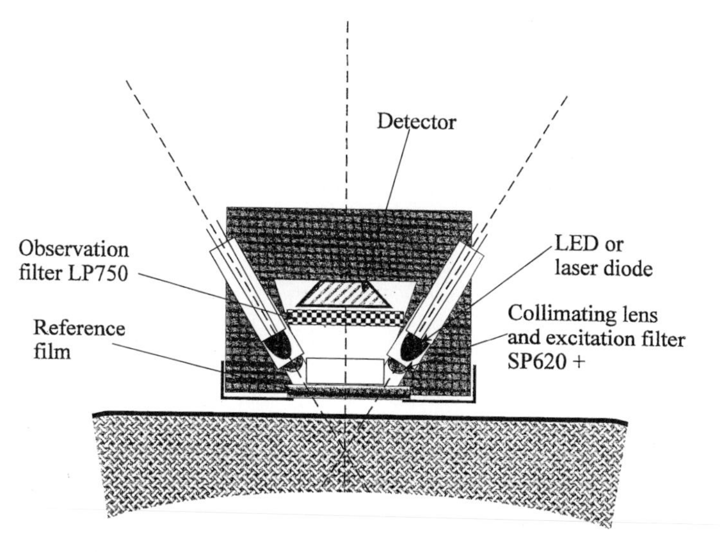

FIG. 10. Prototypic device for drug compliance monitoring with the use of modulation sensing.

There are two main experimental configurations for polarization-based sensing: one uses a self-referenced sample and the second uses a highly polarized oriented film.

Self-Referenced Sample. Self-referenced polarization sensing can be accomplished with steady state measurements and can be used whenever the fluorescence intensity changes in response to the analyte. The concept of the self-referenced method is shown in Fig. 11. The sensor consists of two parts. One side contains

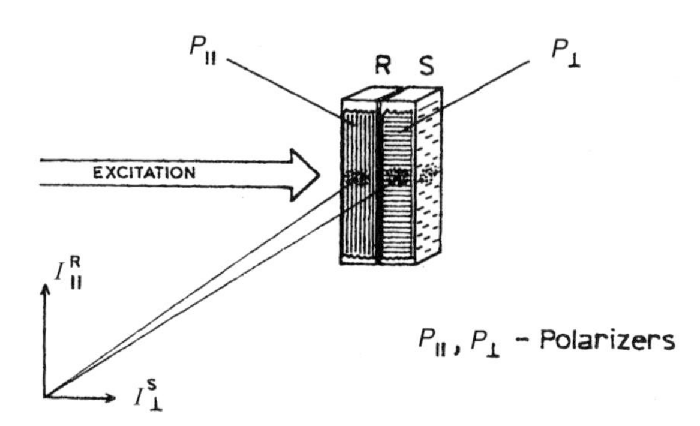

FIG. 11. Concept of self-referenced polarization method.

the sensing fluorophore and a constant concentration of analyte and is regarded as the reference side (R). The second side contains the sensing fluorophore and a variable concentration of analyte and is regarded as the sample side (S). The emission from the reference side passes through the vertically oriented film polarizer and emission from the sample side passes through the horizontal film polarizer. The combined emission from both sides of the sensor reaches the detector. The polarization of the combined emission is given by

$$P = \frac{I_{\parallel}^{\mathrm{T}} - I_{\perp}^{\mathrm{T}}}{I_{\parallel}^{\mathrm{T}} + I_{\perp}^{\mathrm{T}}} \tag{9}$$

where the superscript T indicates total intensity from the probe and the reference of parallel ($\parallel$) and perpendicular ($\perp$) components of the emission, that is, from both sides of the sensor. Because of the polarizers in front of the probe and reference, the polarized intensities are given by

$$\begin{aligned} I_{\parallel}^{\mathrm{T}} &= I_{\parallel}^{\mathrm{R}} \\ I_{\perp}^{\mathrm{T}} &= I_{\perp}^{\mathrm{S}} \end{aligned} \tag{10}$$

where $I_{\parallel}^{\mathrm{R}}$ and $I_{\perp}^{\mathrm{S}}$ are the intensities from the reference and sample, respectively. For isotropic excitation unpolarized emission of the sample and reference, the polarized intensities are directly proportional to the total intensity from each side of the sensor. In general, intensities from sample and reference depend on a number of constant instrumental factors. For simplicity, we have chosen not to indicate these factors explicitly.

Substituting Eq. (10) into Eq. (9), we have

$$P = \frac{I_{\parallel}^{\mathrm{S}} - I_{\perp}^{\mathrm{R}}}{I_{\parallel}^{\mathrm{S}} + I_{\perp}^{\mathrm{R}}} = \frac{I^{\mathrm{S}} - I^{\mathrm{R}}}{I^{\mathrm{S}} + I^{\mathrm{R}}} \tag{11}$$

Here the polarization of the combined emission depends only on the relative intensity of the reference and sample sides of the sensor. To optimize the initial experimental conditions to yield the maximal change in polarization, consider a fluorescence probe that changes intensity n-fold in response to analyte. If the initial ratio of the probe to the reference, k, is as defined in Eq. (3) ($k = I^{\mathrm{S}}/I^{\mathrm{R}}$), it is possible to calculate the observed change in polarization ΔP as a function of values n and k:

$$\Delta P = P_0 - P = \frac{k-1}{k+1} - \frac{nk-1}{nk+1} \tag{12}$$

where P_0 is the initial polarization from the sensor. Figure 12 shows simulations of the polarization change, ΔP, as a function of initial intensity ratio, k, for different values of total probe intensity change, n. From this prediction it is possible to optimize initial experimental conditions to result in the largest change in polarization on analyte addition.

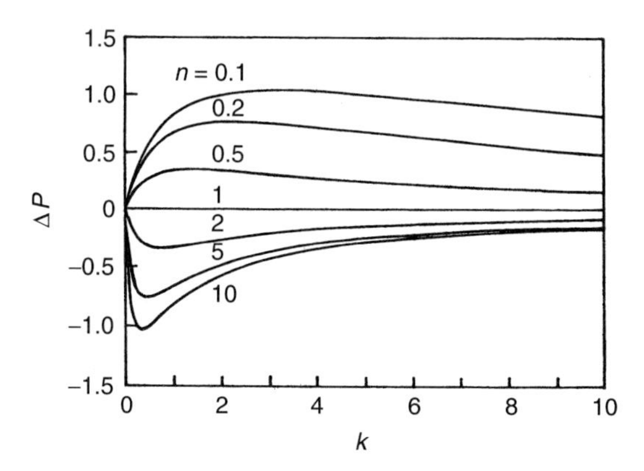

FIG. 12. Expected polarization change, ΔP, as a function of initial intensities ratio, k, for different values of total probe intensity change, n.

Examples of self-referenced polarization-based sensing are presented in Figs. 13 and 14. Figure 13 shows calcium-sensitive fluorophore Fluo-3, which displays a dramatic (~100-fold) increase in fluorescence upon binding calcium. This fluorophore does not show either a spectral or lifetime change and cannot be used as a ratiometric or lifetime probe. Our polarization sensor for calcium contained Fluo-3 in both cuvettes. The reference solution was observed through a vertical polarizer and sample was observed through a horizontal polarizer. On binding of calcium the observed polarization change is large (0.8), resulting in high-precision determination of the calcium concentration (0.01 μM). Figure 14 shows similar data for glucose/galactose-binding protein (GGBP).[41] The polarization sensor contained ANS-Q26C GGBP on both sides of the sensor. The polarization change exceeded 0.3, resulting in submicromolar accuracy of glucose concentration measurement.

Highly Polarized Reference. Another approach to polarization sensing is to use highly polarized films as a reference. Such films can easily be placed on the skin surface for convenient tissue sensing. To maximize the range and sensitivity of the methods it is desirable to use a background with high polarization. A convenient reference signal can be obtained from a stretched thin film that contains an oriented fluorophore. The emission from such films is highly polarized even when the excitation is not polarized.[41–44] Figure 15 shows fluorescence polarization measured for Pyridine-2 oriented in stretched PVA film with polarized and nonpolarized excitation. For nonpolarized excitation we used LED excitation with no excitation polarizer. For high stretching ratios the observed polarization is high and almost independent of the polarization of the excitation source.

Figure 16a shows the concept of polarization sensing with oriented thin film. The oriented film is placed adjacent to the sample, which displays a change in

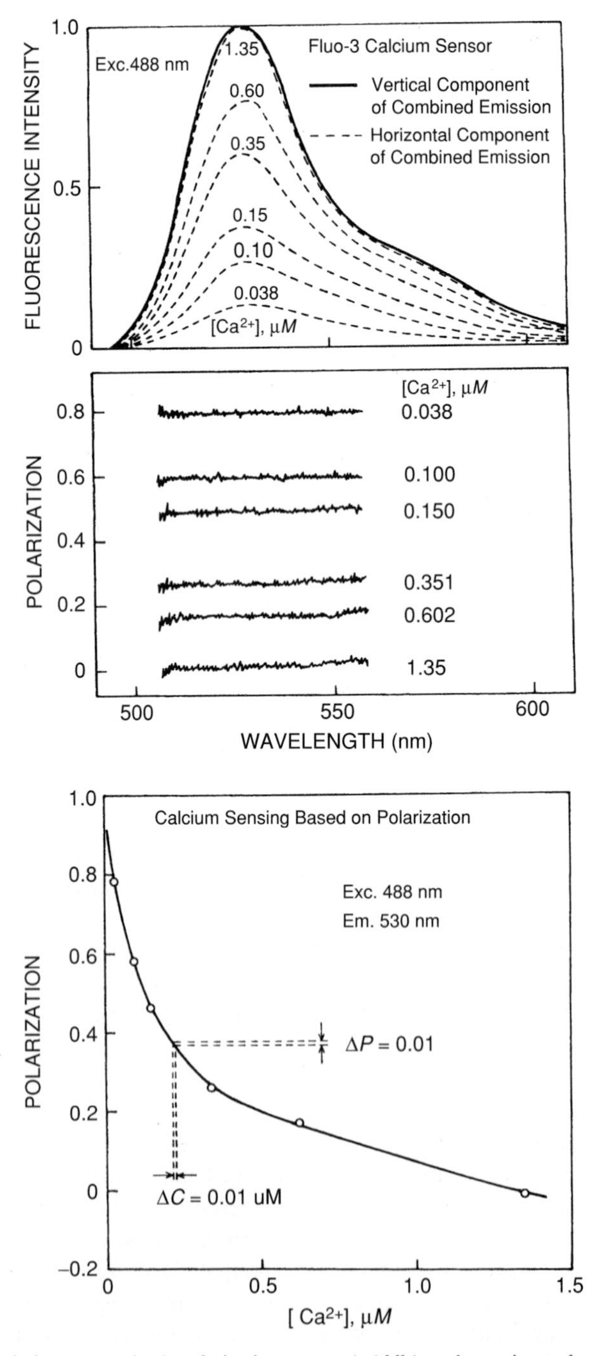

FIG. 13. Emission spectra (*top*), polarization spectra (*middle*), and experimental calibration curve (*bottom*) for polarization-based calcium sensing.

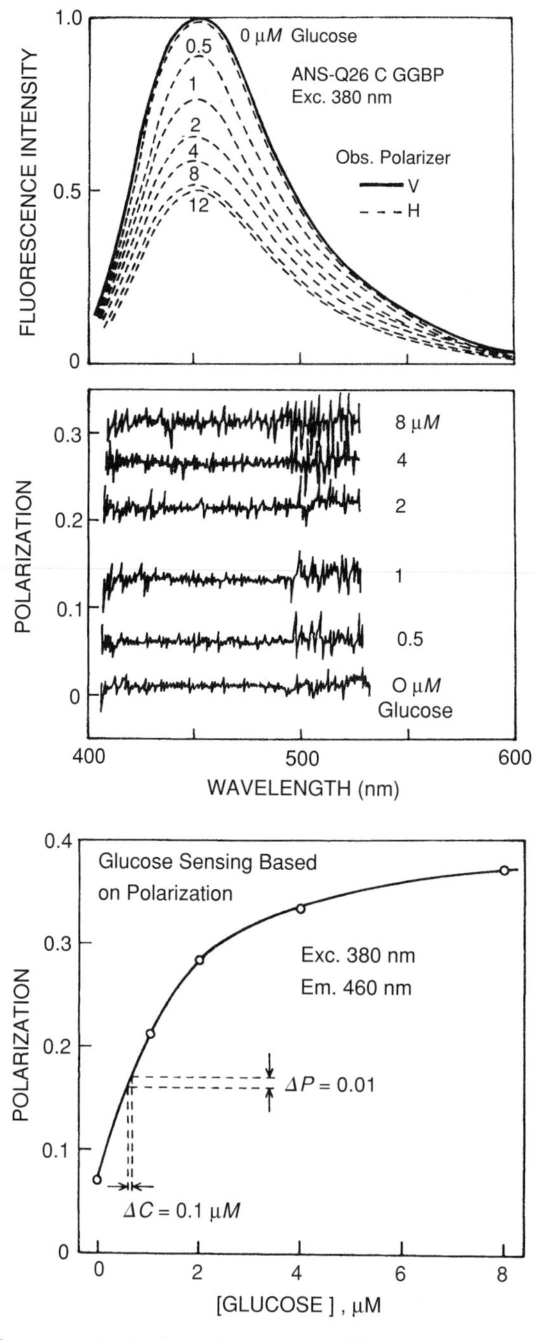

FIG. 14. Emission spectra (*top*), polarization spectra (*middle*), and experimental calibration curve (*bottom*) for polarization-based glucose sensing.

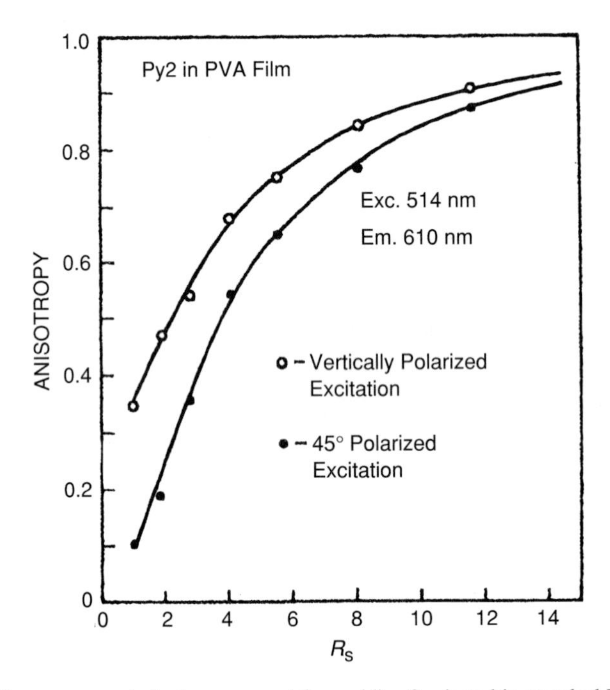

FIG. 15. Fluorescence polarization measured for pyridine 2 oriented in stretched PVA film with polarized and nonpolarized excitation.

intensity in response to the analyte. The reference film is optically thin and does not change polarization of the excitation beam or of the probe emission. The sensing fluorophore only needs to display a change in intensity, not a change in the polarization of its emission. In fact, an isotropic solution of fluorophore with front-face excitation from a nonpolarized light source will display polarization close to 0 in the direction of observation. As shown in Fig. 16 a, observed parallel and orthogonal intensity components are

$$I_{\parallel} = I_{\parallel}^{\mathrm{p}} + I_{\parallel}^{\mathrm{r}} \qquad \text{and} \qquad I_{\perp} = I_{\perp}^{\mathrm{p}} + I_{\perp}^{\mathrm{r}} \tag{13}$$

The measured polarization will be

$$P = \frac{I_{\parallel} - I_{\perp}}{I_{\parallel} + I_{\perp}} = \frac{I_{\parallel}^{\mathrm{p}} - I_{\perp}^{\mathrm{p}} + I_{\parallel}^{\mathrm{r}} - I_{\perp}^{\mathrm{r}}}{I_{\parallel}^{\mathrm{p}} + I_{\perp}^{\mathrm{p}} + I_{\parallel}^{\mathrm{r}} + I_{\perp}^{\mathrm{r}}} \tag{14}$$

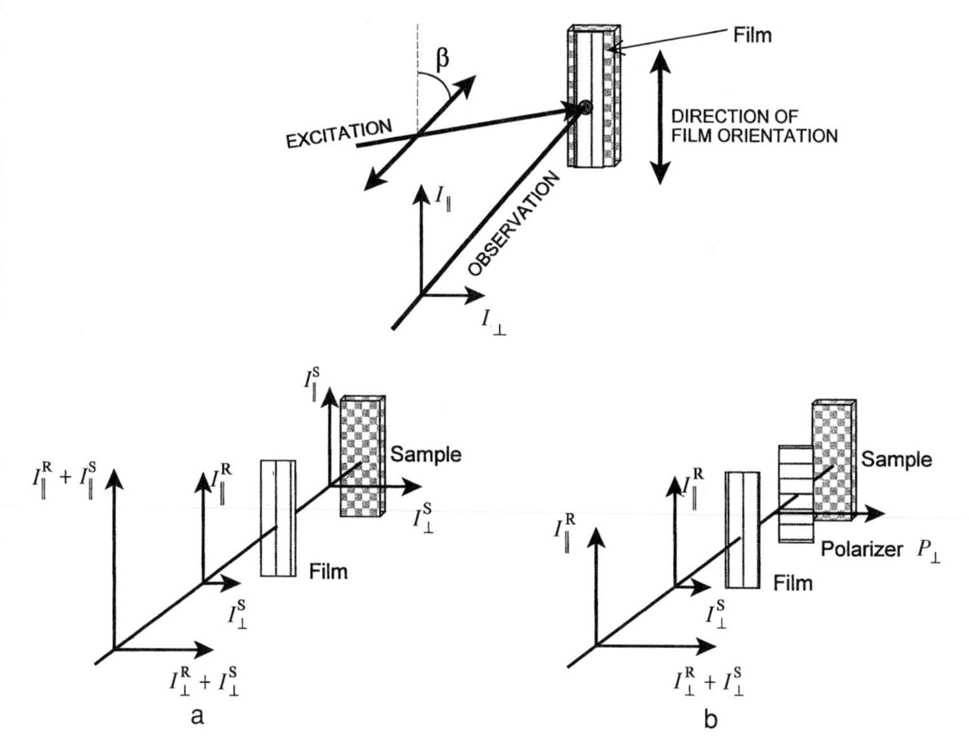

FIG. 16. Optical system for polarization-based sensing with oriented thin film in front. (a) Fluorescence intensity components for oriented film and sample. (b) Fluorescence intensity components for oriented film and sample with the additional polarizer (orthogonally oriented to the film orientation) in front of the sample. Angle β is the polarization of the excitation beam measured from the direction of polymer film orientation.

For nonpolarized emission of the probe $I^{\mathrm{p}}_\perp = I^{\mathrm{p}}_\parallel = I^{\mathrm{p}}$ and Eq. (14) becomes

$$P = \frac{I^{\mathrm{r}}_\parallel - I^{\mathrm{r}}_\perp}{I^{\mathrm{r}}_\parallel + I^{\mathrm{r}}_\perp + I^{\mathrm{p}}} \tag{15}$$

Polarization of stretched polymer film only, P_0, is given by

$$P_0 = \frac{I^{\mathrm{r}}_\parallel - I^{\mathrm{r}}_\perp}{I^{\mathrm{r}}_\parallel + I^{\mathrm{r}}_\perp} = \frac{R_{\mathrm{d}} - 1}{R_{\mathrm{d}} + 1} \tag{16}$$

where $R_{\mathrm{d}} = I^{\mathrm{r}}_\parallel / I^{\mathrm{r}}_\perp$. R_{d}, the ratio of parallel and orthogonal film fluorescence intensities, describes the orientation of embedded fluorophores. Dividing the right-hand

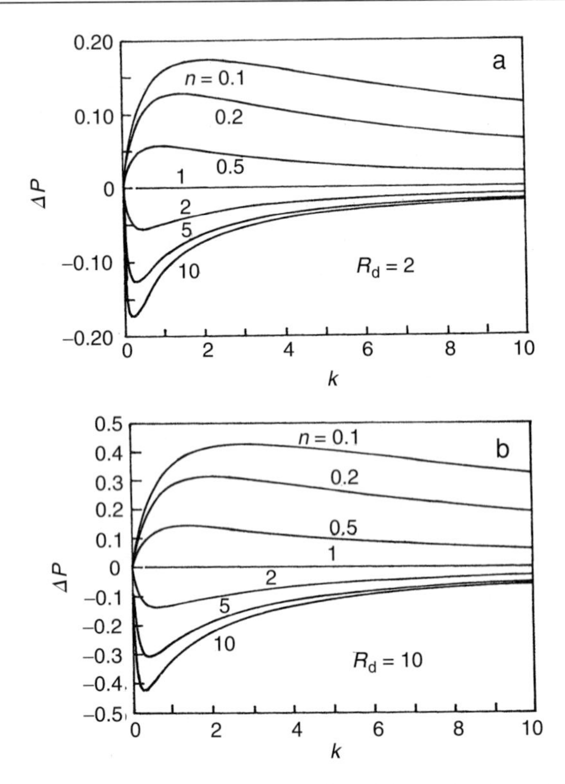

FIG. 17. Polarization change, ΔP, as a function of the initial ratio of sample and background intensities, k, for different values of total probe intensity change, n. (a) Film stretching ratio $R_{\rm d} = 2$, and (b) film stretching ratio $R_{\rm d} = 10$.

side of Eq. (15) by $I^{\rm p}$ yields

$$P = \frac{\frac{I_{\parallel}^{\rm r}}{I^{\rm p}} - \frac{I_{\parallel}^{\rm r}}{R_{\rm d} I^{\rm p}}}{\frac{I_{\parallel}^{\rm r}}{I^{\rm p}} + \frac{I_{\parallel}^{\rm r}}{R_{\rm d} I^{\rm p}} + 1} = \frac{k - \frac{k}{R_{\rm d}}}{k + \frac{k}{R_{\rm d}} + 1} \tag{17}$$

where $k = I_{\parallel}^{\rm r}/I^{\rm p}$ is the ratio between the initial fluorescence intensity of the background, $I_{\parallel}^{\rm r}$, and probe, $I^{\rm p}$. The initial value of k is an important adjustable parameter. If the probe intensity changes n-fold in response to analyte, the expected change in polarization, ΔP, may be calculated:

$$\Delta P = \frac{k - \frac{k}{R_{\rm d}}}{k + \frac{k}{R_{\rm d}} + 1} - \frac{\frac{k}{n} - \frac{k}{nR_{\rm d}}}{\frac{k}{n} + \frac{k}{nR_{\rm d}} + 1} = \frac{k(R_{\rm d} - 1)}{R_{\rm d}(k+1) + k} - \frac{k(R_{\rm d} - 1)}{R_{\rm d}(k+n) + k} \tag{18}$$

Figure 17a and b shows the dependence of ΔP on k for different values of n. Figure 17a and Figure 17b correspond to two different film stretching ratios, $R_{\rm d} = 2$ ($P_0 = 0.333$) and $R_{\rm d} = 10$ ($P_0 = 0.818$), respectively.

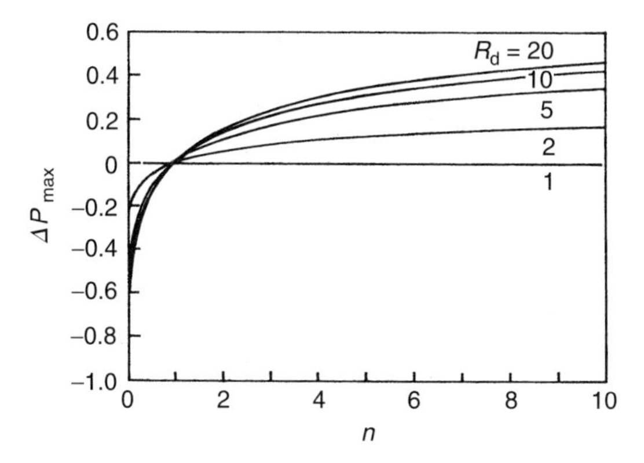

FIG. 18. Dependence of maximal polarization change, ΔP_{max}, as a function of probe intensity change, n, for different film stretching ratios, R_{d}.

Differentiating Eq. (18) yields an analytical expression for the optimal initial ratio between the fluorescence signal and the reference, k,

$$k = \frac{R_{\mathrm{d}}\sqrt{n}}{R_{\mathrm{d}} + 1} \tag{19}$$

The optimal initial conditions that yield the maximum polarization change depend on two values, the value of R_{d} (which characterizes intrinsic polarization of the film) and the total intensity change of the probe in response to the analyte, n.

Substituting Eq. (19) into Eq. (18) allows calculation of the expected maximum polarization change for the sensor:

$$\Delta P_{\mathrm{max}} = \frac{(R_{\mathrm{d}} - 1)(\sqrt{n} - 1)}{(R_{\mathrm{d}} + 1)(\sqrt{n} + 1)} \tag{20}$$

Figure 18 shows the maximal polarization change as a function of the intensity change, n, for different values of R_{d}.

It is interesting to consider limits of possible polarization changes. For an isotropic reference ($R_{\mathrm{d}} = 1$) $\Delta P = 0$, independent of n and sample polarization. For a highly oriented background ($R_{\mathrm{d}} \rightarrow \infty$)$\Delta P$ depends on probe intensity change and for high values of n, $\Delta P \rightarrow 1$. Under common experimental conditions, $R_{\mathrm{d}} \approx 10$ and probe intensity changes of about 5-fold, the expected polarization change, will be more than 0.3. This change in polarization is large and easy to measure.

An example of polarization sensing of pH using 6-carboxyfluorescein and oriented film doped with laser dye Pyridine-2 (Py2) is presented in Fig. 19. The fluorescence intensity of fluorescein increases more than 6-fold for the pH change from pH 5 to pH 7. Fluorescein displays a lifetime near 4 ns and its anisotropy in aqueous solution is expected to be near 0. The polymer film is stretched 5.5-fold

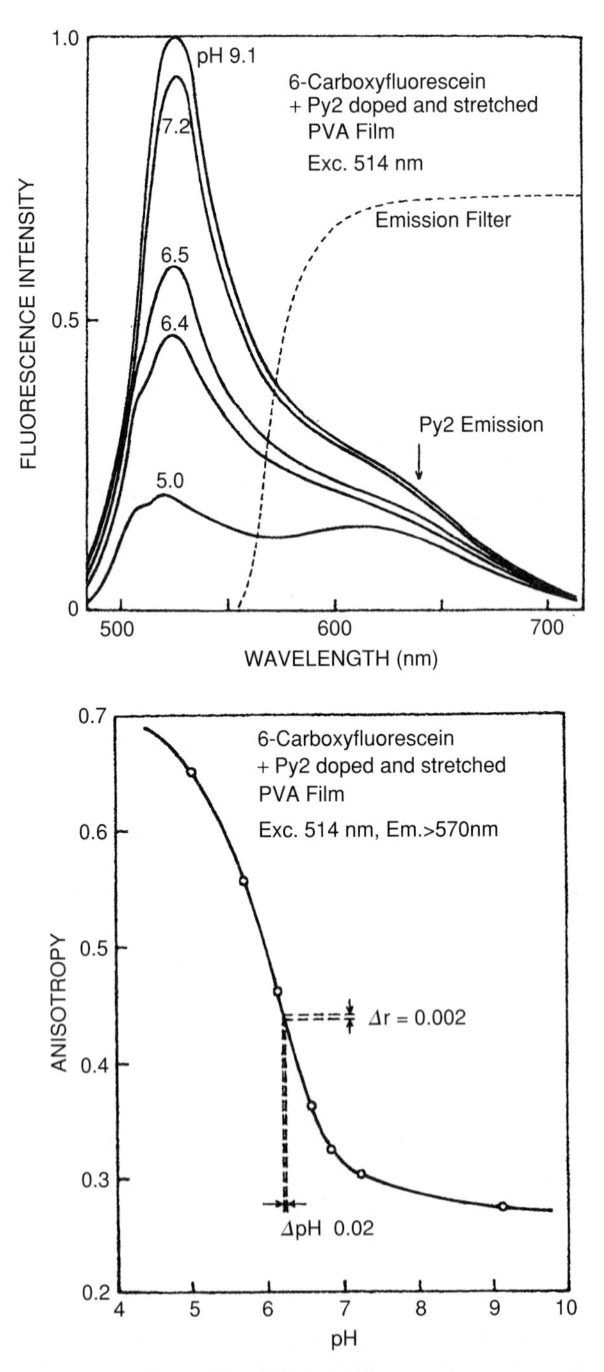

FIG. 19. Emission spectra of 6-carboxyfluorescein with Py2-doped and stretched PVA film (*top*) and experimental calibration curve (*bottom*) for polarization-based pH sensing with oriented polymer film.

and displays polarization near 0.9. The observed polarization change is more than 0.5, resulting in good accuracy at the measured pH (~0.02).

The possibility of using a thin polymer film as a highly polarized reference makes it practical to use polarization sensing to detect analytes through the skin or on the skin. Because steady state fluorescence from highly scattering media such as tissue and skin is intrinsically depolarized a single polarization component may be used to monitor the probe intensity change. This provides a method for significantly enlarging the sensor dynamic range. An extra polarizer may be introduced immediately after the oriented film and in front of the sample. The polarizer must be orthogonal to the film orientation. Figure 16b shows the principles of such an optical system. A horizontal polarizer between the vertically oriented film and isotropic sample does not perturb the excitation or emission of the reference film. The isotropic sample is excited with the horizontal polarization and only horizontal emission is transmitted through the polarizer toward the detector. It is important to mention that the oriented reference film does not change or interrupt the horizontal sample emission. From Fig. 16b it is possible to show that the parallel and orthogonal intensity components are

$$\begin{aligned} I_{\parallel}^{\mathrm{T}} &= I_{\parallel}^{\mathrm{R}} \\ I_{\perp}^{\mathrm{T}} &= I_{\perp}^{\mathrm{R}} + I_{\perp}^{\mathrm{S}} \end{aligned} \tag{21}$$

and observed polarization, P, will be

$$P = \frac{I_{\parallel}^{\mathrm{T}} - I_{\perp}^{\mathrm{T}}}{I_{\parallel}^{\mathrm{T}} + I_{\perp}^{\mathrm{T}}} = \frac{I_{\parallel}^{\mathrm{R}} - I_{\perp}^{\mathrm{R}} - I_{\perp}^{\mathrm{S}}}{I_{\parallel}^{\mathrm{R}} + I_{\perp}^{\mathrm{R}} + I_{\perp}^{\mathrm{S}}} = \frac{R_{\mathrm{d}} - k - 1}{R_{\mathrm{d}} + k + 1} \tag{22}$$

where $R_{\mathrm{d}} = I_{\parallel}^{\mathrm{R}}/I_{\perp}^{\mathrm{R}}$ and k is the ratio between the intensities of the sample fluorescence and the intensity of the orthogonal film component. It is interesting to consider the polarization limits for this configuration. If the total intensity is dominated by the sample emission $k \to \infty$ and $P \to -1$, if the fluorescence signal from the sample is much smaller, $k \to 0$, the polarization approaches polarization of the film, $P \to P_{\mathrm{R}}$. As mentioned previously, the reference films typically display polarization near 0.8 or higher, which results in a large dynamic range for the sensor: $-1 < \Delta P < 0.8$.

One example of a polarization sensing application is drug compliance monitoring,[47] using the sensor configuration shown in Fig. 16b. The essential elements of the sensor are the vertically oriented stretched PVA film (doped with styryl 7) and the horizontally oriented polarizer immediately in front of the sample. This polarization sensor was used to examine rhodamine 800 (Rh800) concentrations in 0.5% Intralipid and in chicken tissue. Figure 20a shows measured fluorescence

[47] Z. Gryczynski, O. Abugo, and J. Lakowicz, *Anal. Biochem.* **273,** 204 (1999).

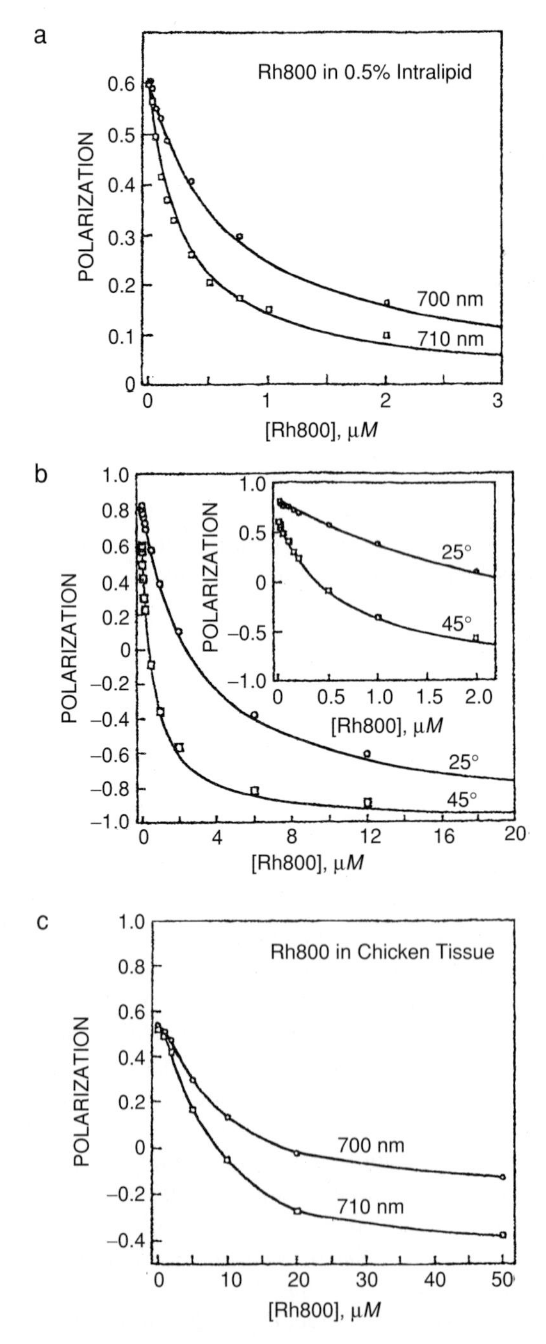

FIG. 20. Measured fluorescence polarization from the sensor shown in Fig. 16b as a function of rhodamine B concentration. (a) Intralipid (0.5%) for two different observation wavelengths; (b) 0.5% Intralipid for two different polarizations of excitation beam (angle β); (c) chicken tissue for two different observation wavelengths.

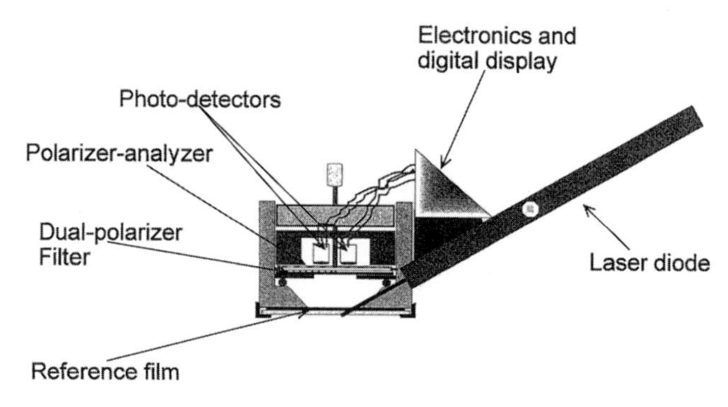

FIG. 21. Schematic of prototype for polarization-based sensing device.

polarization from the sensor for different Rh800 concentrations in 0.5% Intralipid measured at two observation wavelengths (700 and 710 nm) that correspond to different relative sample and reference intensities. It appears that the midpoint of the response can be modified by changing the observation wavelength. Figure 20b shows measurements done for two different polarizations of excitation light. By changing the polarization of excitation light, angle β in Fig. 16, it is possible to regulate the relative contribution of sample and reference signals. Polarization of the excitation light has an angle of 0° when the electric vector of excitation light is parallel to the film stretching direction. For small angles the excitation efficiency of the oriented film is high and excitation of the sample is strongly diminished by the horizontal polarizer. This shifts the midpoint of the sensor response toward higher Rh800 concentrations, allowing accurate measurements of high-range Rh800 concentrations. In contrast, high angles of the excitation polarizer rotation shift the midpoint of the sensor response toward low Rh800 concentrations.

Figure 20c shows measured fluorescence polarization from chicken tissue through the skin for two different observation wavelengths and excitation polarization rotated to 35° from the vertical orientation. Rh800 was injected into the tissue about 1 mm behind the skin.

Figure 21 shows a prototypic device capable of polarization sensing of red dyes on the skin or through the skin. The excitation source is a red laser diode (laser pointer).

Visual Detection

An important opportunity for polarization sensing is visual detection. The concept of visual detection is shown in Fig. 22. The method is based on the visual equalization of the intensity seen through a dual polarizer (DP) and through a

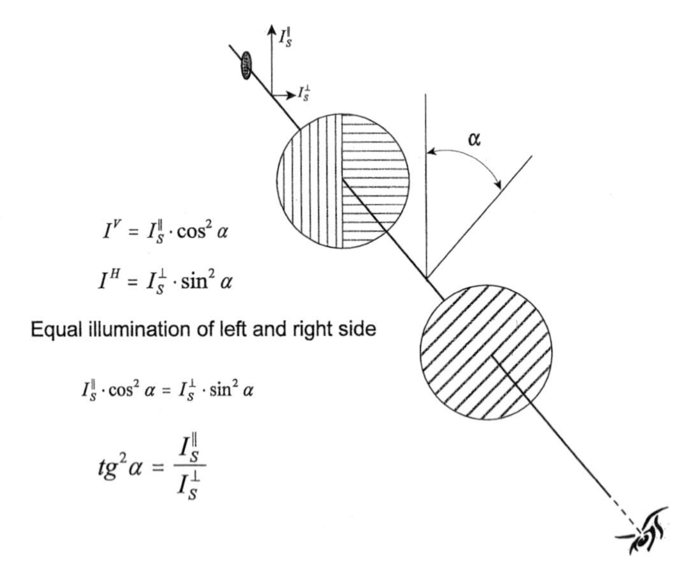

$I_S^{\|}$
$I_S^{\perp}$
α
$I^V = I_S^{\|} \cdot \cos^2 \alpha$
$I^H = I_S^{\perp} \cdot \sin^2 \alpha$
Equal illumination of left and right side
$I_S^{\|} \cdot \cos^2 \alpha = I_S^{\perp} \cdot \sin^2 \alpha$
$tg^2 \alpha = \frac{I_S^{\|}}{I_S^{\perp}}$

FIG. 22. Polarizer configuration for polarization-based sensing with visual detection. Front-face and in-line geometry.

polarizer analyzer (P). The dual polarizer is a crucial element in the visual polarization sensor. This component consists of two adjacent sheet polarizers with the optical axis of one rotated 90° relative to the other polarizer. If the sample is uniformly illuminated, the intensity transmitted by each half of the dual polarizer represents the parallel (||) and perpendicular (⊥) components from the reference film and from the sample, respectively. Visual judgment of absolute intensity is unreliable, but the eye is sensitive to contrast and can match the intensities of two adjunct surfaces with good precision, on the order of 5% for an average tested individual. This fact allows for fairly accurate visual measurement. The fluorescence signal from the sensor includes two parts: (1) unpolarized fluorescence from the probe and (2) highly polarized fluorescence from the reference. As can be determined from Fig. 22, the parallel and orthogonal intensity components from the sensor transmitted through two parts of the dual polarizer are

$$\begin{aligned} I_{\|} &= I_{\|}^{\mathrm{p}} + I_{\|}^{\mathrm{r}} \\ I_{\perp} &= I_{\perp}^{\mathrm{p}} + I_{\perp}^{\mathrm{r}} \end{aligned} \tag{23}$$

By taking into account the isotropic fluorescence of the probe, $I_{\|}^{\mathrm{p}} = I_{\perp}^{\mathrm{p}}$, it is possible to calculate the angle α at which the polarizer analyzer (P) transmits equal intensities for both sides of the DP as

$$\tan \alpha^2 = \frac{I_{\parallel}}{I_{\perp}} \tag{24}$$

There are several experimental configurations that allow polarization sensing with visual detection.[11] Two simple geometries (in-line and front-face) are shown schematically in Fig. 23. For the in-line geometry, the light passes through the reference film and the sample. For the front-face configuration, emission from the sample is viewed from the illuminated surface through the reference film. The absorption of the stretched reference film is adjusted to be comparable to that of the sample fluorescence. An emission filter, F, is used to eliminate excitation and to transmit emission. For in-line geometry the angle of the polarizer, that is, the rotation needed to equalize the intensities on both sides of the dual polarizer in terms of the polarization ratio of the reference, k, and relative starting intensity of the sample to that of the reference, n, is given by

$$\tan^2\alpha = \frac{k/(k+1) + n/2}{1/(k+1) + n/2} \tag{25}$$

A similar relation for the front-face geometry is given by

$$\tan^2\alpha = \frac{k/(k+1)}{1/(k+1) + n} \tag{26}$$

Visual detection is accomplished by rotating the polarizer analyzer until the intensity is equal for both sides of the observation window. The practical readout involves two angle measurements: one with no analyte and another in the presence of analyte. The differential angle $\Delta\alpha$ is called the compensation angle.[11] The compensation angle for in-line geometry is given by

$$\Delta\alpha = \arctan\sqrt{k} - \arctan\sqrt{\frac{k/(k+1) + n/2}{1/(k+1) + n/2}} \tag{27}$$

and for the front-face geometry:

$$\Delta\alpha = \arctan\sqrt{k} - \arctan\sqrt{\frac{k/(k+1)}{1/(k+1) + n}} \tag{28}$$

where $\tan^2\alpha_0 = k$, and α_0 is the angle measured when there is no emission from the sample and all the signal originates from the reference film. If the emission from the film is completely polarized ($k = \infty$), the analyzer should be oriented at 90° to equalize the intensities to 0. For typical values of $k = 12$, the initial angle $\alpha_0 \approx 74^\circ$.

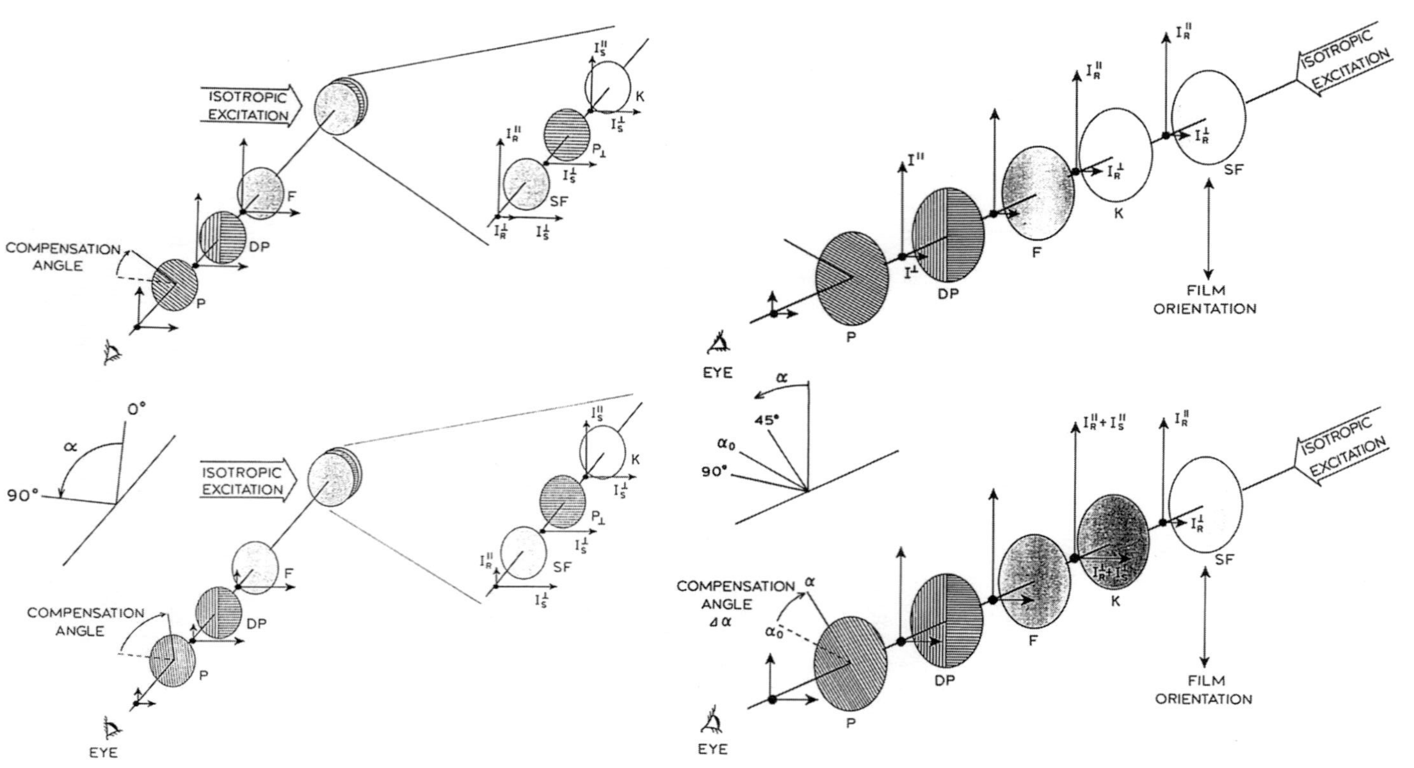

FIG. 23. Concept of visual detection. DP, Dual polarizer; P, polarizer; K, cuvette; SF, stretched PVA film; $P_\perp$, thin film polarizer; F, filter.

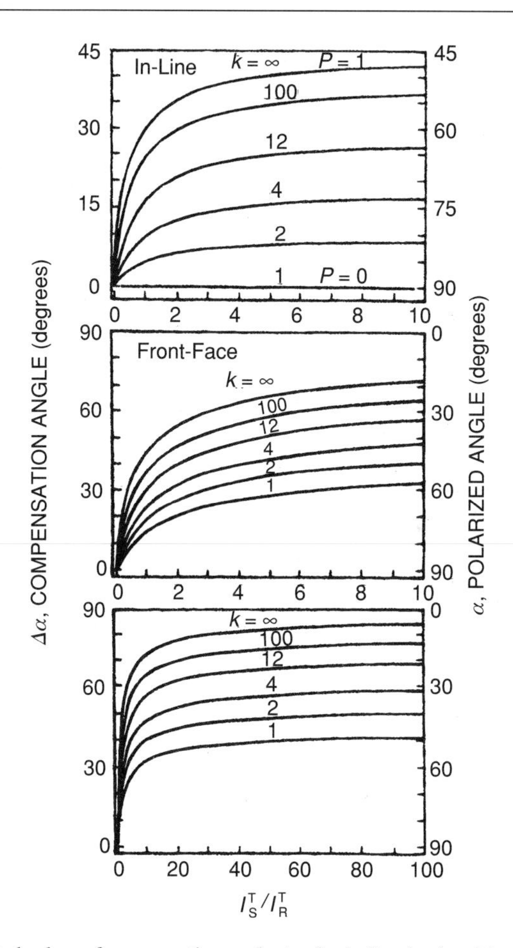

FIG. 24. Simulated values of compensation angle $\Delta\alpha$ for in-line (*top*) and front-face (*middle* and *bottom*) polarization-based sensing with visual detection. The two lower panels are similar, except for the dependent variable scale.

Figure 24 shows the values of $\Delta\alpha$ simulated for various values of k as a function of relative intensities $n = I_S^T/I_R^T$. The total range of polarizer angles is 45° and 90° for in-line and front-face configurations, respectively, as the emission ranges from 100% for the reference film to 100% for the sample. It is possible to detect the position of equal intensity to within 1° of accuracy by visual adjustment of the polarizer analyzer.[11] Figure 25 shows the visual images seen through the polarizer analyzer for increasing concentrations of rhodamine B. The lower part of Fig. 25 shows the dependence of the compensation angle on the concentration of rhodamine B, using in-line geometry. As the rhodamine B concentration increases to 15 μM the compensation angle increases by more than 20°.

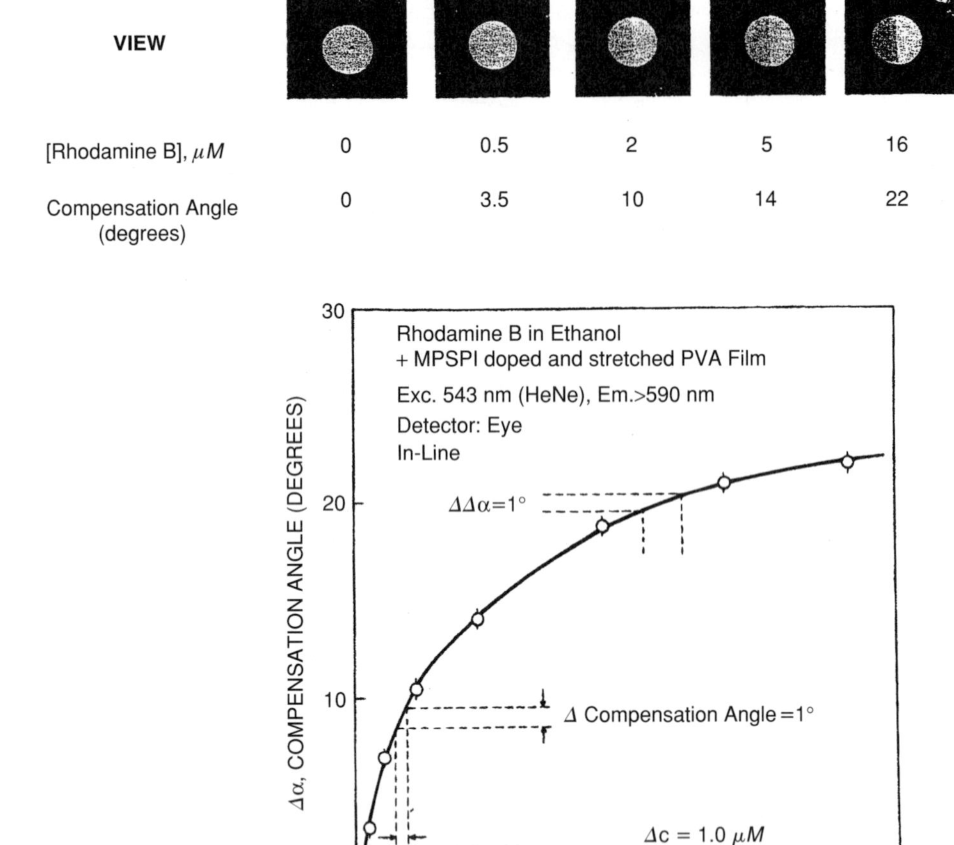

FIG. 25. Emitted light observed through an analyzer polarizer for different concentrations of rhodamine B, using the MPSI reference and in-line geometry (*top*). The listed values are the compensation angles ($\Delta\alpha$) needed to equalize the intensities. Dependence of the compensation angle ($\Delta\alpha$) on the concentration of rhodamine B, using in-line geometry (*bottom*), is measured.

Summary

Novel approaches to sensor design, based on the use of an internal standard with appropriate spectral properties, provide new possibilities for designing simple devices for fluorescence sensing. Detection of combined emission from the reference and an analyte-sensitive fluorophore has been achieved in numerous measurements in cuvettes,[7–10,42,43] tissues,[46,47] and high-throughput formats.[48] These methods have been used with a long-lifetime reference to measure pH, O_2, pCO_2, glucose, and calcium[8,9] by means of modulation-sensing methods as well as by the use of oriented films as the reference for polarization sensing of glucose, pH, oxygen, and lactate.[21–23] Polarization sensing has also been developed with visual detection to measure the concentration of rhodamine B and pH.[24]

Modulation and polarization sensing was found to be effective in highly scattering media such as Intralipid or tissue. The applicability of these technologies to transdermal diagnostics depends on the availability of red fluorophores that can be used *in vivo*. One dye that could possibly be used is indocyanine green (IcG), which absorbs and emits at wavelengths above 700 nm. Furthermore, IcG has already been approved for use in humans for monitoring burn severity[49] and it has been detected through the skin.[50]

It appears likely that modern optics and electronic technology will allow the development of practical devices for biomedical use as shown in Scheme 1.

[48] J. R. Lakowicz, I. Gryczynski, and Z. Gryczynski, *J. Biomol. Screening* **5,** 123 (2000).
[49] J. M. Still, E. J. Law, K. G. Klavuhn, T. C. Island, and J. Z. Holtz, *Burns* **27,** 364 (2001).
[50] R. Boushel, H. Langberg, J. Olesen, M. Nowak, L. Simonsen, J. Bülow, and M. Kjær, *J. Appl. Physiol.* **89,** 1868 (2000).

[3] Bioluminescence and Chemiluminescence

By J. WOODLAND HASTINGS and CARL HIRSCHIE JOHNSON

Introduction

Three earlier volumes of *Methods in Enzymology* have been devoted to bioluminescence and chemiluminescence.[1–3] In this article we present an overview of the biochemistry of several systems that have found analytical and reporter applications, along with examples of such applications.

Researchers studying bioluminescence have long relied on their own ingenuity to construct custom photomultiplier devices for measurements of bioluminescence. Fortunately, a wide range of instruments, both analog and digital (photon counting), are now available commercially.[1–3] Studying bioluminescence and/or using it as a tool is thus more accessible than ever before.

[1] M. DeLuca, ed., *Methods Enzymol.* **57** (1978).
[2] M. A. DeLuca and W. D. McElroy, eds., *Methods Enzymol.* **133** (1986).
[3] M. M. Ziegler and T. O. Baldwin, eds., *Methods Enzymol.* **305** (2000).

0076-6879/03 $35.00

TABLE I
LUCIFERASE GENES AND PROTEINS FROM DIFFERENT ORGANISMS

Organism	Gene	Protein
Bacteria	*luxAB*	LuxAB
Firefly	*fluc*	fLUC
Renilla	*rluc*	rLUC
Gaussia	*gluc*	gLUC
Dinoflagellate	*lcf*	LCF
Aequorea	*aeq*	AEQ

There are many different bioluminescent organisms, from bacteria and fungi to mollusks, crustaceans, insects, fishes, and many more.[4,5] This is matched by a similar diversity of mechanisms, meaning that the substrates (luciferins) and enzymes (luciferases), as well as the genes responsible, are also different (Table I). The available evidence indicates that many or most of the different systems originated independently during evolution,[6,7] so bioluminescence is an excellent case of convergent evolution. However, there is a degree of unity at the chemical level, as indicated by generic reaction below: all are enzyme-catalyzed reactions in which the luciferins are oxidized by molecular oxygen, and in all probability all involve the formation and breakdown of an intermediate luciferase-bound peroxide or hydroperoxide.[8]

Bioluminescence is an enzyme-catalyzed chemiluminescence reaction in which the energy released (40 to 60 kcal) is used to produce an intermediate or product in an electronically excited state, P*, which then emits a photon. The emission does not come from or depend on light absorbed, as in fluorescence or phosphorescence, but the excited state produced is indistinguishable from that produced in fluorescence after the absorption of a photon by the ground state of the molecule concerned (Scheme 1). Thus, there cannot be efficient bioluminescence if the emitter has a low fluorescence quantum yield, and kinetic events happening on a time scale longer than nanoseconds must be due to processes preceding the formation of the excited emitter.[9]

[4] E. N. Harvey, "Bioluminescence." Academic Press, New York, 1952.
[5] P. J. Herring, ed., "Bioluminescence in Action," p. 570. Academic Press, New York, 1978.
[6] J. W. Hastings, *J. Mol. Evol.* **19,** 309 (1983).
[7] J. W. Hastings and K. V. Wood, *in* "Photobiology for the 21st Century" (D. P. Valenzeno and T. P. Coohil, eds.). Valdenmar Publishing, Overland Park, KS, 2001.
[8] T. Wilson and J. W. Hastings, *Annu. Rev. Cell Dev. Biol.* **14,** 197 (1998).
[9] T. Wilson, *Photochem. Photobiol.* **62,** 601 (1995).

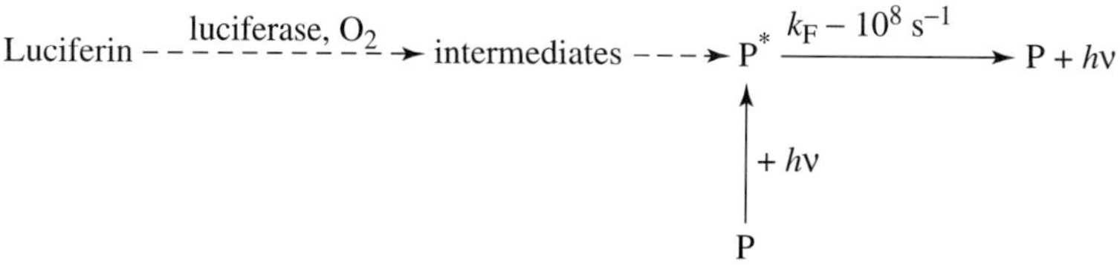

$$\text{Luciferin} \xrightarrow{\text{luciferase, } O_2} \text{intermediates} \longrightarrow P^* \xrightarrow{k_F - 10^8\ s^{-1}} P + h\nu$$

$$P \xrightarrow{+h\nu} P^*$$

SCHEME 1

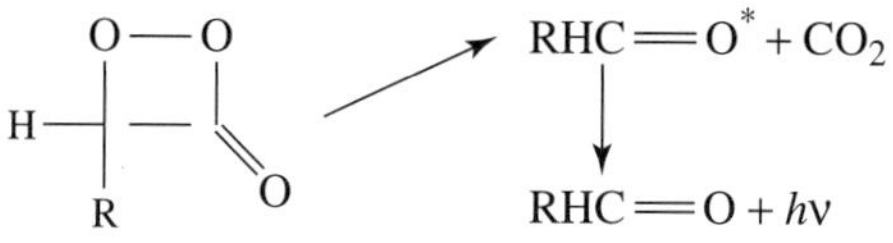

$$\text{dioxetanone (H, R)} \longrightarrow RHC{=}O^* + CO_2 \longrightarrow RHC{=}O + h\nu$$

SCHEME 2

In both bioluminescence and chemiluminescence in solution, two types of intermediate peroxides have been characterized: linear peroxides and those with four-membered energy-rich dioxetanone rings. In bioluminescence, these are exemplified by the bacterial system and the firefly or coelenterate systems, respectively.

Dioxetanone cleavage in bioluminescence forms CO_2 and an excited carbonyl (Scheme 2); it may be catalyzed by an internal electron transfer to the O—O bond from the R moiety if its oxidation potential is appropriately low. The consequence is a fast cascade of processes terminating in the generation of the excited state of RHC=O, possibly via a final charge-annihilation process. Although not proved, even in model systems, this proposal of a chemically induced electron-exchange luminescence (CIEEL)[10,11] is often invoked and discussed in the literature, in both the firefly/coelenterate cases, and also for the bacterial reaction.

In bioluminescence, the expectation that the emission should match spectrally the fluorescence of the reaction product may not be realized. One reason is that the ground state of the excited product may be unstable and quickly break down to a nonfluorescent product, or that the emission may originate from an enzyme-bound intermediate or product, whose intensity and spectrum may differ from that of the free excited product. A more exotic mechanism involves a second fluorophore (F) to which the energy from the primary excited species (P*) is nonradiatively transferred, thereby causing this accessory fluorophore to become excited and emit its own fluorescence (Scheme 3). An example of this was first discovered

[10] G. B. Schuster, *Acc. Chem. Res.* **12,** 366 (1979).

[11] L. H. Catalani and T. Wilson, *J. Am. Chem. Soc.* **111,** 2633 (1989).

$$\begin{array}{l} P^* + F \longrightarrow F^* + P \\ \qquad\qquad\quad \downarrow \\ \qquad\qquad\quad F \;\; + h\nu \end{array}$$

SCHEME 3

in the coelenterate system,[12] with the green fluorescent protein (GFP) identified as the secondary emitter;[13] blue and yellow accessory fluorescent proteins were discovered later in the bacterial system.[14–16]

Radiationless transfer can happen by either the so-called electron-exchange (or Dexter) mechanism, when there is collisional contact between energy donor and acceptor, or via so-called fluorescence resonance energy transfer (FRET), which occurs at distances much greater than molecular diameters if there is a good overlap between the absorption spectrum of the second fluorophore and the emission spectrum of the primary emitter.[17–19] Energy transfer to GFP is considered to involve FRET.

Bacterial accessory emitters, described below, may actually participate in the reaction by binding to an enzyme–substrate intermediate and influencing the kinetics of the reaction, and in this process acquire excitation energy in a different way, bypassing excitation of the primary emitter.[20,21]

Bacterial Luminescence

Luminous bacteria are found ubiquitously in the oceans and can be readily isolated directly from sea water. Colonies typically emit a continuous blue-green light, about 10^4 or 10^5 photons s^{-1} per cell, when strongly expressed; they do not flash or otherwise modulate the intensity. Some but not all species of luminous bacteria grow symbiotically as pure cultures in photogenic organs of many host species, notably fish and squid, and thereby provide the light source for the host.

[12] O. Shimomura, F. H. Johnson, and Y. Saiga, *J. Cell. Comp. Physiol.* **59,** 223 (1962).
[13] J. G. Morin and J. W. Hastings, *J. Cell. Physiol.* **77,** 313 (1971).
[14] P. Koka and J. Lee, *Proc. Natl. Acad. Sci. U.S.A.* **76,** 3068 (1979).
[15] E. Ruby and K. Neason, *Science* **196,** 432 (1977).
[16] P. Macheroux, K. U. Schmidt, P. Steinerstauch, S. Ghisla, P. Colepicolo, R. Buntic, and J. W. Hastings, *Biochem. Biophys. Res. Commun.* **146,** 101 (1987).
[17] A. A. Lamola, "Energy Transfer and Organic Photochemistry." Interscience, New York, 1969.
[18] N. J. Turro, "Modern Molecular Photochemistry," p. 628. Benjamin/Cummings, Menlo Park, CA, 1978.
[19] P. Wu and L. Brand, *Anal. Biochem.* **218,** 1 (1994).
[20] J. W. Eckstein, K. W. Cho, P. Colepicolo, S. Ghisla, J. W. Hastings, and T. Wilson, *Proc. Natl. Acad. Sci. U.S.A.* **87,** 1466 (1990).
[21] G. Sirokmán and J. W. Hastings, *Photochem. Photobiol.* **66,** 198 (1997).

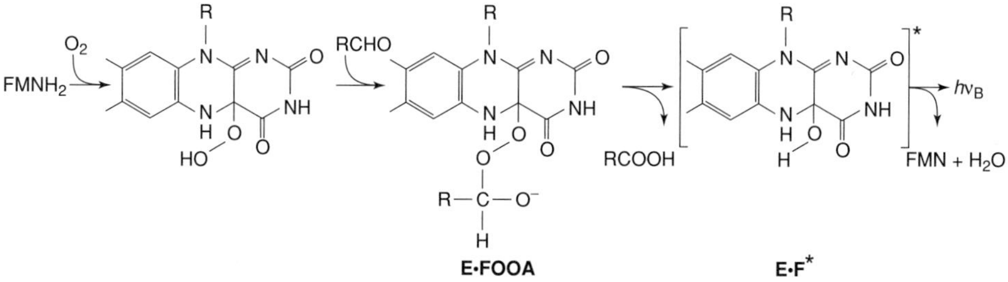

SCHEME 4

Bacterial luciferase catalyzes the mixed function oxidation of a long-chain aldehyde and reduced flavin mononucleotide, $FMNH_2$.[22] The enzyme is an external flavin monoxygenase (EC 1.14.14.3). Curiously, no other enzymes of this general type have been found to emit light, even at low quantum yields. The pathway itself constitutes a shunt of cellular electron transport at the level of flavin, and $FMNH_2$ is called the luciferin, because it gives rise to the emitter. Several intermediates, notably luciferase–peroxyflavin,[23,24] have been characterized, but the most critical, the postulated peroxyhemiacetal, E · FOOA, has so far resisted isolation. An interesting feature of this reaction (Scheme 4) is its inherent slowness: at 20° the time required for a single catalytic cycle is about 20 sec. Although many steps are involved,[25] it is the lifetime of E · FOOA that determines the rate of the reaction.[26] The emitter is the enzyme-bound 4a-hydroxyflavin, E · F*,[27] and the quantum yield is ~0.2.

Electrochemical studies,[28] as well as the effect of 8-position substituents of various oxidation potentials on the flavin ring,[29] indicate that the rate-determining step is influenced by electron transfer (or charge redistribution) toward the peroxide bond, which is estimated to be unusually weak: cleavage of the O—O bond could follow either the transfer of an electron from the N5 lone pair, or the shift of a proton from the aldehyde moiety to the O—O bond. The excitation step is presently regarded as a charge annihilation between two radical ion centers. All electron shifts and bond rearrangements are considered to take place within the constraints of the enzyme pocket, which is assumed to keep the reacting groups in contact

[22] S.-C. Tu and H. I. X. Mager, *Photochem. Photobiol.* **62,** 615 (1995).

[23] J. W. Hastings, C. Balny, C. Le Peuch, and P. Douzou, *Proc. Natl. Acad. Sci. U.S.A.* **70,** 3468 (1973).

[24] J. Vervoort, F. Muller, J. Lee, W. A. M. van den Berg, and C. T. W. Moonen, *Biochemistry* **25,** 8062 (1986).

[25] H. M. Abu-Soud, A. C. Clark, W. A. Francisco, T. O. Baldwin, and F. M. Raushel, *J. Biol. Chem.* **268,** 7699 (1993).

[26] P. Macheroux, S. Ghisla, and J. W. Hastings, *Biochemistry* **32,** 14183 (1993).

[27] M. Kurfürst, P. Macheroux, S. Ghisla, and J. W. Hastings, *Biochim. Biophys. Acta* **924,** 104 (1987).

[28] G. Merenyi, J. Lind, H. I. X. Mager, and S.-C. Tu, *J. Phys. Chem.* **96,** 10528 (1992).

[29] J. W. Eckstein, J. W. Hastings, and S. Ghisla, *Biochemistry* **32,** 404 (1993).

and in proper orientation, a role tentatively attributed to the solvent cage in the examples of "intramolecular CIEEL."

Luciferases from all bioluminescent bacteria studied are heterodimers of the α (~40-kDa) and β (~35-kDa) subunits. The crystal structure of *Vibrio harveyi* luciferase has been determined at both 2.4- and 1.5-Å resolution, but only in the absence of substrates.[30,31] A deep pocket on the α subunit extending to the β subunit may be the catalytic site, consistent with the finding that mutants of catalytic properties are confined to the α subunit.[32]

Although luciferase mutants may have slightly altered emission spectra,[33] major shifts occur in species possessing accessory proteins carrying chromophores: 6,7-dimethyl-8-(1′-D-ribityl)lumazine in the blue-shifted strains,[14] and FMN or riboflavin in the red-shifted strains.[16,34–36] *In vitro,* these two homologous and functionally analogous proteins [lumazine protein (LumP) and bacterial yellow fluorescent protein (YFP), respectively] shift the emission peaks of the corresponding luciferase reactions in a concentration-dependent manner. In the case of YFP, it does more than accept energy; it accelerates by up to 10-fold the rate of intensity decay of the *in vitro* reaction of *Vibrio fischeri* luciferase, in a concentration-dependent manner.[20] It must therefore interact with and destabilize an enzyme-bound intermediate, such as the peroxyhemiacetal, and deviate the reaction course.[21,37]

The *lux* operon includes *luxA* and *luxB* genes encoding the two luciferase subunits. It also contains, among others, three genes (*luxC, D*, and *E*) encoding proteins that make up the fatty acid reductase complex (for aldehyde synthesis), as well as *luxR* and *luxI*, in two divergent operons, involved in the novel autoinduction (quorum-sensing) regulatory mechanism[38,39] (Diagram 1).

The biosynthesis of the entire system is actually autoregulated, and it was the discovery of this mechanism that led to the discovery of quorum sensing, now known to be widespread among bacteria.[40,41] Its biological meaning is that the transcription of certain genes occurs only when the cell density is high enough for

[30] A. J. Fisher, F. M. Raushel, T. O. Baldwin, and L. Rayment, *Biochemistry* **34,** 6581 (1995).
[31] A. J. Fisher, T. B. Thompson, J. B. Thoden, T. O. Baldwin, and I. Rayment, *J. Biol. Chem.* **271,** 21956 (1996).
[32] T. W. Cline and J. W. Hastings, *Biochemistry* **11,** 3359 (1972).
[33] T. W. Cline and J. W. Hastings, *J. Bacteriol.* **118,** 1059 (1974).
[34] S. C. Daubner, A. M. Astorga, G. B. Leisman, and T. O. Baldwin, *Proc. Natl. Acad. Sci. U.S.A.* **84,** 8912 (1987).
[35] H. Karatani and J. W. Hastings, *Photochem. Photobiol.* **18,** 227 (1993).
[36] V. N. Petushkov, B. G. Gibson, and J. Lee, *Biochem. Biophys. Res. Commun.* **211,** 774 (1995).
[37] G. Sirokmán, T. Wilson, and J. W. Hastings, *Biochemistry* **34,** 13074 (1995).
[38] E. A. Meighen, *Annu. Rev. Genet.* **28,** 117 (1994).
[39] S. Ulitzur and P. V. Dunlap, *Photochem. Photobiol.* **62,** 625 (1995).
[40] K. H. Nealson, T. Platt, and J. W. Hastings, *J. Bacteriol.* **104,** 313 (1970).
[41] J. W. Hastings and E. P. Greenberg, *J. Bacteriol.* **181,** 2667 (1999).

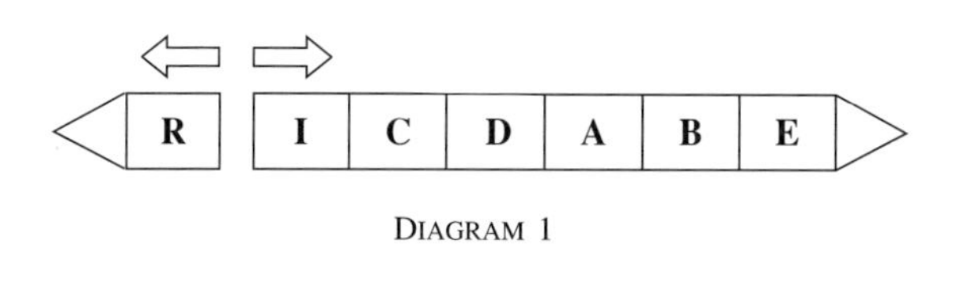

DIAGRAM 1

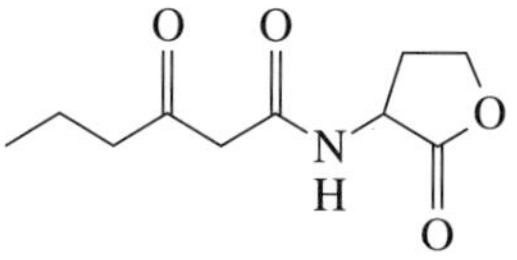

VAI

SCHEME 5

the products of these genes to act optimally. The transcription of the *lux* operon is thus restrained by the autoinduction mechanism and then triggered by a freely diffusible pheromone (autoinducer), which is synthesized by the bacteria and accumulates both intra- and extracellularly as the cells grow, as in the light organ of a host such as a fish or a squid.[42] In the laboratory, a not-yet luminescing, low-density culture can indeed be caused to synthesize luciferase and to emit light by the addition of cell-free medium from a late culture.[40]

In *V. fischeri* the autoinducer (VAI) is an *N*-acylhomoserine lactone (Scheme 5).[43] The product of *luxI* on the rightward operon synthesizes the autoinducer from *S*-adenosylmethionine and an acyl-[acyl-carrier-protein].[44–46] The *luxR* gene on the leftward operon encodes the LuxR protein, whose N-terminal domain binds the autoinducer[47]; this causes the LuxR C-terminal domain to act as a transcriptional regulator, by binding DNA in synergy with RNA polymerase at a palindromic sequence located upstream of the transcription start of the rightward *lux* operon.[48]

When the concentration of VAI in the medium (and therefore within the cell) is below a threshold, transcription of the rightward operon goes on at a steady but low level, allowing for the build-up of autoinducer concentration to such a point

[42] K. J. Boettcher and E. G. Ruby, *J. Bacteriol.* **1995,** 1053 (1995).

[43] A. Eberhard, A. L. Burlingame, C. Eberhard, G. L. Kenyon, K. H. Nealson, and N. J. Oppenheimer, *Biochemistry* **20,** 2444 (1981).

[44] M. I. Moré, L. D. Finger, J. L. Stryker, C. Fuqua, A. Eberhard, and S. C. Winans, *Science* **272,** 1655 (1996).

[45] B. L. Hanzelka and E. P. Greenberg, *J. Bacteriol.* **178,** 5291 (1996).

[46] A. L. Schaefer, D. L. Val, B. L. Hanzelka, J. E. Cronan, and E. P. Greenberg, *Proc. Natl. Acad. Sci. U.S.A.* **93,** 9505 (1996).

[47] B. L. Hanzelka and E. P. Greenberg, *J. Bacteriol.* **177,** 815 (1995).

[48] A. M. Stevens and E. P. Greenberg, *J. Bacteriol.* **179,** 557 (1997).

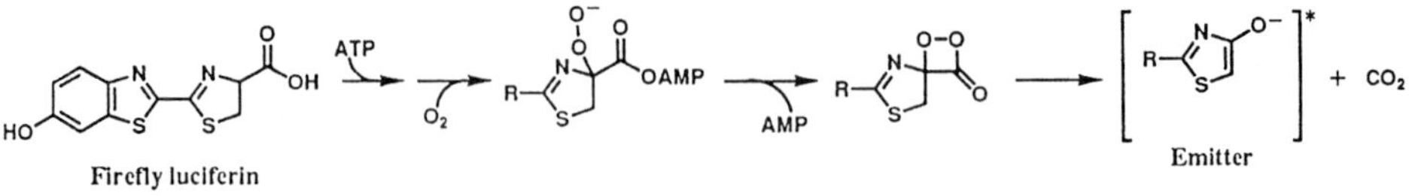

Firefly luciferin

ATP

O_2

OAMP

AMP

Emitter

$+ CO_2$

SCHEME 6

that VAI associates with LuxR and activates transcription of both operons in an autocatalytic feedback loop. Luciferase synthesis is increased by VAI binding to LuxR, as is the production of VAI due to a positive feedback loop.[49,50]

Firefly Bioluminescence

Fireflies are lampyrid beetles, but members of other beetle families are also luminous and utilize similar biochemical systems, notably the railroad worm with its lanterns emitting at different wavelengths,[51] and click beetles. In fireflies the major function of light emission is for communication during courtship, in which one sex emits a flash as a signal, to which the other responds, usually in a species-specific pattern.[52–54] The time delay between the two is a signaling feature in some cases, but the flashing pattern or kinetic features may also be utilized. Firefly flashing in areas of Southeast Asia is particularly striking, especially the synchronous rhythmic flashing of *Pteroptyx* species, which form congregations of many thousands in single trees, where the males produce an all-night-long display, with flashes every few seconds.[55]

Firefly luciferase is a 62-kDa monooxygenase not involving a metal or cofactor, the reaction chemistry being basically the same or similar for all beetles, all reacting with firefly luciferin, a benzothiazoylthiazole (Scheme 6).[56,57] The enzyme first catalyzes the condensation of luciferin with ATP in the presence of Mg^{2+} to form the luciferyl adenylate, with ATP providing AMP as a good leaving group. This is followed by the reaction with oxygen and the cyclization of the peroxide to form the critical energy-rich dioxetanone intermediate. The breakdown of the dioxetanone

[49] A. Eberhard, T. Longin, C. A. Widrig, and S. J. Stranick, *Arch. Microbiol.* **155,** 294 (1991).
[50] P. V. Dunlap, *J. Biolumin. Chemilumin.* **7,** 203 (1992).
[51] V. R. Viviani and E. J. H. Bechara, *Ann. Entomol. Soc. Am.* **90,** 389 (1997).
[52] J. Case, *in* "Insect Communication" (T. Lewis, ed.), p. 195. Harcourt Brace Jovanovich, London, 1984.
[53] J. E. Lloyd, *in* "How Animals Communicate" (T. A. Sebeok, ed.), p. 164. Indiana University Press, Bloomington, IN, 1977.
[54] J. E. Lloyd, *Science* **210,** 669 (1980).
[55] J. Buck, *Q. Rev. Biol.* **63,** 265 (1988).
[56] Y. Ohmiya, N. Ohba, H. Toh, and F. I. Tsuji, *in* "Bioluminescence and Chemiluminescence" (A. K. Campbell, L. J. Kricka, and P. E. Stanley, eds.), p. 572. John Wiley & Sons, Chichester, 1994.
[57] K. V. Wood, *Photochem. Photobiol.* **62,** 662 (1995).

provides energy to form the excited oxyluciferin, with an overall quantum yield of ~0.9 photon per oxidized luciferin.[58] Even though all beetle luciferins are the same, their emissions differ in color, from green to orange and red (railroad worm); this is attributed to the tertiary structure at the catalytic site.[59]

The crystal structure of firefly luciferase without substrates shows a large N-terminal domain and a small C-terminal domain linked by a flexible, four-residue loop, with a cleft exposed to water between the two.[60] The most conserved residues are located on the facing surfaces of this cleft and on the loop connecting the domains, and this area may be the active site. Modeling of the active site containing substrates luciferin and Mg-ATP[61] and crystal structures of luciferase containing inhibitor,[62] as well as a model based on energy minimization,[63] are in general agreement.

The cDNAs of several beetle luciferases have been cloned and expressed in *Escherichia coli* and several eukaryotes.[64–68] Different beetle luciferases have a 40–50% sequence identity at the amino acid level, and there are sequence similarities with all enzymes that also activate the carboxyl group of their substrates via adenylation, such as 4-coumarate:CoA ligases.[57,69] This is reflected in an effect of coenzyme A on the firefly luciferase reaction.[70]

Mutagenesis of the luciferase cDNA of the Japanese firefly (*Luciola cruciata*) resulted in emissions ranging from green to red, as a consequence, in each case, of single amino acid substitutions.[71] In the case of click beetle luciferase isozymes, expression in *E. coli* of four different cDNAs encoding proteins differing in only a few amino acids resulted in peak emissions ranging from green to orange; these substitutions were all in a short region of the the N-terminal domain.[67]

How a single residue substitution can radically affect the emission spectrum requires knowing its exact location in relation to the active site and the identity

[58] H. H. Seliger and W. D. McElroy, *Biochem. Biophys. Res. Comm.* **1,** 21 (1959).

[59] F. McCapra, D. J. Gilfoyle, D. W. Young, N. J. Church, and P. Spencer, *in* "Bioluminescence and Chemiluminescence" (A. K. Campell, L. J. Kricka, and P. E. Stanley, eds.), p. 387. John Wiley & Sons, Chichester, 1994.

[60] E. Conti, N. P. Franks, and P. Brick, *Structure* **4,** 287 (1996).

[61] B. R. Branchini, R. A. Magyar, M. H. Murtiashaw, S. M. Anderson, and M. Zimmer, *Biochemistry* **37,** 15311 (1998).

[62] N. Franks, A. Jenkins, E. Conti, W. Lieb, and P. Brick, *Biophys. J.* **75,** 2205 (1998).

[63] T. P. Sandalova and N. N. Ugarova, *Biochemistry (Moscow)* **64,** 962 (1999).

[64] S. J. Gould, G.-A. Keller, and S. Subramani, *J. Cell Biol.* **105,** 2923 (1987).

[65] J. H. Devine, G. D. Kutuzova, V. A. Green, N. N. Ugarova, and T. O. Baldwin, *Biochim. Biophys. Acta* **1173,** 121 (1993).

[66] T. Masuda, H. Tatsumi, and E. Nakano, *Gene* **77,** 265 (1989).

[67] K. W. Wood, A. Y. Lam, H. H. Seliger, and W. D. McElroy, *Science* **244,** 700 (1989).

[68] D. W. Ow, K. V. Wood, M. DeLuca, J. R. deWet, D. R. Helinski, and S. H. Howell, *Science* **234,** 856 (1986).

[69] J. Schroder, *Nucleic Acids Res.* **17,** 460 (1989).

[70] R. L. Airth, W. C. Rhodes, and W. D. McElroy, *Biochim. Biophys. Acta* **27,** 519 (1958).

[71] N. Kajiyama and E. Nakano, *Protein Eng.* **4,** 691 (1991).

of the excited reaction product. It is assumed that the excited monoanion of the keto form of oxyluciferin emits red light.[57] It had been thought that the wild-type yellow emission resulted from enolization during the lifetime of the excited state, at physiological pH 8. Indeed, the 5,5′-dimethyl analog of luciferin, which cannot enolize, emits only red light. It was also thought that low pH (~pH 6) results in red light because of more extended conjugation. Oxyluciferin is so extremely unstable that it is practically impossible to confirm experimentally these conclusions.[72]

The previous work had thus focused on the identity of two discrete emitters, one emitting yellow, the other red light. But the observation that naturally occurring and mutant luciferases emitted in a whole range of colors, each with an emission spectrum consisting of a narrow, shoulderless band, cannot be accounted for by the superposition of the spectra of these two emitters. In fact, the effect of protein environment on the color of emission may be due to restriction on the conformation of luciferin.[59,73] Although the molecule is probably insensitive in the ground state to the angle of twist between the two rings around the C2–C2′ axis, in the excited state the energy minimum is estimated to correspond to a structure with the rings at a 90° angle. When a twisted excited oxyluciferin anion emits, it ends up in a twisted ground state, because conformation changes are too slow to compete with emission of a photon; the energy of the electronic transition will then correspond to photons toward the red end of the spectrum. But if the luciferin intermediate is constrained by luciferase to be nearly planar, the energy of the emitted photon will be larger, that is, blue shifted. Thus the degree of twist might determine any one of a continuum of emission colors.

The firefly light organ comprises a series of photocytes arranged in a rosette, positioned radially around a central tracheole, which supplies oxygen to the organ.[74] The organ comprises a series of such rosettes, stacked in many dorsoventral columns. Photocyte vesicles containing luciferase colocalize with peroxisomes by immunolabeling,[75] and the C-terminal tripeptide SKL peroxisomal targeting sequence is present in luciferase of American species; expressed in mammalian cells, luciferase localizes to peroxisomes.[76–78] However, the SKL sequence is absent in the Japanese firefly luciferase.[66]

[72] E. H. White and D. F. Roswell, *Photochem. Photobiol.* **53,** 131 (1991).

[73] F. McCapra, *in* "Bioluminescence and Chemiluminescence" (J. W. Hastings, L. J. Kricka, and P. E. Stanley, eds.), p. 7. John Wiley & Sons, Chichester, 1997.

[74] H. Ghiradella, *in* "Microscopic Anatomy of Invertebrates," Vol. 11a: "Insecta," p. 363. Wiley-Liss, New York, 1998.

[75] C. H. Hanna, T. A. Hopkins, and J. Buck, *J. Ultrastruct. Res.* **57,** 150 (1976).

[76] G.-A. Keller, S. J. Gould, M. DeLuca, and S. Subramani, *Proc. Natl. Acad. Sci. U.S.A.* **84,** 3264 (1987).

[77] S. J. Gould, G.-A. Keller, N. Hosken, J. Wilkinson, and S. Subramani, *J. Cell Biol.* **108,** 1657 (1989).

[78] U. Soto, R. Pepperkok, W. Ansorge, and W. W. Just, *Exp. Cell Res.* **205,** 66 (1993).

Nerve impulses triggering flashes travel to the lantern, the neurotransmitter being octopamine.[79–81] However, the nerve termini are not on photocytes, but on the adjacent tracheolar cells, which form the tubes carrying oxygen to the organ. This probably accounts for the time delay between the arrival of the nerve impulse and the flash onset. The favored theory for the control of flashing is that it involves oxygen availability for the luminescence reaction. Biochemical studies support this possibility; the removal of oxygen from a reaction mixture extinguishes luminescence, and the reintroduction of oxygen results in a flash with a peak intensity more than 100 times the level before removal of oxygen, with a rapid decay back to that level without the removal of oxygen.[82,83] This can be reliably attributed to an accumulation of the luciferyl-adenylate intermediate, whose reaction with oxygen and the subsequent steps leading to light emission are extremely rapid compared with the relatively slow formation of that intermediate.

If this is so in the living system, how is the oxygen withheld and then introduced to produce a flash? Photocytes are richly endowed cortically with mitochondria,[84] which are credited with keeping the cell interior anaerobic, surely made easier if the oxygen supply is limited. Timmins *et al.*[85] propose that oxygen entry to photocytes is restricted by fluid in tracheoles, and that a flash is generated by a transient withdrawl of fluid, this caused by an action potential-triggered osmotic change in the tracheolar cells, allowing oxygen to enter the photocytes and give rise to the flash.

An alternative theory proposes instead that mitochondrial oxygen consumption is fully responsible for keeping the cells anaerobic even when oxygen entry is not restricted.[86] They then propose that neurally released octopamine activates nitric oxide synthase in the abutting cells, and that NO then transiently inhibits photocyte mitochondrial respiration and results in increased oxygen levels in the peroxisomes, thus causing flashing. How neurally released octopamine might result in the postulated activation of nitric oxide synthase was not established.

The flashing mechanism, whatever it may be, must account for the rapid kinetics, complex waveforms, multiple flashes, and high-frequency flickering, all of which seem unlikely to be regulated by a mechanism involving synthesis and

[79] J. F. Case and L. G. Strause, *in* "Bioluminescence in Action" (P. J. Herring, ed.), p. 331. Academic Press, New York, 1978.

[80] D. Oertel and J. R. Case, *J. Exp. Biol.* **65,** 213 (1976).

[81] A. D. Carlson and M. Jalenak, *J. Exp. Biol.* **122,** 453 (1986).

[82] J. W. Hastings, W. D. McElroy, and J. Coulombre, *J. Cell Comp. Physiol.* **42,** 137 (1953).

[83] W. D. McElroy and J. W. Hastings, *in* "The Luminescence of Biological Systems" (F. H. Johnson, ed.), p. 161. AAAS Press, Washington, D.C., 1955.

[84] H. Ghiradella, *J. Morphol.* **153,** 187 (1977).

[85] G. S. Timmins, F. J. Robb, C. M. Wilmot, S. K. Jackson, and H. M. Swartz, *J. Exp. Biol.* **204,** 2795 (2001).

[86] B. A. Trimmer, J. R. Aprille, D. M. Dudzinski, C. J. Lagace, S. M. Lewis, T. Michel, S. Oazi, and R. M. Zayas, *Science* **292,** 2486 (2001).

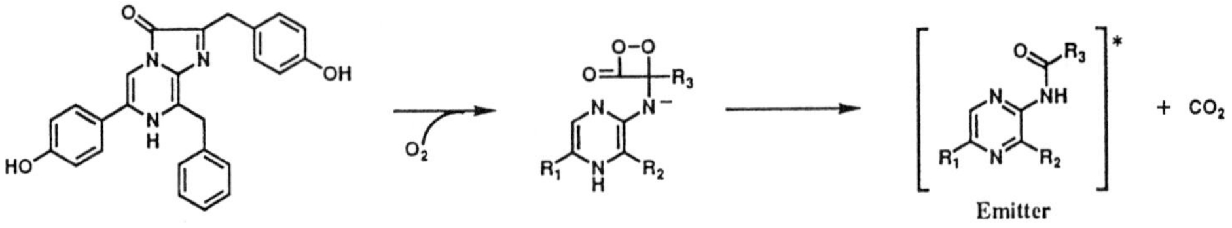

Coelenterate luciferin

Emitter

$+ CO_2$

SCHEME 7

enzymatic inhibition. Gas transfer also seems ill suited for rapid kinetics, but in comparing different species, there is a strong positive relationship between the extent of the tracheal supply system in the adults and the flashing ability, and lantern tracheoles exhibit a uniquely strengthened morphology in flashing adults.[84]

Cnidaria (Coelenterate) and Ctenophore Bioluminescence

The sea pansy *Renilla,* the jellyfish *Aequorea,* the hydroid *Obelia,* and the ctenophore *Mnemiopsis* are among the best studied of many bioluminescent species in this group. They emit light as brief, bright flashes, typically caused by mechanical stimulation and spread by conducted calcium action potentials. An inward calcium current occurs in depolarized support cells, not the photocytes.[87] Flashing is believed to function to deter predation.

The luciferin, coelenterazine, is an imidazolopyrazine, which occurs widely in luminous and nonluminous marine organisms.[88–90] Its reaction with oxygen gives a dioxetanone intermediate, which breaks down to give CO_2 and oxidized luciferin (coelenteramide) in the excited state (Scheme 7).[8,91]

In *Renilla,* an anthozoan, four proteins are involved.[92] The first, a sulfokinase, removes a sulfate from a precursor, and the freed coelenterazine binds to a nonenzymatic protein (18.5 kDa, with three Ca^{2+}-binding sites); calcium triggers its release and light emission ensues, with the reaction catalyzed by *Renilla* luciferase (35 kDa). The fourth protein is the green fluorescent protein, *Renilla* GFP, which (like the sulfokinase and binding protein) is not needed for the light-emitting reaction as such and, indeed, is not present in all anthozoans. In the absence of GFP, luciferase-bound excited coelenteramide emits blue light (λ_{max}, ~480 nm); in its presence, the light emitted is green (λ_{max}, ~509 nm).

[87] K. Dunlap, K. Takeda, and P. Brehm, *Nature (London)* **325,** 60 (1987).

[88] O. Shimomura, *Comp. Biochem. Physiol.* **86B,** 361 (1987).

[89] O. Shimomura, S. Inoue, F. H. Johnson, and Y. Haneda, *Comp. Biochem. Physiol.* **65B,** 435 (1980).

[90] C. M. Thomson, P. J. Herring, and A. K. Campbell, *J. Biolumінesc. Chemiluminesc.* **12,** 87 (1997).

[91] O. Shimomura, *in* "Chemical and Biological Generation of Excited States" (W. Adam and G. Cilento, eds.), p. 249. Academic Press, New York, 1982.

[92] M. J. Cormier, *in* "Bioluminescence in Action" (P. J. Herring, ed.), p. 75. Academic Press, New York, 1978.

When first isolated, the hydrozoan *Aequorea* system appeared not to have the classic luciferase–luciferin system, or to require oxygen for luminescence,[91] as had been discovered earlier for living ctenophores.[4] In the presence of a calcium chelator, they isolated and purified a "photoprotein," aequorin, which requires only calcium for light production. It is now clear that the photoprotein is simply a stable luciferase reaction intermediate to which an oxygenated form of the coelenterazine is already bound, probably as a hydroperoxide. Calcium, for which the protein has three binding sites, triggers the flash by allowing the reaction to go to completion via the dioxetanone intermediate. Thus, instead of triggering at the stage of luciferin availability, as in *Renilla,* calcium acts on a reaction intermediate in *Aequorea.*

The crystal structures of the "photoproteins" aequorin and obelin (from the hydrozoan *Obelia*) have been reported. Aequorin is a globular molecule with a hydrophobic cavity where the coelenterazine 2-hydroperoxide is located and stabilized.[93] It possesses four helix–loop–helix "EF-hand" domains, three of which bind calcium. In the obelin structure only a single oxygen atom occurs at the 2-position.[94,95]

The three (12-amino acid-long) Ca^{2+}-binding sites of the three studied proteins, aequorin, obelin, and luciferin binding protein from *Renilla,* are homologous to the Ca^{2+}-binding sites of other calcium-binding proteins, such as calmodulin. Calmodulin has four such sites, with the spacing between sites 3 and 4 being the same as that between sites 2 and 3 of the three bioluminescence proteins. It is speculated that these genes had a common ancestor and that site 2 in calmodulin might have become the luciferin-binding site in the coelenterate proteins.[96]

As in *Renilla,* the *Aequorea* emission is blue (~486 nm) when the reaction is carried out in extracts, but green (~508 nm) from the living organism. Shimomura and co-workers discovered a protein with green fluorescence associated with the *Aequorea* luminescence system that they recognized could be responsible for converting the blue luminescence of aequorin into a green color.[12,97] Later, Morin and Hastings[13] discovered that such a protein, which they referred to as green fluorescent protein (GFP), was located exclusively in photocytes in the colonial hydroid *Obelia,* and that it occurred also in *Renilla.* This scenario occurs in many coelenterates: green emission *in vivo* due to GFP, but blue in extracts, where it is typically too dilute for energy transfer. The involvement of energy transfer was first inferred in the case of *Obelia,* where, in extracts, photoprotein and GFP are found together in granules.[13] If the cells are mechanically ruptured in sea water containing $MgCl_2$

[93] J. F. Head, S. Inouye, K. Teranishi, and O. Simomora, *Nature (London)* **405,** 372 (2000).

[94] Z.-J. Liu, E. S. Vysotski, C. J. Chen, J. P. Rose, J. Lee, and B. C. Wang, *Protein Sci.* **9,** 2085 (2000).

[95] Z. J. Vysotski, L. Deng, J. Rose, B. C. Wang, and J. Lee, *in* "Bioluminescence and Chemiluminescence 2000" (J. Case, S. Haddock, L. Kricka, and P. Stanley, eds.). World Scientific, London, 2001.

[96] F. I. Tsuji, Y. Ohmiya, T. F. Fagan, H. Toh, and S. Inouye, *Photochem. Photobiol.* **62,** 657 (1995).

[97] F. H. Johnson, O. Shimomura, Y. Saiga, L. C. Gershman, G. T. Reynolds, and J. R. Waters, *J. Cell. Comp. Physiol.* **60,** 85 (1962).

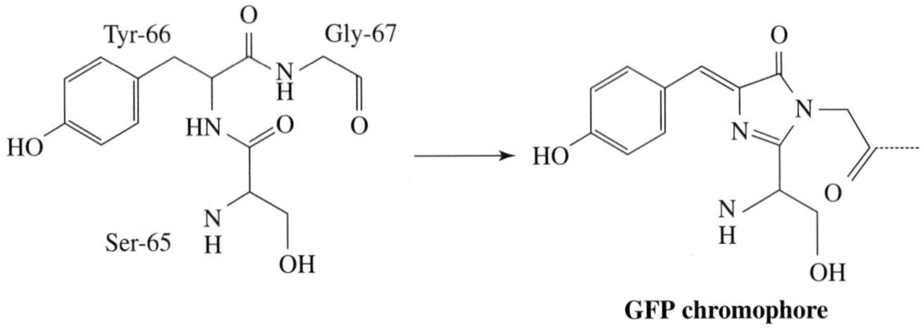

Tyr-66 Gly-67

Ser-65

GFP chromophore

SCHEME 8

(a Ca^{2+} antagonist), and then centrifuged lightly to remove large cell debris, the supernatant contains both intact granules and the soluble "photoprotein" (i.e., the luciferase/luciferin hydroperoxide, Ca-triggerable system). If calcium is added to the supernatant, it causes a flash of blue light by reacting with the photoprotein. If water is now added, the granules osmotically rupture, and green light is produced as the photoprotein and its associated GFP come in contact with calcium. However, if the order of additions is reversed, first water and then calcium, only blue light is emitted, because the photoprotein–GFP complex has dissociated in dilute solution, thus preventing energy transfer.

Later experiments with *Renilla* showed that energy transfer still takes place efficiently in reaction mixtures containing as little as 0.1 μM GFP.[98] At such a low concentration, energy transfer could take place only if the two proteins were preassociated, because the average distance between nonassociated proteins would be an order of magnitude too large to allow for radiationless transfer. Chromatography showed, indeed, that one luciferase complexes with a GFP homodimer.[99]

The most unusual (and valuable) feature of the GFPs is that the chromophore is covalently bound. In *Aequorea* GFP, it results from the posttranslational cyclization, dehydration, and oxidation of residues S65-Y66-G67 in the 238-amino acid protein (Scheme 8).[100] The cloning of a cDNA of this GFP,[101] and the demonstration that its expression in prokaryotic or eukaryotic cells produced a fluorescent protein, thus without the need for any coelenterate-specific enzymes, opened the gates to innumerable applications of GFP as a reporter gene and a marker of cellular localization.[102]

[98] W. W. Ward and M. J. Cormier, *J. Phys. Chem.* **80,** 2289 (1976).

[99] W. W. Ward and M. J. Cormier, *Photochem. Photobiol.* **27,** 389 (1978).

[100] R. Heim, D. C. Prasher, and R. Y. Tsien, *Proc. Natl. Acad. Sci. U.S.A.* **91,** 12501 (1994).

[101] D. C. Prasher, V. K. Eckenrode, W. W. Ward, F. G. Prendergast, and M. J. Cormier, *Gene* **111,** 229 (1992).

[102] M. Chalfie, Y. Tu, G. Euskirchen, W. W. Ward, and D. C. Prasher, *Science* **263,** 802 (1994).

Even though the GFP peptides of *Aequorea* and *Renilla* are not highly similar in sequence,[103] their imidazolinone chromophores are identical, and the fluorescence spectra of both consist of a narrow band at 509 nm. However, the absorption spectrum of wild-type *Aequorea* GFP shows two bands, a major peak at 395 nm and a minor one at 475 nm, whereas that of *Renilla* GFP consists of a single peak at 498 nm, fivefold more intense than the 395-nm peak of *Aequorea* GFP. This in itself demonstrates, as for the visual pigments, how much the protein environment of a chromophore may influence its spectral properties. A large number of mutants of *Aequorea* GFP have now made this abundantly clear, while making available custom-made GFP of specially desirable spectroscopic properties.[104] Mutations in GFP far removed from the chromophore can affect the absorption and fluorescence spectra, but the full-length gene is nearly essential: only one amino acid can be deleted at the N terminus, and at most 15 at the C terminus, without loss of fluorescence.[103]

Several steps are required for the biosynthesis of the chromophore: folding of the peptide chain to bring residues 65 and 67 to the geometry appropriate for cyclization, dehydration, and finally oxidation to form a C=C double bond on the phenolic side chain of tyrosine and thus create the 4-hydroxycinnamyl part of the chromophore. If GFP is expressed in *E. coli* grown anaerobically, the correct molecular weight protein is produced but it is not fluorescent. The subsequent appearance of fluorescence requires only the admission of oxygen, and will occur in extracts, but requires several hours, thereby limiting the use of GFP as a fast reporter of gene expression.[100] No enzyme is involved, because it occurs even in dilute lysates.

Crystal structures show that the chromophores are located at the center of a protein cylinder, a lantern-like structure dubbed a β can, in both wild-type GFP[105,106] and mutants.[107,108] Many of the special properties of GFP, such as its thermal stability and resistance to proteolysis, may be attributed to this structure. The inside cavity is closed to solvent on top and bottom by short segments of α helices, and there is clearly no room for an enzyme to catalyze chromophore formation.

Dinoflagellate Bioluminescence

Dinoflagellates are unicellular marine eukaryotes that emit light as brief ($\sim$100 ms) and bright ($\sim 10^{10}$ photons) flashes. These come from numerous ($\sim$400

[103] A. B. Cubitt, R. Heim, S. R. Adams, A. E. Boyd, L. A. Gross, and R. Y. Tsien, *Trends Biochem. Sci.* **20,** 448 (1995).

[104] R. Heim and R. Y. Tsien, *Curr. Biol.* **6,** 178 (1996).

[105] K. Brejc, T. K. Sixma, P. A. Kitts, S. R. Kain, R. Y. Tsien, M. Ormö, and S. J. Remington, *Proc. Natl. Acad. Sci. U.S.A.* **94,** 2306 (1997).

[106] F. Yang, L. G. Moss, and G. N. Phillips, Jr., *Nat. Biotechnol.* **14,** 1246 (1996).

[107] M. Ormö, A. B. Cubitt, K. Kallio, L. A. Gross, R. Y. Tsien, and S. J. Remington, *Science* **273,** 1392 (1996).

[108] R. M. Wachter, B. A. King, R. Heim, K. Kallio, R. Y. Tsien, S. G. Boxer, and S. J. Remington, *Biochemistry* **36,** 9759 (1997).

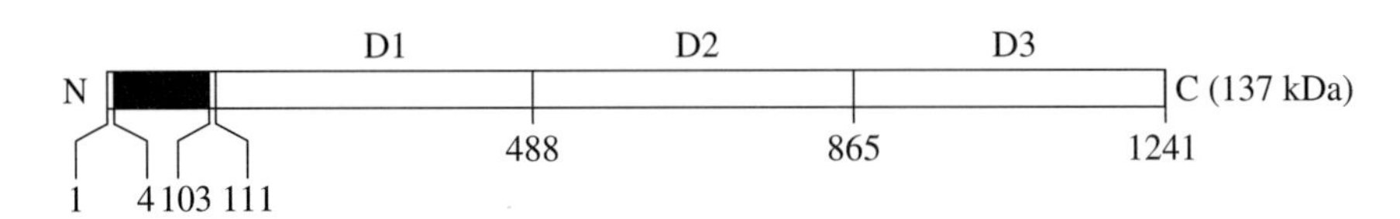

DIAGRAM 2

per cell) small (0.4-μm) organelles called scintillons,[109] occurring as bulbous cytoplasmic projections into the vacuole.[110,111] In the best studied species, *Gonyaulax polyedra,* the scintillons contain both the luciferase (LCF) and the substrate (luciferin, a tetrapyrrole,[112] which is bound to a second and very different protein, the luciferin-binding protein, LBP).[113–115] Two other luminous species of dinoflagellates, *Pyrocystis lunula* and *P. noctiluca,* also possess scintillons[116] and have a similar luciferase structure,[117] but the luciferin in those species does not appear to be associated with a binding protein.[118,119]

The gene encoding luciferase from *G. polyedra* is composed of three intramolecularly homologous domains, each encoding a catalytic site, preceded at the 5′ end by a region with homology to the corresponding region of the LBP gene.[120] In the genome the *lcf* gene is organized as tandem repeats[121] and, like the gene for LBP,[122] does not contain introns. The spacer region between the *lcf* genes has no TATA box or other known conserved eukaryotic promoter elements, but a candidate promoter sequence has been identified by comparison with the dinoflagellate peridinin–chlorophyll binding protein[121,123] (Diagram 2).

The repeated sequences are approximately 75% identical overall at the protein level, and ~95% identical in the more central regions, presumably the locations of

[109] J. W. Hastings and J. C. Dunlap, *Methods Enzymol.* **133,** 307 (1986).

[110] M.-T. Nicolas, G. Nicolas, C. H. Johnson, J.-M. Bassot, and J. W. Hastings, *J. Cell Biol.* **105,** 723 (1987).

[111] L. Fritz, D. Morse, and J. W. Hastings, *J. Cell Sci.* **95,** 321 (1990).

[112] H. Nakamura, Y. Kishi, O. Shimomura, D. Morse, and J. W. Hastings, *J. Am. Chem. Soc.* **111,** 7607 (1989).

[113] M.-T. Nicolas, D. Morse, J.-M. Bassot, and J. W. Hastings, *Protoplasma* **160,** 159 (1991).

[114] D. M. Morse, A. M. Pappenheimer, and J. W. Hastings, *J. Biol. Chem.* **264,** 11822 (1989).

[115] M. Desjardins and D. Morse, *Biochem. Cell Biol.* **71,** 176 (1993).

[116] M.-T. Nicolas, B. M. Sweeney, and J. W. Hastings, *J. Cell Sci.* **87,** 189 (1987).

[117] O. K. Okamoto, L. Liu, D. L. Robertson, and J. W. Hastings, *Biochemistry* **40,** 15682 (2001).

[118] R. E. Schmitter, D. Njus, F. M. Sulzman, V. D. Gooch, and J. W. Hastings, *J. Cell. Physiol.* **87,** 123 (1976).

[119] R. Knaust, T. Urbig, L. Li, W. Taylor, and J. W. Hastings, *J. Phycol.* **34,** 167 (1998).

[120] L. Li, R. Hong, and J. W. Hastings, *Proc. Natl. Acad. Sci. U.S.A.* **94,** 8954 (1997).

[121] L. Li and J. W. Hastings, *Plant Mol. Biol.* **36,** 275 (1998).

[122] D.-H. Lee, M. Mittag, S. Sczekan, D. Morse, and J. W. Hastings, *J. Biol. Chem.* **268,** 8842 (1993).

[123] Q. H. Le, P. Markovic, J. W. Hastings, R. V. M. Jovine, and D. Morse, *Mol. Gen. Genet.* **255,** 595 (1997).

highly conserved catalytic domains. Unexplained is the extremely low frequency of synonymous (silent) nucleotide substitutions in the central regions, but it may be that either the DNA or RNA in those regions has a function in addition to encoding protein, possibly the binding of a regulatory protein.

Searches have yielded no sequences in the database having even the slightest indication of sequence similarity to the three repeat domains of dinoflagellate luciferase.[7] As noted above, the N-terminal region of LCF (amino acids 4 to 103) shares a ~50% sequence similarity with the N-terminal region of LBP (amino acids 6–105), also not found elsewhere in the database. This suggests that this region serves a similar but still unknown function in the two proteins. It does not appear to constitute a signal or transit peptide, and LCF and LBP do not travel across a membrane in the formation of the scintillons.[110] Also, this region is not required for the luciferase reaction, because peptides of each of the domains, in which this portion of the molecule has been removed, are active.[120,124] It seems possible that it could be involved in the association between the two proteins, which associate near the Golgi before migrating together to the vacuolar membrane.[110,113]

Flashing, which is initiated by mechanical stimulation and serves to deter predators, is postulated to be the result of a rapid and transient change in pH within the scintillon, caused by a propagated action potential that opens membrane channels, allowing protons to enter rapidly from the acidic vacuole.[110] Both proteins have pK_a values of about 6.7.[114,125] At pH values above pH 7.5 luciferin is bound to LBP, and LCF is not active, a property shared by peptides of each of its three domains. As a consequence, the reaction occurs maximally at pH 6.3, and virtually not at all at pH 7.5 and above. The N-terminal regions of each of the repeat domains contain conserved histidines identified as being responsible for this effect.[124]

Gonyaulax luciferase, and probably all dinoflagellate luciferases, are mechanistically unusual enzymes. The presence of repeated conserved sequence domains in one enzyme molecule is not unprecedented, but to our knowledge, *Gonyaulax* luciferase is the first in which each of these domains has been shown to be catalytically active by itself.

Other Bioluminescence Systems

Mollusks

Snails (gastropods), clams (bivalves), and cephalopods (squid) have bioluminescent members.[126] The last are by far the most numerous; they are also diverse,

[124] L. Li, L. Liu, R. Hong, R. L. Robertson, and J. W. Hastings, *Biochemistry* **40,** 1844 (2001).

[125] N. Krieger and J. W. Hastings, *Science* **161,** 586 (1968).

[126] R. E. Young and T. M. Bennett, *in* "The Mollusca" (M. R. Clarke and E. R. Trueman, eds.), p. 241. Academic Press, San Diego, CA, 1988.

both in form and function, rivaling the fishes in these respects. As is also true for fishes, some squid utilize symbiotic luminous bacteria, whereas others are self-luminous, thus representing different evolutionary approaches for the acquisition of the ability to emit light.

Many squid possess compound photophores having associated optical elements, used in spawning and other interspecific displays (communication). They may emit different colors of light, and are variously located near the eyeball, on tentacles, body integument, or associated with the ink sac or other viscera. In some species, luminescence intensity is regulated in response to changes in ambient light, indicative of a camouflage function.

Along European coasts there is a clam, *Pholas dactylus,* which inhabits compartments that it makes by boring holes in soft rock. When irritated, it produces a bright cellular luminous secretion, squirted out through the siphon as a blue cloud, presumably to somehow deter or thwart predators. The reaction has been studied, but the structure of the luciferin, which involves a protein-bound chromophore, remains unknown. The luciferase is a copper-containing large (>300-kDa) glycoprotein.[127] It can serve as a peroxidase with several alternative substrates, indicating the involvement of a peroxide in the light-emitting pathway.

Among gastropods, a New Zealand pulmonate limpet, *Latia neritoides,* is notable as the only known truly freshwater luminous species, secreting a green luminous slime. Its luciferin is an enol formate of an aldehyde, but the emitter and products in the reaction are unknown; in addition to its luciferase (molecular mass, ~170 kDa; EC 1.14.99.21), a "purple protein" (molecular mass, ~40 kDa) is also required, but only in catalytic quantities, suggesting that it may be somehow involved as a recycling emitter.

Annelids

There are both marine and terrestrial luminous annelids.[5] *Chaetopterus* is a marine polychaete that constructs and lives in U-shaped tubes in sandy bottoms, exuding luminescence on stimulation, but the chemistry of the reaction has eluded researchers. Other marine polychaetes include members of the family Syllidae, such as the Bermuda fireworm, and the polynoid worms, which shed their luminous scales as decoys. Extracts of the latter have been shown to emit light on the addition of superoxide ion.

Luminous terrestrial earthworms, some of which are quite long (up to 60 cm),[128] exude coelomic fluid. Cells lyse to produce a mucus-emitting blue-green light. In *Diplocardia longa* luminescence in extracts involves a copper-containing luciferase (molecular mass, ~300 kDa) and a luciferin (*N*-isovaleryl-3 amino-1-propanal).

[127] J. P. Henry and J. W. Hastings, *Mol. Cell Biochem.* **3,** 81 (1974).

[128] J. E. Wampler, *Methods Enzymol.* **133,** 249 (1981).

Crustaceans

The small cyprindinid ostracods[129] such as *Vargula* have two glands whose nozzles squirt luciferin and luciferase (EC 1.13.12.6) into sea water, where they react and produce a spot of light, useful either as a decoy or for communication.[130] Cypridinid luciferin is a substituted imidazopyrazine, like coelenterazine, that reacts with oxygen to form an intermediate cyclic peroxide, yielding CO_2 and an excited carbonyl. However, in the cypridinid reaction calcium is not involved, and the luciferase gene has no homologies with the gene for the corresponding coelenterate proteins that might indicate convergent evolution at the molecular level.

Euphausiid shrimp possess compound photophores with accessory optical structures and emit a blue, ventrally directed luminescence. The system is unusual because both luciferase and luciferin cross-react with the dinoflagellate system.[109] This cross-taxon similarity indicates another possible exception to the rule that luminescence in distantly related groups has independent evolutionary origins. The shrimp might obtain luciferin nutritionally, but the explanation for the occurrence of functionally similar proteins is not evident. One possibility is lateral gene transfer; convergent evolution is another. Analyses of gene structures for homologies should provide insight into this question.

Fishes

Bioluminescence in fishes (marine only) is highly diverse, and occurs in both teleost (bony) and elasmobranch (cartilaginous) fishes, but for most relatively little is known about the luminescence.[131] As noted above, luminescence in some is due to luminous bacteria cultured in special organs, but most are self-luminous, including the midshipman fish *Porichthys,* with its array of photophores distributed like buttons on a military uniform. Its luciferin and luciferase cross-react with the cypridinid system described above, an enigma until it was discovered that Puget Sound (Washington) fish have photophores but are unable to luminesce, but can do so if injected with cypridinid luciferin or fed the animals. Luciferin may thus be obtained nutritionally, but did the luciferase in this fish originate independently to make use of the available substrate, or was the ability to synthesize luciferin lost secondarily from a complete system? If the latter, this would be analogous to the loss of the ability to synthesize vitamins in mammals.

Essentially nothing is known of the biochemistry of open sea and midwater species such as sharks, some of which may have several thousand photophores. The teleosts include the gonostomatids such as *Cyclothone,* with simple photophores,

[129] P. J. Herring, *in* "Physiological Adaptations in Marine Animals" (M. S. Laverack, ed.), p. 323. Society of Experimental Biology, Cambridge, 1985.

[130] A. S. Cohen and J. G. Morin, *J. Crustacean Biol.* **9,** 297 (1989).

[131] P. J. Herring, *J. Oceanogr. Mar. Biol.* **20,** 415 (1982).

and the hatchet fishes, having compound photophores; emission is directed exclusively downward, indicating that the light serves to camouflage the silhouette from below, counterilluminating the downwelling ambient light. A clear case of a functional use of luminescence is in *Neoscopelus,* which has photophores on the tongue, allowing it to attract prey to just the right location.

Analytical Applications

Both bioluminescence and chemiluminescence have come into widespread use for quantitative determinations of specific substances in biology and medicine, both in extracts and in living cells, and as a function of time.[2,3,132–136] Such methods have several advantages. Instrumentation for measuring light emission is now commercially available; it is sensitive and relatively free of the background and interference characteristic of many other analytical techniques.[136a] A typical photon-counting instrument can readily detect an emission of about 10^4 photons s^{-1}, which corresponds to the transformation of 10^5 molecules s^{-1} (or 6×10^6 min^{-1}) if the quantum yield is 10% (typically reported values for bioluminescence reactions are this great or greater). Luminescent tags have been developed that are as sensitive as radioactivity, and now replace radioactivity in many assays, notably immunoassays.[137]

Because the different luciferase systems have different specific requirements, many different substances can be detected (Table II). Each of these substances may in turn be used by or are the product of any of many different reactions, so the activities (amounts) of enzymes catalyzing those reactions may thereby be determined quantitatively. There are indeed further kinds of linkages, so that the potential number of different assays capable of being linked to light emission is indeed large.

One of the first bioluminescent assays, which is still widely used, is the measurement of ATP with firefly luciferase.[138] This assay, used in the original

[132] M. DeLuca and L. J. Kricka, *Arch. Biochem. Biophys.* **226,** 285 (1983).

[133] L. J. Kricka, P. E. Stanley, G. H. G. Thorpe, and J. P. Whitehead, "Analytical Applications of Bioluminescence and Chemiluminescence." Academic Press, New York, 1984.

[134] J. Schomerich, R. Andreesen, A. Knapp, M. Ernst, and W. G. Woods, "Bioluminescence and Chemiluminescence." John Wiley & Sons, New York, 1987.

[135] A. Pazzagli, E. Cadenas, L. J. Kricka, A. Roda, and P. E. Stanley, "Bioluminescence and Chemiluminescence." John Wiley & Sons, New York, 1989.

[136] A. Roda, M. Pazzagli, L. J. Kricka, and P. E. Stanley, eds., "Bioluminescence and Chemiluminescence." John Wiley & Sons, Chichester, 1999.

[136a] J. E. Wampler, *in* "Bioluminescence in Action" (P. J. Herring, ed.), p. 1. Academic Press, New York, 1978.

[137] L. Kricka and T. P. Whitehead, *J. Pharmaceut. Biomed.* **5,** 829 (1987).

[138] B. L. Strehler and J. R. Totter, *in* "Methods of Biochemical Analysis," Vol. 1 (D. Glick, ed.). Interscience, New York, 1954.

TABLE II
ANALYTES THAT DIRECTLY AFFECT LUCIFERASE REACTION[a]

Organism	Analyte
Firefly	ATP, ADP, AMP, etc.
	Coenzyme A
	Inorganic pyrophosphate
Bacteria	FMN, $FMNH_2$
	NAD(H), NADP(H)
	Long-chain aldehydes
	Oxygen
Coelenterates	Calcium
Scale worms (polynoidin)	Superoxide

[a] Any enzyme system that consumes or produces such a compound may be determined with the respective luciferase.

work demonstrating photophosphorylation in chloroplasts, was responsible for the serendipitous discovery of delayed light emission in green plants.[139] Because many different enzymes use or produce ATP, their activities may be monitored using this assay, measuring either an increase or decrease in ATP levels. The firefly luciferase reaction is also affected by other substances, such as coenzyme A and inorganic pyrophosphate, so these may similarly be determined with this system.

Bacterial luciferase is specific for reduced flavin mononucleotide ($FMNH_2$), and thus this luciferase may be used to assay FMN and to distinguish it from riboflavin and FAD. Also, any reaction that produces or utilizes NAD(H), NADP(H), or long-chain aldehyde, either directly or indirectly, can be coupled to this light-emitting reaction.[140] Although all bioluminescent reactions can, in principle, be used to detect oxygen, bacterial luciferase is especially useful because its K_m for oxygen is low ($\sim 2 \times 10^{-7} M$). Such determinations can also be made with intact bacteria; indeed, luminous bacteria were used 100 years ago to demonstrate the production of oxygen in photosynthesis. They continue to be used for oxygen detection in special applications,[141] and an oxygen electrode incorporating luminous bacteria has been developed.[142] Intact luminous bacteria have been used for a large number of other analytical purposes, including determinations of lipopolysaccharide, lipases, and less specific identification of toxic components in water.[143]

[139] B. L. Strehler and W. A. Arnold, *J. Gen. Physiol.* **34,** 809 (1951).

[140] H. Watanabe, J. W. Hastings, K. Wulff, G. Michal, and F. Staehler, *J. Appl. Biochem.* **4,** 508 (1982).

[141] B. Chance and R. Ohnishi, *Methods Enzymol.* **57,** 223 (1978).

[142] J. E. Lloyd, *Science* **210,** 669 (1980).

[143] S. Ulitzur, *J. Biolumin. Chemilumin.* **12,** 179 (1997).

The purified photoprotein aequorin has been widely used for the detection of intracellular Ca^{2+} and its changes under various experimental conditions; the protein is relatively small (~20 kDa), nontoxic, and may be readily injected into cells in quantities adequate to detect calcium over the range of 3×10^{-7} to 10^{-4} *M*. It was used to demonstrate directly that calcium is released into the cytoplasm, triggering contraction of muscle cells,[144] and is similarly released in egg cells after fertilization.[145] The stoichiometry and the transient nature of calcium release in such cells means that injected aequorin can continue to serve as a monitor of cellular calcium levels for hours or days.[146] For determinations over longer time periods apoaequorin can be expressed transgenically in cells of interest. With coelenterazine in the medium, which the cells take up, aequorin is regenerated intracellularly, thereby providing a direct measure of the calcium concentration in the cellular compartment where expression occurs.[147–153]

Reporter Applications

Luciferases have proved to be useful indicators of biological processes, even of processes that are not directly related to bioluminescence per se. An early reporter application was the use of endogenous bioluminescence rhythms to monitor the progression of the circadian (daily) clock in the dinoflagellate alga, *Gonyaulax*.[154,155] In that alga, a biological clock controls all the components of its luminescence system: luciferase, luciferin, luciferin-binding protein, and its light-emitting organelle, the scintillon.[156,157] The cells glow rhythmically, providing an endogenous marker for the clock. Another early example of using a luciferase as a

[144] E. B. Ridgeway and C. C. Ashley, *Nature (London)* **219,** 1168 (1968).

[145] J. C. Gilkey, L. F. Jaffe, E. B. Ridgeway, and G. T. Reynolds, *J. Cell Biol.* **76,** 448 (1978).

[146] J. R. Blinks, W. G. Wier, P. Hess, and F. G. Prendergast, *Prog. Biophys. Mol. Biol.* **40,** 1 (1982).

[147] C. H. Johnson, M. R. Knight, T. Kondo, P. Masson, J. Sedbrook, A. Haley, and A. Trewavas, *Science* **269,** 1863 (1995).

[148] M. R. Knight, A. K. Campbell, S. M. Smith, and A. J. Trewavas, *Nature (London)* **352,** 524 (1991).

[149] M. R. Knight, S. M. Smith, and A. J. Trewavas, *Proc. Natl. Acad. Sci. U.S.A.* **89,** 4967 (1992).

[150] M. R. Knight, N. D. Read, A. K. Campbell, and A. J. Trewavas, *J. Cell. Biol.* **121,** 83 (1993).

[151] M. N. Badminton, J. M. Kendall, G. Sala-Newby, and A. K. Campbell, *Exp. Cell Res.* **216,** 236 (1995).

[152] D. Button and A. Eidsath, *Mol. Biol. Cell* **7,** 419 (1996).

[153] R. Rizutto, M. Brini, C. Bastianutto, R. Marsault, and T. Pozzan, *Methods Enzymol.* **260,** 417 (1995).

[154] B. M. Sweeney and J. W. Hastings, *J. Cell. Comp. Physiol.* **49,** 115 (1957).

[155] T. Roenneberg and W. Taylor, *Methods Enzymol.* **305,** 104 (2001).

[156] C. H. Johnson, S. Inoue, A. Flint, and J. W. Hastings, *J. Cell Biol.* **100,** 1435 (1985).

[157] M.-T. Nicolas, G. Nicolas, C. H. Johnson, J.-M. Bassot, and J. W. Hastings, *in* "Bioluminescence and Chemiluminescence" (J. Schölmerich, R. Andressen, A. Kapp, M. Ernst, and W. G. Woods, eds.), p. 413. John Wiley & Sons, London, 1987.

reporter in a heterologous system was the microinjection of the calcium-sensitive luciferase, aequorin, into muscle fibers to monitor Ca^{2+} levels during muscular contraction.[144]

With the cloning of several different luciferase genes and the ability to transform the genes into heterologous hosts to make transgenic luminescent organisms,[68] the number of applications using luciferases has exploded.[158–162] Genetically encoded luciferases in transgenic organisms have been incredibly useful as reporters of single-gene promoters,[158–162] but also in global screens by promoter traps[163] and by luminescent versions of microarrays.[164] Examples from the field of circadian rhythms are discussed below as case studies. Luciferases have been a particularly useful reporter for circadian studies because, unlike GFP, luciferases are generally degraded rapidly in heterologous systems and therefore they turn over, which is essential for monitoring a cyclic process. In addition, they do not require excitation by light, which could directly perturb the circadian system and compromise the study.

Primary tasks to which genetically encoded luciferases have been harnessed include (1) as a reporter of promoter activity in promoter::luciferase fusions,[158–163] (2) as an indicator of ion concentrations, especially of Ca^{2+},[146] (3) as molecular reporters of circadian rhythms, and (4) as the donor system in bioluminescence resonance energy transfer (BRET) to monitor protein–protein interactions, protease activity, and so on.[165,166]

Native Luciferases as Reporters

Several of the luciferases described above have been used extensively as reporters, especially the workhorse luciferases of *Vibrio harveyii* (LuxAB), fireflies (fLUC), *Renilla* (rLUC), and the Ca^{2+}-dependent photoprotein/luciferase of *Aequorea* (AEQ). Other luciferases that have not been used extensively but have attractive characteristics that may encourage use in the future include the luciferases of *Gaussia* (gLUC), *Vargula, Gonyaulax* (LCF), and the Ca^{2+}-dependent

[158] J. Engebrecht, M. Simon, and M. Silverman, *Science* **227,** 1345 (1985).

[159] A. J. Millar, S. R. Short, K. Hiratsuka, N.-H. Chua, and S. A. Kay, *Plant Mol. Biol. Rep.* **10,** 324 (1992).

[160] C. Aflalo, *Int. Rev. Cytol.* **130,** 269 (1991).

[161] J. Alam and J. L. Cook, *Anal. Biochem.* **188,** 245 (1990).

[162] K. V. Wood, *in* "Bioluminescence and Chemiluminescence: Current Status" (P. E. Stanley and L. J. Kricka, eds.), p. 543. John Wiley & Sons, Chichester, 1991.

[163] Y. Liu, N. F. Tsinoremas, C. H. Johnson, N. V. Lebedeva, S. S. Golden, M. Ishiura, and T. Kondo, *Genes Dev.* **9,** 1469 (1995).

[164] T. K. van Dyk, E. J. DeRose, and G. E. Goyne, *J. Bacteriol.* **183,** 5496 (2001).

[165] Y. Xu, D. W. Piston, and C. H. Johnson, *Proc. Natl. Acad. Sci. U.S.A.* **96,** 151 (1999).

[166] Y. Xu, A. Kanauchi, A. G. von Arnim, D. W. Piston, and C. H. Johnson, *Methods Enzymol.* **360,** [12], 2003 (this volume).

photoprotein/luciferase of *Obelia* (OBE). The respective advantages and disadvantages of these luciferases are described below.

Modified Luciferases as Reporters

In addition to the native isoforms of these luciferases, such as those emitting different colors of light,[67] genetically modified versions have been developed that are useful in specific applications. One such modification involves the mutation of fLUC to alter its spectral emission.[167,168] Another is the alteration of codon usage to improve expression in heterologous systems. An example of this is the codon optimization of rLUC for mammalian codon bias (commercially available from BioSignal [Montreal, ON, Canada] and Promega [Madison, WI]); the codon bias of *Renilla* genes is significantly different from that in mammalian cells and the codon-optimized versions of rLUC are brighter when expressed in mammalian cells than is native rLUC. There does not appear to be any effect of codon optimization on fundamental characteristics of rLUC such as kinetics and emission spectrum.

A completely different kind of modification is the genetic fusion of a luciferase to GFP. Such chimeras allow the spatial expression patterns of the fusion proteins to be examined by the fluorescence of GFP, which is usually easier to localize by microscopy than is bioluminescence. One such fusion protein is a GFP::fLUC chimera that allows temporal patterns of gene expression to be monitored by bioluminescence (the half-life of the fusion protein in mammalian cells is about 2 hr), whereas spatial patterns can be visualized by fluorescence.[169] Another useful chimera is that of GFP::AEQ, which allows the visualization of Ca^{2+} flux kinetics in single mammalian cells.[170] The fusion of GFP to AEQ appears to increase the half-life of AEQ in mammalian cells significantly, so that the reporter accumulates to higher levels than with AEQ, enabling the measurement of the aequorin luminescence from single transfected mammalian cells. The authors also claim that the affinity of GFP::AEQ for Ca^{2+} is higher than that of AEQ, but this finding should be independently confirmed.

Reporter Stability

A key feature of any reporter protein that is used for temporal studies is its turnover in terms of synthesis and degradation. If a reporter is unstable, it might not accumulate to detectable levels. On the other hand, if it is too stable (as is usually

[167] Y. Ohmiya, T. Hirano, and M. Ohashi, *FEBS Lett.* **384,** 83 (1996).
[168] V. R. Viviani and Y. Ohmiya, *Photochem. Photobiol.* **72,** 267 (2000).
[169] R. N. Day, M. Mawecki, and D. Berry, *Biotechniques* **25,** 848 (1998).
[170] V. Baubet, H. Le Mouellic, A. K. Campbell, E. Lucar-Meunies, P. Fossier, and P. Brulet, *Proc. Natl. Acad. Sci. U.S.A.* **97,** 7260 (2000).

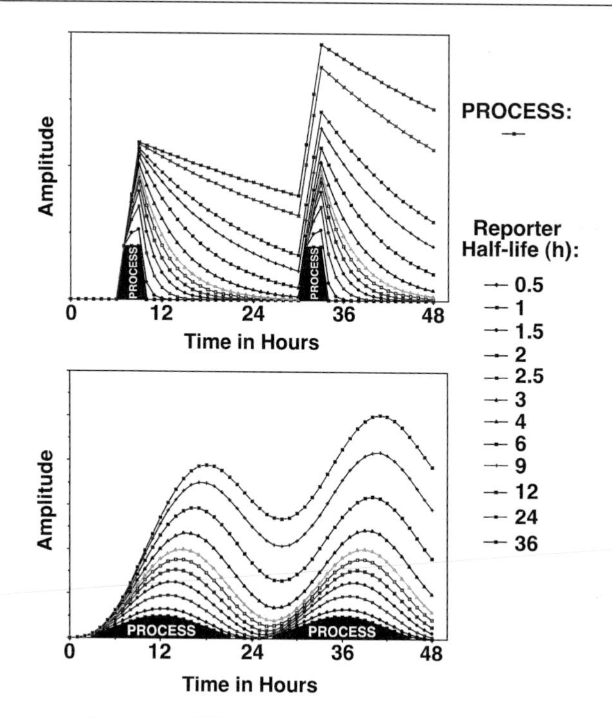

FIG. 1. Importance of reporter stability for accuracy of mimicking the temporal profile of an underlying process. Amplitude of a reporter signal (ordinate) as a function of time (abscissa) is shown for reporter molecules having different half-lives (in hours).

true for native GFP), then it does not give an accurate reflection of the kinetics of the process it is designed to report. In our modeling of luminescence reporters for circadian rhythms (Fig. 1), we find that a reporter with a half-life of 0.5–1 hr will give a luminescence rhythm that reasonably mimicks that of the underlying clock-controlled process (e.g., rhythmic promoter activity). As the half-life of the reporter increases, the luminescence remains rhythmic, but becomes progressively phase-delayed relative to the underlying rhythm, and the peak-to-trough amplitude decreases. With a reporter half-life of 12 hr, the peak of the luminescence rhythm is about 6 hr later and the amplitude is less than 30% of the underlying rhythm. For even more stable reporters, the luminescence rhythm becomes almost undetectable, even if the promoter is being activated rhythmically.

As it happens, luciferases tend to be unstable when expressed in heterologous hosts. For example, fLUC expressed in mammalian cells has a half-life of 3–4 hr at 37°,[169,171] and a destabilized version with a half-life of 0.9 hr has been designed.[172]

[171] J. F. Thompson, L. S. Hayes, and D. B. Lloyd, *Gene* **103,** 171 (1991).

[172] G. M. Leclerc, F. R. Boockfor, W. J. Faught, and L. S. Frawley, *Biotechniques* **29,** 590 (2000).

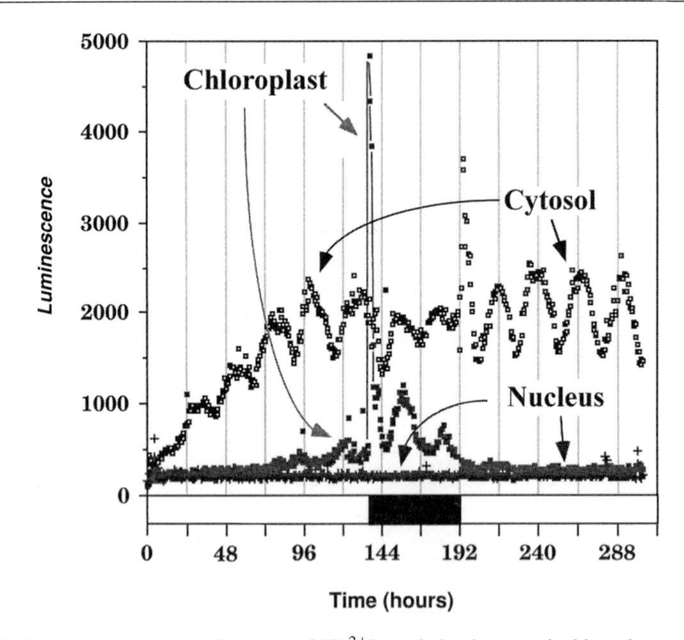

FIG. 2. Time course analyses of patterns of [Ca^{2+}] regulation in cytosol, chloroplast, and nucleus of plants, using apoaequorin reporter expressed as a transgene (white bar represents constant light, black bar represents constant darkness) (see Refs. 147 and 184).

Although some destabilized GFPs are also now commercially available, native luciferases are thus usually better reporters of temporal information than are native GFPs.

Targeted Reporters

One of the benefits of genetically encoded reporter proteins is that they can be targeted to specific tissues if tissue-specific promoters are available, and/or to specific organelles if targeting sequences have been characterized for the host organism. A specific example is the ability to target AEQ to specific organelles to measure subcellular Ca^{2+}. In mammalian cells, aequorin has been targeted to cytosol, nucleus, endoplasmic reticulum, and mitochondria,[151–153] and in plants, to cytosol, chloroplast, and nucleus[147,148] (Fig. 2). For AEQ, targeting capability is an advantage over the other methods for measuring Ca^{2+} levels such as fluorescent dyes or microelectrodes, which can be used only in the cytosol. In addition, AEQ that is targeted to organelles tends to remain at those sites, whereas fluorescent dyes, such as Fura-2, that begin in the cytosol tend to become compartmentalized into unidentified granules after a few hours.

Important Characteristics of Luciferins for Reporter Applications

When selecting a particular luciferase to use as a reporter, many of the characteristics described above (molecular weight, monomeric or multimeric structure, emission spectrum, etc.) are important criteria. However, the other player in the luminescence reaction—the luciferin—is just as important in determining whether a luciferase will work optimally in a particular application. Several key considerations of the luciferins/substrates/cofactors relate to their use in heterologous reporter systems, for example, membrane permeability, solubility, hydrophobicity, toxicity, stability to oxidation or degradation, and endogenous occurrence. In the following paragraphs that describe the various luciferins, the corresponding luciferases are indicated in parentheses.

Beetle luciferin (fLUC): This luciferin is relatively stable in culture media and is permeable to many cell types (but not all). Beetle luciferin is not toxic to the cells we have tested (plants, mammalian cell cultures, mammalian brain slices).

ATP (fLUC): Because the fLUC reaction requires ATP, its use as a reporter of promoter activity is not appropriate in metabolically inhibited cells. However, it might be useful as a reporter of metabolic activity.

Decanal or other long-chain aldehydes (LuxAB): Decanal can be presented in solution or as a vapor, and thus it is excellent for use with colonies on petri dishes.[173] Decanal is also permeable to eukaryotic cell membranes. In some types of cells, decanal can be toxic if the concentration is too high. Instead of adding decanal exogenously, it has also been possible to transform cells with the *luxCDE* genes so that the aldehyde is synthesized endogenously.[174]

$FMNH_2$ (LuxAB): LuxAB needs reduced flavin mononucleotide, which is available inside prokaryotic cells, but not in the cytosol of eukaryotic cells. This problem might be circumvented by targeting LuxA and LuxB to the mitochondrion (and/or the chloroplast in plants) of eukaryotic cells so that $FMNH_2$ will be available for the luminescence reaction.

Coelenterazine (AEQ, OBE, rLUC, gLUC): Coelenterazine is a hydrophobic molecule that is permeable to almost all cell types, but it can be unstable in complex medium and is subject to oxidation. Also, in complex medium, coelenterazine can autoluminesce. If the reporter activity is low and the cells are in complex medium, coelenterazine autoluminescence can confound the estimation of true reporter luminescence. Coelenterazine is not toxic to the cells we have tested (bacteria, plants, mammalian cell cultures).

[173] T. Kondo, N. F. Tsinoremas, S. S. Golden, C. H. Johnson, S. Kutsuna, and M. Ishiura, *Science* **266,** 1233 (1994).

[174] F. Fernandez-Pinas and C. P. Wolk, *Gene* **150,** 169 (1994).

Gonyaulax luciferin (LCF): Like coelenterazine, this luciferin is sensitive to oxidation. Its ability to permeate heterologous cell types has not been tested.

Oxygen (ALL luciferases): Because all luciferases need oxygen, they are not appropriate for anaerobic conditions (except that oxygen is only needed for *charging* of AEQ and OBE).

Luciferases versus Green Fluorescent Proteins

Because the choice of genetically encoded reporters is often between luciferases and GFPs, it is worthwhile to describe their respective drawbacks (and, by contrast, their respective advantages).

Drawbacks of Bioluminescence Techniques

1. The signal is usually dimmer than with fluorescence (therefore longer exposures are needed, as are specialized, highly sensitive instruments that are often less commonly available than fluorimeters and fluorescence microscopes).
2. Poorer spatial resolution in the microscope, confocal microscopy cannot be used (except with the GFP::fLUC and GFP::AEQ chimeras, where this is not a problem).
3. A substrate is needed that might be unstable, impermeable, or toxic.

Drawbacks of Fluorescence Techniques

1. Almost all tissues are autofluorescent to some degree because of NADH, collagen, flavins, and other such compounds, and some are highly so (e.g., photosynthetic cells, partly due to chlorophyll). Autofluorescence can create an unacceptably high background.
2. Fluorescent reporters can bleach under excitation, complicating quantification.
3. Fluorescence excitation can damage tissue (phototoxicity) or stimulate photoresponsive systems (e.g., retina, circadian clocks).

Examples of Luciferases as Reporters: Case Studies from Circadian Rhythms

As mentioned already, the first use of bioluminescence as a reporter of circadian clocks was the monitoring of the endogenous rhythm of bioluminescence from the alga *Gonyaulax*. Both stimulated bioluminescence "flashes" and steady state "glow" are rhythmic.[175,176] From these pioneering studies came the inspiration to introduce luciferase genes into nonluminescent species to monitor circadian

[175] J. W. Hastings, *Cold Spring Harbor Symp. Quant. Biol.* **25,** 131 (1960).

[176] C. H. Johnson and J. W. Hastings, *J. Biol. Rhythms* **4,** 417 (1989).

rhythms. Luminescence reporters are ideal for this application because they do not require excitation by light (as does GFP) that might phase-shift the circadian rhythms. Also, it was determined empirically that the half-lives of luciferases in these systems were amenable to accurate reporting of the underlying clockwork. The first two heterologous systems in which luciferases were used to track circadian clocks were both photosynthetic: the flowering plant *Arabidopsis thaliana*[177] and the cyanobacterium *Synechococcus elongatus*.[178] The plant study used the firefly *fluc* gene and the cyanobacterial study used the bacterial *luxAB* genes, both fused to clock-controlled promoters. Both rhythmically luminescent systems proved to be admirably fitted for use in high-throughput screening for clock mutants.[173,179]

The success of this approach in photosynthetic organisms led to its application to animals, with the construction of transgenic animals expressing luciferase. The first such study was in *Drosophila*, in which a clock gene promoter was fused to the *fluc* gene.[180] This group then extended these studies to show that tissues isolated from different parts of the animal had excellent rhythms *in vitro*![181] For that study, GFP constructs were used to identify the locations of reporter construct expression. This spatial-versus-temporal profiling used independent GFP and fLUC constructs; if the GFP::fLUC reporter introduced later[169] had been available then, it would have aided the cross-correlation of the fluorescence spatial patterns with the luminescence temporal patterns. The latest development in the arena of reporters of animal circadian clocks is the extension of this approach to mammals. Transgenic mice and rats have been developed that express fLUC under the control of a clock gene promoter. Investigations of these remarkable rodents have finally proven what was long suspected: that circadian oscillators operate in both central brain pacemakers and peripheral tissue.[182] Moreover, the rhythmic luminescence rhythm has been monitored from the central brain pacemaker in intact, living animals![183]

The above described cases are examples of using luciferases as reporters of promoter activity. However, another approach has been to use AEQ as a reporter of intracellular Ca^{2+} over a time of several days. Most researchers had believed

[177] A. J. Millar, S. R. Short, N.-H. Chua, and S. A. Kay, *Plant Cell* **4,** 1075 (1992).

[178] T. Kondo, C. A. Strayer, R. D. Kulkorni, W. Taylor, M. Ishiura, S. S. Golden, and C. H. Johnson, *Proc. Natl. Acad. Sci. U.S.A.* **90,** 5672 (1993).

[179] A. J. Millar, I. A. Carre, C. A. Strayer, N. H. Chua, and S. A. Kay, *Science* **267,** 1161 (1995).

[180] C. Brandes, J. D. Plautz, R. Stanewsky, C. F. Jamison, M. Straume, K. V. Wood, S. A. Kay, and J. C. Hall, *Neuron* **16,** 687 (1996).

[181] J. D. Plautz, M. Kancko, J. C. Hall, and S. A. Kay, *Science* **278,** 1632 (1997).

[182] S. Yamazaki, R. Numano, M. Abe, A. Hida, R. Takahashi, M. Ueda, G. D. Block, Y. Sakaki, M. Menaker, and H. Tei, *Science* **288,** 682 (2000).

[183] S. Yamaguchi, M. Kobayashi, S. Mitsui, Y. Ishida, G. T. van der Worst, M. Suzuki, S. Shibata, and H. Okamura, *Nature* (*London*) **409,** 684 (2001).

that the maintenance of constant basal Ca^{2+} levels in unstimulated cells was an incontrovertible fact of biology. The expression of AEQ in plants established for the first time that the average basal level of intracellular Ca^{2+} is not constant over long time intervals, but that the setpoint for calcium levels is under the control of a circadian oscillator.[147] Figure 2 shows data of transgenic tobacco seedlings in which the apoaequorin has been targeted to different compartments (cytosol, chloroplast stroma, nucleus).[184] These time-course data show that Ca^{2+} regulation is not the same in different compartments, again illustrating the utility of targetable reporters.

Finally, a method using a protein splicing-based complementation of a split reporter, either GFP or firefly luciferase, has been developed for reporting protein–protein interactions.[185,186] This is a matter of central importance and great interest in biology, and the method overcomes limitations inherent in the commonly used yeast two-hybrid assay, as well as other methods.

Acknowledgment

We thank Dr. T. Mori for the modeling analysis depicted in Fig. 1.

[184] N. T. Wood, A. Haley, M. Viry-Moussaid, C. H. Johnson, A. H. van der Luit, and A. J. Trewavas, *Plant Physiol.* **125,** 787 (2001).
[185] T. Ozawa, S. Nogami, M. Sato, Y. Ohya, and Y. Umezawa, *Anal. Chem.* **72,** 5151 (2000).
[186] T. Ozawa, A. Kalhara, M. Sato, K. Tachihara, and Y. Umezawa, *Anal. Chem.* **73,** 2516 (2001).

[4] Fluorescent Amino Acid Analogs

By SUSAN M. TWINE and ARTHUR G. SZABO

The use of fluorescence probes as reporters of structure, dynamics, interactions, and local microenvironments is now commonplace in studying biological systems. In this article, we review the use of analogs of tryptophan (Trp) as intrinsic probes of protein structure in protein bimolecular complexes. We first summarize how various Trp analogs can be used in studies of protein complexes. The fluorescence properties of the fluorinated Trp analogs are reviewed and shown to be useful tools in fluorescence studies of proteins. Methods of analog incorporation into proteins and methods of determining the degree of analog incorporation into an expressed protein are then discussed. Finally, we review some of the more recent data concerning the effects of analog incorporation on protein structure and function, some new analogs, and their applications.

0076-6879/03 $35.00

Background

Among biopolymers, proteins are unique in displaying intrinsic fluorescence chromophores that provide valuable information about their structure, dynamics, and segmental molecular details. Trp, as the dominant fluorophore, has a number of well-documented properties including its sensitivity to environmental factors and its usefulness in reporting protein conformational changes, subunit associations, denaturation, and ligand binding.

To study a particular Trp-containing protein in a complex with another Trp-containing protein presents difficulties in resolving and assigning the fluorescence properties to the several Trp components. This problem can be circumvented in two ways. Trp residues may be removed by site-directed mutagenesis, so that one protein can be rendered fluorescently "silent," with respect to Trp fluorescence. This process can perturb protein secondary structure and influence the function of the protein. Alternatively, analogs of Trp can be biosynthetically incorporated into one of the protein partners in place of Trp. Analogs, such as 7-azatryptophan (7AW) and 5-hydroxytryptophan (5HW), have red extended absorbance spectra in relation to Trp. These analogs have structures similar to the indole moiety in Trp, so that there should be a negligible effect on protein structure and function. They present the advantage of fluorescence excitation at wavelengths longer than the absorbance of Trp and thus their fluorescence can be selectively excited.

Historically, amino acid analogs were used in the 1950s to investigate metabolite pathways and mechanisms of protein synthesis.[1,2] Pardee *et al.*[1] described the growth of an *Escherichia coli* auxotroph on medium containing 2-azatryptophan (2AW), 7AW, and 5-methyltryptophan (5MeW). Figure 1 shows the structures of some fluorescent Trp analogs. Bacterial growth was sustained on all but the 5MeW-containing medium. Presumably, the tryptophanyl-tRNA synthetase cannot form 5MeW adenylate and acylate the $tRNA^{Trp}$ with this analog. Later reports investigated the role of the essential enzyme in the process of analog incorporation, tryptophanyl-tRNA synthetase (TrpRS).[1–5] TrpRS aminoacylates the cognate tRNA with Trp, or its analog, which is then incorporated into the growing polypeptide chain. The mechanisms of TrpRS in the presence of Trp and various analogs are under active investigation. This laboratory has developed an assay for prokaryotic TrpRS activity based on the formation of 7AW-adenylate, exploiting the acute environmental sensitivity of its fluorescence.[4,5]

[1] A. B. Pardee, V. G. Shore, and L. S. Prestridge, *Biochim. Biophys. Acta* **21,** 406 (1956).

[2] E. W. Davie, V. V. Koningsberger, and F. Lipman, *Arch. Biochem. Biophys.* **65,** 21 (1956).

[3] N. Sharon and F. Lipmann, *Arch. Biochem. Biophys.* **69,** 219 (1957).

[4] C. W. V. Hogue, Ph.D. Dissertation. University of Ottawa, Ottawa, Ontario, Canada, 1994.

[5] C. V. W. Hogue and A. G. Szabo, *Biophys. Chem.* **48,** 159 (1993).

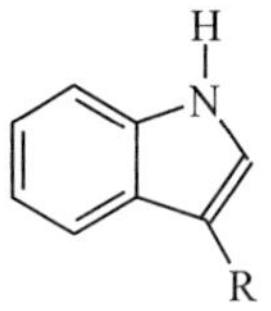

Tryptophan (W) 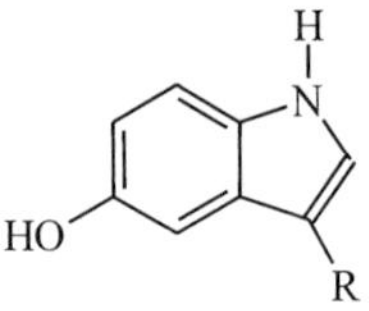 5-hydroxytryptophan (5HW) 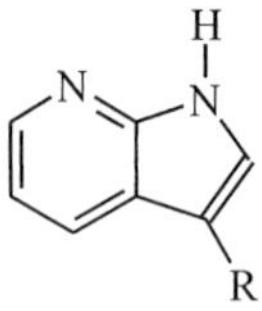 7-azatryptophan (7AW)

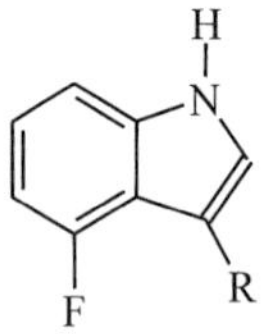

4-fluorotryptophan (4FW) 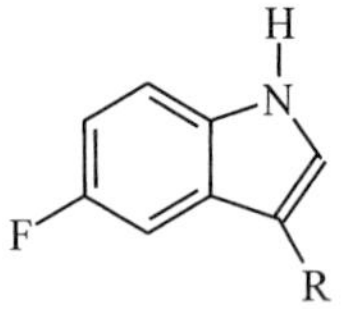 5-fluorotryptophan (5FW) 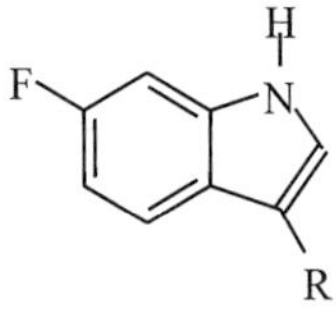 6-fluorotryptophan (6FW)

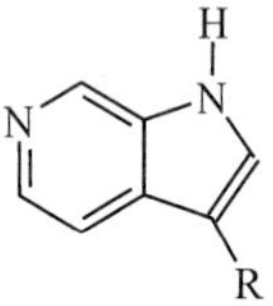

6-azatryptophan (6AW) 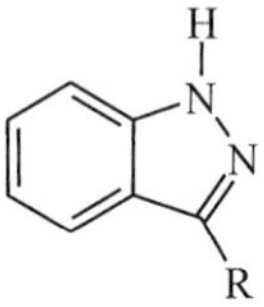 2-azatryptophan (2AW)

FIG. 1. Structures of tryptophan analogs.

The earliest example of a purified analog-containing protein was alkaline phosphatase.[6] The enzyme was expressed in the presence of either 7AW or 2AW. No change in enzyme activity was reported, but large changes in the absorbance and fluorescence spectra of the protein were observed. By the 1970s fluorinated analogs were used for ^{19}F NMR (nuclear magnetic resonance) studies of proteins. Several articles summarize the use of fluorinated Trp and fluorinated phenylalanine (Phe) analogs in NMR,[7,8] the discussion of which are beyond the scope of this article.

Early attempts to incorporate Trp analogs into proteins were hampered by the lack of protein expression technology. In certain instances, when the protein expression was under the control of a natural inducible promoter, a high degree of analog incorporation was achieved.[6] In the absence of an inducible promoter the degree of analog incorporation was observed to be highly unpredictable, as

[6] S. Schlesinger, *J. Biol. Chem.* **243,** 3877 (1968).
[7] B. D. Sykes, H. I. Weinarte, and M. J. Schlesinger, *Proc. Natl. Acad. Sci. U.S.A.* **71,** 469 (1974).
[8] F. A. Pratt and C. Ho, *Biochemistry* **14,** 3035 (1975).

both Trp and the analog were required to sustain cell growth. The development of expression vectors with tightly regulated protein expression has reduced the unpredictability of analog incorporation.

Interest in the use of Trp analogs as spectroscopic probes has been renewed. Petrich *et al.* conducted a number of studies of the photophysics of 7AW, and its parent chromophoric moiety, 7-azaindole (7AI).[9–19] Similar studies have proved to be essential if the fluorescence signal of the analogs is to be correctly interpreted when incorporated into protein systems. In 1992, groups led by Szabo and by Ross independently demonstrated the potential of 5HW as a probe of protein–protein[20] and protein–nucleic acid interactions.[21] In both cases the red extended shoulder of the analog absorbance was retained, allowing a window for selective red-edge excitation of the 5HW chromophore in the presence of proteins and nucleic acids. The use of this, and other Trp analogs, as biosynthetically incorporated protein structural probes has grown significantly since this time.

Spectral Properties of Tryptophan and Its Analogs

The analogs 7AW, 5HW, and 4-fluorotryptophan (4FW) are most commonly selected in fluorescence studies by virtue of their favorable spectral properties. Figure 2 shows the absorbance properties of 7AW, 5HW, 4FW, and Trp, and the photophysical properties of 7AW, 5HW, and Trp are shown in Table I. These spectral properties have been extensively reviewed previously,[22] but are summarized here in order to place them in the context of other Trp analogs used in more recent studies.

The absorbance of the indole side chain of Trp is rationalized in terms of two overlapping, nearly isoenergetic π–π^* transitions denoted ^1L_a and ^1L_b. The

[9] M. Negrerie, S. M. Bellefeuille, S. Whitman, J. W. Petrich, and R. W. Thornburg, *J. Am. Chem. Soc.* **112,** 7419 (1990).

[10] M. Negrerie, M. Gai, S. M. Bellefeuille, and J. W. Petrich, *J. Phys. Chem.* **95,** 8663 (1991).

[11] F. Gai and J. W. Petrich, *J. Am. Chem. Soc.* **114,** 8343 (1992).

[12] M. Negrerie, F. Gai, J. C. Lambry, J. L. Martin, and J. W. Petrich, *J. Phys. Chem.* **97,** 5045 (1993).

[13] Y. Chen, R. L. Rich, F. Gai, and J. W. Petrich, *J. Phys. Chem.* **97,** 1770 (1993).

[14] R. L. Rich, Y. Chen, D. Neven, M. Negrerie, F. Gai, and J. W. Petrich, *J. Phys. Chem.* **97,** 1781 (1993).

[15] R. L. Rich, M. Negrerie, J. Li, S. Elliott, R. W. Thornburg, and J. W. Petrich, *Photochem. Photobiol.* **58,** 28 (1993).

[16] Y. Chen, F. Gai, and J. W. Petrich, *J. Am. Chem. Soc.* **115,** 10158 (1993).

[17] Y. Chen, F. Gai, and J. W. Petrich, *J. Phys. Chem.* **98,** 2203 (1994).

[18] F. Gai, R. L. Rich, and J. W. Petrich, *J. Am. Chem. Soc.* **116,** 735 (1994).

[19] R. L. Rich, F. Gai, Y. Chen, and J. W. Petrich, *Proc. SPIE* **2137,** 435 (1994).

[20] C. W. V. Hogue, I. Rasquina, A. G. Szabo, and J. P. MacManus, *FEBS Lett.* **310,** 269 (1992).

[21] J. B. A. Ross, D. F. Senear, E. Waxman, B. B. Kombo, E. Rusinova, Y. T. Huang, W. R. Laws, and C. A. Hasselbacher, *Proc. Natl. Acad. Sci. U.S.A.* **89,** 12023 (1992).

[22] J. B. A. Ross, A. G. Szabo, and C. W. V. Hogue, *Methods Enzymol.* **273,** 151 (1997).

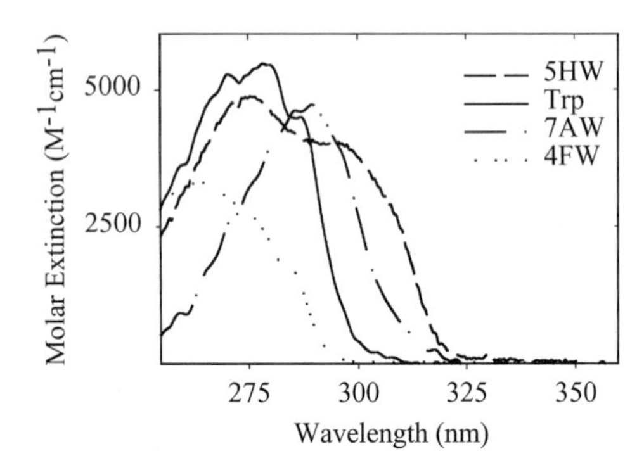

FIG. 2. Absorbance spectra of tryptophan (Trp), 5-hydroxytryptophan (5HW), 7-azatryptophan (7AW), and 4-fluorotryptophan (4FW) in neutral pH buffer, 20°.

fluorescence properties of Trp are highly dependent on the environment of the fluorophore, and this is due to the large change in dipole moment resulting from excitation of the L_a transition. In aqueous solutions at pH 7, Trp has a quantum yield of 0.14 with an emission maximum of 352 nm. In acetonitrile the quantum yield increases to 0.25 and the emission maximum exhibits a blue shift to 334 nm.[22]

The fluorescence properties of 7AW are even more sensitive to local environmental effects than are those of its natural analog. The quantum yield of 7AW increases from 0.01 in aqueous solution, pH 7, to 0.25 in acetonitrile.[22] In a protein matrix, such as alkaline phosphatase, the fluorescence maximum of 7AW is

TABLE I
SPECTROSCOPIC PROPERTIES OF TRYPTOPHAN ANALOGS

Analog	Absorbance		Fluorescence			
	λ_{max} (nm)	ε (M^{-1} cm^{-1})	λ_{max} (nm)	ϕ	$\langle\tau\rangle$ (ns)[a]	Ref.[b]
Tryptophan	280 (288)[c]	5400	353	0.13	2.8	1
5-Hydroxytryptophan	277 (298)[c]	4800	339	0.256	3.6	2
7-Azatryptophan	290	6000	403	0.016	0.75	2
5-Fluorotryptophan	285	5400	360	0.14	2.7	1
6-Fluorotryptophan	281	4900	366	0.14	4.2	1

[a] The mean lifetime $\langle\tau\rangle = \sum \alpha_i \tau_i^2 / \sum \alpha_i \tau_i$, where α_i and τ_i are amplitude and lifetime of the ith component, respectively.

[b] *Key to references:* (1) C. Wong and M. R. Eftink, *Biochemistry* **37,** 8938 (1998); (2) C. W. V. Hogue, Ph.D. Dissertation. University of Ottawa, Ottawa, ON, Canada, 1994.

[c] Absorbance values in parentheses indicate absorbance shoulders.

observed at 370 nm, whereas in aqueous buffer a broad maximum is centered at 412 nm.[6,22] Early denaturation experiments also recognized the dramatic quenching effect of water.[6] Exploitation of the acute environmental sensitivity of 7AW permits its use as a probe of protein structural changes and ligand binding. The fluorescence decay behavior of 7AW is complex after excitation, exhibiting multiexponential decay kinetics[5,10,12,23] similar to Trp in proteins. Contrary to expressed views, this complexity increases the information content, when the 7AW photophysics are well understood. It has been shown that 7AW can be highly valuable as a biological probe, providing spectral information regarding changes in solvation, frequently associated with protein structural perturbations and ligand-binding reactions.

5HW is a naturally occurring amino acid that is not found in any naturally occurring protein, and it is a precursor of the neurotransmitter serotonin. 5HW has a higher quantum yield in water (0.21) in comparison with Trp (0.14), with an emission maximum of 339 nm.[22] The fluorescence maximum of this analog is relatively insensitive to changes in the solvation state of its environment. However, a decrease in 5HW quantum yield in acetonitrile, to 0.18, was observed.[22] The highly structured red-extended absorbance shoulder of 5HW has been shown to be almost entirely ^1L_b in nature. The ^1L_b transition of 5HW has little or no solvent sensitivity, occurs at a lower energy level than the ^1L_a transition, and can be selectively excited at wavelengths longer than 315 nm. The most distinct advantage of 5HW as a fluorescence probe originates from this nonoverlapping nature of the ^1L_a and ^1L_b transitions above 310 nm. Selective excitation of the singlet L_b state in 5HW allows its use as an effective anisotropy probe because its fluorescence is not depolarized via internal conversion processes. Therefore, proteins containing 5HW have significantly higher anisotropy values compared with the same protein containing Trp. This was first demonstrated with studies of the 5HW "model" compound, *t*-Boc(α-amino)5HW,[21] and Y57(5HW)oncomodulin.[20]

4-Fluorotryptophan (4FW) has been reported to be nonfluorescent in aqueous solution, at room temperature,[24] and if biosynthetically incorporated into a protein can be used to eliminate Trp fluorescence, obviating the need to remove the Trp residue from the protein by site-directed mutagenesis.

5-Fluorotryptophan and 6-Fluorotryptophan

These analogs have been used primarily in ^{19}F NMR spectroscopy, as pioneered by the groups of Sykes and Ho,[7,8] and have not been used extensively as spectroscopic probes of protein structure and interactions until more recently. The

[23] J. K. Judice, T. R. Gamble, E. C. Murphy, A. M. de Vos, and P. G. Schultz, *Science* **261,** 1578 (1993).
[24] P. M. Bronskill and J. T. Wong, *Biochem. J.* **249,** 305 (1988).

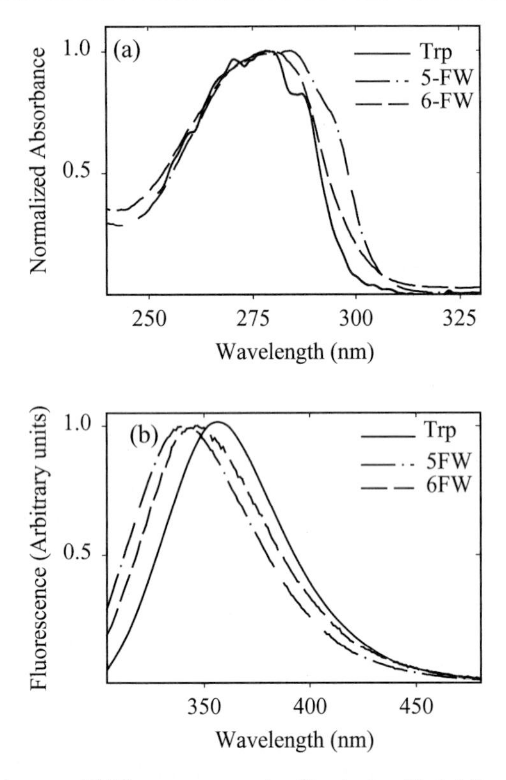

FIG. 3. (a) Absorbance and (b) fluorescence spectra of tryptophan (Trp), 5-fluorotryptophan (5FW), and 6-fluorotryptophan (6FW) in neutral pH buffer, 20°.

absorbance and fluorescence properties of the free analogs at pH 7.3 are shown in Table I and illustrated in Fig. 3. Both 5- and 6-fluorotryptophans have absorbance maxima close to that of Trp and moderate extinction coefficients, but their absorbance bandwidth is broader, with increased extinction on the lower energy side.

Incorporation of Tryptophan Analogs into Proteins

The primary method for the production of analog-incorporated mutant proteins is the technique of *in vivo* protein expression, using auxotrophic hosts. This approach is advantageous, requiring only minor modifications to existing protein expression protocols, and is capable of producing milligram quantities of mutant protein. Other strategies, such as *in vitro* transcription/translation using nonsense tRNA molecules, chemically misacylated with a Trp analog, have been developed and are discussed in the following sections.[23,25–29] Total chemical synthesis of peptides containing analogs has been accomplished and further developed to produce synthetic proteins.

Methods of Analog Incorporation

Previously two methods for the biosynthetic incorporation of analogs have been described.[23] To facilitate the incorporation of the analog of interest the expression system for the protein of interest must be transferred to an organism that is a Trp auxotroph. *Escherichia coli* is the most commonly used organism selected as the host strain for protein expression and a number of *E. coli* Trp auxotrophs are available for use. To maximize the synthesis of analog mutant protein the expression system must be under the control of a stringent, nonleaky promoter. A leaky promoter will lead to high levels of basal expression, causing the accumulation of protein containing natural Trp. Factors influencing the level of analog incorporation are numerous and in our experience depend on the expression system, the protein being expressed, as well as other environmental factors. To maximize incorporation low levels of basal protein expression are desirable. Hence conditions were developed in which intracellular pools of Trp were depleted before protein expression was induced. One concern is the production of proteins essential to expression of the mutant protein, such as T7 RNA polymerase in the T7 promoter system,[30] which is produced postinduction. Such proteins contain Trp in their native sequence and thus postinduction will result in analog incorporation into these proteins as well as into the protein of interest. This may affect their functionality and subsequent levels of protein expression of the protein of interest.

To obtain high yields of the desired protein mutant, two variants of the protein expression protocol are used.

Two-Step Method. Normally, the analogs are toxic to cell growth and attempts to incorporate them into expressed proteins in *E. coli* BL21 cells, for example, by adding analog to the growth medium, have met with only limited success. Furthermore, the extent of analog incorporation is difficult to control under these conditions.

[25] C. Wong and M. R. Eftink, *Biochemistry* **37,** 8938 (1998).

[26] J. D. Bain, D. A. Wacker, C. G. Glabe, T. A. Dix, and A. R. Chamberlin, *J. Am. Chem. Soc.* **111,** 8013 (1989).

[27] J. D. Bain, D. A. Wacker, E. E. Kuo, M. H. Lyttle, and A. R. Chamberlin, *J. Org. Chem.* **56,** 4615 (1991).

[28] C. J. Noren, S. J. Anthony-Cahill, M. C. Griffith, and P. G. Schultz, *Science* **244,** 182 (1989).

[29] D. M. Mendel, J. A. Ellman, Z. Chang, D. L. Veenstra, P. A. Kollman, and P. Schultz, *Science* **256,** 1798 (1992).

[30] F. W. Studier, A. H. Rosenberg, J. J. Dunn, and J. W. Dubendorff, *Methods Enzymol.* **185,** 60 (1990).

The *E. coli* auxotroph containing the expression system for the protein of interest is grown in a rich medium such as LB or TB supplemented with glycerol,[31] to a high cell density ($OD_{600} > 1.0$). Typically, at 37° and 2 liters of culture this will take 6–10 hr. The cells are harvested by centrifugation and washed in minimal medium to remove traces of the original medium. The cells are transferred to 1 liter of M9 minimal medium supplemented with 1% (w/v) casamino acids, 2 m*M* $MgSO_4$, 0.1 m*M* $CaCl_2$, and 100 mg of thiamin. This medium is devoid of Trp. Antibiotics are added to the medium after cooling, to the final concentration normally used for bacterial growth.

The growth step in rich medium is used to attain a high cell density in a short period of time. After washing and resuspension into minimal medium a further growth period of 1 hr is advised to deplete any residual Trp that may be present in the medium or within the cells. The medium is then supplemented with 20–50 mg of L-analog or 40–100 mg of DL-analog and recombinant protein expression induced by the addition of isopropylthiogalactoside (IPTG).

One-Step Method. Trp auxotroph cells containing the expression plasmid are grown in M9 minimal medium as described for the two-step method but supplemented with 2% (w/v) casamino acids, glycerol (20 ml/liter) as a carbon source, and L-Trp (4 mg/liter). The cells are then grown to a constant OD_{600}, usually for 15–18 hr, until the Trp is depleted. The growth medium is not changed and the analog supplement is added directly to the medium immediately before induction of protein expression.

This method has some advantages in comparison with the two-step method. One is the avoidance of the intermediate harvesting step. The second is that in the T7 promoter system T7 lysozyme is produced by the pLysS plasmid. Processes such as harvesting cells may damage cells, liberating T7 lysozyme that is produced by this expression system,[32] and this causes further cell lysis.

Estimating Levels of Incorporation of Tryptophan Analogs into Proteins

The main concern when preparing analog-incorporated protein is the level of incorporation: the relative amounts of analog and L-Trp within the protein. Incorporation of Trp analogs by an *in vivo* method has an inherent degree of unpredictability, which can relate to the growth conditions, protein expression system, and in some cases the position of the Trp residue within the protein. In addition, it is important to remember that the cells will selectively utilize residual Trp in preference to an analog. For these reasons, it is essential that the relative levels of

[31] J. Sambrook, E. F. Fritsch, and T. Maniatis, "Molecular Cloning: A Laboratory Manual," 2nd Ed., Vols. I, II, and III. Cold Spring Harbor Laboratory Press, Cold Spring Harbor, NY, 1989.

[32] C. Wong and M. R. Eftink, *Biochemistry* **37,** 8947 (1998).

analog incorporation must be estimated. A number of methods have been described to achieve this objective. These include spectral analysis, using "basis set spectra" in the linear combination of spectra (LINCS), and amino acid or peptide analysis following protein hydrolysis and subsequent high-performance liquid chromatography (HPLC) and mass spectrometry. In this section, we describe the use of these methods together with their respective advantages and disadvantages.

Before a detailed analysis of the amino acid content of an analog-incorporated protein is undertaken, spectral characteristics of the purified protein may provide a qualitative estimate of the degree of incorporation.

1. Absorption spectra: Analogs such as 5HW and 7AW have a red-extended absorbance shoulder in comparison with Trp. When levels of analog incorporation are high, an absorbance shoulder between 300 and 320 nm is clearly visible. 4-Fluorotryptophan (4FW) does not have a red-extended absorbance shoulder, but the absorbance peak is blue shifted relative to that of Trp, owing to a λ_{max} of 260 nm for 4FW. If incorporation is high, then a shift in the absorbance maximum may be observed.

2. Fluorescence emission spectra will show changes characteristic of the incorporated analog, relative to the native protein. For example, proteins containing 5HW can be excited at 310 nm and a fluorescence maximum at 335 nm is evident. Alternatively, a protein into which 4FW has been incorporated, when excited at 295–300 nm, will show negligible fluorescence at 340 nm.

3. Phosphorescence spectra of the analogs are characteristic for the incorporated analog and may be observed.[25,32] However, this generally requires low-temperature samples.

4. Near-UV circular dichroic (CD) spectra can also be useful, as 5HW and 7AW show signals that extend to 320 nm.

The estimation of the aromatic amino acid content of proteins from their absorbance spectra in comparison with the absorbance spectra of the aromatic amino acids was proposed by Edelhoch[33] and Wetlaufer.[34] This involved measuring the fluorescence and absorbance spectra of the protein under denaturing conditions (6 *M* guanidine hydrochloride). In a similar way the Trp and analog content of the protein can be estimated by comparison of the protein spectrum with the spectra of various model compounds under the same conditions. The spectra of model compounds such as *N*-acetyltryptophanamide (NATA) and *N*-acetyltyrosinamide (NAYA), and *t*-BOC(α-amino)7AW, constitute the basis set spectra. The absorbance spectrum of the analog-containing protein in 6 *M* guanidine hydrochloride is measured and the analog content of the protein is estimated

[33] H. Edelhoch, *Biochemistry* **6,** 1948 (1967).

[34] D. Wetlaufer, *Adv. Protein Chem.* **17,** 303 (1962).

by deconvolving a linear combination of the basis set spectra.[33–36] This method relies heavily on knowing the extinction coefficients of model compounds accurately and determining whether they are good representations of the same aromatic groups when incorporated into proteins and in the denatured proteins.

The general method for determining the amino acid content of proteins and peptides uses acid hydrolysis of the protein, followed by chromatographic separation and detection of a colored or fluorescent adduct. Experience has shown that such traditional methods of amino acid analysis of proteins are not appropriate for tryptophan because it is inherently acid labile. Alternative methods use base hydrolysis[37,38] or nonoxidative acid conditions that do not degrade Trp, such as with methanesulfonic acid.[39] The resulting hydrolysate is lyophilized and resuspended in an appropriate buffer for HPLC analysis using a reversed-phase column. The aromatic amino acids can be identified by their retention time and by monitoring the absorbance at 280 nm. This precludes the necessity of a secondary reaction to label the amino acid with a dye. Zhang *et al.*[39] quantified the incorporation of 4FW into arginyl-tRNA synthetase by performing base hydrolysis and monitoring the elution of aromatic amino acids by absorbance at 280 nm. Concentrations of Trp and 4FW were calculated from the areas under the retention curves, after calibration with a known amount of each sample. A linear relationship was found between the peak area and mass of Trp or 4FW in the 100- to 1000-ng range. 4FW was found to have replaced 95% of the Trp in the arginyl-tRNA synthetase. Another report, using a similar method of alkaline hydrolysis, determined the 5FW content of glutathione transferase to be greater than 95%.[40]

Another method uses electrospray mass spectrometry (ESMS) to quantify the relative amounts of analog versus L-Trp in a recombinant protein. Analysis of rat parvalbumin mutant F102W(5HW) by ESMS gave two peaks, one 16 mass units (m.u.) above that for the F102W protein, corresponding to the addition of a hydroxyl group in the analog-containing protein.[41] Fluorinated Trp-containing rat parvalbumin mutants showed an additional peak at m/e +18 as a result of the addition of a fluorine atom. The unit mass difference between Trp and 7AW is not sufficient to differentiate the two amino acids in the proteins by this method. The fraction of the analog-containing peak in relation to the L-Trp protein peak provides an estimate of the percentage analog incorporation. Using this method, the percentage incorporation of fluorinated analogs into rat parvalbumin F102W was found to be 71–77%.[41] The 5HW-containing protein showed 34.6% incorporation.[41]

[35] W. R. Laws and C. A. Hasselbacher, *Proc. Natl. Acad. Sci. U.S.A.* **89,** 12023 (1992).

[36] E. Waxman, E. Rusinova, C. A. Hasselbacher, G. P. Schwartz, W. R. Laws, and J. B. A. Ross, *Anal. Biochem.* **210,** 425 (1993).

[37] T. E. Hugli and S. Moore, *J. Biol. Chem.* **247,** 2828 (1974).

[38] S. Delhaye and J. Landry, *Anal. Biochem.* **159,** 175 (1986).

[39] Q. S. Zhang, L. Shen, E. D. Wang, and Y. L. Wang, *J. Protein Chem.* **18,** 187 (1999).

[40] J. F. Parsons, G. Xiao, G. L. Gilliland, and R. N. Armstrong, *Biochemistry* **37,** 6286 (1998).

[41] M. Acchione, unpublished data (2001).

TABLE II
MELTING TEMPERATURES FOR THERMAL UNFOLDING OF RAT PARVALBUMIN MUTANT F102W AND STAPHYLOCOCCAL NUCLEASE

	T_m(°C)			
	Rat parvalbumin[a]	Staphylococcal nuclease[b,c]		
Analog	(F102W)	Wild type	V66W	V66W′
Tryptophan	66	53	49	45
5-Hydroxytryptophan	63	54	23	36
7-Azatryptophan	58	45	45	21
4-Fluorotryptophan	67	54	52	44
5-Fluorotryptophan	69	53	53	42
6-Fluorotryptophan	68	54	53	44

[a] M. Acchione, personal, unpublished results (2001).
[b] C. Wong and M. R. Eftink, *Biochemistry* **37,** 8947 (1998).
[c] Data rounded to two significant figures.

Analog Incorporation and Protein Stability

A traditional way of introducing a fluorescent reporter probe, other than Trp, into proteins is by covalent attachment of a chromophore to cysteine or lysine residues. Numerous chromophores have been developed for this purpose, and are coupled with amine, sulfhydryl, or histidine side chains in proteins. Such probes have been of great value, but the effect of their size and properties on the protein structure and function is a continual concern. The incorporation of Trp analogs has proved an attractive alternative because it is usually assumed that they minimally perturb the protein structure. In this section, we review some of the more recent studies investigating the effects of various Trp analogs on protein structure and function.

A series of studies conducted with staphylococcal nuclease have characterized the effects of analog incorporation on the thermal stability of the protein and its fragments.[25,32,42–45] They reported that the effect of analog was dependent on the mutant studied and the position of the Trp residue within the protein. The melting points of the proteins examined are summarized in Table II.

Wild-type staphylococcal nuclease has a single Trp residue at position 140, and the V66W mutant has two Trp residues at positions 140 and 66. V66W′ is a

[42] A. Ozarowski, J. Q. Wu, S. K. Davis, C. Y. Wong, M. R. Eftink, and A. H. Maki, *Biochemistry* **37,** 8954 (1998).
[43] M. R. Eftink, R. Ionesu, G. D. Ramsay, C. Y. Wong, J. Q. Wu, and A. A. Maki, *Biochemistry* **35,** 8084 (1996).
[44] C. Y. Wong and M. R. Eftink, *Protein Sci.* **6,** 689 (1997).
[45] D. Shortle and A. K. Mecker, *Biochemistry* **28,** 938 (1989).

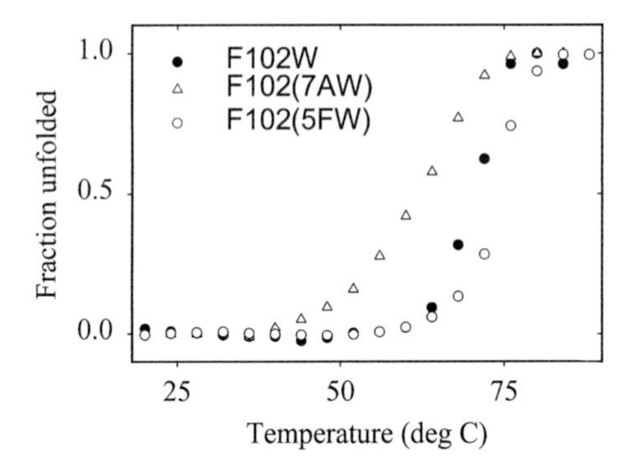

FIG. 4. Thermal unfolding of rat parvalbumin mutants F102W, F102(7AW), and F102(5FW).

C-terminal truncated fragment of the wild-type nuclease containing a single Trp residue at position 66. Previous studies have shown the V66W′ fragment to be a marginally stable structure, such that the degree of secondary and tertiary structure retained by the protein is highly dependent on amino acid substitutions.[45,46] Trp-140 is located within the C-terminal α helix of the protein, whereas Trp-66 is situated within the β barrel hydrophobic core region in the N-terminal region.[32]

5HW was not found to perturb the thermal stability of the wild-type protein, whereas it significantly destabilized both V66W and the V66W′ fragment. Substitution of 5HW at positions 66 and 140 in V66W did not appear to alter the overall thermal stability of the protein, but did result in an increased degree of cooperativity in the unfolding process.

7AW was found to perturb the stability of the wild-type protein and of the V66W and V66W′ mutant proteins. All three proteins were reported to be partially unfolded, retaining only the β barrel structure of the L-Trp-containing proteins. The alteration in protein structure was explained by the potential of the polar imino group at position-7 of the aromatic ring of 7AW to form hydrogen bonds with other amino acids. The natural environment of the indole side chain of the Trp residue would not be expected to accommodate a hydrogen bond acceptor at position-7.

The fluorinated Trp analogs appear to have a small effect on the thermal stability of the proteins. This effect appears to be dependent on the protein studied.

The rat parvalbumin mutant, F102W, has a single Trp situated within the hydrophobic core of the protein. The thermal unfolding curves of F102W, F102(7AW), and F102(5FW) are shown in Fig. 4, and melting points for the analog incorporated proteins are shown in Table II. Substitution with 7AW or 5HW resulted in a dramatic decrease in the melting temperature of the protein, indicative of a change in structural stability. This is evident from the shift in the midpoint of the thermal

[46] D. Shortle and C. Abygunwardana, *Structure* **1,** 121 (1993).

denaturation curve shown in Fig. 4. All fluorinated derivatives were observed to have a stabilizing effect on rat parvalbumin (RP) F102W. Fluorine has a covalent radius only slightly larger than that of hydrogen (1.35 and 1.2 Å, respectively) and therefore should not have any large volume effect. Wong and Eftink studied the partition coefficients of the Trp analogs and the results indicated that the fluorinated derivatives are slightly more hydrophobic than Trp itself.[32] This may be the origin of the stabilizing effect observed in the fluorinated RP mutant proteins, where the Trp residue is located within the hydrophobic core of the protein.

New Applications

Broos *et al.* presented the use of 7AW as an intrinsic chromophore for membrane proteins in vesicles and in live *E. coli* cells.[47] Purification of membrane proteins is inherently difficult, involving solubilization, purification, and reconstitution into proteoliposomes. Such purifications can alter the functional and oligomeric states of the protein and interactions with other membrane proteins. In addition, the reconstituted proteins are not unidirectional, complicating the analysis of ligand interactions in the protein. Using Trp auxotrophic cells, 7AW was incorporated into the *E. coli* membrane protein mannitol permease. The alloprotein was estimated to represent 10% of the total bacterial membrane protein. Production of inside-out (ISO) vesicles gave a functional alloprotein that had no reported differences in mannitol phosphorylation activity. The addition of mannitol to L-Trp protein resulted in an 8% increase in fluorescence intensity. Under the same conditions the 7AW protein gave a 28% reduction. In living cells addition of mannitol resulted in a 5% decrease in 7AW fluorescence, indicating that the procedure may be useful for the study of membrane proteins *in vivo.*

Another application of analog-incorporated proteins was presented by Soumillion *et al.*,[48] in which they investigated the subunit association and dissociation of the T4 bacteriophage gene 45 protein. This protein functions as a sliding clamp in the DNA polymerase holoenzyme, markedly increasing its processivity. The gene 45 protein is a homotrimer and it is unclear whether subunit exchange occurs and what role this may play in the loading and unloading of the enzyme from DNA. The so-called "loading" of gene 45 protein (g45p) on the DNA is an ATP-driven reaction catalyzed by the gene 44/62 protein complex. The authors presented a new example of using Trp analogs in proteins to study protein–protein interactions. They utilized the covalent label *N*-(acetylaminoethyl)-8-napthylamine-1-sulfonic acid (AEDANS), attached to a cysteine residue inserted at position-162, at the g45p subunit interface, in close proximity to Trp-91 within

[47] J. Broos, F. der Veld, and G. T. Robillard, *Biochemistry* **38,** 9798 (1999).
[48] P. Soumillion, D. J. Sexton, and S. J. Benkovic, *Biochemistry* **37,** 1819 (1998).

the associated subunit. Subunit association and dissociation were monitored by fluorescence resonance energy transfer from Trp-91 in one subunit to the AEDANS in the adjacent subunit. Addition of gene 45 protein with 4FW replacing Trp-91 led to a decrease in the fluorescence of AEDANS, suggesting that subunit exchange in solution was occurring, forming trimers containing 4FW-45 and V162C-45. The nonradiative properties of 4FW abolished the resonance energy transfer to AEDANS. The use of both covalently labeled protein and analog-incorporated protein provided evidence of subunit exchange in the trimer. They were able to show that although subunit exchange in gene 45 protein does not participate in the mechanism of DNA loading, it may facilitate gene 45 protein disassembly from DNA.

The photophysical properties of both 7AW and 5HW have been extensively characterized. The group led by Sengupta[49–51] has proposed the use of 5-hydroxyindole (5HI), the chromophoric moiety of 5HW, as an extrinsic probe of membrane and protein structure. The fluorescence maxima of 5HI and 5HW are insensitive to the surrounding environment. This solvent insensitivity has been explained by a significant energy separation between the two low-lying π–π^* singlet electronic states, L_a and L_b, that occur in indoles. The L_b state, a solvent-insensitive transition, lies below L_a in the hydroxyindole, so that fluorescence results exclusively from the L_b state. This separation also results in 5HI having a high intrinsic anisotropy when it is immobilized. The potential of 5HI as an extrinsic anisotropy probe for membranes and proteins has been examined. 5HI was incorporated in unilamellar liposomal membranes, where an r value of 0.16 was reported at 25° in comparison with a solution value of close to zero.[51] The temperature dependence of 5HI anisotropy within the liposomal membrane exhibited a sigmoidal shape characteristic of the thermodynamic transition of the phospholipid from the gel-to-liquid crystalline states. The transition temperature obtained from this data was consistent with previously reported values.

5HI has also been proposed as a probe of protein structure, using the highly characterized protein bovine serum albumin (BSA) as a model protein. It is well known that BSA and its human homolog HSA have at least one indole-binding site. The presence of two Trp residues in BSA has hampered spectroscopic characterization of this site. The anisotropy of 5HI was monitored in the presence of increasing concentrations of BSA, producing a characteristic hyperbolic curve with the plateau region of constant anisotropy corresponding to a 1 : 1 stoichiometry of BSA : 5HI. Use of the analog 5HI has proved a useful alternative. The binding constant was found to be 10^6 M^{-1}.[51]

[49] B. Sengupta, J. Guharay, and P. K. Sengupta, *Spectrochim. Acta A* **56,** 1213 (2000).

[50] J. Guharay, B. Sengupta, and P. K. Sengupta, *Spectrochim. Acta A* **54,** 185 (1998).

[51] B. Sengupta, J. Guharay, and P. K. Sengupta, *J. Mol. Struct.* **559,** 347 (2001).

The technique of *in vitro* translation relies on the suppression of amber codons by tRNAs aminoacylated with nonnatural amino acids.[52–60] Rothschild and Gite[60] reviewed the technique of site-directed nonnative amino acid replacement (SNAAR). Essentially, the concept involves the use of chemically aminoacylated tRNAs that correspond to stop codons inserted at specific sites within the target DNA sequence. The tRNA used has the anticodon corresponding to a specific stop codon and will thus insert the nonnatural amino acid at that point in the mRNA. This "read through" of the stop codon is the process known as suppression and can be accomplished through the use of the corresponding suppressor tRNA.[54,60–63] In *E. coli* the suppressor codons used are TAG (amber), TAA (ochre), and TGA (opal). The overall approach is illustrated in Fig. 5. In brief, the suppressor tRNA is aminoacylated with the nonnatural amino acid. The method that has been used to place a fluorescent nonnatural amino acid on the suppressor tRNA involves the chemical acylation of the 5′-terminal dinucleotide of the tRNA, CA, with the fluorescent analog. This is followed by an enzymatic ligation of the labeled oligonucleotide to a truncated suppressor tRNA. This has been demonstrated with 7AW,[58] 5HW, and an ε-dansyllysine.[58,64] The misacylated suppressor tRNA is then added to the *in vitro* protein translation system, or *in vivo* if the misacylated tRNA can be readily introduced into cells.[65] The suppressor tRNA is then directed during protein synthesis to the inserted stop codon on the mRNA and the nonnatural amino acid is incorporated into the nascent polypeptide chain.

[52] J. D. Bain, E. S. Diala, C. G. Glabe, T. A. Dix, and A. R. Chamberlain, *J. Am. Chem. Soc.* **111,** 8013 (1989).

[53] J. D. Bain, D. A. Wacker, E. E. Juo, M. H. Lyttle, and A. R. Chamberlin, *J. Org. Chem.* **56,** 4615 (1991).

[54] C. J. Noren, S. J. Anthony-Cahill, and P. G. Schultz, *Science* **244,** 182 (1989).

[55] D. M. Mendel, J. A. Ellman, Z. Chang, D. L. Veenstra, P. A. Kollman, and P. G. Schultz, *Science* **256,** 1798 (1992).

[56] J. K. Judice, T. R. Gamble, E. C. Murphy, A. M. de Vos, and P. G. Schultz, *Science* **261,** 1578 (1993).

[57] V. W. Cornish, D. R. Benson, C. A. Altenbach, K. Hideg, W. L. Hubbell, and P. G. Schultz, *Proc. Natl. Acad. Sci. U.S.A.* **91,** 2910 (1994).

[58] D. R. Liu, T. J. Magliery, M. Pastrnak, and P. G. Schultz, *Proc. Natl. Acad. Sci. U.S.A.* **94,** 10092 (1997).

[59] L. Jermutus, L. A. Ryabova, and A. Pluckthun, *Curr. Opin. Biotechnol.* **9,** 534 (1998).

[60] K. J. Rothschild and S. Gite, *Curr. Opin. Biotechnol.* **10,** 64 (1999).

[61] S. J. Anthony-Cahill, M. C. Griffith, C. J. Noren, D. J. Suich, and P. G. Schultz, *Trends Biochem. Sci.* **14,** 400 (1990).

[62] E. Resto, A. Lida, M. D. van Cleve, and S. M. Hecht, *Nucleic Acids Res.* **20,** 5979 (1992).

[63] G. Xiao, J. F. Parsons, K. Tesh, R. N. Armstrong, and G. L. Gilliland, *J. Mol. Biol.* **281,** 323 (1998).

[64] L. E. Steward, C. S. Collins, M. A. Gilmore, J. E. Carlson, J. B. A. Ross, and A. R. Chamberlin, *J. Am. Chem. Soc.* **119,** 6 (1997).

[65] M. W. Nowak, J. P. Gallivan, S. K. Silverman, C. G. Labarca, D. A. Dougherty, and H. A. Lester, *Methods Enzymol.* **293,** 504 (1998).

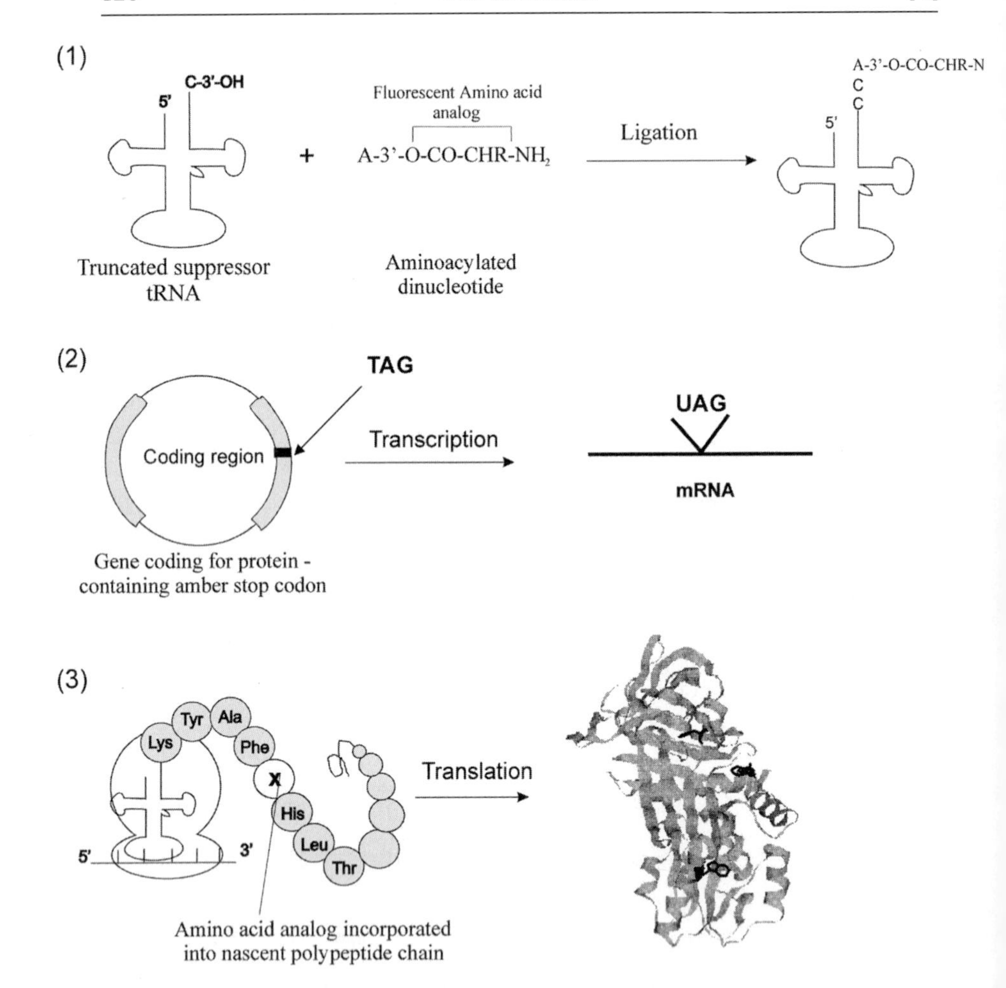

FIG. 5. Incorporation of a fluorescent amino acid analog into a protein, using SNAAR. (1) A truncated tRNA is ligated to an aminoacylated dinucleotide; (2) the TAG amber stop codon is inserted into the gene of the protein of interest; (3) cell-free protein synthesis, using aminoacylated suppressor tRNA, results in the incorporation of the amino acid analog at the desired position. Adapted from K. J. Rothschild and S. Gite, *Curr. Opin. Biotechnol.* **10,** 64 (1999).

The drawback of this system is the inefficiency of the suppression of the stop codon. Release factors within the cell extract used in the cell-free system compete with the suppressor tRNA at the stop codon, resulting in termination of protein synthesis. The groups led by Chamberlin and others[64–66] used a strain of *E. coli* with a faulty release factor 1 (RF1) when producing the S30 transcription/translation

[66] L. E. Steward and A. R. Chamberlin, *Methods Mol. Biol.* **77,** 325 (1998).

TABLE III
PRODUCTION OF β-GALACTOSIDASE USING CELL-FREE PROTEIN TRANSLATION

DNA template	Suppressor	β-Gal[a,b] (μg)
pT7lac-7amb (wild type)		2.71 ± 0.21
pT7lac-7amb	Phe-tRNA-dCA	0.78 ± 0.075
pT7lac-7amb	5HW-tRNA-dCA	0.35 ± 0.0015
pT7lac-7amb	7AW-tRNA-dCA	0.23 ± 0.006
pT7lac-7amb	ε-dnsLys-tRNA-dCA	0.089 ± 0.014

[a] Data from L. E. Steward, C. S. Collins, M. A. Gilmore, J. E. Carlson, J. B. A. Ross, and A. R. Chamberlin, *J. Am. Chem. Soc.* **119,** 6 (1997).

[b] Details the yields from 50-μl lysate reactions. pT7-7amb DNA template contains the *lacZ* gene with an amber stop codon at position-7.

mix for in the *in vitro* synthesis of β-galactosidase. This gave higher yields of full-length protein transcripts for the incorporation of 5HW and 7AW into the enzyme β-galactosidase.[64] Interestingly, the same group reports the successful incorporation of dansyllysine into the same enzyme. All three amino acid analogs were chemically ligated to a suppressor tRNA molecule ($tRNA^{Gly}_{CUA}$). The limited spectral analysis carried out indicates that the dansyl fluorescence in dnsLys-β-galactoside corresponded well with the emission spectra reported for dansyl fluorescence of proteins that have been nonspecifically posttranslationally modified. Table III shows the yields of protein obtained from the cell-free synthesis of β-galactosidase using suppressor tRNA aminacylated with phenylalanine, 5HW, 7AW, or ε-dansyllysine.

Koval and Oliver[67] took a novel approach and used unassigned codons from the organism *Micrococcus luteus*. The organism has six unassigned codons that theoretically could be used to encode nonnatural amino acids. The *M. luteus* unassigned codon AGA was incorporated into a DNA template. When added to an *M. luteus* S30 extract termination of the growing polypeptide occurred. Complete translation of the target protein was achieved when the extract was supplemented with *E. coli* tRNA. This circumvents the need for the use of suppressor tRNAs and the difficulties associated with them.

In vivo systems based on the suppression of stop codons have been used in *Xenopus laevis* oocytes, where misacylated tRNAs were directly injected into the cells.[68] This has the advantage that protein folding and posttranslational modification can be carried out in a natural intracellular environment.

[67] A. K. Koval and J. S. Oliver, *Nucleic Acids Res.* **25,** 4685 (1997).

[68] A. E. Johnson, *Trends Biochem. Sci.* **18,** 235 (1993).

Although this technique has huge potential for site-directed incorporation of nonnatural amino acids, including fluorescently labeled amino acid analogs, the major drawback at the present time is the low yield of protein.

Chemical Synthetic Methods

An alternative to the molecular biological approach is the total chemical synthesis of peptides and proteins. Theoretically, this would allow complete flexibility as to the type and site of residue incorporation. In addition, chemical synthesis of peptides can yield large quantities of protein.[69] In this way small peptides containing 5HW [65,70] and 7AW[12,23] have been produced.

Peptide synthesis using commercially available Trp analogs requires their conversion to the *t*-Boc(α-amino) derivative. Coupling may then be accomplished by standard solid-phase or solution synthesis.[65,69] 5HW is commercially available as the L-form, and therefore synthetic peptides contain pure L-5HW. 7AW is commercially available as a DL-racemic mix, requiring the separation of the diastereomeric peptides. A number of methods have been explored to resolve the enantiomers of 7AW or effect total chemical or enzymatic synthesis of the pure enantiomers.[71,72]

Complete chemical synthesis of a Ras-binding peptide (RBP) containing N^1-methyl-7-azatryptophan has been reported.[73] Drawing on innovations in peptide chemistry[74,75] multimilligram amounts of proteins up to 20 kDa in size have been produced of sufficient quality for structural studies.[76,77] The RBP was synthesized as a series of N-terminal peptide fragments, from which the full-length protein was constructed by native chemical ligation.[73] Structurally, the synthetic protein was not distinguishable from the recombinant protein, using far-UV CD. The chemically synthesized RBP was found to interact with Ras in the same manner as the recombinant protein.

[69] G. Barany and R. B. Merrifield, *in* "The Peptides" (E. Gross and J. Meienhofer, eds.), Vol. 2, p. 3. Academic Press, New York, 1980.

[70] W. R. Laws, G. P. Schwartz, E. Rusinova, G. T. Burke, Y. C. Chi, P. Katsoyannis, and J. B. A. Ross, *J. Protein Chem.* **14,** 225 (1995).

[71] J. D. Brennan, C. W. V. Hogue, B. Rajendran, K. J. Willis, and A. G. Szabo, *Anal. Biochem.* **252,** 260 (1997).

[72] L. Lecointe, V. Rolland-Fulcrand, M. L. Roumestant, P. Viallefont, and J. Martinez, *Tetrahedron Asym.* **9,** 1753 (1998).

[73] J. R. Sydor, C. Herrmann, S. B. H. Kent, R. S. Goody, and M. Engelhard, *Proc. Natl. Acad. Sci. U.S.A.* **96,** 7865 (1999).

[74] P. E. Dawson, T. W. Muir, I. Clark-Lewis, and S. B. H. Kent, *Science* **266,** 776 (1994).

[75] J. P. Tam, Y. Lu, C. Liu, and J. Shao, *Proc. Natl. Acad. Sci. U.S.A.* **92,** 12485 (1995).

[76] L. E. Canne, A. R. Ferré-D'Amaré, S. K. Burley, and S. B. H. Kent, *J. Am. Chem. Soc.* **117,** 2998 (1995).

[77] T. M. Hackeng, C. M. Mounier, C. Bon, P. E. Dawson, J. H. Griffith, and S. B. H. Kent, *Proc. Natl. Acad. Sci. U.S.A.* **94,** 7845 (1997).

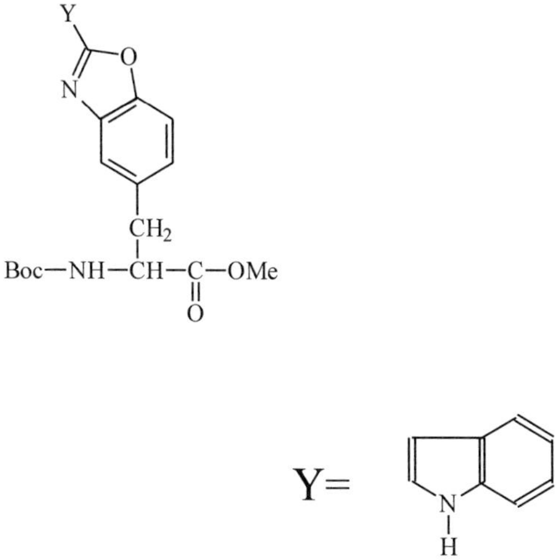

FIG. 6. Structure of *N*-[(tert-butoxy)carbonyl]-3-[2-(1*H*-indol-3-yl)benzoxazol-5-yl]-L-alanine methyl ester.

This method can be used for the incorporation of a diverse array of fluorescent labels, with the prerequisite that they have a reactive group, and are stable to the conditions under which the peptides are synthesized.

Other Fluorescent Amino Acid Analogs

Although 5HW and 7AW are by far the most commonly used amino acid analogs and have certain advantages in their spectral properties, it would still be worthwhile to have fluorescent amino acids with additional unique fluorescent properties. The use of constrained analogs of the aromatic amino acids is reviewed in McLaughlin and Barkley.[78] These have not proved amenable to biosynthetic incorporation into proteins. Other fluorescent probes that are currently available have desirable spectral characteristics, such as long-wavelength fluorescence maximum, high quantum yield, and simple photokinetics, but cannot be incorporated directly into the amino acid chain of proteins because of the lack of an amino acid moiety. The synthesis of a new, nonproteogenic amino acid has been described, based on 2-arylbenzoxazoles, a group of fluorescent compounds with long-wavelength absorbance and higher quantum yield than Trp. The result, [2-(1*H*-indol-3-yl)benzoxazol-5-yl]-L-alanine (Fig. 6) has an absorbance maximum in polar and nonpolar solvent in the range 310–320 nm, with a molar extinction coefficient of 20,000 M^{-1} cm^{-1}.[79] The emission maximum of the fluorophore

[78] M. L. McLaughlin and M. D. Barkley, *Methods Enzymol.* **278,** 190 (1997).

[79] Y. Kanagae, K. Peariso, and S. S. Martinez, *Appl. Spectrom.* **50,** 316 (1996).

in methylcyclohexane was 340 nm, and 360 nm in methanol. The quantum yield is solvent sensitive, 0.77 in methanol and 0.57 in methylcyclohexane. The high quantum yield and its increase with solvent polarity is consistent with the findings for this fluorophore without the amino acid moiety.[80] It is also reported that these probes are photostable. It is predicted that this probe will be incorporated into proteins by virtue of the amino acid moiety. If this is to be performed *in vivo,* studies of the recognition by the cellular protein synthesis machinery are clearly essential. If the chromophore is not recognized by the alaninyl-tRNA synthetase and the cognate tRNA cannot be charged then an alternative method of protein incorporation will be required. In addition, there is no information regarding the cellular toxicity of such compounds to *E. coli* cells, or their uptake by such cells.

A large number of Trp analogs have been synthesized, as shown in Fig. 1. Although a number of these have useful photophysical properties and can be biosynthetically incorporated into proteins as probes of structure and function, a number of analogs remain untested. The Trp analog 6-azatryptophan (6AW) has been shown to have a significant absorbance at 325 nm with a molar extinction coefficient of 9800 M^{-1} cm^{-1} [81]; however, its fluorescence properties had not been previously investigated. The clear spectral maximum of the absorbance of 6AW would provide a window for selective excitation at its absorbance maximum. We have investigated the photophysical properties of 6AW and the parent chromophoric moiety 6-azaindole (6AI).[80]

In a preliminary investigation we have shown that 6AW is readily incorporated into proteins, using a single Trp mutant of the calcium-binding protein, calmodulin, as a model protein.[80] The absorbance spectrum of 6AW, shown in Fig. 7, with a high extinction coefficient maximum at 325 nm, initially interested us in the analog as a spectral probe. In protein systems, when one of the protein partners in a protein–protein complex has a large number of Trp residues the selective excitation of the other partner can be problematic. The sum of the Trp absorbance on the red edge of the spectrum can become significant and make the selective excitation of an analog in the protein partner difficult. The advantage of analog incorporation is then lost. The absorbance spectrum of 6AW initially suggested that it could be used to circumvent this problem. The fluorescence properties of 6AW were also promising, having a large quantum yield in water and an emission maximum considerably shifted from that of Trp, at about 400 nm. The fluorescence and absorbance properties of 6AW and 6AI are shown in Fig. 7.

Initial photophysical studies showed the absorbance and fluorescence properties of both 6AW and 6AI to be acutely pH dependent. Figure 8 shows the absorbance and fluorescence spectra of 6AI in solutions of selected pH. Both exhibited isobestic points in their absorbance spectra, indicative of a ground state

[80] S. M. Twine, L. Murphy, P. Malinowski, R. S. Phillips, and A. G. Szabo, *Biophys. J.* **80,** 1537 (2001).
[81] G. M. Moran, R. S. Philips, and P. F. Fitzpatrick, *Biochemistry* **38,** 16283 (1999).

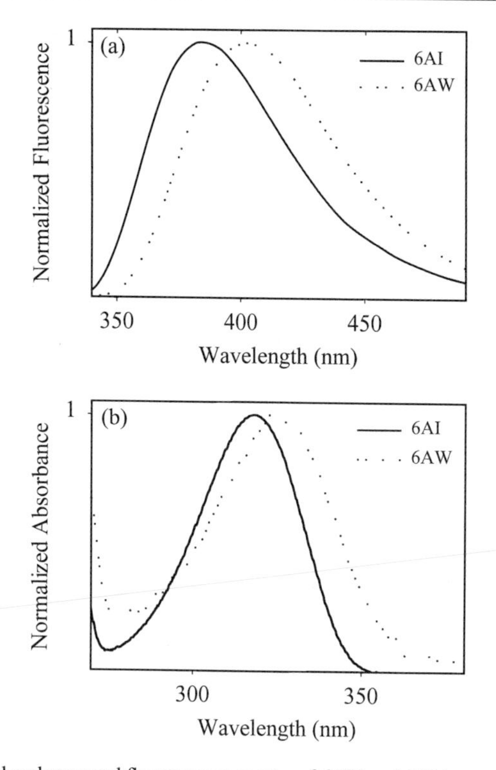

FIG. 7. Absorbance and fluorescence spectra of 6AW and 6AI in water, pH 6.

equilibrium. The isosbestic point in the absorbance spectra is clearly observed at 302 nm. The fluorescence of 6AI decreases markedly at higher pH values. We have proposed the equilibrium to be between the N6 protonated and unprotonated forms, as shown in Fig. 9. More detailed studies have shown the photophysics of 6AI to be complex, with possible excited state proton transfer reactions. 6AI has some potential as a sensitive probe of localized protein structure, reporting on the presence of proton-accepting or -donating groups.

In Summary

In vivo methods of incorporating nonnatural amino acids into proteins have several advantages: (1) the technique may be readily applied to an existing *E. coli* expression system by simple transformation of the expression plasmid into a Trp auxotrophic strain of *E. coli;* (2) protein yields are usually high, although generally slightly lower in comparison with expression of the L-Trp protein; (3) little or no modification of the protein purification techniques is required; and (4) specialized

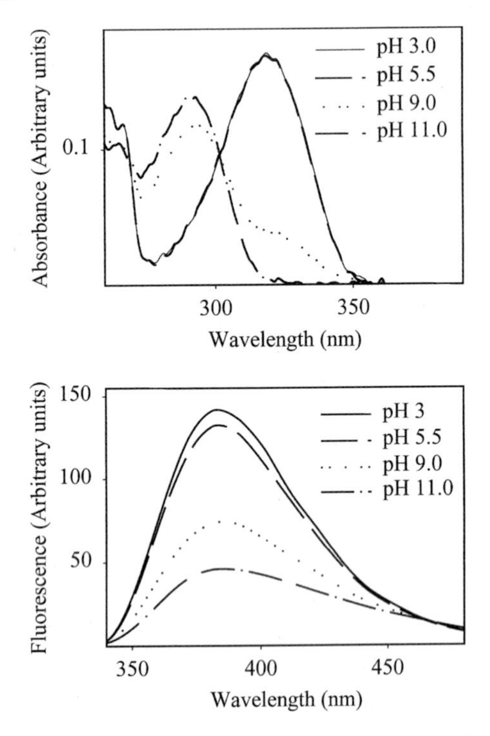

FIG. 8. Absorbance and fluorescence spectra of 6AI in solutions of selected pH.

equipment is not required in a laboratory that routinely uses recombinant protein expression.

The biosynthetic approach does, however, have certain disadvantages: (1) 100% incorporation of the Trp analog at a particular site is not ensured; (2) the method is generally restricted to proteins containing only a few Trp residues because all Trps are replaced; and (3) the method is restricted to the few analogs that can successfully replace Trp in the cellular biosynthetic machinery.

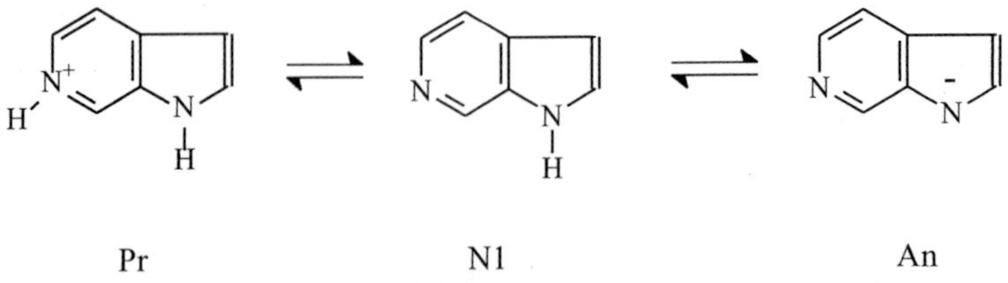

Pr N1 An

FIG. 9. Proposed relationship between the various forms of 6AI.

TABLE IV
ANALOG-CONTAINING PROTEINS

Protein	Percent incorporation					Method of estimation	Ref.[a]
	5HW	7AW	4FW	5FW	6FW		
Rat parvalbumin	35			71–77		LINCS or mass spectrometry	1
Glutathione transferase				95		HPLC	2
Staphylococcal nuclease	95	98	97	>100		LINCS	3
Tropomyosin	79					LINCS	4
Arginyl-tRNA synthetase			>95			HPLC	5
Mannitol permease		ND				—	6
Sigma factor (σ^N)		70–90				LINCS	7
T45 clamp protein			~95			LINCS	8
Endotoxin A			100			LINCS	9
Sporulation repressor		95				LINCS	10

[a] *Key to references:* (1) M. Acchione, unpublished data (2001); (2) J. F. Parsons, G. Xiao, G. L. Gilliard, and R. N. Armstrong, *Biochemistry* **37,** 6282 (1998); (3) C. Wong and M. R. Eftink, *Biochemistry* **37,** 8947 (1998); (4) C. S. Farah and F. C. Reinach, *Biochemistry* **38,** 10543 (1999); (5) Q. S. Zhang, L. Shen, E. D. Wang, and Y. L. Wang, *J. Protein Chem.* **18,** 187 (1999); (6) J. Broos, F. der Veld, and G. T. Robillard, *Biochemistry* **38,** 9798 (1999); (7) D. J. Scott, A. L. Ferguson, M. Gallegos, M. Pitt, M. Buck, and J. G. Hoggett, *Biochem. J.* **352,** 539 (2000); (8) P. Soumillion, D. J. Sexton, and S. J. Benkovic, *Biochemistry* **37,** 1819 (1998); (9) F. Mohammadi, G. A. Prentice, and R. A. Merrill, *Biochemistry* **40,** 10273 (2001); (10) D. J. Scott, S. Leejeerajumnean, J. A. Brannigan, R. J. Lewis, A. J. Wilkinson, and J. G. Hoggett, *J. Mol. Biol.* **293,** 997 (1999).

Concluding Remarks

Advances in genomic sequencing, including the sequencing of the human genome, has led to expanded activity in the area known as proteomics. New protein–protein interactions are being uncovered and their role in cellular processes are of extensive interest. The use of biosynthetically incorporated Trp analogs in the study of interaction of new protein partners would be a distinct advantage.

The laboratories of Szabo and Ross first demonstrated the incorporation of a 5HW into recombinant proteins *in vivo,* using *E. coli* Trp auxotroph cells.[20,21] Since that time a number of advances have been made in the incorporation of Trp analogs into proteins and their use in the study of protein structure, function, and dynamics on protein-protein interactions. The number of researchers using this concept to study the systems of interest is increasing. A summary of studies of using analog-incorporated proteins is shown in Table IV. The search for new fluorescent Trp analogs and the characterization of their photophysical properties is a research area of merit.

[5] Fluorescent Nucleotides: Synthesis and Characterization

By CHRISTINE R. CREMO

Introduction

Fluorescent nucleotides have been widely used to study the structure and function of proteins and nucleic acids. With the advent of methods to study macromolecules at the single-molecule level with fluorescence imaging techniques, and the advent of two-photon excitation technologies, the interest in new nucleotide derivatives with appropriate fluorescent properties has increased.

The goal of this article is to provide the biochemist, who may be an initiate to synthetic organic methods, with the tools to synthesize and characterize fluorescent nucleotides. Basic information about the general approaches to synthesis such as reagent preparation, monitoring of synthetic reactions, chromatographic methods, spectral characterization, and compositional analysis are provided to assist the reader to replicate published syntheses and/or to prepare new compounds.

The general classes of nucleotide derivatives are summarized and general synthetic routes are described. Specific protocols are provided only when the synthesis demonstrates a route to a general class of compounds. Modifications or improvements to published protocols for common derivatives are mentioned. Spectral properties of the analogs mentioned are summarized.

The literature on synthetic nucleotide analogs, including those without fluorescent properties, is vast and pertinent to the present topic. Readers are directed to references describing methods to incorporate nucleotides into oligonucleotides. Dinucleotides and nucleosides are mentioned only occasionally. Readers are directed to excellent reviews concerning ATP analogs,[1] nucleotide analogs used in antiviral therapy,[2] cyclic nucleotide analogs,[3] α-phosphorothioate-tagged triphosphates for RNA function/structure analysis[4] by nucleotide analog interference mapping,[5] cyclic ADP-ribose,[6] ribose-modified fluorescent nucleotides,[7] fluorescent pteridine nucleoside analogs,[8] and nucleotides designed for affinity labeling.[9,10]

[1] C. Bagshaw, *J. Cell Sci.* **114,** 459 (2001).
[2] A. Van Rompay, M. Johansson, and A. Karlsson, *Pharmacol. Ther.* **87,** 189 (2000).
[3] F. Schwede, E. Maronde, H. Genieser, and B. Jastorff, *Pharmacol. Ther.* **87,** 199 (2000).
[4] S. A. Strobel, *Curr. Opin. Struct. Biol.* **9,** 346 (1999).
[5] S. P. Ryder and S. A. Strobel, *Methods* **18,** 38 (1999).
[6] F. J. Zhang, Q. M. Gu, and C. J. Sih, *Bioorg. Med. Chem.* **7,** 653 (1999).
[7] D. M. Jameson and J. F. Eccleston, *Methods Enzymol.* **278,** 363 (1997).
[8] M. E. Hawkins, *Cell Biochem. Biophys.* **34,** 257 (2001).
[9] M. Hohenegger, M. Freissmuth, and C. Nanoff, *Methods Mol. Biol.* **83,** 179 (1997).
[10] R. F. Colman, *Subcell. Biochem.* **24,** 177 (1995).

0076-6879/03 $35.00

TABLE I
NOMENCLATURE OF NATURALLY OCCURRING PYRIMIDINE AND PURINE BASES AND RESPECTIVE NUCLEOSIDES AND NUCLEOTIDES[a]

	Pyrimidines		Purines		
	Adenine	Guanine	Cytosine	Thymine	Uracil
Abbreviation	A or Ade	G or Gua	C or Cyt	T or Thy	U or Ura
Nucleoside	Adenosine	Guanosine	Cytidine	Thymidine	Uridine
Abbreviation	Ado	Guo	Cyd	dThd	Urd
Nucleotide	Adenosine 5′-phosphate or 5′-adenylic acid	Guanosine 5′-phosphate or 5′-guanylic acid	Cytidine 5′ phosphate or 5′-cytidylic acid	Thymidine 5′-phosphate or 5′-thymidylic acid	Uridine 5′-phosphate or 5′-uridylic acid
Abbreviation	AMP	GMP	CMP	dTMP	UMP
Sugar	Ribose	Ribose	Ribose	2′-Deoxyribose	Ribose

[a] Modified from Ref. 11 with permission of Elsevier Science. Nucleotides are phosphorylated at the 5′ position unless otherwise indicated. All nucleotides of DNA have 2′-deoxyribose instead of ribose, as for the RNA nucleotides. They are designated with the prefix "d," such as dNMP, where N stands for nucleoside, or as dA, etc. Diphosphate derivatives are NDPs and triphosphates are NTPs.

Nomenclature and Structure of Nucleotides

The nomenclature of nucleotides and nucleosides can be confusing and is thus summarized here (Table I[11] and Fig. 1[12–14]). The nucleotides of RNA (ribonucleotides) contain a ribose ring, whereas those of DNA contain 2′-deoxyribose (deoxyribonucleotides). Nucleosides do not have phosphates and nucleotides have phosphate attached to the 5′-O unless otherwise indicated.

Structures of the common physiological substrates from which most fluorescent nucleotides are derived are shown in Fig. 1. The numbering system for purines and pyrimidines (Fig. 1) begins at the base and continues to the sugar, where numbers in combination with primes are used.

The phosphate groups are designated α, β, and γ as shown. Phosphate groups at other positions on the sugar are designated 2′-PO_4 or 3′-PO_4. Structure **1** in Fig. 1 is nucleotide triphosphate (NTP); the three common nonhydrolyzable or slowly hydrolyzable modifications of the triphosphate moiety are also shown (Fig. 1, structures **2**, **3**, and **4**), along with the trivial names. Many of the syntheses described herein can be extended to these phosphate derivatives. This is often also true for cyclic nucleotides.

[11] D. E. Metzler, "Biochemistry: The Chemical Reactions of Living Cells." Academic Press, New York, 1977.
[12] R. S. Goody and F. Eckstein, *J. Am. Chem. Soc.* **93,** 6252 (1971).
[13] R. G. Yount, D. Babcock, W. Ballantyne, and D. Ojala, *Biochemistry* **10,** 2484 (1971).
[14] T. C. Myers, K. Nakamura, and J. W. Flesher, *J. Am. Chem. Soc.* **85,** 3292 (1963).

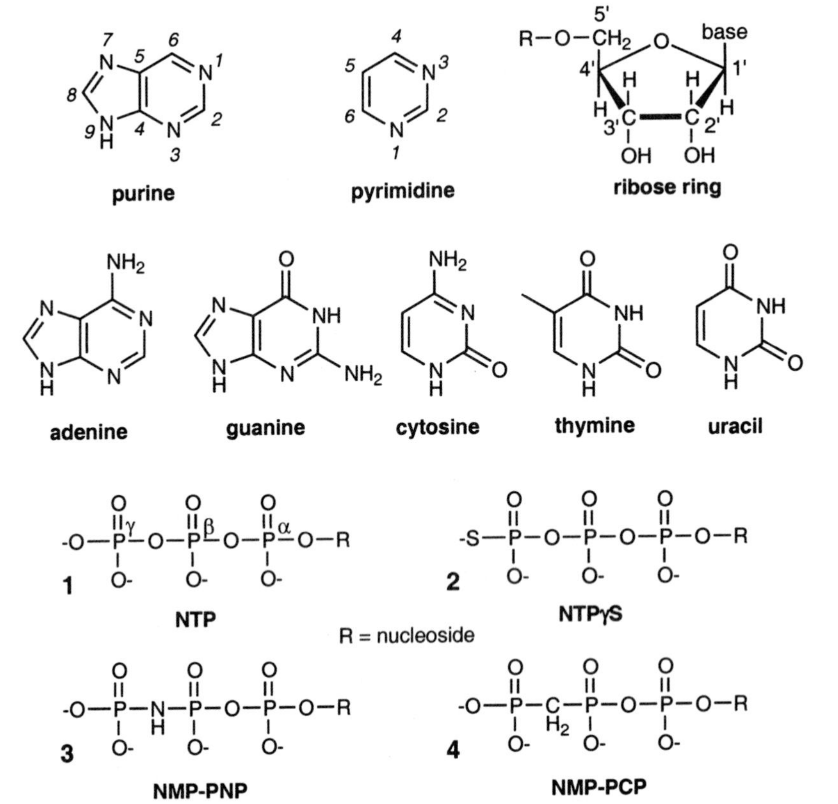

purine pyrimidine ribose ring

adenine guanine cytosine thymine uracil

NTP NTPγS R = nucleoside NMP-PNP NMP-PCP

FIG. 1. Structures, numbering, and naming of nucleotide moieties. First (*top*) row: International Union of Pure and Applied Chemistry (IUPAC; http://www.chem.qmw.ac.uk/iupac/) numbering for the two parent bases, purine and pyrimidine, and for the ribose ring. R, Phosphates. Second row: The five major purines and pyrimidines of nucleic acids in their dominant tautomeric form. Third and fourth (*bottom*) rows: Naming convention for the triphosphate moiety and other triphosphate analogs commonly used in fluorescent nucleotides. R_1, Nucleoside; nucleoside triphosphate (**1**); nucleoside 3′-thiotriphosphate (**2**), as in ATPγS [R. S. Goody and F. Eckstein, *J. Am. Chem. Soc.* **93,** 6252 (1971)]; nucleosidyl imidodiphosphate (**3**), as in AMPPNP [R. G. Yount, D. Babcock, W. Ballantyne, and D. Ojala, *Biochemistry* **10,** 2484 (1971)]; nucleosidyl methylenediphosphonate (**4**), as in AMPCMP (see Refs. 13 and 14).

Synthesis

Equipment

A rotary evaporator with a good vacuum pump is used to remove solvents. A lyophilizer can be used for aqueous solutions. A source of dry ice should be available to cool the traps for the rotary evaporator (dry ice in 2-propanol is useful) and to freeze solutions for lyophilization. A fast protein liquid chromatography (FPLC), high-performance liquid chromatography (HPLC), or automated low-pressure chromatography unit with a variable UV–visible (UV–VIS) detector is advised. For thin-layer chromatography (TLC), a hand-held UV light is needed to visualize compounds. A jar with a watchglass or petri dish cover works well to develop TLC plates. A household hairdryer is used to dry the plates. A distillation apparatus is useful to prepare solvents. For some applications a vacuum line is useful, but is often not essential. A CO_2 tank is needed to bubble CO_2 into triethylamine (TEA) solutions and an argon or N_2 tank is useful to flush reactions to minimize exposure to air. Fluorescent compounds are often sensitive to light, so room lights should be turned off, or vessels covered with aluminum foil. A low-speed table top centrifuge or microcentrifuge is often needed. Positive pressure pipettes (Microman, Rainin) are useful to quantitatively transfer organic solvents.

Finding Compounds

One of the rate-limiting steps in synthesis is to find the reactants for the synthesis. For a small fee, the availability and sources of any compound can be searched through Chem Sources (http://www.chemsources.com/). Structures and sources can be searched via Chemical Abstracts, which is available online at most university libraries. The following companies are among those that currently specialize in marketing fluorescent nucleotides and their precursors: Molecular Probes (Eugene, OR; http://www.probes.com/), Jena Bioscience (Jena, Germany; http://www.jenabioscience.com/nucleotides.html), Biolog Life Science Institute (Bremen, Germany; http://www.biolog.de/index.html), and TriLink BioTechnologies (San Diego, CA; http://www.trilinkbiotech.com/). Alt specializes in nucleotide photoaffinity labels (Lexington, KY; http://www.altcorp.com/). Three-dimensional structural information about many nucleotides and nucleosides is available at http://www.nyu.edu/pages/mathmol/library.

Solvent and Chemical Preparation

Presented below is a brief description of commonly used, commercially available chemicals that may require treatments or modifications before use. A compendium of procedures for reagents for organic synthesis is recommended.[15]

Trialkylamines. Trialkylamines are subject to air-catalyzed oxidation. Often, a yellow color indicates the presence of oxidized compounds. Tributylamine (TBA) should be redistilled under vacuum. TEA can be refluxed over *p*-toluene sulfonyl chloride to react with any secondary amines, and then distilled at atmospheric

[15] M. Fieser, ed., "Fieser and Fieser's Reagents for Organic Synthesis," Vol. 19. John Wiley & Sons, New York, 1999.

pressure. Storage of both TEA and TBA should be in a dark bottle at 4°. Alternatively, small bottles of the highest purity commercial grade can be stored under a vacuum seal and the reagent can be removed with N_2 pressure to avoid exposure of the bulk solvent to air. Aldrich (Milwaukee, WI) supplies a technical bulletin describing how to handle air-sensitive reagents.

Triethylammonium-Bicarbonate Buffer. TEA–bicarbonate buffers are prepared immediately before use by bubbling CO_2 into a 1–2 *M* aqueous solution of TEA (two phases) at 4°, using a sintered glass diffuser, until the pH reaches pH 7.6–7.8. This process can take many hours for a 4-liter volume. The buffer is a single phase at this pH. TEA–bicarbonate buffers release CO_2 over time and the pH will rise. The CO_2 is more soluble at lower temperature and therefore the lowest pH values can be obtained at 0–4°. Any increase in temperature will result in outgassing, which can cause problems during chromatography.

Other Solvents. Dimethylformamide (DMF) should be freshly redistilled and stored over 3-Å molecular sieves (pellets) to maintain an anhydrous condition. Any solvent that is required to be anhydrous should be shaken with and stored over molecular sieves to adsorb water. Filter after settling if necessary.

N,N′-Carbonyldiimidazole. *N,N′*-Carbonyldiimidazole (CDI) is hygroscopic and should be stored over desiccant. Reaction with water inactivates the compound. Dry CDI in a vacuum desiccator over P_2O_5 before use. Dissolution of CDI in a solvent should not result in bubbling if the solvent is dry.

Preparation of Triethyl- and Tributylammonium Salts of Nucleotides. Many syntheses of nucleotides are performed under anhydrous or nearly anhydrous conditions. Therefore, the first step is often to prepare the TEA or TBA salts from commercially available sodium, lithium, or other water-soluble salts. The methods described here can also be used to prepare the TEA, pyridinium, and TBA salts of phosphate and pyrophosphate.

Strong cation exchangers (Dowex 50; Dow Chemical, Midland, MI) are used for these purposes. These resins are sold in either the H^+ or Na^+ forms in different mesh sizes. Many investigators use the 100–200 mesh size. The H^+ and Na^+ forms can be interconverted with NaOH and HCl, respectively. However, it is best to begin with the correct form. When trying to remove Na^+ from a nucleotide it is best to avoid any use of the Na^+ form of Dowex. New resins should be washed according to the manufacturer's recommendations to remove colored impurities. As a rule of thumb, the grams of resin needed is calculated as [(equivalents of charge in nucleotide) $\times 10 \times$ (g/equivalents of resin)]. Resins can be reused.

1. Direct conversion to the TBA salt: A Dowex 50W (strong cation exchanger) column is prepared in the free acid form (H^+) by treatment with HCl and exhaustive flushing with water. The sodium salt of the nucleotide in water is applied to the column and, as the compound exchanges sodium for protons, the eluting

nucleotide will be acidic, a condition that promotes hydrolysis of the phosphoanhydride linkages. Therefore in this case it is essential to run the column at 4° and to ensure that the eluate is dripped into rapidly stirred cold tributylamine (in excess over the number of phosphates) to neutralize the solution. The elution of the acidic nucleotide from the column can be monitored with pH paper. Excess TBA can be removed by repetitive evaporation from methanol.

2. Conversion to the TEA salt before conversion to the TBA salt: Water-soluble salts of nucleotides are applied to a Dowex 50 column (H^+ form) that has been prepared in the TEA form by equilibration with 1 *M* TEA until the eluant is basic and subsequent flushing with water until the eluant is neutral. Nucleotide elution can be monitored by spotting samples onto a TLC plate with fluorescent indicator. Reduce the solvent volume, add excess TBA, and dry by rotary evaporation. The product is often a clear solid or an oil.

It is important to convert all of the compound to the correct salt. An unsuccessful transformation will be evident, as the putative TEA or TBA salt will not be soluble in an organic solvent. Precipitate indicates the presence of water-soluble ions and the procedure should be repeated.

Preparation of Na^+ Salts of Nucleotides. Nucleotides are often synthesized as the TEA or TBA salts. TEA and TBA salts are usually oils and must be dissolved in organic solvents, such as methanol or DMF. Because these amines can be oxidized during handling, they may be sources of UV-absorbing or other contaminating or reactive material generated by degradation. TEA and TBA can also complicate proton and ^{13}C- NMR (nuclear magnetic resonance) spectra. In the author's experience, long-term storage of TEA and TBA salts can promote degradation of nucleotides, whereas Na^+ or Li^+ salts appear to be more stable, either in solution or as solids. Nucleotides are usually commercially available as sodium, lithium, or other water-soluble salts. For these reasons, it is useful to convert these alkylammonium salts to sodium salts for analysis and storage. There are two easy methods.

A fresh solution of saturated NaI (protected from light) in acetone is cooled to −20°. The nucleotide is dried repetitively from methanol or other dry organic solvent to remove residual excess amines and dissolved in the minimal volume of dry organic solvent in which the alkylammonium salt is soluble. It is critical that this volume be kept to the minimum. The nucleotide solution is cooled on ice and the cold NaI solution is added dropwise until no more precipitation can be observed. Nucleotides with good solubility in the organic solvent used will have difficulty precipitating if the volume is too large. The flask can be triturated as needed and the contents transferred to a centrifuge tube. The precipitate is the Na^+ salt of the nucleotide, which can be recovered by centrifugation. The amorphous solid can be washed repetitively with dry acetone or ether, and dried with a stream of N_2.

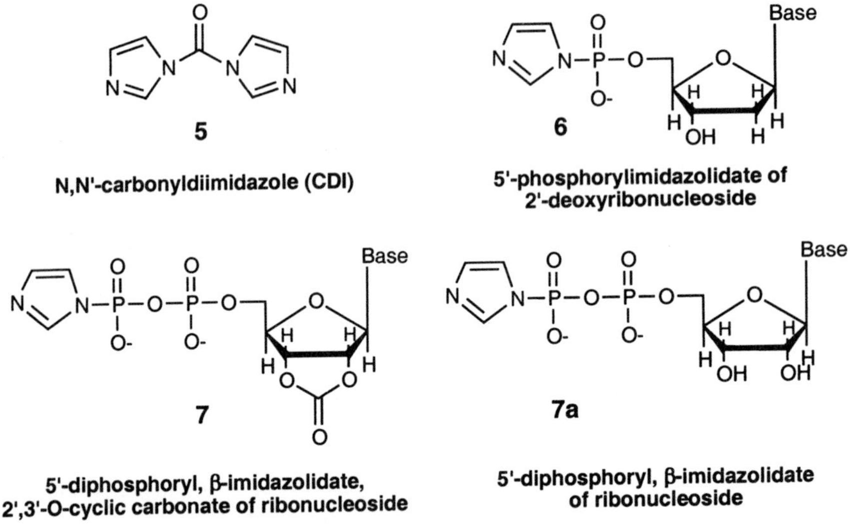

5

N,N'-carbonyldiimidazole (CDI)

6

5'-phosphorylimidazolidate of 2'-deoxyribonucleoside

7

5'-diphosphoryl, β-imidazolidate, 2',3'-O-cyclic carbonate of ribonucleoside

7a

5'-diphosphoryl, β-imidazolidate of ribonucleoside

FIG. 2. Reaction of CDI (**5**) with deoxyribonucleotides gives structure **6**. Reaction with ribonucleotides (NDP shown here) gives **7**, which can be hydrolyzed to **7a**.

Alternatively, the alkylammonium salts can be passed through a column of Dowex 50W-X8 in the sodium form and eluted with water.

Phosphorylation of Nucleotide Monophosphate Derivatives

Chemical Methods

Preparation of the tri- and diphosphates from monophosphates of deoxyribonucleotides can be achieved chemically by the method of Hoard and Ott.[16] CDI (**5**; Fig. 2) is reacted with the tributylammonium salt of a nucleotide monophosphate under anhydrous conditions to give the 5′-phosphorylimidazolidate (**6**; Fig. 2). From this intermediate, the diphosphate and triphosphate may be synthesized by adding the tributylammonium salt of phosphate or pyrophosphate, respectively. Although this method has been applied to ribonucleotides, it has the complication that CDI will modify the ribose 2′- and 3′-OH to the cyclic carbonate,[17] resulting in structure **7** (Fig. 2) for the diphosphate case. A mildly basic treatment is sufficient to hydrolyze the cyclic carbonate to **7a** (Fig. 2). The half-life of the cyclic carbonate is 5–7 hr at pH 7.6 (22°). Therefore, if ion-exchange chromatography in TEA–bicarbonate is used for purification, it is likely that the cyclic carbonate

[16] D. Hoard and D. Ott, *J. Am. Chem. Soc.* **87,** 1785 (1965).

[17] M. Maeda, A. Patel, and A. Hampton, *Nucleic Acids Res.* **4,** 2843 (1977).

will be hydrolyzed during the chromatography. Alternatively, the reverse synthesis can be performed,[17] whereby CDI is added to phosphate or pyrophosphate to form the respective imidazolidate. After quenching excess CDI with methanol, the imidazolidate is added to AMP. This modification does not allow the cyclic carbonate to form. An interesting modification of the Hoard and Ott[16] procedure has been reported to be faster and to give higher yields.[18] Preparation of the 5′-monophosphates from nucleosides is often accomplished by the method of Yoshikawa *et al.*[19]

Enzymatic Methods

Enzymatic methods are often preferable to chemical methods because reactions can go to completion, given the optimum conditions. They are usually easier than chemical methods, requiring less manipulation, and are therefore often preferred for radioactive syntheses. The following protocols are a few examples of the use of enzymes in fluorescent nucleotide synthesis. A complete treatment of this topic is not attempted here.

Phosphorylation of Nucleoside Monophosphate to Nucleoside Diphosphate and Nucleoside Triphosphate Derivatives. It is often possible to convert nucleoside monophosphate (NMP) derivatives to the diphosphate (NDP) and triphosphate (NTP) derivatives by enzymatic methods, depending on enzyme availability and specificity. An example of this approach is given for the conversion of formycin A monophosphate (FMP) to formycin A triphosphate (FTP).[20,21] This procedure also works for the respective etheno derivatives. Adenylate kinase converts FMP plus ATP to FDP plus ADP and pyruvate kinase converts FDP plus phosphoenolpyruvate (PEP) to FTP plus pyruvate.

The following mixture is incubated for 60 min at 30°:

TEA (0.1 *M*, pH 7.6), 0.1 *M* KCl, 0.002 *M* $MgSO_4$	100 μl
FMP	1 μmol
ATP (to prime synthesis)	0.1 μmol
Phosphoenolpyruvate (Sigma, St. Louis, MO)	3 μmol
Adenylate kinase (Sigma)	2 units
Pyruvate kinase (Boehringer, Mannheim, Germany)	2 units

Another example of this approach is found in Eccleston *et al.*[22] The 5′-monophosphate of 2′-amino-2′-deoxyguanosine was converted to the triphosphate.

[18] I. Yanachkov, J. Y. Pan, M. Wessling-Resnick, and G. E. Wright, *Mol. Pharmacol.* **51,** 47 (1997).
[19] M. Yoshikawa, T. Kata, and H. Miki, *Bull. Chem. Soc. Jpn.* **42,** 3505 (1969).
[20] B. Schobert, *Anal. Biochem.* **226,** 288 (1995).
[21] E. F. Rossomando, J. H. Jahngen, and J. F. Eccleston, *Proc. Natl. Acad. Sci. U.S.A.* **78,** 2278 (1981).
[22] J. F. Eccleston, E. Gratton, and D. M. Jameson, *Biochemistry* **26,** 3902 (1987).

The following reaction mixture was incubated at 30° and additional enzyme aliquots were added at 6 and 24 hr. After 48 hr, the reaction was adjusted to pH 2 with Dowex 50W resin in the H^+ form, filtered to remove denatured enzymes, and adjusted to pH 7.6 before purification by diethylaminoethyl (DEAE) chromatography. The final reaction volume is 12 ml:

2′-Amino-2′-deoxyguanosine-5′-monophosphate	2 m*M*
Phosphoenolpyruvate	67 m*M*
Tris (pH 7.6, adjusted after mixing)	50 m*M*
$MgCl_2$	5 m*M*
ATP	17 μM
Pyruvate kinase (rabbit muscle)	2500 units
Guanosine 5′-monophosphate kinase (hog brain)	10 units

Dephosphorylation of Nucleoside Triphosphate to Nucleoside Diphosphate Derivatives. If both the diphosphate and triphosphate derivatives of the analog are needed, it is best to synthesize the triphosphate and hydrolyze a portion of it to the diphosphate, using chicken or rabbit skeletal myosin subfragment 1[23] or myosin at 1–2 mg/ml. The enzyme is generally tolerant of nucleotide modifications.

The reaction is performed at 25° at pH 7.0 to 8.0 in morpholinepropane sulfonic acid (MOPS) or Tris buffer (50 m*M*) with 1 m*M* $MgCl_2$ (or 1 m*M* $CaCl_2$) and 0.5 m*M* EDTA. For myosin, add NaCl to 200 m*M*. Protein can be pelleted by adding 2 volumes of cold acetone or ethanol, incubating the mixture at 4° for 10 min, and centrifuging it at ~15,000–20,000*g*.

Preparation of [γ-^{32}P]ATP Analogs. Preparation of [γ-^{32}P]ATP analogs from the unlabeled ATP analogs can be accomplished by the method of Glynn and Chappell.[24] This method was used for the preparation of [γ-^{32}P]DNS-ATP (**32**).

General Synthetic Approaches

There are two general synthetic approaches to the preparation of fluorescent nucleotides. First, the base can be modified to give a fluorescent base that has properties similar to those of the unmodified base. Second, fluorescent groups can be appended to either a modified or unmodified version of the base, the sugar, or the phosphate(s). Fluorescence properties of the nucleotides mentioned in this review are shown in Table II.[22,25–60]

[23] A. G. Weeds and R. S. Taylor, *Nature* (*London*) **257,** 54 (1975).
[24] I. M. Glynn and J. B. Chappell, *Biochem. J.* **90,** 147 (1964).

TABLE II
SPECTROSCOPIC PROPERTIES OF FLUORESCENT NUCLEOTIDES[a]

Structure number	Trivial name or class of compound	Analog of:	Excitation or absorption wavelength (nm)[b]	Molar extinction coefficient (M^{-1} cm^{-1})[c]	Emission maximum (nm)[d]	Quantum yield[e]	Lifetime (ns)[f]	Refs.[g]
			Base-modified analogs					
8	1,N^6-Ethenoadenosine derivatives	Adenosine 5′-phosphates	294 abs 305 ex	2,900	415	0.59	26.5	25, 26
	$Bz_2\varepsilon$ADP	ADP	264 abs	32,250	410	0.053	5.1	26
	$Bz_2\varepsilon$dADP	dADP	264 abs	32,250	410	0.10	6.1	
8b	8-Azido-1,N^6-etheno-ATP	ATP	288 abs 340 ex	7,370	402	—	—	27
8c	8-Azido-1,N^6-etheno-cAMP	cAMP	290 abs 360 ex	3,500	402	—	—	28
	5′-[*p*-(Fluorosulfonyl)benzoyl]-1,N^6-Ethenoadenosine	Adenosine	275 abs 308 ex	6,210	412	0.008	—	29
9	1,N^6-Etheno-2-azaadenosine derivatives	Adenosine 5′-phosphates	290 abs 354 abs 354 ex	5,000 1,530	494	—	—	30–32
10	Formycin A nucleotides	Adenosine 5′-phosphates	295 abs 295 ex	9,500	340	0.064	<1	33
13	2-Aminopurine nucleotides	Adenosine 5′-phosphates	303 abs 303 ex	7,100	370	0.86	7.0	33
12a	6MAP	2′-Deoxyadenosine	329 abs 320 ex	8,510 (CH_3OH)	430	0.39	3.8	34
12b	DMAP	2′-Deoxyadenosine	333 abs 310 ex	8,910 (CH_3OH)	430	0.48	4.8	34
11	Linear benzoadenosine derivatives	Adenosine 5′-phosphates	331 abs 348 ex	9,750	372	0.44	3.7	35, 36
15	Linear benzoguanosine derivatives	Guanosine 5′-phosphates	323 abs 323 ex	5,600	385	0.39	6	36
14a	3-MI	2′-Deoxyguanosine	~350 abs	13,490[h] (CH_3OH)	~420	0.88	—	37

continued

TABLE II (*continued*)

Structure number	Trivial name or class of compound	Analog of:	Excitation or absorption wavelength (nm)[b]	Molar extinction coefficient (M^{-1} cm^{-1})[c]	Emission maximum (nm)[d]	Quantum yield[e]	Lifetime (ns)[f]	Refs.[g]
14b	6-MI	2′-Deoxyguanosine	340 ex	—	431	0.7	6.3	38
16	1-Methyl-2-azido-6-oxopurine ribosides	Guanosine 5′-phosphates	307 abs	8,850	400	0.25	—	39
			307 ex					
			Phosphate-modified analogs					
19	(γ-AmNS)ATP		315 abs	5,580	460	0.8	20	40–42
	(γ-AmNS)dATP		315 ex					
			323 ex	4,200				
20a	(γ-1,5-EDANS)ATP	ATP	255 abs	3.2×10^4	495	0.11	—	43
			366 ex					
21	(PB)MABA-CDP	CDP	325 abs	2,700	428	—	—	44
22	Cm-GTP	GTP	342 abs	6,000	440	0.007	—	45
			Ribose-modified analogs					
32	DNS-ATP	ATP	334 abs	3,950	554	0.052	—	46
			345 ex					
26	DEDA-ADP	ADP	330 abs	4,350	555	0.089	3.6	47
42a	1,5-DAN-ATP	ATP	333 ex		609	—	—	48
42b	2,5-DAN-ATP	ATP	345 ex		550	—	7.8[i]	48
38a	2′-*O*-Dansyl-ATP	ATP	325 ex		501	—	—	49
38b	8-Azido-2′-*O*-dansyl-ATP	ATP	339 abs	—	—	—	—	50
31	Fluorescamine-GDP	GDP	380 abs	2,400	484[j]	—	7.74	22
			385 ex					
41a	TNP-adenosine nucleotides	Adenosine 5′-phosphates	408 abs	26,400	552[j]	0.0002	190 ps	51, 52
			470 abs	18,500			50 ps[k]	
41b	TNP-guanosine nucleotides	Guanosine 5′-phosphates	408 abs	26,500	552[j]	—	—	53
			470 abs	18,300				
39a	Mant-nucleotides	Adenosine 5′-phosphates	356 abs	5,800	446	0.22	4.0	54, 55
			350 abs	5,700	442	0.24	—	

		Guanosine 5′-	355 abs	5,400	445	0.21		
		phosphates	350 abs	5,300	441	0.26		
		cAMP						
		cGMP						
39b	Ant-nucleotides	Adenosine 5′-phosphates	332	4,700	428	0.14	—	54, 55
		Guanosine 5′-phosphates	332	4,600	428	0.14	—	
		cAMP	333	4,500	430	0.11	—	
		cGMP	332	4,400	427	0.12	—	
			Unusual analogs					
43	DNS-triphosphate	Adenosine 5-phosphates	320 abs	5,410	545	—		56
44	BzAF	ATP	490 abs 490 ex	81,000	—	—		57, 58

[a] For nucleotides mentioned in this article. All values are for approximately neutral pH in aqueous buffer unless otherwise indicated. Structures can be found in Figs. 3–13.

[b] Some wavelengths are the maxima of the absorption spectrum, or the excitation spectrum as indicated. Other wavelengths are peaks of the lowest energy absorbance band that are most relevant to fluorescence measurements, but are not the only maxima in the spectrum.

[c] The molar extinction coefficient is for the absorption wavelength given.

[d] The emission maximum is for corrected emission spectra unless otherwise indicated. The excitation wavelength used is often not quoted in the original publications. Usually it is close to the excitation maximum.

[e] The quantum yield is relative to quinine sulfate = 0.7. Values are adjusted if another value for quinine sulfate was used.

[f] Most lifetimes are single exponentials. For double exponentials, the amplitude-weighted average is reported.

[g] These are the primary references for most of the data for a compound. There may be other references containing data that are not quoted.

[h] This is a correction from the original manuscript (Ref. 59).

[i] Ethanol–water (1 : 1, v/v).

[j] Uncorrected emission spectrum.

[k] Biexponential decay, 87% (v/v) glycerol–water; see Ref. 60.

Modifications to Base

For base-modified derivatives, there are two classes of compounds. The first class includes modifications to the native structure of the base itself that render the base fluorescent. The focus here is to direct the reader to the original syntheses and subsequent improvements for this type of compound. The second class is derived from a common reactive precursor that is not fluorescent, to which fluorescent compounds are attached. This class of derivative is not mentioned in this review.

[25] J. A. I. Secrist, J. R. Barrio, N. J. Leonard, and G. Weber, *Biochemistry* **11,** 3499 (1972).
[26] C. R. Cremo and R. G. Yount, *Biochemistry* **26,** 7524 (1987).
[27] H. J. Schafer, P. Scheurich, G. Rathgeber, and K. Dose, *Nucleic Acids Res.* **5,** 1345 (1978).
[28] E. K. Keeler and P. Campbell, *Biochem. Biophys. Res. Commun.* **72,** 575 (1976).
[29] J. J. Likos and R. F. Colman, *Biochemistry* **20,** 491 (1981).
[30] K. C. Tsou, K. F. Yip, E. E. Miller, and K. W. Lo, *Nucleic Acids Res.* **1,** 531 (1974).
[31] H. Miyata and H. Asai, *Biochem. Biophys. Res. Commun.* **105,** 296 (1982).
[32] K. F. Yip and K. C. Tsou, *Tetrahedron Lett.* **33,** 3087 (1973).
[33] D. C. Ward, E. Reich, and L. Stryer, *J. Biol. Chem.* **244,** 1228 (1969).
[34] M. E. Hawkins, W. Pfleiderer, O. Jungmann, and F. M. Balis, *Anal. Biochem.* **298,** 231 (2001).
[35] N. J. Leonard, M. A. Sprecker, and A. G. Morrice, *J. Am. Chem. Soc.* **98,** 3987 (1976).
[36] N. J. Leonard and G. E. Keyser, *Proc. Natl. Acad. Sci. U.S.A.* **76,** 4262 (1979).
[37] M. E. Hawkins, W. Pfleiderer, A. Mazumder, Y. G. Pommier, and F. M. Balis, *Nucleic Acids Res.* **23,** 2872 (1995).
[38] M. E. Hawkins, W. Pfleiderer, F. M. Balis, D. Porter, and J. R. Knutson, *Anal. Biochem.* **244,** 86 (1997).
[39] G. Weigand and R. Kaleja, *Eur. J. Biochem.* **65,** 473 (1976).
[40] L. R. Yarbrough, J. G. Schlageck, and M. Baughman, *J. Biol. Chem.* **254,** 12069 (1979).
[41] J. G. Schlageck, M. Baughman, and L. R. Yarbrough, *J. Biol. Chem.* **254,** 12074 (1979).
[42] L. R. Yarbrough, *Biochem. Biophys. Res. Commun.* **81,** 35 (1978).
[43] F. Y. Wu, A. W. Abdulwajid, and D. Solaiman, *Arch. Biochem. Biophys.* **246,** 564 (1986).
[44] M. G. Rudolph, T. J. Veit, and J. Reinstein, *Protein Sci.* **8,** 2697 (1999).
[45] S. Sastry, *Biophys. Chem.* **91,** 191 (2001).
[46] T. Watanabe, A. Inoue, Y. Tonomura, S. Uesugi, and M. Ikehara, *J. Biochem.* (*Tokyo*) **90,** 957 (1981).
[47] C. R. Cremo, J. M. Neuron, and R. G. Yount, *Biochemistry* **29,** 3309 (1990).
[48] I. Mayer, A. S. Dahms, W. Riezler, and M. Klingenberg, *Biochemistry* **23,** 2436 (1984).
[49] D. Thoenges and W. Schoner, *J. Biol. Chem.* **272,** 16315 (1997).
[50] H. Chuan, J. Lin, and J. H. Wang, *J. Biol. Chem.* **264,** 7981 (1989).
[51] T. Hiratsuka and K. Uchida, *Biochim. Biophys. Acta* **453,** 293 (1973).
[52] T. Hiratsuka, *Biochim. Biophys. Acta* **453,** 293 (1976).
[53] T. Hiratsuka, *J. Biol. Chem.* **260,** 4784 (1985).
[54] T. Hiratsuka, *Biochim. Biophys. Acta* **742,** 496 (1983).
[55] T. Hiratsuka, *J. Biol. Chem.* **257,** 13354 (1982).
[56] M. Onodera, and K. Yagi, *Biochim. Biophys. Acta* **253,** 254 (1971).
[57] P. Pal and P. Coleman, *J. Biol. Chem.* **265,** 14996 (1990).
[58] J. E. Rosen, *SPIE* **1885,** 349 (1993).
[59] M. E. Hawkins, personal communication.
[60] J. Y. Ye, M. Yamauchi, O. Yogi, and M. Ishikawa, *J. Phys. Chem.* **103,** 2812 (1999).

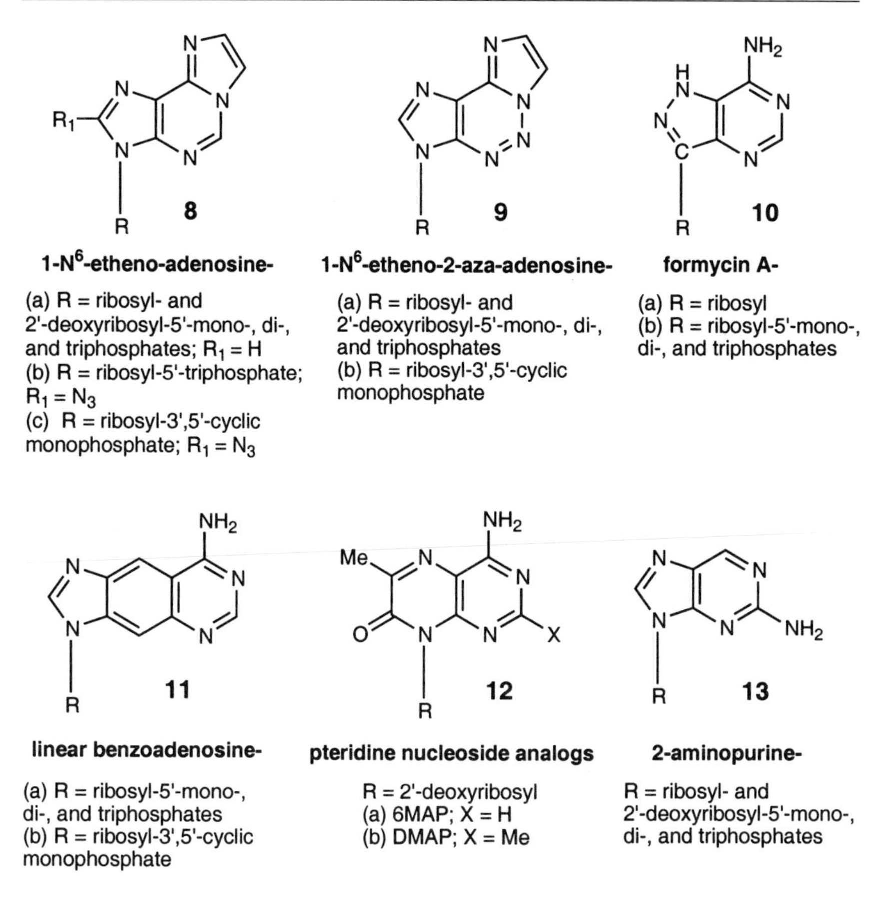

FIG. 3. Fluorescent analogs of the adenine ring.

MODIFICATION TO ADENINE RING

1,N⁶-Ethenoadenosine derivatives. One of the most exploited approaches to the preparation of base-modified fluorescent adenosine analogs is the highly cited $1,N^6$-etheno modification of the adenine ring[25] (**8**; Fig. 3), which is based on the original work of Kochetkov *et al.*[61] The modified adenine ring generally mimicks the actions of the parent derivatives. In addition, its spectral properties are excellent, with a high quantum yield and long lifetime (~20 ns).

Many other derivatives containing the etheno modification have been synthesized and a few are mentioned here. 3′(2′)-*O*-(4-Benzoylbenzoyl)-1,

[61] N. K. Kochetkov, V. N. Shibaev, and A. A. Kost, *Tetrahedron Lett.* **22,** 1993 (1971).

N^6-ethenoadenosine diphosphate (Bz$_2\varepsilon$ADP) has the photoreactive benzoylbenzoyl modification by an ester linkage to the ribose ring that has proved useful for myosin active site studies.[26,62] 8-Azido-1,N^6-etheno-ATP (**8b**; Fig. 3) carries the photoreactive azido group along with the fluorescent etheno modification on the adenine ring.[27]

Many etheno-modified adenine nucleotides are commercially available. In the single-step, high-yield synthesis the nucleotide is reacted with chloroacetaldehyde (2-chloro-1-ethanal). Chloroacetaldehyde can be purchased as an aqueous solution (~50%) and should be distilled under reduced pressure immediately before use.[25,63] Great care should be taken when using chloroacetaldehyde, as it has an acrid penetrating odor; is a severe eye, skin, and upper respiratory tract irritant; and is a mutagen. It is recommended that the products of the synthesis be purified by ion-exchange chromatography, a step not mentioned in the original synthesis.

1,N^6-Etheno-2-azaadenosine 5′-phosphates. 1,N^6-Etheno-2-azaadenosine derivatives (**9** in Fig. 3) have not been as widely used as the 1,N^6-ethenoadenosine derivatives. However, they do have favorable spectral properties, in particular the long emission wavelength. In general, the fluorescence properties of these probes have not been well characterized. The nucleoside,[30] the 5′-tri- and 5′-diphosphate,[64] and the 5′,3′-cyclic phosphate[65] can be prepared from the respective 1,N^6-ethenoadenosine nucleotides (**8**) in two steps. The synthesis of the 2′,3′-cyclic monophosphate has been reported.[66] Muscle researchers have found the 5′-triphosphate to be useful, as it binds to myosin with a large increase in fluorescence.[31,64,67–70] The 5′-monophosphate can replace the covalently attached AMP moiety of adenylylated glutamine synthetase and has proved useful in the analysis of the kinetic cycle.[71,72] 5′-(*p*-Fluorosulfonylbenzoyl)-2-aza-1,N^6-ethenoadenosine has been used to chemically modify phosphofructokinase.[73]

Formycin A 5′-phosphates. Formycin A (**10a** in Fig. 3; 7-amino-3-[β-D-ribofuranosyl]pyrazolo[4,3-*d*]pyrimidine) is a naturally occurring nucleoside antibiotic that mimicks adenosine. The carbon at position-8 of the purine ring is reversed

[62] Y. Luo, D. Wang, C. R. Cremo, E. Pate, R. Cooke, and R. G. Yount, *Biochemistry* **34,** 1978 (1995).
[63] J. E. Baker, I. Brust-Mascher, S. Ramachandran, L. E. LaConte, and D. D. Thomas, *Proc. Natl. Acad. Sci. U.S.A.* **95,** 2944 (1998).
[64] S. J. Smith and H. D. White, *J. Biol. Chem.* **260,** 15156 (1985).
[65] N. Yamaji and M. Kato, *Chem. Lett.* **4,** 311 (1975).
[66] K. W. Lo, K. F. Yip, and K. C. Tsou, *J. Neurochem.* **25,** 181 (1975).
[67] H. D. White and I. Rayment, *Biochemistry* **32,** 9859 (1993).
[68] E. Pate, K. Franks-Skiba, H. White, and R. Cooke, *J. Biol. Chem.* **268,** 10046 (1993).
[69] H. D. White, B. Belknap, and W. Jiang, *J. Biol. Chem.* **268,** 10039 (1993).
[70] H. Nagano and T. Yanagida, *J. Mol. Biol.* **177,** 769 (1984).
[71] S. G. Rhee, G. A. Ubom, J. B. Hunt, and P. B. Chock, *J. Biol. Chem.* **257,** 289 (1982).
[72] S. G. Rhee, G. A. Ubom, J. B. Hunt, and P. B. Chock, *J. Biol. Chem.* **256,** 6010 (1981).
[73] D. W. Craig and G. G. Hammes, *Biochemistry* **19,** 330 (1980).

with the nitrogen at position-9 and it is therefore a C-riboside. Formycin A 5′-phosphates (**10b**) have received wide use[33,74,75] because they generally mimic the action of adenosine 5′-phosphates. However, the nucleoside formycin A, from which the 5′-phosphates are synthesized, is currently not commercially available. It can be prepared from cultures of *Nocardia interforma* (American Type Culture Collection [ATCC], Manassas, VA)[76] and purified by ion-exchange chromatography.[77] A multistep chemical synthesis[78] has been shortened and improved.[79] Formycin A 5′-phosphates can be prepared by conventional synthetic and enzymatic methods.[16,20,21,74,80,81]

Linear benzoadenosine derivatives. The linear benzoadenine ring (**11** in Fig. 3; 8-aminoimidazo[4,5-*g*]quinazoline) is an analog of adenine that is stretched out (by 2.4 Å) in a linear manner by addition of a benzene ring between the two terminal rings. These analogs have been referred to as dimensional probes,[82] used to test the dimensions of adenine-binding sites. The original synthesis of the base[83] has been extended to the nucleoside[35] and nucleotides[82] including the 3′,5′-cyclic nucleotide.[84] This approach to dimensional nucleotides has been extended to the guanosine (see structure **15** in Fig. 4), inosine, and xanthosine derivatives.[36]

Pteridine-based adenosine analogs. Pteridine-based adenosine analogs are based on the pteridine ring (**12**; Fig. 3). 4-Amino-6-methyl-8-(2′-deoxy-β-D-ribofuranosyl)-7(8*H*)-pteridone (6MAP; **12a** in Fig. 3) and 4-amino-2,6-dimethyl-8-(2′-deoxy-β-D-ribofuranosyl)-7(8*H*)-pteridone (DMAP; **12b** in Fig. 3) have been synthesized also as the phosphoramidates for automated incorporation into DNA.[34] The relatively high quantum yields of the monomers are quenched by interactions with neighboring based in the oligomers. The authors report little disruption to the DNA structure on the basis of melting point data. Applications have been reviewed.[8] To date there is no report of the preparation of the 5′-phosphate derivatives of these bases.

[74] D. C. Ward, A. Cerami, E. Reich, G. Acs, and L. Altwerger, *J. Biol. Chem.* **244,** 3243 (1969).

[75] R. J. Ankrett, A. J. Rowe, R. A. Cross, J. Kendrick-Jones, and C. Bagshaw, *J. Mol. Biol.* **217,** 323 (1991).

[76] T. Kunimoto, T. Sawa, T. Wakashiro, M. Hori, and H. Umezawa, *J. Antibiot.* (*Tokyo*) **24,** 253 (1971).

[77] H. Umezawa and S. Kondo, *Zaidan Hojin Biseibutsu Kagaku Kenkyu Kai,* U.S. Patent 4,151,347, 1979.

[78] J. G. Buchanan, A. R. Edgar, R. J. Hutchison, A. Stobie, and R. H. Wightman, *J. Chem. Soc. Chem. Commun.* **5** (1980).

[79] J. G. Buchanan, A. O. Jumaah, G. Kerr, R. R. Talekar, and R. H. Wightman, *J. Chem. Soc.* **1,** 1077 (1991).

[80] M. Ikehara and T. Tezuka, *Nucleic Acids Res.* **1,** 907 (1974).

[81] E. De Clercq, J. Balzarini, D. Madej, F. Hansske, and M. J. Robins, *J. Med. Chem.* **30,** 481 (1987).

[82] N. J. Leonard, D. I. Scopes, P. VanDerLijn, and J. R. Barrio, *Biochemistry* **17,** 3677 (1978).

[83] N. J. Leonard, A. G. Morrice, and M. A. Sprecker, *J. Org. Chem.* **40,** 356 (1975).

[84] M. J. Schmidt, L. L. Truex, N. J. Leonard, D. I. Scopes, and J. R. Barrio, *J. Cyclic Nucleotide Res.* **4,** 201 (1978).

14

methylisoxanthopterins

(a) 3-MI; R_1= Me, R_2 = H, R = ribosyl
(b) 6 -MI; R_1 = H; R_2 = Me, R = 2'-deoxyribosyl

15

linear benzoguanosine-

R = ribosyl 5'-phosphates

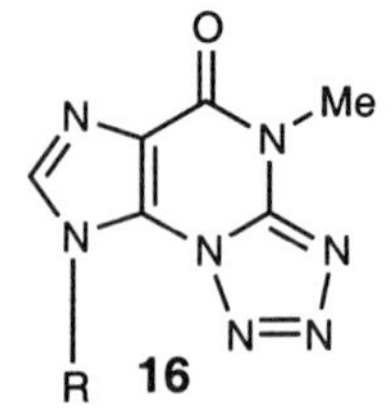

1-methyl-2-azido-6-oxo-purine-

(a) R = ribosyl 5'-monophosphate
(b) R = ribosyl 5'-triphosphate

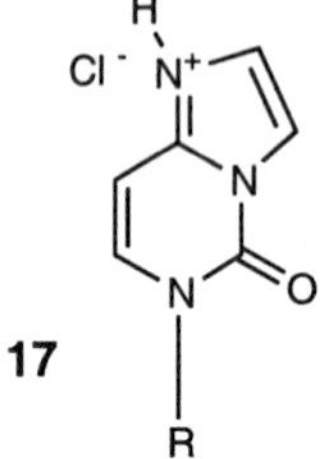

3,N⁴-ethenocytidine

R = ribosyl 5'-phosphates

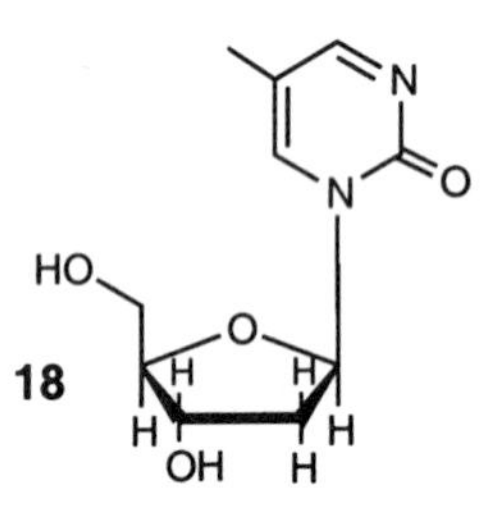

5-methylpyrimidin-2-one nucleoside

FIG. 4. Fluorescent analogs of the guanine (*top row*) and cytosine (*bottom row*) rings.

2-Aminopurine (isoadenine) nucleoside 5′-phosphates. 2-Aminopurine (**13**; Fig. 3), an analog of adenosine and guanosine, can be incorporated into DNA and RNA.[85,86] Therefore, it is widely used as a site-specific probe of nucleic acid structure and dynamics. 2-Aminopurine can base pair with thymine in a Watson–Crick configuration and with cytosine in a wobble configuration.[87] The high quantum yield of the monomeric nucleotide is quenched on incorporation into DNA/RNA and the emission intensity and lifetime are sensitive to the chemical environment. It is interesting that the molecular interactions that give rise to spectroscopic changes are still under careful study.[88,89] The synthesis of stable 2-aminopurine phosphoramidites that couple with high efficiency under standard automated synthesis conditions is reported.[90]

[85] I. Zagorowska and R. W. Adamiak, *Biochimie* **78,** 123 (1996).
[86] J. A. Doudna, J. W. Szostak, A. Rich, and N. Usman, *J. Org. Chem.* **55,** 5547 (1990).
[87] L. C. Sowers, Y. Boulard, and G. V. Fazakerley, *Biochemistry* **39,** 7613 (2000).
[88] E. L. Rachofsky, E. Seibert, J. T. Stivers, R. Osman, and J. B. Ross, *Biochemistry* **40,** 957 (2001).
[89] J. M. Jean and K. B. Hall, *Proc. Natl. Acad. Sci. U.S.A.* **98,** 37 (2001).
[90] J. Fujimoto, Z. Nuesca, M. Mazurek, and L. C. Sowers, *Nucleic Acids Res.* **24,** 754 (1996).

The base, 2-aminopurine, is an inhibitor of serine-threonine protein kinases and is currently used as a valuable tool to understand the role of kinases in gene expression. Although studies have examined the specificity of various enzymes for 2-aminopurine nucleotides (e.g., McClure and Scheit[91]), the current protein literature is not reporting extensive use of the nucleotides.

MODIFICATION TO GUANINE RING

Pteridine-based guanosine analogs or methylisoxanthopterins. 3-Methyl-8-(2′-deoxy-β-D-ribofuranosyl) isoxanthopterin [or 3-MI (**14a**); Fig. 4] and 6-methyl-8-(2′-deoxy-β-D-ribofuranosyl) isoxanthopterin [or 6-MI (**14b**); Fig. 4] are members of a new class of pteridine-based fluorophores. These analogs have been synthesized along with the 3′-*O*-(bβ)-cyanoethyl-*N*-diisopropyl)phosphoramidite derivatives for use in a DNA synthesizer and are commercially available (Trilink). However, as of yet, the 5′-mono-, 5′-di-, or 5′-triphosphates have not been reported. It is likely that they would be useful in studies of guanosine nucleotide-binding proteins.

3-MI (**14a**)[37] and 6-MI (**14b**)[38] can replace guanosine in an oligonucleotide by incorporation through a 3′,5′-phosphodiester linkage. The relatively large quantum yields of the nucleotides are an advantage and incorporation into oligonucleotides results in quenching, depending on the position in the strand and the neighboring bases. A major advantage of these derivatives is stability toward photochemical decomposition[63] and the evidence that they only minimally disrupt DNA structure.[34] There have been many applications, all of which involve incorporation into oligonucleotides.[34,37,92,93]

Linear benzoguanosine nucleotides. The linear benzoguanosine 5′-mono-, 5′-di-, and 5′-triphosphates (**15**; Fig. 4) have been synthesized[36] as derivatives of guanosine with a spacer phenyl ring between the pyrimidine and imidazole rings of guanosine.

1-Methyl-2-azido-6-oxopurine ribosyl 5′-phosphates. The interesting guanosine derivative 1-methyl-2-azido-6-oxo-purine ribosyl 5′-phosphate (**16a**; Fig. 4) has had little use in the literature but has interesting properties.[39] At equilibrium, the ring is essentially all in the tetrazolo form and not in the azido form. The authors report that this probe is not stable above pH 8. The triphosphate (**16b**) has been shown to be a substrate for tubulin. Nonmethylated derivatives are also fluorescent and can be used as photoreactive probes because the azido form is more predominant.

MODIFICATION TO CYTOSINE RING

3,N^4-Ethenocytidine derivatives. The synthesis of 3,N^4-ethenocytidine (**17**; Fig. 4) was reported in the initial publication of the more widely used 1,N^6-ethenoadenosine derivatives (**8**).[25] Unlike the adenosine counterpart,

[91] W. R. McClure and K. H. Scheit, *FEBS Lett.* **32,** 267 (1973).

[92] A. M. Moser, M. Patel, H. Yoo, F. M. Balis, and M. E. Hawkins, *Anal. Biochem.* **281,** 216 (2000).

[93] K. Wojtuszewski, M. E. Hawkins, J. L. Cole, and I. Mukerji, *Biochemistry* **40,** 2588 (2001).

3,N^4-ethenocytidine did not receive wide use because only the protonated form is fluorescent (pK_a 4.0). This limiting feature can be overcome by N^1-alkylation but the quantum yields of these and the unalkylated compounds are low (<0.01).[94]

5-Methylpyrimidin-2-one-nucleoside. Singleton *et al.*[95] have reported an improved synthesis of this fluorescent 2′-deoxycytidine analog (**18**; Fig. 4). It is useful in studies of DNA structure[96–100] but does not base pair with dA.

Modifications to Phosphates

PHOSPHORAMIDATES. Phosphoramidates, such as **19–22** (Fig. 5), are generally synthesized through the condensation of a fluorescent primary amine and the terminal phosphate of a nucleotide by use of the water-soluble carbodiimide EDC [*N*-ethyl-*N*′-(3-dimethylaminopropyl)carbodiimide] to activate the terminal phosphate.

Many reports[43,101,102] use modifications of the method of Yarbrough *et al.*[40] for the synthesis of (γ-AmNS)ATP (**19a**). 1-Aminonaphthalene 5-sulfonate (2 mmol) is added to 10 ml of water and the pH is adjusted to pH 5.8 with 1 *N* NaOH. The saturated solution is centrifuged to remove insoluble material. A 12.5 m*M* aqueous solution of ATP (4 ml) is adjusted to pH 5.8 and 2 ml of 1 *M* EDC (dissolved in water) is added. The mixture is equilibrated to 20° and the reaction is initiated by adding the saturated solution of aminonaphthalene 5-sulfonate. The pH is maintained between pH 5.65 and 5.75 by adding 0.1 *N* HCl. After 2.5 hr, sufficient TEA–bicarbonate (pH 7.5) is added to bring the solution to 0.05 *M* in 50 ml and the product is isolated by DEAE chromatography.

The UDP derivative (**19b**) has been synthesized to characterize stacking interactions by NMR.[101] The synthesis of a similar aminonaphthalene 5-sulfonate derivative of ATP, (γ-1,5-EDANS)ATP (**20a**), containing an ethylenediamine linker between the fluorophore and the phosphates, has been described.[43] Dunkak *et al.*[102] have reported a similar synthesis, but with 5-amino-2-naphthalenesulfonic acid to form uridine-5′-triphosphoro-γ-5-(2-sulfonic acid)naphthylamidate (γ-ANS-UTP); **20b**) used to measure RNA transcription rates.

Similar chemical methods have been used to synthesize a CDP analog that contains the Mant group, (Pβ)MABA-CDP (**21**), which is a probe for the NMP-binding site of UMP/CMP kinase.[44] The oxazine ring of *N*-methylisatoic anhydride

[94] J. R. Barrio, P. D. Sattsangi, B. A. Gruber, L. G. Dammann, and N. J. Leonard, *J. Am. Chem. Soc.* **98,** 7408 (1976).

[95] S. F. Singleton, F. Shan, M. W. Kanan, C. M. McIntosh, C. J. Stearman, J. S. Helm, and K. J. Webb, *Org. Lett.* **3,** 3919 (2001).

[96] B. Gildea and L. W. McLaughlin, *Nucleic Acids Res.* **17,** 2261 (1989).

[97] P. G. Wu, T. M. Nordlund, B. Gildea, and L. W. McLaughlin, *Biochemistry* **29,** 6508 (1990).

[98] B. A. Connolly and P. C. Newman, *Nucleic Acids Res.* **17,** 4957 (1989).

[99] H. P. Rappaport, *Nucleic Acids Res.* **16,** 7253 (1988).

[100] H. P. Rappaport, *Biochemistry* **32,** 3047 (1993).

[101] G. Dhar and A. Bhaduri, *J. Biol. Chem.* **274,** 14568 (1999).

[102] K. S. Dunkak, M. R. Otto, and J. M. Beechem, *Anal. Biochem.* **243,** 234 (1996).

19

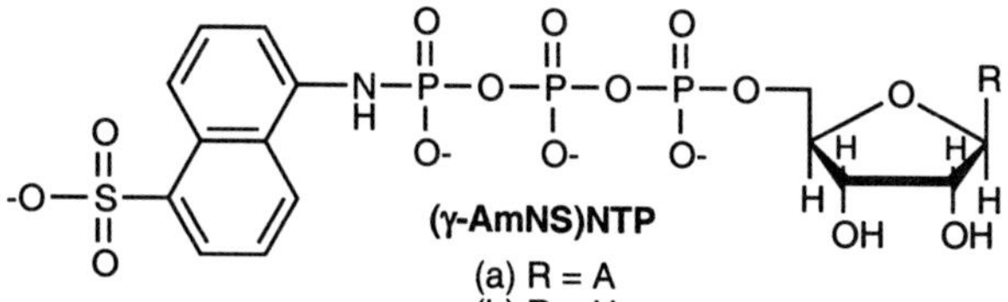

(γ-AmNS)NTP

(a) R = A
(b) R = U

20

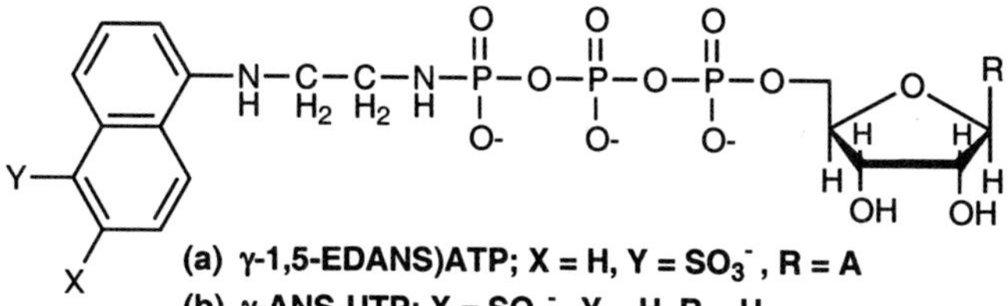

(a) γ-1,5-EDANS)ATP; X = H, Y = SO_3^-, R = A
(b) γ-ANS-UTP; X = SO_3^-, Y = H, R = U

21

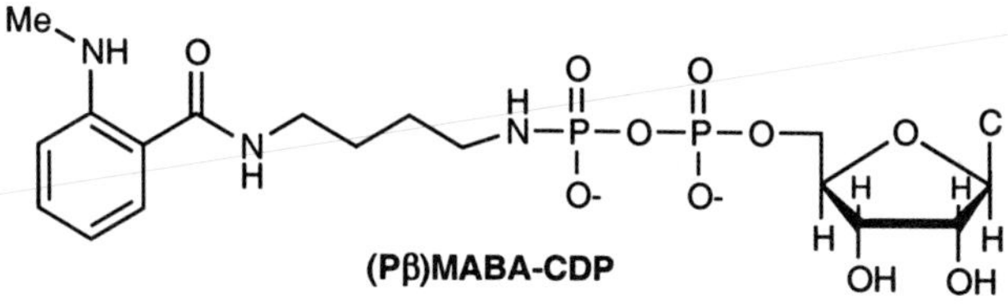

(Pβ)MABA-CDP

22

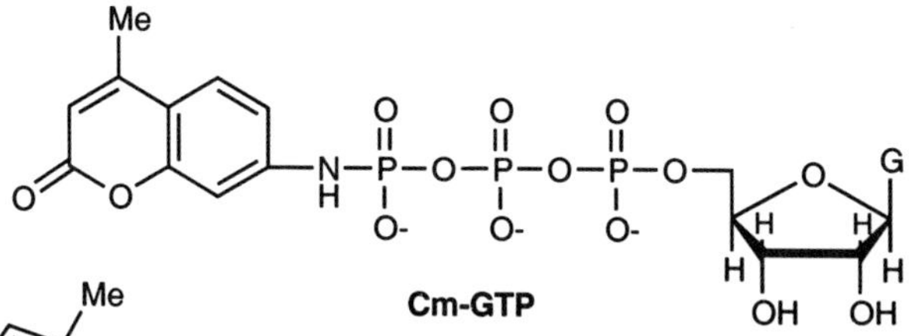

Cm-GTP

23

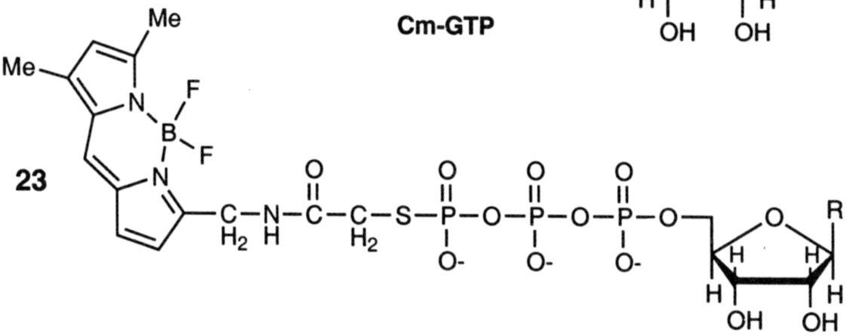

(a) ApppγsBODIPY; R = A
(b) GpppγSBODIPY; R = G

FIG. 5. Examples of fluorescent probes conjugated to nucleotides by phosphoramidate (**19–22**) and phosphorothioate (**23**) linkages.

is opened with diaminobutane and the free aliphatic amine is coupled to CDP by use of EDC.

A coumarin-derivatized GTP (**22**; Cm-GTP) has been synthesized in 2-(*N*-morpholino)-ethanesulfonic acid (MES) buffer, pH 5.2, using the coupling agent 1-cyclohexyl-3,2-(morpholinoethyl)carbodiimide metho-*p*-toluene sulfonate.[45] This compound has been used in an assay for transcription using T7 RNA polymerase.

PHOSPHOROTHIOATES. The sulfur of phosphorothioates (**2**; Fig. 1) is a good nucleophile that reacts with iodoacetates and maleimides to form the phosphorothioesters. This is a route to the synthesis of stable fluorescent nucleotides modified selectively at the thiophosphate (e.g., **23**; Fig. 5).

Typically, the sodium or lithium salt of the nucleotide (e.g., ATPγS[12] is added to water (5–30 m*M*) and the pH is adjusted to pH 9.0 with sodium bicarbonate. An equimolar amount of a thiol-reactive fluorophore (e.g., iodoacetates or maleimides), dissolved in an appropriate organic solvent (5–30 m*M*), is added and the mixture is stirred at ambient temperature for several hours or until complete as judged by TLC. BODIPY FL (Molecular Probes) triphosphorothioates have been used as diadenosine (ApppA; **23a**; Fig. 5) and diguanosine (GpppG; **23b**; Fig. 5) analogs as have the respective coumarin derivatized probes.[103] Compound **23b** has also proved useful to measure real-time binding of nucleotides to G proteins.[104]

Modifications to Ribose or Modified Ribose Ring

There are many choices of analogs for experiments with macromolecules that can tolerate modifications to the ribose ring. An important distinction between them is whether they can be isolated as pure isomers (either 2′-O or 3′-O modified) versus mixtures of isomers.

If it is necessary to utilize a single isomer of a ribose-modified nucleotide, there are four general approaches: (1) 2′(3′)-*O*-carbamates (**25**; Fig. 6) are mixtures of isomers, but individual isomers can often be isolated; (2) 2′-deoxy or 3′-deoxy derivatives are not mixtures of isomers; (3) amino sugars can be modified at the 2′- or 3′-amine, yielding pure isomers; and (4) amines can be attached to a periodate-oxidized ribose ring (**28**; Fig. 8). All four of these approaches are addressed in this article.

Ribose esters are an equilibrium mixture of isomers (**24**; Fig. 6), as are carbamates (**25**; Fig. 6). It has been shown that 3′-O- and 2′-O-isomers of ribose esters can be separated chromatographically at low pH, but the transesterification rates at neutral pH are so fast (on the order of minutes at pH 7) that a mixture of isomers will be regenerated within the time frame of most biological experiments.[7,105,106]

[103] A. Draganescu, S. C. Hodawadekar, K. R. Gee, and C. Brenner, *J. Biol. Chem.* **275,** 4555 (2000).
[104] D. P. McEwen, K. R. Gee, H. C. Kang, and R. R. Neubig, *Anal. Biochem.* **291,** 109 (2001).
[105] H. Rensland, A. Lautwein, A. Wittinghofer, and R. S. Goody, *Biochemistry* **30,** 11181 (1991).
[106] J. F. Eccleston, K. J. Moore, G. G. Brownbridge, M. R. Webb, and P. N. Lowe, *Biochem. Soc. Trans.* **19,** 432 (1991).

24

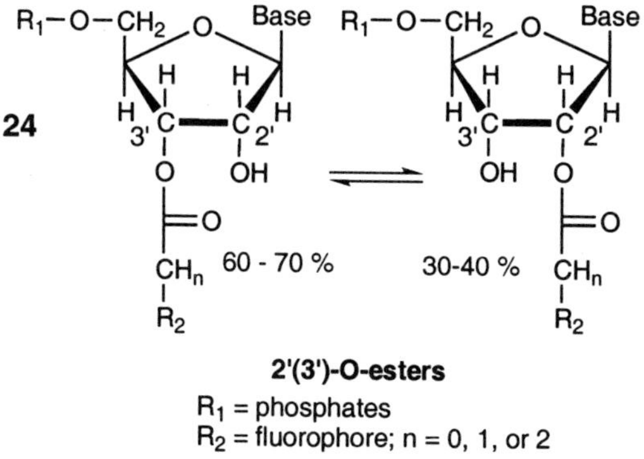

60 - 70 % ⇌ 30-40 %

2'(3')-O-esters

R_1 = phosphates
R_2 = fluorophore; n = 0, 1, or 2

25

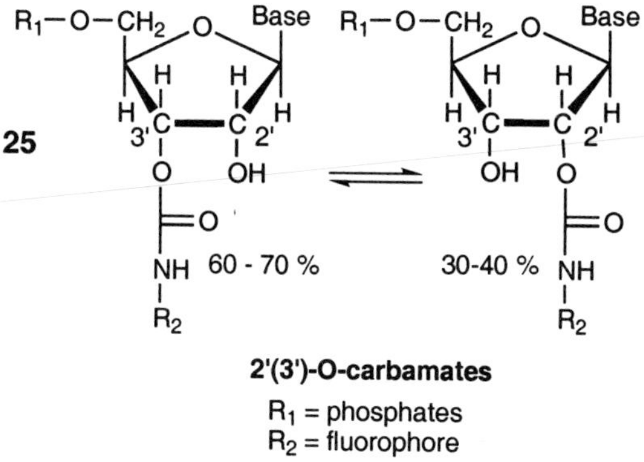

60 - 70 % ⇌ 30-40 %

2'(3')-O-carbamates

R_1 = phosphates
R_2 = fluorophore

FIG. 6. At equilibrium, esters and carbamates of the ribose ring are mixtures of the 2′-O-isomer (~30–40%) and 3′-O-isomer (~60–70%).

In contrast, transcarbamoylation of carbamates is much slower and it is possible to obtain single isomers and use them at physiological pH without significant reequilibration.[7,107–109]

The use of pure isomers solves two general problems. First, depending on the macromolecule, it is possible that the different isomers will interact or turn over with different rate constants, thus complicating kinetic analyses. Second, the different isomers often have different fluorescence properties. For example, by comparing the spectral properties of benzoylbenzoyl esters of etheno-ADP,

[107] K. Oiwa, J. F. Eccleston, M. Anson, M. Kikumoto, C. T. Davis, G. P. Reid, M. A. Ferenczi, J. E. T. Corrie, A. Yamada, H. Nakayama, and D. R. Trentham, *Biophys. J.* **78,** 3048 (2000).

[108] P. B. Conibear, D. S. Jeffreys, C. K. Seehra, R. J. Eaton, and C. R. Bagshaw, *Biochemistry* **35,** 2299 (1996).

[109] M. R. Webb and J. E. T. Corrie, *Biophys. J.* **81,** 1562 (2001).

$3'(Bz_2)2'$ dεADP, and $3'(2')Bz_2\varepsilon$ADP, it was found that fluorescence emission is significant only from the 3′-*O*-isomer.[26] Different fluorescence properties from the two isomers are not a general rule however. It may be that a longer linker between the nucleotide base and the fluorophore will allow sufficient conformational flexibility to decrease the differences between the two isomers. For example, studies of Cy3 (Amersham Life Science, Piscataway, NJ) derivatives of ADP/ATP with an ethylenediamine linker between the adenine base and the Cy3 fluorophore showed that the emission of the 3′-*O*-isomer was ~90% of the 2′-*O*-isomer.[107]

PREPARATION OF 2′(3′)-*O*-RIBOSE ESTERS FROM IMIDAZOLIDES OF CARBOXYLIC ACIDS. The general route to synthesis of 2′(3′)-*O*-ribose esters (**24**) is described by Gottikh *et al.*[110] Gottikh *et al.*[110] advise that the composition of the final reaction mixture should be <30% strongly solvating aprotic solvents (dimethyl sulfoxide, DMF, formamide, pyridine), with the remainder of the solution being water mixed with weakly solvating aprotic solvents [acetonitrile (AcN), tetrahydrofuran (THF), acetone]. These general conditions will avoid aminoacylation of the amino groups of the adenine, guanine, and cytidine rings, leading to exclusive acylation at the ribose ring.

CDI (**5**; Fig. 2) is stirred for 15 min at room temperature with the fluorescent carboxylic acid under anhydrous conditions to form the respective imidazolide. This reaction is typically done in dry DMF or THF. Often, CDI is added in excess (1.5- to 4-fold) of the carboxylic acid to drive the reaction to completion. If the final acylation reaction mixture is to contain little water, it may be prudent to quench the excess CDI with methanol before nucleotide addition to avoid any potential for the formation of the 2′,3′-*O*-cyclic carbonate (**7**). An aqueous solution of the nucleotide (equimolar or up to 10-fold less than the imidazolide) is then added to the imidazolide (or vice versa) and left to stir for 3 hr to overnight at room temperature. THF or other weakly solvating aprotic solvent may be mixed with water to improve solubility if required. Most protocols use neutral solutions[111] of the sodium salts of the nucleotides, and other salts such as the TEA salts have been reported. This may not be critical, as the imidazole will strongly buffer the solution. Many protocols use an excess of imidazolide over nucleotide because the imidazolide simultaneously hydrolyzes under the aqueous acylation conditions. This should improve yields, but can lead to formation of the 2′,3′-*O*-disubstituted compound[111,112] especially if the reaction is left for more than 3 hr. Using an excess of nucleotide over imidazolide will favor the formation of the α-phosphate acylated product.[111] Schafer *et al.*[111] also advise that the nucleotide should be no less than 0.5 *M* (acylation reaction), but this is a misprint and should read 0.05 *M*.

[110] B. P. Gottikh, A. A. Krayevsky, N. B. Tarussova, P. P. Purygin, and T. L. Tsilevich, *Tetrahedron* **26,** 4419 (1970).

[111] G. Schafer, U. Lucken, and M. Lubben, *Methods Enzymol.* **126,** 682 (1986).

[112] S. J. Jeng and R. J. Guillory, *J. Supramol. Struct.* **3,** 448 (1975).

On the basis of the above-described considerations, the following can be described as a generic large-scale synthesis.

Formation of imidazolide of carboxylic acid

1. Dissolve 1.5 mmol of CDI (dried) in 1.0 ml of dry DMF or THF (there should be no bubbling).
2. Add 1.0 mmol of carboxylic acid (dried).
3. React at room temperature with stirring for 15–30 min. This gives ~1.0 mmol of activated carboxylic acid.

Preparation of nucleotide. Dissolve 0.5 mmol of nucleotide in 4 ml of H_2O.

Acylation reaction

1. Add the imidazolide to the nucleotide in one addition (or vice versa).
2. Stir for 3 hr to overnight at room temperature.
3. Monitor the reaction by TLC.

Workup. The majority of unreacted carboxylic acid and imidazole can be removed by repeated washes with acetone or other solvent mixture in which the nucleotide is insoluble, followed by centrifugation. The mixture can then be further purified by chromatography.

Stability. Esters are not stable under basic conditions. Therefore, DEAE ion-exchange chromatography in TEA–bicarbonate (TEAB) may allow for hydrolysis, especially during solvent evaporation after chromatography. Many esters have been purified by Sephadex LH-20 chromatography in H_2O or 0.1 *M* ammonium formate. Keep products cold whenever possible. Store at $-80°$ as the dry sodium salt or in solution at pH 5.

General small-scale synthesis: 5–20 μmol of nucleotide. Prepare more imidazolide than needed to ensure a successful reaction. Then use only the amount as needed per the instructions given above.

PREPARATION OF 2′-DEOXY-3′-*O*-ESTERS. The general considerations for preparation of the 2′(3′)-*O*-esters apply to preparation of 2′-deoxy-3′-*O*-esters. An example of synthesis of a 2′-deoxy-3′-*O*-ester is found in Cremo and Yount.[26]

PREPARATION OF 2′-*O*,3′-*O*-CARBAMATES BY REACTION OF 2′,3′-*O*-CYCLIC CARBONATE OF RIBOSE WITH AMINES

Overview. Carbamates of the ribose ring (**25**; Fig. 6) are widely used because they offer several advantages over esters, as discussed above. In addition, carbamates are more stable to base-catalyzed hydrolysis than esters. This is important because hydrolysis products often have higher quantum yields than the respective esters or carbamates.

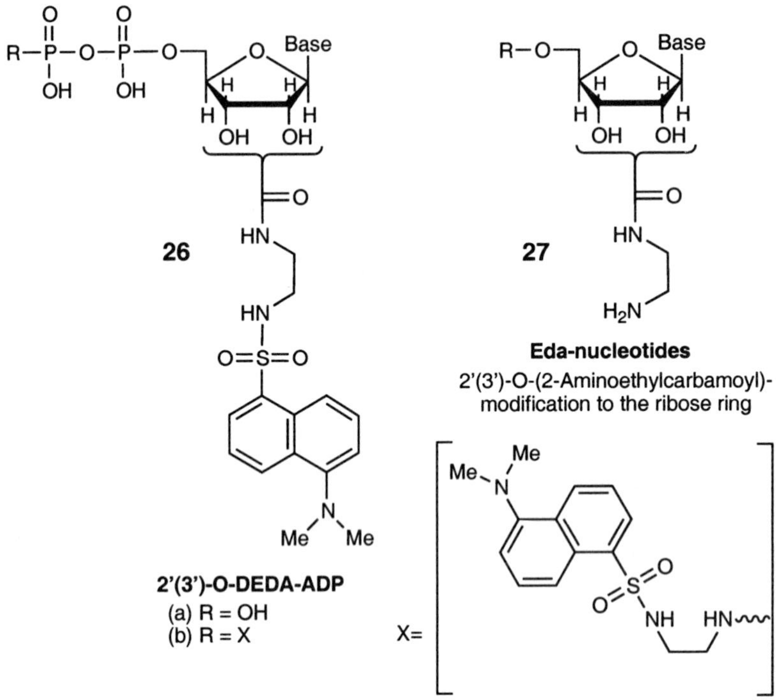

FIG. 7. Examples of 2′(3′)-*O*-carbamates of the ribose ring.

Cremo *et al.*[47] reported the first synthesis of a nucleotide with a carbamate linkage to the ribose ring [**26a** in Fig. 7; 2′(3′)-*O*-DEDA-ADP]. The approach is based on the work of Maeda *et al.*,[17] who characterized the 5′-phosphorylimidazolidate 2′,3′-*O*-cyclic carbonate of an NMP (**7**; Fig. 2) formed during treatment of ribonucleotides with CDI. Primary amines react with **7** (Fig. 2) at the modified ring to give the respective 2′(3′)-*O*-carbamate, and at the modified terminal phosphate to give the respective phosphoramide (e.g., **26b**; Fig. 7). The phosphoramide can be selectively cleaved under mildly acidic conditions to give the unmodified terminal phosphate (e.g., **26a**; Fig. 7). This mild acid treatment may produce some diphosphates from triphosphates.

Preparation of 5′-diphosphoryl, β-imidazolidate, 2′,3′-O-cyclic carbonate of nucleotide diphosphate. The example is for ADP[47] but can be applied to any nucleotide 5′-mono-, 5′-di-, or 5′-triphosphate. Dry CDI (4 mmol) in 18 ml of dry DMF is added to a freshly dried preparation of the TBA salt of ADP (1 mmol; see Preparation of Triethylamine and Tributylamine Salts of Nucleotides, above). The reaction is stirred at 20° for 2 hr or until reaction reaches completion as monitored by TLC [R_f 0.37; 1-butanol–acetic acid–H_2O (5 : 2 : 3, v/v/v); ADP

0.08]. Essentially all of the material is converted to **7** (Fig. 2). The remaining CDI is then quenched with absolute methanol (1.52 mmol) and left to stand for 0.5 hr. This material is ready for reaction with the free base of an amine.

Preparation of 2′(3′)-O-DEDA-ADP. The free base of dansylethylenediamine[113] (DEDA; 1 mmol in 3 ml of dry DMF) is added to 0.4 mmol of **7** and stirred at 25° for 4 days. The product migrates as a closely spaced doublet in the above-described solvent system corresponding to **26b**. Water is added (20 ml; less can be added), the pH is adjusted to pH 2.5–3.5 with glacial acetic acid (HCl can also be used), and the reaction is allowed to proceed at 25° overnight to remove dansyl from the terminal phosphate to generate **26a**. Another example of a synthesis of this type is for a coumarin derivative.[7]

The general purification scheme for such a derivative depends on the properties of the fluorophore. In the original DEDA-ADP synthesis, the solvent was evaporated and the residue was dissolved in a small volume of methanol and the Na^+ salt was prepared with NaI (see Preparation of Na^+ Salts of Nucleotides, above). Unreacted DEDA was removed by repeated acetone washes followed by centrifugation. It is time-consuming to evaporate the solution as described above and it may be possible to apply the reaction directly to a DEAE column as described for a coumarin derivative[7] if the excess free amine will not comigrate with the final product.

2′(3′)-O-(2-Aminoethylcarbamoyl) modification to the ribose ring. The modification, with ethylenediamine coupled to the ribose ring (**27**; Fig. 7) by a carbamoyl linkage, has been given the trivial name of Eda-nucleotide. Eda-nucleotides are universal precursors that will react with any amine-reactive fluorescent compound.[7,114]

Since the original publication of this synthesis,[114] Jameson and Eccleston[7] have elaborated details and corrections,[115] and there have been a few modifications,[109] which are included here.

The following protocol has been demonstrated for adenine or guanine nucleotides, but the protocol should be applicable to other bases. This example is for ATP. A 3-fold excess of 1,2-ethylenediamine (81 μl, 1.5 mmol) is added to the tributylammonium salt of **7** (Fig. 2; triphosphate form; 0.5 mmol prepared as described above, in 2.25–5.0 ml of dry DMF) while the solution is stirred rapidly. A white precipitate forms immediately, and the solution is stirred for 3 hr at 5°. At this point it is possible to collect the nucleotide by centrifugation. This will leave much of the imidazole and ethylenediamine in the supernatant, and the subsequent pH adjustment (see below) will thus require less acid. Alternatively, 80 ml of water can be added directly.

[113] M. E. DePecol and D. B. McCormick, *Anal. Biochem.* **101,** 435 (1980).

[114] T. L. Hazlett, K. J. Moore, P. N. Lowe, D. M. Jameson, and D. F. Eccleston, *Biochemistry* **32,** 13575 (1993).

[115] B. S. Watson, T. L. Hazlett, J. F. Eccleston, C. Davis, D. M. Jameson, and A. E. Johnson, *Biochemistry* **34,** 7904 (1995).

The clear solution is taken to pH 2.5 with concentrated HCl and stirred for ~20 hr at 5° with occasional pH adjustment. Reaction progress is checked by HPLC with a Partisil SAX column (0.4 × 25 cm; Whatman, Clifton, NJ) with 0.4 *M* $(NH_4)_2HPO_4$ as the mobile phase, adjusted to pH 4.0 with concentrated HCl, and running at 2 ml/min. For the ATP derivative, acid treatment takes the main peak from 2.5 to 7 min, similar to that for ATP itself. The solution is then raised to pH 9.0 with 5 *M* NaOH and stirred at 20° for 90 min to remove an unknown modification, presumably on the purine. Guanosine 5′-phosphates may need a longer treatment. The progress of reaction is monitored by HPLC as described above, with the formation of two overlapping peaks centered at 5.5 min, corresponding to 2′-*O*- and 3′-*O*-EdaATP. The EdaATP can be purified at 5° on a DEAE-cellulose column (200 ml) preequilibrated in 10 m*M* TEAB, pH 7.6. The product is eluted by a linear gradient of this buffer from 10 to 450 m*M* (1.5 liter). EdaATP (mixed isomers) elutes at 240 m*M* TEAB in ~50% yield from ATP. The isomers are not well separated by this procedure. The buffer is removed by repeated rotary evaporation from methanol and EdaATP is stored at 20° as a concentrated solution at pH 7.0.

Preparation of fluorescent derivatives of Eda-nucleotides. Fluorescent amine-reactive compounds such as sulfonyl chlorides,[116] isothiocyanates,[108] and succinimidyl esters[107] (~1 m*M* or higher) can be conjugated to Eda-nucleotides (~1 m*M* or higher) within a few hours at room temperature in water–acetone or water–DMF mixtures buffered with $NaHCO_3$ (20–100 m*M*; pH 8.6). Carboxylic acids can be activated with isobutylchloroformate.[109] These conditions are selective for the highly reactive primary amine of the Eda-nucleotides. Products can be purified by DEAE chromatography.

The number of fluorescent nucleotides that can be prepared by this approach is limited only by the availability of fluorescent amine-reactive compounds. Table III[7,107–109,114–122] shows a list of nucleotides based on **27** (Fig. 7). To date, various coumarin,[109,114] Cy3,[107,123,124] Cy5,[107] rhodamine,[115,116,118] and fluorescein,[108,114,117] derivatives of adenine and guanine nucleotides have been

[116] J. E. T. Corrie, C. T. Davis, and J. F. Eccleston, *Bioconjug. Chem.* **12,** 186 (2001).

[117] A. J. Sowerby, C. K. Seehra, M. Lee, and C. R. Bagshaw, *J. Mol. Biol.* **234,** 114 (1993).

[118] E. M. Zera, D. P. Molloy, J. K. Angleson, J. B. Lamture, T. G. Wensel, and J. A. Malinski, *J. Biol. Chem.* **271,** 12925 (1996).

[119] S. Chaen, I. Shirakawa, C. Bagshaw, and H. Sugi, *Biophys. J.* **73,** 2033 (1997).

[120] M. Xiao, H. Li, G. E. Snyder, R. Cooke, R. G. Yount, and P. R. Selvin, *Proc. Natl. Acad. Sci. U.S.A.* **95,** 15309 (1998).

[121] R. B. Mujumdar, L. A. Ernst, S. R. Mujumdar, C. J. Lewis, and A. S. Waggoner, *Bioconjug. Chem.* **4,** 105 (1993).

[122] J. E. T. Corrie, V. R. N. Munasinghe, and W. Rettig, *J. Heterocyclic Chem.* **37,** 1447 (2000).

[123] S. Chaen, I. Shirakawa, C. R. Bagshaw, and H. Sugi, *Biophys. J.* **73,** 2033 (1997).

[124] I. Shirakawa, S. Chaen, C. R. Bagshaw, and H. Sugi, *Biophys. J.* **78,** 918 (2000).

TABLE III
FLUORESCENT NUCLEOTIDES SYNTHESIZED FROM Eda-NUCLEOTIDES[a]

Trivial name	Fluorescent amine-reactive compound used for synthesis	Biological system studied	Refs.
FEDA-ATP	Fluorescein 5-isothiocyanate (FITC, isomer 1)	Myosin ATPase	108, 117
FEDA-GTP		p21^{ras} GTPase	114[b]
FEDA-GDP		Transducin	118
REDA-ATP	Rhodamine B isothiocyanate [5(6) mixed isomers]	Myosin ATPase	108
		Transducin	118
Rhodamine-Eda-ATP	Sulforhodamine chlorides (isomeric monosulfonyl chlorides of sulforhodamine B)	Myosin ATPase	116
GTP-Rh	Lissamine rhodamine B sulfonyl chloride	Aminoacyl-tRNA elongation factor Tu complex	115
Cy3-Eda-ATP, Cy3-Eda-ADP	Cy3.29-OH succinimidyl ester[c] (Cy3.29-OSu)	Skeletal muscle myofibrils	7, 107, 119
Cy5-Eda-ATP, Cy5-Eda-ADP	Cy5.29-OH succinimidyl ester[c] (Cy5.29-OSu)	Myosin	7, 107
Deac-Eda-ATP	7-Diethylaminocoumarin-3-carboxylic acid[c] activated with isobutyl chloroformate	Actin–myosin	109
But-Eda-ATP or coumarin 343-Eda-ATP	Butterfly coumarin carboxylic acid (Coumarin 343; Acros, Loughborough, UK) activated with isobutyl chloroformate	Actin–myosin	109
Mbc-Eda-ATP	7-Ethylamino-8-bromocoumarin-3-carboxylic acid[d] activated with isobutyl chloroformate	Actin–myosin	109
2-Azido-ATP, terbium chelate[e]	Novel multidentate terbium chelate with antenna	Myosin	120

[a] Compound **27**; "Eda-nucleotides" is the name used by the original authors.
[b] Note a correction for this synthesis. "Ethylenediamine (167 μmol)..." should read "Ethylenediamine (167 μl)...."
[c] See Ref. 121.
[d] See Ref. 122.
[e] Note derived from ethylenediamine linkage but a similar butylenediamine linkage.

synthesized. Generally, the fluorescence properties of the modified nucleotides are similar to those of the original fluorophore and are therefore not included in Table III. Many reports to date have made use of fluorescent carbamates of nucleotides for single-molecule, or single molecular assembly, measurements.[107,117,124,125]

[125] T. Funatsu, Y. Harada, H. Higuchi, M. Tokunaga, K. Saito, Y. Ishii, R. D. Vale, and T. Yanagida, *Biophys. Chem.* **68,** 63 (1997).

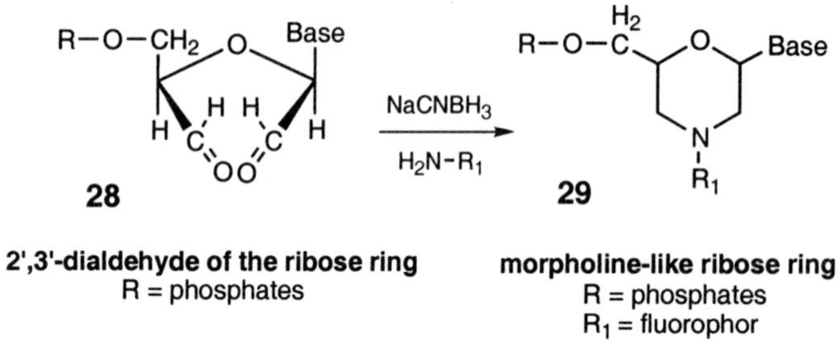

28

2',3'-dialdehyde of the ribose ring
R = phosphates

$NaCNBH_3$
H_2N-R_1

29

morpholine-like ribose ring
R = phosphates
R_1 = fluorophor

FIG. 8. Modification of the 2′,3′-dialdehyde of the ribose ring (**28**) with a fluorescent amine to form a morpholidate-like ribose ring (**29**).

REACTION OF PERIODATE-OXIDIZED RIBOSE WITH AMINES. Periodate treatment of ribose-containing nucleotides results in cleavage of the 2′,3′ carbon–carbon bond and concomitant oxidation to the 2′,3′-dialdehyde (**28**; Fig. 8). These compounds have been extensively used as an affinity labels for the nucleotide-binding sites of many proteins[126–131] as they readily react to form Schiff bases with proximal lysine residues. These can be converted to stable adducts by reduction with $NaBH_4$ or $NaCNBH_3$. RNA can be selectively labeled at the 3′ end by periodate oxidation, followed by reaction with aldehyde-reactive fluorophores.[132]

Hileman *et al.*[133] have characterized the products of the reaction of the dialdehydes of GDP and ATP (ox-GDP and ox-ATP) (**28**; Fig. 8) with fluorescent amines and hydrazines with subsequent reduction with $NaCNBH_4$ or $NaBH_4$ or both. The most useful product (**29**) was obtained by using only $NaCNBH_3$ as the reductant. The ribose ring has been converted to a six-membered morpholine-like ring with no hydroxyls attached. The fluorophore is attached to the nitrogen of the ring via a stable C–N bond. The authors showed that a fluorescein derivative and a similar rhodamine derivative bound to to eIF-2. A similar synthetic approach has been used to prepare affinity columns of nucleotides.[134]

This class of nucleotide derivative could be generally useful, but little is currently known about their ability to bind to macromolecules as substrate analogs.

[126] S. B. Easterbrook-Smith, J. C. Wallace, and D. B. Keech, *Eur. J. Biochem.* **62,** 125 (1976).

[127] M. E. Peter, B. Wittmann-Liebold, and M. Sprinzl, *Biochemistry* **27,** 9132 (1988).

[128] M. E. Peter, J. She, L. A. Huber, and C. Terhorst, *Anal. Biochem.* **210,** 77 (1993).

[129] G. G. Chang, M. S. Shiao, K. R. Lee, and J. J. Wu, *Biochem. J.* **272,** 683 (1990).

[130] L. A. Huber and M. E. Peter, *Electrophoresis* **15,** 283 (1994).

[131] P. Clertant, P. Gaudray, E. May, and F. Cuzin, *J. Biol. Chem.* **259,** 15196 (1984).

[132] O. W. Odom, Jr., D. J. Robbins, J. Lynch, D. Dottavio-Martin, G. Kramer, and B. Hardesty, *Biochemistry* **19,** 5947 (1980).

[133] R. E. Hileman, K. M. Parkhurst, N. K. Gupta, and L. J. Parkhurst, *Bioconjug. Chem.* **5,** 436 (1994).

[134] R. Rayford, D. D. Anthony, Jr., R. E. O'Neill, Jr., and W. C. Merrick, *J. Biol. Chem.* **260,** 15708 (1985).

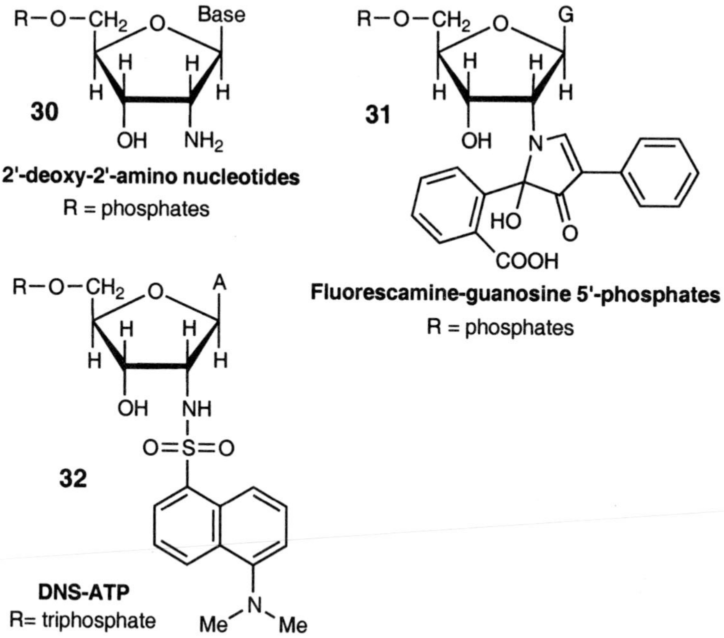

FIG. 9. Examples of fluorescent nucleotides (**31**, **32**) derived from 2′-deoxy-2′-amino nucleotides (**30**).

They are relatively easy to synthesize because the 2′,3′-dialdehyde precursors[126] are commercially available (Sigma). In addition, the products are not mixtures of isomers and they are relatively stable.

DERIVATIVES OF 2′-AMINO-2′-DEOXYRIBOSE NUCLEOTIDES BY REACTION WITH 2′-AMINO GROUP. The uridine, cytidine, and adenosine derivatives containing the 2′-amino-2′-deoxyribose ring (**30**; Fig. 9) are commercially available (http://www.trilinkbiotech.com/) and the synthesis of the adenosine triphosphate derivative[46] and of the guanosine 5′-mono- and 5′-triphosphate have been reported from commercially available precursors (http://www.trilinkbiotech.com/). This precursor can be modified with an amine-reactive fluorophore. Examples of this are found in the reaction with fluorescamine[22] to form fluorescamine-GMP, -GDP, and -GTP (**31**) and with dansyl chloride to form the sulfonamide DNS-ATP [**32**; 2′-(5-*N*-dimethylnaphthalene-1-sulfonyl)amino-2′-deoxy-ATP]. DNS-ATP (**32**) has been used to study the kinetics of myosin[46] and of the mitochondrial F_1-ATPase.[135]

[135] I. Matsuoka, T. Watanabe, and Y. Tonomura, *J. Biochem.* (*Tokyo*) **90,** 967 (1981).

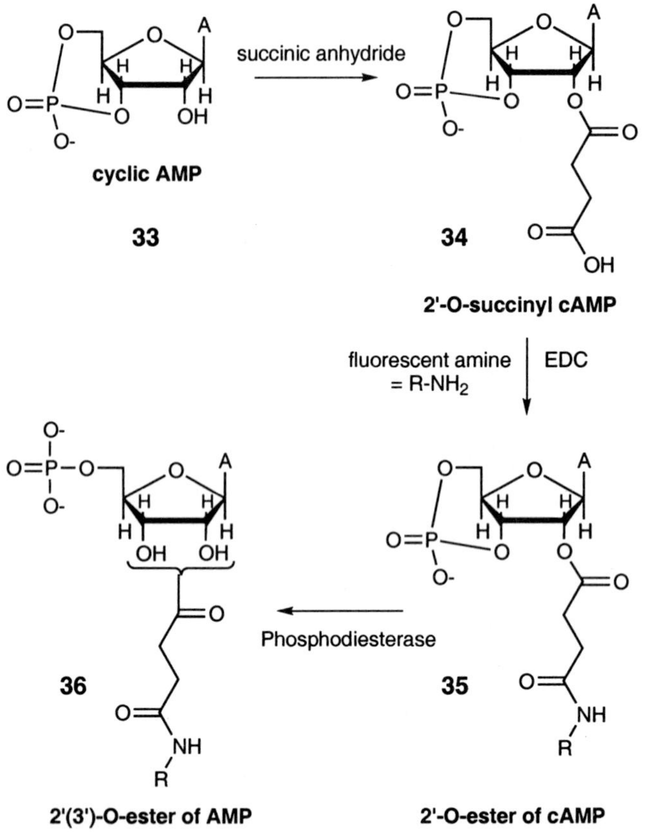

FIG. 10. Potentially general approach to the synthesis of a family of fluorescent cAMP (**35**) and AMP (**36**) derivatives.

GENERAL APPROACH TO PREPARATION OF 2′-*O*-ESTERS OF cAMP AND 2′(3′)-*O*-ESTERS OF AMP. Figure 10 shows a potentially general approach to the synthesis of a family of fluorescent cAMP and AMP derivatives. This approach was demonstrated by Ueda *et al.*[136] in the synthesis of a Cy3-modified cAMP used for single-molecule analysis of chemotactic signaling in cells. Succinic anhydride reacts with the ribose to give 2′-*O*-succinyl-cAMP (**34**; Fig. 10). The free carboxylate is available for reaction with fluorescent amines to give **35**. Although not demonstrated, it should often be possible to generate the AMP derivatives, **36**, from **35** by treatment with phosphodiesterase. It is likely that this approach gives higher yields than the preparation of 2′(3′)-*O*-ribose esters from imidazolides

[136] M. Ueda, Y. Sako, T. Tanaka, P. Devreotes, and T. Yanagida, *Science* **294,** 864 (2001).

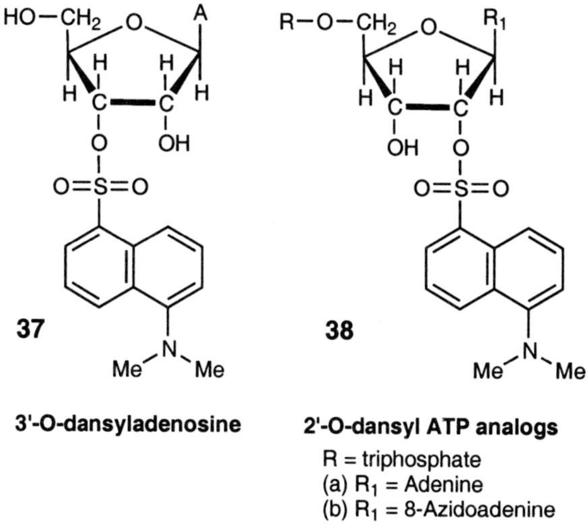

37 3'-O-dansyladenosine

38 2'-O-dansyl ATP analogs
R = triphosphate
(a) R_1 = Adenine
(b) R_1 = 8-Azidoadenine

FIG. 11. Examples of sulfonyl esters of the ribose ring.

of carboxylic acids (see above). Caution is advised in this approach as succinic anhydride may also modify N^6 of the adenine ring.

SULFONYL ESTERS OF RIBOSE RING. Dansyl chloride has been reacted with adenosine (**37**; Fig. 11),[137] ATP (**38a**),[49,138] and 8-azido-ATP (**38b**)[50] to from the respective sulfonyl esters. The adenosine derivative was synthesized in dry pyridine and the authors suggested that the product was solely the 3′-*O*-sulfonyl ester, based on results of a methylation reaction and the lack of evidence for isomers in the ^{1}H NMR spectrum. The 8-azido-ATP derivative (**38b**),[50] synthesized in a bicarbonate buffer at pH 9.8, was characterized as the 2′-*O*-sulfonyl ester on the basis of the ^{1}H NMR spectrum. The dansylated ATP derivative[49] has been synthesized under anhydrous conditions in DMF and assumed to be the 2′-*O*-sulfonyl ester. Another report, synthesizing the ATP derivative in buffer at pH 10, provided ^{1}H NMR evidence that the modification was solely on the 2′-OH. A similar derivative of ATP containing a sulfonyl ester of pyrene was reported to be a mixture of the 2′-*O*- and 3′-*O*-isomers, but without a reported characterization.[139]

EXAMPLES OF FLUORESCENT RIBOSE-MODIFIED NUCLEOTIDES

N-Methylanthraniloyl (Mant) and anthraniloyl (Ant) esters of ribonucleotides. The mono-, di-, and triphosphates of Mant and Ant derivatives of adenine and guanine nucleotides (**39**; Fig. 12) can be synthesized as described by Hiratsuka.[54]

[137] G. Skorka, P. Shuker, D. Gill, J. Zabicky, and A. H. Parola, *Biochemistry* **20,** 3103 (1981).

[138] S. G. Huang and M. Klingenberg, *Biochemistry* **34,** 349 (1995).

[139] D. Thoenges, E. Amler, T. Eckert, and W. Schoner, *J. Biol. Chem.* **274,** 1971 (1999).

39

Mant- and Ant-ribose esters

(a) 2'(3')-O-(N-methyl)anthraniloyl-nucleotide (Mant); X = Me
(b) 2'(3')-O-anthraniloyl-nucleotide (Ant); X = H
R = mono-,di- or triphosphate
R1 = adenine, guanine or cytidine

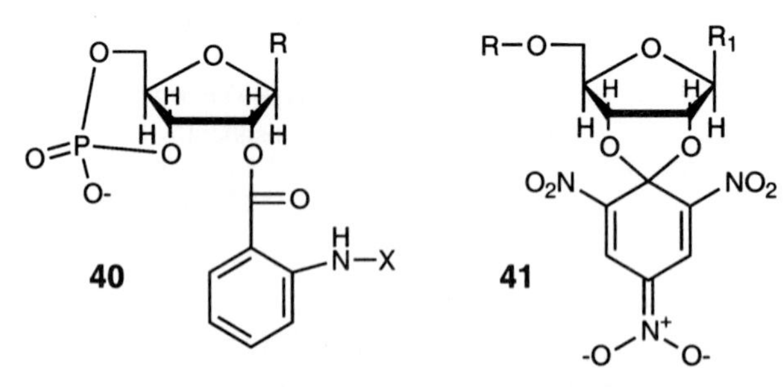

5',3'-O-cyclic-2'-O-Mant and 2'-O-Ant esters

Mant; X = Me
Ant; X = H
R = Adenine or guanine

TNP-nucleotides

R = mono-,di- or triphosphate
(a) R_1 = adenine
(b) R_1 = guanine

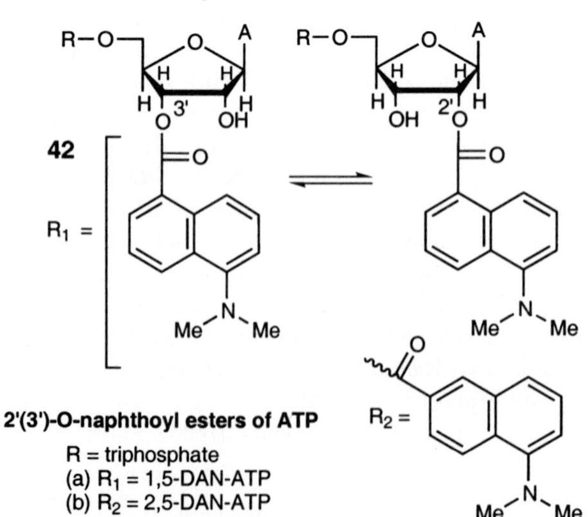

2'(3')-O-naphthoyl esters of ATP

R = triphosphate
(a) R_1 = 1,5-DAN-ATP
(b) R_2 = 2,5-DAN-ATP

FIG. 12. Examples of fluorescent ribose-modified nucleotides.

These probes are mixtures of the 3′-O- and 2′-O-isomers[47] and have been widely used. Eccleston *et al.*[106] have shown that the half-time of base-catalyzed transesterification is 7 min at pH 7.4 for Mant-GDP.

Mant- and Ant-nucleotides are easy to synthesize and are now commercially available (Molecular Probes). These syntheses are reliable and are not further elaborated here, except to report improvements or additional information.

The syntheses of the cAMP and cGMP derivatives with the fluorophore attached at the 2′-O position of the ribose (**40**; Fig. 12) have also been reported.[55] Neal *et al.*[140] report the synthesis of Mant-GTP, Mant-GDP, and Mantp[NH]ppG using DEAE-cellulose instead of LH-20 as described in the original synthesis, for purification. Others[118,141–143] report the use of DEAE or QAE-Sephadex[144] to purify other Mant-nucleotides including Mant-ATPγS. This suggests that these probes are relatively stable to base hydrolysis. The nucleotides could be analyzed for purity on anion-exchange HPLC (Partisil-10 SAX column, 250 × 4.6 mm; Whatman) with 85% 0.6 *M* $(NH_4)H_2PO_4$ (pH 4.0; adjusted with HCl)–15% methanol.[140] Zera *et al.*[118] found it useful to futher purify Mant-GTP on preparative polyethyleneimine (PEI)-cellulose TLC developed in 1.5 *M* KH_2PO_4 and extracted with 0.8 *M* TEA–bicarbonate.

Remmers[145] describes a detailed synthesis and purification of Mant-GTPγS using high-performance anion-exchange and perfusion chromatography columns.

3′-Anthraniloyl-2′-deoxy-ATP has been synthesized from isatoic anhydride and purified by reversed-phase chromatography (LiChroprep RP-18, 25–40 μm), using 1 m*M* TEA–acetate to elute the starting materials and 90% 1 m*M* TEA–acetate/10% AcN to elute the product.[146] Mant-Gpp(NH)p was synthesized with modifications to the original synthesis.[53] The 5′-Mant-ester of adenosine has been prepared to study the adenosine kinase.[147]

Mant-cytidine derivatives are more difficult to synthesize than the adenosine or guanosine derivatives. The syntheses of 5′-*O*-Mant-cytidine, 2′(3′)-*O*-Mant-cytidine, 5′-*O*-Mant-2′-deoxycytidine, and 3′-*O*-Mant-2′-deoxycytidine have been reported to require protecting groups of the appropriate hydroxyl and amino groups, although a detailed synthesis was not available.[148]

[140] S. E. Neal, J. F. Eccleston, and M. R. Webb, *Proc. Natl. Acad. Sci. U.S.A.* **87,** 3562 (1990).

[141] S. K. Woodward, J. F. Eccleston, and M. A. Geeves, *Biochemistry* **30,** 422 (1991).

[142] K. J. Moore, M. R. Webb, and J. F. Eccleston, *Biochemistry* **32,** 7451 (1993).

[143] K. J. Moore and T. M. Lohman, *Biochemistry* **33,** 14550 (1994).

[144] J. John, R. Sohmen, J. Feuerstein, R. Linke, A. Wittinghofer, and R. S. Goody, *Biochemistry* **29,** 6058 (1990).

[145] A. E. Remmers, *Anal. Biochem.* **257,** 89 (1998).

[146] R. Sarfati, V. Kansal, H. Munier, P. Glaser, A. Gilles, E. Labruyere, M. Mock, A. Danchin, and O. Barzu, *J. Biol. Chem.* **265,** 18902 (1990).

[147] H. Pelicano, G. Maury, A. Elalaoui, M. Shafiee, J. L. Imbach, R. S. Goody, and G. Divita, *Eur. J. Biochem.* **248,** 930 (1997).

[148] M. Shafiee, G. Gosselin, J. L. Imbach, S. Eriksson, and G. Maury, *Nucleosides Nucleotides* **18,** 717 (1999).

2′,3′-O-(2,4,6-Trinitrocyclohexadienylidene)-modified ATP and GTP. TNP-modified nucleotides (TNP-ATP, **41a**; and TNP-GTP, **41b**; Fig. 12) contain a Meisenheimer complex moiety at neutral pH in water, which fluoresces weakly, but fluoresces strongly on binding many proteins. This property makes them attractive for single-molecule imaging studies.[149] TNP-modified nucleotides are currently in wide use in the literature and are commercially available. One drawback of these probes is the short lifetime (<200 ps).

TNP-ATP[51] is prepared at 30° by dropwise addition over 3 hr of a 3.5 *M* excess of 2,4,6-trinitrobenzene 1-sulfonate to an aqueous solution of the sodium salt of ATP adjusted to pH 9.5 with 4 *M* LiOH (1.65 mmol in 10 ml). The reaction is allowed to continue at room temperature in the dark for 4 days, after which it is evaporated to dryness at 30°. The residue is dissolved in acetone–methanol (3 : 1, v/v) and filtered with repeated washes of the solid with acetone. The solid is dissolved in water and purified on a 2.2 × 60 cm Sephadex LH-20 column. The synthesis of the TNP-GTP (**41b**) and TNP-Gpp(NH)p derivatives are prepared with 2,4,6-trinitrochlorobenzene as described.[53]

2′(3′)-O-Naphthoyl esters of ATP. Mayer *et al.*[48] have synthesized a series of 2′(3′)-*O*-(*N*-dimethyl)naphthoyl esters of ATP using the methods described herein (see Preparation of 2′(3′)-*O*-Ribose Esters from Imidazolides of Carboxylic Acids, above). Figure 12 shows two of these isomers. 1,5-DAN-ATP (**42a**) has been used as a probe for an uncoupling protein[138] as it has a different interaction with this protein than DNS-ATP (**32**),[46] which has a sulfonamide linkage to the ribose ring.

Unusual Fluorescent Nucleotide Analogs. Onadera and Yagi[56] report the synthesis of **43** (Fig. 13), which is a substrate for the myosin ATPase. The dansyl group appears to mimic the adenine ring and the ethanolamine linker between the dansyl and the phosphates appears to provide spacing equivalent to the ribose ring. Nakamaye and co-workers[150] have developed photoaffinity labels with a similar structure, with the 4-azido-2-nitrophenyl moiety replacing the dansyl group (NANTP). This class of analog is also a substrate for the myosin ATPase[151] and has proved useful in the understanding of the structure and function of the myosin active site.[152]

Pal and Coleman[57] and Rosen[58] have described a purine nucleotide triphosphate analog, 4-benzoyl(benzoyl)-1-amidofluorescein (**44**), that is a fluorescent photoaffinity label. It has been used to study the submitochondrial ATPase, creatine phosphokinase,[58] and the F_1-ATPase.[57]

[149] J. Y. Ye, Y. Yamane, M. Yamauchi, H. Natatsuka, and M. Ishikawa, *Chem. Phys. Lett.* **14,** 607 (2000).

[150] K. L. Nakamaye, J. A. Wells, R. L. Bridenbaugh, Y. Okamoto, and R. G. Yount, *Biochemistry* **24,** 5226 (1985).

[151] E. Pate, K. L. Nakamaye, K. Franks-Skiba, R. G. Yount, and R. Cooke, *Biophys. J.* **59,** 598 (1991).

[152] A. M. Gulick, C. B. Bauer, J. B. Thoden, E. Pate, R. G. Yount, and I. Rayment, *J. Biol. Chem.* **275,** 398 (2000).

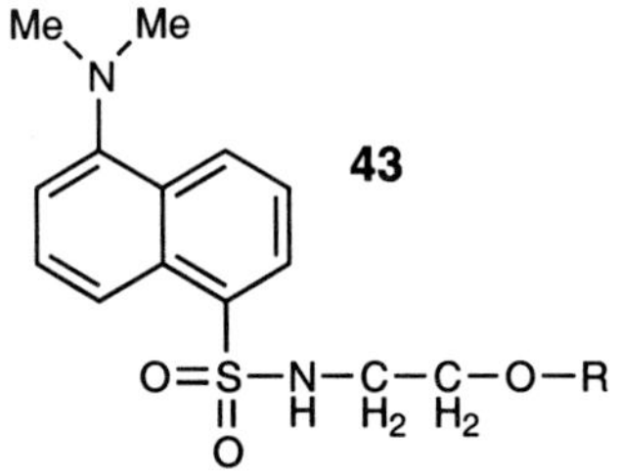

43

2-(dansylamino)ethyl-nucleotide analogs

(a) R = H
(b) R = phosphate
(c) R = triphosphate

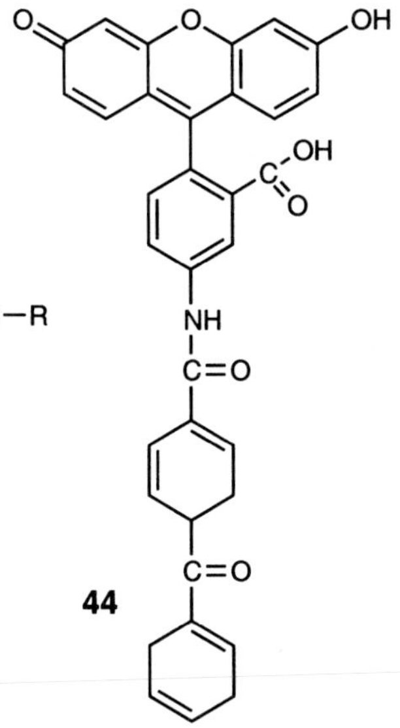

44

4-benzoyl(benzoyl)-1-amidofluorescein (BzAF)

FIG. 13. Examples of unusual fluorescent nucleotide analogs.

Fluorophore Selection

There are many fluorophores available that can be appended to nucleotides and modified nucleotides. The selection of which fluorophore to use and where to place it on the nucleotide is often based on what has worked for that class of protein in the past.

Another consideration is the spectral properties of the fluorophore. For example, the FEDA-ATP Cy3-ATP and Cy5-ATP derivatives have proved useful for single-myofibril[124] and single-molecule detection[107] because of their high extinction coefficients and quantum yields. With the advent of two-photon excitation, the problem of the small Stokes shift of these larger fluorophores can be overcome. Now excitation can be performed at longer wavelengths than emission and thus the two are easily spectrally separated.

For single-molecule detection, the introduction of the evanescent field[125,153] has decreased the problem of emission from fluorophores not bound to target sites. However, it is useful to choose fluorophores with low quantum yields free in solution and higher quantum yields bound to the macromolecule.

[153] Y. Harada, T. Funatsu, M. Tokunaga, K. Saito, H. Higuchi, Y. Ishii, and T. Yanagida, *Methods Cell Biol.* **55,** 117 (1998).

For energy transfer studies, the absorption of the acceptor must overlap the emission of the donor fluorophore. Fluorophores with larger extinction coefficients can measure longer distances. Consideration of the flexibility of linker groups can affect the orientation factor, κ^2.

Most suppliers of fluorophores will give spectral properties, solubilities, and references of uses. Do not assume that all fluorophores are provided in a pure form. If more than one compound is observed by analytical TLC, for example, reactivity tests with model compounds (amines, thiols, etc.) can be performed to determine whether the contaminants are reactive. If they do not react, it may not be necessary to purify the fluorophore mixture before attachment to the nucleotide. Also, the spectral properties of the contaminant should be considered. If it fluoresces or absorbs strongly, it may complicate the analysis of the experimental results.

Characterization

General Considerations

The proper characterization of synthetic nucleotides usually involves chromatographic, chemical, and spectral techniques. The characterization usually has two goals. First, the identity of the product must be established, which usually involves typical structure verification tools available to the organic chemist. These tools are mentioned with specific reference to their use in the special case of nucleotides and nucleosides. Second, the level of purity must be established depending on the application. Rigorous applications such a crystallography or kinetics can often require purity better than 99%. It should be verified, for commercial and other preparations, that by-products, hydrolytic breakdown products, inhibitors/activators, and other interfering compounds are absent from the preparation. An example of a case in which extreme purity is required is for studies of the substrate specificity of nucleotide analogs. If the analogs do not interact with the macromolecule in question, but the parent nucleotides strongly interact, trace levels of the parent nucleotides can give inconsistent results. A good example of this is reported for the interaction of Mant-esters and various carbamates of GTP with transducin.[118]

Chromatographic Methods

Analytical Thin-Layer Chromatography Techniques. Analytical TLC is an essential tool to follow the course of a synthesis and to characterize the final products for publication and/or verification of structure. Standards are prepared and samples are retained at each step of the synthesis and subsequently compared on TLC plates. It is always advisable, as in any synthesis, to analyze the starting material by TLC before starting the synthesis. Many fluorescent compounds are not provided in pure form by the manufacturer, and nucleotides are often contaminated with hydrolysis products (i.e., ADP in ATP). TLC is a fast and easy way to develop an initial idea of the purity. If purification is required (when in doubt, purify), preparative TLC or HPLC methods should be used. TLC is also a rapid method to determine whether the nucleotide is a substrate for an enzyme.

THIN-LAYER CHROMATOGRAPHY PLATES. Most investigators favor the use of silica gel or cellulose plastic- or aluminum-backed analytical TLC plates (0.1 mm thick; Merck silica 60 F 254). PEI-cellulose (polyethyleneimine anion exchange; Brinkmann, Westbury, NY) plates are also used. These are often sold as precut plates (5×20 cm) that can be cut in half. Each 5×10 cm plate has enough room for application of five to eight spots. Plates can be purchased with or without fluorescent indicator (designated F) dissolved in the silica or cellulose.

APPLICATION OF SAMPLES. It is best to apply compounds with capillary applicators (Microcaps; Drummond, Broomall, PA) in the smallest possible spots (1–2 μl) in readily volatile solvents and to allow spots to completely dry before developing the plate. Heat-stable compounds can be dried onto the plate with a hair dryer. Nonfluorescent samples of approximately 1 m*M* typically give excellent quenching of the plates and less is required for fluorescent compounds. Less concentrated samples should be applied repetitively in small aliquots. Low-volatility solvents such as DMF should be avoided.

DEVELOPING THIN-LAYER CHROMATOGRAPHY PLATES. The solvent chamber (a glass jar with a watchglass or petri dish cover) should contain only a sufficient amount of solvent to wet the bottom of the plate, avoiding initial contact with the applied spots. It is best to allow the solvent chamber to come to equilibrium before developing the plate. Better separation is achieved if solvents are allowed to migrate the full length of the plate.

Common solvent recipes are listed in Table IV[103,154] and often must be tested by trial and error for the particular synthesis. R_f values for new compounds should be reported, along with appropriate standards, in more than one solvent system for the convenience of the reader. Typically, the more highly charged nucleotides migrate more slowly (triphosphates) and the effect of hydrophobic appended groups is to increase the mobility. Fresh solvents are essential as some solvent recipes will phase separate after storage and should no longer be used.

VISUALIZATION OF COMPOUNDS. Nonfluorescent nucleotides will quench the fluorescent indicator on the plates and appear as a dark spot under UV illumination. Small hand-held UV illuminators (Ultra-Violet Products, Cambridge, UK; http://www.uvp.com/) are sufficient to observe plates. Many illuminators (UVGL-15; Ultra-Violet Products) have two lamps for short-wave length (254 nm) and long-wavelength (365 nm) excitation. The short wavelength will illuminate the

[154] H. Wada, H. Nakamura, and K. Miyatake, *J. Carbohydrates Nucleosides Nucleotides* **4,** 231 (1977).

TABLE IV
COMMON THIN-LAYER CHROMATOGRAPHY RECIPES FOR ANALYSIS OF NUCLEOTIDES AND NUCLEOSIDES[a]

Solvent system	Reagents	Relative amounts in volume	Ref.
1	Isobutyric acid[b]	66	
	Concentrated ammonium hydroxide	1	
	Water	33	
2	Isobutyric acid	75	
	Concentrated ammonium hydroxide	1	
	Water	24	
3	*n*-Butanol	5	
	Glacial acetic acid	2	
	Water	3	
4	Ethanol	7	
	Ammonium acetate, 1 *M*	3	
5	2-Propanol	7	
	Concentrated ammonium hydroxide	1	
	Water	2	
6	Dioxane	40	103
	2-Propanol	20	
	Water	35	
	Concentrated ammonium hydroxide	35	
7	Acetone	4	154
	Chloroform	1	
	Triethylamine	1	
	Water	1	
8	LiCl (0.5 *M*), 2 *M* formic acid, or 0–0.5 *M* LiCl for use with PEI-cellulose		

[a] All recipes are to be used with silica gel plates unless otherwise indicated.
[b] Isobutyric acid has a bad odor and should be used in the hood. Avoid any contact with skin or clothing and avoid contact with vapors. Gloves should be worn at all times, and gloves should be disposed of in the hood to evaporate before further disposal.

fluorescent indicator on the plate and the long wavelength is useful for fluorescent probes that absorb light near this wavelength. UV-protective glasses should always be worn. Fluorescent compounds do not require the use of the fluorescent indicator on the plates, but can be observed on them as fluorescent spots. Amine-containing derivatives can be reacted directly with ninhydrin reagent to give a colored spot after heating. Phosphorus-containing compounds can be detected with the Dittmer–Lester reagent.[155]

[155] J. C. Dittmer and R. L. Lester, *J. Lipid Res.* **5,** 126 (1964).

ANALYSIS OF ENZYMATIC REACTIONS WITH NUCLEOTIDES BY THIN-LAYER CHROMATOGRAPHY. Enzymatic reactions with nucleotides can be quickly analyzed by TLC. This can be a useful method to verify nucleotide structure with enzymes of known specificity (i.e., alkaline phosphate; see below), to determine enzyme specificity for fluorescent nucleotides, and to identify products of reaction of enzymes with nucleotide derivatives. The following are a few applications of TLC in these modes.

Alkaline phosphatase hydrolysis of triphosphates. Alkaline phosphatase can degrade a triphosphate derivative to the nucleoside. The time course of the enzymatic reaction can be monitored by TLC. A progression from the triphosphate to the di- and monophosphates and ultimately to the nucleoside should be apparent with the appropriate standards for comparison. The following is a procedure for this assay.

Nucleotide is mixed with 10× buffer [0.5 *M* Tris-HCl (pH 9–10), 10 m*M* $MgCl_2$, 1 m*M* $ZnCl_2$] to a final concentration of ~0.1–1 m*M* nucleotide at 1× buffer. An aliquot of the reaction is applied to a TLC plate along with an aliquot of the buffer alone. The reaction is begun by adding an aliquot (1/20 of the reaction volume) of alkaline phosphatase orthophosphoric monoester phosphohydrolase (EC 3.1.3.1 from *E. coli* type III; Sigma). Time points are removed and immediately applied to the TLC plate. The reaction is quenched by drying the spot with a hair dryer. A positive control reaction with ATP or other nucleotide is highly recommended to establish optimal reaction conditions.

Quantitative imaging of fluorescence on thin-layer chromatography plates. With the advent of highly sensitive cooled charge-coupled device (CCD) cameras attached to imaging systems, TLC can now be used to quantitatively monitor the rates of enzymatic reactions of fluorescent nucleotides. Plates are imaged in the epi-UV mode, that is, illumination with UV light from above. An example of this method[103] determines both k_{cat}/k_m and the products of the reactions of fluorescent ApppA analogs (coumarin and BODIPY labeled) with a member of the histidine triad superfamily of nucleotide binding proteins.

Anion-Exchange Chromatography. Anion-exchange chromatography is the most widely used preparative chromatographic technique to purify nucleotides. In the most commonly used preparative method, samples are applied to a weak anion exchanger such as a diethylaminoethyl-modified resin (e.g., DEAE-cellulose or DEAE-Sephadex; bicarbonate form washed with water) at low ionic strength in water or weakly buffered pH ~7.5 TEA–bicarbonate (TEAB, <0.1 *M*) and eluted with a pump at 4° with linear gradients up to 1 *M* TEAB. This solvent is useful because it is volatile and excess TEA can be removed from the pooled peaks by rotary evaporation at low temperature. One drawback of this solvent is that it degasses CO_2 if the temperature is increased or as it ages. This can cause problems with bubbles in the chromatographic apparatus. In addition, the pH of the buffer increases as CO_2 evolves, which generates basic conditions. This is known to be a problem in the synthesis of ribose-modified esters as esters can hydrolyze under these basic conditions.

It is highly recommended, but not essential, that columns be attached to a variable-wavelength UV–VIS detector with a flow cell. High optical densities can be avoided with a short path length flow cell and a wavelength setting that is off the absorbance maximum.

It should be noted that ion-exchange chromatography as described above often cannot separate nucleotides from negatively charged side products of a synthetic reaction. For example, Mahoney and Yount[156] report that 5′-adenosine di- and triphosphate analogs cannot be resolved from pyrophosphate and imidodiphosphate (PNP_i), using TEAB solvents with ion-exchange chromatography. This is a problem in radioactive syntheses and for determination of extinction coefficients.

For analytical purposes, a strong anion-exchange HPLC column can be used. Jameson and Eccleston[7] used ion-exchange HPLC to purify Mant-GDP and Mant-GTP (**39**) from GDP and GTP, using an ammonium phosphate–methanol solvent system. The 2′-O- and 3′-O-isomers of FEDA-ATP (Table III[117]) could be separated with an SAX column (Synchropak Q 300 A) with an isocratic elution with 0.6 *M* ammonium dihydrogen phosphate at pH 4.0 with 20% methanol.[108] Webb *et al.*[109] have used a Partisphere SAX (Whatman) to isocratically separate analytical amounts of 2′-*O*- and 3′-*O*-coumarin-modified ATP derivatives in 75% 0.35 *M* ammonium phosphate (pH 4.0) and 25% methanol.

Many authors have used AG MP-1 (Bio-Rad, Hercules, CA), which is also a strong anion exchanger, to analyze cyclic nucleotides. In these procedures, the sample is applied in water and eluted with a gradient of trifluoroacetic acid.

A disadvantage of anion-exchange chromatography is the nonspecific adsorption of hydrophobic large fluorophores onto the resin.[7] For these compounds, reversed-phase HPLC techniques or gel filtration on Sephadex LH-20 is often preferable.

Gel Filtration

SEPHADEX LH-20. Sephadex LH-20 (Amersham Biosciences) is a gel-filtration medium that can be used in both aqueous and organic solvents. It is a hydroxypropylated cross-linked dextran with both hydrophilic and lipophilic properties. Sephadex LH-20 is versatile because it can separate on the basis of size alone and/or on the extent of partitioning between stationary and mobile phases. One advantage of this method is that eluting solvents are often volatile. A comprehensive book on the use of this resin for preparative purifications is available.[157]

LH-20 chromatography is useful in the preparative mode and can be performed with a low-pressure chromatography unit. LH-20 can be used to separate Mant- and

[156] C. W. Mahoney and R. G. Yount, *Anal. Biochem.* **138,** 246 (1984).

[157] H. Henke, "Preparative Gel Chromatography on Sephadex LH-20." University of Uppsala, Uppsala, Sweden, 1995.

Ant-modified nucleotides (**39a** and **39b**) from their parent compounds.[54,55] Mant-AMP, Mant-GMP, and the respective cyclic nucleotides were chromatographed in 30% ethanol and the respective NDP and NTP derivatives were eluted with water. Mant-derivatized products eluted later than their parent compounds.

As in any gel-filtration chromatography, it is important that the sample be applied to the column in a minimal volume (preferably <5% of the total bed volume). Cremo and Yount[26] found that separation of εADP from $Bz_2\varepsilon$ADP was improved by adding NaCl (~0.1 *M*) to the sample before chromatography in water. Williams and Coleman have used LH-20 with an ammonium formate solvent for the synthesis of Bz_2ADP.[158]

BIO-GEL P. Many investigators use Bio-Gel P (Bio-Rad), an uncharged polyacrylamide-based support, for the desalting of nucleotides than have been chromatographed in nonvolatile salts, and/or cleanup of nucleotides isolated by paper or preparative thin-layer chromatography. Typically the nucleotide (5–10 mg) is applied to a 1 × 60 cm column and eluted with water or dilute acid.

Reversed-Phase High-Performance Liquid Chromatography. Reversed-phase HPLC is another widely used method to purify nucleotides and has been reviewed.[159–161] The choice of the protocol depends on the application, preparative versus analytical, and whether the product can be isolated in the presence of nonvolatile salts. Table V[20,107,156,162–164] shows examples of reversed-phase protocols from the literature. Most procedures rely on silica-based columns and TEA, Na^+, or NH_4^+ phosphate salts mixed with alcohols.

Many applications of fluorescent nucleotides cannot tolerate high concentrations of salts in the purified nucleotide. Therefore, it may be desirable to use a volatile buffer. Mahoney and Yount[156] report the use of TEAB–ethanol mixtures at pH 6.7 to separate 5′-adenosine di- and triphosphate analogs from pyrophosphate and imidodiphosphate (PNP_i). This method has the advantage that the buffer system is volatile, but a drawback is that the buffers must be kept under CO_2 pressure to avoid degassing and thus pH values above 7, which degrade silica-based columns. Schobert[20] and Oiwa *et al.*[107] describe similar procedures that do not require CO_2 pressure on the solvents, at higher pH (values pH 7–7.5), using columns that are more stable to these slightly basic conditions.

[158] N. Williams and P. S. Coleman, *J. Biol. Chem.* **257,** 2834 (1982).

[159] P. R. Brown, A. M. Krstulovic, and R. A. Jartwick, *Adv. Chromatogr.* **18,** 101 (1980).

[160] M. Zakaria and P. R. Brown, *J. Chromatogr.* **226,** 267 (1981).

[161] H. A. Scoble and P. R. Brown, "High Performance Liquid Chromatography" (C. Horvath, ed.), p. 3. Academic Press, London, 1983.

[162] A. M. Gilles, I. Cristea, N. Palibroda, I. Hilden, K. F. Jensen, R. S. Sarfati, A. Namane, J. Ughetto-Monfrin, and O. Barzu, *Anal. Biochem.* **232,** 197 (1995).

[163] T. Uesugi, K. Sano, Y. Uesawa, Y. Ikegami, and K. Mohri, *J. Chromatogr. B Biomed. Sci. Appl.* **703,** 63 (1997).

[164] H. Ashihara, N. Yabuki, and K. Mitsui, *J. Biochem. Biophys. Methods* **21,** 59 (1990).

TABLE V
SELECTED SEPARATIONS OF NUCLEOTIDES BY REVERSED-PHASE HIGH-PERFORMANCE LIQUID CHROMATOGRAPHY

Column (source type)	Column size	Solvent A	Solvent B	Nucleotides separated	Ref.
Nucleosil 5-μm C_{18} (Alltech)	150 × 4.6 mm	50 m*M* K_2HPO_4 (pH 5.5)	AcN	UTP, UDP, UMP, uracil, cAMP	162
PRP-1 (Hamilton; polymeric)	150 × 4.1 mm	100 m*M* TEAB (pH 6.7)	100 m*M* TEAB, pH 6.7 in ethanol	8-N_3ATP, phosphate, PP_i	156
Spherisorb ODS (Waters)	250 × 4.6 mm	100 m*M* TEAB (pH 6.7)	100 m*M* TEAB, pH 6.7 in ethanol	AMP, ADP, ATP, 8-N_3ATP, εAMP-PNP, PNP, PO_4, εADP	156
Nucleosil C_{18} AB 5 μm (Alltech)	4.6 × 150 mm or 1 × 25 cm	2 m*M* TBA, 0.05–0.09 ammonium bicarbonate, 2–6% AcN (pH 7)	Isocratic	Mono-, di-, and triphosphates of formycin A and ε-adenosine	20
Capcell Pak-C_{18} SG 5 μm (Shiseido, Japan; silica)	150 × 6.0 mm; analytical	0.1 *M* TEA-phosphate (pH 6.5 and 8.0)	5% 0.1 *M* TEA-phosphate 95% methanol	Adenosine 3′- and 2′-monophosphates, AMP, ADP, ATP, adenosine 2′,5′- and 3′,5′-diphosphate, adenosine sulfate derivatives	163
Nova-Pak C_{18} (Waters; 4 μm silica)	10-nmol scale	10% (v) AcN, 90% (v) 100 m*M* TEAB (pH 7.4)	Isocratic	Separation of 2′-*O*-Cy3-Eda-ATP from the 3′-isomer	107
Asahipak GS-320 H (Shodex, Thompson Instruments; mixed-mode)	250 × 76 mm	10 m*M* Sodium phosphate (pH 4.4)	Isocratic	Many nucleotides and nucleosides	164

Mixed-mode is a column type that has not had extensive use to date, but has promise to effectively separate nucleotides with large fluorophores. Depending on the solvent used, it is capable of working in size-exclusion, ion-exchange, and reversed-phase modes simultaneously. An example of this approach is given by Ashihara *et al.*[164] as summarized in Table V.

Capillary Electrophoresis. Capillary electrophoresis (CE) is an emerging new analytical tool for nucleotide analyses. Nucleotide CE relies upon charge differences for separation. It is rapid and requires trace amounts of sample and is thus a useful approach to monitoring synthetic reactions. Since peak areas can reflect the amounts of compounds it can be a powerful tool for purity determinations. The ability to use laser-induced fluorescence detection makes this method a perfect match for fluorescent nucleotide synthesis. The methodologies currently developed are beyond the scope of this review but are covered in excellent recent articles[165–167] and reviews.[168]

Determination of Total and Acid-Labile Phosphate

If the nature of the phosphate groups on a compound is known (i.e., triphosphate vs. diphosphate) then the molar extinction coefficient of the compound can be determined (or vice versa) by reading the absorbance and performing a chemical assay of the total or acid-labile phosphate content.

There are two types of phosphates present in nucleotides with respect to determination of phosphate. Acid-labile groups, which are the phosphoanhydride linkages, are cleaved by boiling in 1 *N* HCl whereas acid-stable groups or the phosphoester linkages require ashing with $Mg(NO_3)_2$. Total phosphate is determined by combining the two procedures. Care must be taken to avoid spurious phosphate from detergents used to clean glassware. It is recommended that new glass test tubes be used and that a phosphate standard curve be prepared, using dried K_2HPO_4.

Preparation of Nucleotide Sample for Total Phosphate Analysis; Hydrolysis of Phophoester Linkages. An aliquot of the sample and appropriate phosphate standards containing ~0.01–0.1 μmol of total PO_4 are added to 0.1 ml of 10% $Mg(NO_3)_2 \cdot 6H_2O$ (prepared in absolute ethanol) in a glass test tube and mixed well. The samples are dried in an oven at 80° and then heated in a hood over a strong Bunsen burner flame with rapid shaking and further heated until brown fumes have disappeared. The sample is then treated as described below to hydrolyze phosphoanhydrides (if present) and the phosphate content is determined spectrophotometrically (see below).

Hydrolysis of Acid-Labile Phosphoanhydrides in Nucleotides. An aliquot of the sample and appropriate phosphate standards containing ~0.01–0.1 μmol of acid-labile phosphate is prepared to a final HCl concentration of 1 *N* in glass test tubes. The recommended volume is 0.1 ml. The tube is capped with a clean marble and boiled for 15 min (works for the PCP linkage, but the PNP linkage requires 30 min[169]) in a water bath.

Colorimetric Determination of Phosphate in the 0.01 to 0.1 Micromolar Range. The following method, based on the work of Fiske and Subbarow[170] with modifications by Rockstein and Herron,[171] detects a phosphomolybdate complex, which has a molar extinction coefficient of about 4000–5000 M^{-1} cm^{-1} at 700–720 nm and therefore is useful for samples with 0.01 to 0.1 μmol of phosphate.

[165] M. Uhrova, Z. Deyl, and M. Suchanek, *J. Chromatogr. B Biomed. Appl.* **681,** 99 (1996).
[166] K. A. Cruickshank, *Anal. Biochem.* **269,** 21 (1999).
[167] Y. Zhang, Z. Shao, A. P. Somlyo, and A. V. Somlyo, *Biophys. J.* **72,** 1308 (1997).
[168] S. E. Geldart and P. R. Brown, *J. Chromatogr. A* **828,** 317 (1998).
[169] R. G. Yount, *Adv. Enzymol.* **43,** 1 (1975).
[170] C. H. Fiske and Y. Subbarow, *J. Biol. Chem.* **66,** 375 (1925).
[171] M. Rockstein and P. W. Herron, *Anal. Chem.* **23,** 1500 (1951).

To the sample (0.01 to 0.1 μmol; prepared either for total or acid-labile analysis), 0.5 ml of an acidic molybdate solution (6.6 g of ammonium molybdate $[(NH_4)_6Mo_7O_{24}\cdot 4H_2O]$ in 100 ml of 7.5 N H_2SO_4; add distilled H_2O to an 800-ml final volume) is added, followed by 0.05 ml of $FeSO_4$ (1 g of $FeSO_4$ to 10 ml of 0.15 N H_2SO_4). The solutions are adjusted to the same final volume with the minimal amount of distilled H_2O and mixed well. After 10 min the optical density is read at 702 nm. Residual TBA but not TEA ions interfere with this assay.[156]

Colorimetric Determination of Phosphate in the 0.001 to 0.01 Micromolar Range. Many investigators use the method of Lanzetta *et al.*,[172] which is based on the method of Hess and Derr.[173] This method generates a complex between phosphomolybdate and malachite green at low pH that has a molar extinction coefficient of ~80,000 M^{-1} cm^{-1} [172] at 660 nm. This method is useful to analyze for contaminating phosphate in nucleotides because the conditions do not promote the hydrolysis of tri- or dinucleotides. A "control" sample of the nucleotide without acid hydrolysis should be included to test for the presence of free phosphate.

A great advantage of this method is that it is approximately 10- to 20-fold more sensitive than the procedure mentioned above. However, more care must be taken to avoid spurious phosphate contamination. Preparation of the reagents is tedious and they are not as stable as the reagents for the procedure described above. Malachite green should be filtered with Whatman 50 filter paper. Sterox (nonionic detergent) is not readily available. An alternative detergent is Tergitol NP-9 (nonylphenol polyethylene glycol ether) or Tween 20 from Sigma. Phosphate contamination of detergents can be a problem and batches should be checked. The assay requires the use of a detergent, as standard curves will not be linear up to 8 nmol without the detergent.

Nuclear Magnetic Resonance Techniques

^{1}H and ^{31}P NMR spectra are essential to structural analysis of nucleotides. ^{13}C NMR spectra are also useful but are not as commonly reported and are not discussed here. Spectral parameters [including some Raman and infrared (IR) spectra] of many reference bases and nucleosides and other relevant compounds are available online at the Integrated Spectral Data Base System for Organic Compounds (SDBS) currently served at http://www.aist.go.jp/RIODB/SDBS/menu-e.html. However, at this time the ^{31}P NMR data for nucleotides are not presented. Spectra are commercially available online, for example, at http://www.chemicalconcepts.com/.

^{1}H Nuclear Magnetic Resonance. It is a common practice to report the one-dimensional (1-D) ^{1}H NMR spectrum of new nucleotide derivatives. Sufficient information can often be obtained from a 1-D ^{1}H NMR spectrum if spectral parameters of reference compounds are available. Assignments of ^{1}H NMR peaks can

[172] P. A. Lanzetta, L. J. Alvarez, P. S. Reinach, and O. A. Candia, *Anal. Biochem.* **100,** 95 (1979).
[173] H. H. Hess and J. E. Derr, *Anal. Biochem.* **63,** 607 (1975).

TABLE VI
^{1}H NMR CHEMICAL SHIFT AND J-COUPLING VALUES FOR ATP IN D_2O^a

Source	Proton	Shift (ppm)	Multiplicity	J (Hz)	Connectivity
Ribose ring	1′CH	6.126	d	5.7	1′–2′
	2′CH	4.796	t	5.3	2′–3′
	3′CH	4.616	dd	3.8	3′–4′
	4′CH	4.396	qu	3.0	4′–5′
	5′CH_2	4.295	m	3.1	4′–5′
		4.206	m	−11.8	5′–5′
Adenine base	2 CH	8.234	s		
	8 CH	8.522	s		

[a] Data from Ref. 174 after Ref. 175. Data acquired at 35°, pH between pH 7 and 8. Multiplicity given here was observed at 250 MHz. Multiplicity definitions: s, singlet; d, doublet; t, triplet; qu, quintet; m, other multiplet. The sodium salt of 2,2-dimethyl-2-silapentane-5-sulfonate (DSS) was added to the solutions as a chemical shift reference. All chemical shifts are reported with reference to the trimethyl hydrogen resonance of DSS set at 0.0 ppm. [Reprinted from T. Son and C. Chachaty, *Biochim. Biophys. Acta* **500,** 405 (1977) with permission of Elsevier Science.]

also be achieved by ^{1}H–^{1}H shift-correlated 2-D NMR (correlation spectroscopy, COSY), and other methods.

SAMPLE PREPARATION. ^{1}H NMR unfortunately requires more than analytical quantities of compounds. For example, a typical ^{1}H NMR spectrum requires about 0.6–0.8 ml of 1–10 m*M* nucleotide, or ~1–10 μmol depending on field strength. Nucleotide samples are usually analyzed as the sodium salts in neutral D_2O or D_2O–CH_3OD mixtures, using tetramethylsilane as an internal reference. The pH can be adjusted with NaOD or with DCl. Nucleosides are often dissolved in CH_3OD, DMSO-*d,* acetone-*d,* or mixtures thereof. TEA or TBA salts are to be avoided as these spectra may overlap selected upfield resonances of the nucleotide. Samples must be repetitively evaporated/lyophilized from D_2O to remove H_2O, which can overlap sugar resonances. Spectra can be obtained at 35–40° to avoid the superposition of the residual H_2O signal with those of the ribose ring.[174]

If line broadening is observed it may be useful to add EDTA to ~1 m*M* to chelate trace amounts of paramagnetic ions. Alternatively, paramagnetic metal ion impurities can be removed with a Chelex 100 column (Bio-Rad)[18] or by shaking the solutions with Chelex 100 followed by filtration.[174]

REFERENCE SPECTRA. A useful reference for standard ^{1}H chemical shifts and coupling constants of nucleotides is presented by Son and Chachaty.[174] Table VI[174,175] shows a tabulated ^{1}H–^{31}P noise-decoupled ^{1}H NMR spectrum of ATP, giving the peak multiplicities with coupling constants taken from Son and

[174] T. Son and C. Chachaty, *Biochim. Biophys. Acta* **500,** 405 (1977).
[175] V. Govindaraju, K. Young, and A. A. Maudsley, *NMR Biomed.* **13,** 129 (2000).

TABLE VII
^{1}H NMR CHEMICAL SHIFTS OF NUCLEOSIDE MONO-, DI-, AND TRIPHOSPHATES[a]

Nucleotide	H_8	H_2	H_1'	H_2'	H_3'	H_4'	H_5'	H_5''
ATP	8.522	8.234	6.129	4.796	4.616	4.396	4.295	4.206
ADP	8.151	8.234	6.128	4.751	4.604	4.375	4.249	4.198
AMP	8.594	8.228	6.115	4.761	4.487	4.331	3.971	3.971
ITP	8.478	8.205	6.124	4.809	4.630	4.393	4.289	4.211
IDP	8.471	8.205	6.128	4.751	4.617	4.378	4.254	4.207
IMP	8.564	8.193	6.124	4.783	4.512	4.368	2.034	4.010
GTP	8.129	—	59.20	4.781	4.586	4.354	4.262	4.216
GDP	8.104	—	5.914	4.731	4.602	4.325	4.225	4.180
GMP	8.172	—	5.902	4.748	4.484	4.320	4.012	3.992

[a] In D_2O at neutral pD at 35°. Reproduced with permission from T. Son and C. Chachaty, *Biochim. Biophys. Acta* **500,** 405 (1977) with permission of Elsevier Science.

Chachaty.[174] From Table VI it should be possible to understand the data shown in Table VII[174] (chemical shifts) and Table VIII[174] (coupling constants) for various nucleotides.[174] The di- and triphosphates have distinct resonances for the 5′ and 5″ protons, whereas the monophosphates do not. If the problem at hand requires fine structural details for various protons, H–^{31}P noise decoupling should be applied. Nondecoupled spectra show a complicated multiplet for $H_{5'}$ and $H_{5''}$. Otherwise, this noise decoupling may have little effect on the spectra.

RESOLUTION OF 2′-O- AND 3′-O-ISOMERS OF RIBOSE ESTERS AND CARBAMATES. ^{1}H NMR has been a useful tool to understand the chemical equilibrium between 2′-*O* and 3′-*O*-esters and carbamates of the ribose ring. Although it is now firmly in the literature that these compounds exist as mixtures of isomers at equilibrium (60–70% 3′-O-isomer and 30–40% 2′-O-isomer), there have been confusing reports of esters described as 3′-O-isomers,[48,54,158,176] despite earlier reports[110,112,177] that equilibrium mixtures of the two isomers would be expected (see Cremo and Yount[26] for discussion). Some of this confusion stemmed from the outdated convention to name the ester mixture as 3′-*O*-ester with the understanding that the 2′-*O*-ester was also present.

Mahmood *et al.*[178] and Cremo and Yount[26] reported 1-D ^{1}H NMR spectra showing that benzoyl-benzoyl and *N*-methylanthraniloyl esters of ADP and ATP were mixtures containing approximately 60–70% 3′-O-isomer and 30–40% 2′-O-isomers. These reports were in agreement with NMR spectra of naphthoyl esters of ADP.[111] Cremo *et al.*[47] also reported that carbamates of the ribose ring were mixtures of the 2′-O- and 3′-O-isomers in similar ratios to the esters. The assignments

[176] N. G. Kambouris and G. G. Hammes, *Proc. Natl. Acad. Sci. U.S.A.* **82,** 1950 (1985).
[177] G. Schafer and G. Onur, *Eur. J. Biochem.* **97,** 415 (1979).
[178] R. Mahmood, C. Cremo, K. L. Nakamaye, and R. G. Yount, *J. Biol. Chem.* **262,** 14479 (1987).

TABLE VIII
VICINAL COUPLING CONSTANTS OF NUCLEOSIDE DI- AND TRIPHOSPHATES[a]

	Coupling constant (Hz)								
Nucleotide	$J_{1'2'}$	$J_{2'3'}$	$J_{3'4'}$	$J_{4'5'}$	$J_{4'5''}$	$J_{5'5''}$	$J_{4'P}$	$J_{5'P}$	$J_{5''P}$
ATP	5.7	5.3	3.8	3.0	3.1	−11.8	1.9	6.5	4.9
ADP	5.3	5.3	4.2	3.0	3.3	−11.8	2.0	6.5	5.0
ITP	5.7	5.3	3.8	3.2	3.4	−11.8	2.0	6.5	5.0
IDP	5.3	5.0	4.4	3.2	3.4	−11.7	1.8	6.2	5.2
GTP	5.8	5.2	3.4	3.2	3.7	−11.7	1.9	6.2	5.3
GDP	5.3	5.3	4.2	3.2	3.7	−11.7	1.9	6.5	5.5

[a] In D_2O at neutral pH at 35°. Reproduced with permission from T. Son and C. Chachaty, *Biochim. Biophys. Acta* **500,** 405 (1977) with permission of Elsevier Science.

were made on the basis of coupling constants and comparison with standard spectra. Protons on the ribose ring that are attached to carbons carrying the electron-withdrawing fluorophores (e.g., $C_{3'}$, 3′-isomer) are found downfield of the signal for the same proton but of the other isomer (e.g., $C_{3'}$, 2′-isomer). The proportions of the 2′-O and 3′-O-isomers were determined by measuring the relative peak areas of the resonance signals for H_8, $H_{1'}$, $H_{2'}$, and $H_{3'}$ at pH 5. It is now generally recognized that esters and carbamates of the ribose ring are mixtures of isomers.[7]

^{31}P Nuclear Magnetic Resonance. For ^{31}P NMR spectra, proton-decoupled chemical shifts for the α, β, γ, and other phosphorus nuclei are easily interpreted and fall into defined ranges. Of course, proton-nondecoupled spectra are also of great value, but are more difficult to interpret and are not discussed here. Because the chemical shift frequently identifies the class of phosphorus nuclei, the presence or absence of such shifts can be valuable evidence of purity and identity. A reference collection of proton-decoupled chemical shifts is available[179] and other ^{31}P databases are commercially available (e.g., http://www.acdlabs.com/products/spec_lab/predict_nmr/pnmr/). A theoretical discussion of the conformational analysis of nucleotides by ^{31}P NMR, with emphasis on nucleic acids, is presented by Gorenstein.[180]

Table IX[18,162,181–183] shows the spin–spin ^{1}H–^{31}P-decoupled ^{31}P NMR spectral parameters for selected nucleotides and common contaminates of nucleotide syntheses. Data for many other nucleotide derivatives are summarized in Lebedev

[179] J. Tebby, ed., "Handbook of Phosphorus-31 Nuclear Magnetic Resonance Data." CRC Press, Boca Raton, FL, 1991.

[180] D. G. Gorenstein, *Annu. Rev. Biophys. Bioeng.* **10,** 355 (1981).

[181] A. V. Lebedev and A. I. Rezvukhin, *Nucleic Acids Res.* **12,** 5547 (1984).

[182] J. R. Barrio, M. C. Barrio, N. J. Leonard, T. E. England, and O. C. Uhlenbeck, *Biochemistry* **17,** 2077 (1978).

[183] P. Rosch, H. R. Kalbitzer, and R. S. Goody, *FEBS Lett.* **121,** 211 (1980).

TABLE IX

SPIN–SPIN ^{1}H–^{31}P-DECOUPLED ^{31}P NMR SPECTRAL PARAMETERS FOR SELECTED NUCLEOTIDES

	Chemical shift (ppm)[a]			Coupling constant (Hz)		
Compound	P_α	P_β	P_γ	$J_{\alpha\beta}$	$J_{\beta\gamma}$	Ref.
dATP	−11.43	−22.83	−7.71	20.5	20.5	181
ATP	−11.45	−22.66	−7.33	19.8	19.8	181
ADP	−11.13	−6.96		23.1		181
AMP	3.882					182[b]
3′-PO$_4$5′-AMP	3.902	4.175 (3′-P)				182[b]
2′-PO$_4$5′-AMP	3.962	3.696 (2′-P)				182[b]
ADPβS	−11.1	33.9		31.2		181
ATPγS	−10.6	−22.0	35.0	19.6	29.0	181
GTPβS	−11.31	29.50	−5.95	—	—	183
GTPαS	43.66	−22.40	−5.66	—	—	183
AMPCP	21	9.2				181[c]
XMPPNP[d]	−9.94	−6.95	−0.08	20.9	4.4	18
XMPPCP	−10.12	14.15	12.19	27.0	7.4	18
dTDP	−11.26	−7.02		23.6		181
dTMP (5′-imidazolide)	−10.3					181
UTP	−10.72	−22.34	−9.92	19.92	19.19	162

[a] Chemical shifts are reported as parts per million downfield from the reference (85% H_3PO_4). Therefore, negative parts per million indicate that the signal is upfield from the reference. The chemical shift is calibrated to zero for the reference. Conditions: D_2O or H_2O, pH neutral unless otherwise indicated.

[b] pH 10.

[c] pH 6.

[d] X, Xanthosine.

and Rezvukhin.[181] At neutral pH, in water, the resonances for the common nucleotides fall upfield from H_3PO_4 at 0 ppm, and thus have negative chemical shifts. Phosphorothioate derivatives and nucleotides with a P–C bond are exceptions as ^{31}P nuclei adjacent to sulfur and carbon resonate downfield. Because hydrolysis of the phosphoanhydride linkages can be a problem for certain synthetic protocols, it is useful to use ^{31}P NMR to probe for the presence of the degradation products such as phosphate and pyrophosphate.

Typically, nucleotide solutions for ^{31}P NMR are in the millimolar range or less. Rosch *et al.* prepared their nucleotide solutions at <12 m*M* to avoid intermolecular interactions.[183] Samples should be treated to remove paramagnetic ions as described for ^{1}H NMR. Integrated signal areas are often a good measure of the concentration of the nuclei.

Mass Spectrometry

Mass spectral information of many nucleosides is available online at the Integrated Spectral Data Base System for Organic Compounds (SDBS) currently served at http://www.aist.go.jp/RIODB/SDBS/menu-e.html. Samples can be analyzed in both the positive and negative ion mode with a fast atom bombardment (FAB) source. Samples can be dissolved in water at 1 μg/ml and a 1-μl aliquot added to a matrix of dithiothreitrol–dithioerythritol (DTT–DTE), or glycerol–water (1 : 1, v/v),[184] for positive ion spectra or triethanolamine for negative ion spectra.[185] Walton *et al.*[184] provide a useful analysis of the relative abundances of the protonated ions, sodium salts, and respective adducts of a variety of cyclic AMP analogs. In addition, the characteristics of the collision-induced dissociation (CID) spectra are discussed. Zhang *et al.*[167] provide an example of the use of on-line capillary electrophoresis–electrospray ionization mass spectrometry of nucleotides in the positive ion mode.

Absorption and Fluorescence

Absorption Spectra and Extinction Coefficients. A proper characterization of fluorescent nucleotides requires reporting of the absorption spectrum and molar extinction coefficient. Molar extinction coefficients are useful because the exact mass of many nucleotide preparations is not known. For example, the number of counterions and water molecules associated with the phosphate groups is often uncertain. Molar extinction coefficients can be calculated from the acid-labile and total phosphate content (see Determination of Total and Acid-Labile Phosphate, above) and the absorption spectrum. The nucleotide must not be contaminated with free phosphate or other nonnucleotide phosphate compounds.

It is often a good assumption that the extinction coefficient of a fluorescent nucleotide is the sum of the extinction coefficients of the unmodified base and the appended fluorophore. Therefore, the ratio of the appended fluorophore to the base can be calculated as a check on structure. Often the fluorophore will absorb at the wavelength maximum of the base and this contribution to the absorbance must be accounted for. In contrast, the fluorophore usually has an absorbance maximum separate from the base.

Fluorescence Spectra, Excitation Spectra, Quantum Yields, Lifetimes, and Polarization Spectra. It is common to report the fluorescence spectra, excitation spectra, quantum yields, lifetimes, and polarization spectra of new fluorescent nucleotides and to use these data to verify structure. Methods for determining these parameters are beyond the scope of this review. An excellent text covering these topics is recommended.[186]

[184] T. J. Walton, M. A. Bayliss, M. L. Pereira, D. E. Games, H. G. Genieser, A. G. Brenton, F. M. Harris, and R. P. Newton, *Rapid Commun. Mass. Spectrom.* **12,** 449 (1998).

[185] R. M. Graeff, T. F. Walseth, K. Fryxell, W. D. Branton, and H. C. Lee, *J. Biol. Chem.* **269,** 30260 (1994).

[186] J. R. Lakowicz, "Principles of Fluorescence Spectroscopy," 2nd Ed. Kluwer Academic/Plenum, New York, 1991.

[6] Photophysics of Green and Red Fluorescent Proteins: Implications for Quantitative Microscopy

By VINOD SUBRAMANIAM, QUENTIN S. HANLEY, ANDREW H. A. CLAYTON, and THOMAS M. JOVIN

Introduction

The green fluorescent protein (GFP) from the jellyfish *Aequorea victoria* and its mutants constitute a class of fluorophores that has revolutionized the evaluation of molecular interactions and visualization of biological systems. GFP and a more recently discovered red fluorescent protein, drFP583 (commercially available as DsRed; Clontech, Palo Alto, CA), a distant homolog from a coral *Discosoma* species,[1] have attracted enormous attention as important reporters in cell, developmental, and molecular biology (for comprehensive overviews, see Refs. 2–5). When fused to proteins of interest and expressed *in vivo,* fluorescent proteins (FPs) act as versatile indicators of structure and function within cells[6–8] and can be imaged with the full repertoire of fluorescence microscopy techniques. FPs and their constructs are finding increasing use in fluorescence lifetime imaging microscopy (FLIM) and fluorescence resonance energy transfer (FRET) modes of microspectroscopy for the elucidation of protein–protein interactions, signaling, and trafficking in cellular systems.[9–16] FPs exhibit a host of photophysical features that

[1] M. V. Matz, A. F. Fradkov, Y. A. Labas, A. P. Savitsky, A. G. Zaraisky, M. L. Markelov, and S. A. Lukyanov, *Nat. Biotechnol.* **17,** 969 (1999).
[2] R. Y. Tsien, *Annu. Rev. Biochem.* **67,** 509 (1998).
[3] P. M. Conn, ed., *Methods Enzymol.* **302** (1999).
[4] K. F. Sullivan and S. A. Kay, ed., *Methods Cell Biol.* **58** (1999).
[5] M. Chalfie and S. Kain, "Green Fluorescent Protein: Properties, Applications and Protocols." Wiley-Liss, New York, 1998.
[6] T. Misteli and D. L. Spector, *Nat. Biotechnol.* **15,** 961 (1997).
[7] A. Miyawaki and R. Y. Tsien, *Methods Enzymol.* **327,** 472 (2000).
[8] J. Lippincott-Schwartz, T. H. Roberts, and K. Hirschberg, *Annu. Rev. Cell Dev. Biol.* **16,** 557 (2000).
[9] A. Honda, S. R. Adams, C. L. Sawyer, V. Lev-Ram, R. Y. Tsien, and W. R. G. Dostmann, *Proc. Natl. Acad. Sci. U.S.A.* **98,** 2437 (2001).
[10] M. Zaccolo and T. Pozzan, *IUBMB Life* **49,** 375 (2000).
[11] J. Llopis, S. Westin, M. Ricote, J. H. Wang, C. Y. Cho, R. Kurokawa, T. M. Mullen, D. W. Rose, M. G. Rosenfeld, R. Y. Tsien, and C. K. Glass, *Proc. Natl. Acad. Sci. U.S.A.* **97,** 4363 (2000).
[12] R. M. Siegel, J. K. Frederiksen, D. A. Zacharias, F. K. M. Chan, M. Johnson, D. Lynch, R. Y. Tsien, and M. J. Lenardo, *Science* **288,** 2354 (2000).
[13] S. Matsuyama, J. Llopis, Q. L. Deveraux, R. Y. Tsien, and J. C. Reed, *Nat. Cell Biol.* **2,** 318 (2000).
[14] F. S. Wouters, P. J. Verveer, and P. I. Bastiaens, *Trends Cell Biol.* **11,** 203 (2001).
[15] P. J. Verveer, F. S. Wouters, A. R. Reynolds, and P. I. Bastiaens, *Science* **290,** 1567 (2000).
[16] T. Ng, D. Shima, A. Squire, P. I. Bastiaens, S. Gschmeissner, M. J. Humphries, and P. J. Parker, *EMBO J.* **18,** 3909 (1999).

0076-6879/03 $35.00

have been studied extensively by high-resolution ensemble and single-molecule spectroscopic techniques. These properties differ significantly between the variant molecules, either by design or by happenstance, and are subject to perturbation by the biological system under investigation. Thus, although the FPs can be exploited to gain insight into molecular processes on and within cells, an accurate and quantitative interpretation of the data necessitates appropriate regard for the intrinsic complexity of these unique probes.

The simplest, and by far the most common, application of FPs is as a passive marker fused to a target protein of interest for visualizing its spatiotemporal distribution. A large number of such applications in cell and developmental biology have been reported. We do not focus on this type of investigation, however, but rather on situations in which the photophysical properties of FPs are exploited to secure information about the particular cellular milieu and molecular state of the FP fusion protein. Examples might be pH- or FRET-dependent lifetimes and/or spectral perturbations reflecting protein–protein interactions such as homo- or heteroassociation.

GFP variants have been classified according to their spectral properties.[2,17] Mutations in and around the chromophore have systematic effects on the spectra. This knowledge has been exploited to generate variants with specific properties, such as significantly red-shifted excitation and emission peaks that minimize interference from cellular autofluorescence and potentiate a greater number of donor–acceptor FRET pairs. The most red-shifted GFP so far has been the yellow fluorescent protein class, featuring a phenolate anion with a stacked π-electron system and exhibiting excitation and emission peaks at ~514 and ~527 nm, respectively. This spectral palette was extended significantly by the discovery and cloning of DsRed, which has absorption and emission spectra peaking at ~559 and ~583 nm, respectively, rendering this molecule a likely acceptor candidate for FRET studies based on conventional GFP donors.[18,19] However, the utility of DsRed in such applications is somewhat limited by its maturation behavior,[19–21] which is depicted in Fig. 1. In fact, the complex photophysics of DsRed[22–25] remain controversial

[17] G. J. Palm and A. Wlodawer, *Methods Enzymol.* **302,** 378 (1999).
[18] D. L. Moon, M. G. Erickson, and D. T. Yue, *Biophys. J.* **80,** 362a (2001).
[19] H. Mizuno, A. Sawano, P. Eli, H. Hama, and A. Miyawaki, *Biochemistry* **40,** 2502 (2001).
[20] G. S. Baird, D. A. Zacharias, and R. Y. Tsien, *Proc. Natl. Acad. Sci. U.S.A.* **97,** 11984 (2000).
[21] S. Jakobs, V. Subramaniam, A. Schonle, T. M. Jovin, and S. W. Hell, *FEBS Lett.* **479,** 131 (2000).
[22] V. Subramaniam, *Biophys. J.* **80,** 7a (2001).
[23] F. Malvezzi-Campeggi, M. Jahnz, K. G. Heinze, P. Dittrich, and P. Schwille, *Biophys. J.* **81,** 1776 (2001).
[24] A. A. Heikal, S. T. Hess, G. S. Baird, R. Y. Tsien, and W. W. Webb, *Proc. Natl. Acad. Sci. U.S.A.* **97,** 11996 (2000).
[25] B. Lounis, J. Deich, F. I. Rosell, S. G. Boxer, and W. E. Moerner, *J. Phys. Chem. B* **105,** 5048 (2001).

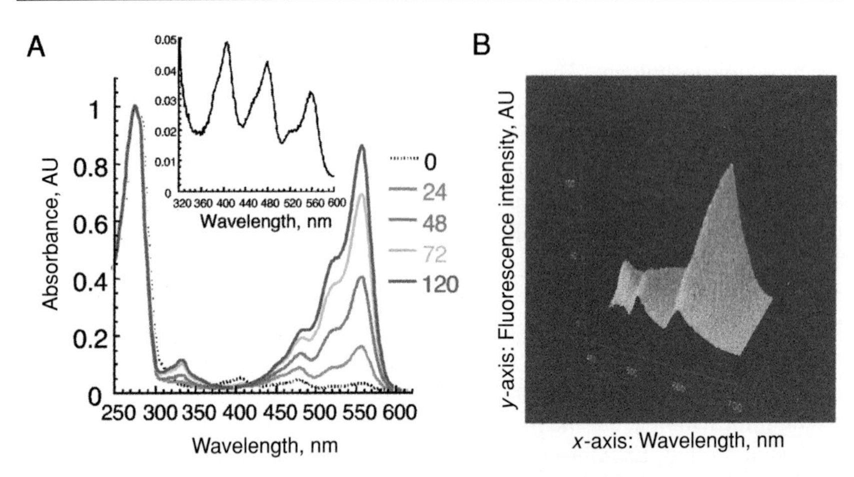

FIG. 1. Maturation of DsRed monitored by different parameters. (A) Temporal evolution of DsRed absorbance spectra. DsRed production in *Escherichia coli* was induced by addition of IPTG and allowed to proceed for 3 hr. The cells were then pelleted and frozen. DsRed was purified and flash frozen. An aliquot of the purified protein was sealed in a cuvette and incubated at 20°. Absorbance spectra were measured at 24-hr intervals for 5 days. Curves have been normalized to tryptophan absorbance at 277 nm. *Inset:* Three absorbing species (406-, 478-, and 558-nm peaks) in the immature state. (B) Temporal evolution of DsRed fluorescence spectra. Excitation was at 400 nm, and emission spectra were acquired at 410–700 nm. DsRed was expressed and purified as described above, and incubated at 20°. Spectra were acquired every 15 min for 25 hr (data from Ref. 28).

despite progress in elucidating the structure of the protein.[26,27] DsRed is presumed to form tetramers in solution,[20] and matures into the red fluorescent form (which may also appear as a dimer[28]) via a GFP-like green fluorescent intermediate.[19–21] If a significant residual green component intrinsic to the DsRed acceptor persists, a clear differentiation of this emission from that of the donor may be problematical.

Green and red fluorescent proteins exhibit photophysical properties that are generally more complex than those of traditional fluorophores, for example, fluorescein and rhodamine. The susceptibility to the microenvironment of the FP chromophore embedded in the interior of the protein varies greatly, depending on specific amino acid substitutions or more dramatic changes (e.g., permutations) in the primary sequence. FPs undergo both static and dynamic quenching, are sensitive to proton (pH), Ca^{2+}, and other ion concentrations, exhibit photochromism, form dimers and higher order oligomers, undergo time-dependent maturation, exhibit blinking behavior, and engage in hetero- and homo-energy transfer and other

[26] M. A. Wall, M. Socolich, and R. Ranganathan, *Nat. Struct. Biol.* **7,** 1133 (2000).

[27] D. Yarbrough, R. M. Wachter, K. Kallio, M. V. Matz, and S. J. Remington, *Proc. Natl. Acad. Sci. U.S.A.* **98,** 462 (2001).

[28] A. Sacchetti, V. Subramaniam, T. M. Jovin, and S. Alberti, *FEBS Lett.* **525,** 13 (2002).

excited state processes leading to lifetime alterations. These differences can be exploited to yield information beyond that yielded merely by the fluorescence intensity integrated over a given spectral window, the parameter most commonly measured in optical microscopy. As a consequence, a broad-based effort is in progress to improve the spectroscopic capabilities of fluorescent microscopes so as to better exploit the spectral, temporal, and polarization characteristics of fluorescence emission. The following treatment reflects this circumstance and addresses the interplay between the photophysics of the FPs, experimentally observable phenomena, instrumentation, and methods of data presentation and analysis. Simple additions to the fluorescence microscope, inspired by a fundamental appreciation of the underlying photophysics of FP probes targeted to biological structures, can serve to maximize the utility of extracted information.

Exploiting Photophysical Properties of Fluorescent Proteins

Fluorescence Resonance Energy Transfer (FRET)

Perhaps the most important use of FPs other than in mere imaging experiments is the probing of proximity relationships between proteins and the dynamics of protein–protein interactions in cells. This goal is generally achieved by FRET, a physical process by which energy is transferred nonradiatively, that is, via long-range dipole–dipole coupling, from an excited donor fluorophore to an acceptor chromophore (which may or may not be a fluorophore). The acceptor may also be the same type of molecule as the donor, in which case the process is referred to as homotransfer (or, more specifically, energy migration), or a different type of molecule, in which case the process is denoted as heterotransfer.

FRET has been reviewed extensively in previous volumes of *Methods in Enzymology*.[29,30] From an experimental standpoint, we can cite at least eight observable phenomena diagnostic of FRET: (1) dynamic quenching of the donor emission (reduced intensity), (2) sensitized emission of the acceptor molecule (if it is fluorescent), (3) reduction in the photobleaching rate of the donor, (4) donor-mediated photobleaching of the acceptor, (5) increased anisotropy (polarization) of donor fluorescence in the case of heterotransfer (distinct donor and acceptor species), (6) reversal of donor quenching and reduction of donor anisotropy after acceptor photobleaching, (7) depolarization of donor fluorescence in the case of homotransfer (energy migration between identical molecules), and (8) increased anisotropy after (partial) photobleaching of a species undergoing homotransfer. These phenomena are explored in the following sections of this article.

[29] R. M. Clegg, *Methods Enzymol.* **211,** 353 (1992).
[30] P. R. Selvin, *Methods Enzymol.* **246,** 300 (1995).

Donor Quenching

In FRET experiments, donor quenching is typically interpreted as an indicator of energy transfer and assessed by recording intensity changes in the donor fluorescence, depression of the lifetime of the donor, and/or variations in the ratio of donor and acceptor fluorescence.[31,32] However, it is important to keep in mind that quenching may arise from many processes other than FRET, some of which involve interactions in the ground state and thus static mechanisms.

In the case of FPs, static quenching can result from distortion of the fluorophore, for example, through interaction with antibodies targeted to the FP itself or via cellular mechanisms induced by the process under study. Jayaraman *et al.*[33] observed anion-dependent static quenching of a yellow fluorescent protein (YFP) mutant (YFP-H148Q); the fluorescence intensity decreased without a corresponding change in the fluorescent lifetime. The effect was ascribed to binding of the anion to the ground state at a site close to the triamino acid chromophore. In imaging situations, similar processes can occur, resulting in the loss of donor intensity without sensitization of the acceptor or a change in donor lifetime. It is advisable to verify the attribution to FRET of a diminished emission intensity by confirming the presence of a corresponding decrease in donor lifetime, an increase in the sensitized emission of the acceptor, enhanced donor photostability, and/or recovery (increase) of both the fluorescence lifetime and intensity upon photobleaching the acceptor.[34,35]

Dynamic quenching of FPs by aqueous phase quenchers (other than protons and other ions) is not generally observed. However, lifetime analysis frequently reveals differences in FP lifetimes obtained *in vivo* versus *in vitro*. The mechanism of this effect is unclear, although one can propose the intervention of a number of potential factors: self-quenching, pH, temperature, viscosity, and solution constituents.

Photoconversion, Photochromic Effects, and Photobleaching

Extensive mutagenesis of GFP has yielded a range of spectral variants optimized for different purposes, of which FRET-based analysis is of paramount importance.[36] However, as a consequence of the variability and complexity of the photophysics of each variant, there is ample room for error in interpreting results.

[31] R. M. Clegg, *in* "Fluorescence Resonance Energy Transfer" (X. F. Wang and B. Herman, eds.), p. 179. John Wiley & Sons, New York, 1996.

[32] G. W. Gordon, G. Berry, X. H. Liang, B. Levine, and B. Herman, *Biophys. J.* **74,** 2702 (1998).

[33] S. Jayaraman, P. Haggie, R. M. Wachter, S. J. Remington, and A. S. Verkman, *J. Biol. Chem.* **275,** 6047 (2000).

[34] P. I. H. Bastiaens and T. M. Jovin, *in* "Fluorescence Resonance Energy Transfer Microscopy" (J. E. Celis, ed.), Vol. 3, p. 136. Academic Press, New York, 1998.

[35] F. S. Wouters, P. I. H. Bastiaens, K. W. A. Wirtz, and T. M. Jovin, *EMBO J.* **17,** 7179 (1998).

[36] R. Heim, *Methods Enzymol.* **302,** 408 (1999).

For example, high-resolution hole-burning experiments show that wild-type and mutant GFPs can occur in three optically interconvertible conformations.[37,38] Thus a change in color of a variant used as an acceptor would usually be interpreted as evidence of FRET, but may also arise from a photoinduced conversion or optical switching between conformations within a specific GFP mutant.

Dramatic photoconversion effects are exhibited by wild-type GFP, which undergoes a photochromic transition between a protonated and deprotonated species on excitation at 400 nm.[39–42] A very important development (from a cell biological perspective) is that of an FP mutant that exhibits a two-orders-of-magnitude increase in emission intensity on exposure to blue light.[43] A photoactivation of this magnitude is reminiscent of conventional photodecaging but has the inestimable advantage of being operative in the living cell using genetically expressed proteins. The ability to generate molecular populations synchronized in time and space within a cell by such means would/will greatly facilitate measurements of protein diffusion, translocation, binding, and conformational transition.

A further important light-induced reaction is irreversible photobleaching. The photobleaching properties of FPs are not only of intrinsic photophysical interest[39,44] but are also significant in relation to acceptor photobleaching protocols.[34] In one such experiment, we combined spectrally resolved FLIM with photobleaching protocols so as to image the red chromophore of DsRed and its variants and determine whether energy transfer occurs between "green" and "red" chromophores. For example, the E8 mutant of DsRed (with the N42H substitution, provided by BD Biosciences Clontech, Palo Alto, CA) possesses both chromophores but exhibits little or no FRET; that is, there is no increase in donor fluorescence intensity or lifetime upon photobleaching the putative (red) acceptor (Fig. 2).

In the following sections we deal with other salient photophysical properties of FPs, emphasizing possible ambiguities arising in commonly used experimental designs as well as novel methodologies capable of yielding decisive, relatively unequivocal information. In all cases, major attention is devoted to imaging living or fixed cells.

[37] T. M. H. Creemers, A. J. Lock, V. Subramaniam, T. M. Jovin, and S. Volker, *Nat. Struct. Biol.* **6,** 557 (1999).

[38] T. M. H. Creemers, A. J. Lock, V. Subramaniam, T. M. Jovin, and S. Volker, *Proc. Natl. Acad. Sci. U.S.A.* **97,** 2974 (2000).

[39] G. H. Patterson, S. M. Knobel, W. D. Sharif, S. R. Kain, and D. W. Piston, *Biophys. J.* **73,** 2782 (1997).

[40] A. B. Cubitt, R. Heim, S. R. Adams, A. E. Boyd, L. A. Gross, and R. Y. Tsien, *Trends Biochem. Sci.* **20,** 448 (1995).

[41] M. Chattoraj, B. A. King, G. U. Bublitz, and S. G. Boxer, *Proc. Natl. Acad. Sci. U.S.A.* **93,** 8362 (1996).

[42] G. Striker, V. Subramaniam, C. A. M. Seidel, and A. Volkmer, *J. Phys. Chem. B* **103,** 8612 (1999).

[43] G. H. Patterson and J. Lippincott-Schwartz, *Science* **297,** 1873 (2002).

[44] G. S. Harms, L. Cognet, P. H. M. Lommerse, G. A. Blab, and T. Schmidt, *Biophys. J.* **80,** 2396 (2001).

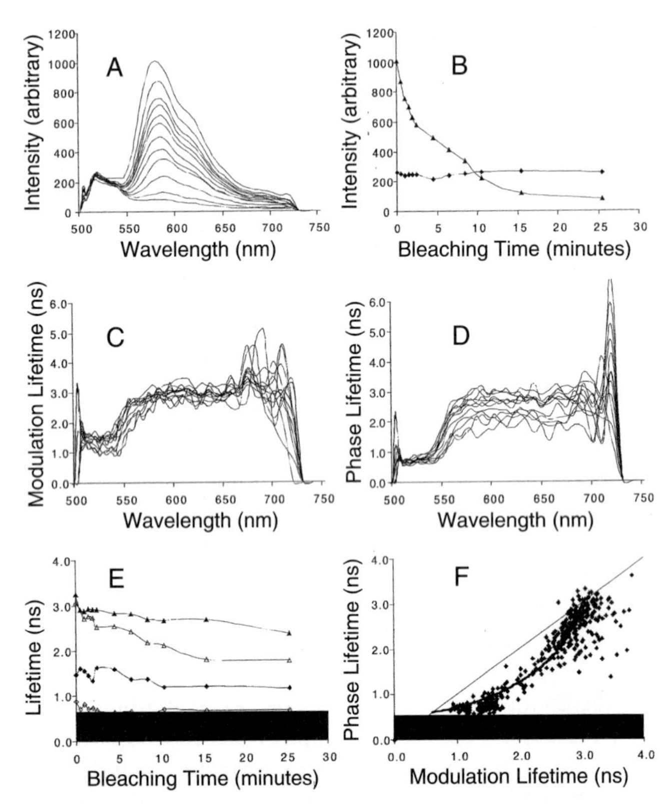

FIG. 2. Spectral FLIM. Photobleaching behavior of the DsRed E8 mutant expressed in *E. coli.* (A) Exposure to light at 546 nm selectively photobleached the red emission. (B) The red portion of the spectrum at 580 nm showed approximately log-linear bleaching (▲) while the green portion monitored at 518 nm (◆) remained unchanged. The modulation (C) and phase (D) lifetime spectra showed a shorter lifetime in the green region and a longer lifetime in the red. (E) During photobleaching both the modulation and phase lifetimes in the red (τ_m, ▲; τ_ϕ, ◆; 580 nm) decreased slowly while little or no change was observed in the equivalent parameters in the green (τ_m, △; τ_ϕ, ◇; 518 nm). (F) Plot of τ_m versus τ_ϕ for all the data in the range of 510–640 nm, showing heterogeneous mixing of two noninteracting fluorophores with lifetimes of ~0.6 ns (green form) and 3 ns (red form). Noisy data at longer wavelengths were excluded [Q. S. Hanley, D. J. Arndt-Jovin, and T. M. Jovin, *Appl. Spectrosc.* **56,** 155 (2002)].

Polarization and Oligomerization: Energy Migration Fluorescence Resonance Energy Transfer (emFRET)

Although great attention has been devoted to the use of FPs as heterotransfer FRET partners, their photophysical properties render them suitable as probes based on homotransfer FRET. The latter process is facilitated when the excitation (absorption) and emission spectra overlap to a significant degree (manifested as a small Stoke's shift). Most FPs fall in this category, as is reflected in the large values for the Förster critical self-transfer distance.[45] For example, the computed homotransfer Förster radii (R_0) for enhanced green fluorescent protein (EGFP) and for YFP are 4.7 and 5.6 nm, respectively. These values are comparable to an R_0 of $\sim$5 nm for the "classic" fluorescein–rhodamine heterotransfer pair. Homotransfer does not alter the ensemble excited state lifetime (and thus intensity), but causes a significant depolarization of fluorescence because the secondarily excited molecules do not retain the orientational distribution established by photoselection during the primary excitation by light.

Perturbations of the emission anisotropy also result from rotational diffusion of the probe (depolarization), changes in lifetime (depolarization or hyperpolarization), and effect(s) of the microenvironment on the relative orientations of the excitation and emission transition moments. Table I summarizes dynamic anisotropy data for various FPs and FP complexes. The FPs have been used as probes of (micro)viscosity of cellular compartments and to detect complex formation via changes in the hydrodynamic volume of the rotational unit or via homotransfer FRET [which we denote here and elsewhere[46] as energy migration FRET (emFRET)]. It is fortunate that the chromophore of FPs generally resides within the β-barrel structure of the protein, exhibiting negligible local motion (exceptions are discussed by Volkmer *et al.*[47]) and thus a rotational correlation time (ϕ) characteristic of the entire protein (typically $\sim$17 ns for monomeric FP in buffer). This value is large in comparison with the fluorescence lifetime (2.5–3 ns), as expressed by the relative parameter $\sigma = \tau/\phi$ of 0.1–0.2. It follows that the contribution of rotational diffusion to depolarization of FPs is limited quantitatively (e.g., to $\sim r_0/10$) under any and all circumstances, particularly in the case of fusion proteins restricting the rotational motion of the attached FP. In contrast, emFRET can be very effective in depolarizing FP emission, that is, if the transfer reaction competes efficiently with the other modes of singlet state deactivation. This condition is readily met in the case of molecular complexes[48] or at high concentrations achieved in solution or in expressing bacteria.[46] Thus, with FPs, a clear separation of the time scales operative for rotational diffusion and emFRET can and does arise, in contrast to the small dye molecules exhibiting emFRET (either free or loosely conjugated to macromolecules) in numerous classic studies of this phenomenon.[49] Another advantage of FPs is their high initial anisotropy

[45] G. H. Patterson, D. W. Piston, and B. G. Barisas, *Anal. Biochem.* **284,** 438 (2000).

[46] A. H. A. Clayton, Q. S. Hanley, D. J. Arndt-Jovin, V. Subramaniam, and T. M. Jovin, *Biophys. J.* **83,** 1681 (2002).

[47] A. Volkmer, V. Subramaniam, D. J. S. Birch, and T. M. Jovin, *Biophys. J.* **78,** 1589 (2000).

[48] I. Gautier, M. Tramier, C. Durieux, J. Coppey, R. B. Pansu, J. C. Nicolas, K. Kemnitz, and M. Coppey-Moisan, *Biophys. J.* **80,** 3000 (2001).

[49] C. Bojarski and K. Sienicki, *in* "Energy Transfer and Migration in Fluorescent Solutions" (J. F. Rabek, ed.), Vol. I, p. 1. CRC Press, Boca Raton, FL, 1989.

TABLE I
ANISOTROPY DECAY PARAMETERS OF FLUORESCENT PROTEINS

Construct, conditions	r_0	ϕ (ns)	$\bar{r}$	Notes	Ref.
Wild-type GFP					
Pure protein, pH 8.0,	0.36 ± 0.01	16 ± 1		400-nm exc	47
0.1 *M* NaCl	0.51 ± 0.01	17 ± 1		800-nm exc; time domain	
Pure protein, pH 8.0,	0.389	14.9	0.306	470-nm exc; time domain	50
+8.9 *M* CsCl	0.380	16.0	0.310		
GFP in T4 phage	0.282	46	0.298	reduced rotation	
$IPIII_TGFP$ in T4 phage	0.247	45	0.266	reduced rotation; emFRET	
S65T-GFP					
Pure protein, pH 8.0,	0.34 ± 0.01	16 ± 1		400-nm exc	47
0.1 *M* NaCl	0.51 ± 0.01	15 ± 1		800-nm exc; time domain	
Pure protein, pH 7.4		20 ± 1	0.325	492-nm exc; frequency domain	92
Free FP in cytoplasm		36 ± 3		viscosity probe	48
Free FP in CHO cells	0.26 ± 0.01	23 ± 2		confocal; time domain	
$TK_{27}GFP$	0.26 ± 0.02	35 ± 9			
$TK_{366}GFP$	0.26 ± 0.01	81 ± 15			
Aggregates	0.23 ± 0.01	2.4, 200		emFRET	
EGFP					
Pure protein, pH 8.2	0.385	13.5		450-nm exc; time domain	93
Pure protein, pH 9.0	0.34, 0.39	14 ± 1		407-, 488-nm exc; time domain	94
Pure protein, pH 7.5, 2°	0.37 ± 0.1	10.6		480-nm exc; time domain	62
GFP–Fv fusion	0.38 ± 0.1	15.8		hinge motion GFP–Fv	
CHO cells; ER lumen		39 ± 5		488-nm exc; time domain	95
Mitochondria	0.4	23 ± 1		488-nm exc; rFLIM	96
Pure protein, pH 8.0	0.37	17 ± 1			46
DsRed					
Pure protein, pH 9.0		0.2, 53 ± 8^a		490 exc; time domain	24

Abbreviations: exc, Excitation; ER, endoplasmic reticulum.

[a] This value is compatible with the assumption of a tetrameric form, compared with $\phi < 20$ ns characteristic of a monomeric 27-kDa fluorescent protein.

(r_0), approaching the limit value of 0.4 (e.g., for EGFP[24,46]). However, the greatest benefit of emFRET in studies of homoassociation is the simplified requirement for establishing a suitable cellular construct: only a single protein species need be expressed, as opposed to the dual expression of different donor and acceptor fusion proteins required for heterotransfer FRET, with all the attendant problems of establishing and maintaining specific relative expression levels after transfection and cloning.

Examples of Energy Migration Fluorescence Resonance Energy Transfer with Fluorescent Proteins

There have been relatively few applications of emFRET in studies of protein oligomerization (Table I). Using steady state and time-resolved fluorescence spectroscopy, Mullaney *et al.*[50] used various cleavable constructs to probe the environment of packaged bacteriophage T4. The GFP was rotationally restricted in all cases, and emFRET was proposed as the cause of enhanced depolarization in two of the proteins. The evidence was the low steady state anisotropy compared with that of free GFP in solution, and a reduced initial anisotropy (from time-resolved measurements). The latter effect presumably resulted from energy transfer faster than the temporal resolution of the instrumentation.

In another study[48] confocal point time-correlated single-photon counting was used to determine the intensity and anisotropy decays of *herpes simplex* virus thymidine kinase in living cells. In cells containing punctate fluorescence the anisotropy decays of the kinase were interpreted in terms of a rapid depolarization event (with an apparent ϕ of $\sim$2 ns) and a long unresolved component ($\phi > 200$ ns).

We have reported anisotropy decay experiments using an adaptation of classic difference phase spectroscopy[51] to wide-field frequency-domain lifetime imaging microscopy.[46] The advantages of such an anisotropy FLIM (rFLIM) system are the acquisition of anisotropy decay and intensity decay information as a function of position over the entire image (as opposed to single-point determinations by time-domain techniques) and the rapid rate of data acquisition. We have used the rFLIM instrument to measure the rotational diffusion of EGFP in solution and in cells (bacterial and mammalian), characterized by initial and limiting anisotropies (r_0, r_∞) and rotational correlation times. There was good agreement with values reported for other monomeric FPs (Table I).

Images and derived data recorded from EGFP-expressing bacteria in the presence of a free EGFP background are shown in Fig. 3. The following lines of evidence suggested the existence of emFRET for the EGFP inside (but not outside) the bacteria: (1) the steady state anisotropy and the AC ratio of the polarized emission components (defined in Ref. 46) from intracellular EGFP are lower than those of the free EGFP; (2) the two-dimensional histograms representing the anisotropy as a function of the relative pixel intensity demonstrate an inverse relationship between the fluorophore concentration and the anisotropy, consistent with concentration depolarization; (3) the lifetime–intensity histograms show a lack of correlation between the intensity and lifetime distributions. Thus, changes in lifetime were not responsible for the decreased anisotropy inside the bacteria; and (4) high-resolution confocal photobleaching experiments led to an increase

[50] J. M. Mullaney, R. B. Thompson, Z. Gryczynski, and L. W. Black, *J. Virol. Methods* **88,** 35 (2000).
[51] G. Weber, S. L. Helgerson, W. A. Cramer, and G. W. Mitchell, *Biochemistry* **15,** 4429 (1976).

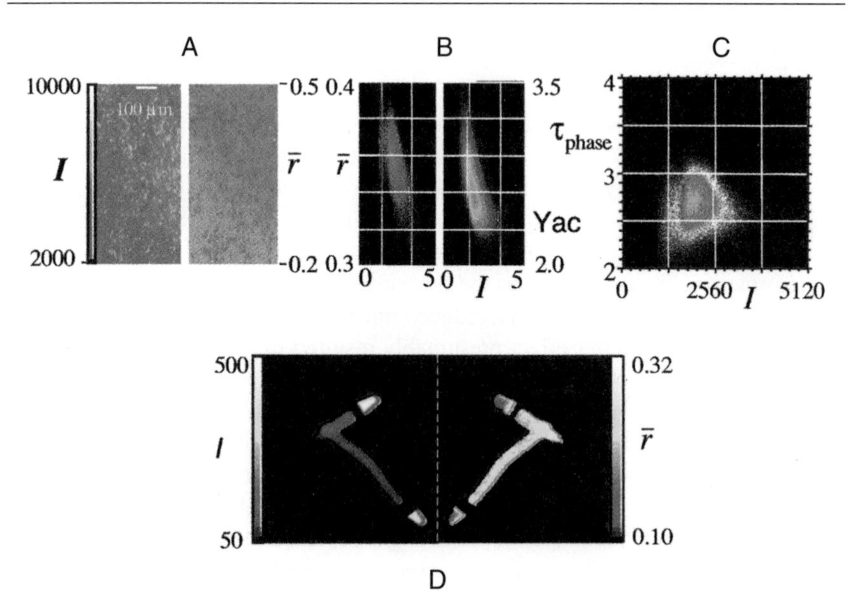

FIG. 3. FLIM–rFLIM images and derived parameter histograms for EGFP expressed in *E. coli* bacteria and free in solution. (A) Wide-field intensity (*I*, *left*) and steady state anisotropy (*r*, *right*) images of the same selected field. (B) Two-dimensional histograms: ($\bar{r}$ vs *I*, *left*) and AC ratio of the polarized emission components (Y_{ac}, *I*, *right*). (C) Two-dimensional phase-lifetime versus intensity (τ_{phase} vs *I*) histogram. τ and ϕ values in nanoseconds; intensities in arbitrary units. The objective was a Nikon Plan Fluor ×20, air (NA 0.5). (D) Confocal laser scanning microscope images of isolated *E. coli* bacteria expressing EGFP, demonstrating emFRET by anisotropy enhancement after photobleaching. Intensity (*left*) and anisotropy (*right*) of bacteria after local photobleaching of the main cell body. For compactness of presentation the intensity and anisotropy images were merged as mirror images. The objective was a Zeiss C-Apochromat water immersion ×63 (NA 1.2) (data from Ref. 46).

in the anisotropy upon partial photobleaching of the bacteria, as also expected for emFRET (see below).

Determination of Local Concentration and Aggregation States of Proteins by Energy Migration Fluorescence Resonance Energy Transfer

A distinct advantage of genetically encoded fluorescent tags such as the FPs is the 100% labeling efficiency of the protein of interest. The use of light to photobleach (but not photoconvert) these proteins allows the experimenter to control the relative surface density (concentration) of fluorophores in a stoichiometric manner, a possibility not generally available in the context of conventional chemical labeling protocols. An example of anisotropy enhancement after photobleaching is shown in Fig. 3 for a bacterium expressing the EGFP molecule. A clear increase in anisotropy upon gradual photobleaching of the sample was observed, consistent with an emFRET mechanism responsible for the depolarization of the EGFP. An analysis of the dependence of the measured homotransfer on surface or voxel density can yield information about protein concentration, association, and clustering according to theories developed for two- or three-dimensional systems.[52–57]

We have extended the existing theory of difference phase fluorimetry to account for the superposition of rotational diffusion and energy migration, and applied the new formalism to determine the concentration of EGFP inside bacteria.[46] Figure 4A illustrates the observed concentration depolarization of pure EGFP in buffer. The formalism predicts a lack of quenching (the intensity should be linear with concentration in the absence of protein self-quenching) but a progressive depolarization. Such behavior was observed. From the parameters obtained by application of appropriate analytical equations, we estimated an apparent emFRET Förster radius of 7.3 nm, an unexpectedly and anomalously high value that is under further investigation. Figure 4B shows a two-dimensional histogram representation of extracellular and intracellular dynamic depolarization of EGFP in bacteria. The upper cloud corresponds to data from free EGFP in the medium and the mean rotational parameters ($r_0 = 0.39$; $\phi = 19$–20 ns) are consistent with an isotropic rotation model of EGFP, close to the values listed in Table I for dilute solutions. This homogeneous mode of relaxation was confirmed by the good agreement between the modulation- and phase-derived isotropic rotational correlation times (Fig. 4D). The lower clouds of Fig. 4B and D correspond to depolarization parameters of the EGFP in the intracellular compartment. These values are consistent with a rotation and concentration depolarization model corresponding to an intracellular EGFP concentration extending up to 0.25 m*M*, that is, a range in which emFRET is operative. The lack of equivalence of the two apparent rotational correlation times rules out trivial changes in local viscosity or rotational diffusion as the source of the reduced anisotropy of intracellular EGFP, and the tight lifetime distribution in Fig. 4C excludes changes in decay rates from denaturation or dequenching. The formalism is being extended to treat protein association and applied to studies of the oligomerization of receptor tyrosine kinases.

Combined Heterotransfer Fluorescence Resonance Energy Transfer and Energy Migration Fluorescence Resonance Energy Transfer

Many natural and synthetic photosystems utilize the phenomenon of energy migration to funnel the initially absorbed energy at one specific location to a

[52] J. Baumann and M. D. Fayer, *J. Chem. Phys.* **85,** 4087 (1986).
[53] T. G. Dewey and G. G. Hammes, *Biophys. J.* **32,** 1023 (1980).
[54] P. K. Wolber and B. S. Hudson, *Biophys. J.* **28,** 197 (1979).
[55] A. K. Kenworthy and M. Edidin, *J. Cell Biol.* **142,** 69 (1998).
[56] A. K. Kenworthy, N. Petranova, and M. Edidin, *Mol. Biol. Cell* **11,** 1645 (2000).
[57] O. Tcherkasskaya, L. Klushin, and A. M. Gronenborn, *Biophys. J.* **82,** 988 (2002).

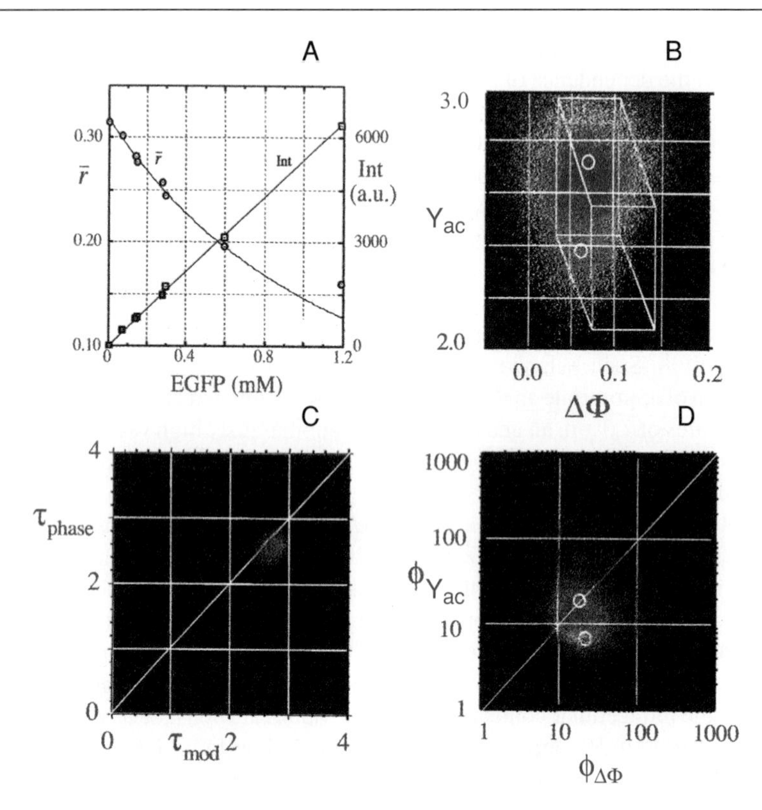

FIG. 4. Analysis of rotational diffusion and concentration depolarization (emFRET) of EGFP in solution and in *E. coli* bacteria. (A) EGFP solutions were measured in microcuvettes (pathlength, 0.25 mm) at 20°. Values for anisotropy (○, $\bar{r}$) averaged over an emission range of 520–560 nm, and intensities (□, Int) corrected for inner filter effects on excitation) are plotted as a function of EGFP concentration. (B) rFLIM 2-D (Y_{ac} vs $\Delta\Phi$) histogram of free extracellular EGFP (upper cloud) and masked bacteria (lower cloud). Circles indicate mean values. Solid lines: range of values predicted for EGFP as an isotropic rotator undergoing emFRET as a function of concentration (see text for details). (C) FLIM 2-D τ_{phase}–τ_{mod} histogram of EGFP in *E. coli* bacteria. (D) Two-dimensional rFLIM $\phi_{\Delta\Phi}$–$\phi_{Y_{ac}}$ histogram of free extracellular EGFP (upper orange cloud) and masked bacteria (lower green cloud). Circles indicate mean values calculated from the mean values of (B) (data from Ref. 46).

distant acceptor or trap site.[58,59] These systems contain multiple coupled donor chromophores that efficiently relay the absorbed energy to the acceptor. The possibility of energy migration is an important consideration when interpreting hetero-FRET experiments with FPs as energy donors. This eventuality will be accentuated

[58] V. Sundstrom, T. Pullerits, and R. van Grondelle, *J. Phys. Chem. B* **103,** 2327 (1999).
[59] K. P. Ghiggino and T. A. Smith, *Prog. React. Kinet.* **18,** 375 (1993).

if the protein of interest is overexpressed, resulting in high local concentrations, and/or the formation of oligomers. The occurrence of emFRET invalidates an implicit assumption in most heterotransfer FRET measurements, that is, that donor–donor transfer is negligible. Moreover, the simultaneous occurrence of emFRET and heterotransfer FRET in a strongly interacting system will lead to erroneous (generally biased to lower values) estimations of the true proximity (or pair distribution function of the donors and acceptors). This caveat not only applies to the assessment of complex formation between two partners, but is also of particular relevance in studies of membrane proteins and domains. Heterotransfer measurements are often used to examine whether proteins are randomly or nonrandomly distributed in membrane-associated structures. Strong donor–donor interactions leading to emFRET necessarily bias quantitative interpretations of the data. Therefore, the indiscriminate use of FPs as fluorescent donors in heterotransfer FRET experiments is not advisable unless the existence of emFRET can be ruled out or taken into account. Measurement of the dynamic anisotropy (rFLIM) or steady state anisotropy (easier but potentially ambiguous) of the sample in the absence of a heterotransfer FRET acceptor provides a valuable diagnostic tool (see above).

Fluorescence Lifetimes

FPs exhibit fluorescence lifetimes that depend on the nature of the construct, the excitation wavelength, the solution conditions, and the temperature. A selection of frequency-domain FLIM data for FPs is given in Table II and a set of FLIM images

TABLE II
SELECTED FLUORESCENT PROTEIN LIFETIME MEASUREMENTS BY FLUORESCENT LIFETIME IMAGING MICROSCOPY

Construct	τ_m (ns)	τ_ϕ (ns)	λ (nm)	Ref.
EGFP, pure protein	2.96	2.79	488	86
ErbB1–EGFP	2.40	2.17	488	
EGFP in *E. coli*	2.75	2.63	488	46
EGFP in cells	2.42	2.36	488	97[a]
ErbB1–EGFP	2.43	2.11	488	
EGFP–p110α	2.44	2.13	488	
Wild-type GFP, pure protein	3.37	3.32	488	86
CFP in cells	2.68	2.07	458	97
	2.23	1.32	488	
NLS–CFP	2.74	2.18	458	97

[a] Additional microscopy references: see Refs. 15, 16, 48, 83, and 98–104.

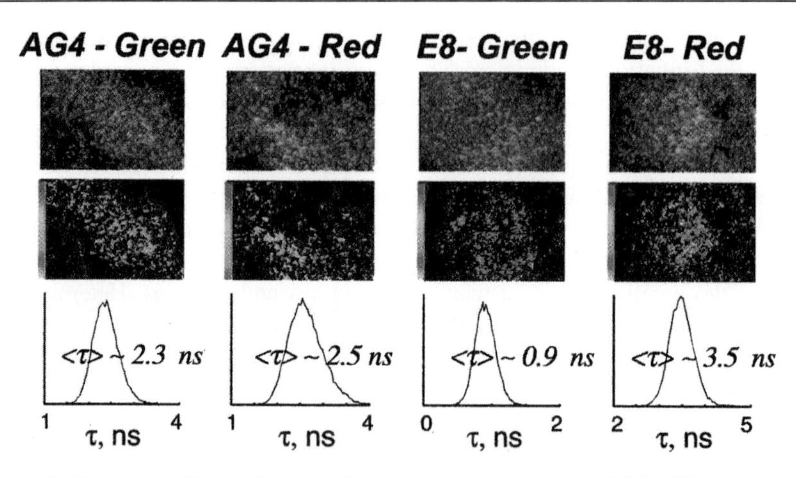

FIG. 5. Fluorescence lifetime imaging microscopy (frequency-domain FLIM) of DsRed mutants. Total intensity (*top*) and fluorescence lifetime (*middle*) images, and calculated lifetime histograms (*bottom*) of *E. coli* expressing the predominantly green-emitting AG4 (see Ref. 91) (amino acid replacements V71M, V105A, and S197T) and the dual green- and red-emitting E8 (N42H) mutants of DsRed in the green (520- to 560-nm bandpass) and red (580-nm long-pass) portions of the spectrum. The green and the red forms of AG4 showed nearly equal lifetimes. E8 exhibited a much shorter lifetime in the green than in the red. The DsRed mutants were provided by BD Biosciences Clontech.

and derived lifetime distributions of bacteria expressing DsRed mutants is given in Fig. 5. FP lifetimes have also been characterized extensively by time-domain techniques.[24,60–63] The observed emission of a fluorophore reflects the competition between radiative and nonradiative processes. Thus, the ratio of the radiative rate to the sum of all radiative and nonradiative rates defines the quantum efficiency. The intervention of additional physical or photophysical pathways, for example, FRET, will necessarily affect the observed lifetime.

Experimental Approaches

In the preceding sections, we explored photophysical phenomena of FPs that can shed light on the biological process under investigation. In this section, we describe in more detail experimental approaches combining spectroscopic techniques and microscopy and designed to exploit the unique characteristics of FPs

[60] M. Cotlet, J. Hofkens, M. Maus, T. Gensch, M. Van der Auweraer, J. Michiels, G. Dirix, M. Van Guyse, J. Vanderleyden, A. Visser, and F. C. De Schryver, *J. Phys. Chem. B* **105,** 4999 (2001).

[61] M. Cotlet, J. Hofkens, S. Habuchi, G. Dirix, M. Van Guyse, J. Michiels, J. Vanderleyden, and F. C. De Schryver, *Proc. Natl. Acad. Sci. U.S.A.* **98,** 14398 (2001).

[62] M. A. Hink, R. A. Griep, J. W. Borst, A. van Hoek, M. H. M. Eppink, A. Schots, and A. Visser, *J. Biol. Chem.* **275,** 17556 (2000).

[63] H. Lossau, A. Kummer, R. Heinecke, F. Pollingerdammer, C. Kompa, G. Bieser, T. Jonsson, C. M. Silva, M. M. Yang, D. C. Youvan, and M. E. Michelbeyerle, *Chem. Phys.* **213,** 1 (1996).

in the cellular milieu. Particular issues related to instrument design and to data analysis are considered.

Emission Spectroscopic Imaging

The goal of emission spectroscopy imaging is to obtain spectra from individual locations in a microscopic field. There are five primary approaches for achieving this end: (1) point measurements,[64] (2) slit or line measurements,[65,66] (3) wavelength scanning,[67–70] (4) Fourier encoding of the wavelength domain,[71] and (5) Hadamard encoding of one[72,73] or two[74] spatial dimensions. A simple system can be implemented by interposing an imaging spectrograph between the microscope image plane and a charge-coupled device (CCD) camera. We have constructed such an imaging spectroscopy instrument using both prism-based and grating spectrographs (Fig. 6). The entrance slit of the imaging spectrograph is mounted at the primary image plane and a CCD camera is placed at the focal plane of the spectrograph. In comparison with FLIM systems incorporating image intensifiers (see below), a minimization of the number of intervening optical elements is sought in order to optimize spectral sensitivity and resolution. We routinely use an Hg-Ne lamp (or, alternatively, an attenuated Hg arc lamp) for wavelength calibration and evaluation of resolution.

The system depicted schematically in Fig. 6 is well suited for the simultaneous observation of multiple FPs and for assessing the intensity distributions of the FRET donor–acceptor pairs, that is, by quantitating the relationship between donor quenching and acceptor sensitization.

Fluorescent Lifetime Imaging Microscopy (FLIM)

A large number of FLIM systems have been described for operation extending from the picosecond[75] to the millisecond regions.[76] Recent technological

[64] G. J. Puppels, F. F. M. de Mul, C. Otto, J. Greve, M. Robert-Nicoud, D. J. Arndt-Jovin, and T. M. Jovin, *Nature (London)* **347,** 301 (1990).

[65] J. A. Timlin, A. Carden, M. D. Morris, J. A. Bonadio, C. E. Hoffler, K. M. Kozloff, and S. A. Goldstein, *J. Biomed. Optics* **4,** 28 (1999).

[66] Q. S. Hanley, P. J. Verveer, and T. M. Jovin, *Appl. Spectrosc.* **52,** 783 (1998).

[67] J. F. Turner and P. J. Treado, *Appl. Spectrosc.* **50,** 277 (1996).

[68] H. R. Morris, C. C. Hoyt, and P. J. Treado, *Appl. Spectrosc.* **48,** 857 (1994).

[69] M. D. Schaeberle, H. R. Morris, J. F. Turner, and P. J. Treado, *Anal. Chem.* **5,** 175A (1999).

[70] D. C. Youvan, *Nature (London)* **369,** 79 (1994).

[71] Z. Malik, R. A. Buckwald, A. Talmi, Y. Garini, and S. G. Lipson, *J. Microsc.* **182,** 133 (1996).

[72] Q. S. Hanley, P. J. Verveer, and T. M. Jovin, *Appl. Spectrosc.* **53,** 1 (1999).

[73] P. J. Treado, A. Govil, M. D. Morris, K. D. Sternitzke, and R. L. McCreery, *Appl. Spectrosc.* **44,** 1270 (1990).

[74] G. Chen, E. Mei, W. Gu, X. Zeng, and Y. Zeng, *Anal. Chim. Acta* **300,** 261 (1995).

[75] K. Dowling, M. J. Dayel, M. J. Lever, P. M. W. French, J. D. Hares, and A. K. L. Dymoke-Bradshaw, *Opt. Lett.* **23,** 810 (1998).

[76] G. Marriott, R. M. Clegg, D. J. Arndt-Jovin, and T. M. Jovin, *Biophys. J.* **60,** 1374 (1991).

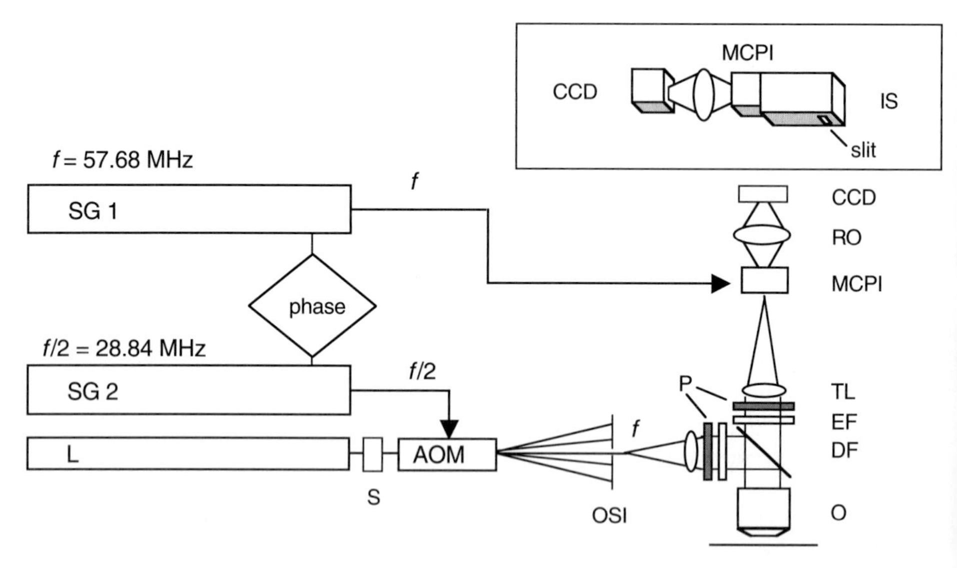

FIG. 6. Multidimensional imaging microscopy. Schematic of a modular wide-field imaging microscope incorporating emission spectroscopy, anisotropy decay, and lifetime modalities. Emission spectroscopic imaging utilizes the laser light source, microscope, a spectrograph (*inset*), and CCD detector (L, O, DF, EF, TL, IS, and CCD). Fluorescence lifetime imaging microscopy (FLIM) employs modulated excitation and modulated detection via an intensified CCD camera (L, SG1, SG2, S, AOM, OSI, O, DF, EF, TL, MCPI, RO, and CCD). In spectrally resolved FLIM (sFLIM; *inset*) a spectograph is placed between the tube lens and an image intensifier of a standard FLIM apparatus. For measurements of static and dynamic depolarization polarizers are introduced into the excitation and emission paths of the microscope (FLIM + P). Symbols: SG1, SG2, signal generators; L, argon ion laser; S, shutter; AOM, acousto-optic modulator; OSI, order selection iris; O, objective; DF, dichroic filter; EM, emission filter; TL, tube lens; MCPI, microchannel plate intensifier; RO, relay optics; CCD, charge-coupled device camera; P, polarizers; IS, imaging spectograph (see Refs. 46 and 87).

extensions include optical sectioning,[77–79] two-photon techniques,[80–82] and multi-frequency excitation.[83] FLIM instruments typically fall into two classes, depending on whether they are operated in the time domain[84] or the frequency domain.[76,85]

[77] M. J. Cole, J. Siegel, S. E. D. Webb, R. Jones, K. Dowling, P. M. W. French, M. J. Lever, L. O. D. Sucharov, M. A. A. Neil, R. Juskaitis, and T. Wilson, *Opt. Lett.* **25,** 1361 (2000).

[78] K. Carlsson and A. Liljeborg, *J. Microsc.* **191,** 119 (1998).

[79] E. P. Buurman, R. Sanders, A. Draaijer, H. C. Gerritsen, J. J. F. Vanveen, P. M. Houpt, and Y. K. Levine, *Scanning* **14,** 155 (1992).

[80] M. Straub and S. W. Hell, *Appl. Phys. Lett.* **73,** 1769 (1998).

[81] T. French, P. T. C. So, D. J. Weaver, T. Coelho-Sampaio, E. Gratton, E. W. Voss, and J. Carrero, *J. Microsc.* **185,** 339 (1997).

[82] P. T. C. So, W. N. Yu, K. Berland, C. Y. Dong, and E. Gratton, *Bioimaging* **3,** 1 (1995).

[83] A. Squire, P. J. Verveer, and P. I. H. Bastiaens, *J. Microsc.* **197,** 136 (2000).

[84] X. F. Wang, T. Uchida, D. M. Coleman, and S. Minami, *Appl. Spectrosc.* **45,** 360 (1991).

[85] R. M. Clegg, G. Marriott, B. A. Feddersen, E. Gratton, and T. M. Jovin, *Biophys. J.* **57,** A375 (1990).

In our estimation, the most versatile and efficient implementation of wide-field FLIM is by phase-modulation techniques based on homodyne detection (Fig. 6). Such a system is relatively easy to implement in terms of both data acquisition and processing and is very efficient in its utilization of the fluorescent emission. The method is based on the sinusoidal modulation of the excitation at a fixed frequency (f_0) and measurement of the steady state intensity of the emission with a two-dimensional (CCD) detector, the gain of which is modulated at f_0 but with a variable phase shift relative to the excitation. A series of images is recorded for different relative phase between the modulated illumination and detection, resulting in a stack of N images designated as $g(x, y, n)$, where x and y represent the pixel position in the image and n is the index of the N evenly spaced phase samples. The data are Fourier transformed over n and both the relative modulation depth, m, and the phase shift, ϕ (not to be confused with the rotational correlation time introduced above), of the emitted light are computed for every pixel in the sample image from the discrete Fourier coefficients $G(x, y, f)$, where f is the frequency [Eqs. (1)–(3), in which correction factors for instrumental phase delay and demodulation, derived from a calibrated reference compound,[86] have been omitted].

$$G(x, y, f) = \sum_{n=0}^{N-1} g(x, y, n) e^{2\pi ni/N} \tag{1}$$

$$m_{\text{samp}}(x, y) = \frac{|G(x, y, f_0)|}{|G(x, y, 0)|} \tag{2}$$

$$\phi_{\text{samp}}(x, y) = \tan^{-1}\left(\frac{\text{Im}[Gx, y, f_0]}{\text{Re}[G(x, y, f_0)]}\right) \tag{3}$$

From these parameters, two apparent lifetimes are computed: a phase lifetime (τ_ϕ) and a modulation lifetime (τ_m):

$$\tau_\phi = \frac{\tan\phi}{2\pi f_0} \qquad \tau_m = \frac{\sqrt{m^{-2} - 1}}{2\pi f_0} \tag{4}$$

A frequency-domain fluorescence lifetime system may be constructed as an add-on to virtually any fluorescence microscope. The photocathode of a modulatable image intensifier is placed at the primary image plane and the signal formed at the output phosphor plate is relayed to the CCD camera via a lens system (e.g., tandem, back-to-back camera lenses) or an optical fiber bundle. The modulation signal is typically in the range of 1–300 MHz, generated by a computer-controlled programmable signal generator. A second signal generator is frequency locked to

[86] Q. S. Hanley, V. Subramaniam, D. J. Arndt-Jovin, and T. M. Jovin, *Cytometry* **43,** 248 (2001).

the first and used to modulate the excitation source, for example, a multiline argon ion laser or a mixed argon–krypton ion laser. Alternative light sources suitable for full-field FLIM include laser diodes, light-emitting diodes (LEDs), and lamps. These can be modulated directly, whereas continuous wave (CW) ion lasers require the use of an acousto-optic modulator (AOM; temperature sensitive) or an electro-optical modulator; these devices have distinct advantages and disadvantages. An AOM results in optical modulation at twice the driving frequency, and its zero-order diffracted output is selected with an iris and relayed to the microscope illumination port with a multimode optical fiber. The latter is usually agitated mechanically to scramble modes and reduce laser speckle. Further homogenization can be achieved with opal glass or light-shaping diffusers.

We have found it expeditious to define three types of FLIM experiments[86]: type I, in which the lifetime differences are *within* a single image; type II, in which the variation in lifetime is *between* images; and type III, in which lifetime heterogeneity is measured. These distinctions were made because it has been our experience that random and systematic errors between images are generally far greater than expected from the statistics computed within a single image. Random errors between lifetime images and within a single lifetime image are dominated by different factors. For example, within a single image, factors such as instrumental drift and temporal variations in laser intensity affect all pixels equally, a situation that does not apply between images.[86] Type I experiments require some knowledge of the system, that is, for distinguishing different regions in an image based on FRET efficiencies (see below). In type II FLIM experiments, replicate measurements are required and significance tests need to be performed on the set of image means rather than on the statistics of the means themselves.[86]

Spectrally Resolved Fluorescent Lifetime Imaging Microscopy (sFLIM)

The FLIM system described in Fig. 6 may be readily adapted for spectrally resolved measurements by placing an imaging spectrograph at the microscope image plane and the intensifier-camera assembly at the focal plane of the spectrograph (Fig. 6, inset). It is worth noting that the image plane of a typical fluorescence microscope rarely requires a spectrograph with an $f/\# < 10$. Certain compact prism spectrographs are useful in this application because they have low focal plane curvature and a high (~80%) throughput over the range of 400–800 nm. Data collection in sFLIM is essentially the same as for conventional FLIM measurements. A two-dimensional (λ vs τ) implementation of sFLIM has been achieved in a Programmable Array Microscope (PAM) by application of Hadamard encoding schemes.[87]

[87] Q. S. Hanley, D. J. Arndt-Jovin, and T. M. Jovin, *Appl. Spectrosc.* **56,** 155 (2002).

Static and Dynamic Measurements of Polarized Fluorescence Emission

Modifying a FLIM microscope for polarization-dependent measurements is relatively straightforward. Applied to FPs in cells, anisotropy FLIM (rFLIM) can confirm whether observed depolarizations are consistent with rotational diffusion models or whether other mechanisms must be invoked.

A fixed linear polarizer is introduced into the path of the epi-illuminator and fluorescence emission is detected through a second linear polarizer arranged alternatively in a parallel or perpendicular relation to the illumination polarizer. For determinations of static emission anisotropy, two images are acquired (with the parallel and perpendicular configurations), from which the anisotropy image is derived by computation. Corrections are required for the depolarization effects of large numerical aperture objectives, the differential transmittance of the parallel and perpendicular polarized emission components through the entire system, and possible birefringence in the optical components (objective, dichroic filter, etc.). In our experience, uncorrected anisotropies calculated from measurements in the microscope differ by as much as 20% from those obtained with a reference spectrofluorometer (e.g., a difference of 0.05 for an anisotropy of 0.25). True (corrected) anisotropies can be computed from the experimental ratios of the parallel and perpendicular emission components (R) by applying three correction factors, A, B, and G, according to the semiempirical relationship[88]

$$r = A\left(\frac{RG - 1}{RG + 2B}\right) \tag{5}$$

where G accounts for differences in system response for the two polarized emission components and is computed at each (ith) pixel in images acquired using a reference solution of known anisotropy (usually ~0),

$$G_i = \left(\frac{1 + 2\bar{r}_{\text{ref}}}{1 - \bar{r}_{\text{ref}}}\right)\frac{1}{R_i} \tag{6}$$

where $\bar{r}_{\text{ref}}$ is the steady state anisotropy of the reference solution measured in a spectrofluorometer and R_i is the DC intensity ratio ($I_{\parallel,i}/I_{\perp,i}$) of the polarized emission measured in the microscope.[46,89] We have determined these parameters for several objectives (Table III) using as references 2.5-μm fluorescent microspheres differing in anisotropy (calibrated independently in a spectrofluorometer). This procedure has the virtue of limiting the axial extent of the fluorescent sample and thereby eliminating interference from out-of-focus signals. The values obtained

[88] T. M. Jovin, *in* "Fluorescence Polarization and Energy Transfer: Theory and Application" (M. R. Melamed, P. F. Mullaney, and M. L. Mendelsohn, eds.), p. 137. John Wiley & Sons, New York, 1979.

[89] J. A. Dix and A. S. Verkman, *Biophys. J.* **57,** 231 (1990).

TABLE III
CORRECTION FACTORS FOR APERTURE DEPOLARIZATION IN MICROSCOPE[a]

Objective	NA	A	B	G
Nikon Plan Fluor ×20, air	0.5	0.81	0.94	1.07
Nikon Plan Fluor ×40, air	0.75	0.75	0.55	1.07
Nikon PlanApo ×60, water	1.2	0.94	0.69	1.07
Nikon Plan Apo ×100, oil	1.3	0.67	0.13	1.07

[a] Microscope: Nikon Eclipse E600. In the absence of aperture depolarization, $A = B = 1$. See Eqs. (5) and (6) in text.

from spectroscopic and microscopic anisotropies in this manner differ by $<3\%$. Alternative correction procedures may be needed depending on the applicable orientational distribution.

Dynamic rFLIM experiments are analogous to standard FLIM except that the phase angles and the amplitudes (AC value) of the sample are acquired for the two polarized emission components (Fig. 6). The difference between the two measured phases, and the ratio of the modulation amplitudes, are combined with the lifetime data so as to calculate apparent rotational correlation times and other parameters defining rotational diffusion (r_0, r_∞) and/or other depolarizing processes such as energy migration (emFRET).

Multicomponent Systems

In dealing with complex cellular systems, the resolution of an unknown number of lifetime components arising from multiple species (e.g., a system consisting of molecules undergoing association and thus exhibiting FRET and free molecules with no FRET) is a desirable but potentially difficult undertaking. Global analysis of single-frequency FLIM data with an *a priori* assumption of two spatially invariant lifetime species has been proposed as a method for deriving the population distribution of lifetime states within an image.[90] The authors have used this strategy for quantitating the fractional populations of molecules, assuming a strictly bimodal FRET distribution (quenched, unquenched). However, in the cellular environment, lifetimes and FRET efficiencies may vary locally in an arbitrary manner, thereby leading to erroneous interpretations based on inappropriate

[90] P. J. Verveer, A. Squire, and P. I. H. Bastiaens, *Biophys. J.* **78,** 2127 (2000).

assumptions. This point is made more transparent by the mathematical exposition presented below.

We assume the existence of two donor species, an unquenched species A, with a fractional concentration $1 - \alpha$ and a fluorescence lifetime τ_0, and a FRET-quenched species B with fractional concentration α and a relative fluorescence lifetime $\beta = \tau/\tau_0$ (range, 0–1). Thus, the FRET efficiency is given by $E = 1 - \beta$. A sinusoidally modulated excitation light source evokes a fluorescence emission given by the sum of two sinusoidal functions corresponding to species A and B, each with a characteristic phase lag $\Delta\Phi_i$ and modulation m_i relative to that the excitation driving function (m_{ex}). The normalized fluorescence emission is given by

$$\text{fluor}(t) = \frac{\alpha\beta[1 + m_{\text{ex}}m_{\text{B}}\sin(\omega t - \Delta\Phi_{\text{B}})] + (1-\alpha)[1 + m_{\text{ex}}m_{\text{A}}\sin(\omega t - \Delta\Phi_{\text{A}})]}{\alpha\beta + (1-\alpha)} \tag{7}$$

where $\Delta\Phi_{\text{A}} = \tan^{-1}[\omega\tau_0]$, $\Delta\Phi_{\text{B}} = \tan^{-1}[\beta\omega\tau_0]$, $m_{\text{A}} = [1 + (\omega\tau_0)^2]^{-1/2}$, $m_{\text{B}} = [1 + (\beta\omega\tau_0)^2]^{1/2}$, and $\omega = 2\pi f_0$.

Using the heterodyne or homodyne detection strategy, we obtain a composite phase lag $\Delta\Phi$ and modulation m, from which we derive an apparent phase lifetime τ_ϕ and apparent modulation lifetime τ_{m}. For the system in question, they are given as normalized (relative to the unquenched lifetime τ_0) quantities by

$$\frac{\tau_\phi}{\tau_0} = \frac{\tan\Delta\Phi}{\omega\tau_0} = \left[\frac{\dfrac{\alpha\beta^2}{1+(\beta\omega\tau_0)^2} + \dfrac{(1-\alpha)}{1+(\omega\tau_0)^2}}{\dfrac{\alpha\beta}{1+(\beta\omega\tau_0)^2} + \dfrac{(1-\alpha)}{1+(\omega\tau_0)^2}}\right] \tag{8}$$

$$\frac{\tau_{\text{m}}}{\tau_0} = \frac{\sqrt{m^{-2}-1}}{\omega\tau_0} = \frac{1}{\omega\tau_0} \times \left(\frac{[1-\alpha(1-\beta)]^2}{\left[\dfrac{\alpha\beta^2}{1+(\beta\omega\tau_0)^2} + \dfrac{(1-\alpha)}{1+(\omega\tau_0)^2}\right]^2 + \left[\dfrac{\alpha\beta}{1+(\beta\omega\tau_0)^2} + \dfrac{(1-\alpha)}{1+(\omega\tau_0)^2}\right]^2} - 1\right)^{1/2} \tag{9}$$

These expressions are functions of the normalized variables α, β (or equivalently E), and $\omega\tau_0$. They can be inverted so as to obtain α, and β as functions of τ_ϕ/τ_0 and τ_{m}/τ_0 (and $\omega\tau_0$); the equations are involved and are not given here. However,

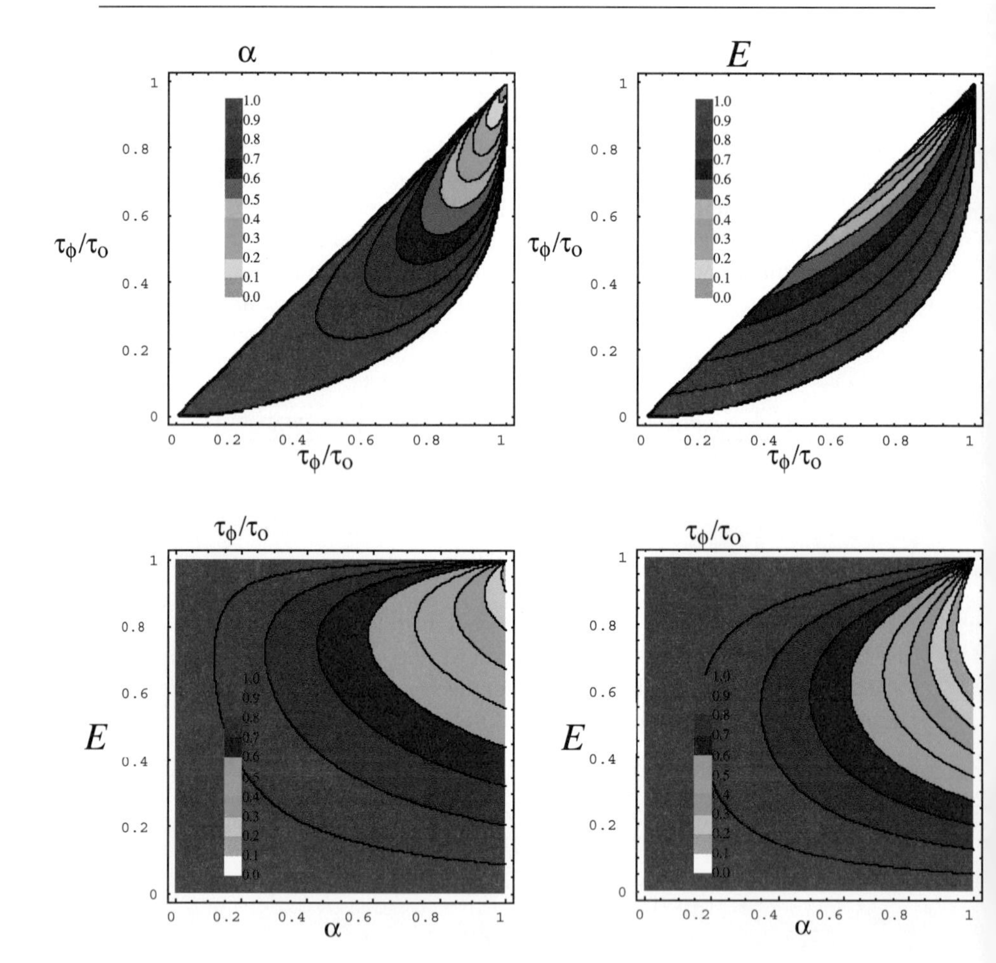

FIG. 7. Parameter correlations in phase-modulation FLIM applied to FRET analysis. Model system of two components, one unquenched with fractional concentration $1-\alpha$, and the second quenched to an extent given by the FRET efficiency E. Dependence of each parameter (α, E, τ_ϕ/τ_0, and τ_m/τ_0) on two other parameters, represented as a 2-D histogram. *Insets:* Color tables showing discrete intervals of ranges of the represented quantity. See text for details and equations. In these simulations, $\omega\tau_0 = 2$.

all four functions can be represented in convenient contour plots (a representative set of which is shown in Fig. 7; E is depicted rather than β), from which several conclusions can be derived: (1) all combinations of feasible (0–1) values of α and E yield a corresponding pair of unique values for τ_ϕ/τ_0 and τ_m/τ_0; (2) the paired values of τ_ϕ/τ_0 and τ_m/τ_0 that correspond to feasible values of α and E, are restricted to certain regions; (3) τ_ϕ/τ_0 is always $<\tau_m/\tau_0$; (4) a given value of τ_ϕ/τ_0

can generally be paired with more than one value of τ_m/τ_0. The converse is also true. The corresponding values of α and E can vary greatly; (5) the value of α can vary greatly, depending on E. Thus, in a heterogeneous population (varying E), the interpretation of given values of τ_ϕ/τ_0 and τ_m/τ_0 can be ambiguous; and (6) the magnitude and resolution of the various parameters depends greatly on the (normalized) modulation frequency. Multifrequency determinations[83] are almost essential for removing or at least reducing ambiguity of data interpretation in the case of complex systems.

Acknowledgments

The authors are indebted to numerous colleagues who have contributed to the concepts, instrumentation, and applications featured in this article. The generous supply of FP vectors from scientific sources and Clontech is gratefully acknowledged. The work was supported by the Max Planck Society. V.S. and A.H.A.C. were recipients of postgraduate fellowships from the Human Frontier Science Program.

[91] A. V. Terskikh, A. F. Fradkov, A. G. Zaraisky, A. V. Kajava, and B. Angres, *J. Biol. Chem.* **277,** 7633 (2002).

[92] R. Swaminathan, C. P. Hoang, and A. S. Verkman, *Biophys. J.* **72,** 1900 (1997).

[93] M. A. Uskova, J.-W. Borst, M. A. Hink, A. van Hoek, A. Schots, N. L. Klyachko, and A. J. W. G. Visser, *Biophys. Chem.* **87,** 73 (2000).

[94] A. A. Heikal, S. T. Hess, and W. W. Webb, *Chem. Phys.* **274,** 37 (2001).

[95] M. J. Dayel, E. F. Horn, and A. S. Verkman, *Biophys. J.* **76,** 2843 (1999).

[96] A. Partikian, B. Olveczky, R. Swaminathan, Y. X. Li, and A. S. Verkman, *J. Cell Biol.* **140,** 821 (1998).

[97] R. Pepperkok, A. Squire, S. Geley, and P. I. Bastiaens, *Curr. Biol.* **9,** 269 (1999).

[98] R. A. G. Cinelli, A. Ferrari, V. Pellegrini, A. Signorelli, M. Tyagi, M. Giacca, and F. Beltram, *Austr. J. Chem.* **54,** 107 (2001).

[99] A. G. Harpur, F. S. Wouters, and P. I. Bastiaens, *Nat. Biotechnol.* **19,** 167 (2001).

[100] P. J. Verveer, A. Squire, and P. I. H. Bastiaens, *J. Microsc.* **202,** 451 (2001).

[101] S. Jakobs, V. Subramaniam, A. Schonle, T. M. Jovin, and S. W. Hell, *FEBS Lett.* **479,** 131 (2000).

[102] G. Jung, S. Mais, A. Zumbusch, and C. Brauschle, *J. Phys. Chem.* **104,** 873 (2000).

[103] A. Squire and P. I. H. Bastiaens, *J. Microsc.* **193,** 36 (1999).

[104] J. Widengren, U. Mets, and R. Rigler, *Chem. Phys.* **250,** 171 (1999); F. S. Wouters and P. I. Bastiaens, *Curr. Biol.* **9,** 1127 (1999).

[7] Development of Genetically Encoded Fluorescent Indicators for Calcium

By ATSUSHI MIYAWAKI, HIDEAKI MIZUNO, TAKEHARU NAGAI, and ASAKO SAWANO

Introduction

Green fluorescent protein (GFP)-based fluorescent indicators for Ca^{2+} offer significant promise for monitoring Ca^{2+} in previously unexplored organisms, tissues, organelles, and submicroscopic environments because they are genetically encoded, function without cofactors, can be targeted to any intracellular location, and are bright enough for single-cell imaging.[1–3] Two general approaches have been followed to develop these indicators.

1. Fluorescence resonance energy transfer (FRET) between two GFPs of different colors[4–9] is highly sensitive to the relative orientation and distance between the two fluorophores. FRET is amenable to emission ratioing, which is more quantitative than single-wavelength monitoring, and is also an ideal readout for fast imaging using laser-scanning confocal microscopy. Cameleons are chimeric proteins composed of a short-wavelength mutant of GFP, calmodulin (CaM), a glycylglycine linker, the CaM-binding peptide of myosin light chain kinase (M13), and a long-wavelength mutant of GFP.[10,11] Ca^{2+} binding to CaM initiates an intramolecular interaction between CaM and M13,[12] which changes the chimeric protein from an extended to a more compact conformation, thereby increasing the efficiency of FRET from the shorter to the longer wavelength mutant GFP. We developed a red cameleon that contains a red fluorescent protein from a *Discosoma* species (DsRed; Clontech, Palo Alto, CA) as the FRET acceptor.[13]

[1] R. Y. Tsien, *Annu. Rev. Biochem.* **67,** 509 (1998).
[2] A. Miyawaki and R. Y. Tsien, *Methods Enzymol.* **327,** 472 (2000).
[3] R. Y. Tsien, *in* "Imaging Neurons" (R. Yuste, F. Lanni, and A. Konnerth, eds.), p. 55.1. Cold Spring Harbor Laboratory Press, New York, 1999.
[4] L. Stryer, *Annu. Rev. Biochem.* **47,** 819 (1978).
[5] B. Herman, *Methods Cell Biol.* **30,** 219 (1989).
[6] T. M. Jovin and D. J. Arndt-Jovin, *Annu. Rev. Biophys. Chem.* **18,** 271 (1989).
[7] R. Y. Tsien, B. J. Bacskai, and S. R. Adams, *Trends Cell Biol.* **3,** 242 (1993).
[8] P. I. Bastiaens and A. Squire, *Trends Cell Biol.* **9,** 48 (1999).
[9] R. Y. Tsien and A. Miyawaki, *Science* **280,** 1954 (1998).
[10] A. Miyawaki, J. Llopis, R. Heim, J. M. McCaffery, J. A. Adams, and R. Y. Tsien, *Nature* (*London*) **388,** 882 (1997).
[11] A. Miyawaki, O. Griesbeck, R. Heim, and R. Y. Tsien, *Proc. Natl. Acad. Sci. U.S.A.* **96,** 2135 (1999).
[12] T. Porumb, P. Yau, T. S. Harvey, and M. Ikura, *Protein Eng.* **7,** 109 (1994).
[13] H. Mizuno, A. Sawano, P. Eli, H. Hama, and A. Miyawaki, *Biochemistry* **40,** 2502 (2001).

0076-6879/03 $35.00

2. A second approach is to engineer single GFPs so that their fluorescence properties are sensitive to Ca^{2+}. To diminish the rigidity of the β-can structure of GFP, we generate a circularly permuted GFP (cpGFP), in which the amino and carboxyl portions have been interchanged and reconnected by a short spacer between the original termini.[14] The resulting new amino and carboxyl termini of cpGFP are fused to calmodulin and its target peptide, M13.[15] The chimeric protein, named pericam, is fluorescent and has spectral properties, which change reversibly with the Ca^{2+} concentration, probably because the interaction between calmodulin and M13 leads to an alteration in the environment surrounding the chromophore.

This article describes the design and construction of the red cameleons and pericams, as well as the potential uses and limitations of these two types of GFP-based Ca^{2+} indicators. The first section suggests factors that should be considered when designing or troubleshooting FRET experiments employing DsRed. The second section explains the rationale behind the use of cpGFPs, and details a method in use in our laboratory for optimization of these indicators.

Construction of Red-Shifted Cameleons: Genetically Encodable Indicators for Ca^{2+}

Complex Features of Red Fluorescent Protein from Discosoma: DsRed

The GFP from *Aequorea victoria* (*Aequorea* GFP) is now widely used in molecular and cellular biology studies.[1] Several other GFP-like fluorescent proteins have been isolated from fluorescent but nonbioluminescent Anthozoa species, especially coral.[16] Among them is a red emitter peaking at 583 nm, called DsRed (Clontech). Although the GFP-like fluorescent proteins share only 26–30% sequence identity with *Aequorea* GFP, they possess several features of GFP structure, including the 11-stranded "β-can" fold. Because DsRed has longer wavelengths of excitation and emission than are currently available from *Aequorea* GFP, it has attracted tremendous interest as a resonance energy transfer acceptor. However, complicated features in its spectra and structure[13,17] limit the usefulness of DsRed as a FRET acceptor: (1) its absorption spectrum is broad, with several shoulders and peaks other than the main peak at 558 nm. In addition, during maturation of the chromophore there is a green component peaking at 500 nm in the emission spectrum. DsRed would not be an ideal acceptor for FRET, if it is easily excited

[14] G. S. Baird, D. A. Zacharias, and R. Y. Tsien, *Proc. Natl. Acad. Sci. U.S.A.* **96,** 11241 (1999).

[15] T. Nagai, A. Sawano, P. E. Sun, and A. Miyawaki, *Proc. Natl. Acad. Sci. U.S.A.* **98,** 3197 (2001).

[16] M. V. Matz, A. F. Fradkov, Y. A. Labas, A. P. Savitsky, A. G. Zaraisky, M. L. Markelov, and S. A. Lukyanov, *Nat. Biotechnol.* **17,** 969 (1999).

[17] G. S. Baird, D. A. Zacharias, and R. Y. Tsien, *Proc. Natl. Acad. Sci. U.S.A.* **97,** 11984 (2000).

directly by the light that should excite a donor selectively; (2) its ability to fluoresce depends on the formation of a tetrameric complex, which may result in unwanted associations with proteins to which DsRed is covalently linked; (3) when DsRed is expressed in eukaryotic cells, a spotty pattern of red fluorescence is often observed, suggesting that the protein has a tendency to aggregate.

Despite these drawbacks, DsRed does have a number of favorable features, including (4) a high molar extinction coefficient and fluorescence quantum yield[17]: the fluorescence brightness of the fully matured protein is comparable to that of rhodamine dyes; (5) excellent resistance to pH extremes: both the molar extinction coefficient and fluorescence quantum yield are constant between pH 5 and 11. A pH-resistant FRET could be obtained if DsRed is combined with a donor GFP that has a pH-insensitive quantum yield[2]; (6) less susceptibility to photobleaching than GFP and yellow fluorescent protein (YFP); and (7) high thermostability: DsRed matures faster and more efficiently at 37° than at room temperature.[13]

Fluorescence Resonance Energy Transfer from Aequorea Green Fluorescent Protein Variants to DsRed

We have examined how well three *Aequorea* GFP variants, EYFP.1, enhanced cyan fluorescent protein (ECFP), and Sapphire, work as donors when paired with DsRed. EYFP.1 (same as EYFP-V68L/Q69K) is a less pH-sensitive YFP.[11] Sapphire (same as H9-90) is a GFP variant containing a mutation of Thr-203 to Ile, which results in stabilization of the neutral form of the chromophore.[1] Use of the three pairs, EYFP.1–DsRed, ECFP–DsRed, and Sapphire–DsRed, as well as a popular FRET pair, ECFP–EYFP.1, is discussed from both a theoretical and a practical standpoint. First, we suggest that DsRed is a desirable acceptor because of its superb resistance to pH and light and its excellent energy-absorbing ability. We then point out practical problems in the use of DsRed and, on the basis of the merits and demerits, discuss the performance of three red cameleons that contain DsRed as the acceptor.

Advantages of DsRed as Fluorescence Resonance Energy Transfer Acceptor

Intracellular pH (pH_i) changes dynamically in relation to cellular events such as Ca^{2+} mobilization. In hippocampal neurons, for example, glutamate and/or depolarization stimuli decrease the intracellular pH by 0.2–0.5. Effective use of FRET for studies in such environments requires that both the quantum yield of the donor and the molar extinction coefficient of the acceptor should be indifferent to physiological changes in pH.[2] Most, although not all, *Aequorea* GFP variants are quenched by acidic pH. It is unfortunate that CFP and YFP, frequently used as a donor and acceptor pair, have a quantum yield and a molar extinction coefficient, respectively, that are acid sensitive (Fig. 1A and C). When the pH decreases, these two negative effects reinforce each other, resulting in a loss of FRET between

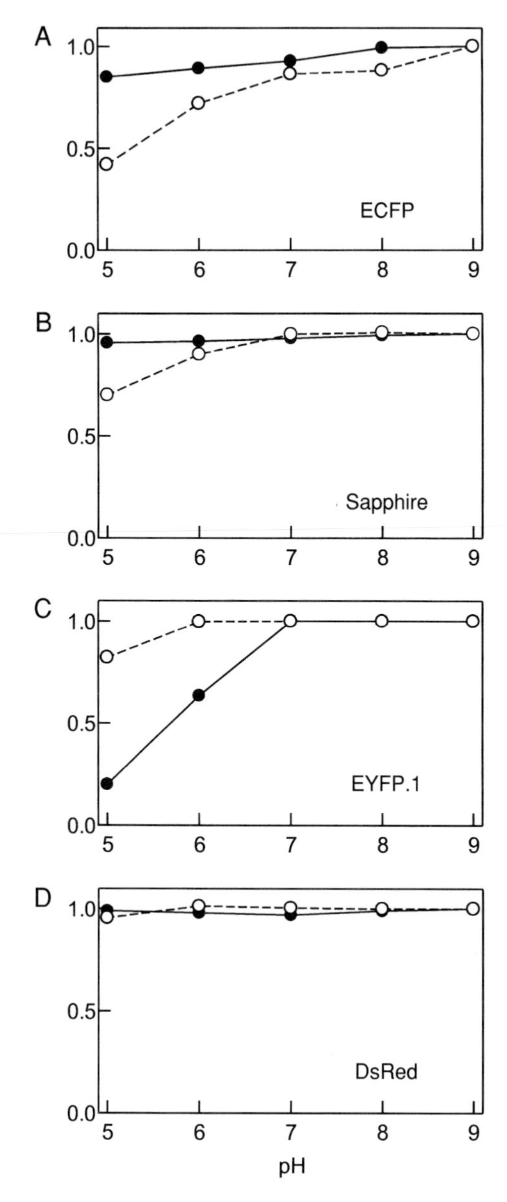

FIG. 1. pH dependency of normalized exctinction coefficients (closed circles connected with solid lines) and quantum yields (open circles connected dashed lines) of ECFP (A), Sapphire (B), EYFP.1 (C), and DsRed (D).

CFP and YFP. Below pH 6.5, for instance, the FRET from ECFP to EYFP.1 is significantly disturbed. DsRed makes a more desirable acceptor because it has a pH-resistant molar extinction coefficient (Fig. 1D). On the other hand, EYFP.1 makes an ideal donor because of its pH-resistant quantum yield (Fig. 1C). Although the quantum yields of Sapphire and ECFP are slightly pH sensitive, they are rather constant within a physiological pH range (Fig. 1A and B). Whereas the pH sensitivity of the molar extinction coefficient of the donor does not matter for pH-resistant FRET, a pH-resistant quantum yield of an acceptor would be preferable because it would allow us to measure the pH-resistant FRET efficiency by observing the ratio of donor and acceptor emissions. In this respect, DsRed also makes an ideal acceptor because its quantum yield is completely indifferent to changes in pH (Fig. 1D).[13,17]

The optical properties of DsRed have been reexamined, using the fully matured protein.[17] Although its molar extinction coefficient and fluorescence quantum yield were originally reported to be 22,500 M^{-1} cm^{-1} and 0.23,[16] one article has reported higher values of 75,000 M^{-1} cm^{-1} and 0.7, respectively.[17] Of particular importance is its outstanding ability to absorb energy. We have utilized highly matured DsRed prepared in bacteria incubated at 37°, and have obtained a slightly higher extinction coefficient of 78,000 M^{-1} cm^{-1}. Using this value, the R_0 (the distance at which FRET efficiency is 50%) values for the pairs EYFP.1–DsRed, ECFP–DsRed, and Sapphire–DsRed are calculated to be 6.1, 4.9, and 5.7 nm, respectively. It is therefore expected that DsRed can be used as an efficient acceptor for FRET. In our hands the R_0 value for the ECFP–EYFP.1 pair is 5.1 nm.

For quantitative FRET measurements, the photostability of the acceptor is important. If acceptor molecules are bleached more easily than donor molecules, the stoichiometry of donors to acceptors increases, resulting in a dilution of the FRET signal by the emission of donors that are accompanied by dead acceptors. In principle, photobleaching of donor molecules does not matter. The high photosensitivity of YFP such as EYFP.1, when used in combination with DsRed, would thus not be a problem for donor lifetime measurement. However, the ratio of acceptor to donor emissions is considerably affected by the bleaching of EYFP.1 owing to cross-excitation of DsRed (see below).

Factors That Perturb Measurements of Emission Ratio of DsRed to Aequorea Green Fluorescent Protein Variants

For FRET measurements, it is usual to monitor the ratio of acceptor-to-donor emissions or, more precisely, the ratio of the signal through the FRET channel to that through the donor channel.[2] These two ratios are not necessarily the same. Although the ratiometric measurement is easy to carry out and can cancel out variations in sample thickness, excitation intensity, and overall indicator concentration, the ratioing is affected by some optical factors. In our studies, selective excitation of the donor GFP variant over the acceptor DsRed was attempted, using a conventional epifluorescence microscope equipped with excitation filters 480DF10, 440DF20, and 400DF15. Normalized excitation and emission spectra of the donors and DsRed along with the wavelengths that pass through the excitation filter and two emission filters (donor and FRET channels) are shown in Fig. 2A–C. As a reference, the wavelengths that pass through the filters used for FRET between ECFP and EYFP.1 are shown with the spectra (Fig. 2D).

Generally, the emission spectrum rises steeply and falls gradually as a function of the wavelength, and the excitation spectrum is a mirror image of the emission spectrum. In each of these four FRET pairs, there is thus no leak of the acceptor emission into the donor channel. On the other hand, the donor emission does spill over into the FRET channel to a certain degree. Such cross-detection is most prominent for the ECFP–EYFP.1 pair, and is not negligible for the EYFP.1–DsRed pair (Fig. 2A). To diminish the relative contribution of the donor emission to the signal through the FRET channel, a long pass filter (565EFLP) was used to collect the emission of DsRed (Fig. 2A–C). When the FRET channel collects a significant quantity of the donor emission, it is not possible to obtain the true ratio of acceptor to donor emissions.

Direct excitation of the acceptor by the illumination light prevents our obtaining the FRET efficiency from the ratio of acceptor to donor emissions. Cross-excitation of the acceptor may be a serious problem when DsRed is used as the acceptor because its absorption spectrum is broad enough to include short wavelengths. The combination of EYFP.1 and DsRed suffers seriously from this problem. Illumination with light from a xenon lamp through a 480DF10 filter directly excites DsRed (Fig. 2A). Cross-excitation of DsRed also occurs with illumination through a 440DF20 filter for excitation of ECFP (Fig. 2B). However, it should be noted that DsRed is not excited by light of about 400 nm, which is the most effective wavelength for exciting Sapphire. Thus, the pairing of Sapphire–DsRed is free from significant cross-excitation (Fig. 2C).

The signal through the donor channel is the combined emission from both donor molecules that are involved in FRET and those that are not. The signal through the FRET channel consists of the emission from donors, the emission from acceptors that are directly excited, as well as the emission from acceptors that are given energy from donors. Therefore, the perturbation of the ratio value that we actually monitor depends on the relative sensitivity of the signal components to pH and light.

Improvements of Cameleons

The original version of cameleon had blue and green mutants (BFP and GFP) as the donor and acceptor, respectively.[10] Cameleons have subsequently been improved to be (1) better expressed in mammalian cells at 37°, (2) shifted to

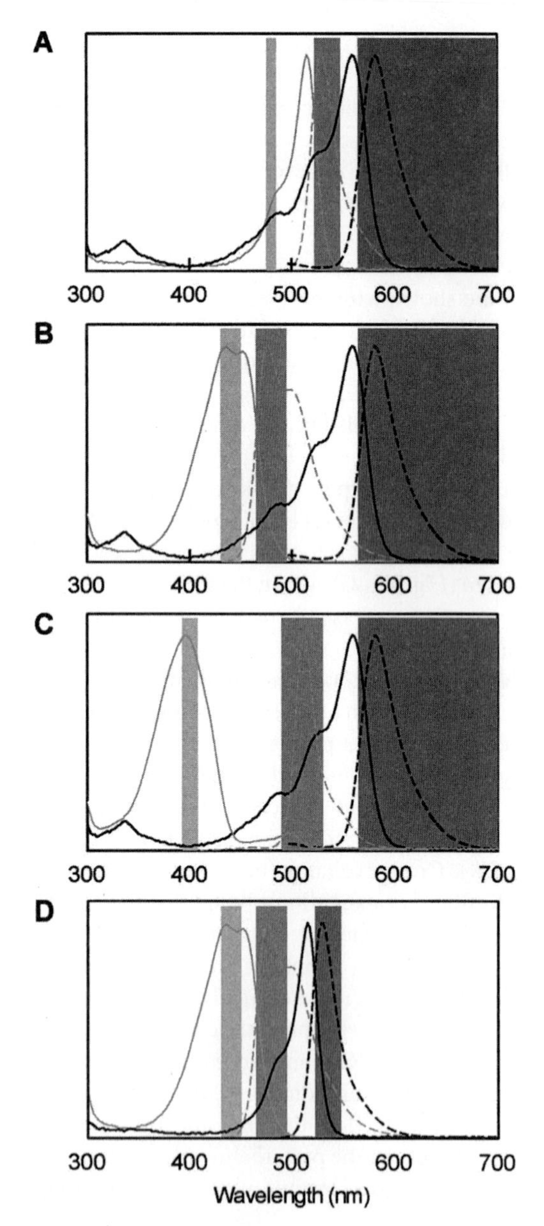

FIG. 2. Normalized excitation (solid lines) and emission (dashed lines) spectra of donor (light lines) and acceptor (dark lines) for EYFP.1–DsRed (A), ECFP–DsRed (B), Sapphire–DsRed (C), and ECFP–EYFP.1 (D) pairs. The pass bands of the excitation filter and two emission filters (donor channel and FRET channel) are indicated by light, medium, and dark gray boxes, respectively. They are 480DF10/535DF25/565EFLP for the EYFP.1–DsRed pair (A), 440DF20/480DF30/565EFLP for the ECFP–DsRed pair (B), 400DF15/510WB40/565EFLP for the Sapphire–DsRed pair (C), and 440DF20/480DF30/535DF25 for the ECFP–EYFP.1 pair (D).

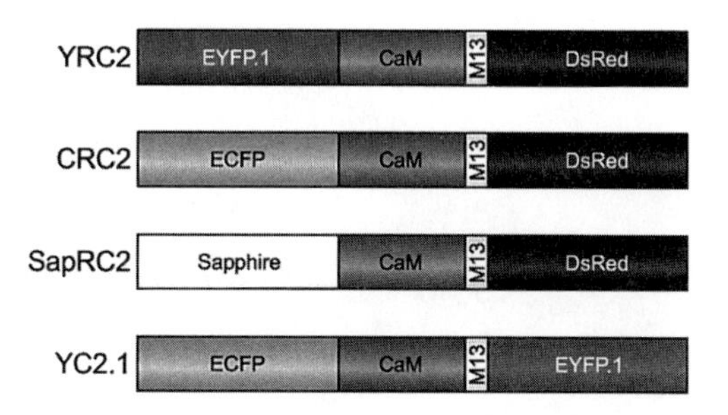

FIG. 3. Schematic structures of three red cameleons, YRC2, CRC2, and SapRC2, and of yellow cameleon-2.1.

longer wavelengths, and (3) resistant to acidic pH.[11] First, enhanced genes with mammalian codon usage and mutations for improved folding of the protein at 37° were introduced. Second, the blue mutant (BFP) proved to be the dimmest and most bleachable of the GFPs. It also required ultraviolet excitation, which is potentially injurious, results in a great deal of cellular autofluorescence, and could interfere with the use of caged compounds. Therefore enhanced cyan and yellow mutants, ECFP and EYFP, were substituted for the enhanced blue and green mutants, respectively, to make yellow cameleons. Third, despite the considerable promise of yellow cameleons, they still have problems that need amelioration. One of these problems is that EYFP is quenched by acidification. This perturbed the signals of yellow cameleons, mimicking a decrease in [Ca^{2+}] when the cellular environment became more acidic. The pH sensitivity of yellow cameleons has been greatly reduced by introducing mutations V68L and Q69K into EYFP (EYFP.1).[11] The improved yellow cameleons, including yellow cameleon-2.1 (Fig. 3), permit Ca^{2+} measurements without perturbation by pH changes between pH 6.5 and 8.0.

Red Cameleons

Three red cameleons, which have DsRed as the acceptor, have been constructed (Fig. 3).[13] Yellow red cameleon-2 (YRC2), cyan red cameleon-2 (CRC2) and Sapphire red cameleon-2 (SapRC2) contain EYFP.1, ECFP, and Sapphire, respectively, as the donor. All the red cameleons carry intact CaMs, and thus exhibit a relatively high affinity for Ca^{2+} (K'_d, 0.2–0.4 μM; data not shown). One unfavorable biochemical feature of DsRed is its obligatory tetrameric structure. Our multiangle light scattering (MALS) analysis reveals a high molecular mass (approximately 300 kDa) for the red cameleons, indicating that they form homotetramers. Figure 4 shows an imaginary depiction of the structure of a red cameleon complex. Despite its oligomerization, red cameleon has been demonstrated to work as a Ca^{2+} indicator.[13]

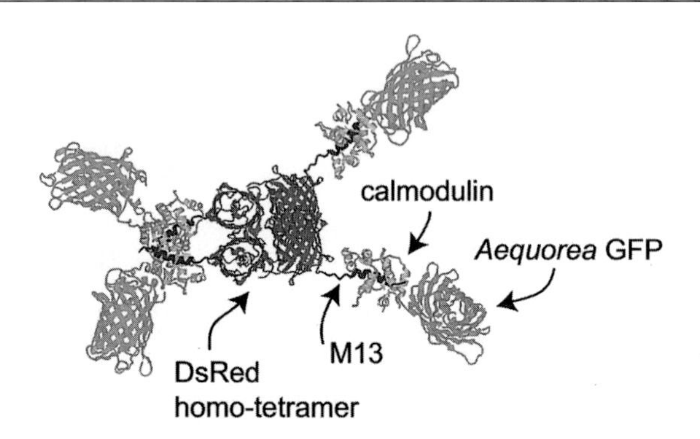

FIG. 4. A homotetrameric complex of Ca^{2+}-saturated red cameleon. Structures of the *Aequorea* GFPs and homotetrameric DsRed are derived from crystallography. Structure of the Ca^{2+}-calmodulin–M13 complex is from NMR. However, the relative distance and orientation between the *Aequorea* GFP and DsRed are unknown.

YRC2 and SapRC2 have been expressed in dissociated hippocampal neurons, in which not only an elevation of $[Ca^{2+}]_c$ but also a decrease in intracellular pH occur as a result of glutamate and/or depolarization stimuli. Figure 5A shows the spontaneous oscillation in $[Ca^{2+}]_c$ observed in a neuron expressing YRC2. The neuron is connected with several other neurons, and oscillatory changes in $[Ca^{2+}]_c$ occur synchronously in all the neurons. The oscillations are augmented when the neurons are stimulated with 1 μM glutamate and suppressed by the application of 1 μM tetrodotoxin. The gradual increase in the ratio before the stimulation may result from the vulnerability of EYFP.1 to light. Also a considerable drop in the baseline of the donor signal (the fluorescence of EYFP.1) is observed after glutamate stimulation (Fig. 5B), indicating that only EYFP.1 is quenched by acidification. These perturbations are explained by the differential sensitivity of EYFP.1 and DsRed to pH and light, and by a significant amount of cross-excitation of DsRed. Transient changes in $[Ca^{2+}]_c$ induced by depolarization are compared in YRC2-expressing and SapRC2-expressing neurons (Fig. 6).[13] The application of 20 mM KCl decreases the intracellular pH by 0.2–0.4. In addition to the Ca^{2+} response, the fluorescence of EYFP.1 (the donor in YRC2) is suppressed for more than 300 sec (Fig. 6B), whereas little long-term effect on the signal through the FRET channel is observed. Therefore, the decay in the $[Ca^{2+}]_c$ transient appears to be slow (Fig. 6A). In contrast, the fluorescent signals through the donor and the FRET channel change reciprocally in a neuron expressing SapRC2 (Fig. 6D). Consequently, the emission ratio quickly returns to the basal level (Fig. 6C). SapRC2 seems to be tolerant of acidosis, which is consistent with the fact that both Sapphire and DsRed are practically indifferent to pH changes.[13]

However, SapRC2 has been found to have an enigmatic pH sensitivity. Recombinant SapRC2 protein is prepared in bacteria, and its pH sensitivity is investigated

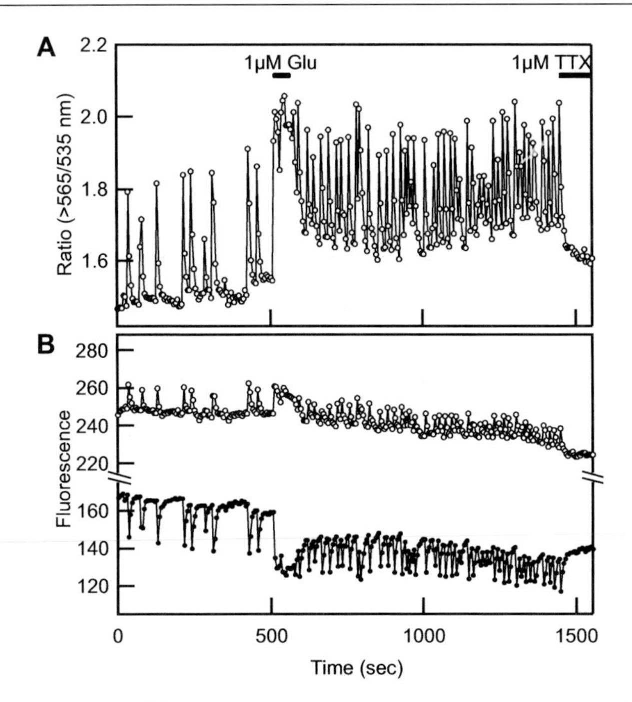

FIG. 5. Oscillations in $[Ca^{2+}]_c$ observed in a hippocampal neuron expressing YRC2. (A) Ratio of >565-nm to 535-nm emissions; (B) >565-nm (open circles) and 535-nm (closed circles) emissions. Perfusate was changed to buffer containing 1 μM glutamate (Glu) or 1 μM tetrodoxin (TTX) during the time indicated in (A). Reprinted with permission from H. Mizuno, A. Sawano, P. Eli, H. Hama, and A. Miyawaki, *Biochemistry* **40,** 2502 (2001). Copyright © 2001 American Chemical Society.

by *in vitro* experiments. Contrary to our expectations, the emission ratio of DsRed to Sapphire decreases as a function of pH in both the presence and absence of Ca^{2+} (Fig. 7A). Similar pH sensitivity is observed for Ca^{2+}-saturated SapRC2 (Fig. 7B), YRC2, and CRC2 (data not shown) by *in situ* pH titration experiments. At present we have no plausible explanation for this pH sensitivity. Because lowering the pH decreases the emission of the donor *Aequorea* GFP variants while increasing that of DsRed (data not shown), it is likely that there is a pH-dependent change in the proximity and relative angular orientation between DsRed and the *Aequorea* GFPs. This may be due to an alteration in some electrostatic interaction between the two fluorescent proteins from different species. This should be a caveat to those interested in using DsRed as a partner for FRET. Checking and clamping of ambient pH is still desirable to prevent artifacts. Aside from the pH sensitivity, Ca^{2+} measurements using SapRC2 appear to be free of other optical problems; there is no cross-excitation of DsRed and a negligible amount of cross-detection (in the detection of Sapphire emission in the FRET channel). As a whole, therefore, SapRC2 is the most reliable Ca^{2+} indicator among the three red cameleons.

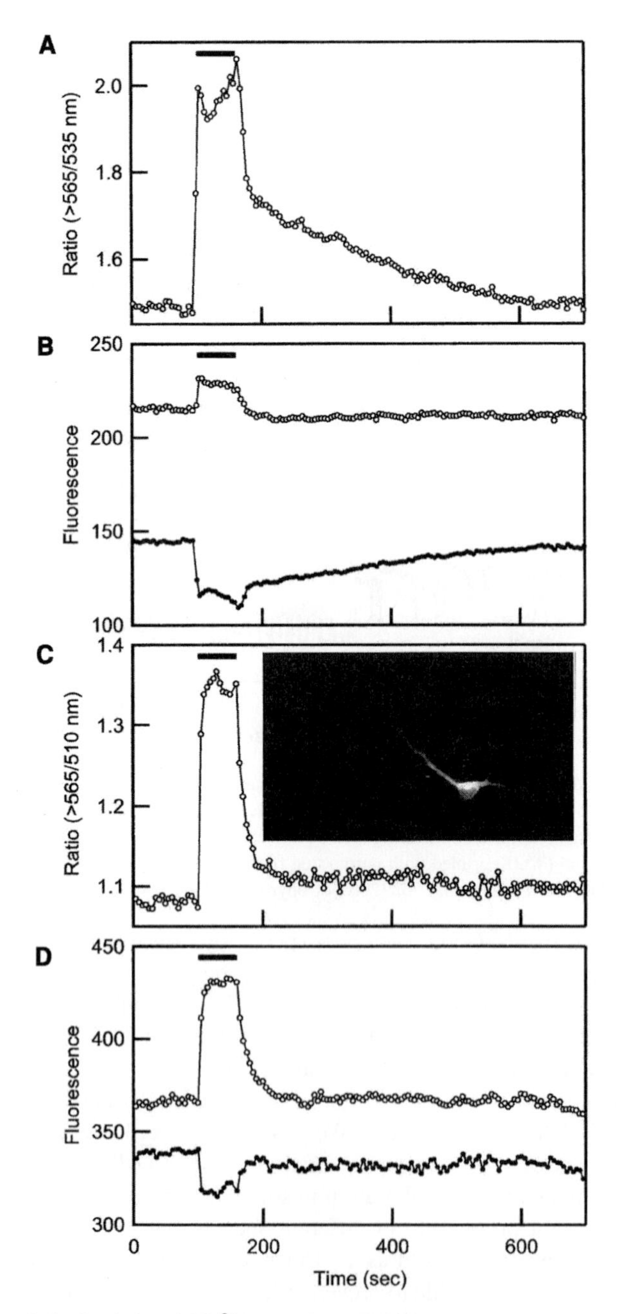

FIG. 6. Depolarization-induced $[Ca^{2+}]_c$ transients in hippocampal neurons expressing YRC2 (A and B) and SapRC2 (C and D). *Inset* in (C) shows a fluorescence image of the neuron expressing SapRC2 acquired through the donor channel. (A) Ratio of >565-nm to 535-nm emissions. (C) Ratio of >565-nm to 510-nm emissions. (B and D) Individual emissions of acceptor (open circles) and donor (closed circles). Perfusate was changed to the buffer containing 20 m*M* KCl during the time indicated with bars. Reprinted with permission from Ref. 13.

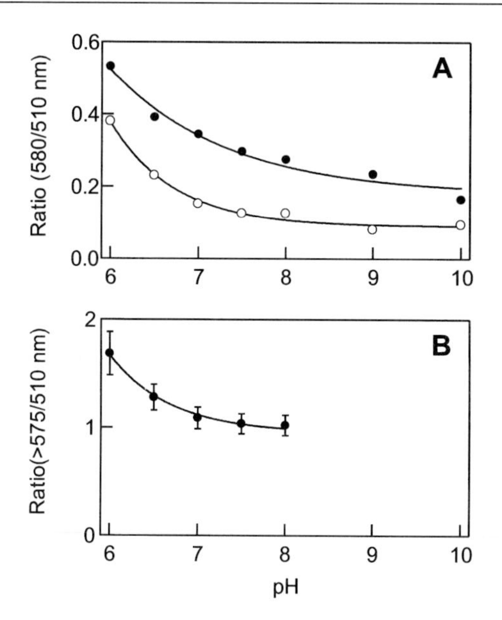

FIG. 7. pH dependency of the ratio of DsRed (acceptor) to Sapphire (donor) emissions from SapRC2 *in vitro* (A) and *in vivo* (B). (A) Emission spectra of the bacterially expressed SapRC2 were measured in the presence of 100 μM EGTA (open circles) and then 100 μM EGTA plus 1 mM $CaCl_2$ (closed circles). (B) SapRC2 expressed in HeLa cells was saturated with Ca^{2+} by adding 1 μM ionomycin to the medium. The intracellular pH was then varied with different pH buffers containing 20 μM nigericin and 20 μM monensin.

Confocal Imaging of $[Ca^{2+}]_c$ *Using SapRC2*

The fluorescence signal from SapRC2 can be visualized with a confocal laser scanning microscope (CLSM). For selective excitation of Sapphire (the donor in SapRC2), a new blue laser diode emitting at 405 nm (Nichia, Tokyo, Japan) seems to work well (H.M. and A.M., unpublished results, 2001). We have also used a system that contains a femtosecond pulsed Ti:sapphire laser (Tsunami; Spectra-Physics, Mountain View, CA) for two-photon excitation microscopy. This system consists of an inverted microscope (IX70; Olympus, Melville, NY), a scanning unit (Fluoview FV500; Olympus), a prechirper unit to create negative dispersion using a prism pair for compensation of group velocity dispersion, a wavelength-tunable Ti:sapphire laser that is pumped by a 5-W green laser (Millennia; Spectra-Physics), and a direct detector box accommodating a beam splitter, two emission filters, and two photomultiplier tubes (PMTs). The laser light is attenuated with neutral density (ND) filters and passed through a dichroic mirror (DM650; Olympus). The objective lens used is a UPlanAPO 60 × W/IR (Olympus). The fluorescent emission is passed though an infrared (IR) cut filter

A650RIF; Olympus) and divided with a beam splitter (560DRLP; Omega Optical, Brattleboro, NY). Two emission filters, 510WB40 (Omega Optical) and 580DF30 (Omega Optical) are used to collect the emissions of Sapphire and DsRed, respectively. An outline of an imaging experiment using SapRC2 is as ws.

Plate HeLa cells on coverslips in a Petri dish and allow them to attach for Transfect cells in the dish with 1 μg of cDNA (the SapRC2 gene in pcDNA3; vitrogen, Carlsbad, CA) using Lipofectin (GIBCO-BRL, Gaithersburg, MD).

Between 2 and 4 days posttransfection, image the HeLa cells on an inverted microscope. Expose the cells to reagents at room temperature in Hanks' balanced solution (HBSS) containing 1.26 mM $CaCl_2$.

Define several variables for image acquisition. They include (a) wavelength intensity of the laser, (b) scanning speed, (c) interval of image sampling, and oltage of the PMT. In our experience, the intensity of the laser should be attenuated to 3–6% using ND filters for Ca^{2+} imaging on a time scale of sub-seconds. Cells expressing SapRC2 are excited at 770 nm. The average power of output laser beam is 0.7–0.8 W at this wavelength. The wavelength is opti- with an emphasis on efficient excitation of Sapphire. The femtosecond-pulsed laser at short wavelength (λ < 800 nm) has proved to selectively the red fluorescence component of DsRed through a three-photon excita- mechanism.[18] Such bleaching does not happen with a sufficiently attenuated

Choose moderately bright cells in which the fluorescence is uniformly dis- uted in the cytosolic compartment but excluded from the nucleus, as would be xpected for a 300-kDa protein without targeting signals.

At the end of an experiment, convert the fluorescence signal into $[Ca^{2+}]_c$. R_{max} $_{min}$ can be obtained as follows. To saturate intracellular SapRC2 with Ca^{2+}, increase the extracellular $[Ca^{2+}]$ to 10–20 mM in the presence of 1–5 μM iono- Wait until the fluorescence intensity reaches a plateau. Then, to deplete the indicator of Ca^{2+}, wash the cells with Ca^{2+}-free medium (1 μM ionomycin, 1 mM and 5 mM $MgCl_2$ in nominally Ca^{2+}-free HBSS). The *in situ* calibration $^{2+}$] is based on the equation

$$[Ca^{2+}] = K'_d[(R - R_{min})/(R_{max} - R)]^{(1/n)}$$

K'_d is the apparent dissociation constant corresponding to the Ca^{2+} concen- at which R is midway between R_{max} and R_{min}, and n is the Hill coefficient. SapRC2, use $K'_d = 0.2$ μM and $n = 0.62$.

Marchant, G. E. Stutzmann, M. A. Leissring, F. M. LaFerla, and I. Parker, *Nat. Biotechnol.* **19,** (2001).

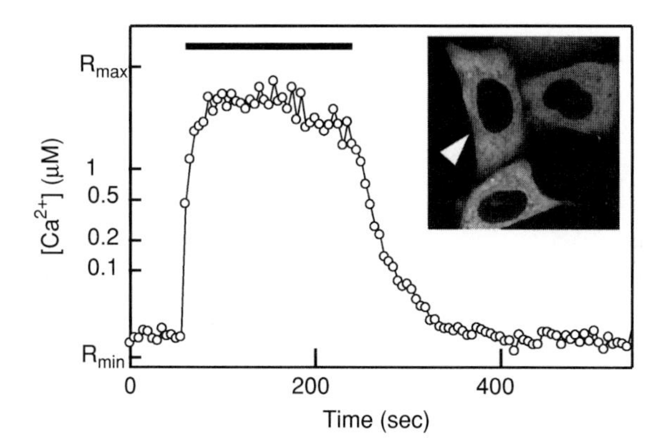

The use of two-photon excitation microscopy for observation of histamine-induced $[Ca^{2+}]_c$ HeLa cells expressing SapRC2. The trace was obtained from the cell indicated by a whead (*inset*). Perfusate was changed to buffer containing 10 μM histamine during the time with a bar. Calibrated $[Ca^{2+}]_c$ is shown on the right with R_{max} and R_{min} values. Reprinted permission from H. Mizuno, A. Sawano, P. Eli, H. Hama, and A. Miyawaki, *Biochemistry* **40,** (2001). Copyright © 2001 American Chemical Society.

Figure 8 shows a histamine (10 μM)-induced $[Ca^{2+}]_c$ transient in a HeLa xpressing SapRC2.

When incubated for longer than 4 days after transfection, a spotty fluo- pattern is observed in the cytosolic compartment, indicative of aggre- SapRC2. A new version of DsRed (DsRed2), released by Clontech, supposedly optimized for high solubility and low aggregation. In our hands, substitution of DsRed2 for the old DsRed does not improve the solubility SapRC2.

Construction of Circularly Permuted Green Fluorescent Protein Sensitive to Ca^{2+}

cularly Permuted Yellow Fluorescent Protein

Although cameleons utilize two fluorescent proteins of different colors, we eloped single GFPs sensitive to Ca^{2+} ions. Wild-type GFP (WT-GFP) bimodal absorption spectrum with two peak maxima, at 395 and 475 nm, corresponding to the protonated (neutral) and the deprotonated (anionic) states of chromophore, respectively.[1] The ionization state is modulated by a hydrogen network, comprising an intricate network of polar interactions between the chromophore and several surrounding amino acids. The chromophore of most ariants titrates with single pK_a values, indicating that the internal proton equilibrium is disrupted by external pH. One variant, the yellow fluorescent protein

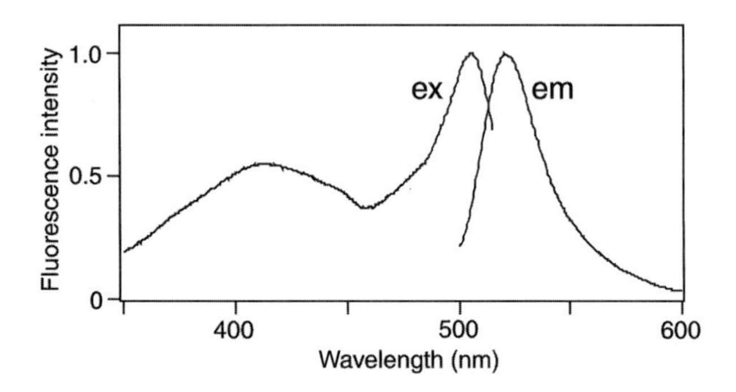

9. Fluorescence excitation (ex) and emission (em) spectra of cpEYFP.1, recorded at 530 and respectively. Spectra were normalized to a maximum value of 1.0. Reprinted with permission Nagai, A. Sawano, P. E. Sun, and A. Miyawaki, *Proc. Natl. Acad. Sci. U.S.A.* **98,** 3197 (2001). yright © 2001 National Academy of Sciences, U.S.A.

has a T203Y substitution that is responsible for the red-shift emission at nm. Ormö *et al.* predicted that the tyrosine introduced at position 203 would olved in a π-stacking interaction with the chromophore.[19] This idea was confirmed by X-ray crystallography.[20]

ithin the rigid "β-can" structure of GFP variants, Baird *et al.* found a site ould tolerate circular permutations—where two portions of the polypeptide ipped around the central site.[14] With obvious clefts in the β can, the chromophore of circularly permuted GFPs (cpGFPs) seems to be more accessible to protons from outside the protein. The cpGFPs might be used to convert changes interaction between two protein domains into a change in the electrostatic potential of the chromophore, in other words, to transduce information about the interaction into a fluorescent signal. A YFP variant, EYFP.1,[11] is subjected to circular permutation. The original N and C termini are fused via a pentapeptide GGSGG, and Y145 and N144 are made the new N and C termini, re- ely. The resulting chimeric protein is called cpEYFP.1. Figure 9 shows the xcitation and emission spectra of bacterially expressed cpEYFP.1. The excitation spectrum has two peaks at 417 and 506 nm, which is reminiscent of the bimodal excitation spectrum of WT-GFP. Unlike the protonated form of most other YFPs, the protonated form of cpEYFP.1 absorbing at about 420 nm fluoresces. The gene for cpEYFP.1 is cloned in-frame into the *Pst*I and *Kpn*I sites of pRSET$_B$ (Invitrogen), yielding cpEYFP.1/pRSET$_B$ (Fig. 10). The *Pst*I and *Kpn*I sites are chosen because

Ormö, A. B. Cubitt, K. Kallio, L. A. Gross, R. Y. Tsien, and S. J. Remington, *Science* **273,** 1392 (1996).

Wachter, M. A. Elsliger, K. Kallio, G. T. Hanson, and S. J. Remington, *Structure* **6,** 1267 (1998).

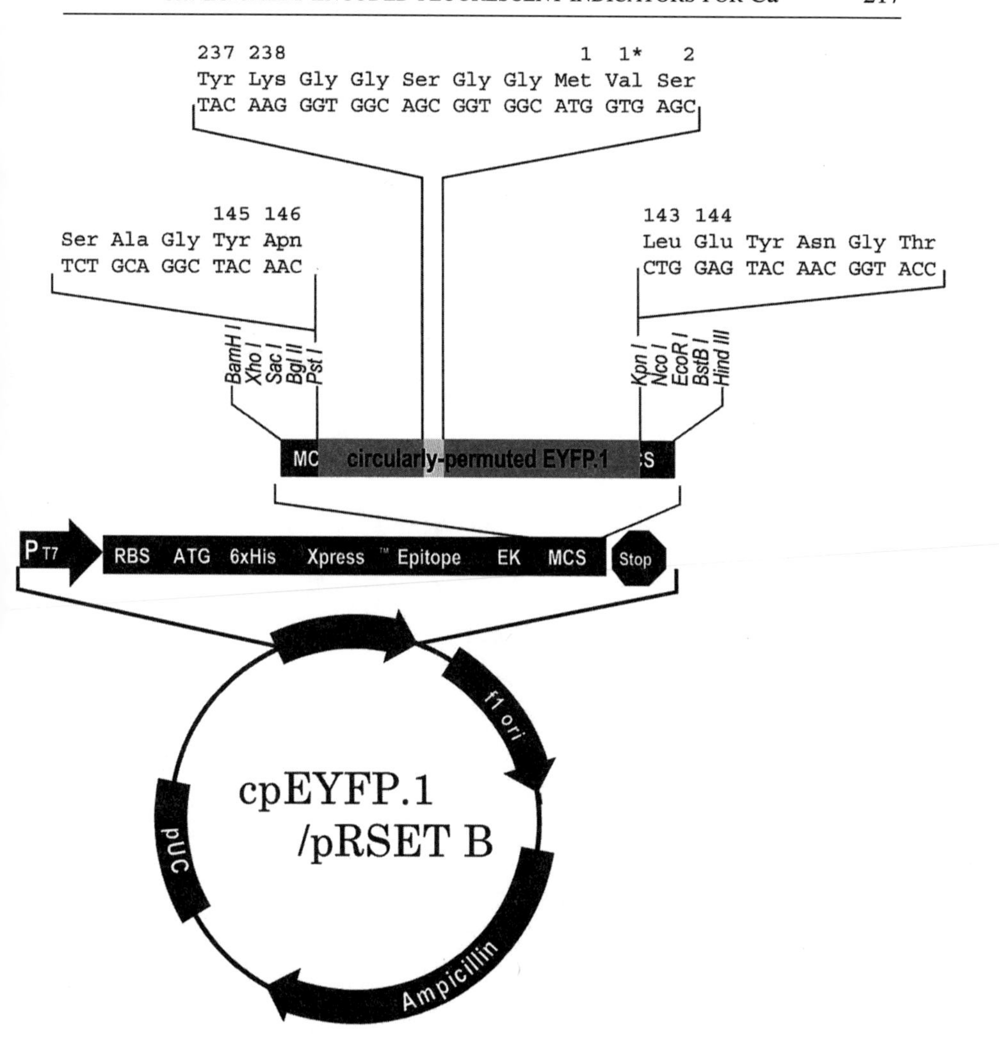

Map of plasmid cpEYFP.1/pRSET$_B$, showing the sequences of the boundaries between multicloning site and the cpEYFP.1 gene and as well as of the linker connecting the original N and EYFP.1.

located in the center of the multicloning site of the vector. Thus, there are *Bam*HI, *Sac*I, *Xho*I, *Bgl*II, and *Pst*I sites at the 5′ end of the cpEYFP.1 gene, as *Kpn*I, *Nco*I, *Eco*RI, *Bst*BI, and *Hin*dIII sites at the 3′ end of the gene. plasmid allows us to easily fuse any two proteins whose heteromerization is interest to Y145 and N144 of cpEYFP.1.

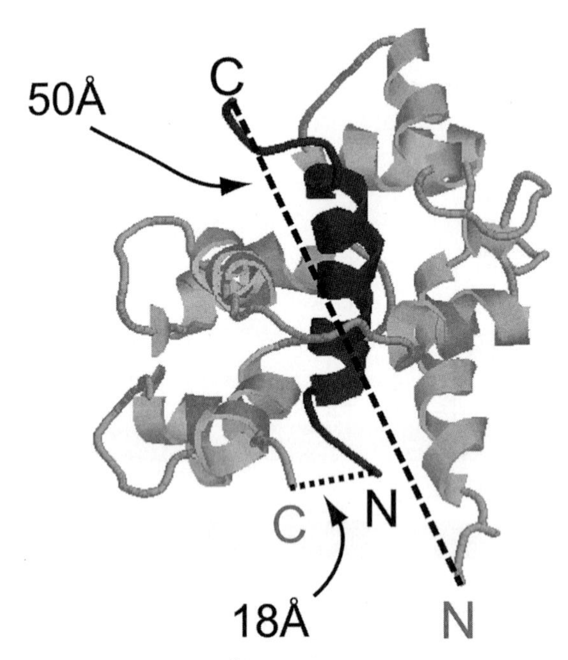

11. Structure of the complex of Ca^{2+}-bound CaM (light gray) and M13 (dark gray) derived NMR (Ref. 21). The N terminus of CaM and the C terminus of M13 are connected with a dashed N terminus of M13 and the C terminus of CaM are connected with a dotted line. Distances calculated using PDB.

Construction of Pericams

Figure 11 shows the nuclear magnetic resonance (NMR) structure of the com- Ca^{2+}-bound CaM and M13 in solution.[21] The C terminus of CaM and the terminus of M13 are relatively close (18 Å apart). To maintain the close prox- between Y145 and N144 of cpEYFP.1, we first connected the C terminus of and the N terminus of M13 to the N and C termini of cpEYFP.1, respectively. Although the resulting conservatively designed chimera is fluorescent, it does not ny response to Ca^{2+}. We next tried interchanging CaM and M13. In this construct, cpEYFP.1 is fused to the C terminus of M13 through a tripeptide linker and through a GTG linker to the N terminus of the E104Q CaM mutant 12). We have utilized a variant of CaM, in which the conserved bidentate glutamate at position 104 in the third Ca^{2+} binding loop has been changed to glutamine.[10] Because the N terminus of CaM and C terminus of M13 are rather apart (50 Å when the complex has formed), the β barrel of cpEYFP.1 might

Ikura, G. M. Glore, A. M. Gronenborn, G. Zhu, C. B. Clee, and A. Bax, *Science* **256,** 632 (1992).

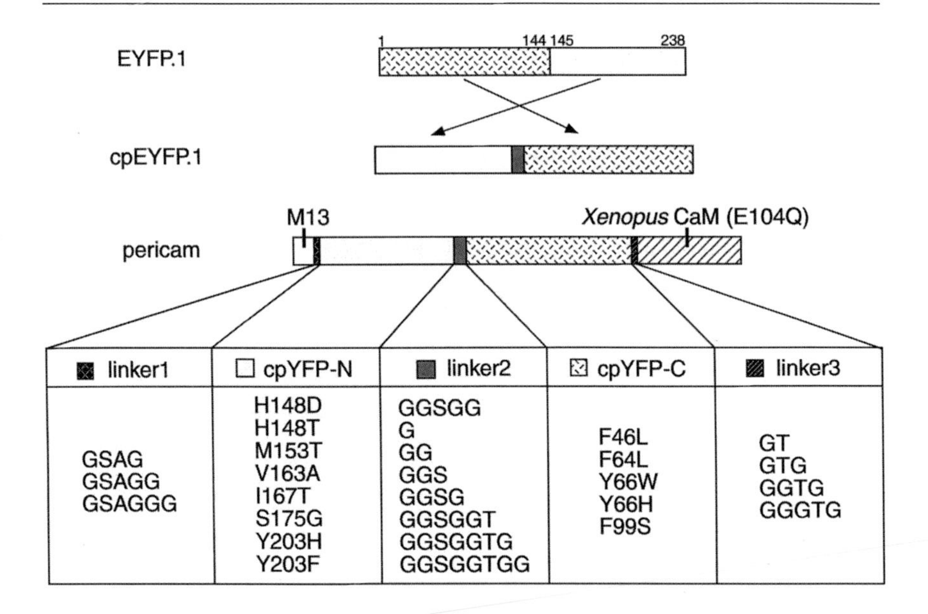

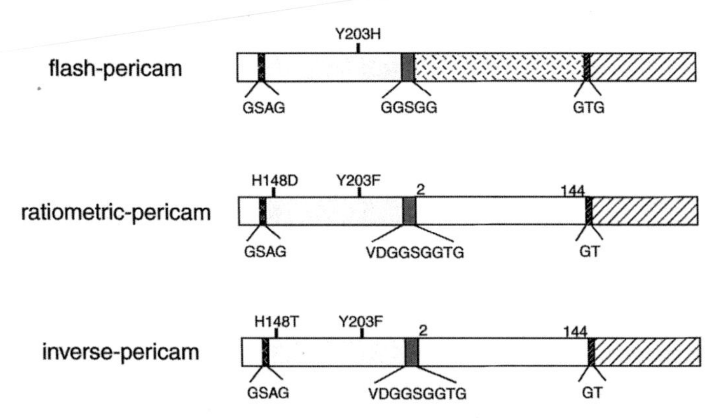

Strategy for the construction of pericams. On the basis of pericam, intensive optimization and amino acids was performed to create flash pericam, ratiometric pericam, and inverse The optimized amino acid sequences of linkers and amino acid substitutions for the three are shown below and above the bars, respectively.

considerably twisted. However, this radically designed chimeric protein is fluo- and, as we hoped, shows Ca^{2+} sensitivity. The protein, having a circularly EYFP.1 and a CaM is named "pericam." Also, the CaM and M13 pro- from cpEYFP.1 remind us of the bill of a pelican. When excited at 485 nm, -bound pericam shows an emission peak at 520 nm, three times brighter than $^{2+}$-free pericam (data not shown).

Pericam

obtain pericams with a larger dynamic range, we have optimized several acids involved in the proton coordinating network. Substitution of His-203 improves the dynamic range significantly. The new pericam, called "flash pericam" (Fig. 12), exhibits an eightfold increase in fluorescence in the presence $^{+}$ (Fig. 13D), suggesting it can be used as a single-wavelength indicator of]. In the absence of Ca^{2+}, flash pericam exhibits an absorbance spectrum to that of cpEYFP.1 (Fig. 13A, broken line). On saturation with Ca^{2+}, the 490-nm absorbance peak increases at the expense of the 400-nm peak (Fig. 13A, line), indicating that the association of Ca^{2+}–CaM with the M13 peptide deprotonates the chromophore, resulting in a leftward shift of the pH titration (Fig. 13G). Note that the Ca^{2+}-bound flash pericam with the ionized chromophore at pH 9 is about twice as bright as the Ca^{2+}-free pericam (at pH > 10). Therefore, the interaction between CaM and M13 might have direct steric effects chromophore that are separate from the pH effect, and change its ionization or reduce its out-of-plane distortions in a way that enhances radiationless The latter possibility is likely, because Ca^{2+} binding to flash pericam in- the quantum yield severalfold as well as the molar extinction coefficient 490 nm (data not shown). Figure 13G shows that Ca^{2+}-bound flash peri- alkaline quenched (pH > 10), suggesting that the incomplete β-can structure collapses.

Ratiometric Pericam

Most YFPs have a tyrosine or a histidine at position 203 and contain nonfluo- protonated species that absorb around 400 nm.[1] Similarly, fluorescence hardly detectable when flash pericam is excited around 400 nm. On the other replacing the amino acid at position 203 in YFP with phenylalanine makes protonated species fluorescent and allows it, on excitation at 400 nm, to give a predominant emission peak at 455 nm,[22] suggesting that proton transfer in xcited state is inhibited. The 455-nm emission indicates the existence of the protonated excited state species. Hoping to create a ratiometric Ca^{2+} indicator, we introduced Phe-203 in flash pericam. Again, the linkers and several amino acids to be critical for the optimization of protein folding and Ca^{2+} sensitivity. numerous constructs were tested, "ratiometric pericam" was derived from pericam by introducing the mutations H203F, H148D, and F46L; deleting glycine before CaM; and replacing the GGSGG linker between the original C termini with VDGGSGGTG (Fig. 12). As seen with flash pericam, Ca^{2+} binding promotes ionization of the chromophore in ratiometric pericam. Therefore, xhibits a Ca^{2+}-dependent change in the absorbance spectrum similar to flash

Dickson, A. B. Cubitt, R. Y. Tsien, and W. E. Moerner, *Nature* **388,** 355 (1997).

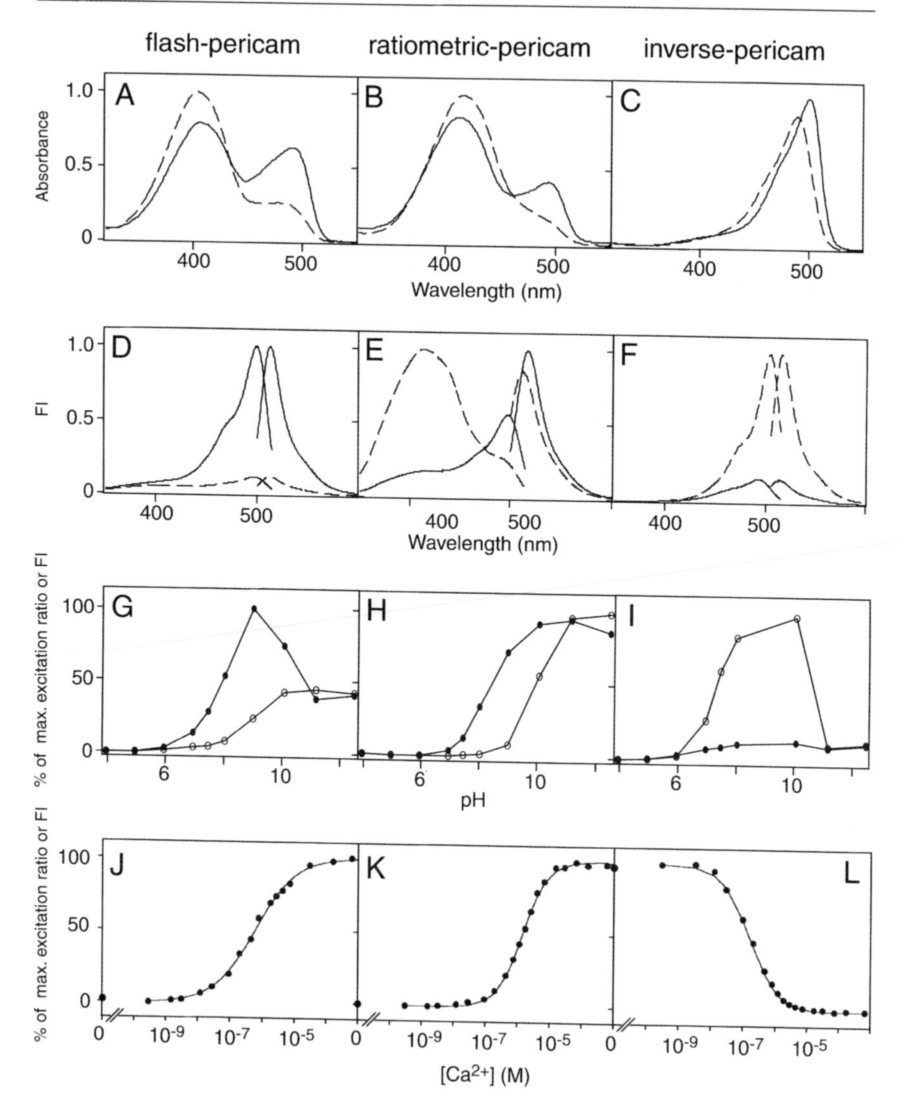

In vitro properties of flash pericam (A, D, G, and J), ratiometric pericam (B, E, H, and inverse pericam (C, F, I, and L). Absorbance (A–C) and fluorescence excitation and –F) spectra of pericams. (A–F) The spectra and data points were obtained in the presence or absence (broken line) of Ca^{2+}. (G–I) pH dependence of normalized amplitudes in the emission peak (G) and in the 516-nm emission peak (I) as well as the excitation ratio of (H) in the presence (closed circles) and absence (open circles) of Ca^{2+}. (J–L) Ca^{2+} titration pericams. FI, Fluorescence intensity. Reprinted with permission from T. Nagai, A. Sawano, and A. Miyawaki, *Proc. Natl. Acad. Sci. U.S.A.* **98,** 3197 (2001). Copyright © 2001 National Sciences, U.S.A.

pericam (Fig. 13B), and the pH titration curve is shifted leftward in the presence $^{+}$ (Fig. 13H). In contrast to flash pericam, however, ratiometric pericam bimodal excitation spectrum with peaks at 415 and 494 nm (Fig. 13E), the relative intensities of green fluorescence (511–517 nm) emitted when with 494- and 415-nm light are changed by about 10-fold between the -saturated and the Ca^{2+}-free forms (Fig. 13E). The excitation ratio (494/415) a monophasic Ca^{2+} dependence with an apparent dissociation constant (K'_d) μM and a Hill constant of 1.1 (Fig. 13K).

Pericam

During the semirandom mutagenesis of ratiometric pericam, we found an interesting protein with the substitution D148T, whose green fluorescence 515 nm) emitted when excited at 500 nm decreases to 15% in the presence $^{+}$ (Fig. 13F)—essentially the opposite of what happens with flash pericam. the protein has been named "inverse pericam" (Fig. 12). One possibility is the binding of Ca^{2+} to inverse pericam may have promoted protonation of chromophore. However, the following results indicate this is not the case. At 7.4, Ca^{2+} only red-shifts the peak of the absorbance spectrum from 490 to nm, with no change in the tiny hump around 400 nm (Fig. 13C); Ca^{2+} binding appears not to have affected the protonation state of the chromophore. Also, the Ca^{2+}-bound and Ca^{2+}-free inverse pericams are pH titrated with similar alues (Fig. 13I). In fact, we have found that the quantum yield is decreased Ca^{2+} binding (data not shown). These findings suggest that the change in uorescence intensity is mostly due to a direct effect of the Ca^{2+}-related structural on the chromophore.

Efficient Strategy for Site-Directed and Semirandom Mutagenesis

the above described studies, we demonstrated that it is possible to drama- change the Ca^{2+}-dependent behavior of pericam just by introducing subtle mutations in the amino acids close to the chromophore. Although there are some rationally introduced mutations, such as the substitution of phenylalanine for tyro- residue 203 (Y203F) to create ratiometric pericam, in most cases pericam been randomly mutated. The hydrogen bond network around the chromophore complex for the effects of mutations on the behavior of the chromophore to predicted. The Quik Change site-directed mutagenesis kit (Stratagene, La Jolla, a widely used method to achieve efficient site-directed mutagenesis. This method begins with a supercoiled, double-stranded DNA (dsDNA) plasmid con- the gene of interest as a template and two complementary synthetic oligonu- cleotide primers containing the desired mutation. The oligonucleotide primers are xtended during temperature cycling by the high-fidelity *Pfu Turbo* DNA poly- merase. The product is then treated with *Dpn*I endonuclease, which cuts only fully

hemimethylated 5′-GATC-3′ sequences in duplex DNA, resulting in the selec-
digestion of the template DNA. The *in vitro*-synthesized and nicked plasmid
containing the desired mutation, is then transformed into *Escherichia coli*.
this protocol is simple, rapid, and efficient, it suffers from the following
antages. (1) Mutations are introduced at only one site at a time. Introduc-
multiple mutations at different sites is time consuming because it requires
transformation and DNA preparation step between each consecutive round of
mutagenesis; (2) two complementary mutagenic oligonucleotide primers are
for each mutation site; and (3) the protocol does not allow random muta-
using degenerative primers.

new protocol, presented below, bypasses the above-described limitations
preserving the simplicity and efficiency of the original protocol.[23] Figure 14
overview of our protocol, which is composed of the following three steps:
polymerase chain reaction (PCR), (2) a *Dpn*I digestion, and (3) the synthesis
double-stranded plasmid DNA.

The PCR includes a thermostable DNA ligase to join the multiple extended
on the circular template in every cycle, generating circular single-stranded
(ssDNA) molecules that carry mutations. The mutant strands dominate as
molecules in the reaction mixture. A small fraction of the species anneals
sense strand of parental DNA, making a hemimethylated dsDNA. Although
shows two phosphorylated mutagenic primers on an antisense strand, more
primers on either strand can be utilized.

ild-type template DNA is eliminated by *Dpn*I digestion. The methylated
(wild-type plasmid DNA) and hemimethylated (a hybrid of the mutant strand and
wild-type antisense strand) dsDNAs are digested.

The nonmutant strands of some DNA fragments produced by *Dpn*I digestion
the ssDNA synthesized in the PCR and serve as megaprimers to complete
synthesis of DNA plasmids that are able to replicate in bacteria. This final re-
utilizes the *Pfu* polymerase, *Taq* DNA ligase, dNTPs, and NAD carried over
PCR. The versatility of the protocol has been proved in our mutagenesis
in which pericams have been improved and diversified. Mutations have
introduced simultaneously to optimize amino acids at multiple sites. Mutated
acids at residues 66, 148, 167, and 203 increase Ca^{2+} sensitivity whereas
residues 46, 64, 99, 147, 153, 163, and 175 allow better maturation and
(see Fig. 12).

Optimization of Amino Acids at Multiple Sites

provide an example, a protocol for inventing a ratiometric Ca^{2+} indicator
ash pericam as the starting material is described below. We know from

ano and A. Miyawaki, *Nucleic Acids Res.* **28,** e78 (2000).

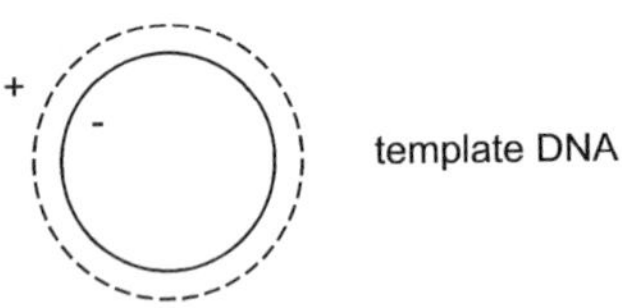

14. Schematic diagram of the multiple-site mutagenesis protocol. The two mutagenic primers designed on a minus strand. The methylated template DNA strands are represented by thin lines, *in vitro*-synthesized DNA strands are represented by thick lines. The plus and minus strands wn as dotted and solid lines, respectively. X, Introduced mutation; ⋙, the species on the left dominant in the reaction mixture. Reprinted from A. Sawano and A. Miyawaki, *Nucleic Acids Res.* (2000) by permission of Oxford University Press.

pericam that Phe-203 needs to be replaced with His. At the same time a mutation will be introduced at residue 148. As mentioned above, H148D turned out to be among the mutations required to make ratiometric pericam ash pericam.

Synthesize the antisense primers for Y203F and a random mutation at posi- They are 5′-GGCGGACTG**GAA**GCTCAGGTA-3′ and 5′-CATGATATA CGTT**NNN**GCTGTTGTA-3′, respectively (N = A or T or C or G).
Phosphorylate the primers at the 5′ end with T4 polynucleotide kinase.
Carry out a thermal cycling reaction in a 50-μl total volume, using 50 ng of plasmid DNA (pRSET$_B$/flash pericam), 14 pmol of each primer, dNTPs each), 2.5 U of cloned *Pfu* DNA polymerase (Stratagene) in 0.5× *Pfu* polymerase reaction buffer, and 20 U of *Taq* DNA ligase (New England BioLabs, MA) in 0.5× *Taq* DNA ligase buffer containing 50 nmol of NAD. The ycler is programmed as follows: preincubation at 65° for 5 min to allow to repair any nicks in the template; initial denaturation at 95° for 2 min; at 95° for 30 sec, 55° for 30 sec, and 65° for 7 min; and postincubation at min. The time at 65° is made relatively long so that the extended primers fully ligated.
Add 1 μl (20 U) of *Dpn*I (New England BioLabs) to the sample (50 μl), incubate at 37° for 1 hr.
Subject the sample (51 μl) to denaturation at 95° for 30 sec, followed by ycles at 95° for 30 sec, 55° for 1 min, and 70° for 7 min.
Use 2 μl of the final sample to transform competent *E. coli* cells [(JM109 by the Ca^{2+}-coprecipitation technique.

interesting approach to improve and diversify the fluorescence and bio- properties of pericam is to systematically test all known mutations that been identified as changing the properties of GFP. When the second-strand is primed by the standard sense primer (T7 primer in the case of pRSET$_B$), simultaneous mutation of up to eight different sites can be achieved efficiently Also, a decrease in the molar ratio of primers to the template DNA en- another semirandom mutagenesis approach in which at each position either mutation is introduced or the position is left unchanged.

[8] Development and Application of Caged Calcium

By GRAHAM C. R. ELLIS-DAVIES

Why Cage Calcium?

Changes in intracellular calcium (Ca) concentration control a myriad of vital physiological processes including muscle contraction, secretion, mitosis, channel gating, chemotaxis, and stomatal pore closure. In the late nineteenth century Ringer provided the first hint that Ca was important for muscle contraction, and Locke suggested a role for Ca in neurotransmission. In the 1950s and 1960s the pivotal role of calcium in these processes was firmly established, and definitively demonstrated in nerve terminals in 1973.[1]

Innovation in measurement techniques has often gone hand-in-hand with important discoveries in physiology and neurobiology. Historically, revolutions in pipette fabrication have had a profound effect on our understanding of electrical signaling in cells.[1–3] More recently, breakthroughs in optical and DNA-based technologies have provided an array of new tools with which to study cells.[3] Rapid changes in the concentration of intracellular constituents involved in signaling processes using solution exchange are impossible because of the diffusional delays associated with dialysis; however, use of caged compounds circumvents this problem, and provides many other additional advantages. "Caged" here designates that the biological signaling molecule has been rendered inert by covalent modification of its active functionality with a photoremovable protecting group.[4–6] Irradiation with a brief pulse of light frees the caged substrate. In a cellular context uncaging provides the means to activate the biological process controlled by the caged substrate. There are two major advantages to the use of such photorelease technology, compared with traditional rapid mixing techniques.

1. Kinetic resolution: Many physiological processes take place over a duration of 1–100 ms. Caged substrates can be photoreleased in the submillisecond time domain, and thus they can be used to activate these events; traditional rapid mixing

[1] J.-M. Jeng, *Nat. Rev. Neurosci.* **3,** 71 (2002).
[2] B. Sakmann and E. Neher, eds., "Single-Channel Recording," 2nd Ed. Plenum, New York, 1995.
[3] E. Neher, *Neuron* **20,** 389 (1998).
[4] S. R. Adams and R. Y. Tsien, *Annu. Rev. Physiol.* **55,** 755 (1993).
[5] J. M. Nerbonne, *Curr. Opin. Neurobiol.* **6,** 379 (1996).
[6] G. C. R. Ellis-Davies, *in* "Imaging Living Cells: A Laboratory Manual" (R. Yuste, F. Lanni, and A. Konnerth, eds.). Cold Spring Harbor Laboratory Press, Cold Spring Harbor, NY, 2000.

METHODS IN ENZYMOLOGY, VOL. 360
0076-6879/03 $35.00

techniques, on the other hand, do not provide true kinetic resolution for such processes.

2. Noninvasive release: The use of light to change the substrate concentration means that the release can be spatially defined, nonperturbing, and intracellular. (a) Spatially defined substrate release from the caged molecule occurs as it is released only where the light is incident. A corollary of this is that uniform illumination of a preparation produces a uniform concentration increase in the substrate, thus circumventing any problems that could be encountered owing to inhomogeneity resulting from mixing techniques. On the other hand, localized release can be realized by focusing the incident light to a precisely defined area; (b) physically nonperturbing concentration jumps are achieved by the very nature of photorelease technology (the wavelength of light is chosen to be nondestructive to proteins); (c) rapid *intracellular* effector release is a unique property of photorelease technology. The biologically inert caged compounds can be loaded into intracellular compartments by a variety of techniques, and concentration jumps can be effected in an environment that is otherwise inaccessible.

Caging calcium would provide the means to produce homogeneous changes in intracellular concentrations of calcium. In principle, these changes could be of a known amount, so that quantitative correlations between [Ca] and biological activity can be discovered. Ideally, such changes should also be extremely fast (submillisecond time domain), as it is known that Ca channel opening produces rapid fluctuations in cellular [Ca]. Photolysis of caged Ca should produce a step increase in calcium concentration, enabling the experimenter to "switch on" the Ca-dependent process under study, by "clamping" the [Ca] to a known, higher value.

Because it is now apparent that calcium is the most important and ubiquitous signaling molecule, the reasons to cage it are self-evident. Two groups have developed two different approaches to caging calcium. These are discussed, along with a representative selection of the hundreds of experiments that have been performed with these optical tools.

How to Cage Calcium

Almost all the compounds that have been caged are organic molecules, and so the photochemical masking group is attached directly to the molecule itself, blocking its biological activity.[4–6] The inorganic cation calcium cannot form such covalent bonds; therefore a new strategy had to be developed for caging this second messenger. To this end, photolabile derivatives of known high-affinity calcium chelators [*O*,*O*′-bis(2-aminophenyl)ethyleneglycol-*N*,*N*,*N*′,*N*′-tetraacetic acid (BAPTA), ethylenediaminetetraacetic acid (EDTA), and ethylene

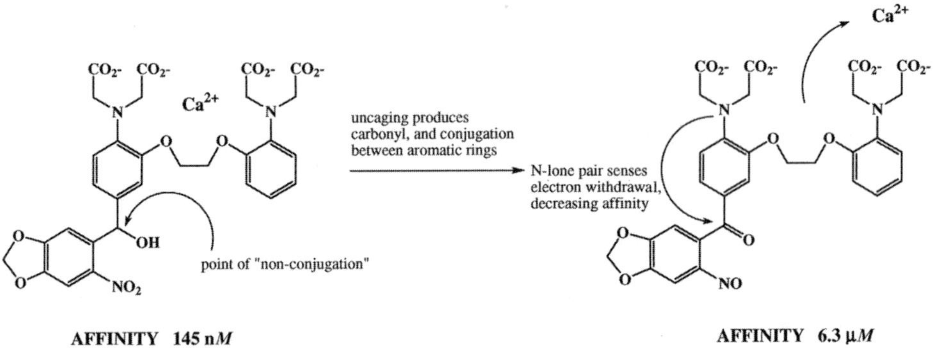

FIG. 1. Photochemical reaction of nitr-5.

glycol-bis(β-aminoethyl ether)-*N,N,N′,N′*-tetraacetic acid (EGTA)] have been synthesized.[7–13] These molecules decrease their affinity for calcium on irradiation, thus uncaging some of the bound calcium. Two different applications of this strategy have been developed, one based on photochemical modification of the buffering capacity of BAPTA derivatives, the other on photochemical scission of the backbone of either EDTA or EGTA.

BAPTA-Based Calcium Cages

Several photolabile derivatives of BAPTA have been synthesized.[7–9] The commercially available nitr-5 is representative of this approach.[8] Photochemical elimination of water changes the buffering capacity of BAPTA 54-fold (see Fig. 1). Creation of the benzylic ketone functionality introduces electron-withdrawing capacity into the BAPTA Ca coordination sphere, which is sensed by one of the nitrogen lone pairs, and thus the K_d of nitr-5 decreases from 145 n*M* to 6.3 μM.[8] The distinct advantages of this approach to (un)caging Ca are (1) the K_d for divalent cations of the cage is pH independent above ~6.5, as BAPTA derivatives are unprotonated in this range; and (2) the cage is Ca selective. The disadvantages are (1) the rate of reaction is 2500 s^{-1} (approximately equal to many Ca-dependent reactions); (2) the quantum yield of photolysis is 0.035 (Ca-loaded nitr-5) or 0.012 (Ca-free nitr-5), and thus the overall use of absorbed light is relatively

[7] R. Y. Tsien and R. S. Zucker, *Biophys. J.* **50,** 843 (1996).
[8] S. R. Adams, J. P. Y. Kao, G. Grynkiewicz, A. Minta, and R. Y. Tsien, *J. Am. Chem. Soc.* **110,** 3212 (1988).
[9] S. R. Adams, V. Lev-Ram, and R. Y. Tsien, *Chem. Biol.* **4,** 867 (1997).
[10] G. C. R. Ellis-Davies and J. H. Kaplan, *J. Org. Chem.* **53,** 1966 (1988).
[11] J. H. Kaplan and G. C. R. Ellis-Davies, *Proc. Natl. Acad. Sci. U.S.A.* **85,** 6571 (1988).
[12] G. C. R. Ellis-Davies and J. H. Kaplan, *Proc. Natl. Acad. Sci. U.S.A.* **91,** 187 (1994).
[13] G. C. R. Ellis-Davies, *Tetrahedron Lett.* **39,** 953 (1998).

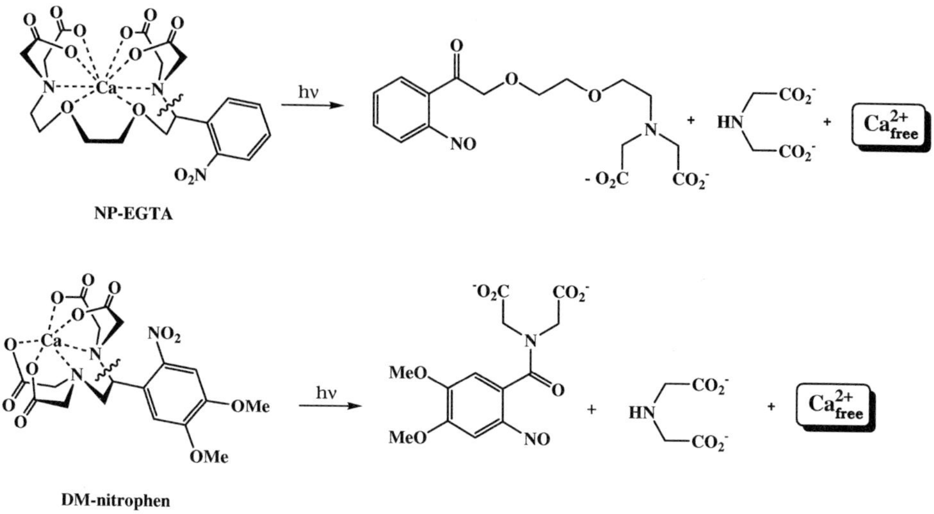

NP-EGTA

DM-nitrophen

FIG. 2. Photochemistry of NP-EGTA and DM-nitrophen.

inefficient; (3) the affinity for Ca before photolysis is 145 n*M*, and thus only a small fraction of the cage can be loaded with Ca (~40%) before $[Ca]_{free}$ reaches activating concentrations (see Fig. 3); and (4) the affinity for Ca changes 54-fold, and thus only a small fraction of the bound Ca is released on photolysis (~0.25%). Recently, azid-1 has been synthesized,[9] but this is not yet commercially available. In some respects this is a much better Ca cage than nitr-5, principally it does make *very* efficient use of incident light (it has a large extinction coefficient, and its quantum yield of photolysis is 1). Further, it undergoes a slightly larger (500-fold) decrease in affinity upon photolysis. However, it does have a lower prephotolysis affinity than nitr-5 for Ca of 230 n*M*, so that only a small fraction of the cage can be Ca loaded before activating levels of $[Ca]_{free}$ are reached.

EDTA- and EGTA-Based Calcium Cages

A wide range of photolabile EDTA, EGTA, and EGTA analogs have been synthesized.[10,12,13] The commercially available DM-nitrophen[10,11] and NP-EGTA[12] are representative of this approach. In contrast to the photochemistry of BAPTA-based Ca cages (Fig. 1), irradiation of DM-nitrophen and NP-EGTA *cuts a covalent bond in the chelator backbone.*[10,14] Uncaging results in severe disruption of the Ca coordination sphere, producing "hemichelators" of known, very low affinity for Ca (see Fig. 2). The advantages of this approach to (un)caging Ca are: (1) the rates

[14] G. C. R. Ellis-Davies, J. H. Kaplan, and R. J. Barsotti, *Biophys. J.* **70,** 1006 (1996).

TABLE I
PHYSICOCHEMICAL PROPERTIES OF CALCIUM CAGES AND PARENT CHELATORS

Source	K_d for Ca (nM)	K_d for products (mM)	Affinity change (x-fold)	K_d for Mg (mM)	Quantum yield	Extinction coefficient (M^{-1} cm^{-1})	Rate of photolysis (s^{-1})	Rate of Ca release (s^{-1})
EDTA	32			0.005				
EGTA	150			12				
DM-nitrophen	5	3	600,000	0.0025	0.18	4,300	8×10^4	3.8×10^4
NP-EGTA	80	1	12,500	9	0.20–0.23	975	5×10^5	6.8×10^4
BAPTA	110			17				
nitr-5	145	0.0063	54	8.5	0.012–0.035	5,500	2.5×10^3	ND
nitr-7	54	0.003	42	5.4	0.011–0.042	5,500	2.5×10^3	ND
azid-1	230	0.12	520	8	1.0	3.3×10^4	ND	

Abbreviation: ND, Not determined.

of chelator fragmentation and of Ca release are fast (DM-nitrophen, 80,000–500,000 s^{-1}; NP-EGTA, 38,000–68,000 s^{-1}); (2) the quantum yields of release are in the range of 0.18–0.23 (Ca-loaded and Ca-free chelators uncage with essentially equal efficiency); (3) affinities before photolysis are high (DM-nitrophen, 5 nM; NP-EGTA, 80 nM); and (4) large changes in Ca affinities are produced by uncaging, and thus much of the bound Ca is released. The disadvantages of this approach are: (1) K_d values are pH sensitive in the physiological range; and (2) DM-nitrophen is not Ca selective (K_d for Mg is 2.5 μM). However, NP-EGTA *is* a Ca-selective cage as it is based, of course, on EGTA. Physicochemical properties of these three widely used, commercially available cages are summarized in Table I.

Design of "Perfect" Calcium Cage

The "ideal" calcium cage has not yet been synthesized. It would combine the best properties of nitr-5, azid-1, DM-nitrophen, and NP-EGTA in one molecule: (1) very high affinity before photolysis, (2) large K_d decrease on uncaging, (3) Ca-selective chelation, (4) rapid chemical reaction and release of Ca, (5) pK_a values below 6; (6) efficient use of absorbed light (large quantum yield), and (7) large extinction coefficient (the extinction coefficient is the measure of the efficiency of light absorption). The final design criterion requires more detailed consideration. For practical purposes optical densities (ODs) of the cage in a cell should be less than 20% or else inhomogeneous uncaging will result from inner filtering effects across the cell. For example, if [cage] = 2 mM, then the extinction coefficient of the cage must be less than 10/mM/cm for a 100-μm path length/cell.

In chromaffin cells (~20 μm in diameter) Neher and co-workers have often used [DM-nitrophen] = 10 mM. This solution has an OD of ~0.09. If azid-1 were substituted then the OD would rise to 0.66, resulting in uneven release of Ca across the cell, because of the inner filtering effect mentioned above. Thus, for most one-photon uncaging experiments, the moderate extinction coefficient of the dimethoxynitro[8,10,11,13] cage is advantageous. For two-photon uncaging of Ca a much larger extinction coefficient is desirable (see discussion below).[9,15]

In spite of the "compromises" in the properties of nitr-5, DM-nitrophen, and NP-EGTA, many useful experiments can be (and have been) performed with these cages. Cage selection depends on how much Ca is required for a particular experiment, how quickly it must be delivered, and whether $[\mathrm{Mg}]_{\mathrm{free}}$ needs to be in the physiological range.

How to Use Caged Calcium

The choice of which of the three commercially available cages should be used will depend on two factors: (1) how much Ca is required and (2) what level of [Mg] is necessary for the experiment.

DM-nitrophen, nitr-5, and NP-EGTA as Calcium Cages

Each of three commercially available cages, and their photoproducts, have different divalent cation-binding properties (Table I). These K_{d} values determine (1) the percentage loading by Ca of each cage (before photolysis) before an activating level of $[\mathrm{Ca}]_{\mathrm{free}}$ is reached, (2) the percentage of cage that must be photolyzed before any net change in [Ca] is effected, and (3) the amount of $[\mathrm{Ca}]_{\mathrm{free}}$ that can be generated after complete (or partial) photolysis of the cage.

DM-nitrophen has both the highest and lowest affinities before and after photolysis, respectively, permitting the largest changes in levels of $[\mathrm{Ca}]_{\mathrm{free}}$. Nitr-5 has the lowest affinity before photolysis and the highest after uncaging, allowing the smallest changes in $[\mathrm{Ca}]_{\mathrm{free}}$. The chelation properties of NP-EGTA place it somewhere between these two cages in its ability to control [Ca]. In Fig. 3, calculations of $[\mathrm{Ca}]_{\mathrm{free}}$ versus percent photolysis at pH 7.2 are presented, using $[\mathrm{cage}]_{\mathrm{total}} = 2$ mM and a $[\mathrm{Ca}]_{\mathrm{total}}$ that gives a $[\mathrm{Ca}]_{\mathrm{free}} = 100$ nM for each cage (no other divalent cations or buffers are included). Complete photolysis of DM-nitrophen gives ~1 mM $[\mathrm{Ca}]_{\mathrm{free}}$, and only 5–10% photolysis is required before Ca is released effectively. About 0.1 mM [Ca] can be released from NP-EGTA, and photolysis of 45% of the cage is required before very much Ca is released. In the case of nitr-5 about 3 μM $[\mathrm{Ca}]_{\mathrm{free}}$ can be generated when all the cage is photolyzed. These data

[15] F. DelPrincipe, M. Egger, G. C. R. Ellis-Davies, and E. Niggli, *Cell Calcium* **25,** 85 (1999).

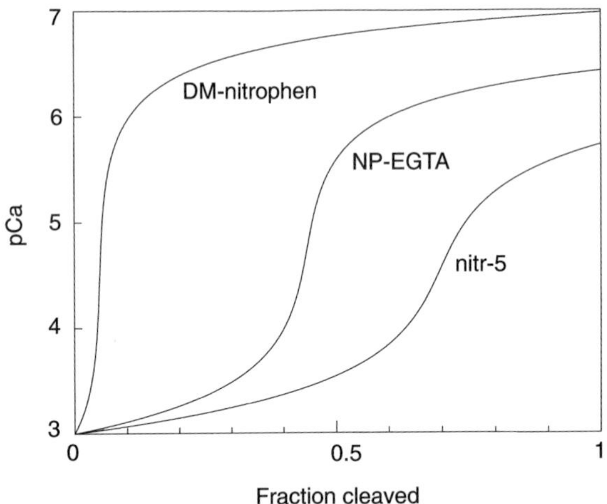

FIG. 3. Calculated changes in $[Ca]_{free}$ versus percent photolysis for three calcium cages [2 m*M*], and $[Ca]_{total}$ such that $[Ca]_{initial}$ is 100 n*M* in every case. $[Ca]_{total}$ values: for DM-nitrophen, 1.905 m*M*; NP-EGTA, 1.108 m*M*; nitr-5, 0.825 m*M*.

illustrate the relative strengths of DM-nitrophen, nitr-5, and NP-EGTA as calcium cages, suggesting that cage selection could depend on the Ca requirements of the envisioned experiment. If millimolar [Ca] is required, then DM-nitrophen must be used,[16–19] in spite of its Mg-binding properties. If the "normal intracellular milieu" is necessary, then either NP-EGTA or nitr-5 could be used. The latter can provide a predictable "ramplike" change in [Ca] on photolysis, whereas the former produces fairly large [Ca] jumps (see Fig. 3). Of course, all cells have endogenous buffers, varying in their concentrations and affinities, that will decrease (perhaps significantly) the value for $[Ca]_{free}$ generated by uncaging from DM-nitrophen, nitr-5, and NP-EGTA.

Experiments with Low [Mg]

If cells/experiments can tolerate a $[Mg]_{free}$ of 0–60 μM, then DM-nitrophen is the cage of choice. Neher and co-workers have used DM-nitrophen to perform many elegant experiments in secretory cells. In the absence of any Mg or MgATP, photolysis of the DM-nitrophen: Ca complex has been used to stimulate secretion in chromaffin cells,[16,17] goldfish bipolar retina,[18] calyx of Held,[19] and inner

[16] C. Heinemann, R. H. Chow, E. Neher, and R. S. Zucker, *Biophys. J.* **67,** 2546 (1994).
[17] T. Xu, M. Naraghi, H. Kang, and E. Neher, *Biophys. J.* **67,** 2546 (1997).
[18] R. Heidelberger, C. Heinemann, E. Neher, and G. Matthews, *Nature* (*London*) **371,** 513 (1994).
[19] R. Schneggenburger and E. Neher, *Nature* (*London*) **406,** 889 (2000).

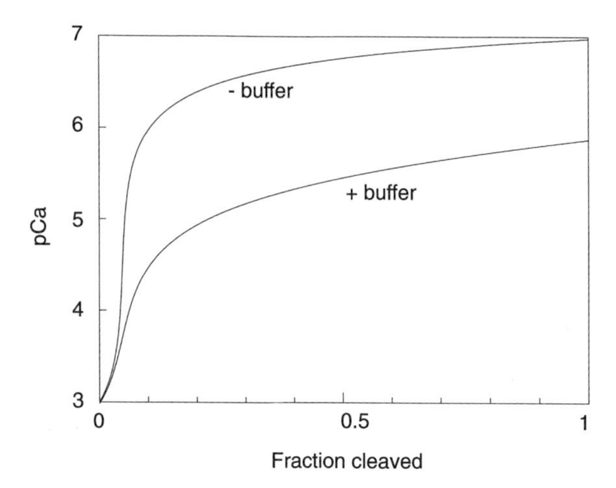

FIG. 4. Effect of immobile cellular buffers on the calculated changes in $[Ca]_{free}$ versus percent photolysis for DM-nitrophen. [Cage] = 2 mM, $[Ca]_{total}$ = 1.905 mM, and minus/plus 135 μM buffers that have a K_d = 4.5 mM.[43,49]

hair cells.[20] Typically, in chromaffin cells, inner hair cells, and bipolar neurons the [DM-nitrophen] and [Ca] used are 10 mM, and uncaging produces from 3 to 600 μM $[Ca]_{free}$. DM-nitrophen is the only calcium cage that could evoke such a wide range of changes in [Ca]. The Neher group has also been able to sustain MgATP-dependent steps in secretion in chromaffin cells by judicious choice of [DM-nitrophen], [Ca], [Mg], and [ATP].[21] Figure 4 illustrates the calculated change in $[Ca]_{free}$ on photolysis of DM-nitrophen under conditions assuming the K_d values shown in Table I, without and with the estimated buffering capacity of a chromaffin cell.[17] These calculations are essentially consistent with the measurements of [Ca] by Neher and co-workers in their many experiments. DM-nitrophen has also been used under similar conditions in PC12 cells,[22,23] pancreatic acinar cells,[24] β cells,[25] and Chinese hamster ovary (CHO) cells[26] by Kasai and co-workers (reviewed in Kasai[27]). Zucker and co-workers have also performed many creative experiments

[20] T. Voets, R. F. Toonen, E. C. Brian, H. de Wit, T. Moser, J. Rettig, T. C. Sudhof, E. Neher, and M. Verhage, *Neuron* **31,** 581 (2001).

[21] K. D. Gillis, R. Mossner, and E. Neher, *Neuron* **16,** 1209 (1996).

[22] H. Kasai, H. Takagi, Y. Ninomiya, T. Kishimoto, A. Yosida, T. Yoshioka, and Y. Miyashita, *J. Physiol.* **494,** 53 (1996).

[23] T. Kishimoto, T. T. Liu, Y. Ninomiya, H. Takagi, T. Yoshioka, G. C. R. Ellis-Davies, Y. W. Miyashita, and H. Kasai, *J. Physiol.* **533,** 627 (2001).

[24] K. Ito, Y. Miyashita, and H. Kasai, *EMBO J.* **16,** 242 (1997).

[25] N. Takahashi, T. Kadowaki, Y. Yazaki, Y. Miyashita, and H. Kasai, *J. Cell Biol.* **138,** 55 (1997).

[26] Y. Ninomiya, T. Kishimoto, Y. Miyashita, and H. Kasai, *J. Biol. Chem.* **271,** 17751 (1996).

[27] H. Kasai, *Trends Neurosci.* **22,** 88 (1999).

with all three commercially available caged Ca.[28] Significantly, uncaging from DM-nitrophen in squid giant synapse[29] produced large jumps in measured [Ca] that were in accordance with modeled concentration changes (1–300 μM range). Zucker and co-workers have concluded from the size of photoevoked biological phenomena that DM-nitrophen releases more Ca than does nitr-5 (see, e.g., Fryer and Zucker[30]), and that such differences are consistent with the physicochemical properties of the two cages (Table I).

Experiments with Physiological [Mg]

Many cellular processes, of course, require millimolar concentrations of MgATP, MgGTP, or Mg for effective operation. If $[Mg]_{free}$ is in this range, DM-nitrophen becomes caged Mg. NP-EGTA and Nitr-5 are the two commercially available cages that have low prephotolysis affinities for Mg, and high affinities for Ca (see Table I). They have been successfully loaded into cells by the AM-ester technique, and then used as caged calciums.[31–34] For example, Lagnado *et al.* loaded fura-2 along with NP-EGTA into neurons[31] so that they could monitor the amount of Ca uncaged. The percentage loading of the cage with Ca was probably determined by intracellular buffering (in contrast to whole-cell dialysis through patch pipettes with 10 m*M* DM-nitrophen/Ca, in which the cage is the dominant Ca buffer, and so sets $[Ca]_{rest}$); nevertheless, micromolar increases in [Ca] were photoevoked in the presence of all the normal intracellular constituents. Such experiments illustrate that under normal intracellular conditions both NP-EGTA and nitr-5 are Ca-selective photolabile chelators.

Use of Calcium Indicators with Caged Calcium

The development of many Ca-selective fluorescent dyes has enabled the quantitative use of caged Ca by several groups (see above). Concomitant with the introduction of fluo-3, the group headed by R. Tsien demonstrated that it could be used to quantify Ca uncaging inside cells.[35] Irradiation wavelengths for uncaging and imaging were designed by the Tsien group to be spectrally separate: nitr-5 is

[28] Reviews: R. S. Zucker, *J. Physiol. (Paris)* **87,** 25 (1993); R. S. Zucker, *Methods Cell Biol.* **40,** 31 (1994); R. S. Zucker, *Neuron* **17,** 1049 (1996); R. S. Zucker, *Curr. Opin. Neurobiol.* **9,** 305 (1999).

[29] K. R. Delaney and R. S. Zucker, *J. Physiol.* **426,** 473 (1990).

[30] M. W. Fryer and R. S. Zucker, *J. Physiol.* **464,** 501 (1993).

[31] L. Lagnado, A. Gomis, and C. Job, *Neuron* **17,** 957 (1996).

[32] M. Kress and S. Guenther, *J. Neurophysiol.* **81,** 2612 (1999).

[33] T. Kimitsuki and H. Ohmori, *J. Physiol.* **458,** 27 (1992).

[34] T. M. Gomez and N. C. Spitzer, *Nature (London)* **397,** 350 (1999); T. M. Gomez, E. Robles, M.-M. Poo, and N. Spitzer, *Science* **291,** 1983 (2001); T. M. Gomez and N. C. Spitzer, *J. Neurobiol.* **44,** 174 (2000).

[35] J. P. Y. Kao, A. T. Harootunian, and R. Y. Tsien, *J. Biol. Chem.* **264,** 8179 (1989).

excited at 350 nm and fluo-3 is excited at 490 nm. Thus, uncaging did not bleach fluo-3, and nitr-5 (and all other caged compounds) is transparent at 490 nm.[35]

Ratiometric Ca indicators such as fura-2 and BTC are easier to use for the quantitative measurement of [Ca] than fluo-3, as the ratio technique eliminates [dye] as a factor from the analysis. (The Bolsover group has introduced a solution to this problem by including a yellow dye along with fluo-3 in the intracellular solution.[36]) However, quantitative use of Ca indicators such as fura-2, furaptra, and fura-2FF that are excited in the same wavelength range as uncaging requires careful control experiments. The Almers group,[37–39] the Neher group,[16–21,40–52] and the Zucker group[29,30,53–57] have outlined these procedures in great detail. There are several concerns about using fura-2 with caged Ca that they address: (1) "cross-bleaching" of chromophores, (2) the effects of cage photoproducts and starting materials having different absorption spectra on indicator excitation, (3) photoproduct fluorescence, (4) collisional quenching of fura-2 by cage,[58] and (5) effects of the intracellular milieu on the apparent affinity of fura-2 for Ca.[59]

Kasai and co-workers have used an alternative ratiometric Ca indicator, BTC (and its low-affinity version, BTC-5N), for [Ca] measurement when uncaging Ca

[36] A. Fleet, G. C. R. Ellis-Davies, and S. R. Bolsover, *Biochem. Biophys. Res. Commun.* **250,** 786 (1998).

[37] P. Thomas, J. G. Wong, and W. Almers, *EMBO J.* **12,** 303 (1993).

[38] P. Thomas, J. G. Wong, A. Lee, and W. Almers, *Neuron* **11,** 93 (1993).

[39] T. D. Parsons, G. C. R. Ellis-Davies, and W. Almers, *Cell Calcium* **19,** 185 (1996).

[40] E. Neher and R. S. Zucker, *Neuron* **10,** 21 (1993).

[41] L. Y. Huang and E. Neher, *Neuron* **17,** 135 (1996).

[42] R. H. Chow, J. Klingauf, C. Heinemann, R. S. Zucker, and E. Neher, *Neuron* **16,** 369 (1996).

[43] T. Xu, M. Naraghi, H. Kang, and E. Neher, *Biophys. J.* **73,** 532 (1997).

[44] T. Moser and E. Neher, *J. Neurosci.* **17,** 2314 (1997).

[45] J. Rettig, C. Heinemann, U. Ashery, Z. H. Sheng, C. T. Yokoyama, W. A. Catterall, and E. Neher, *J. Neurosci.* **17,** 6647 (1997).

[46] M. Haller, C. Heinemann, R. H. Chow, R. Heidelberger, and E. Neher, *Biophys. J.* **74,** 2100 (1998).

[47] C. Smith, T. Moser, T. Xu, and E. Neher, *Neuron* **20,** 1243 (1998).

[48] A. Betz, U. Ashery, M. Rickmann, I. Augustin, E. Neher, T. C. Sudhof, J. Rettig, and N. Brose, *Neuron* **21,** 123 (1998).

[49] T. Xu, T. Binz, H. Niemann, and E. Neher, *Nat. Neurosci.* **1,** 192 (1998).

[50] T. Voets, E. Neher, and T. Moser, *Neuron* **23,** 607 (1999).

[51] R. Schneggenburger, A. C. Meyer, and E. Neher, *Neuron* **23,** 399 (1999).

[52] D. Beutner, T. Voets, E. Neher, and T. Moser, *Neuron* **29,** 681 (2001).

[53] N. V. Marrion, R. S. Zucker, S. J. Marsh, and P. R. Adams, *Neuron* **6,** 533 (1991).

[54] P.-M. Lau, R. S. Zucker, and D. Bentley, *J. Cell Biol.* **145,** 1265 (1999).

[55] R. K. Ayer and R. S. Zucker, *Biophys. J.* **77,** 3384 (1999).

[56] J. Wang and R. S. Zucker, *J. Physiol.* **533,** 757 (2001).

[57] K. M. Ohnuma, M. D. Whim, R. Fetter, L. K. Kaczmarek, and R. S. Zucker, *J. Physiol.* **535,** 647 (2001).

[58] R. S. Zucker, *Cell Calcium* **13,** 29 (1992).

[59] M. Zhao, S. M. Hollingsworth, and S. M. Baylor, *Biophys. J.* **70,** 896 (1996).

from DM-nitrophen, NP-EGTA, and DMNPE-4. Because the BTC chromophore is excited at different wavelengths than caged Ca (emission ratio of 430- and 480-nm excitation, with uncaging at 350 nm), Ca is not uncaged during imaging, as it is with fura-based dyes. They have described in detail their careful method of calibration of BTC when using caged Ca.[22,23] Thus, the literature contains descriptions of the quantitative use of caged Ca in many published protocols, which use either ratiometric or nonratiometric fluorescent Ca dyes.

Uncaging Light Pulse Width, Chemical Yield, and [Ca] Waveform

Pulsed[60–66] and continuous wave[67] (CW) light sources have both been used effectively to uncage Ca in many biological experiments. Pulsed light sources used are Nd:YAG,[62] excimer,[61] ruby lasers,[60,63] and flash lamps. YAG and ruby lasers have pulse widths of ~3 and 35 ns, respectively, and flash lamps have pulse widths in the range of 0.5–1.5 ms. The rate of reaction of DM-nitrophen is about 85,000 s^{-1}.[14] Thus, if a YAG laser is used for uncaging, the rate of reaction is the kinetically rate-limiting step for Ca release, but if a flash lamp is used, the light pulse width becomes the rate-limiting factor. Consequently, the pulse width of the light source can have a profound effect on the [Ca] waveform, as shown in Fig. 5. The [Ca] is modeled for light pulses of 3 ns (YAG laser) and 1 ms (flash lamp), assuming 5% photolysis to give a $[Ca]_{free}$ jump from 100 n*M* ([DM-nitrophen] = 2 m*M*, [Ca] = 1.905 m*M*) to about 10 μM. A laser pulse produces a considerably larger spike in [Ca] than the flash lamp. The size and shape of the [Ca] waveform reflects the relatively slow Ca-binding kinetics of DM-nitrophen and its photoproducts, and obviously the exact nature of this pulse is acutely sensitive to percentage photolysis, [cage], [Ca], and pH.

The light source pulse width also affects the chemical yield for uncaging for an individual flash. Because nitroaromatic chromophores are nonfluorescent, there are only two modes of deexcitation of the singlet state: chemical reaction (uncaging) or vibronic relaxation (nonradiative electronic transition from the singlet to the

[60] G. Rapp, K. J. V. Poole, Y. Meada, G. C. R. Ellis-Davies, J. McCray, and R. S. Goody, *Ber. Bunsenges. Phys. Chem.* **93,** 410 (1989).

[61] A. Eisenrauch, M. Juhaszova, G. C. R. Ellis-Davies, E. Bamberg, and M. P. Blaustein, *J. Membr. Biol.* **145,** 151 (1995).

[62] H. H. Valdivia, J. H. Kaplan, G. C. R. Ellis-Davies, and W. J. Lederer, *Science* **267,** 1997 (1995).

[63] B. Zimmermann, G. C. R. Ellis-Davies, A. V. Somlyo, and A. P. Somlyo, *J. Biol. Chem.* **270,** 23966 (1996).

[64] M. Wilding, E. W. Wright, R. Patel, G. C. R. Ellis-Davies, and M. Whitaker, *J. Cell Biol.* **135,** 1991 (1996).

[65] N. Takahashi, T. Kadowaki, Y. Yazaki, G. C. R. Ellis-Davies, Y. Miyashita, and H. Kasai, *Proc. Natl. Acad. Sci. U.S.A.* **96,** 760 (1999).

[66] J. Xu, Y. Xu, G. C. R. Ellis-Davies, G. A. Augustine, and F. W. Tse, *J. Neurosci.* **22,** 53 (2002).

[67] S.-H. Wang and G. J. Augustine, *Neuron* **15,** 755 (1995).

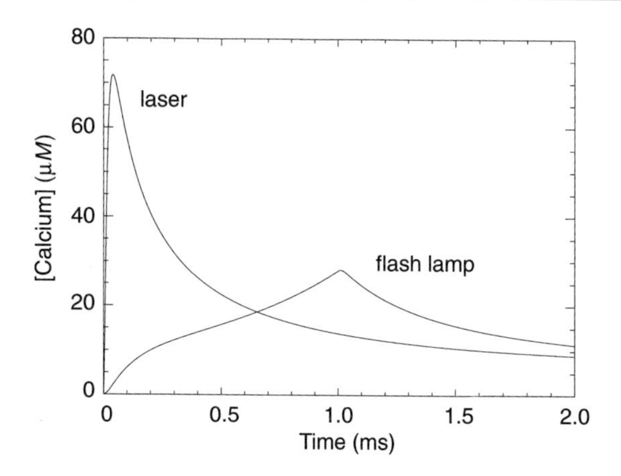

FIG. 5. Effect of uncaging light pulse width on the calculated changes in $[Ca]_{free}$ versus time for DM-nitrophen. [Cage] = 2 mM, and $[Ca]_{total}$ = 1.905 mM.

ground state). The later period is typically a few nanoseconds for aromatic chromophores. Thus, uncaging using a flash lamp can produce multiple rounds of excitation, deexcitation, and reexcitation of the same chromophore, giving chemical yields that are much higher than quantum yields, and so producing large changes in [Ca]. Use of short pulse-width lasers (e.g., YAG), in contrast, limits the chemical yield to the quantum yield for the uncaging reaction. Ruby lasers (which we use for our flash photolysis experiments[14]) have pulse widths about 10 times longer than singlet lifetimes, and so can give good chemical yields for uncaging, along with the useful properties normally associated with coherent light. Unfortunately, ruby lasers are no longer commercially available. Thus, for high chemical yield in uncaging experiments either flash lamps or shuttered CW light sources must be used.

Two-Photon Uncaging of Calcium

Attempts have been made to use two-photon excitation to release Ca in living cells. Niggli and co-workers have had some success with DM-nitrophen[68] and DMNPE-4.[15] However, to produce modest increases in [Ca], high laser energies had to be used (~10–20 mW in the image plane). Azid-1 has a two-photon cross-section (i.e., the efficiency of the simultaneous absorption of two red photons, giving the first excited singlet state) of 1 GM,[69] but no reports of applications in cellular physiology have appeared, perhaps because it has a low affinity before uncaging (see Table I), making the release of large quantities of Ca difficult.

[68] P. Lipp and E. Niggli, *J. Physiol.* **508,** 801 (1998).

[69] E. B. Brown, J. B. Shear, S. R. Adams, R. Y. Tsien, and W. W. Webb, *Biophys. J.* **76,** 489 (1999).

Because two-photon photolysis of glutamate is now practical,[70,71] it gives hope that when the "perfect" cage Ca is developed (see above), highly localized release of Ca at reasonable power levels will become technically feasible.

Summary

Several caged calciums have been synthesized since 1986, and three are commercially available: DM-nitrophen, NP-EGTA, and nitr-5. Each of these caged compounds has uniquely useful properties, making the choice of which cage to use dependent on the specific experiment (i.e., the cell type and divalent cation requirements of the experiments within purview). Significantly, methods have been developed for all three cages that permit their quantitative use inside many cell types, including those with some of the most demanding of requirements for experiments with caged calcium, namely, in relating presynaptic [Ca] to postsynaptic function.[19,72] The success of such experiments using DM-nitrophen and NP-EGTA suggests that caged calcium is now a mature tool for cellular physiology and neurobiology.

Acknowledgment

I thank Stephen Hollingworth for help with the modeling in Figs. 3–5.

[70] G. C. R. Ellis-Davies, *J. Gen. Physiol.* **114,** 1a (1999).
[71] M. Matsuzaki, A. Tachikawa, G. C. R. Ellis-Davies, Y. Miyashita, M. Iino, and H. Kasai, *Nat. Neurosci.* **4,** 1086 (2001).
[72] J. H. Bollman, B. Sakmann, and J. G. G. Borst, *Science* **289,** 953 (2000).

[9] Application of Fluorescent Probes to Study Mechanics and Dynamics of Ca^{2+}-Triggered Synaptotagmin C2 Domain–Membrane Interactions

By JIHONG BAI and EDWIN R. CHAPMAN

Introduction

The application of fluorescence spectroscopy to address biological problems, including the structure, function, and dynamics of molecular machines, is a well-established approach that is currently enjoying a renaissance as a "biophotonics" tool. Here, we describe the application of fluorescence spectroscopy to gain unique insights into a step proposed to couple Ca^{2+} influx to the secretion of

0076-6879/03 $35.00

neurotransmitters from neurons. This step, the Ca^{2+}-triggered interaction of a protein domain (the C2A domain of synaptotagmin, described below) with lipid bilayers, is just one example of how fluorescence techniques afford the detailed analysis of protein–membrane interactions, including the mechanism of protein docking onto membranes and the dynamics and affinities of these interactions.

Visualizing Rapid Ca^{2+}-Triggered Rearrangements between Macromolecules That Mediate Neuronal Exocytosis

Neurotransmitters are stored in specialized secretory organelles and are released into the synapse by the abrupt opening of a preassembled fusion pore that connects the lumen of the vesicle to the extracellular space.[1] The release process is extremely rapid[2–4] and is precisely regulated by Ca^{2+}.[5] These observations dictate that Ca^{2+} triggers rapid changes in the protein–protein and protein–membrane interactions that mediate the opening (and subsequent dilation) of fusion pores. To understand this mechanism, time-resolved methods that are applicable to both protein–membrane and protein–protein interactions are required. Here, we describe the use of fluorescence spectroscopy to determine the immediate consequences of Ca^{2+} binding to a putative Ca^{2+} sensor for rapid exocytosis, synaptotagmin I (reviewed by Augustine[6] and Misura *et al.*[7]). Key questions addressed by these methods are whether synaptotagmin I exhibits the Ca^{2+} sensitivity, divalent cation specificity, and speed of response consistent with physiological measurements for secretion.

Synaptotagmin I spans the secretory vesicle membrane once and has a large cytoplasmic region composed of two C2 domains, designated C2A and C2B. C2 domains are widespread structural motifs found in more than 120 proteins in the human genome[8]; these motifs often function as Ca^{2+}-sensing domains. In many cases (e.g., lipid kinases and phospholipases), Ca^{2+} binding to C2 domains drives their translocation to membranes, thereby juxtaposing separate catalytic domains onto the bilayer, where they can efficiently interact with substrate. The C2A domain of synaptotagmin I interacts with membranes in response to Ca^{2+} and serves as a model system to understand C2 domain–membrane interactions.[9,10]

[1] M. Lindau and W. Almers, *Curr. Biol.* **5,** 509 (1995).
[2] R. Llinas, I. Z. Steinberg, and K. Walton, *Biophys. J.* **33,** 323 (1981).
[3] B. L. Sabatini and W. G. Regehr, *Nature (London)* **384,** 170 (1996).
[4] S. Mennerick and G. Matthews, *Neuron* **17,** 1241 (1996).
[5] B. Katz, "The Release of Neurotransmitter Substances." Liverpool University Press, Liverpool, UK, 1969.
[6] G. J. Augustine, *Curr. Opin. Neurol.* **11,** 320 (2001).
[7] K. M. S. Misura, A. P. May, and W. I. Weis, *Curr. Opin. Struct. Biol.* **10,** 662 (2000).
[8] International Human Genome Sequencing Consortium, *Nature (London)* **409,** 860 (2001).
[9] B. Davletov and T. Südhof, *J. Biol. Chem.* **268,** 26386 (1993).
[10] E. R. Chapman and R. Jahn, *J. Biol. Chem.* **269,** 5735 (1994).

Ca^{2+}-triggered C2A–lipid bilayer interactions are likely to be a key to the *in vivo* function of synaptotagmin, potentially accelerating fusion in response to Ca^{2+},[11] as supported by a genetic study.[12] Thus, an understanding of the mechanism by which Ca^{2+} triggers the interaction of C2A with membranes should provide general insights into C2 domain function and specific insights into the mechanics of the Ca^{2+}-triggered fusion reaction.

Scanning Optical Probe Mutagenesis

The C2A domain of synaptotagmin I is the first C2 domain whose structure has been solved,[13] making it possible to place reporters at site-directed locations on the surface of the domain. At present, a number of biophysical approaches [e.g., nuclear magnetic resonance (NMR) and crystallography] cannot be readily applied to this problem, because of the difficulties inherent in using phospholipid bilayers, and the static nature of the measurements. We therefore have adopted fluorescence methods to determine the mechanism and dynamics of synaptotagmin C2A–membrane interactions [note that electron paramagnetic resonance studies have also proved useful in the study of protein–membrane interactions (reviewed by Hubbell *et al.*[14–15]].

The C2A domain of synaptotagmin I forms a compact eight-strand β-sandwich structure (Fig. 1[13]). Three flexible loops that protrude from one end of the domain mediate the binding of two or three Ca^{2+} ions.[13,16] To identify the C2A–membrane interface, we "scan" the surface of the recombinant domain by placing single tryptophan (Trp) residues at the positions indicated in Fig. 1, using the overlapping polymerase chain reaction (PCR)–primer method described by Higuchi.[17] Trp residues are suitable fluorescence reporters for a number of reasons: (1) the sensitivity of indole side-chain fluorescence to local environment can be exploited to monitor local conformational or environmental changes, for example, as a consequence of binding Ca^{2+} and/or lipids (e.g., Meers[18]); (2) Trp residues are relatively rare in protein sequences, and thus it is often possible for a protein or domain to possess a lone Trp reporter that can be readily moved around within the structure via mutagenesis; (3) in some cases, naturally occurring Trp residues can be exploited to study interactions with effectors (see, e.g., Davis *et al.*[11]), obviating the

[11] A. F. Davis, J. Bai, D. Fasshauer, M. J. Wolowick, J. L. Lewis, and E. R. Chapman, *Neuron* **24,** 363 (1999).

[12] R. Fernandez-Chacon, A. Konigstorfer, S. H. Gerber, J. Garcia, M. F. Matos, C. F. Stevens, N. Brose, J. Rizo, C. Rosenmund, and T. C. Südhof, *Nature (London)* **410,** 41 (2001).

[13] R. B. Sutton, B. A. Davletov, A. M. Berghuis, T. C. Südhof, and S. R. Sprang, *Cell* **80,** 929 (1995).

[14] W. L. Hubbell, D. S. Cafiso, and C. Altenbach, *Nat. Struct. Biol.* **7,** 735 (2000).

[15] W. L. Hubbell, A. Gross, R. Langen, and M. A. Lietzow, *Curr. Opin. Struct. Biol.* **8,** 649 (1998).

[16] J. Ubach, X. Zhang, X. Shao, T. C. Südhof, and J. Rizo, *EMBO J.* **17,** 3921 (1998).

[17] R. Higuchi, *in* "PCR Protocols: A Guide to Methods and Applications" (M. A. Innis, D. H. Gelfand, J. J. Sninsky, and T. J. White, eds.), p. 177. Academic Press, New York, 1990.

[18] P. Meers, *Biochemistry* **29,** 3325 (1990).

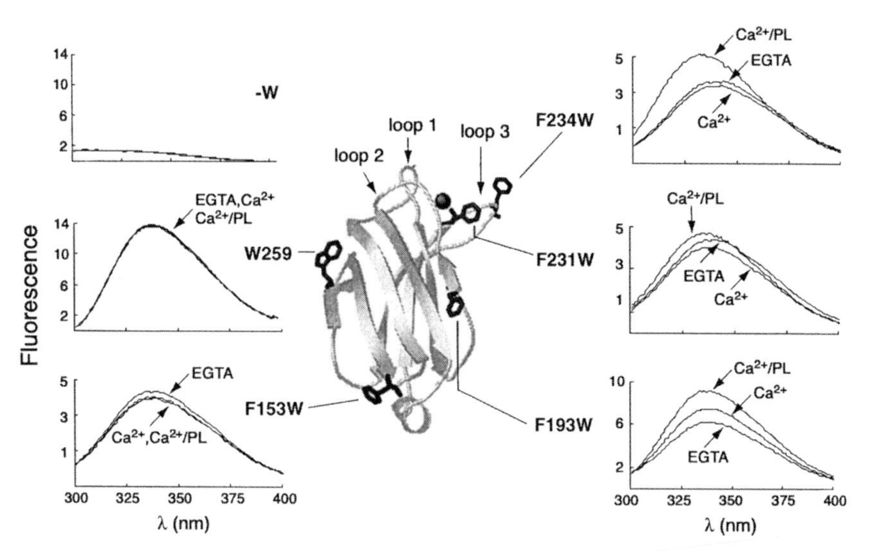

FIG. 1. An example of "scanning optical probe mutagenesis," in which single Trp residues are placed on the surface of the C2A domain of synaptotagmin I. The crystal structure of C2A was rendered using Molscript [P. J. Kraulis, *J. Appl. Crystallogr.* **24,** 946 (1991)] and Raster3D [E. A. Merritt and D. J. Bacon, *Methods Enzymol.* **277,** 505 (1997)] according to a description file modified from Sutton *et al.* (Ref. 13). The sole naturally occurring Trp at position 259 is indicated in black, as are the native phenylalanine (Phe) residues (positions 234, 231, 193, and 153) that were individually changed to Trp reporters (in a background lacking Trp-259). The Ca^{2+}-binding loops (loops 1–3) are indicated with arrows and the sphere represents a water molecule that occupies the position of one of the Ca^{2+} ions. Proteins were adjusted to 5 μM in TBS buffer [20 m*M* Tris (pH 7.4), 150 m*M* NaCl], and aromatic residues were excited at 288 nm. The Trp emission spectra were collected from 300 to 450 nm, using a PTI QM-1 fluorometer. Spectra were first obtained in the presence of 2 m*M* EGTA, then after the addition of Ca^{2+} to a free concentration of 1 m*M*, and finally after subsequent addition of liposomes composed of 50% PS/50% PC (250 μM total lipid). Fluorescence intensity is given in arbitrary units. [Modified and reproduced from E. R. Chapman and A. F. Davis, *J. Biol. Chem.* **273,** 13995 (1998), with permission.]

need for mutations; and (4) Trp substitutions are often conservative, particularly when replacing Phe or Tyr residues, decreasing the chances of perturbing protein function. However, to rule out effects of the reporter mutations on the function of the protein, steady state radioligand-binding assays should be carried out. For example, the Ca^{2+} requirements for C2 domain–membrane interactions, and the absolute level of binding, can be readily compared between wild-type and mutant proteins, using radiolabeled liposomes[9,19,20a] as follows.

³H-Labeled Liposome-Binding Assays. Small unilamellar liposomes (~50 nm) are prepared by sonication. Brain-derived phosphatidylserine (PS) and

[19] N. Brose, A. G. Petrenko, T. C. Südhof, and R. Jahn, *Science* **256,** 1021 (1992).

[20a] E. R. Chapman and A. F. Davis, *J. Biol. Chem.* **273,** 13995 (1998).

phosphatidylcholine (PC) (dissolved in chloroform) are obtained from Avanti Polar Lipids (Alabaster, AL). L-3-Phosphatidyl[*N-methyl*-^{3}H]choline-1,2-dipalmitoyl ([^{3}H]PC) is purchased from Amersham Pharmacia Biotech (Piscataway, NJ) (specific activity, 84.0 Ci/mmol). A mixture of lipids containing 0.44 mg of PS, 1.31 mg of PC, and 20 μCi of [^{3}H]PC is dried as a thin film in a glass test tube under a stream of nitrogen; trace solvent is then removed by placing the test tube in a vacuum for at least 30 min. Dried lipids are hydrated in 1 ml of HEPES buffer [50 m*M* HEPES–NaOH (pH 7.4), 0.1 *M* NaCl] for 30 min, and then resuspended by vortexing for 2 min. The test tube containing the lipid suspension is placed in a W-220 bath sonicator (Heat Systems-Ultrasonics (now Misonix), Farmingdale, NY], and the lipid suspension is sonicated (20 kHz, 200 W) for 8 min. The residual large particles remaining in the suspension are removed by centrifugation at 21,000*g* for 10 min at room temperature. Liposomes are stored at 4° and used within 1 week.

The liposome-binding assays are carried out by using 15 μg of recombinant proteins bound to 10 μl of glutathione–Sepharose (Amersham Pharmacia Biotech) and 10 μl of ^{3}H-labeled liposomes (per data point) in HEPES buffer with either 2 m*M* EGTA or 1 m*M* Ca^{2+}. The samples are incubated for 15 min at room temperature in Micro Bio-Spin chromatography columns (Bio-Rad, Hercules, CA) with continuous mixing, using a VXR IKA-Vibrax with a VX 2E Eppendorf attachment (IKA Works, Wilmington, NC). The samples are then washed three times with binding buffer plus 2 m*M* EGTA or 1 m*M* Ca^{2+}. Rapid washing is carried out by removing the plug in the base of the Micro Bio-Spin column and loading the column onto a vacuum manifold. After washing, the columns are placed in 20-ml scintillation vials filled with 10 ml of scintillation fluid. Radioactivity is quantified by liquid scintillation counting.

Once it is confirmed that the mutant versions of the protein under study retain Ca^{2+}- and lipid-binding activity, fluoroescence studies can be carried out. To selectively study the site-directed fluorophore, native Trp residues present in the protein or domain under study should be removed via mutagenesis (generally by substituting them with Phe or Tyr residues). In some cases, removal of "background" Trp residues by substitution mutagenesis can adversely affect the function of a protein. Indeed, this is observed when the fluorescence studies described here for C2A are extended to the intact cytoplasmic domain of synaptotagmin (C2A–C2B), which contains Trp residues in the C2B domain that cannot be substituted by other amino acids because they are critical for function.[20b] In such cases, it is often possible to place a lone cysteine (Cys) residue at the site of interest, without altering protein function. This Cys can then be easily labeled with thiol-reactive fluorescent probes. In the case of the cytoplasmic domain of synaptotagmin, site-directed Cys residues

[20b] J. Bai, P. Wang, and E. R. Chapman, *Proc. Natl. Acad. Sci. U.S.A.* **99,** 1665 (2002).

are labeled with AEDANS [acetyl-*N'*-(5-sulfo-1-naphthyl)ethylenediamine] reporters without disrupting any known functions of the protein.[20b] This alternative approach is described below in Stoichiometry of Protein–Liposome Complex. Finally, we note that if neither the Trp nor Cys approach is tenable, inteins can be used to splice protein fragments, that harbor exogenous fluorophores, into the protein or protein domain under study (reviewed by Xu *et al.*[21] and Perler[22]).

The Trp "sensors" shown in Fig. 1 can be used to search for potential C2A–lipid interfaces by fluorescence spectroscopy. Each Trp reporter version of C2A is expressed in *Escherichia coli* as a glutathione transferase (GST) fusion protein, using pGEX vectors (Amersham Pharmacia Biotech). The C2A domains are cleaved from the GST tag, using thrombin, and used for spectroscopic studies.[20a] The reporter emission spectra are collected for each Trp mutant protein in the presence of EGTA or Ca^{2+} in the presence or absence of liposomes.

Large (~100 nm) unilamellar liposomes[23] are prepared with an Avanti Polar Lipids extruder according to the manufacturer instructions. This method is used to generate liposomes of uniform size, which is critical for the kinetics and stoichiometry experiments described further below. Briefly, lipids (dissolved in chloroform) are mixed to the desired compositions and dried in a film as described above. Dried lipids are hydrated and suspended in Tris-buffered saline [20 m*M* Tris (pH 7.4), 150 m*M* NaCl; TBS] for 30 min. The fully hydrated samples are vortexed for 1 min and loaded into one end of the Avanti Polar Lipids extruder. The samples are extruded through 100-nm pore filters (Whatman Nuclepore Process Business, Newton, MA) 20 times to produce large unilamellar liposomes (LUVs).

Fluorescence measurements are carried out as follows. Fluorescence measurements are made at 24°, using a PTI QM-1 fluorometer and Felix software. Protein samples are adjusted to 5 μM in HEPES buffer and excited at 288 nm with 2-nm resolution. Emission spectra are collected from 300 to 450 nm and are corrected for blank samples (containing everything except for the fluorophore), dilution (due to the addition of EGTA, Ca^{2+}, or liposomes), and instrument response.

Two issues should be noted when carrying out these measurements: the inner filter effect and photobleaching. The inner filter effect occurs in samples with high absorbance (the absorbance can be due to the fluorophore itself or due to other absorbing components of the sample). Under these conditions, the incident excitation light is absorbed at the front face of the sample and does not properly penetrate the sample. The result is that the detected fluorescence decreases as the sample absorbance increases. As implied, the magnitude of the inner filter effect depends on the geometric relationship between the excitation and emission

[21] M. Q. Xu, H. Paulus, and S. R. Chong, *Methods Enzymol.* **326,** 376 (2000).

[22] F. B. Perler, *Cell* **92,** 1 (1998).

[23] R. C. MacDonald, R. I. MacDonald, B. P. M. Menco, K. Takeshita, N. K. Subbarao, and L. R. Hu, *Biochim. Biophys. Acta* **1061,** 297 (1991).

detection paths and on the thickness of the sample. To avoid inner filter effects, it is generally advisable that the sample absorbance measured at the excitation wavelength not exceed 0.1. In our studies using liposomes, which present a barrier to incident light, we titrate Trp (free in solution) with increasing concentrations of the liposome mixture. This control titration shows that at liposome concentrations as high as 22 n*M*, only a small (~2%) decrease in fluorescence occurs.

Photobleaching (i.e., photodestruction) of the fluorophore occurs because the excited state is generally much more chemically reactive than the ground state. A small fraction of the excited fluorophore molecules participate in chemical reactions that reduce their fluorescence. Photobleaching often results from the reaction of molecular oxygen with the excited triplet state of dyes, producing highly reactive singlet oxygen. The rate of these reactions depends on a number of parameters, including the chemical environment, the excitation light intensity, the dwell time of the excitation beam, and the number of repeat scans. In practice, it is important to wait until the baseline fluorescence becomes stable before assaying for changes in the emission of the reporter as a function of a manipulation (e.g., the addition of Ca^{2+}).

The fluorescent reporters in C2A fall into three categories: (1) the fluorescence of Trps placed at positions 259 and 153 is not affected by Ca^{2+} and/or liposomes. Under the conditions of these experiments, each construct binds similar levels of radiolabeled liposomes.[20a] Thus, the lack of a response is not secondary to disruption of activity within the mutant C2A domain; rather, the environment of these reporters is not affected by Ca^{2+} and liposome binding. The Trp at position 193 falls into category (2): Ca^{2+} binding results in an increase in Trp fluorescence intensity that is further increased by addition of liposomes. However, this increase is not associated with a shift in the emission maxima, as would be expected if Trp-193 shifted to a hydrophobic environment (i.e., the interior of the lipid bilayer). Thus, the effect of liposomes may be to increase the binding of Ca^{2+}; synaptotagmin binds Ca^{2+} poorly in the absence of lipids (Brose *et al.*[19]; the data described below provide a mechanistic explanation for the coupling of lipid and Ca^{2+} binding, and suggest that the Ca^{2+}-induced change in Trp-193 results from subtle conformational changes that do not involve direct contact with bilayers). Finally, Trp residues at positions 231 and 234 comprise category (3): Ca^{2+} plus liposomes induce increases in Trp fluorescence intensity and also cause a blue shift in the emission spectra. These changes are characteristic of fluorophores that have moved into a more hydrophobic environment, either within the C2A domain, or within the lipid bilayer in the C2A–liposome complex (this issue is directly addressed below). The change in fluorescence is completely dependent on the presence of both Ca^{2+} and liposomes. These data provide the first indication that a Ca^{2+}-binding loop of a C2 domain interacts with membranes. Furthermore, the stronger fluorescence changes exhibited by Trp-234 versus Trp-231 can be explained in light of this model: Trp-234 lies at the distal tip of loop 3, whereas Trp lies near the body of C2A and would be expected to penetrate target membranes to a lesser degree.

It is also possible that the fluorescence changes of Trp-234 and Trp-231 are due to conformational changes within the protein and are not due to penetration into membranes. In the next section we describe a method that can be used to assay directly whether any of the Trp reporters penetrate into the lipid bilayer. Furthermore, this method can be used to measure the depth of penetration.

Direct Measurement of Protein Penetration into Lipid Bilayers

Lipid-embedded quenchers are used to determine whether the fluorescent reporters described in Scanning Optical Probe Mutagenesis (above) penetrate membranes. One series of quenchers, doxyl spin labels attached to the acyl chain of phosphatidylcholine (PC), are shown in Fig. 2 (left).[11,24] If a Trp residue comes in direct contact with the doxyl spin label, its fluorescence will be efficiently quenched (as compared with the fluorescence of samples containing liposomes that lack the membrane-embedded quencher), providing a direct means to assay for penetration of the reporters described above. These experiments have been reported by Chapman and Davis[20a]: Trp-234 and Trp-231 are quenched by the doxyl spin labels, and thereby directly penetrate into the bilayer in response to binding Ca^{2+}. In contrast, Trp-193, -153, and -259 are unaffected by the quenchers, strongly indicating that the Ca^{2+}-induced fluorescence changes exhibited by Trp-193 reflect Ca^{2+}-induced conformational changes in C2A.[20a] We note that the initial scanning study described here made use of a truncated C2 domain that ends at residue 258 (to avoid the lone naturally occurring Trp at position 259); C2A ends at roughly residue 265.[13] At 1 m*M* Ca^{2+} (the concentrations used in the fluorescence studies), the position 258 truncation binds the same level of radiolabeled liposomes as the full-length domain. However, subsequent analysis revealed that the truncation adversely affects liposome binding at lower [Ca^{2+}]. Therefore, all these experiments were repeated with full-length C2A (residues 96–265), which harbors a W259F mutation to remove the naturally occurring Trp. The properties of the W259F mutant C2A domain are indistinguishable from the wild-type domain, and all the optical reporters, except for Trp-193, yield the same results. In the W259F background, Trp-193 does not exhibit marked fluorescence changes as it does in the position 258-truncated background. This finding suggests that removal of a short C-terminal segment of the C2A domain may result in a "looser" structure that is affected by the binding of Ca^{2+}. All subsequent experiments were carried out with reporters placed in the full-length C2A W259F background.

[24] J. Bai, C. Earles, J. Lewis, and E. R. Chapman, *J. Biol. Chem.* **275,** 25427 (2000).

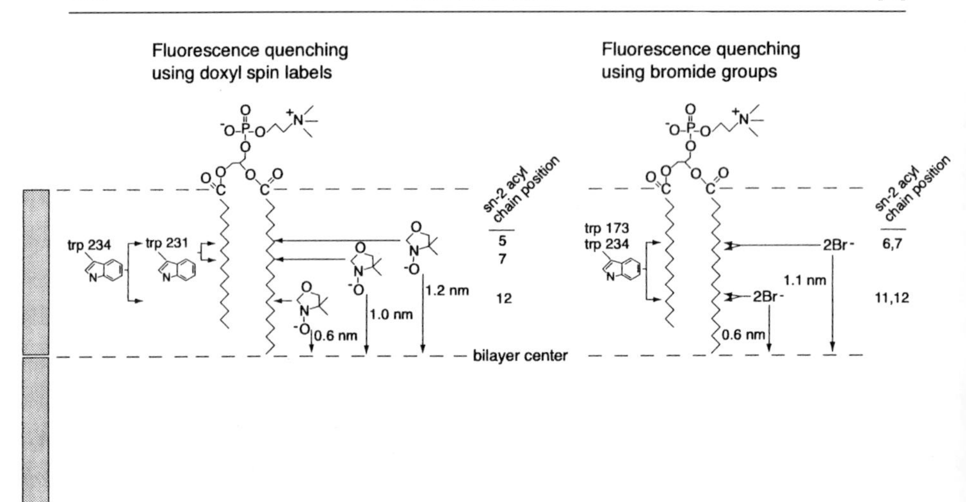

FIG. 2. Diagrammatic representation of direct measurement of C2A penetration into lipid bilayers, using membrane-embedded quenchers. On the left-hand side, the doxyl spin labels are indicated at the 5-, 7-, and 12-positions of the *sn*-2 acyl chain of PC. On the right-hand side, the bromide quenchers at the 6,7- and 11,12-positions of the acyl chain of PC are indicated. The relative positions of the quenchers from the center of the lipid bilayer are also indicated [L. A. Chung, J. D. Lear, and W. F. Degrado, *Biochemistry* **31,** 6608 (1992)]. Only those Trp residues that penetrate into the lipid bilayer will make contact with the quenchers and become quenched. The Trp reporters described in Fig. 1 were analyzed with the doxyl spin labels, and Trp-231 and Trp-234 were the only reporters that were quenched by the spin labels; these residues directly penetrate into lipid bilayers and are indicated [modified and reproduced from E. R. Chapman and A. F. Davis, *J. Biol. Chem.* **273,** 13995 (1998) with permission]. The precise depth of penetration was measured with the bromide quenchers and Trp reporters placed at position 234 (the distal tip of loop 3) and position 173 (the distal tip of loop 1). Both loops penetrated membranes and were quenched to similar degrees (Refs. 11 and 24). The quenchers used for these studies were obtained from Avanti Polar Lipids. The doxyl quenchers were synthetic 1-palmitoyl-2-stearoyl (*n*-DOXYL)-*sn*-glycero-3-phosphocholine (doxyl-PC), with the spin labels at the 5-, 7-, or 12-position of the *sn*-2 acyl chain, as well as 1-palmitoyl-2-stearoyl (6,7) dibromo-*sn*-glycero-3 phosphocholine (6,7-Br_2-PC) and 1-palmitoyl-2-stearoyl (11,12) dibromo-*sn*-glycero-3 phosphocholine (11,12-Br_2-PC). Bromide quenchers were synthetic 1-palmitoyl-2-stearoyl (*n*-DOXYL)-*sn*-glycero-3-phosphocholine (doxyl-PC), with the spin labels at the 5-, 7-, or 12-position of the *sn*-2 acyl chain. The brominated quenchers were 1-palmitoyl-2-stearoyl (6,7) dibromo-*sn*-glycero-3 phosphocholine (6,7-Br_2-PC) and 1-palmitoyl-2-stearoyl (11,12) dibromo-*sn*-glycero-3 phosphocholine (11,12-Br_2-PC).

To measure the depth of penetration of Trp-234 in the distal tip of Ca^{2+}-binding loop 3, we have applied a parallax method.[25,26a] In these experiments, membrane-embedded bromide quenching groups are used (shown diagrammatically in Fig. 2, right). Also, the possibility that Ca^{2+}-binding loop 1 of C2A might also penetrate into bilayers is addressed by placing a Trp reporter at position 173, which

[25] A. Chattopadhyay and E. London, *Biochemistry* **26,** 39 (1987).
[26a] F. S. Abrams and E. London, *Biochemistry* **31,** 5312 (1992).

is the distal tip of Ca^{2+}-binding loop 1 (Fig. 3B[11,20a,24]). As the mole fraction of brominated lipid is systematically increased, the degree of quenching of Trp-234 increases (Fig. 3A). Similar data are observed for Trp-173, demonstrating that, like loop 3, loop 1 also penetrates into membranes.[11] Qualitatively, the more shallow quencher (Br at the 6,7-positions) is more effective than the deeper quencher (Br at the 11,12-positions), indicating that both Trp reporters penetrate to a depth closer to the 6,7-quencher (which lies 1.1 nm from the bilayer center) than to the deeper quencher. To more precisely determine the depth of penetration, parallax analysis is applied. The distance of the Trp reporter, from the bilayer center (Z_{CF}), is given by Eq. (1):

$$Z_{CF} = L_{C_1} + [-\ln(F_1/F_2)/\pi C - L^2]/2L \quad (1)$$

where L_{C_1} is the distance from the bilayer center to the shallow quencher (11 Å for 6,7-Br_2-PC), C is the mole fraction of the quencher divided by the lipid area (70 $Å^2$), F_1 and F_2 are the relative fluorescence intensities of the fluorophore in the presence of the shallow quencher (6,7-Br_2-PC) and deep quencher (11,12-Br_2-PC), respectively, and L is the difference in the depth of the two quenchers (0.9 Å per CH_2 or CBr_2 group). For our calculations, the thickness of the hydrophobic region is taken to be 29–32 Å. Using these values and Eq. (1), the depth of penetration of Trp-173 and Trp-234 is ~9.5 and ~10.4 Å, respectively, away from the center of the bilayer. If one-half of the hydrophobic core of the bilayer is 14–15 Å, these data indicate that both reporters penetrate ~4–5 Å into the hydrophobic core, or roughly one-sixth into the bilayer. These findings are modeled in Fig. 3B; the equal degree of penetration of loops 1 and 3 suggests that C2A docks onto membranes in a perpendicular orientation.[24]

These data provide insights into a long-standing question regarding the biochemistry of C2 domains: why do they bind Ca^{2+} relatively poorly in the absence of membranes?[19,27] The finding that the Ca^{2+}- and membrane-binding sites correspond to the same region of the domain indicates that the lipid head groups may complete the Ca^{2+} coordination sites in C2A, thereby facilitating Ca^{2+} binding. This model sharply contrasts with other Ca^{2+}-binding domains in which Ca^{2+} drives large conformational changes/structural rearrangements that underlie the regulation of effectors. For example, Ca^{2+} binding drives the exposure of a myristoyl group from recoverin, which in turn drives membrane targeting.[28] The distinct Ca^{2+}/lipid-binding mechanism of the C2A domain of synaptotagmin does not require large-scale structural rearrangements, and therefore may be specialized for

[26b] H. Heller, K. Schaefer, and K. Schulten, *J. Phys. Chem.* **97,** 8343 (1993).

[26c] T. J. McIntosh and P. W. Holloway, *Biochemistry* **26,** 1783 (1987).

[27] M. D. Bazzi and G. L. Nelsestuen, *Biochemistry* **29,** 7624 (1990).

[28] A. M. Dizhoor, C. K. Chen, E. Olshevskaya, V. V. Sinelnikova, P. Phillipov, and J. B. Hurley, *Science* **259,** 829 (1993).

A

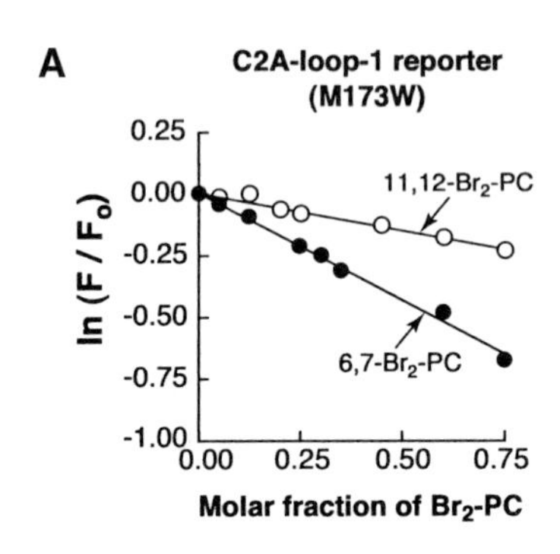

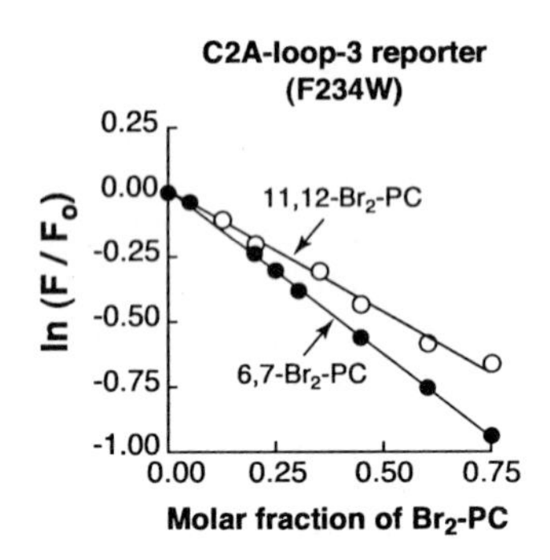

B

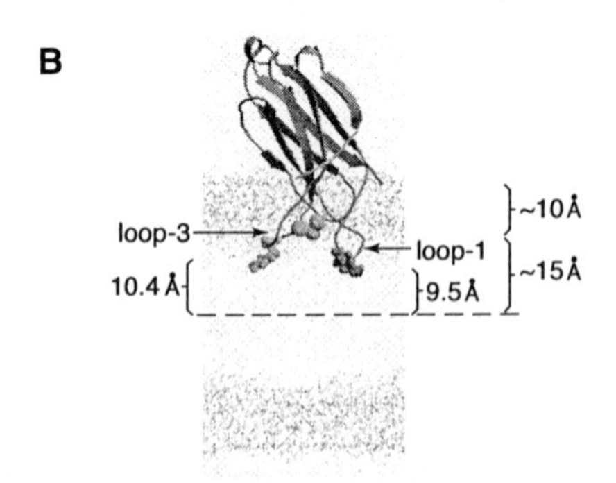

FIG. 3. Application of membrane-embedded bromine quenchers and the parallax method to measure the depth of penetration of Ca^{2+}-binding loops 1 and 3 of C2A. (A) On Ca^{2+}-triggered binding to membranes, Trp residues in Ca^{2+}-binding loops 1 (M173W, *top*) and 3 (F234W, *bottom*) of C2A are efficiently quenched by membrane-embedded quenchers. For these experiments, C2A · Ca^{2+} · liposome complexes were assembled using 5 μM C2A-Trp-reporter, 1 mM total lipid (25% PS and 75% PC or Br_2-PC as indicated), and 1 mM Ca^{2+}. Samples were excited at 285 nm and the emission spectra were collected from 300 to 380 nm and were corrected for blank, dilution, and instrument response. The fluorescence intensities in the absence (F_0) and presence of quencher in the liposomes (F) were determined by integrating the spectra. The degree of quenching was determined as a function of the

speed of response, as dictated by the rapid kinetics of exocytosis. In the following section we describe experiments in which the Trp reporters in C2A are exploited to determine the speed of response of this domain.

Time-Resolved Measurements of Protein–Lipid Interactions

C2A–membrane interactions are resolved in time by a stopped-flow rapid mixing approach. This method has advantages over other techniques such as plasmon resonance, temperature jump, or uncaging methods. Plasmon resonance suffers from the limitation that one of the binding partners must be immobilized on a surface, potentially changing the properties of the interaction under study, whereas stopped-flow experiments are carried out in solution and offer much faster mixing times. Temperature jump experiments provide excellent time resolution, but the extraction of rate constants from these experiments, in which an electrical discharge is used to heat the samples, is less straightforward than in stopped-flow mixing experiments. Finally, flash photolysis of caged Ca^{2+}, in conjunction with monitoring the fluorescence of the Trp reporters as they penetrate bilayers, could in principle afford excellent time resolution. However, the presence of free cage, coupled with incomplete photolysis, results in a period of time in which a new equilibrium must be established, and the time course of this process is similar to the "dead time" inherent in stopped-flow experiments.

For stopped-flow experiments, an optical reporter that can be used to monitor the interaction under study must be established. The Trp reporters described above provide an ideal means to monitor protein–membrane interactions because of the short lifetime of the Trp fluorophore. In this case, steady state equilibrium measurements (i.e., the radioligand-binding assay described above) demonstrated that the Ca^{2+} dependence for C2A–lipid interactions is not affected by the reporter mutations.[11] These findings strongly suggest that the reporter mutations do not significantly affect the kinetics of Ca^{2+}-triggered C2A–membrane interactions. We point out that the Trp reporters provide a readout for penetration. However, fluorescence resonance energy transfer (FRET) experiments, between the Trp reporter

molar fraction of brominated PC (Br_2-PC) with the bromine quenchers at either the 6,7-position (closed circles) or 11,12-positions (open circles). The depth of penetration was calculated by parallax analysis (see text for details). For our calculations we used the values obtained at a 0.75 molar fraction of Br_2-PC. According to this analysis, C2A penetrates about one-sixth into the hydrophobic core of the lipid bilayer. (B) Scale model depicting the Ca^{2+}-triggered docking of C2A onto a lipid bilayer. The crystal structure of C2A is rendered as described in Fig. 1; the simulated lipid bilayer is modified from H. Heller, K. Schaefer, and K. Schulten, *J. Phys. Chem.* **97,** 8343 (1993). The thickness of the hydrophilic (10 Å) and hydrophobic (15 Å) regions of one leaflet of the bilayer (Ref. 26c) are indicated in the model. The distances of the Trp reporters from the center of the bilayer (shown with a dashed line) are also indicated. [Modified and reproduced from J. Bai, C. Earles, J. Lewis, and E. R. Chapman, *J. Biol. Chem.* **275,** 25427 (2000), with permission.]

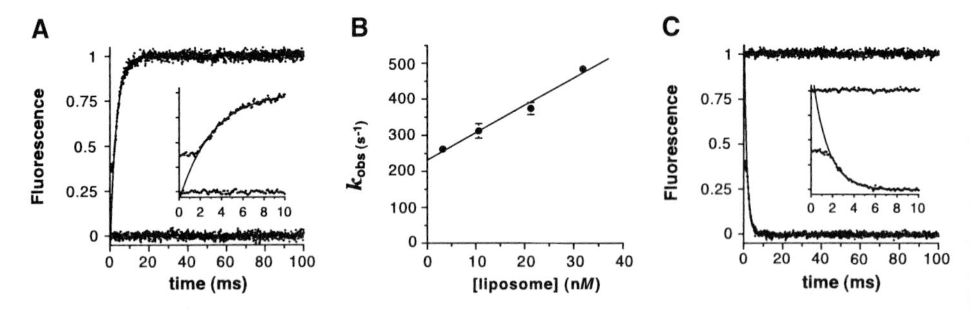

FIG. 4. Monitoring the dynamics of Ca^{2+}-regulated C2 domain–membrane interactions, using stopped-flow rapid mixing techniques. The fluorescence changes exhibited by C2A-F234W on Ca^{2+}-triggered binding to liposomes were used to monitor the kinetics of C2A–Ca^{2+}–liposome interactions. C2A-F234W was excited at 285 nm and emitted light was collected with a 326.1-nm bandpass filter (10-nm width; obtained from Melles Griot). All kinetics data are shown referenced to an arbitrary time = 0. The 2-ms plateau phase at the beginning of each trace corresponds to the signal that is collected during the "flow" and before the "stop"; the dead time was 0.7–1.2 ms. (A) Binding of C2A-F234W–Ca^{2+} to liposomes in real time. Liposomes (25% PS/75% PC; 11 n*M* final [liposome]) were premixed with Ca^{2+} (100 μM final concentration) and then rapidly mixed with C2A-F234W (5 μM final concentration). As a control, mixing experiments were also carried out with 2 m*M* EGTA instead of Ca^{2+} (lower trace), providing a true minimum reference point. Data (1000 points) were collected for 100 ms and are plotted with a best-fit single exponential function ($k_{obs} = 312\ s^{-1}$). The first 10 ms is shown on an expanded time scale in the inset. (B) Determination of k_{on} and k_{off} for the C2A-F234W–liposome complex in the presence of Ca^{2+}. Stopped-flow experiments were carried out as in (A) and k_{obs} was determined by fitting the data with single exponential functions and plotted versus [liposome]. The *y* intercept yields k_{off} for the C2A-F234W–liposome complex in the presence of Ca^{2+} (240 s^{-1}) and the slope yields k_{on} ($0.8 \times 10^{10}\ M^{-1}\ s^{-1}$) for binding of C2A-F234W–Ca^{2+} to liposomes (k_{obs} = [liposome]$k_{on} + k_{off}$). Error bars represent standard deviations from four independent experiments. (C) Disassembly kinetics of the C2A-F234W–Ca^{2+}–liposome complex on chelation of Ca^{2+}. C2A-F234W–Ca^{2+}–liposome complexes were assembled at 100 μM Ca^{2+}, using liposomes composed of 25% PS/75% PC. Disassembly reactions were carried out by rapidly mixing these complexes with 5 m*M* EGTA (final concentration). As a control, samples were mixed with buffer lacking EGTA, thus providing a maximum signal reference. Disassembly data were collected for 100 ms; the plot includes a best-fit single exponential function. The first 10 ms is shown on an expanded time scale in the inset. [Reprinted from A. F. Davis, J. Bai, D. Fasshauer, M. J. Wolowick, J. L. Lewis, and E. R. Chapman, *Neuron* **24,** 363 (1999), with permission from Elsevier Science.]

and an acceptor attached to the liposomes (e.g., dansyl-PC), can be readily applied to assay for binding. In our kinetic studies, the FRET and penetration signals have identical profiles, indicating that binding and penetration are closely coupled in time.[11] In principle, such an approach could be used to dissect apart steps in other protein–membrane interactions; in some cases, a slower penetration step may be preceded by a rapid initial binding step.

An example of a stopped-flow rapid mixing experiment is shown in Fig. 4A. In the upper trace, C2A harboring Trp-234 is premixed with Ca^{2+}, and this sample is then rapidly mixed with liposomes containing acidic phospholipids (25% PS/75% PC). Mixing results in a rapid increase in the fluorescence intensity as the Trp reporters penetrate into lipid bilayers. The same kinetics are observed when Ca^{2+} is premixed with liposomes and then rapidly mixed with C2A; this is because at [Ca^{2+}] levels required to trigger binding (micromolar range), Ca^{2+} binding occurs within the dead time of the instrument (~1 ms[11]). As a control (Fig. 4A, lower trace), Ca^{2+} is replaced by EGTA and membrane penetration is not observed. The increase in fluorescence shown in Fig. 4A is well fitted by a single exponential function, yielding k_{obs}. To determine the on and off rates for C2A–liposome interactions in the presence of Ca^{2+}, k_{obs} is plotted as a function of the [liposome] (*Note:* A stoichiometry of 97,000 lipids per 100-nm liposome is used to calculate the liposome concentration[29]). From this plot, the *y* intercept yields k_{off} (240 s^{-1}) and the slope yields k_{on} [0.8×10^{10} M^{-1} s^{-1}; Eq. (2), which assumes pseudo first-order kinetics]:

$$k_{obs} = [\text{liposome}]k_{on} + k_{off} \tag{2}$$

The value for k_{on} closely approaches the collisional limit calculated for this reaction (1.9×10^{10} M^{-1} s^{-1} for liposomes with a diameter of 100 nm[11,30]). These data yield a K_d of ~30 nM (in 100 μM Ca^{2+} with liposomes composed of 25% PS/75% PC), and indicate that anionic lipids and Ca^{2+} cooperate to form a highly stable complex, as predicted by the overlapping Ca^{2+} and lipid-binding sites within C2A.[11,20a,24]

We next address the rate at which the C2A-Ca^{2+}-liposome complex responds to decreases in [Ca^{2+}] by rapidly mixing the assembled complex with excess Ca^{2+} chelator (Fig. 4C, lower trace). As a control, complexes are also mixed with Ca^{2+} and no change in fluorescence is observed (Fig. 4C, upper trace). When mixed with chelator, the disassembly kinetics are extremely rapid (k_{diss} = 700 s^{-1} for complexes containing 100 μM Ca^{2+} and liposomes harboring 25% PS/75% PC), demonstrating that C2A can rapidly respond to both increases as well as decreases in [Ca^{2+}]. Comparisons with other C2 domains reveal that the C2A domain of synaptotagmin is "built for speed" and exhibits the fastest kinetics yet measured for a C2 domain.[11,31,32]

The mechanism of C2A–Ca^{2+}–membrane interactions, and the kinetics data described above, are summarized in the model shown in Fig. 5. A key point here is that the Ca^{2+}-triggered penetration of C2A into lipid bilayers fulfills the kinetic constraints of rapid exocytosis. In the presence of Ca^{2+}, the free energy associated with high-affinity C2A–membrane interactions may accelerate exocytosis by reducing the energy of activation for the membrane fusion reaction.

[29] P. R. Cullis, D. B. Fenske, and M. J. Hope, *in* "Biochemistry of Lipids, Lipoproteins and Membranes" (D. E. Vance and J. E. Vance, eds.), 3rd Ed., p. 1. Elsevier Science, Amsterdam, 1996.

[30] Y. Lu, M. D. Bazzi, and G. L. Nelsestuen, *Biochemistry* **34,** 10777 (1995).

[31] E. A. Nalefski, M. A. Wisner, J. Z. Chen, S. R. Sprang, M. Fukuda, K. Mikoshiba, and J. J. Falke, *Biochemistry* **40,** 3089 (2001).

[32] E. A. Nalefski, M. M. Slazas, and J. J. Falke, *Biochemistry* **36,** 12011 (1997).

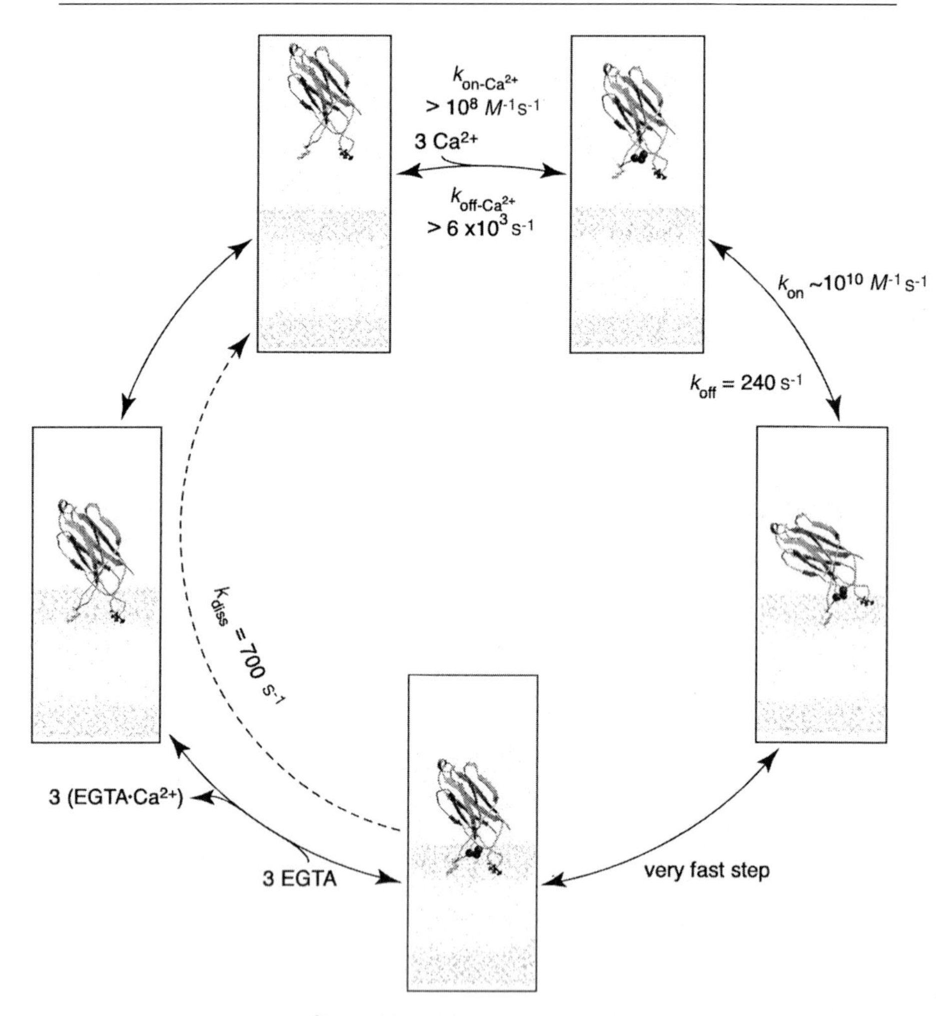

FIG. 5. Model for C2A–Ca^{2+}–liposome dynamics. Summary of the rate constants: $k_{on\text{-}Ca^{2+}} \geq 10^8\ M^{-1}\ s^{-1}$; $k_{off\text{-}Ca^{2+}} \geq 6 \times 10^3\ s^{-1}$ [determined from $k_{on\text{-}Ca^{2+}}$ and a K_d of 61 μM (Refs. 11 and 53)]; $k_{on} \approx 10^{10}\ M^{-1}\ s^{-1}$; $k_{off} = 240\ s^{-1}$ (in the presence of 100 μM Ca^{2+} with liposomes composed of 25% PS/75% PC); $k_{diss} = 700\ s^{-1}$ (for complexes assembled at 100 μM Ca^{2+}, using liposomes composed of 25% PS/75% PC). A detailed discussion of these constants is described in Davis *et al.* (Ref. 11).

Other Applications

The optical reporter system can also be used to study a number of additional fundamental aspects of protein–membrane complexes. Our example, the C2A–membrane interaction, is regulated by Ca^{2+}. A common means to measure the

Ca^{2+} dependence for complex formation is by immobilizing C2A onto beads and "pulling down" radiolabeled liposomes as a function of the [Ca^{2+}] (described in detail above). However, the pull-down approach suffers two limitations; first, it is not a true equilibrium binding assay because the beads must be washed to remove unbound liposomes, and second, the interaction of beads coated with C2A with the surface of a liposome is polyvalent, giving rise to errors in the apparent cooperativity of the Ca^{2+} dependence. Both of these problems are solved by using the optical reporters to monitor C2A–membrane interactions in solution. Furthermore, this assay can be readily exploited to measure the ability of other cations to drive C2A penetration into membranes. These experiments will make it possible to relate the divalent cation dependence of membrane penetration to the ability of divalent cations to trigger secretion.[33] We have initiated these studies, and these data are described in the next section. Finally, we have used this system to determine the stoichiometry of C2A–liposome complexes, an experiment made possible by the development of a solution-based binding assay, described below (Stoichiometry of Protein–Liposome Complex).

Divalent Cation Specificity of a C2-Domain–Membrane Interaction. Exocytosis is selectively triggered by Ca^{2+} versus other divalent cations; however, other metals are capable of triggering release to some extent.[33–36] Studies have begun to characterize, in detail, the ability of different metals to drive secretion of secretory organelles.[33] Thus, an important question to address among the 13 known isoforms of synaptotagmin concerns the metal requirements of synaptotagmin C2-domain effector interactions. We have begun to address this issue for the lipid penetration activity of the C2A domain of synaptotagmin, by taking advantage of the Trp-234 reporter. For these experiments, the degree of penetration is determined by monitoring increases in Trp-234 fluorescence in the presence of liposomes at increasing metal concentrations. As shown in Fig. 6, Ca^{2+} drove the most extensive degree of C2A–membrane penetration, as evidenced by the largest increase in fluorescence intensity using this metal. Ba^{2+} and Sr^{2+} were also able to drive penetration, albeit to a lesser degree (42 and 40%, respectively, versus Ca^{2+}, at saturating metal concentrations). Cd^{2+} and Mn^{2+} drove slight penetration of C2A (14 and 8.3%, respectively, versus Ca^{2+}), whereas Co^{2+} and Mg^{2+} had no effect. This parameter reports the relative extent of Trp penetration. We also used this assay to determine the effective concentrations for half-maximal penetration (designated as the EC_{50}). The EC_{50} values followed a similar pattern: $Ca^{2+} < Ba^{2+} < Sr^{2+}$ (the precise values and Hill coefficients are given in the caption to Fig. 6). The Ca^{2+}

[33] G. J. Augustine, M. P. Charlton, and S. J. Smith, *Annu. Rev. Neurosci.* **10,** 633 (1987).

[34] T. Kishimoto, T. T. Liu, Y. Ninomiya, H. Takagi, T. Yoshioka, G. C. R. Ellis-Davies, Y. W. Miyashita, and H. Kasai, *J. Physiol.* **533,** 627 (2001).

[35] Y. Goda and C. F. Stevens, *Proc. Natl. Acad. Sci. U.S.A.* **91,** 12942 (1994).

[36] J. L. Tomsig and J. B. Suszkiw, *J. Neurochem.* **66,** 644 (1996).

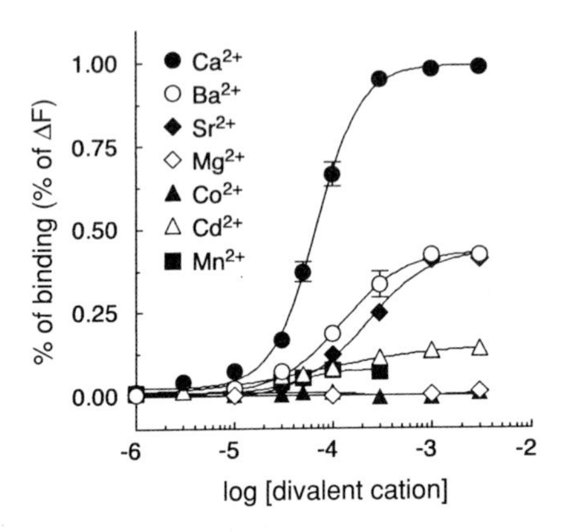

FIG. 6. Using fluorescent reporters to study the ion specificity of C2A–membrane interactions. C2A-F234W (5 μM), in the presence of 22 n*M* liposomes, was excited at 285 nm and spectra were collected from 300 to 400 nm as a function of [divalent cation]. Spectra were integrated, corrected, normalized, and plotted (closed circles) versus [divalent cation]. Data were fitted with GraphPad Prism 2.0 software. These results show that the C2A–membrane interaction was selectively promoted by Ca^{2+} versus other divalent ions ($EC_{50} = 69\ \mu M$; Hill coefficient = 1.93). Ba^{2+} and Sr^{2+} were less efficacious (Ba^{2+}: $EC_{50} = 124.3\ \mu M$, Hill coefficient = 1.37; Sr^{2+}: $EC_{50} = 223.1\ \mu M$, Hill coefficient = 1.38). Cd^{2+} and Mn^{2+} triggered only slight increases in fluorescence intensity. Mg^{2+} and Co^{2+} did not cause any changes in fluorescence. In all experiments, error bars represent the standard deviations from a minimum of three independent determinations. None of the metals had a significant effect on reporter fluorescence in the absence of liposomes.

requirement for penetration (69 μM) lies within the range of Ca^{2+} dependencies observed *in vivo* (range, from 10 to 200 μM Ca^{2+}).[37–39]

In future studies, this analysis will be applied to each C2 domain of each synaptotagmin isoform. These data will then be compared with metal requirements for fusion in various cell types.[33–36] This comparison should reveal insights into which synaptotagmin isoforms govern fusion in distinct secretory cells.

Stoichiometry of Protein–Liposome Complex. In this example, we used the intact cytoplasmic domain of synaptotagmin (residues 96–421, designated C2A–C2B) because it occupies twice the volume of the isolated C2A domain and will yield a more accurate determination of the number of synaptotagmins that can bind to a 100-nm liposome. For these studies we place a single reporter in position 234 of C2A–C2B (i.e., Ca^{2+}-binding loop 3 in the C2A domain). We do not use a

[37] R. Heidelberger, C. Heinemann, and G. Matthews, *Nature (London)* **371,** 513 (1994).

[38] J. H. Bollman, B. Sakmann, and J. G. G. Borst, *Science* **289,** 953 (2000).

[39] R. Schneggenburger and E. Neher, *Nature (London)* **406,** 889 (2000).

Trp residue, because removal of background Trp residues within the C2B domain adversely affects the protein. Instead, we use an AEDANS label placed on Cys-234 as follows.

Labeling of Single-Cysteine Mutant of C2A–C2B with IAEDANS. Purified recombinant C2A–C2B mutant (C2A–C2B F234C) is incubated with a 10-fold molar excess of 1,5-IAEDANS (Molecular Probes, Eugene, OR) at 25° for 1 hr in HEPES buffer [50 m*M* HEPES-NaOH (pH 7.4), 0.1 *M* NaCl]. Modified protein is then separated from unreacted free fluorophore, using Sephadex G-25 desalting columns (Amersham Pharmacia Biotech). Trace residual free probe is removed by dialyzing against 4 L of HEPES buffer for at least 4 hr at 4°. The AEDANS absorption spectrum is obtained with a Shimadzu (Kyoto, Japan) Biospec-1601 spectrophotometer, and an extinction coefficient of $6.0 \times 10^3 M^{-1}$ cm^{-1}, at 337 nm,[40] is used to calculate the AEDANS concentration. The protein concentration is determined by Coomassie blue staining of proteins separated by sodium dodecyl sulfate–polyacrylamide gel electrophoresis (SDS–PAGE), using bovine serum albumin (BSA) as a standard. From these values, the labeling ratios may be determined and are found to be 0.85–0.95 mol of label per mole of protein.

We then take advantage of the high affinity of Ca^{2+}-triggered C2A–C2B–liposome complexes (K_d in the nanomolar range)[20b] to determine the stoichiometry of liposomes "coated" with C2A–C2B F234C-AEDANS (designated as C2A*–C2B). Like Trp, AEDANS reports membrane penetration as a marked increase in fluorescence intensity and a blue shift in the emission spectra. Figure 7A shows the emission spectra of the AEDANS reporter as a function of [liposome]. The emission spectra are corrected, integrated, and the relative changes in fluorescence are plotted versus [liposome]. Due to the affinity of the interaction, the rise in the plot is linear until saturation is observed; the inflection point reveals the concentration of liposomes required to bind 2 μM C2A*–C2B; this value is ~10 n*M* liposomes, yielding a stoichiometry of ~200 C2A*–C2B domains per 100-nm liposome. This experiment has also been carried out in reverse: the liposome concentration is fixed at 4 n*M* and the [C2A*–C2B] is increased (Fig. 7B). Again, the relative change in fluorescence is plotted and saturation occurs at ~0.9 μM C2A*–C2B, yielding a stoichiometry of ~225 C2A*–C2B molecules per 100-nm liposome.

The lipid-binding surface area of C2A*–C2B is $\sim 2.8 \times 10^3$ Å^2 (estimated from the crystal structure reported by Sutton *et al.*[41]) and the surface area of a 100-nm liposome is $\sim 4.2 \times 10^6$ Å^2. From our stoichiometry experiments, ~15% of the liposome surface is coated with C2A*–C2B at saturation. Because synaptotagmin does not bind PC, these data suggest it binds to patches of PS. The fraction of the

[40] E. N. Hudson and G. Weber, *Biochemistry* **12,** 4154 (1973).

[41] R. B. Sutton, J. A. Ernst, and A. T. Brunger, *J. Cell Biol.* **147,** 589 (1999).

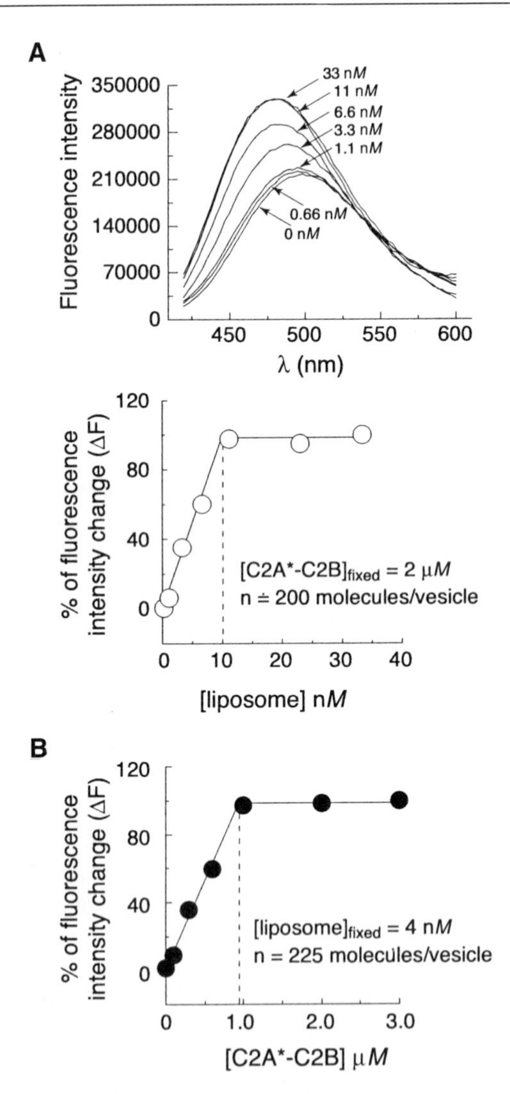

FIG. 7. Titration experiments reveal the stoichiometry of synaptotagmin–liposome complexes. In these experiments, a lone Cys residue was engineered into the distal tip of Ca^{2+}-binding loop 3 of the first C2 domain of the cytoplasmic domain of synaptotagmin I (designated as C2A–C2B F234C). *Note:* In this construct, the native Cys at position 277 was also replaced with an alanine residue. C2A–C2B F234C was labeled with an environmentally sensitive fluorescent reporter, AEDANS. The labeled protein is designated C2A*–C2B. This fluorescent group was used to report the membrane interaction of Ca^{2+}-binding loop 3 as an increase in intensity and a blue shift in the emission spectrum. (A) Liposome titration: C2A*–C2B was adjusted to 2 μM in TBS buffer and liposomes (~100 nm) were added at the indicated concentrations by using a liposome stock (20 mM total lipids, ~208 nM liposomes; 25% PS/75% PC). *Top:* Emission spectra obtained in the presence of 0.5 mM Ca^{2+} and

liposome that is coated with synaptotagmin is in reasonable agreement with the PS content of the liposomes (25%).

Conclusions

The human genome contains more than 120 different genes encoding proteins that harbor C2 domains; C2 domains are the twenty-fourth most common motif in the genome.[8] In many (but not all) cases, C2 domains bind membranes in response to Ca^{2+} signals.[42] However, the mechanism of "docking" onto membranes was not understood. The crystal structure of cytosolic phospholipase C held the first clue.[43,44] Molecular modeling suggested that in order for the catalytic domain of this protein to bind substrate, the Ca^{2+}-binding loops of C2A would be juxtaposed to the lipid bilayer. By "scanning" the surface of the crystal structure of the C2A domain of synaptotagmin, using Trp residues as optical reporters, the region that directly interacts with, and partially inserts into, the lipid bilayer was subsequently identified.[11,20a,24] The reporter mutations did not affect the equilibrium binding properties of C2A–Ca^{2+}–liposome complexes and were therefore used to determine the kinetics of the assembly and disassembly reactions. These kinetic studies revealed that C2A is "built for speed" and readily fulfills the kinetic requirements of rapid exocytosis.[11] The optical reporters used to monitor Ca^{2+}-triggered C2A–membrane interactions are now being exploited to study the metal requirements for C2 domain–membrane interactions. These studies should make it possible to further relate synaptotagmin–membrane interaction to physiological studies of the metal requirements for fusion in different cell types.

Subsequent studies of other C2 domains, using fluorescence and electron spin reporters, have confirmed that the Ca^{2+}-binding loops of other C2 domains also

[42] E. A. Nalefski and J. J. Falke, *Protein Sci.* **5,** 2375 (1996).
[43] L. Essen, O. Perisic, R. Cheung, M. Katan, and R. L. Williams, *Nature* (*London*) **380,** 595 (1996).
[44] L. Essen, O. Perisic, D. E. Lynch, M. Katan, and R. L. Williams, *Biochemistry* **36,** 2753 (1997).

the indicated [liposome]. AEDANS reporters were excited at 336 nm and the emission spectra were collected from 420 to 600 nm, using a PTI QM-1 fluorometer. All spectra were corrected for blank, dilution, and instrument response. *Bottom:* Corrected spectra (*top*) were integrated and the changes in fluorescence intensity were plotted versus [liposome]. The fluorescence intensity increased linearly during the liposome titration and reached a maximum at ~10 n*M* liposomes. Thus, approximately 200 C2A*–C2B molecules were bound to a 100-nm liposome composed of 25% PS/75% PC. (B) Protein titration: In these experiments, C2A*–C2B was adjusted to the desired concentrations, using an 80 μM stock. AEDANS fluorescence spectra were first obtained in the presence of 0.1 m*M* EGTA and 4-nm liposomes, and then after by the addition of Ca^{2+} to 0.5 m*M*. The Ca^{2+}-induced fluorescence intensity changes were plotted versus the protein concentration. The fluorescence intensity increased linearly and reached a maximum at ~0.9 μM protein. Thus, approximately 225 C2A*–C2B molecules were bound to a 100-nm liposome composed of 25% PS/75% PC.

form the C2 domain–membrane interface.[45,46] In addition, in the crystal structure of the C2 domain of protein kinase Cα, a glycerolphosphoserine molecule was localized between Ca^{2+}-binding loops 1 and 3.[47] Thus, among those C2 domains that dock onto membranes in response to Ca^{2+}, docking appears to occur via a conserved mechanism.

Acknowledgments

We thank Cynthia Earles, Ward Tucker, and Ralf Langen for critical comments on this manuscript and David Gaston for molecular modeling. This study was supported by grants from the NIH (GM 56827-01 and AHA 9750326N), and by the Milwaukee Foundation. E.R.C. is a Pew Scholar in the Biomedical Sciences. J.B. is supported by an AHA Pre-doctoral Fellowship.

[45] A. Ball, R. Nielsen, M. H. Gelb, and B. Robinson, *Proc. Natl. Acad. Sci. U.S.A.* **96,** 6637 (1999).
[46] E. A. Nalefski and J. J. Falke, *Biochemistry* **37,** 17642 (1998).
[47] N. Verdaguer, S. Corbalan-Garcia, W. F. Ochoa, I. Fita, and J. C. Gomez-Fernandez, *EMBO J.* **18,** 6329 (1999).

[10] Caging Proteins through Unnatural Amino Acid Mutagenesis

By E. JAMES PETERSSON, GABRIEL S. BRANDT, NIKI M. ZACHARIAS, DENNIS A. DOUGHERTY, and HENRY A. LESTER

Introduction

Caged compounds provide unparalleled temporal and spatial control over cellular processes. Although a large number of photoreleasable small molecules and peptides have been generated,[1–4] relatively few full-length caged proteins have been prepared or utilized.[5] Caging groups can be incorporated in peptides generated through chemical synthesis, but this places a limit on the size of the protein.[6,7] Bayley, Marriot, and others have used cysteine- and lysine-modifying reagents

[1] S. R. Adams and R. Y. Tsien, *Annu. Rev. Physiol.* **55,** 755 (1993).
[2] G. Dorman and G. D. Prestwich, *Trends Biotechnol.* **18,** 64 (2000).
[3] G. Marriott, ed., *Methods Enzymol.* **291** (1998).
[4] J. A. McCray and D. R. Trentham, *Annu. Rev. Biophys. Biophys. Chem.* **18,** 239 (1989).
[5] K. Curley and D. S. Lawrence, *Curr. Opin. Chem. Biol.* **3,** 84 (1999).
[6] V. Borisenko, D. C. Burns, Z. Zhang, and G. A. Woolley, *J. Am. Chem. Soc.* **122,** 6364 (2000).
[7] K. Curley and D. S. Lawrence, *J. Am. Chem. Soc.* **120,** 8573 (1998).

0076-6879/03 $35.00

with photoreactive groups to cage proteins posttranslationally.[6,8–13] However, this technique suffers from an inability to target the protecting group to a specific residue and a limit on the kinds of functionality that can be revealed by deprotection. Both of these problems can be eliminated by using the nonsense suppression method for incorporating unnatural amino acids site specifically.[14–17] This permits almost complete freedom in choosing a caged moiety and its location in the protein. A number of caged proteins have been generated in this manner, using both *in vitro* and *in vivo* translation systems. Figure 1 (structures **1–12**) shows the amino acids that have been incorporated through nonsense suppression.

Incorporating caged unnatural amino acids makes it possible to photochemically control a broad range of processes, from protein folding, to protein–protein interactions, to ligand binding. For example, Short *et al.* were able to control dimerization of human immunodeficiency virus type 1 (HIV-1) protease *in vitro* by incorporating a nitrobenzyl (NB)- or nitroveratryl (NV)-derivatized aspartate (**1**, **2**).[18] Mendel and co-workers also used an Asp(NB) in caging the active site of T4 lysozyme.[19] In addition, the Schultz laboratory has used the decaging of a serine residue [Ser(NB), **3**] to initiate intein splicing from the *Thermococcus litoralis* DNA Vent polymerase.[20]

In our laboratory, we have expressed several receptors containing photodeprotectable unnatural amino acids in *Xenopus* oocytes, permitting *in vivo* electrophysiological studies. Tyrosine (NB) (**4**) was incorporated in place of an endogenous tyrosine to cage the binding site of the mouse muscle nicotinic acetylcholine receptor (nAChR).[21] Flash photolysis permitted recovery of the response of nAChR

[8] H. Bayley, *Bioorg. Chem.* **23,** 340 (1995).

[9] C. Chang, B. Niblack, B. Walker, and H. Bayley, *Chem. Biol.* **2,** 391 (1995).

[10] C. Chang, T. Fernandez, R. Panchal, and H. Bayley, *J. Am. Chem. Soc.* **120,** 7661 (1998).

[11] R. Golan, U. Zehavi, M. Naim, A. Patchornick, and P. Smirnoff, *Biochim. Biophys. Acta* **1293,** 238 (1996).

[12] G. Marriott, *Biochemistry* **33,** 9092 (1994).

[13] S. Thompson, J. A. Spoors, M.-C. Fawcett, and C. H. Self, *Biochem. Biophys. Res. Commun.* **201,** 1213 (1994).

[14] J. A. Ellman, D. Mendel, S. J. Anthony-Cahill, C. J. Noren, and P. G. Schultz, *Methods Enzymol.* **202,** 301 (1991).

[15] C. J. Noren, S. J. Anthony-Cahill, M. C. Griffith, and P. G. Schultz, *Science* **244,** 182 (1989).

[16] M. W. Nowak, J. P. Gallivan, S. K. Silverman, C. G. Labarca, D. A. Dougherty, and H. A. Lester, *Methods Enzymol.* **293,** 504 (1998).

[17] M. E. Saks, J. R. Sampson, M. W. Nowak, P. C. Kearney, F. Du, J. N. Abelson, H. A. Lester, and D. A. Dougherty, *J. Biol. Chem.* **271,** 23169 (1996).

[18] G. F. Short, M. Lodder, A. L. Laikhter, T. Arslan, and S. M. Hecht, *J. Am. Chem. Soc.* **121,** 478 (1999).

[19] D. Mendel, J. A. Ellman, and P. G. Schultz, *J. Am. Chem. Soc.* **113,** 2758 (1991).

[20] S. N. Cook, W. E. Jack, X. Xiong, L. E. Danley, J. A. Ellman, P. G. Schultz, and C. J. Noren, *Angew. Chem. Int. Ed. Engl.* **34,** 1629 (1995).

[21] J. C. Miller, S. K. Silverman, P. M. England, D. A. Dougherty, and H. A. Lester, *Neuron* **20,** 619 (1998).

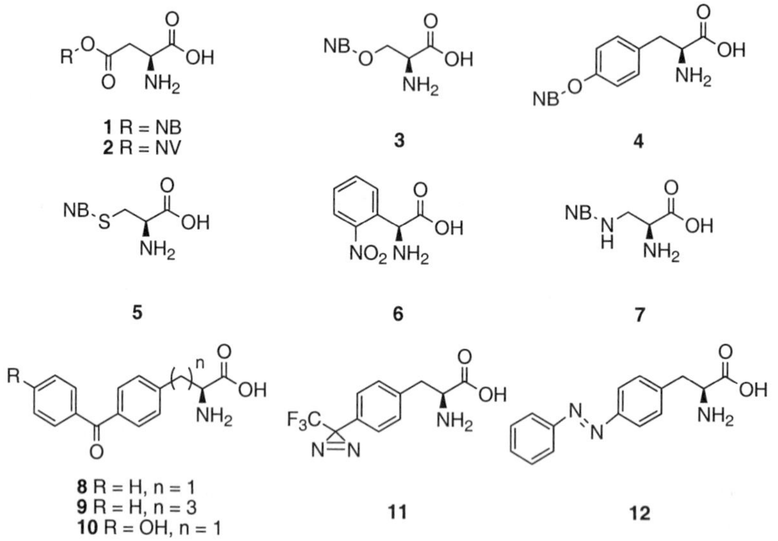

FIG. 1. Photoactive unnatural amino acids incorporated through nonsense suppression. caged compounds: aspartate (**1** and **2**), serine (**3**), tyrosine (**4**), cysteine (**5**), nitrophenylglycine (Npg, **6**), and β-aminoalanine (**7**). Other photoreactive amino acids: *p*-benzoylphenylalanines (**8–10**), trifluoromethyldiazirinylphenylalanine (TmdPhe, **11**), and *p*-phenylazophenylalanine (**12**). NB, Nitrobenzyl; NV, nitroveratryl.

to acetylcholine. The pore of the same ion channel has been caged by incorporating *o*-nitrobenzylcysteine (**5**) and *o*-nitrobenzyltyrosine in pore-lining segments of the receptor.[22] Photochemical initiation of protein trafficking was demonstrated by decaging a Tyr(NB) on the inward rectifier potassium channel Kir2.1.[23] Revealing the tyrosine hydroxyl group permitted recognition by endocytotic machinery.

The caged residue need not be a natural amino acid. We have used the unnatural amino acid 2-nitrophenylglycine (Npg, **6**) to initiate protein splicing in Shaker B K^+ channels and muscle nAChRs.[24] Caged β-aminoalanine (**7**) has also been incorporated.[25] In fact, the residue need not even be an amino acid; there is

[22] K. P. Philipson, J. P. Gallivan, G. S. Brandt, D. A. Dougherty, and H. A. Lester, *Am. J. Physiol. Cell Physiol.* **281,** C195 (2001).

[23] Y. Tong, G. Brandt, M. Li, G. Shapovalov, E. Slimko, A. Karschin, D. Dougherty, and H. Lester, *J. Gen. Physiol.* **117,** 103 (2001).

[24] P. M. England, H. A. Lester, N. Davidson, and D. A. Dougherty, *Proc. Natl. Acad. Sci. U.S.A.* **94,** 11025 (1997).

[25] V. W. Cornish, D. Mendel, and P. G. Schultz, *Angew. Chem. Int. Ed. Engl.* **34,** 621 (1995).

precedent for the use of noncaged hydroxy[26–29] and hydrazino acids.[30] Although they are not precisely caged compounds, it should be noted that amino acids that photocrosslink (**8–10**, **11**)[31–34] or photoisomerize (**12**)[35,36] have also been introduced.

The above-described examples are provided to give an idea of the scope of experiments that are possible with caged unnatural amino acids. We focus here on receptor-based expression methodology and assays. Once a protein of interest and a residue within that protein have been chosen, a photolabile protecting group for the side chain (Fig. 2A, structures **13–20**) and an orthogonal (i.e., nonphotochemical) protecting group for the α-amine of the amino acid (Fig. 2B, structures **21–28**) must be picked. This is necessary for coupling the amino acid to the pdCpA dinucleotide before enzymatic ligation to form the full-length suppressor tRNA. After expression of the protein in *Xenopus* oocytes, the decaging can be done in a variety of ways; we describe two, one of which permits real-time electrophysiological monitoring. Thorough descriptions of *in vivo* nonsense suppression methodology are available elsewhere,[16,17] and therefore we focus on the techniques that apply specifically to using caged unnatural amino acids in this context.

Experimental Protocols

We describe here minimal protocols relevant to nonsense suppression. The following section then deals with aspects of the suppression methodology that are specific to introducing photolabile amino acids. The reader should consult earlier methods papers for guidelines on implementing unnatural amino acid mutagenesis in general.[16,36–39] Here we outline a procedure for suppression of *amber* codons

[26] P. M. England, H. A. Lester, and D. A. Dougherty, *Biochemistry* **38,** 14409 (1999).

[27] P. M. England, H. A. Lester, and D. A. Dougherty, *Tetrahedron Lett.* **40,** 6189 (1999).

[28] P. M. England, Y. Zhang, D. A. Dougherty, and H. A. Lester, *Cell* **96,** 89 (1999).

[29] T. Lu, A. Y. Ting, J. Mainland, L. Y. Jan, P. G. Schultz, and J. Yang, *Nat. Neurosci.* **4,** 239 (2001).

[30] J. A. Killian, M. D. Van Cleve, Y. F. Shayo, and S. M. Hecht, *J. Am. Chem. Soc.* **120,** 3032 (1998).

[31] T. Kanamori, S. Nishikawa, I. Shin, P. G. Schultz, and T. Endo, *Proc. Natl. Acad. Sci. U.S.A.* **94,** 485 (1997).

[32] T. Kanamori, S. Nishikawa, M. Nakai, I. Shin, P. G. Schultz, and T. Endo, *Proc. Natl. Acad. Sci. U.S.A.* **96,** 3634 (1999).

[33] B. Martoglio, M. W. Hofmann, J. Brunner, and B. Dobberstein, *Cell* **81,** 207 (1995).

[34] W. Mothes, S. U. Heinrich, R. Graf, I. Nilsson, G. von Heijne, J. Brunner, and T. A. Rapoport, *Cell* **89,** 523 (1997).

[35] T. Hohsaka, D. Kajihara, Y. Ashizuka, H. Murakami, and M. Sisido, *J. Am. Chem. Soc.* **121,** 34 (1999).

[36] M. Sisido and T. Hohsaka, *Bull. Chem. Soc. Jpn.* **72,** 1409 (1999).

[37] M. A. Gilmore, L. E. Steward, and A. R. Chamberlin, *Top. Curr. Chem.* **202,** 77 (1999).

[38] L. E. Steward and A. R. Chamberlin, *Methods Mol. Biol.* **77,** 325 (1998).

[39] J. S. Thorson, V. W. Cornish, J. E. Barrett, S. T. Cload, T. Yano, and P. G. Schultz, *Methods Mol. Biol.* **77,** 43 (1998).

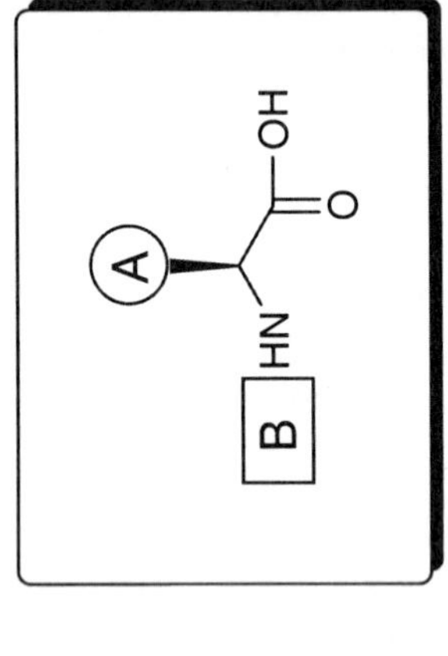
A
13a NB
R1 = H, R2 = H
13b NV
R1 = H, R2 = OMe
14a NVOC
R1 = H, R2 = OMe
14b NMOC
R1 = CO2H, R2 = H
15 Dinitrobenzyl
16 3-Nitrophenyl
17 Phenacyl
18 Desyl
19 Cinnamoyl
R = H or N(Et)2
20 Coumarinyl
R = H, Me, Ac, or
CH3CH2CO
B
21a 4-PO R = H
21b Dimethyl-4-PO R = Me
22 Alloc
23 N-Phenylfluorenyl-S-
2-amino-4-pentenoyl
24 Pyroglutamoyl
25 Bpoc
26 Fmoc
27 Boc
28 Cbz

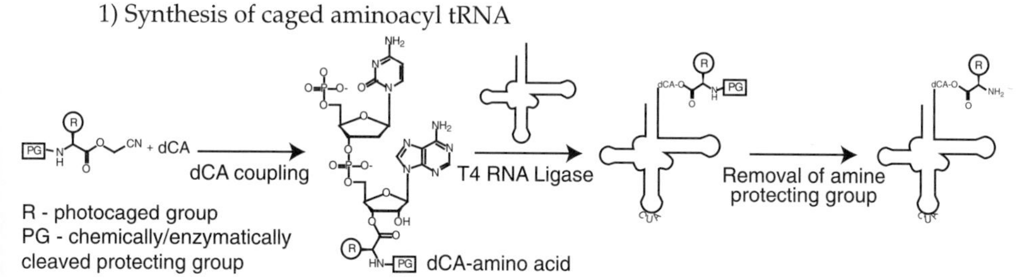

2) Introduction of stop codon into mRNA at site of interest

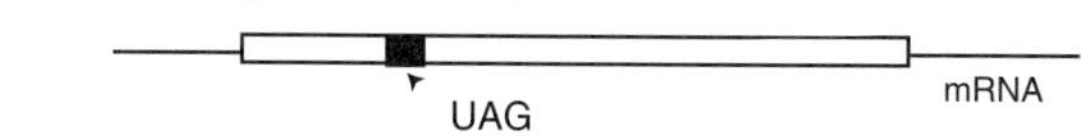

3) Injection of mRNA and charged tRNA to express caged protein

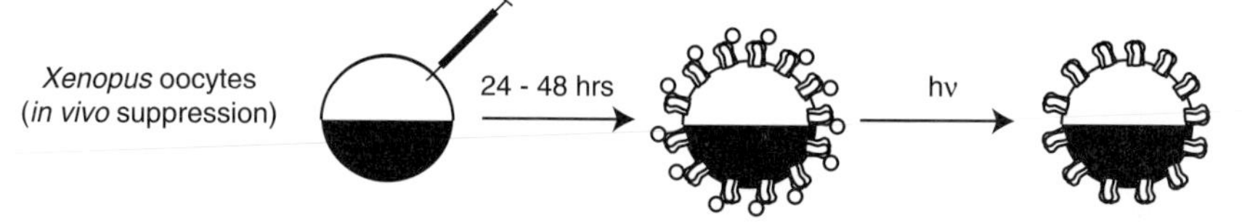

FIG. 3. *In vivo* nonsense suppression of caged unnatural amino acids. Schematic diagram representing the steps described in Experimental Protocols.

in ion channels expressed in *Xenopus* oocytes. Figure 3 gives a schematic view of the process.

Stop Codon Introduction

A stop codon is introduced into the gene of interest by standard techniques of site-directed mutagenesis. The base immediately 5′ of the TAG codon is also changed to a C to facilitate screening by providing a cleavage site for the restriction enzyme *Bfa*I. However, this is not done if this C mutation changes which amino acid is encoded. Typical oocyte expression vectors include a T7 site for *in vitro* transcription, an untranslated sequence (such as that from alfalfa mosaic virus)

FIG. 2. Protecting groups. (**A**) Photolytic side-chain caging groups: **13a**, nitrobenzyl (NB; Ref. 64); **13b**, nitroveratryl (NV; Ref. 64); **14a**, nitroveratryloxycarbonyl (NVOC; Refs. 14 and 64); **14b**, nitromandelyloxycarbonyl (NMOC; Refs. 64 and 65); **15**, dinitrobenzyl (Refs. 66 and 67); **16**, 3-nitrobenzyl (Refs. 68 and 69); **17**, phenacyl (Refs. 70 and 71); **18**, desyl (Refs. 66, 70, and 72); **19**, cinnamoyl (Refs. 73 and 74); **20**, coumarinyl (Refs. 75 and 76). (**B**) Nonphotolytic α-amino protecting groups: **21a**, 4-pentenoyl (4-PO; Refs. 18 and 58); **21b**, 2,2-dimethyl-4-pentenoyl (Ref. 58); **22**, allyloxycarbonyl (Alloc; Refs. 31 and 32); **23**, *N*-benzyl-*N*-phenylfluorenyl-2-amino-4-pentenoyl (Ref. 59); **24**, pyroglutamoyl (Ref. 56); **25**, biphenylisopropoxycarbonyl (Bpoc; Refs. 25, 77, and 78); **26**, (9-fluorenylmethyloxy)carbonyl (Fmoc; Refs. 77 and 78); **27**, *tert*-butyloxycarbonyl (Boc; Refs. 77 and 78) **28**, benzyloxycarbonyl (Cbz; Refs. 77 and 78).

to provide a binding site for the ribosome, and a reliable restriction site 3′ of the poly(A) tail (typically, we employ *Not*I).

mRNA Transcription

The mutant cDNA is linearized with *Not*I and precipitated. mRNA is run off from the linearized DNA template, using a T7 RNA polymerase. The mMESSAGE mMACHINE kit from Ambion (Austin, TX) works well for this purpose. The resulting mRNA is resuspended in RNase-free water and stored at −80°.

tRNA 74-mer Transcription

Truncated (74-nucleotide) tRNA is also generated via runoff *in vitro* transcription. Because of the short length of the transcript, methods specially designed for small RNAs are desirable. We have found the Ambion MEGAshortscript kit to be a convenient way to generate truncated tRNA.

Synthesis of dCA Amino Acid

The actual methods for chemical synthesis of the properly protected and caged amino acid are outside the scope of this review, as is the condensation of the protected, activated amino acid with 5′-*O*-phosphoryl-2′-deoxycytidyl-adenosine (pdCpA). These procedures are detailed in a number of reviews.[16,36,38–40] The side chain can be caged with any of the caging groups shown in Fig. 2A and the α-amine can be caged with a group from Fig. 2B. It must be noted that only nitrobenzyl-based groups **13a** and **13b** have been used in suppression experiments thus far; a discussion of this is found in Caging Groups, below.[18–24,41] Aminoacyl dinucleotides are typically stored as 3 m*M* solutions in dimethyl sulfoxide (DMSO) at −80°. It is desirable to have not only caged unnatural amino acid but also the wild-type residue at the intended position of insertion as a positive control.

Ligation of dCA Amino Acid to tRNA

Full-length charged tRNA is generated by enzymatic ligation of the dCA-amino acid to the 74-nucleotide tRNA, using T4 RNA ligase.[16,36,38–40] Subsequent to ligation, tRNA charged with the requisite caged amino acid is precipitated and resuspended in water buffered to pH 4.5. It is stored at −80° and can be aliquoted to avoid repeated freeze–thaw cycles. When it is thawed, it should be stored on ice. However, charged tRNA is relatively stable at −80°; certain tRNAs have been used successfully in suppression experiments even after several years.

[40] J. M. Humphrey and A. R. Chamberlin, *Chem. Rev.* **97,** 2243 (1997).

[41] M. Lodder, C. F. Crasto, A. L. Laikhter, H. Y. An, T. Arslan, V. A. Karginov, G. F. Short, and S. M. Hecht, *Can. J. Chem.* **78,** 884 (2000).

Injection of tRNA and mRNA into Oocytes

Oocytes are coinjected with N-deprotected aminoacyl-tRNA plus mRNA as follows. Typically, tRNA charged with an enantiomerically pure amino acid protected with the iodine-labile 4-PO group (**21a**) is resuspended at 2 μg/μl. Immediately before injection, it is combined for 10 min with an equal volume (typically 0.5 μl) of saturated aqueous iodine. The 1 μl of deprotected aminoacyl-tRNA is then mixed with an equal volume (1 μl) of mRNA in water. For a multisubunit receptor, such as the nAChR, the suppressed subunit is usually present in some excess over the wild-type subunits. For instance, suppression in the α subunit is often carried out at a subunit ratio of $10:1:1:1$ ($\alpha:\beta:\gamma:\delta$) and a total mRNA concentration of 4.0 ng in 25 nl. Given a 50-nl oocyte injection, each cell thus receives 25 ng of tRNA and 25 ng of mRNA. Positive and negative control oocytes from the same batch are also injected. Negative controls include injection of mRNA without tRNA and injection of mRNA with tRNA that has not been synthetically aminoacylated. As a positive control, tRNA aminoacylated with the wild-type residue is injected along with the mRNA. The oocytes are incubated at 18° with shaking in ND96 medium supplemented with sodium pyruvate, theophylline, gentamicin sulfate, and 5% heat-inactivated horse serum. Typically, expression may be observed after 24 hr and is maximal by 48 hr postinjection.

Other Expression Systems

For biochemical assays, mutant proteins can be produced by *in vitro* translation. Many commercially available extracts may be used.[36,42,43] We have found Promega (Madison, WI) wheat germ extract to be convenient for the detection of full-length proteins by Western blotting.[26–28] Receptor subunits that contain the hemagglutinin epitope and C-terminal hexahistidine (His_6) tags have been prepared for this purpose.

Electrophysiology and Other Assays

Oocytes are assayed in an electrophysiological apparatus designed for simultaneous irradiation and recording (Fig. 4).[21] In a typical experiment, the oocyte is held at the desired membrane potential in a standard two-electrode voltage clamp setup and assayed for expression of the receptor.[23] The prephotolysis record is used to determine the effect, if any, of the presence of the unnatural amino acid and its caging group on receptor function. Optimally, recording continues during the irradiation, so that rate-limiting steps can be monitored. In favorable cases, the decaging proceeds so efficiently that the light is delivered as a flash, briefer than

[42] L. Jermutus, L. A. Ryabova, and A. Pluckthun, *Curr. Opin. Biotechnol.* **9,** 534 (1998).

[43] N. Budisa, C. Minks, S. Alefelder, W. Wenger, F. Dong, L. Moroder, and R. Huber, *FASEB J.* **13,** 41 (1999).

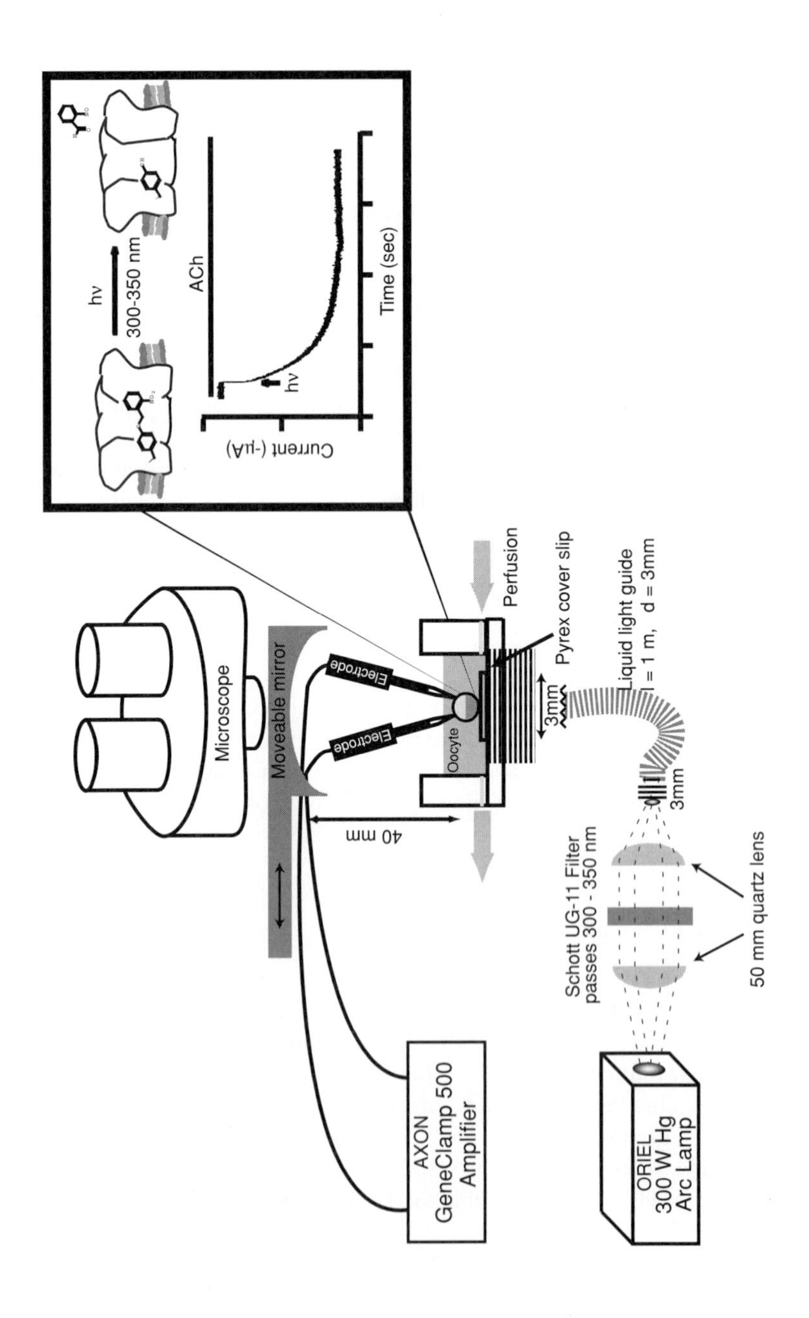
hν
300-350 nm
ACh
Current (-μA)
hν
Time (sec)
Microscope
Moveable mirror
Electrode
Electrode
Oocyte
40 mm
3mm
Perfusion
Pyrex cover slip
Liquid light guide
l = 1 m, d = 3mm
3mm
Schott UG-11 Filter
passes 300 - 350 nm
50 mm quartz lens
AXON
GeneClamp 500
Amplifier
ORIEL
300 W Hg
Arc Lamp

activation of the channels.[21,22] In other cases, the rate-limiting step is delivery of the light itself (several seconds to several hours). It is not practical to record voltage–clamp currents from an oocyte for periods longer than tens of minutes. Therefore, for longer irradiations (as in the case of the 4-hr Npg deprotection), photocleavage of the caging group is not done on the electrophysiology rig. Naturally, the control oocytes injected above are subject to the identical procedure.

When desired, oocytes may be utilized in biochemical assays subsequent to or in addition to electrophysiological recording. Oocytes have been subjected to a full range of modern physiological measurements, such as simultaneous fluorescence and voltage–clamp, modification by cysteine reagents, and electrochemistry; and these techniques are also available for oocytes expressing proteins with caged side chains.

Experimental Considerations

Expression Systems

The first requirement for introducing an unnatural amino acid into a receptor is a viable expression system for that receptor. The majority of experiments employing nonsense suppression as a means of incorporating unnatural amino acids into proteins have relied on bacterial extracts to produce the proteins of interest. Unfortunately, many receptors cannot be functionally reconstituted subsequent to expression in a bacterial system. More recent efforts have employed bacterial extracts modified to increase suppression efficiency and extracts from eukaryotic systems.[37,44] *In vitro* systems such as wheat germ and rabbit reticulocyte extracts have been utilized for nonsense suppression of functional soluble proteins.[36,45,46] In our laboratory, wheat germ extracts are routinely used

[44] A. V. Karginov, M. Lodder, and S. M. Hecht, *Nucleic Acids Res.* **27,** 3283 (1999).
[45] V. A. Karginov, S. V. Mamaev, and S. M. Hecht, *Nucleic Acids Res.* **25,** 3912 (1997).
[46] J. D. Bain, C. Switzer, A. R. Chamberlin, and S. A. Benner, *Nature* (*London*) **356,** 537 (1992).

FIG. 4. Real-time electrophysiological monitoring of protein decaging. Shown is an apparatus for simultaneous irradiation and electrophysiological recording from *Xenopus* oocytes. The output of the arc lamp (in this case, a flashlamp) is passed through a 300- to 350-nm bandpass filter and focused onto a fiber optic liquid light guide. When continuous illumination is used, the beam is also passed through a water filter to eliminate IR. The light guide directs the beam onto the oocyte, which is clamped in a standard two-electrode configuration. A concave mirror can be moved into position above the bath, reflecting some of the beam that has passed around the oocyte back to the shadowed upper surface. The mirror increases the overall flash intensity by ~55%. *Inset:* The millisecond photolytic decaging of a Tyr(NB) incorporated into the pore-lining M2 region of the mouse nAChR γ subunit. The electrophysiological trace below shows that removal of the caging group increases acetylcholine (ACh)-induced current (negative by convention) in the ion channel. Electrophysiological data are reproduced with permission from Philipson *et al.* (Ref. 22).

to generate full-length membrane proteins for polyacrylamide gel electrophoresis (PAGE) analysis.[26,27] In principle, receptors containing unnatural amino acids could be reconstituted into bilayers by using such a system supplemented with microsomes.[47] However, there is no precedent for unnatural amino acid mutagenesis of an integral membrane protein, using a cellular extract. Expression in *Xenopus* oocytes is the only method to date that has been shown to be effective for incorporating unnatural amino acids into functional receptors.[16,48–51]

Suppressor tRNA

It is likely that suppression efficiency in different expression systems is largely dependent on the suppressor tRNA employed. Amber suppressor glutamine tRNA from *Tetrahymena* has given good results in oocytes, and yeast $tRNA^{Phe}$ has occasionally been useful.[17,49,50] The performance of a number of tRNAs in bacterial extracts has been reviewed, although their utility for nonsense suppression in eukaryotic systems is largely untested.[52] Orthogonal tRNAs for mammalian expression have been developed, but their compatibility for use in incorporating unnatural amino acids via nonsense suppression is also unknown.[53] Some practitioners of unnatural amino acid suppression in cell extracts have employed four-base codons and their complementary engineered tRNAs.[36,46] This will be an interesting strategy to pursue in eukaryotic systems, as it may address some of the issues of site variability, readthrough, and truncation discussed below.

Receptor

Suppression has been demonstrated in the major classes of neuroreceptors and ion channels. Unnatural amino acids have been incorporated into a number of ligand-gated channels, such as the nAChR (numerous subunits) and 5-HT_3 receptors, a G protein-coupled receptor (NK1), GIRKs (G protein inward rectifier K^+ channels), and several potassium channels including Kir2.1 and

[47] L. K. Lyford and R. L. Rosenberg, *J. Biol. Chem.* **274,** 25675 (1999).

[48] A. Chollet and G. Turcatti, *Lett. Peptide. Sci.* **5,** 79 (1998).

[49] G. Turcatti, K. Nemeth, M. D. Edgerton, U. Meseth, F. Talabot, M. Peitsch, J. Knowles, H. Vogel, and A. Chollet, *J. Biol. Chem.* **271,** 19991 (1996).

[50] G. Turcatti, K. Nemeth, M. D. Edgerton, J. Knowles, H. Vogel, and A. Chollet, *Receptors Channels* **5,** 201 (1997).

[51] M. W. Nowak, P. C. Kearney, J. R. Sampson, M. E. Saks, C. G. Labarca, S. K. Silverman, W. Zhong, J. Thorson, J. N. Abelson, N. Davidson, P. G. Schultz, D. A. Dougherty, and H. A. Lester, *Science* **268,** 439 (1995).

[52] S. T. Cload, D. R. Liu, W. A. Froland, and P. G. Schultz, *Chem. Biol.* **3,** 1033 (1996).

[53] H. J. Drabkin, H. J. Park, and U. L. RajBhandary, *Mol. Cell. Biol.* **16,** 907 (1996).

Shaker.[21,23,24,29,48,50,54,55] In unpublished work from our laboratory, we have incorporated unnatural amino acids into the cystic fibrosis transmembrane conductance regulator (CFTR), a P2X receptor, and a neurotransmitter transporter. Among these are monomeric and multimeric receptors, both homomeric and heteromeric versions of the latter. We emphasize that heterologous expression of any novel protein is by no means guaranteed, because it is uncertain whether the protein will be folded, assembled, or transported properly. However, our experience suggests that any protein that can be expressed effectively in *Xenopus* oocytes will be amenable to incorporation of unnatural amino acids by nonsense suppression.

Site of Incorporation

On the basis of our experience to date, there is considerable freedom in choosing a suppression site. Unnatural amino acids have been incorporated in extracellular domains,[21,49,50] in transmembrane regions,[22,27] and in intracellular loops.[23,24] Sites have been suppressed successfully near both the N and C termini and in regions with a variety of secondary structures. However, suppression efficiency is variable, and clear rules have yet to be established for predicting whether suppression will work well at any given site. In general, the standard practice has been to attempt suppression of the desired residue at the desired site accompanied by a set of three controls, two negative and one positive.

mRNA Only. When suppression at a new site is attempted, mutant mRNA without exogenous tRNA as a control must be injected. The expression of functional receptors suggests either that there is significant readthrough of the stop codon or that the protein truncated at the proposed insertion site remains functional.

mRNA Plus Uncharged Suppressor tRNA. Another essential control is coinjection of mutant mRNA with suppressor tRNA ligated to dCA. There is a formal possibility that the suppressor tRNA may be recognized by endogenous aminoaocyl-tRNA syntheses. If this is the case, any suppressor tRNA that is uncharged, either from incomplete ligation to the dCA-amino acid or by subsequent hydrolysis of the aminoacyl linkage, may be charged with a natural amino acid. It is important to determine the extent to which functional proteins generated in a nonsense suppression experiment arise as a result of incorporation of the unnatural amino acid or because of the undesired incorporation of a natural amino acid supplied by the host cell. Caged amino acids have the desirable property that they confer novel photosensitivity on the protein, so that a functional distinction may be made between these two populations of receptors.

[54] H. Dang, P. M. England, S. S. Farivar, D. A. Dougherty, and H. A. Lester, *Mol. Pharmacol.* **57,** 1114 (2000).

[55] S. K. Silverman, Ph.D. Thesis. California Institute of Technology, Pasadena, CA (1998).

mRNA Plus Suppressor tRNA with Native Amino Acid. Finally, a positive control is obligatory. Recovery of wild-type activity of the protein *via* nonsense suppression is a valuable positive control. This control will reveal cases in which there is a systemic problem of translation or folding and assembly of the suppressed protein. If wild-type recovery is possible at a given site, unnatural amino acids can usually be incorporated. It may be that expression levels will be low, such that an extremely sensitive technique (typically electrophysiology) is necessary to detect the mutant protein. Low expression may compromise experiments that require somewhat greater amounts of protein. In some cases, expression can be increased by injecting the oocytes multiple times over the course of several days. These subsequent injections may include mRNA only, tRNA only, or both. Although we have observed the best results when injecting a 1 : 1 mixture of mRNA and tRNA, the conditions for optimal expression have not been conclusively determined.

Caging Groups

The choice of caged side chain depends, of course, on the exact nature of the information to be gained from the experiment. The literature contains precedents for caged side-chain hydroxyls [Tyr (**4**)[21–23] and Ser (**3**)[20]], thiols [Cys (**5**)[22]], acids [Asp (**1**)[18,19]], amines [β-aminoalanine (**7**)[25]], and amides [Npg (**6**),[24] in which backbone cleavage can be seen as photolytic "decaging" of the peptide bond]. As mentioned above, the suppression efficiency of an unnatural amino acid is related to the nature of its side chain, but in a complex fashion. Most investigators have employed relatively small caging groups, because the charged tRNA must pass through the ribosome in order for the amino acid to be incorporated into the receptor. As is discussed further below, it is difficult to predict whether problems will be encountered in incorporating a particular amino acid. Investigators have generally reported that steric and especially charge conservation of the native side chain help promote efficient expression. In practice, there is no substitute for simply attempting the suppression at the desired position with the desired unnatural amino acid.

The only caging groups precedented for unnatural amino acid suppression thus far are nitrobenzyl (NB, **13a**)[18–25] and nitroveratryl (NV, **13b**).[18] Other groups also have attractive photochemical properties (Fig. 2A). It remains to be seen whether these groups can be incorporated by nonsense suppression. There are several considerations in choosing a caging group: (1) synthetic characteristics: the degree to which it is synthetically accessible and compatible with coupling to pdCpA; (2) photochemical characteristics: its action spectrum, quantum yield for photolysis, and speed of the dark reactions that complete photolysis; and 3) reactivity in the system of interest: its stability in water, and the possible reactivity of its photoproducts.

The α-amino group of the unnatural amino acid must also be protected for the coupling to pdCpA. Conventionally, this is done with a nitroveratryloxycarbonyl [NVOC (**14a**), Fig. 2A] group. However, because this group requires photochemical deprotection, NVOC is not compatible with the introduction of caged side chains. Although it is theoretically possible to use photocleavable protecting groups with nonoverlapping action spectra, this is not recommended as several α-amino-protecting groups that are cleavable chemically or enzymatically (pyroglutamoyl, **24**)[56] are available. These are illustrated in Fig. 2B. We have used primarily the 4-PO (**21a**)[22–24,57–59] group, but all of the others have been used successfully in suppression experiments. 4-PO is removed by mixing the aminoacyl-tRNA with 25% (v/v) saturated I_2 in water before oocyte injection.

Photolysis Apparatus

Various types of photolysis apparatus have been employed for side-chain decaging. Experiments involving photoreactive unnatural amino acids such as diazirine, aryl azides, and benzophenone (**8–12**)[31–36] are also relevant here, as are experiments with caged small molecules in *Xenopus* oocytes.[60,61] In general, methods of photolysis can be divided into those performed on single oocytes and those performed on batches of oocytes.

An apparatus for real-time decaging in single oocytes (Fig. 4)[21–23,62] is designed for the particular needs of these unusually large cells ($\sim$1 mm in diameter). In most cases, inexpensive arc lamps (either continuous irradiation or pulsed) are preferred over the much more costly lasers if illumination of the entire oocyte is desired.[61] A "point-source" arc has a diameter (2–3 mm) on the order of an oocyte; therefore relatively little light is lost in focusing onto the oocyte. The coherence of the laser becomes an advantage only when the intention is to focus the laser down to a spot $<100\ \mu m$ in diameter, for example, for patch recording. The briefer pulse of the laser can rarely be exploited, because voltage-clamp circuits typically have time resolution on the order of the flash lamp ($\sim$1 ms).

The voltage-clamp electronics must be carefully shielded from the 14-kV trigger pulse of the flash-lamp circuit. Also, metal surfaces in the microelectrodes

[56] J. R. Roesser, C. Xu, R. C. Payne, C. K. Surratt, and S. M. Hecht, *Biochemistry* **28,** 5185 (1989).

[57] M. Lodder, S. Golovine, and S. M. Hecht, *J. Org. Chem.* **62,** 778 (1997).

[58] M. Lodder, S. Golovine, A. L. Laikhter, V. A. Karginov, and S. M. Hecht, *J. Org. Chem.* **63,** 794 (1998).

[59] M. Lodder, B. Wang, and S. M. Hecht, *Tetrahedron* **56,** 9421 (2000).

[60] L. Niu, R. W. Vazquez, G. Nagel, T. Friedrich, E. Bamberg, R. E. Oswald, and G. P. Hess, *Proc. Natl. Acad. Sci. U.S.A.* **93,** 12964 (1996).

[61] I. Parker, N. Callamaras, and W. G. Wier, *Cell Calcium* **21,** 441 (1997).

[62] J. Nargeot, H. A. Lester, N. J. Birdsall, J. Stockton, N. H. Wassermann, and B. F. Erlanger, *J. Gen. Physiol.* **79,** 657 (1982).

must be shielded from the flash itself, to avoid photoelectric effects. Although the flash lamps in our laboratory are custom-built,[62] excellent devices can be bought from Rapp OptoElectronic (Hamburg, Germany; http://www.rapp-opto.com/) and T.I.L.L. Photonics (Gräfelfing, Germany; http://www.till-photonics.de). Our setup employs an Oriel 66011 lamp (ThermoOriel, Stratford, CT) housing with a high-pressure Hg/Xe lamp run at 300 W by an Oriel 68810 power supply. The output is filtered with a (Schott Glas, Mainz, Germany) UG-11 filter [and a water IR filter (ThermoOriel) if the lamp is run in continuous irradiation mode] and focused with 50-mm quartz lenses (ThermoOriel). An Ealing 22-8411 electronic shutter is employed to run the lamp in flash mode. The focused beam is directed onto the oocyte with a liquid light guide (1 m long, 3 mm in diameter; ThermoOriel) aimed through a Pyrex coverslip. Finally, a concave first-surface mirror (20-mm focal length, 50 mm in diameter; Rolyn Optics, Covina, CA) is positioned ~40 mm above the oocyte.

Methods for decaging proteins expressed in many oocytes at once are simpler.[24] In the particular case of Npg decaging, oocytes were irradiated for 4 hr at 4° in Pyrex vials with a 288-W Hg lamp (BLAK-RAY longwave ultraviolet lamp; Ultraviolet Products, San Gabriel, CA) equipped with a 360-nm band pass filter at a distance of 15–30 cm.

Assay

Finally, a fundamental methodological requirement for unnatural amino acid mutagenesis is an assay capable of detecting the effects of photolyzing a caged side chain. In fact, one of the reasons that receptors are such an attractive target for studies of this nature is that electrophysiology may be employed for this purpose. Electrophysiology is exquisitely sensitive and therefore capable of detecting the small amounts of protein generated by nonsense suppression. To date, only two-electrode voltage clamp recordings have been made with receptors containing caged amino acids; but single-channel recordings are in principle possible. In addition, electrophysiology has been successfully coupled with irradiation in a number of experiments.[60,61] (See other articles in this volume as well as *Methods in Enzymology,* Vol. 291.[3])

Other methods may be employed to detect downstream effects of side-chain decaging. There is ample precedent for the application of most biochemical methods to receptors expressed in oocytes.[63a] However, given that small amounts of protein are produced in nonsense suppression experiments, these methods often require some adaptation for use. The literature is not terribly extensive, but there are examples of nonelectrophysiological techniques that have been successful in the

[63a] H. Soreq and S. Seidman, *Methods Enzymol.* **207,** 225 (1992).

[63b] *Methods Enzymol.* **360,** [1], [2], [17], [21], [23], [24], [25], this volume (2003).

analysis of the effects of single side-chain decaging.[23] For example, endocytosis of Kir2.1 containing caged tyrosine has been tracked by both capacitance and surface fluorescence measurements.

Summary

The caging of specific residues of proteins is a powerful tool. This discussion attempts to alert the reader to the considerations that must be made in preparing and analyzing a caged protein through nonsense suppression. Although the suppression methodology is conceptually straightforward, it not possible to provide a failsafe "cook book" method for using caged unnaturals. We have emphasized the preparation of caged receptors expressed in *Xenopus* oocytes, but these approaches can clearly be adapted to many other systems.

Acknowledgment

This work was supported by the National Institutes of Health (Grants GM-29836, NS-11756, NS-34407, and NS-11756).

[64] C. P. Holmes, *J. Org. Chem.* **1997,** 2370 (1997).
[65] F. M. Rossi, M. Margulis, R. E. Hoesch, C. M. Tang, and J. P. Kao, *Methods Enzymol.* **291,** 431 (1998).
[66] K. R. Gee, B. K. Carpenter, and G. P. Hess, *Methods Enzymol.* **291,** 30 (1998).
[67] K. R. Gee, L. Niu, K. Schaper, V. Jayaraman, and G. P. Hess, *Biochemistry* **38,** 3140 (1999).
[68] G. P. Hess and C. Grewer, *Methods Enzymol.* **291,** 443 (1998).
[69] P. Kuzmic, L. Pavlickova, and M. Soucek, *Coll. Czech. Chem. Commun.* **51,** 1293 (1986).
[70] R. S. Givens, J. F. Weber, A. H. Jung, and C. H. Park, *Methods Enzymol.* **291,** 1 (1998).
[71] J. C. Sheehan, K. Umezawa, *J. Am. Chem. Soc.* **38,** 3771 (1973).
[72] J. C. Sheehan, R. M. Wilson, and A. W. Oxford, *J. Am. Chem. Soc.* **93,** 7222 (1971).
[73] B. L. Stoddard, P. Koenigs, N. Porter, K. Petratos, G. A. Petsko, and D. Ringe, *Proc. Natl. Acad. Sci. U.S.A.* **88,** 5503 (1991).
[74] A. D. Turner, S. V. Pizzo, G. Rozakis, and N. A. Porter, *J. Am. Chem. Soc.* **110,** 244 (1998).
[75] T. Furuta and M. Iwamura, *Methods Enzymol.* **291,** 50 (1998).
[76] R. S. Givens and B. Matuszewski, *J. Am. Chem. Soc.* **106,** 6860 (1984).
[77] J. D. Bain, D. A. Wacker, E. E. Kuo, M. H. Lyttle, and A. R. Chamberlin, *J. Org. Chem.* **56,** 4615 (1991).
[78] J. D. Bain, D. A. Wacker, E. E. Kuo, and A. R. Chamberlin, *Tetrahedron* **47,** 2389 (1991).

[11] Preparation and Light-Directed Activation of Caged Proteins

By GERARD MARRIOTT, PARTHA ROY, and KENNETH JACOBSON

Introduction

A major challenge in cell biology is to understand the regulation of cell motility in terms of the multiple protein-mediated reactions that drive this complex process. These mechanism-based studies require knowledge of the intracellular function and regulation of the individual reactions that collectively drive motility. In principle, this information can be obtained by chemical relaxation techniques. In this approach a rapid perturbation such as a temperature jump is used to perturb the reaction of interest from its equilibrium state. The rate of relaxation back to the equilibrium state is continuously recorded, using a signal associated with the reaction, for example, a fluorescence signal from a reactant. The rate constants for the reaction can be calculated from a series of relaxation experiments and used to establish the reaction mechanism.

In this article we focus on methods for perturbing actin-binding protein (ABP) activity in complex molecular environments and for monitoring the state of ABP reactions, using a fluorescent actin conjugate. An emerging practice in cell biology is to study specific reactions of a process within the context of the biological system, for example, within a living cell.[1] Unfortunately, physical perturbations, such as a T-jump, are nonspecific in these complex systems and are often incompatible with cell viability. Kaplan *et al.*,[2] on the other hand, showed that light-directed activation of "caged" substrates can generate protein activity jumps within complex biological environments. Caged derivatives of various enzyme substrates and ligands (reviewed by Corrie and Trentham[3]) including peptides[4] have also been used to study specific reactions and functions of myosin in muscle and nonmuscle cells. On the other hand, we introduced an approach to directly cage ABP activity by modifying essential amino acid residues—protein activity is restored to the caged protein by removing the caged groups, using near-UV light.[5,6] This direct approach to cage protein activity offers several advantages over methods employing caged

[1] J. A. Theriot and T. J. Mitchison, *Nature* (*London*) **352,** 126 (1991).

[2] J. H. Kaplan, B. Forbush, and J. F. Hoffman, *Biochemistry* **17,** 1929 (1978).

[3] J. E. T. Corrie and D. R. Trentham, *in* "Bioorganic Photochemistry" (H. Morrison, ed.). Wiley, New York, 1993.

[4] J. W. Walker, S. H. Gilbert, R. M. Drummond, M. Yamada, R. Sreekumar, R. E. Carraway, M. Ikebe, and F. S. Fay, *Proc. Natl. Acad. Sci. U.S.A* **95,** 1568 (1998).

[5] G. Marriott, *Biochemistry* **33,** 9092 (1994).

[6] G. Marriott and M. Heidecker, *Biochemistry* **35,** 3170 (1996).

0076-6879/03 $35.00

substrates and ligands, including (1) specificity—irradiation of a caged ligand such as ATP within a cell triggers the activity of a host of proteins, whereas activity is generated directly by irradiating a caged protein; (2) speed—an uncaged protein is immediately active at the site of irradiation, but with photoactivated ligands the rate of protein activation depends on the diffusion rate of the ligand; and (3) caged proteins can be used to investigate protein structure–function relationships from the perspective of a single amino acid residue within a complex environment. In this article we describe methods used in our laboratories for perturbing protein activity from caged proteins and present select applications in the areas of protein structure–activity relationships and protein function within living cells. Although the protein modification approach is suitable for caging soluble proteins, it is not suitable for intrinsic membrane proteins such as ion channels and receptors. However the Lester group[7] has described *in vivo* translation approaches to introduce caged amino acids into membrane receptors. The interested reader should consult the excellent review in [10] in this volume[7a] for details of this technique.[7a]

Proteins

The protein preparation should be highly purified according to sodium dodecyl sulfate–polyacrylamide gel electrophoresis (SDS–PAGE) analysis and exhibit maximal activity. Protein activity is assayed before the caging reaction.

Caging Reagents

The most commonly used reagents for caging proteins fall into three categories:

1. Lysine-directed monofunctional caged reagents[5]
2. Cysteine-directed monofunctional caged reagents[6]
3. Heterobifunctional, photocleavable reagents[8,9]

Caged proteins are prepared under precisely defined labeling conditions, including the protein-to-caged reagent ratio and pH. The optimized protocol should lead to the inactivation of >90% protein activity with the fewest number of caged groups. It is important to avoid reaction conditions that result in overlabeling of the protein,[10] because these heterogeneous caged conjugates usually exhibit a low yield of photoactivation.[5]

[7] J. C. Miller, S. K. Silverman, P. M. Englan, D. A. Dougherty, and H. A. Lester, *Neuron* **20,** 619 (1998).

[7a] E. J. Petersson, G. S. Brandt, N. M. Zacharias, D. A. Dougherty, and H. A. Lester, *Methods Enzymol.* **360,** [10], 2003 (this volume).

[8] G. Marriott, H. Kmiyata, and K. Kinosita, *Biochem. Int.* **26,** 943 (1992).

[9] J. Ottl, D. Gabriel, and G. Marriott, *Bioconjug. Chem.* **9,** 143 (1998).

[10] C. H. Self and S. Thompson, *Nat. Med.* **2,** 817 (1996).

Caging Chemistry

Lysine-Directed Monofunctional Caged Reagents

The amino-directed protein-caging reactions[5] are performed with ~20 μM protein in 2–50 mM sodium borate, pH 8–9, for 30 min. Essential cofactors and ions may be added to the labeling buffer so long as they are free of amino groups. 4,5-Dimethoxy-2-nitrobenzyl chloroformate (NVOC-Cl; Fluka, Ronkonkoma, NY) is freshly prepared to 0.1 M in acetone and is rapidly added to the protein to a final concentration of 1 mM. The protein solution may turn cloudy on the addition of NVOC-Cl. The reaction is complete within 30 min and the solution is centrifuged at 10,000g for 10 min at 4° and then dialyzed against a desired buffer at 4°. A smaller volume of the protein solution is used as a control and is subjected to the same conditions described above, with the exception that pure acetone is used instead of NVOC-Cl–acetone. The absorption spectra and activity of the control and caged proteins are measured after centrifugation at 10,000g for 10 min at 4°. The NVOC : protein labeling ratio is calculated from the value of the NVOC absorption at 350 nm, using an extinction coefficient of 5000 M^{-1} cm^{-1}, whereas the protein concentration is determined by the Bradford assay or by SDS–PAGE analysis. In practice, the optimal labeling ratio is determined by varying the reagent concentration between 0.5 and 2 mM at a constant protein concentration (20 μM). Because NVOC-Cl randomly labels lysine residues, caged proteins often exhibit some residual activity; we define a caged protein as a conjugate with $<$10% of the activity of the control protein.

Cysteine-Directed Monofunctional Caged Reagents

An alternative approach for caging proteins is to target essential cysteine residues, using thiol-reactive caged reagents.[6,11] For example, the activity of heavy meromyosin (HMM), luciferase, or papain is completely inhibited by labeling one or more cysteine residues. Caged proteins can be engineered from other proteins by cysteine-scanning mutagenesis; the ideal mutant will be one that is inactive only after labeling with a thiol-reactive caged reagent. Because only a single cage group is removed from cysteine-directed caged proteins the photoactivation yield is usually higher compared with NVOC-caged proteins. The main disadvantage of this approach, however, is the need to generate and screen many single-cysteine mutants.[12]

Bromomethyl-3,4-dimethoxy-2-nitrobenzene (BMDNB) is a simple alkylating reagent for caging proteins on cysteine.[6] The caging reaction involves dialyzing the protein at ~20 μM in a thiol-free buffer within a pH range of pH 6–8. BMDNB

[11] P. Pan and H. Bayley, *FEBS Lett.* **405,** 81 (1997).
[12] C. Y. Chang, B. Niblack, B. Walker, and H. Bayley, *Chem. Biol.* **2,** 391 (1995).

is added to a final concentration of 20–100 μM from a 20 mM dimethylformamide (DMF) solution. The reaction is complete within 60 min at room temperature. The labeled protein is centrifuged at 10,000g for 10 min at 4° and dialyzed against a desired buffer containing 1 mM dithiothreitol (DTT) at 4° in the dark. A smaller volume of protein serves as a control and is subjected to the same conditions, with the exception that pure DMF is used instead of BMDNB–DMF. The absorption spectra and activity of both control and caged proteins are recorded after a second centrifugation run at 10,000g for 10 min at 4°. The labeling ratio of BMDNB to protein is calculated from the absorption value at 350 nm (BMDNB), using an extinction coefficient of 5000 M^{-1} cm^{-1}. The protein concentration is determined by the Bradford assay or SDS–PAGE analysis.

Heterobifunctional, Photocleavable Cross-Linking Reagents

In this caging approach a heterobifunctional, photocleavable cross-linking reagent is used to form a cross-link between a large dextran molecule and the protein of interest; the dextran molecule should inhibit the activity of the protein by physically masking a binding site. This approach can be used to cage almost any wild-type protein by labeling a limited number of lysine residues. The dextran may also incorporate either a fluorescent probe for imaging the caged protein within a cell or a specific antibody for targeting the caged protein to a specific site or cell. We have described two different classes of heterobifunctional, photocleavable reagent for caging proteins: (1) Bromomethyl-2-nitro benzoic acid–N-hydroxy succinamide (BNBA–NHS) for preparing a caged actin dimer[8] and (2) various activated forms of the N-1-(3,4-dimethoxy-6-nitrophenyl)-2,3-epoxypropyl group (DMNEP) for preparing caged actin–dextran complexes.[9] In this review we consider the properties and applications only of DMNEP reagents.

N-1-(3,4-Dimethoxy-6-nitrophenyl)-2,3-epoxypropyl-Based Reagents. DM NEP-based heterobifunctional, photocleavable reagents exhibit novel design features for the cross-linking and uncaging reactions.[9] These include (1) separation of the cross-linking unit from the photochemical unit and (2) incorporation of the dimethoxy-2-nitrophenyl group, which exhibits a high extinction coefficient (5000 M^{-1} cm^{-1} at 350 nm) and a red-shifted action spectrum. Interestingly, we find that the amino group reactivity of the chloroformate group in DMNEP is far lower compared with NVOC-Cl, although this reactivity is improved in several reagents having an activated carbonate or carbamate. The dimethyl aminopyridine (DMAP) salt of DMNEP chloroformate (Scheme 1) is the most reactive reagent for caging proteins in aqueous buffers. Scheme 1 summarizes a DMNEP-mediated protein-caging protocol, which involves reacting the protein (20 μM) with an excess of the DMAP salt of DMNEP (100–500 μM) in thiol-free, 50 mM borate buffer, pH 8.5, for 5 min. The labeling ratio and caging efficiency of the DMNEP-labeled protein are determined by absorption spectrophotometry and

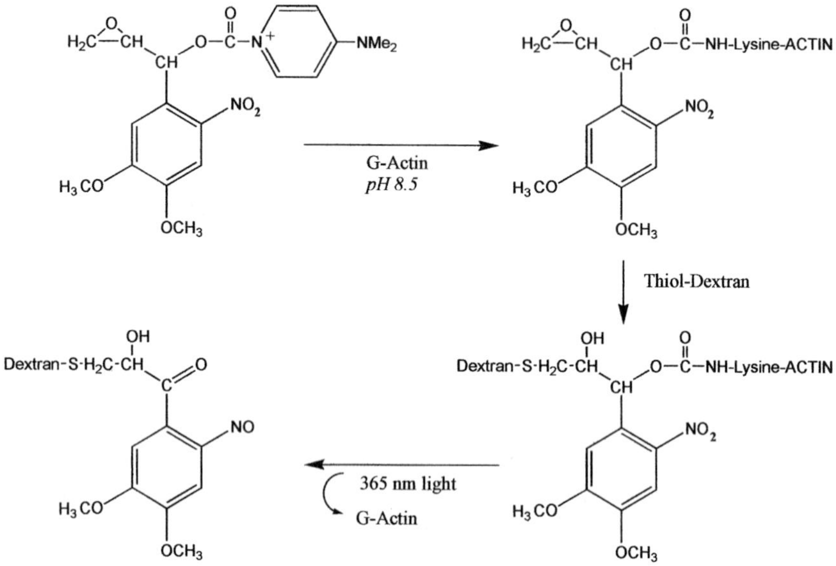

SCHEME 1. Preparation and photoactivation of caged actin, using the DMAP salt of DMNEP.

protein activity measurements. Aminodextran (Molecular Probes, Eugene, OR) is thiolated with iminothiolane (Sigma, St. Louis, MO) according to Ottl *et al.*[9] Essentially, aminodextran dissolved to 50–100 μM in nitrogen-purged, 50 mM borate buffer, pH 8.5, is treated with an excess of freshly prepared iminothiolane (~1 mM, in water). The reaction is left overnight at 4° and excess iminothiolane is removed by dialysis in nitrogen-purged, thiol-free borate buffer, pH 8.5. The thiol content of the dextran is determined by absorption analysis after labeling a small fraction of the conjugate with excess Acrylodan (6-acryloyl-2-dimethylaminonaphthalene, 500 μM; Ottl *et al.*[9]). The thiolated dextran is added to the DMNEP conjugate for 1 hr at 20°. DTT is added to quench the reaction and, if necessary, excess thiolated dextran is removed by gel-filtration or affinity chromatography, for example, using nitrilotriacetate (NTA)–Sepharose for His-tagged proteins. Quantitative analysis of the caging efficiency is determined by measuring the protein activity and/or the amount of non-cross-linked protein as seen by SDS–PAGE.

Rapid and Continuous Measurements of Relaxation Process

The activity of ABPs is measured directly or indirectly, using the fluorescence of 6-propionyl-2-dimethylaminonaphthalene (Prodan; Molecular Probes, Eugene, OR)–actin. This environmentally sensitive fluorescent actin conjugate exhibits a dramatic shift in its emission maximum (496 to 465 nm) with an 8-fold increase in intensity at 465 nm during actin polymerization.[13] Importantly, by irradiating the sample with light between 400 and 420 nm it is possible to selectively excite Prodan–actin in the presence of caged groups. ABPs exert different effects on the polymerization rate of actin, which can be used to determine ABP activity in the unmodified, caged, and uncaged states. The most commonly used assay formats are as follows: (1) filament (+)-end capping: the addition of CapG, a (+)-end capping protein, to Prodan-labeled F-actin at 0.1 μM in the presence of Ca^{2+} prevents depolymerization from the (+)-end, so that the rate of depolymerization is slower compared with control F-actin; (2) filament severing: the addition of actin filament-severing proteins such as gelsolin to 0.1 μM Prodan-labeled F-actin in the presence of Ca^{2+} increases the rate of severing and (+)-end capping. Actin filaments are disassembled much faster in the presence of gelsolin (gelsolin : F-actin, 1 : 10) compared with control filaments; (3) monomer sequestration: the sequestering protein thymosin $\beta 4$ ($T\beta 4$) forms a 1 : 1 complex with G-actin. The rate of actin polymerization is therefore much slower in the presence of $T\beta 4$ compared with control actin; and (4) in addition to these indirect polymerization assays, the direct binding of Prodan–action to ABPs can be quantified by monitoring associated changes in the emission properties of Prodan.[14]

Microscope-Based Photoactivation of Caged Proteins

Cell motility is driven by the spatial and temporal regulation of multiple ABP activities and interactions. It could be argued therefore that mechanism-based investigations of ABP function should be studied within the context of a motile cell. The feasibility of this approach is demonstrated using light-directed activation of caged ABPs. In these studies the relaxation process that follows the photoactivation of a caged ABP can be monitored through time and spatially resolved images of a fluorescent actin[15] or by phase-contrast microscopy with subsequent analyses to determine changes in (1) protein interactions or (2) cell behavior and morphology, for example, protrusive activity at the leading edge.[6,16]

Light-Directed Activation of Caged Proteins

Irradiation of a caged group on a protein leads to the rapid photoisomerization of the 2-nitrophenyl group with subsequent cleavage of the carbamate bond (for NVOC and DMNEP) or thioether bond (for BMDNB). This reaction proceeds

[13] G. Marriott, K. Zechel, and T. M. Jovin, *Biochemistry* **27,** 6214 (1988).

[14] K. Zechel, *Biochem. J.* **290,** 411 (1993).

[15] A. Choidas, A. Jungbluth, A. Sechi, A. Ullrich, and G. Marriott, *Eur. J. Cell Biol.* **77,** 81 (1998).

[16] P. Roy, Z. Rajfur, D. Jones, G. Marriott, and K. Jacobson, *J. Cell Biol.* **153,** 1035 (2001).

efficiently for both types of caged groups at neutral pH. However, the 2-nitroso photoproducts can further react in an irreversible manner with thiol groups; these secondary reactions are quenched *in vitro* with 5 m*M* DTT, whereas *in vivo* the scavenger is thought to be glutathione. Here we describe two irradiation conditions for uncaging proteins, using either continuous, steady state illumination for large-scale analysis of uncaging or pulsed illumination for *in vivo* uncaging of proteins.

Steady State Irradiation

The efficiency of the uncaging reaction is different for each protein and depends on several factors including pH and the molecular environment of the caging group. Quantitative analysis of the uncaging reaction is performed by SDS–PAGE and absorption spectroscopic analyses. In these studies 500 μl of a 20 μM caged protein solution is irradiated in a microcuvette, using the near-ultraviolet emission (340–400 nm) from a 50-W Hg arc lamp. The UV light is selected from this white light source by using a combination of UG-11 and WG340 filters while the intense infrared light is removed with a 1-cm water-filled cuvette. The near-UV light is defocused onto the sample to achieve homogeneous irradiation. The progress of the reaction is determined by SDS–PAGE, protein activity, and absorption spectral analysis and usually the uncaging reaction is complete within 8 min.

Time-Resolved Analysis of Uncaging Reaction

The rate of the photoisomerization of the 2-nitrobenzyl caged group is determined by time-resolved absorption measurements of the aci–nitro intermediates of NVOC and BMNBB, which have transient absorption maxima at 440 nm. The caged protein at $\sim$100 μM is irradiated with a nanosecond pulse of near-ultraviolet laser light and transient absorption spectra are recorded after 0.1, 10, and 50 ms.[9,17] The decay rate of the aci–nitro intermediate is measured by recording the absorption value at 440 nm, from 0.1 to 100 ms following the UV pulse. The decay of the 440-nm absorption value is fitted by a single exponential decay; the decay of a multiply labeled caged protein would probably be multiexponential. The photocleavage rate for 4,5-dimethoxy-2-nitrobenzene-based caged groups is on the order of 50 s^{-1} and somewhat faster for 2-nitrobenzene.[18] Interestingly, Walker *et al.*[4] and Milburn *et al.*[19] have shown that this rate can increase by $\sim 10^4$, using a caged group with a carboxyl group at the benzylic carbon.

[17] R. Uhl, B. Meyer, and H. J. Desel, *Biochem. Biophys. Methods* **10,** 35 (1984).

[18] J. A. McCray and D. R. Trentham, *Annu. Rev. Biophys.* **18,** 239 (1989).

[19] T. Milburn, N. Matsubara, A. P. Billington, J. B. Udgaonkar, J. W. Walker, B. K. Carpenter, W. W. Webb, J. Marque, W. Denk, J. A. McCray, and G. P. Hess, *Biochemistry* **29,** 49 (1989).

Preparation and Applications of Caged Actin-Binding Proteins

Lysine-Directed Monofunctional Caged Reagents

The NVOC-Cl reagent has been used to label amino groups in caged actin,[5] caged transcription factor (Gal-418; Cambridge *et al.*[20]), caged profilin,[21] and caged cofilin.[22] In this article we describe an application of the NVOC-Cl reagent for preparing and uncaging thymosin $\beta4$ (T$\beta4$). T$\beta4$ is a simple G-actin-sequestering protein that inhibits actin polymerization in eukaryotic cells.[23] It has been hypothesized that during cell protrusion G-actin is released from the T$\beta4$ complex in competition with profilin; profilactin is then used to accelerate actin polymerization.[24] If this mechanism is correct then a local release of T$\beta4$ from a caged T$\beta4$ within a motile cell should lead to a decrease in the amount of free profilactin–actin polymerization in this region and this should halt motility. T$\beta4$ has sensitive lysine residues in a nine-residue segment (LKKTETQEK; residues 17–25) that constitute an actin-binding motif.[25] Specifically, electrostatic contacts between actin and Lys-18 and Lys-19 of T$\beta4$ are critical for correct binding.[26–28] The goal of the labeling experiments is to target one or both of these lysine residues with a minimal number of NVOC groups. A reproducible protocol to cage T$\beta4$ with two or three NVOC groups is established by recording T$\beta4$ activity at varying NVOC-Cl:T$\beta4$ ratios. The rate of actin polymerization in the presence of caged T$\beta4$ is almost identical to that of control actin (Fig. 1B), which confirms the inactivation of T$\beta4$ by the covalently attached NVOC groups. Light-directed activation of caged T$\beta4$ under steady state conditions of irradiation leads to a loss of the 350-nm NVOC absorption band and concomitant activation of the G-actin-sequestering activity such that actin polymerization is inhibited in a manner similar to that with pure T$\beta4$.

Light-Directed Activation of Caged Thymosin $\beta4$ in Motile Keratocytes. In control keratocytes (Fig. 2A) the cytoskeleton is characterized by a meshwork of actin filaments in the lamella and a striated actin assembly in the perinuclear region. Increasing the amount of intracellular T$\beta4$ results in a disassembly of the

[20] S. B. Cambridge, R. L. Davis, and J. S. Minden, *Science* **277,** 825 (1997).

[21] G. Marriott, J. Ottl, M. Heidecker, and D. Gabriel, *Methods Enzymol.* **291,** 95 (1998).

[22] G. Marriott, in preparation (2002).

[23] P. J. Goldschmidt-Clermont, M. I. Furman, D. Wachsstock, D. Safer, V. T. Nachmias, and T. D. Pollard, *Mol. Biol. Cell.* **3,** 1015 (1992).

[24] D. Pantaloni and M. F. Carlier, *Cell* **75,** 1007 (1993).

[25] D. Heintz, A. Reichart, M. Mihelic, W. Voelter, and H. Faulstich, *FEBS Lett.* **329,** 9 (1993).

[26] K. Vancompernolle, J. Vandekerckhove, M. R. Bubb, and E. D. Korn, *J. Biol. Chem.* **266,** 15427 (1991).

[27] K. Vancompernolle, M. Goethals, C. Huet, D. Louvard, and J. Vandekerckhove, *EMBO J.* **11,** 4739 (1992).

[28] M. VanTroys, D. Dewitte, M. Goethals, M. F. Carlier, J. Vandekerckhove, and C. Ampe, *EMBO J.* **15,** 201 (1996).

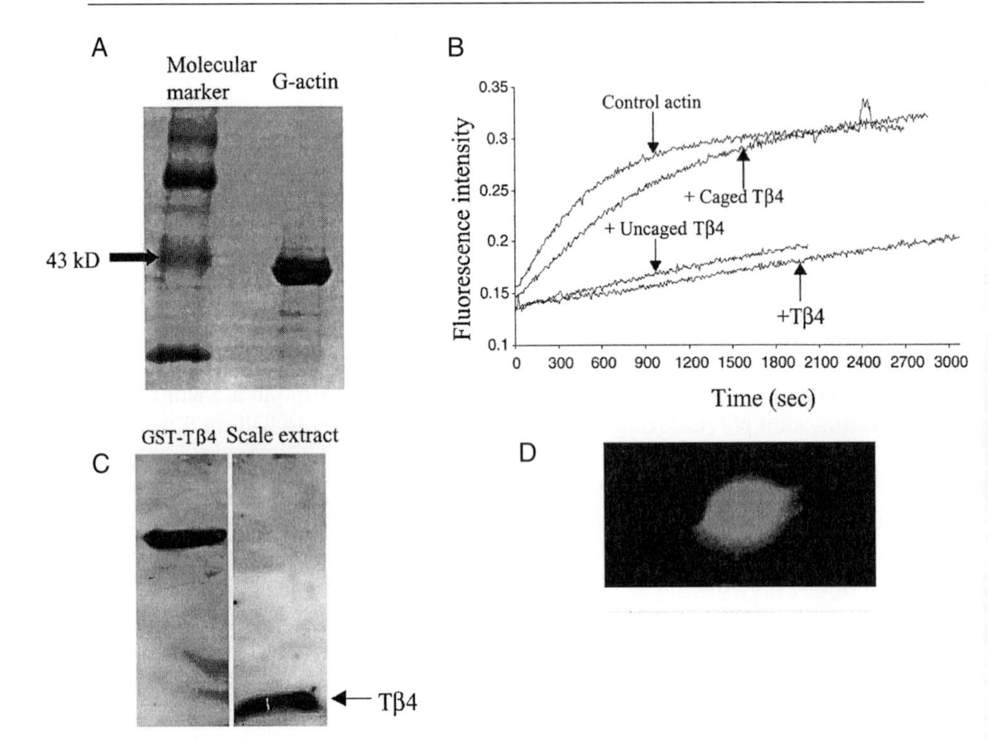

FIG. 1. Spectrofluorophotometric data from *in vitro* Acrylodan–actin polymerization assay performed under various conditions as indicated, demonstrating the loss of actin-sequestering capability of Tβ4 as a result of caging with NVOC (molar ratio of caged Tβ4 to actin, 4 : 1). Uncaging restored the biochemical activity of caged Tβ4, comparable to the efficiency of pure Tβ4.

actin cytoskeleton with accompanying changes in contractility and cell shape and a cessation of cell motility (Fig. 2B). Keratocytes loaded with rhodamine–dextran alone do not show effects on the structure of the actin cytoskeleton (data not shown). Given the dramatic effect of higher levels of Tβ4 on the actin cytoskeleton, we hypothesize that local activation of Tβ4 leads to the disassembly of actin filaments in that region with subsequent inhibition of both cell contractility and actin-based protrusion.

Caged Tβ4 is bead loaded from a 10-mg/ml stock solution to achieve an estimated concentration of 200 μM.[16] The experimental setup for photoactivating caged Tβ4 within motile cells is essentially the same as that described by Ishihara *et al.*[29] Other uncaging systems for microscopes have been described that employ near-UV light delivered from a 100-W Hg arc lamp[6] in place of the He–Cd laser. Caged Tβ4, and caged fluorescein isothiocyanate (FITC)–dextran, are uncaged by near-UV light that is focused onto an ~3-μm-diameter spot on the specimen, using

[29] A. Ishihara, K. Gee, S. Schwartz, K. Jacobson, and J. Lee, *Biotechniques* **23,** 268 (1997).

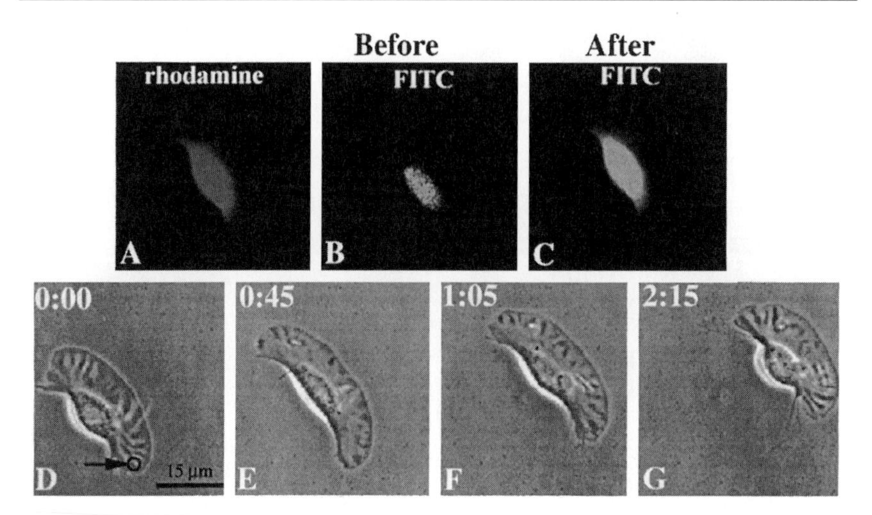

FIG. 2. Control experiment: photoactivation of caged FITC–dextran in the wing of a keratocyte. (A) Positive rhodamine fluorescence identifies a loaded cell. (B) Before uncaging, the fluorescence in the FITC channel was negligible, as expected. (C) After photoactivation, caged FITC–dextran was uncaged successfully as demonstrated by FITC fluorescence throughout the cell. (D–G) Time-lapse microscopy of the same cell shows unaltered locomotion as a result of such photorelease (the 3-μm zone of photoactivation is indicated by a circle).

a ×100 objective. The power of the He–Cd laser beam after passing through the objective is about 10 μW, and cells are irradiated for 100 ms. Control experiments show that keratocyte motility is not affected by the short UV pulse or by the release of the photoproducts generated from irradiated caged fluorescein–dextran or bleaching photoproducts from irradiated rhodamine–dextran.[16] However, we observe significant changes in keratocyte motility after irradiating cells loaded with caged Tβ4. The locomotory characteristics of a motile cell irradiated at the wing change approximately 1 min after the laser pulse (marked by a circle in Fig. 3). This localized irradiation causes cells to turn toward the direction of photoactivation, after pivoting around the tethered, irradiated region. The cell eventually breaks from its tether and is able to locomote in its usual gliding manner. This turning phenomenon is observed in 19 of a total of 23 photoactivated cells, although these cells exhibited a large variation in the degree of turning (15–103°; mean turning, 56 + 24°; see Roy *et al.*[16] for details).

The main limitation of localized photoactivation of caged proteins within a cell is rapid diffusion, which lowers the effective concentration of the protein below the threshold level of activity.[4] Interestingly, the effective diffusion constant of Tβ4 in the cytosol is two orders of magnitude less than that predicted for free diffusion in solution. This property suggests that the rate of Tβ4 diffusion inside the cytoplasm is severely hindered by several factors, which might include transient interactions of Tβ4 with G- and F-actin and/or obstructions imposed

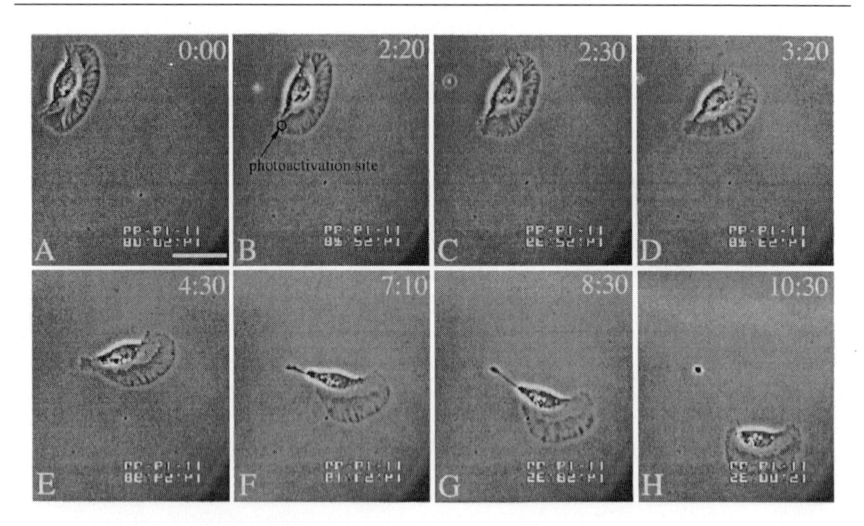

FIG. 3. Uncaging of caged Tβ4 in a motile keratocyte. Time-lapse video microscopy of a locomoting keratocyte after photorelease of Tβ4 at the wing (marked by a circle) shows dramatic turning of the cell toward the direction of photoactivation. (B and C) Frames acquired right before and after photoactivation, respectively. Note the pivoting of the irradiated zone to the substrate as a result of photorelease of Tβ4. About 8 min after photoactivation of Tβ4, normal locomotion is resumed (H). Bar: 15 μm.

by the dense cytoskeletal meshwork.[30,31] Furthermore, the results of a virtual cell simulation presented by Roy *et al.*[16] predict that photoactivated Tβ4 will undergo rapid binding and unbinding to G-actin, which will serve to localize Tβ4. Together, the theoretical and experimental data confirm our hypothesis that a decrease in the free G-actin concentration near the site of Tβ4 uncaging causes (1) a local shift in equilibrium from F-actin to G-actin, which results in a local depolymerization of actin filaments, and (2) a severe reduction of existing filament growth in the photoactivated region. These combined effects result in a change in cell motility as a result of (1) local changes in cell contractility due to the loss of actin filaments, and (2) protrusive activity due to the lower local concentration of free G-actin.

Cysteine-Directed Monofunctional Caged Reagents

Conformational Transitions in Myosin II. Alkylating reagents such as TMR-IA are highly reactive toward Cys-707 of myosin II.[32,33] This has been selectively

[30] K. Jacobson and J. Wojcieszyn, *Proc. Natl. Acad. Sci. U.S.A.* **81,** 6747 (1984).

[31] K. Luby-Phelps, *Curr. Opin. Cell Biol.* **6,** 3 (1994).

[32] E. Reisler, *Methods Enzymol.* **85,** 84 (1982).

[33] D. D. Root and E. Reisler, *Biophys. J.* **63,** 730 (1992).

exploited in many biophysical studies that employ spectroscopic probes attached to Cys-707 to describe functional motions within myosin. However, most Cys-707 conjugates of myosin and heavy meromyosin (HMM) including TMR-IA-labeled HMM are unable to support the motility of actin filaments within *in vitro* assays of muscle contraction.[6,33,34] Furthermore, the ATPase activity of these conjugates is elevated compared with native myosin II. These two results led us to hypothesize that Cys-707, which is located close to the fulcrum of the converter domain, plays an essential role in coupling conformational transition(s) associated with the myosin-binding–hydrolysis-unbinding reactions of ATP and its products to the swinging of its lever arm. We believe that this uncoupling is caused by the chemical group attached to the cysteine, which physically obstructs a conformational transition at the hinge. To test this hypothesis we have described experiments, summarized here, to (1) inhibit HMM function by using the thiol-reactive caging group DMNBB, and (2) restore the coupling of conformational transitions and HMM function by removing the caged group by UV irradiation.

The activities of native HMM, caged HMM, and decaged HMM are analyzed by an *in vitro* motility assay of muscle contraction. Details of this work have been published[6] and therefore only the essential elements of the technique are described here. The reaction conditions for caging HMM through Cys-707 have already been described. Quantitative analysis of the labeling efficiency is performed by recording the absorption value of the HMM conjugate at 350 nm, which shows a single cage group per myosin head.[6] Competition-labeling studies using the well-characterized Cys-707-directed reagent *N*-ethylmaleimide (NEM[32]) in the absence and presence of 1 m*M* ATP show that BMDNB specifically labels Cys-707. This result is confirmed by Malachite green-based measurements of the Ca^{2+}-activated ATPase of HMM, which show that caged HMM exhibits a fivefold higher ATPase activity compared with native HMM (data not shown; and Reisler[32]). Consistent with these results, HMM caged on Cys-707 does not support the motility of actin filaments in an *in vitro* assay whereas actin filaments on unlabeled HMM move at 5 μm/sec.[6]

The presence of a single, specifically labeled caged group on HMM simplified kinetic investigations of the uncaging reaction and the rate of HMM activation after flash photolysis. Time-resolved analysis of the photoisomerization reaction is performed as described earlier, using transient absorption measurements of the 440-nm absorption of the aci–nitro intermediate (Fig. 4A and B). Curve-fitting routines show that the aci–nitro intermediate decays according to a single exponential at a rate of 45 s^{-1}. This rate, which is directly related to the rate of myosin activation, is rapid and similar to that obtained for other BMDNB groups in solution. This is an interesting result because it suggests that in spite of its complexity the protein matrix does not impede the rate of the photocleavage reaction. Indeed,

[34] S. J. Kron and J. A. Spudich, *Proc. Natl. Acad. Sci. U.S.A.* **83,** 6272 (1985).

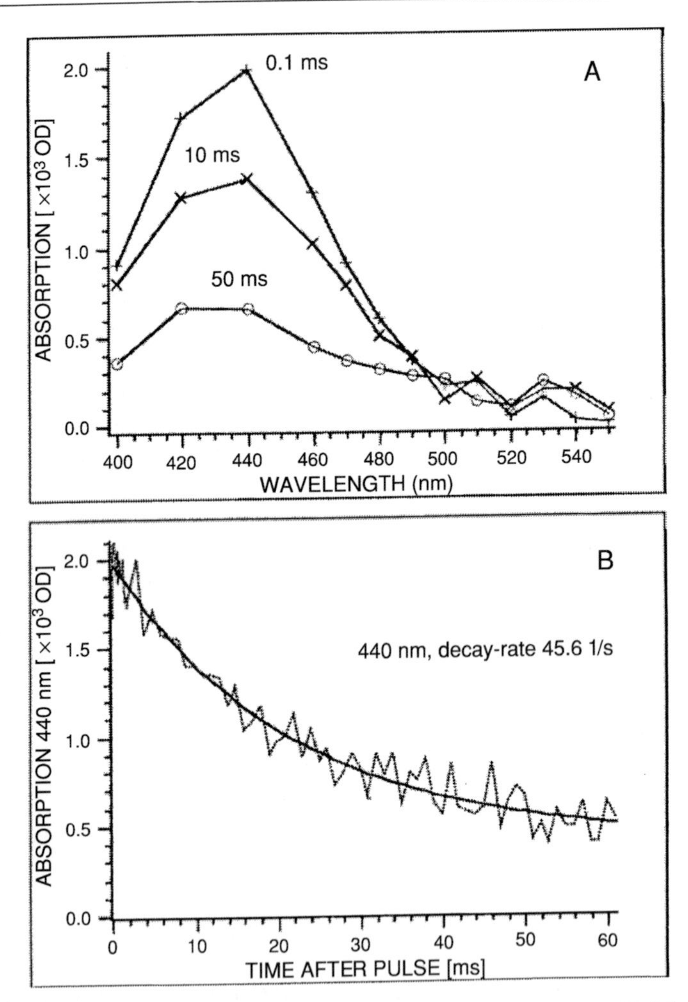

FIG. 4. Selected time-resolved absorption spectra of a 21.4 μM solution of BMDNB-labeled HMM in AB-buffer containing 5 mM DTT after irradiation with a 10-ns pulse of 380-nm light: (A) 0.1 ms (*top*), 10 ms (*middle*), and 50 ms (*bottom*). (B) Time-resolved absorption decay of the sample described in (A), recorded at 440 nm.

caged HMM is activated at a rate comparable to other perturbation methods that use stopped flow and pressure jumps, and is significantly faster than the rate of force development during muscle contraction.[18]

To prove that the photocleavage of the BMDNB group from caged myosin generates fully functional HMM, quantitative measurements of HMM-based motility of actin filaments are performed by an *in vitro* motility assay. Control experiments show that a functional monolayer of fully functional HMM propels actin filaments in the presence of 1 mM ATP at 5 μm/sec. No filament motility is observed for HMM caged on Cys-707. However, irradiation of a caged HMM monolayer in the image field with a 0.5-sec pulse of near-UV light delivered from a 100-W Hg arc lamp initiates the photoisomerization reaction and leads to the movement of most actin filaments in the irradiation field at a rate of 2 μm/sec, although some filament severing is observed. However, all filaments move at maximal velocity after irradiating the same field with a second, 0.5-sec UV flash. These data suggest that the yield of uncaging from a single flash is close to or greater than 90% and reaches 100% after the second flash.[6] This estimate is based on other data showing that a mixture of 5% NEM–HMM and 95% functional HMM reduces the actin filament sliding velocity and leads to frequent stalling and severing of actin filaments. Caged HMM can therefore be used to trigger muscle contraction and provides unique kinetic information about conformational coupling around the hinge region in a physiologically relevant environment.

Investigations of protein structure–function relationships that rely on chemical modification of specific amino acid are often criticized for their lack of suitable controls; for example, it is difficult to prove that the inhibition is due only to the modification of a specific residue and not to secondary, irreversible changes in protein structure that accompany the modification. This issue is clearly resolved by the approach based on light-directed activation of caged protein because the chemical modifier is removed from the protein by modest irradiation, using near-UV light, with full recovery of protein activity.

Heterobifunctional, Photocleavable Reagents

Photocleavable cross-linking reagents can be used as an alternative strategy to cage wild-type proteins.[8,9] We have described novel photocleavable cross-linking reagents and labeling protocols to cage almost any protein.[9] Furthermore, the additional functionality of this class of reagent may be used to generate caged proteins with built-in fluorescent probes for imaging, or antibodies for cell targeting. In this article we have restricted our discussion to the DMNEP-based reagents and a description of the preparation and uncaging of an actin–dextran complex (Scheme 1).

G-actin is first labeled at Cys-374 with TMR-IA according to Marriott[5] to avoid generating actin dimers with the heterobifunctional reagent. The DMAP salt of DMNEP is highly reactive toward amino groups in proteins. However, because it is rapidly hydrolyzed by water, the labeling reaction is performed at a relatively high protein concentration (>20 μM) for a short period of time (<30 min). TMR–actin (30 μM) is treated with the DMAP salt of DMNEP, delivered from either a stock solution in acetonitrile or as a solid, at a final concentration of 100–500 μM in a thiol-free, borate-based G-buffer, pH 8.5. The actin conjugate is centrifuged at

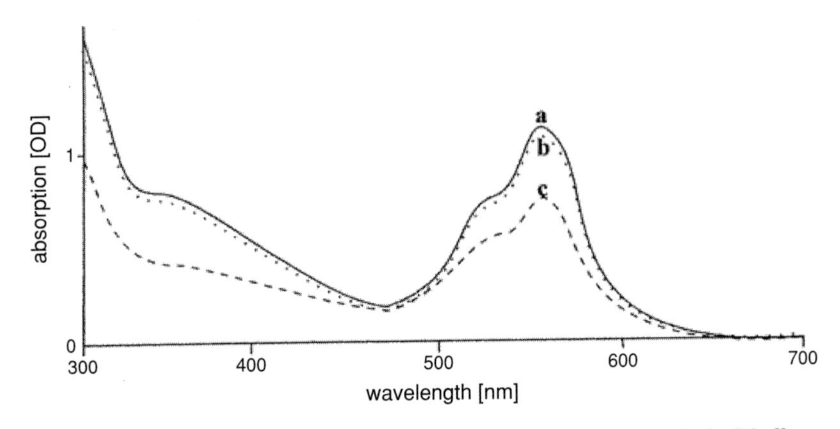

FIG. 5. (a) Absorption spectra of the IA-TMR-labeled G-actin–dextran complex in G buffer containing 2 m*M* DTT. After addition of $MgCl_2$ and KCl to 2 and 100 m*M*, respectively, and 2 hr at room temperature, the sample was centrifuged at 100,000*g* for 60 min. (b) Absorption spectrum of the supernatant fraction. No pellet fraction was found after the centrifugation; that is, the actin was caged. The supernatant fraction was then irradiated with near-ultraviolet light for 10 min, and after 2 hr at room temperature the sample was centrifuged at 100,000*g* for 60 min. (c) Absorption spectrum of the pellet fraction resuspended in an equal volume of F buffer.

10,000*g* for 20 min followed by overnight dialysis in a thiol-free buffer. The final conditions should yield an actin conjugate with two to four caged groups, which retains the majority of its polymerization activity as determined by absorption spectroscopic analysis of TMR–F-actin in the supernatant and pellet fractions of a high-speed centrifugation (100,000*g* for 60 min; Fig. 5). The actin conjugate is then treated with an excess of iminothiolane labeled aminodextran in thiol free G-buffer, pH 8.5. Any residual polymerization activity is inhibited within the actin–dextran complex (Fig. 5). Steady state irradiation of the G-actin–dextran complex with near-UV light for 10 min leads to the cleavage of the cross-link with the formation of native, polymerization-competent G-actin (Fig. 5), carbon dioxide, and the photoproduct, which remains attached to dextran. We expect that this approach will be even more effective when using even larger dextran molecules (>70,000 molecular weight), because fewer cross-linking groups are required for caging and this would translate into higher efficiencies for the uncaging reaction.

[12] Bioluminescence Resonance Energy Transfer: Monitoring Protein–Protein Interactions in Living Cells

By YAO XU, AKIHITO KANAUCHI, ALBRECHT G. VON ARNIM, DAVID W. PISTON, and CARL HIRSCHIE JOHNSON

Introduction

Protein–protein interactions are known to play an important role in a variety of biochemical systems. To date, thousands of protein–protein interactions have been identified by using the conventional two-hybrid system, but this method is limited in that the interaction must occur in the yeast nucleus. This means interactions that strictly depend on cell type-specific processing and/or compartmentalization will not be detected. Therefore, a number of new methods have been developed that rely on reconstitution of biochemical function *in vivo,* such as fluorescence resonance energy transfer (FRET), protein mass spectrometry, or evanescent wave spectroscopy.[1] Among those methods, the resonance energy transfer techniques have potential advantages for assaying protein–protein interactions in living cells and in real time. In this article, we describe a resonance energy transfer method based on bioluminescence and update previously published reviews.[2,3]

Fluorescence Resonance Energy Transfer

Fluorescence resonance energy transfer (FRET)[4,5] is a well-established phenomenon that has been useful in cellular microscopy. When two fluorophores (the "donor" and the "acceptor") with overlapping emission/absorption spectra are within ~50 Å of one another and their transition dipoles are appropriately oriented, the donor fluorophore is able to transfer its excited state energy to the acceptor fluorophore. Therefore, if appropriate fluorophores are linked to proteins that might interact with each other, the proximity of these candidate interactors could be measured by determining whether fluorescence resonance energy is transferred from the donor to the acceptor. Thus, the presence or absence of FRET acts as a "molecular yardstick."

[1] A. R. Mendelsohn and R. Brent, *Science* **284,** 1948 (1999).
[2] Y. Xu, D. W. Piston, and C. H. Johnson, *Spectrum* **12,** 9 (1999).
[3] Y. Xu, A. Kanauchi, D. W. Piston, and C. H. Johnson, *in* "Luminescence BioTechnology" (K. Van Dyke, C. Van Dyke, and K. Woodfork, eds.), p. 529. CRC Press, Boca Raton, FL, 2002.
[4] P. Wu and L. Brand, *Anal. Biochem.* **218,** 1 (1994).
[5] R. M. Clegg, *Curr. Opin. Biotechnol.* **6,** 103 (1995).

METHODS IN ENZYMOLOGY, VOL. 360
0076-6879/03 $35.00

The discovery and development of green fluorescent protein (GFP) and its mutants made possible their use as FRET donors and acceptors.[6–12] Genetically fusing GFP derivatives to the candidate proteins enabled the detection of protein–protein proximity in real time in living cells of the organisms from which the proteins were originally obtained.[9,10] In those studies, blue fluorescent protein (BFP) was used as the donor fluorophore and GFP was the acceptor. As mentioned above, the efficiency of the resonance transfer depends on the spectral overlap of the fluorophores, their relative orientation, as well as the distance between the donor and acceptor fluorophores. By targeting the fusion proteins to specific compartments, this FRET-based assay can also allow protein interactions to be observed within cellular compartments *in vivo,* as has been shown for mitochondria and nuclei.[9,10] However, because FRET demands that the donor fluorophore be excited by illumination, the practical usefulness of FRET can be limited because of the concomitant results of excitation: photobleaching, autofluorescence, and direct excitation of the acceptor fluorophore (see Bioluminescence Resonance Energy Transfer versus Fluorescence Resonance Energy Transfer, below). Furthermore, some tissues might be easily damaged by the excitation light or might be photoresponsive (e.g., retina and most plant tissues).

Bioluminescence Resonance Energy Transfer

In nature, GFP is a resonance energy transducer of the luminescence from the photoprotein aequorin.[13] We have developed a bioluminescence resonance energy transfer (BRET) system for assaying protein–protein interactions that incorporates the attractive advantages of the FRET assay while avoiding the problems associated with fluorescence excitation.[14] In BRET, the donor fluorophore of the FRET pair is replaced by a luciferase, in which bioluminescence from the luciferase in the presence of a substrate excites the acceptor fluorophore through the same resonance energy transfer mechanisms as FRET.

The bioluminescent *Renilla* luciferase (RLUC; molecular mass, 35 kDa) was originally chosen as the donor luciferase in our BRET assay because its emission

[6] R. Heim, D. C. Prasher, and R. Y. Tsien, *Proc. Natl. Acad. Sci. U.S.A.* **91,** 12501 (1994).

[7] R. Heim and R. Y. Tsien, *Curr. Biol.* **6,** 178 (1996).

[8] A. Miyawaki, J. Llopis, R. Heim, J. M. McCaffery, J. A. Adams, M. Ikura, and R. Y. Tsien, *Nature (London)* **388,** 882 (1997).

[9] N. P. Mahajan, K. Linder, G. Berry, G. W. Gordon, R. Heim, and B. Herman, *Nat. Biotechnol.* **16,** 547 (1998).

[10] A. Periasamy and R. N. Day, *J. Biomed. Opt.* **3,** 154 (1998).

[11] X. Xu, A. L. Gerard, B. C. Huang, D. C. Anderson, D. G. Payan, and Y. Luo, *Nucleic Acids Res.* **26,** 2034 (1998).

[12] T. W. Gadella, Jr., G. N. van der Krogt, and T. Bisseling, *Trends Plant Sci.* **4,** 287 (1999).

[13] J. G. Morin and J. W. Hastings, *J. Cell Physiol.* **77,** 313 (1971).

[14] Y. Xu, D. W. Piston, and C. H. Johnson, *Proc. Natl. Acad. Sci. U.S.A.* **96,** 151 (1999).

spectrum is similar to that of the cyan mutant of *Aequorea* GFP ($\lambda_{max} \approx 480$ nm), which has been shown to exhibit FRET with the acceptor fluorophore EYFP, which is an enhanced yellow-emitting GFP mutant.[8] The excitation peak of EYFP (513 nm) does not perfectly match the emission peak of RLUC, but the emission spectrum of RLUC is sufficiently broad that it provides good excitation of EYFP. The spectral overlap between RLUC and EYFP is similar to that of EYFP and the enhanced cyan mutant of GFP, ECFP, which yields a critical Förster radius (R_0) for FRET of ~50 Å.[9] Thus, we would expect significant BRET between RLUC and EYFP, with an R_0 for BRET of ~50 Å. The fluorescence emission of EYFP is yellow, peaking at 527 nm, which is distinct from the RLUC emission peak. Furthermore, RLUC and EYFP do not naturally interact with each other. Fortunately, the substrate for RLUC, coelenterazine, is a hydrophobic molecule that easily permeates cell membranes.

As depicted in Fig. 1, in the BRET assay of protein interactions, RLUC is genetically fused to one candidate protein, and EYFP is fused to another protein of interest that perhaps interacts with the first protein. If RLUC and EYFP are

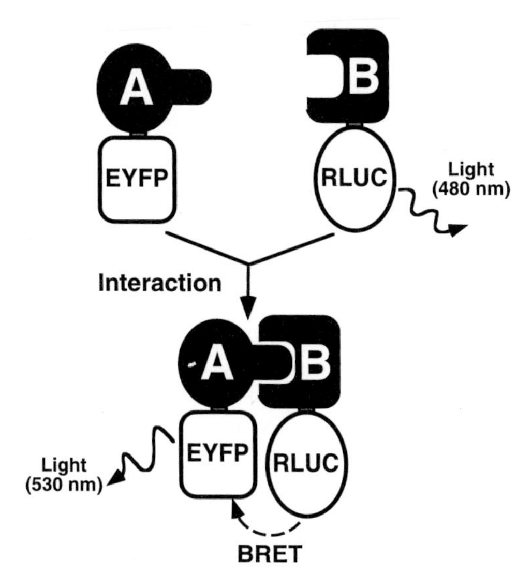

FIG. 1. A diagram of bioluminescence resonance energy transfer (BRET) used for a protein–protein interaction assay. One protein of interest (B) is genetically fused to the donor luciferase RLUC and the other candidate protein (A) is fused to the acceptor fluorophore EYFP. In the presence of the substrate, coelenterazine, RLUC emits luminescence (peak at 480 nm). Interaction between the two fusion proteins can bring RLUC and EYFP close enough for BRET to occur, with an additional emission at a longer wavelength (e.g., peak at 530 nm). The diagram shows the ideal case of 100% resonance transfer; under most experimental circumstances the amount of resonance transfer would be expected to be less than 100%, resulting in emission of light at both 480 and 530 nm. Adapted and modified with permission from Refs. 2 and 3.

brought close enough for resonance energy transfer to occur, the bioluminescence energy generated by RLUC can be transferred to EYFP, which then emits yellow light (Fig. 1). In the BRET assay for protein interaction, this resonance transfer can occur between RLUC–EYFP fusion proteins that interact. If there is no interaction between the two proteins of interest, RLUC and EYFP will be too far apart for significant transfer and only the blue-emitting spectrum of RLUC will be detected. Thus, protein–protein interactions can be monitored both *in vivo* and *in vitro* by detecting the emission spectrum and quantifying the emission ratio (530 nm to 480 nm).

BRET between RLUC and EYFP was first demonstrated in control experiments in which RLUC was fused directly to EYFP through a linkage of 11 amino acids.[14] The luminescence profile of the *Escherichia coli* cells expressing this RLUC::EYFP fusion construct yielded a bimodal spectrum, with one peak centered at 480 nm (as for RLUC), and a new peak centered at 527 nm (as for EYFP fluorescence).[14] This result suggests that a significant proportion of the energy from RLUC is transferred to EYFP and emitted at the characteristic wavelength of EYFP. We concluded that RLUC and EYFP could be an effective combination to apply in a protein–protein interaction assay.

Application of Bioluminescence Resonance Energy Transfer to Clock Proteins

To test BRET as a protein–protein assay, we chose the proteins encoded by circadian (daily) clock genes from cyanobacteria and fused them to RLUC or EYFP. In cyanobacteria, the *kaiABC* gene cluster encodes three proteins, KaiA (molecular mass, 32.6 kDa), KaiB (molecular mass, 11.4 kDa), and KaiC (molecular mass, 58 kDa), that are essential for circadian clock function.[15] Iwasaki *et al.* have used the yeast two-hybrid and *in vitro* binding assays to discover that Kai proteins interact in various ways, such as in the formation of KaiB–KaiB homodimers.[16] First, we tried N-terminal fusions of KaiB to RLUC and to EYFP. The luminescence spectra of *E. coli* expressing these fusions showed a second peak in cells expressing both RLUC::KaiB and EYFP::KaiB (Fig. 2A). This spectrum is similar to that depicted for the fusion protein RLUC::EYFP.[14]

We further tested all possible combinations of KaiB fusions with RLUC or EYFP, including N- vs. N-terminal, N- vs. C-terminal, C- vs. N-terminal, as well as C- vs. C-terminal fusions. All these combinations of the KaiB fusion proteins showed BRET (our unpublished data, 2001). KaiB interactions were also observed

[15] M. Ishiura, S. Kutsuna, S. Aoki, H. Iwasaki, C. R. Andersson, A. Tanabe, S. S. Golden, C. H. Johnson, and T. Kondo, *Science* **281,** 1519 (1998).
[16] H. Iwasaki, Y. Taniguchi, M. Ishiura, and T. Kondo, *EMBO J.* **18,** 1137 (1999).

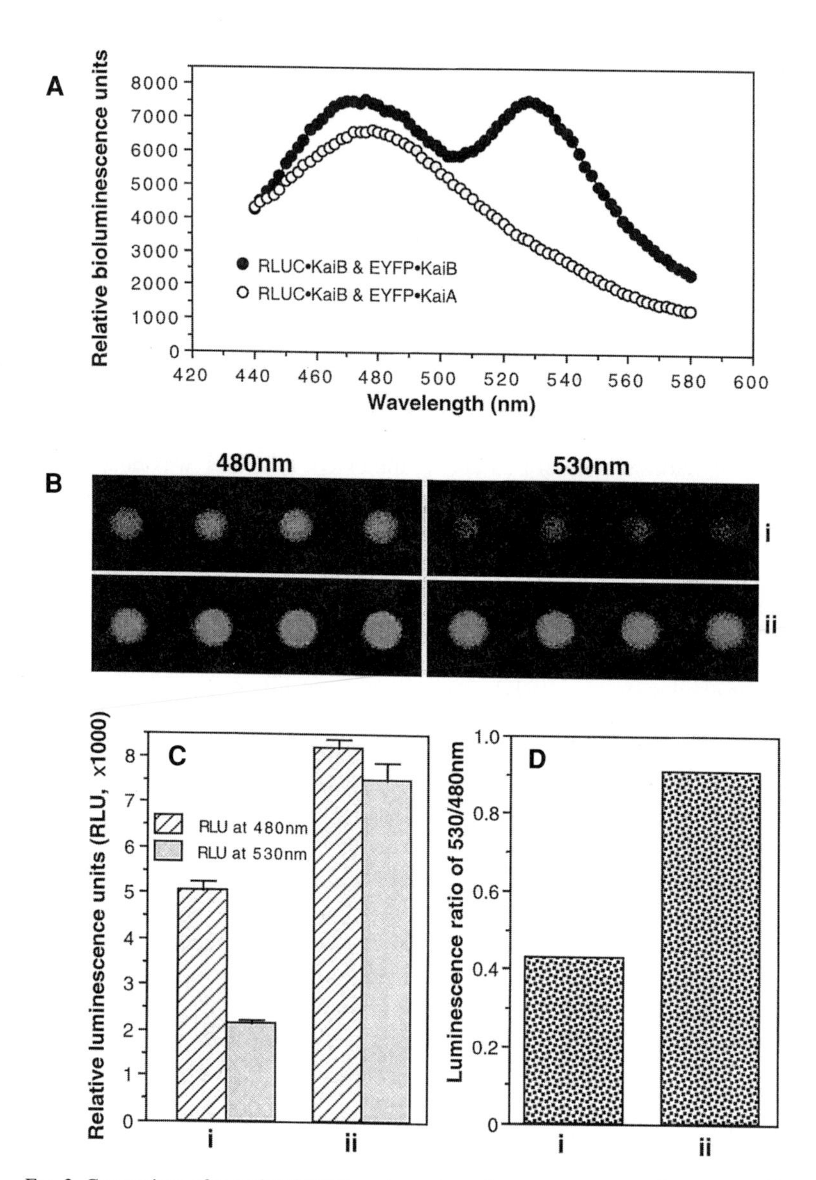

FIG. 2. Comparison of complete BRET spectra, using a fluorescence spectrophotometer with camera images of *E. coli* cells. (A) Luminescence emission spectra (measured with an SPEX fluorescence spectrophotometer) from transformed *E. coli* strains coexpressing fusion proteins exhibiting BRET (RLUC::KaiB and EYFP::KaiB) or fusion proteins that are not exhibiting BRET (RLUC::KaiB and EYFP::KaiA). (B) Luminescence of *E. coli* colonies imaged with a CCD camera through filters transmitting light at 480 or 530 nm. Row i: RLUC::KaiB and EYFP::KaiA combination (no BRET). Row ii: RLUC::KaiB and EYFP::KaiB combination (BRET). (C) Quantification of the luminescence at 480 versus 530 nm for the colonies shown in (B). (D) BRET ratio for the data from (C). (B–D) Comparisons of luminescence from the RLUC::KaiB and EYFP::KaiB combination (column ii) or from the RLUC::KaiB and EYFP::KaiA combination (column i).

in vitro by BRET.[14] To demonstrate that this bimodal spectrum does not occur nonspecifically, we used KaiA as a control, in which EYFP was fused to a slightly truncated KaiA. The luminescence spectra of *E. coli* coexpressing EYFP::KaiA with RLUC::KaiB did not exhibit the second luminescence peak, indicating that no interaction occurred between KaiA and KaiB (Fig. 2A). Our results, therefore, strongly suggest that interaction among KaiB molecules, either in N-terminal or C-terminal fusions to the donor luciferase or the acceptor fluorophore, have brought the RLUC and EYFP into close proximity such that energy transfer occurs for ~50% of the RLUC luminescence. Thus, BRET supports the data from the yeast two-hybrid assay,[16] demonstrating that the clock protein KaiB self-associates to form oligomers.

In the experiments described above, the extent of BRET was determined by measuring emission spectra.[14] For applications such as microscopic imaging and high-throughput screening, it would be more convenient to measure the ratio of luminescence intensities at two fixed wavelengths, for example, 480 and 530 nm. Ratio imaging has the advantage of automatically correcting for differences in overall levels of expression of RLUC and EYFP fusion proteins. Figure 2B shows the images of *E. coli* cultures expressing fusion proteins that either exhibit (row ii) or do not exhibit (row i) BRET. These images of liquid *E. coli* cultures (5-μl cultures) were collected with a charge-coupled device (CCD) camera through interference bandpass filters centered at 480 and 530 nm, respectively. In the cultures coexpressing the interacting combination of RLUC::KaiB with EYFP::KaiB, the amounts of light emitted at 480 and 530 nm are roughly equal, as would be predicted from the spectra depicted (row ii in Fig. 2B, column ii in Fig. 2C). In contrast, in the cultures containing a noninteracting combination of RLUC::KaiB with EYFP::KaiA, there is much less light emitted at 530 nm than at 480 nm (row i in Fig. 2B, column i in Fig. 2C). As we reported previously,[14] the extent of BRET can be quantified according to the 530 nm : 480 nm ratios of luminescence intensity in the image (Fig. 2D). Thus, the 530 nm : 480 nm ratios can apparently be used to evaluate BRET and thereby infer whether protein–protein interaction has occurred.

Bioluminescence Resonance Energy Transfer in Mammalian Cells

The BRET technique has now been successfully extended to other cell types, including plant cells (see below and Fig. 4) and mammalian cells. Figure 3 shows spectra of RLUC and the RLUC::EYFP fusion protein expressed in mammalian cells (COS7 cells). The clear bimodal spectrum of the RLUC::EYFP construct indicates resonance transfer in mammalian cells, as we already reported for *E. coli*.[14] In mammalian cells, Wang *et al.* used BRET (they call it "LRET," but it is the same phenomenon) to demonstrate interaction between insulin-like growth

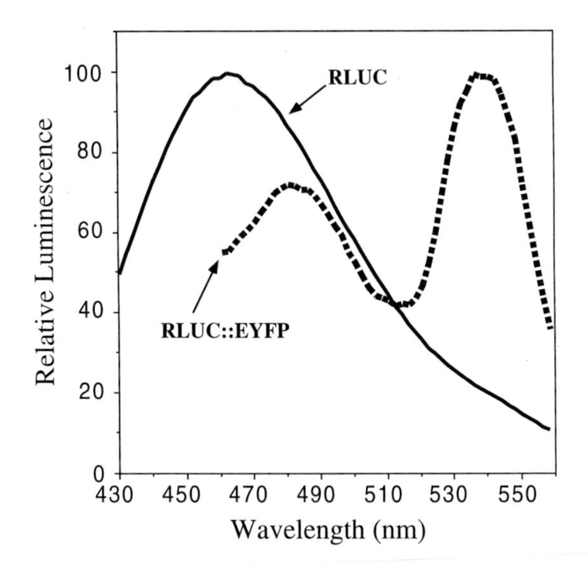

FIG. 3. Luminescence emission spectra measured with an SPEX fluorescence spectrophotometer from COS7 cells transfected with constructs expressing either RLUC or RLUC::EYFP, using the hRLUC version of *Renilla* luciferase.

factor II (IGF-II) and its binding protein, IGFBP-6.[17,18] BRET has been particularly successful in studies involving dimer and/or oligomer formation among receptors *in vivo*. The first such study was that of Angers *et al.*,[19] who used BRET to demonstrate that human β_2-adrenergic receptors form constitutive homodimers in HEK-293 cells. Treatment with the agonist isoproterenol increased the BRET signal, indicating that the agonist interacts with receptor dimers at the cell surface. Since that ground-breaking publication, BRET assays of receptor–receptor interactions have been extended to the thyrotropin-releasing hormone receptor, insulin receptors, opioid receptors, and the cholecystokinin receptor.[20–23]

[17] Y. Wang, G. Wang, D. J. O'Kane, and A. A. Szalay, *Mol. Gen. Genet.* **264,** 578 (2001).

[18] Y. Wang, G. Wang, D. J. O'Kane, and A. A. Szalay, *in* "Bioluminescence and Chemiluminescence: Perspectives for the 21st Century" (A. Roda, L. J. Kricka, and P. Stanley, eds.), p. 475. Wiley, Chichester, 1999.

[19] S. Angers, A. Salahpour, E. Joly, S. Hilairet, D. Chelsky, M. Dennis, and M. Bouvier, *Proc. Natl. Acad. Sci. U.S.A.* **97,** 3684 (2000).

[20] M. McVey, D. Ramsay, E. Kellett, S. Rees, S. Wilson, A. J. Pope, and G. Milligan, *J. Biol. Chem.* **276,** 14092 (2001).

[21] K. M. Kroeger, A. C. Hanyaloglu, R. M. Seeber, L. E. C. Miles, and K. A. Eidne, *J. Biol. Chem.* **276,** 12736 (2001).

[22] N. K. Boute, K. Pernet, and T. Issad, *Mol. Pharm.* **60,** 640 (2001).

[23] Z.-J. Cheng and L. J. Miller, *J. Biol. Chem.* **276,** 48040 (2001).

New Tools/Applications for Bioluminescence Resonance Energy Transfer

Several new tools for BRET have appeared that may prove useful. The first tool is two RLUCs that are codon optimized for mammalian expression. One, hRluc, is available from BioSignal (now part of PerkinElmer Life Sciences, Boston, MA). Transfection of the BioSignal hRluc construct into mammalian cells results in significantly higher luminescence levels than with the original RLUC.[3] The other optimized RLUC, hRL, is available from Promega (Madison, WI; www.Promega.com), and is reported to be much more highly expressed than native RLUC in mammalian cells. The second new tool is a luciferase, isolated from *Gaussia,* that has an emission spectrum like that of RLUC but whose molecular mass is only 20 kDa (available from Prolume, Pinetop, AZ; www.prolume.com).[24] Like RLUC, this luciferase uses coelenterazine as a substrate. By virtue of its smaller size, this luciferase may better allow native interactions without steric hindrance in fusion proteins. Transfection of the humanized version of *Gaussia* luciferase (hGluc) also allows strong luminescence signals in mammalian cells that are somewhat more stable over time than with hRluc.[3]

Another tool of potential advantage is a new fluorescent protein isolated from anthozoans.[25] These proteins, which are remote homologs of GFP, form a new group of fluorescent tags. One of these proteins has a much longer wavelength than any other fluorescent protein yet isolated, with an excitation spectrum peaking at 558 nm and a sharp emission spectrum peaking at 583 nm. The excitation spectrum is broad enough that a luciferase like RLUC might excite it. The advantage of this fluorescent protein is that its emission spectrum is sufficiently red shifted that the separation between BRET and non-BRET luminescence is much greater than with YFP, and hence quantification of BRET could be more accurate. This red fluorescent protein is now available from BD Biosciences Clontech (Palo Alto, CA) as DsRed, and it has been used in a FRET assay of protein–protein interactions in plants.[26] DsRed needs further development as a resonance tool, however, because it is a green fluorescent protein when first synthesized and matures to the red form over time. This means that it can undergo FRET with itself and its use could lead to misinterpretations of FRET/BRET signals.[27] Further, the natural tetramerization of DsRed makes its use in energy transfer studies problematic.[27] Hopefully, a useful mutant form of DsRed can be developed that is naturally a monomer and synthesized immediately into a stable red fluorescent form (see note added in proof).

[24] C. Szent-Gyorgyi, B. T. Ballou, E. Dagnal, and B. Bryan, *Prog. Biomed. Optics* **3600,** 4 (1999).

[25] V. M. Matz, A. F. Fradkov, Y. A. Labas, A. P. Savitsky, A. G. Zaraisky, M. L. Markelov, and S. A. Lukyanov, *Nat. Biotechnol.* **17,** 969 (1999).

[26] P. Más, P. F. Devlin, S. Panda, and S. A. Kay, *Nature* (*London*) **408,** 207 (2000).

[27] G. S. Baird, D. A. Zacharias, and R. Y. Tsien, *Proc. Natl. Acad. Sci. U.S.A.* **97,** 11984 (2000).

New substrates for the luciferase are also available. In their study with mammalian cells, Angers *et al.*[19] used the coelenterazine analog, h coelenterazine, to increase luminescence intensity. This coelenterazine analog and others are available from Molecular Probes (Eugene, OR; www.probes.com). BioSignal markets another coelenterazine analog in which the spectrum of emission is shifted to shorter wavelengths. When used with a GFP mutant adapted to this emission wavelength, this BRET pair results in a higher sensitivity and wider dynamic range. This system is now available under the trademark BRET2. Finally, a new application of BRET is its use for *in vitro* assays, including a homogeneous noncompetitive immunoassay[28] and a homogeneous BRET assay for biotin.[29] It is likely that many other applications will emerge as instrumentation for its assay becomes more common.[30]

Potential Screening System

On the basis of these data, we have proposed a relatively simple scheme for designing an *in vivo* library-screening system for protein–protein interaction through BRET.[2,3,14] By measuring the light emission collected through interference filters, the 530 nm : 480 nm luminescence ratio of *E. coli* (or yeast) colonies expressing a "bait" protein fused to RLUC and a library of "prey" molecules fused to EYFP (or vice versa) could be measured. It would be possible to screen colonies of bacteria or yeast on agar plates, using a camera imaging system. On the other hand, a photomultiplier-based instrument designed to measure luminescence of liquid cultures in 96-well plates could be adapted to high-throughput BRET screening by insertion of switchable 480- or 530-nm filters in front of the photomultiplier tube. Colonies that show high light intensity (i.e., bright colonies) at 530 nm or exhibit an above-background 530 nm : 480 nm ratio could be selected and the "prey" DNA sequence further characterized. Thus, an efficient BRET screening system could be practical by use of an appropriate instrument.

Advantages of Resonance Energy Transfer Techniques: Bioluminescence Resonance Energy Transfer and Fluorescence Resonance Energy Transfer

Features of BRET and FRET techniques offer some attractive advantages over other current assays for protein–protein interactions, especially the yeast

[28] R. Arai, H. Nakagawa, K. Tsumoto, W. Mahoney, I. Kumagai, H. Ueda, and T. Nagamune, *Anal. Biochem.* **289,** 77 (2001).

[29] M. Adamczyk, J. A. Moore, and K. Shreder, *Org. Lett.* **3,** 1797 (2001).

[30] Y. Xu, D. Piston, and C. H. Johnson, *in* "Green Fluorescent Protein: Methods and Protocols" (H. W. Hicks, ed.), p. 131. Humana Press, Totowa, NJ, 2001.

two-hybrid method, which is currently the most widely used. For instance, BRET and FRET can be applied to determine whether the interaction changes with time because the measurement is noninvasive. BRET and FRET are suitable to assay protein–protein interactions in different subcellular compartments or specific organelles of native cells; this has already been shown to work for FRET.[9,10] In particular, the yeast nucleus may be a poor place for some compatible proteins to meet. This advantage of BRET and FRET could be particularly useful in the case of interacting membrane proteins, for which assays are limited with other traditional methods. BRET and FRET may also be used to reveal interactions that depend on cell type-specific posttranslational modifications that do not occur in yeast and therefore cannot be assayed by the yeast-two hybrid method. By using cell type-specific promoters and/or fusion to targeting sequences, GFP-based BRET and FRET indicators can be observed specifically in the cell type and subcellular location of choice. Moreover, BRET and FRET assays could be adapted to monitor the dynamic processes of protein–protein interactions *in vivo,* such as intracellular signaling.

Limitations of Technique

As with any technique, however, the resonance energy transfer methods have some limitations. For example, the efficiency of both BRET and FRET is dependent on proper orientation of the donor and acceptor dipoles. Conformational states of the fusion proteins may fix the dipoles into a geometry that is unfavorable for energy transfer. Further, because the fluorophore/luciferase tags are fused to ends of the potentially interacting molecules, it is possible that some parts of the candidate molecules are interacting without allowing the fluorophore/luciferase tags to be close enough for energy transfer to occur. Consequently, two proteins might interact in a way that is blind to the FRET/BRET technique. In other words, a negative result with a resonance transfer technique does not prove noninteraction. In such a case, testing different combinations of N-terminal and C-terminal fusions in BRET/FRET assays could help to determine the optimal orientation in which candidate proteins interact.

The luciferase/fluorescent protein tags that are fused to the candidate interacting proteins could interfere with the interaction by steric hinderance (this problem is true for the yeast two-hybrid assay as well). Therefore, the smaller the tags, the less likely will be the hindrance. This is a reason why the *Gaussia* luciferase might prove to be superior to RLUC. These luciferase/fluorescent tags might cause inactive or incorrectly folded fusion proteins. For example, the bulkiness of the GFP (and its derivatives) cylinders (20 × 30 Å) have been shown to impede correct folding of some fusion proteins.[12]

Another consideration in the use of GFP variants as fluorophore tags is that the slow kinetics of GFP turnover may hamper measuring the kinetics of interaction (whereas *Renilla* luciferase does not suffer these same disadvantages in turnover rate). New GFPs are available that have been engineered to be less stable (the BD Biosciences Clontech d2EGFP),[31] and the reengineering of BRET fluorophores to be less stable could be useful in temporal studies. Moreover, the acid sensitivity of some of the GFPs might restrict their application to subcellular areas with neutral or higher pH values. However, this limitation can be overcome by utilizing mutants that are less sensitive to pH.[32]

Bioluminescence Resonance Energy Transfer versus Fluorescence Resonance Energy Transfer

BRET has potential advantages over FRET because it does not require the use of excitation illumination. BRET should be superior for cells that are either photoresponsive (e.g., retina or any photoreceptive tissue) or damaged by the wavelength of light used to excite FRET. Moreover, photobleaching of the fluorophores can be a serious limitation of FRET, but it is irrelevant to BRET. Cells that have significant autofluorescence would also be better assayed by BRET than by FRET. This is particularly true for highly autofluorescent tissue, but all cells are autofluorescent to a degree because of ubiquitous fluorescent molecules such as NADH, collagen, and flavins. Plant cells have particularly high autofluorescence, primarily due to photosynthetic pigments. Adaptation of BRET to plant cells is shown in Fig. 4, where fusion proteins between RLUC and various spectral variants of GFP (GFPS65T,[6] GFP5,[33] and EYFP) were tested for BRET after expression in onion epidermal cells (not highly pigmented) and transgenic *Arabidopsis* seedlings (highly pigmented). In onion cells, all these fusion constructs displayed BRET, but EYFP was the optimal BRET acceptor in this cell type (Fig. 4A). As control, coexpression of unfused RLUC and YFP did not result in a significant BRET signal. Moreover, RLUC and RLUC::YFP fusion proteins were stably expressed in transgenic *Arabidopsis* seedlings and were spectrally distinguishable despite the presence of photosynthetic pigments (Fig. 4B). BRET is particularly promising in plant cells because the highly fluorescent photosynthetic pigments and cell wall compounds that are prevalent in plants interfere with FRET-based assays.

In addition, FRET may be prone to complications due to simultaneous excitation of both donor and acceptor fluorophores. Specifically, even with monochromatic laser excitation, it is impossible with the current generation of fluorescent

[31] J. B. Andersen, C. Sternberg, L. K. Poulsen, S. P. Bjorn, M. Givskov, and S. Molin, *Appl. Environ. Microbiol.* **64,** 2240 (1998).

[32] A. Miyawaki, O. Griesbeck, R. Heim, and R. Y. Tsien, *Proc. Natl. Acad. Sci. U.S.A.* **96,** 2135 (1999).

[33] J. Haseloff, K. R. Siemering, D. C. Prasher, and S. Hodge, *Proc. Natl. Acad. Sci. U.S.A.* **94,** 2122 (1997).

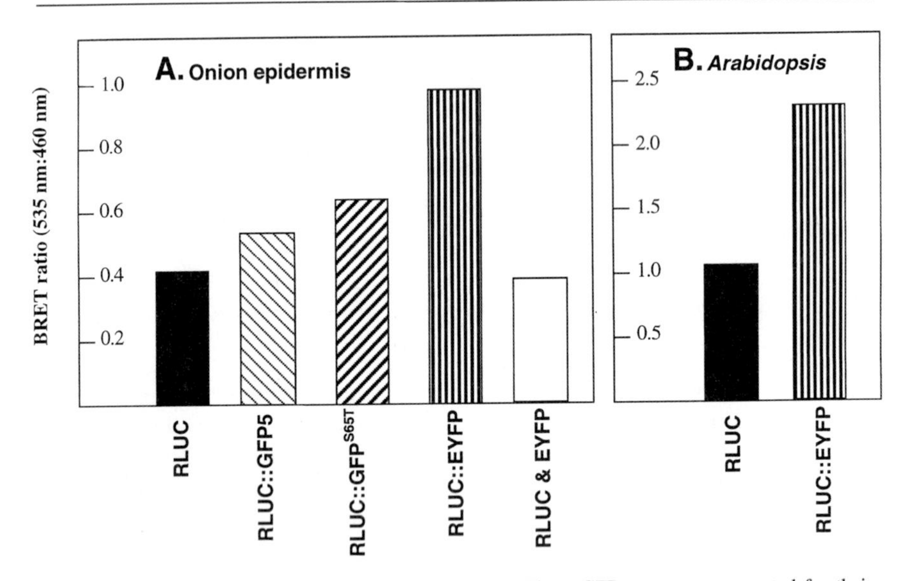

FIG. 4. *In vivo* detection of BRET in plant cells. (A) Three GFP mutants were tested for their suitability as BRET acceptors by fusion to RLUC. Fusion proteins were expressed by transient transformation of onion (*Allium cepa*) epidermal cells, using a Bio-Rad (Hercules, CA) PDS-He particle gun. As a control, unfused RLUC and YFP were coexpressed (RLUC and EYFP). BRET ratios were acquired under *in vivo* conditions through green (535 ± 20 nm) and blue (460 ± 25 nm) interference filters in a Wallac Victor II microplate reader. (B) RLUC and RLUC::YFP were expressed in stably transformed *Arabidopsis thaliana* under the control of the CaMV 35S promoter. BRET ratios were determined in 4-day-old light-grown seedlings, using a Turner Designs TD20/20 luminometer with BRET accessory (blue filter, 380–420 nm; yellow long-pass filter, 520+ nm).

proteins to excite only the donor without exciting the acceptor fluorophore to some degree. In contrast, because BRET does not involve optical excitation, all the light emitted by the fluorophore must result from resonance transfer. Therefore, BRET is theoretically superior to FRET for quantifying resonance transfer. Related to this point is one of the most important advantages of BRET over FRET—namely, that the relative levels of expression of the donor and acceptor partners can be quantified independently: the donor by luminescence and the acceptor by fluorescence. This is difficult with FRET because the acceptor is generally excited to some extent by the excitation wavelength used to excite the donor. With BRET, measuring the fluorescence of the system gives the relative level of the acceptor (YFP/GFP fusion partner) and when coelenterazine is added, the total luminescence of the system measured in darkness gives the relative level of the donor (luciferase fusion partner). Knowledge of the relative levels of the fusion partners is crucial when comparing the results of one experiment with those of another.

A BRET assay requires a substrate for the luciferase. In the case of RLUC and Gluc, coelenterazine is the substrate. Coelenterazine is hydrophobic and can permeate all the cell types we have tested, including bacteria (*E. coli* and cyanobacteria), yeast, *Chlamydomonas,*[34] plant seedlings and calli (Fig. 4[35]), and animal cells in culture (Fig. 3[19]). The major limitation that BRET suffers in comparison with FRET is that the luminescence may sometimes be too dim to measure accurately without a sensitive light-measuring apparatus. With FRET, dim signals can be amplified simply by increasing the intensity or duration of excitation (possibly at the cost of light-induced damage to the cells), whereas with BRET, the only option to improve low signal levels is to integrate the signal for a longer time. New instruments designed for BRET measurements have been introduced: the Fusion from Packard (now part of PerkinElmer Life Sciences, Boston, MA), and the Mithras from Berthold Technologies (Bad Wildbad, Germany) (both are plate-reading luminometers). Other instruments are capable of BRET measurements as well,[30] including the single-channel Turner TD20/20 luminometer (Turner Designs, Sunnyvale, CA) with BRET accessory (used in the measurements in Fig. 4). Manufacturers are continuously developing improved instrumentation for measuring low-light levels, and these improvements in technology will undoubtedly aid the further development of BRET assays of real-time protein–protein interactions in living organisms.

Note added in proof: A new paper reports the development of a red fluorescent protein that is monomeric and matures rapidly to a red fluorescent form: R. E. Campbell, O. Tour, A. E. Palmer, P. A. Steinbach, G. S. Baird, D. A. Zacharias, and R. Y. Tsien, *Proc. Natl. Acad. Sci. U.S.A.* **99,** 7877 (2002).

Acknowledgments

This research was supported by the National Institute of Mental Health (MH 43836 and MH 01179 to C.H.J.), the National Institutes of Health (GM 59984 to C.H.J., DK534343 and CA86283 to D.W.P.), and the National Science Foundation (MCB-9874371 to C.H.J. and MCB 0114653 to A.G.V. and C.H.J.); spectral data were acquired at the Vanderbilt Cell Imaging Shared Resource, supported in part by the NIH through the Vanderbilt Cancer Center (CA68485) and the Vanderbilt Diabetes Center (DK20593).

[34] I. Minko, S. P. Holloway, S. Nikaido, O. W. Odom, M. Carter, C. H. Johnson, and D. L. Herrin, *Mol. Gen. Genet.* **262,** 421 (1999).

[35] J. Sai and C. H. Johnson, *Proc. Natl. Acad. Sci. U.S.A.* **96,** 11659 (1999).

[13] Structure–Function Relationships in Metalloproteins

By JAROSLAVA MIKŠOVSKÁ and RANDY W. LARSEN

Metalloproteins

Metalloproteins represent a diverse class of biological macromolecules that perform a wide array of physiologic functions. These functions include the conversion of light into chemical energy in photoreceptors, electron transport, small molecule transport and storage, structural organization of protein matrices, small molecule sensing, and the synthesis/degradation of various metabolites. The most abundant metals found in human plasma are Fe (2×10^{-5} *M*), Zn (1.7×10^{-5} *M*), Cu (1.7×10^{-5} *M*), Mo (1.0×10^{-5} *M*), Cr (5.5×10^{-8} *M*), V (1.77×10^{-7} *M*), Mn (1.1×10^{-7} *M*), Ni (4.4×10^{-8} *M*), and Co (2.0×10^{-11} *M*).[1] A majority of these metals have a variety of oxidation states and relatively low reduction potentials under physiological conditions, making them ideal for electron transfer reactions or catalytic reactions requiring electron shuttling. In addition, nature takes full advantage of the variety of coordination geometries available to transitions metals to optimize metalloprotein functionality (see Fig. 1). Of the transition metals found in living systems, Fe and Cu are the most prevalent because of their redox lability under physiological conditions and because they are the most abundant of the transition metals in seawater (i.e., the wide belief that life originated in the oceans suggests that metal abundance in living organisms should parallel the abundance in seawater).[1]

As iron and copper are the most abundant of the transition metals in biological molecules, Fe/Cu-containing proteins and enzymes provide a good cross-section of the diversity of function of metalloproteins. Iron-containing proteins and enzymes generally fall into two categories: heme (discussed in the next section) and nonheme iron proteins. Nonheme iron proteins include oxygen carriers (hemerythrins) and electron transport proteins (ferredoxins and iron–sulfur proteins containing Fe_4S_4, Fe_2S_2, and Fe_3S_4 centers).[1,2] A number of other proteins bind Fe for transport and storage. The iron–sulfur proteins are good examples of Nature's use of coordination geometry of transition metals to modulate physical function of metalloproteins, in this case reduction potential of various iron–sulfur proteins containing differing coordination geometries (usually tetrahedral and distorted tetrahedral),

[1] W. W. Porterfield, *in* "Inorganic Chemistry: A Unified Approach," 2nd Ed. Academic Press, San Diego, CA, 1993.

[2] E. I. Stiefel and G. N. George, *in* "Bioinorganic Chemistry" (I. Bertini, H. B. Gray, S. J. Lippard, and J. S. Valentine, eds.), p. 365. University Science Books, Sausalito, CA, 1994.

0076-6879/03 $35.00

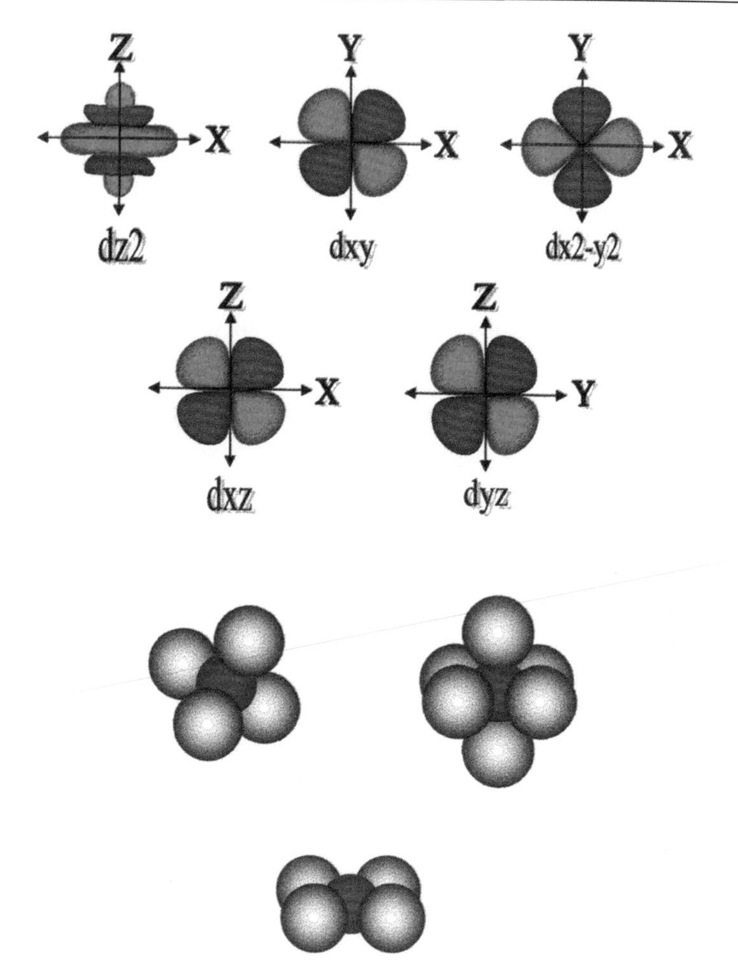

FIG. 1. *Top:* Diagram showing spatial orientation of atomic d-orbitals. *Bottom:* Common metal coordination geometries.

can span a reduction potential range of nearly 1 V ($\varepsilon^0 = -645$ mV for the Fe_4S_4 protein from *Azotobacter vinellandii* ferredoxin I (FdI) to $\varepsilon^\circ = +356$ mV for a high-potential iron–sulfur protein from *Chromatium vinosum*) (Stiefel and George[2] and references therein). Oxygen storage proteins also make use of coordination geometry to facilitate ligand binding. These proteins contain iron centers with more octahedral-like geometry. Thus, understanding structure–function relationships requires a detailed knowledge of the coordination sphere of the iron centers.

Copper proteins are also widely distributed in nature. The copper ion is utilized primarily in oxygen transport and electron transport, much like iron.[3] Electron transfer proteins containing copper active sites include the plastocyanin and azurin classes. Both classes of proteins are found primarily in plant tissue. Azurin-type proteins contain a single copper atom in a trigonal planar coordination geometry with two histidines and one cysteine ligand tightly coordinated. A weakly bound carbonyl group of one glycine residue and one weakly bound methionine make up the remaining ligands. These proteins are characterized by an intense S(Cys) $\rightarrow$ Cu charge transfer transition near 600 nm that has been a convenient spectroscopic probe for structure–function studies. The reduction potentials of these proteins fall in the range of $\sim$+200 to $\sim$+400 mV. Plastocyanins serve as electron shuttles between cytochrome *f* and P700 of photosystem I in higher plants and algae. The Cu ion in these proteins contains two histidines, one cysteine, and one methionine ligand in a trigonally distorted tetrahedral geometry. The reduction potentials of these proteins are in the +245 mV range. Copper proteins also participate in oxygen storage/transport. Hemocyanins are large, multisubunit proteins that transport oxygen in mollusks and anthropods. The active site of these proteins consists of a binuclear copper center, with each copper ion exhibiting a square planar geometry. Each of the copper ions is coordinated to two histidine residues and a bridging oxygen molecule. The nature of the fourth ligand is still not clear. Hemocyanins generally have a lower affinity for oxygen than the iron porphyrin-containing hemoglobins (Hb). For example, hemocyanin from *Leirus quinquestris* R. exhibits a P_{50} of 1 torr whereas human HbA exhibits a P_{50} of 0.7 torr.

Heme Proteins

Heme proteins are one of the most widely distributed metalloproteins in nature.[4–6] Heme proteins participate in electron transfer (cytochromes), oxygenation (monooxygenases), hydrogen peroxide degradation (peroxidases, catalases), small molecule sensing [FixL, PAS (*p*eriod circadian protein, *A*h receptor nuclear translocator protein, *s*ingle-minded protein) domain sensors, and HemAT (aerotactic heme) sensors], transcription regulation [CooA (CO-oxidizing system activator)-type proteins], energy transduction (heme/copper oxidases, cytochrome bc_1, etc.), oxygen transport and storage (hemoglobins and myoglobins), and polymer synthesis/degradation (lignan peroxidase, etc.). Heme proteins contain an

[3] R. Lontie, *in* "Copper Proteins and Copper Enzymes." CRC Press, Boca Raton, FL, 1984.

[4] M. Gouterman, *in* "The Porphyrins" (D. Dolphin, ed.), Vol. III. Academic Press, New York, 1978.

[5] T. G. Spiro, *in* "Iron Porphyrins" (A. B. P. Lever and H. B. Gray, eds.). Addison-Wesley, Reading, MA, 1983.

[6] K. Smith, *in* "Porphyrins and Metalloporphyrins." Elsevier, Amsterdam, 1975.

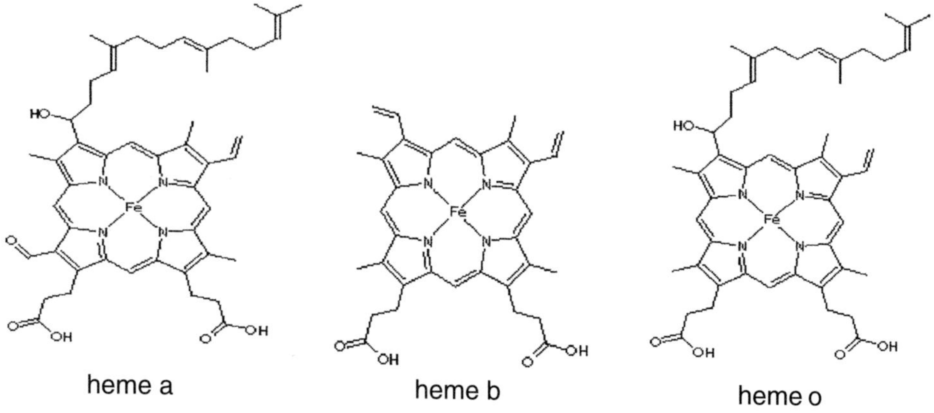

FIG. 2. Structural diagram of various heme groups found in nature.

iron protoporphyrin IX active site (or a derivative of this macrocycle) and are coordinated to the protein via amino acids containing lone pairs of electrons such as histidine, methionine, tyrosine, or cysteine (some π-type interactions may be involved in histidine coordination). The structures of the various hemes found in nature are shown in Fig. 2.

The optical properties of the heme group as well as the fact that the heme group is either the sole active site or one of the active sites of the protein have proven to be an extraordinarily useful probe for monitoring structure–function relationships in a wide variety of heme proteins. The absorption spectrum of the heme macrocycle has been shown to be sensitive to Fe oxidation and spin state, axial ligation, and solvent environment (Fig. 3). This spectrum is dominated by an intense absorption band in the near-UV region (Soret band, $\varepsilon \approx 1 \times 10^5\ M^{-1}\ \mathrm{cm}^{-1}$) and typically two less intense bands in the visible region ($\varepsilon \approx 1 \times 10^4\ M^{-1}\ \mathrm{cm}^{-1}$). The Soret band arises from a $\pi \rightarrow \pi^*$ transition that is a mixture of $a_{1u} \rightarrow e_g^*$ and $a_{2u} \rightarrow e_g^*$ electronic configurations (D_{4h} symmetry).[4–6] The longest wavelength visible band (α band) arises from the lowest energy (0,0) vibronic transition of the Soret transition whereas the second visible band (β band) is attributed to an absorption band with one mode of vibrational excitation (0,1). These absorption bands have been extensively utilized to probe the kinetics of physiological reactions in heme proteins.

Optical Methods to Study Heme Proteins

By far the most widely used method for structure–function studies of heme proteins has been optical absorption spectroscopy. Equilibrium optical absorption

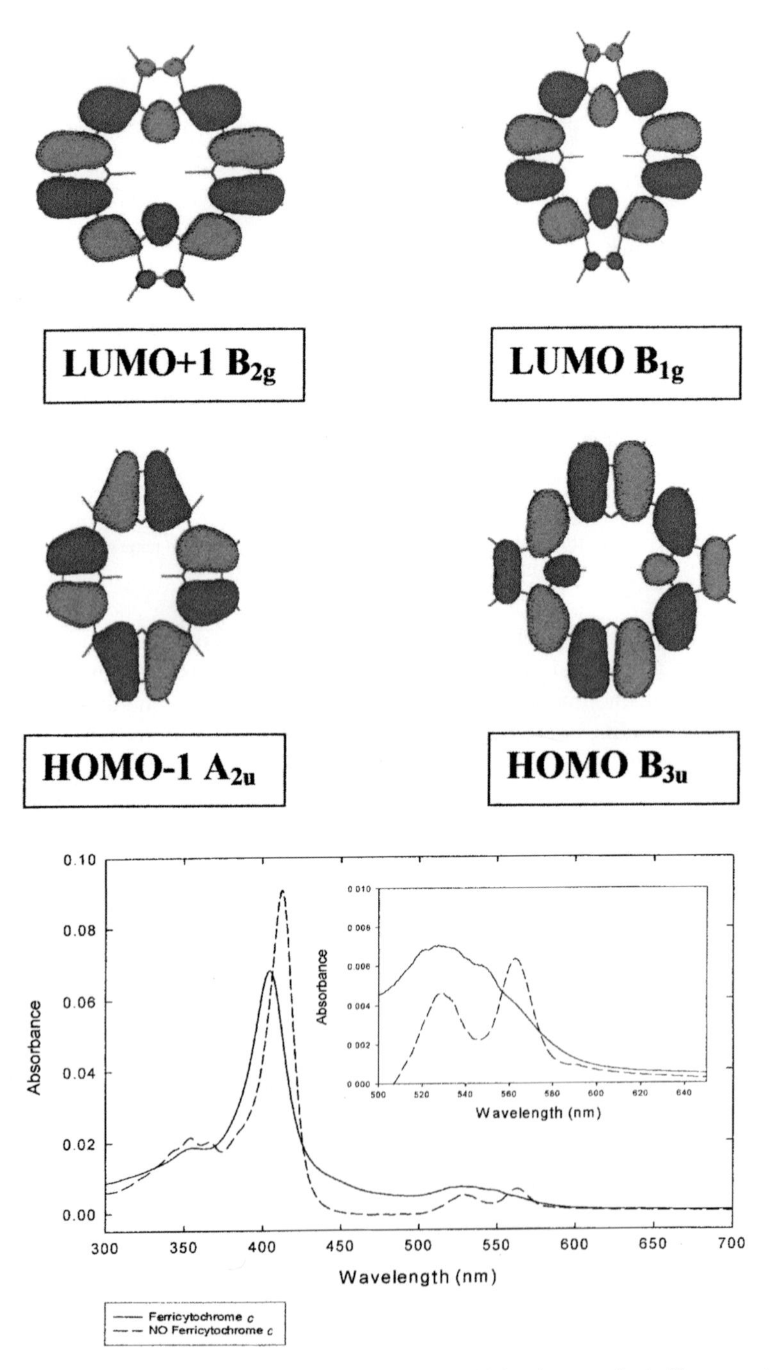

FIG. 3. *Top:* Representative molecular orbitals associated with free-base porphyrin (D_{2h} symmetry). Porphyrin molecular orbitals are only slightly different for the metal-substituted macrocycle (symmetry increased to D_{4h}). *Bottom:* Typical heme protein absorption spectra, in this case, NO-bound ferricytochrome *c* (dash line) and ferricytochrome *c* (solid line).

is routinely utilized to determine the oxidation state and/or the nature of the axial ligation in heme proteins and is especially useful in determining these properties in newly discovered heme proteins (such as FixL and the HemAT class) as well as in heme protein mutants. Of greater utility is the use of time-resolved optical spectroscopy to probe heme protein kinetics under more physiological conditions (see Chen *et al.*[7] for a review). In this type of experiment a perturbation is made to the protein (photolysis of a ligand, photoinjection of an electron, rapid solution heating, rapid pH jump, etc.) and the subsequent relaxation of the protein is monitored by monitoring the changes in the heme spectra as a function of time. A classic example of this type of experiment is the photolysis of CO from CO-bound human hemoglobin (HbA).[7–10] DeoxyHbA exhibits a Soret maximum at 430 nm whereas that of COHbA is found at 415 nm. Early studies monitoring the changes in absorbance as a function of time after full photolysis of CO-bound HbA revealed five kinetics steps of ~40 ns (geminate recombination), ~1 μs (tertiary structural change), ~20 μs [quaternary structural changes associated with the transition between high affinity (R) state and low affinity (T) state], ~200 μs (quaternary structural transition together with CO binding), and ~4 ms (CO recombination to the T-like quaternary structure). Further monitoring of the isosbestic point between the equilibrium deoxy- and CO-bound spectra as a function of time subsequent to photolysis has allowed for the kinetic complexity of the R–T transition to begin to be understood. In a similar way, the photolysis of CO-bound myoglobin (COMb) has revealed only two phases of CO recombination: a geminate recombination phase ($k_{gem} = 3 \times 10^8\ s^{-1}$) and a diffusion-controlled rebinding phase ($k_1 = 7.6 \times 10^5\ M^{-1}\ s^{-1}$). Photolysis experiments involving a large number of mutant Mbs have now provided important details about the structural subtleties of the distal heme pocket in modulating ligand affinities.[11]

Kinetic ligand-binding studies have proved enormously successful in probing structure–function relationships in heme proteins. Fast kinetics studies have also been extremely useful in probing other reactions, such as electron transfer, as well. For example, these studies have begun to illuminate the complex mechanism of dioxygen reduction and active proton transport in heme/copper oxidases found in nearly all aerobic organisms. In higher organisms these enzymes contain two heme chromophores and at least one copper ion (see below for a more detailed description of this class of enzyme). One heme and one copper ion together constitute a dioxygen reduction site, while the remaining heme (and Cu ion in heme/copper oxidases from higher organisms) catalyze electron transfer from the substrate to dioxygen

[7] E. Chen, R. A. Goldbeck, and D. S. Kliger, *Annu. Rev. Biophys. Biomol. Struct.* **26,** 327 (1997).
[8] C. A. Sawicki and Q. H. Gibson, *J. Biol. Chem.* **251,** 1533 (1976).
[9] J. Hofrichter, J. H. Sommer, E. R. Henry, and W. A. Eaton, *Proc. Natl. Acad. Sci. U.S.A.* **80,** 2235 (1983).
[10] A. Bellelli and M. Brunori, *Methods Enzymol.* **232,** 56 (1994).
[11] B. A. Springer, S. G. Sligar, J. S. Olson, and G. N. Phillips, *Chem. Rev.* **94,** 699 (1994).

bound at the heme/copper binuclear center. One of the most successful strategies for probing the pathways of electron transfer as well as the kinetics/dynamics of dioxygen chemistry has been to use the so-called flow–flash method coupled with absorption spectroscopy.[12,13] In this experiment the enzyme is reduced with four electrons (in the bovine case) and CO is bound to the dioxygen reduction site. Under these conditions dioxygen binds to the binuclear center with a rate constant equal to that of the CO dissociation rate from the heme ($0.03\ s^{-1}$).[14] The fully reduced CO bound form of the enzyme is then rapidly mixed with oxygen-saturated buffer and exposed to a short laser pulse that photodissociates the CO on a femtosecond time scale. The subsequent oxygen binding and intramolecular electron transfer within the enzyme are then monitored by probing changes in the heme optical absorption spectra. These studies have shown that the initial two-electron reduction of dioxygen (formation of the so-called P-intermediate) occurs with a rate constant of roughly $3 \times 10^4\ s^{-1}$, which is followed by proton uptake with a rate constant of $\sim 1 \times 10^4\ s^{-1}$.[15–17] Further electron transfer, forming an oxyferryl species at the heme of the binuclear center, occurs with a rate constant of $\sim 5 \times 10^3\ s^{-1}$. The fourth electron is then transferred with a rate constant of $\sim 500\ s^{-1}$. The use of enzyme reconstituted into phospholipid vesicles together with pH-sensitive dyes has further allowed the kinetics of active proton transport to be determined.[16]

These two examples demonstrate the utility of optical spectroscopy in unraveling kinetic complexity in heme-containing proteins and enzymes. By coupling these methods with site-directed mutagenesis a more detailed picture of structure–function relationships in these proteins can be obtained.

Activation Parameters and Thermodynamics

Although the use of optical spectroscopy has been primarily focused on unraveling the often complex kinetics involved with heme protein chemistry, similar methods can also be employed to obtain important information concerning the thermodynamics associated with heme protein reactions. Probing the thermodynamics of metalloprotein function is a critical component in understanding structure–function relationships. In the case of heme proteins, ligand binding,

[12] Q. H. Gibson and L. Milnes, *Biochem. J.* **91,** 161 (1964).

[13] M. I. Verkhovsky, N. Belevich, J. E. Morgan, and M. Wikstrom, *Biochim. Biophys. Acta* **1412,** 184 (1999).

[14] O. Einarsdottir, *Biochim. Biophys. Acta* **1229,** 129 (1995).

[15] H. Michel, J. Behr, A. Harrenga, and A. Kannt, *Annu. Rev. Biophys. Biomol. Struct.* **27,** 329 (1998).

[16] S. Hallen and T. Nilsson, *Biochemistry* **31,** 11853 (1992).

[17] B. C. Hill and C. Greenwood, *Biochem. J.* **218,** 913 (1984).

electron transfer, and so on, provide at least part of the free energy required for catalysis and/or relevant conformational changes.

Ideally, to have a firm understanding of the reaction pathway in heme proteins a complete thermodynamic profile is required. Such a profile would map the magnitudes and time scales of thermodynamic changes (including enthalpy, volume, and entropy changes). There are a number of methods available for determining reaction enthalpy and entropy changes, but these are typically equilibrium techniques [such as equilibrium binding titrations as a function of temperature (reaction enthalpies/entropies) and pressure (reaction volumes)]. Reaction kinetics can also reveal information concerning the thermodynamic profiles. Knowledge of the thermodynamic properties of transition states between various intermediates within an enzymatic or ligand-binding cycle can be utilized to construct the appropriate thermodynamic profile. This can be accomplished by monitoring the various reaction rates as a function of temperature or pressure and then fitting the data to the following expressions:

$$\ln[k_{obs}/(k_b h/T)] = \Delta H^{\ddagger}/RT + \Delta S^{\ddagger}/R \tag{1}$$

(where k_b is Boltzmann's constant, h is Planck's constant, k_{obs} is the observed rate constant, and T is the absolute temperature) and

$$-RT[\partial \ln(k_{obs})/\partial P]_P = \Delta V^{\ddagger} \tag{2}$$

(where R is the universal gas constant, T is the temperature, and P is the applied pressure).[18,19] The slope of the line in Eq. (1) gives the activation enthalpy and the intercept provides the entropy of activation. Likewise, the slope in Eq. (2) gives the activation volume. Determining the activation parameters for all the pathways involved in a chemical reaction allows for the construction of a complete thermodynamic profile (see Fig. 4). The difficulty, however, arises from the fact that the activation parameters cannot always be determined in both the forward and reverse directions. Thus, to obtain the complete profile requires the knowledge of $\Delta H^{\ddagger}/\Delta V^{\ddagger}$ in one reaction direction as well as the overall $\Delta H/\Delta V$ for the given reaction step, which requires determining $\Delta H/\Delta V$ for transient species along the reaction coordinate.

Photothermal Methods

During the last decay calorimetric techniques such as photoacoustic calorimetry (PAC) and photothermal beam deflection (PBD) became popular techniques

[18] R. van Eldik, T. Asano, and W. J. le Noble, *Chem. Rev.* **89,** 549 (1989).

[19] K. Hiromi, *in* "Kinetics of Fast Enzyme Reactions: Theory and Practice." Kodanshi Scientific Books, Kodanshi, Japan, 1979.

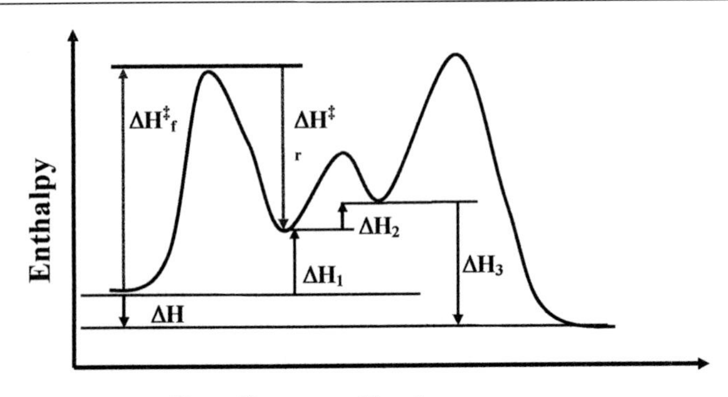

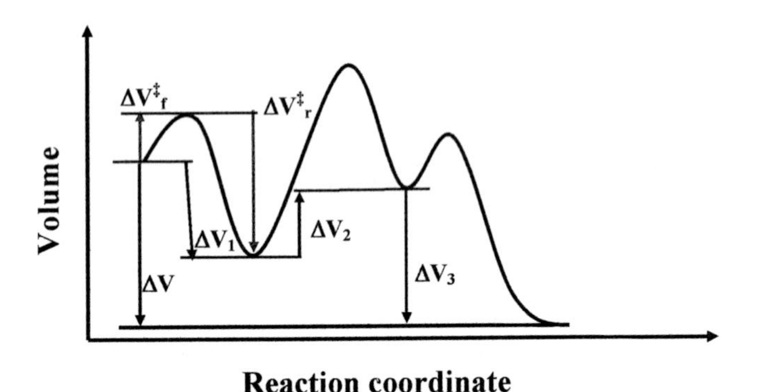

FIG. 4. Examples of enthalpy and volume profiles describing chemical/biochemical processes.

because of their ability to determine the magnitudes and time scales of enthalpy and volume changes resulting from photoinitiated processes. Photoacoustic spectroscopy has been applied to the study of electron transfer reactions in bacterial reaction centers,[20–22] the proton-translocating photocycle of bacteriorhodopsin,[23–25] and conformational changes associated with the ligand dissociation from carboxyhemoglobin, carboxymyoglobin, and cytochrome P450.[26–28] More recently,

[20] J. M. Hou, V. A. Boichenko, B. A. Diner, and D. Mauzerall, *Biochemistry* **40,** 7117 (2001).

[21] J. M. Hou, V. A. Boichenko, Y. C. Wang, P. R. Chitnis, and D. Mauzerall, *Biochemistry* **40,** 7109 (2001).

[22] G. J. Edens, M. R. Gunner, Q. Xu, and D. Mauzerall, *J. Am. Chem. Soc.* **122,** 1479 (2000).

[23] D. Zhang and D. Mauzerall, *Biophys. J.* **71,** 381 (1996).

[24] P. J. Schulenberg, M. Rohr, W. Gartner, and S. E. Braslavsky, *Biophys. J.* **66,** 838 (1994).

[25] S. L. Logunov, M. A. El-Sayed, L. Song, and J. K. Lanyi, *J. Phys. Chem.* **100,** 2391 (1996).

[26] K. S. Peters, T. Watson, and T. Logan, *J. Am. Chem. Soc.* **114,** 4276 (1992).

[27] C. Di Primo, G. H. B. Hoa, E. Deprez, P. Douzou, and S. G. Sligar, *Biochemistry* **32,** 3671 (1993).

[28] J. A. Westrick, K. S. Peters, J. D. Ropp, and S. G. Sligar, *Biochemistry* **29,** 6741 (1990).

photothermal beam deflection has been used to probe long time conformational changes during the bacteriorhodopsin photocycle.[29,30]

In general, both photoacoustic calorimetry (PAC) and photothermal beam deflection (PBD) are based on the same principle. A nonradiative relaxation of a photoexcited molecule (heme, chlorophyll, retinal, etc.) to the ground state causes thermal heating of the surrounding solvent. This results in a rapid volume expansion that generates both pressure changes (acoustic wave) and density changes (change in refractive index) within the illuminated volume. The pressure change can be easily detected by a pressure transducer or a sensitive microphone (PAC). The change in density is primarily responsible for the refractive index change that can be probed by a variety of optical methods. A gradient in the refractive index results in a change in direction of an impinging light ray and the magnitude of the change in direction is directly proportional to the amount of sample heating and/or molar volume change. The photothermal beam deflection technique uses a position-sensitive bicell detector to probe the magnitude of deflection of a laser beam passing through an absorbing sample subsequent to the initiation of solution photochemistry. The theory and application of PAC and PBD have been reviewed in detail elsewhere.[31–33] In this article we summarize the theory of both techniques and describe their implementation and application in our laboratory.

Photoacoustic Calorimetry

General Theory

As mentioned above, acoustic waves are produced when a photoexcited molecule transfers excess energy to the solvent bath. In addition to the solvent volume changes caused by the thermal relaxation process, volume changes in the system of interest resulting from a photo-initiated reaction also contribute to the overall solvent volume change. Thus at least two processes contribute to the overall acoustic wave according to Eq. (3):

$$S = K E_a \Phi(\Delta V_{th} + \Delta V_{con}) \tag{3}$$

where S is the acoustic signal, K is an instrument response parameter, E_a is the number of einsteins absorbed, Φ is the quantum yield of the process, ΔV_{th} is the volume change caused by the thermal heating of the solvent, and ΔV_{con} is the nonthermal volume change. The thermal component of solvent volume change

[29] P. J. Schulenberg, W. Gartner, and S. E. Braslavsky, *J. Phys. Chem.* **99,** 9617 (1995).
[30] A. Losi, I. Michler, W. Gartner, and S. E. Braslavsky, *Photochem. Photobiol.* **72,** 590 (2000).
[31] S. E. Braslavsky and G. E. Heibel, *Chem. Rev.* **92,** 1381 (1992).
[32] K. S. Peters, T. Watson, and K. Mar, *Annu. Rev. Biophys. Biophys. Chem.* **20,** 343 (1991).
[33] D. E. Falvey, *Photochem. Photobiol.* **65,** 4 (1997).

is related to the heat released to the surrounding solvent, Q, according to Eq. (4):

$$\Delta V_{th} = (\beta / C_p \rho) Q \quad (4)$$

where β is the thermal expansion coefficient of the solvent, C_p is the solvent heat capacity, and ρ is the solvent density. For aqueous solutions $\beta / C_p \rho$ is temperature dependent, and therefore ΔV_{th} is a function of temperature. Assuming ΔV_{con} is temperature independent, ΔV_{th} and ΔV_{con} can be separated by examining the temperature dependence of the acoustic wave amplitudes. The instrument response parameter can be eliminated by employing a reference compound that promptly converts energy of an absorbed photon, $E_{h\nu}$, into heat, with a quantum yield of unity, thus giving $\Delta V_{con} = 0$. The amplitude of the acoustic signal for the reference compound, S_{ref}, is then described by

$$S_{ref} = K E_a (\beta / C_p \rho) E_{h\nu} \quad (5)$$

A ratio of the sample signal to the reference signal gives the following expression for the signal ϕ scaled to the energy of an absorbed photon:

$$\phi E_{h\nu} = (S / S_{ref}) E_{h\nu} = \Phi\{Q + [\Delta V_{con} / (\beta / C_p \rho)]\} \quad (6)$$

A plot of $\phi E_{h\nu}$ versus $\beta / C_p \rho$ gives a straight line with a slope equal to $\Phi \Delta V_{con}$ and the intercept equal to the released heat (ΦQ). Subtracting Q from $E_{h\nu}$ gives ΔH for the reaction.

The transducers used in PAC are sensitive not only to the amplitude of the acoustic waves but also to their temporal profile. For example, if the photoinitiated process involves two sequential decay processes on the time scale of the transducer resolution (~100 ns–5 μs), the individual contribution to ΔV_{con} and Q for each decay process can be resolved. The observed time-dependent acoustic signal $E(t)_{obs}$ is produced by the convolution of an instrument response function $T(t)$ with a time-dependent function of the decay processes, $H(t)$:

$$E(t)_{obs} = H(t) * T(t) \quad (7)$$

and

$$H(t) = \phi_1 \exp(-\tau / t_1) + [\phi_2 k_2 / (k_2 - k_1)][\exp(-\tau / t_1) - \exp(-\tau / t_2)] \quad (8)$$

$T(t)$ can be independently determined from the reference compound. The deconvolution of the signal involves estimating of the parameters ϕ_i and k_i for $H(t)$ and a convolution of the estimated $H(t)$ function with the $T(t)$ function. The parameters are varied until the estimated $E(t)$ fits the $E(t)_{obs}$ (based on the χ^2 and residuals of the fit). Processes occurring faster than roughly 100 ns cannot be resolved but the net enthalpy and volume changes can be quantified from the amplitude of the acoustic wave.

Experimental Design

Photoacoustic calorimetry instruments vary in their design and a description of the instrument currently used in our laboratory is given (Fig. 5, top). Acoustic waves are generated by excitation of a sample by a 532-nm laser pulse [Continuum (Santa Clara, CA) Minilite I, Q-switched Nd:YAG (neodymium-doped yttrium aluminum garnet) laser, 6-ns pulse, $\sim$100 μJ]. A 1 $\times$ 1 cm quartz cuvette containing 1 ml of a sample is placed at the center of a Panametrics (Waltham, MA) V103 detector or a home-built detector based on a Transducer Products (New London, CT) PZT-2H piezoelectric crystal (2 mm in diameter) that is silvered on each end. The crystal is housed in an aluminum case such that one side of the crystal makes contact with the top of the cylinder. The crystal is held in place by a steel spring, which also serves as the signal carrier to a BNC connector. Contact between the cuvette and either detector is facilitated by a thin layer of vacuum grease. The acoustic waves created by photochemical reactions in the sample result in compression of the crystal and generation of a voltage that is amplified (ultrasonic preamplifier from Panametrics) and recorded by an NI 5102 15-MHz oscilloscope controlled by VirtualBench-Scope software (National Instruments, Austin, TX).

Photothermal Beam Deflection

Theory

Excitation of an absorbing sample results in the release of heat in the small volume illuminated by the excitation source, creating a virtual lens within the solution. This is due to the change in refractive index of the heated sample volume relative to the cooler surrounding medium. The lens created behaves in such a manner as to cause parallel light rays to diverge through the transient lens. The degree of refractive index change can be monitored by a change in the deflection angle of a probe laser beam passing near the edge of the transient lens. The corresponding deflection signal arising from changes in the refractive index of the sample can be described as

$$S = KE_a\Phi[(dn/dt)(1/\rho C_p)Q + \rho(dn/d\rho)\Delta V + B\Delta n_{abs}] \tag{9}$$

where K represents an instrument response parameter (equivalent to K in PAC), E_a is the number of einsteins absorbed, Φ is the quantum yield, ρ is the solvent density, C_p is the solvent heat capacity, Q is the heat released to the solvent by the sample on excitation, n is the refraction index of solution, and $B\Delta n_{abs}$ is the change in the refractive index due to the absorption changes in the sample.[29] As in

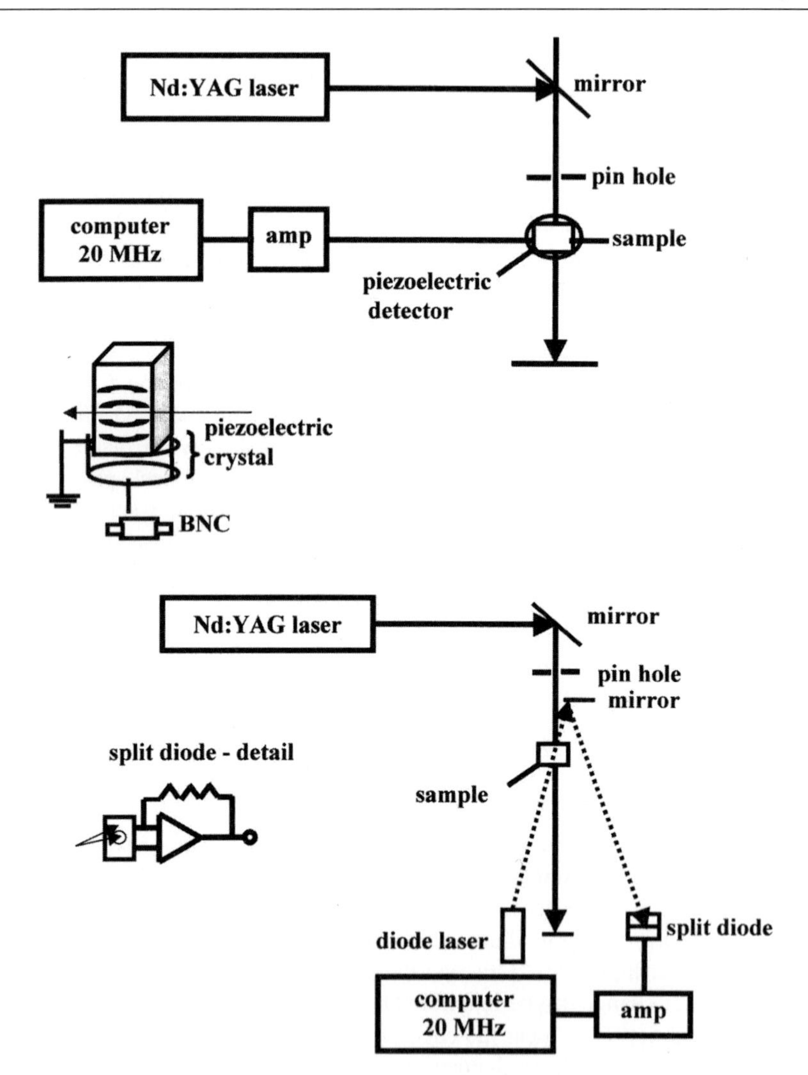

FIG. 5. *Top:* Diagram of the photoacoustic calorimetry instrument. *Bottom:* Corresponding diagram for the photothermal beam deflection instrument. Both instruments utilize a Continuum Minilite I, Q-switched Nd:YAG laser (6-ns pulse, ~100 μJ) to initiate photochemistry and, in the case of PAC, generate an acoustic wave. For the PAC instrument we currently use a Panametrics V103 detector with a corresponding Panametrics amplifier. In the case of the PBD instrument the intensity of a continuous wave probe laser (typically an Edmond Scientific LDL 175, 820 nm, 3 mW) is monitored by a Centronic LD2-5T coupled with an Analog Devices AD844 amplifier operating in a difference mode. For both instruments the signals are detected with a National Instruments NI5102 15-MHz digitizer card. See text for additional details.

the case of PAC we use a reference compound to eliminate the instrument response parameter. The deflection signal of a reference compound is described by

$$S_{\mathrm{ref}} = K E_{\mathrm{a}} E_{h\nu}[(dn/dt)(1/\rho C_{\mathrm{p}})] \tag{10}$$

and taking the ratio of the sample signal to the reference signal gives

$$(S/S_{\mathrm{ref}})E_{h\nu} = \Phi Q + \Phi[\rho(dn/d\rho)\Delta V + B\Delta n_{\mathrm{abs}}]/(dn/dt)(1/\rho C_{\mathrm{p}}) \tag{11}$$

From the plot of $(S/S_{\mathrm{ref}})E_{h\nu}$ versus $\rho C_{\mathrm{p}}/(dn/dt)$ the ΔV (slope of the line) and the amount of the heat released to the solvent (the intercept) can be separated. The $B\Delta n_{\mathrm{abs}}$ term can be evaluated experimentally by evaluating the anisotropy of the deflection data together with the anisotropy obtained from transient absorption measurements.[29] Alternatively, the magnitude of this term can be estimated theoretically by the Kramers–Kronig transform method.[34] As long as the probe laser wavelength is far from any absorption band of the molecule (including any transient absorption bands that may appear during the photochemical process) $B\Delta n_{\mathrm{abs}}$ will be small compared with Δn_{heat} and $\Delta n_{\mathrm{volume}}$. The $\rho(dn/d\rho)$ term is a unitless constant that is temperature independent in the temperature range of our measurement (0–35°). We typically use (dn/dT) values experimentally determined by Abbate *et al.*[35]

Volume and enthalpy parameters of individual reaction steps can be obtained by fitting PBD data to Eq. (12):

$$F = \alpha_0 + \sum \alpha_i[1 - \exp(-t/\tau_i)] \tag{12}$$

where α_i is equivalent to S in Eq. (9) for each intermediate step.

Experimental Design

The experimental setup of our PBD instrument is shown in Fig. 5 (bottom). Nearly colinear pump and probe beams are counterpropagated through the center of the sample cuvette. The pump beam is the 532-nm output of a Nd:YAG laser (Continuum Minilite I, 6-ns pulse, ~100 μJ) which is aligned with a mirror. The probe beam is a continuous wave diode laser [Edmund Scientific (Tonawanda, NY) LDL 175, 820 nm, 3 mW] and its position is fixed with a pin hole (0.5 mm) placed before the sample cuvette. The mirror behind the sample focuses the probe beam to fall equally on both cells of a split photodiode bicell detector [Centronic (Croydon, Surrey, UK) LD2-5T coupled with an Analog Devices (Norwood, MA) AD844 amplifier operating in a difference mode], which is fed into a 200-MHz amplifier of our own design. The generated deflection signal is then digitized by a National Instruments NI 5102 15 MHz oscilloscope controlled by VirtualBench-Scope software (National Instruments).

[34] M. Terazima and N. Hirota, *J. Phys. Chem.* **96,** 7147 (1992).

[35] G. Abbate, U. Bernini, E. Ragozzino, and F. Somma, *J. Phys. D Appl. Phys.* **11,** 1167 (1978).

Applications to Heme Proteins: CO-Bound Heme Model Compounds

The critical importance of heme axial ligation in modulating the biochemistry of heme proteins as well as the potential for iron porphyrins in biomimetic chemistry has stimulated efforts to understand mechanisms of ligand binding and activation in a wide range of iron porphyrins. This has resulted in the synthesis and reactivity studies of a variety of heme model complexes.[4–6,36] A majority of these complexes are designed to mimic the reversible oxygen binding of hemoglobin and the oxygen activation chemistry of peroxidases and cytochrome P450. These models are typically constructed to contain various imidazole or thiolate derivatives as ligands to the heme iron.

Structural changes associated with both the protein and heme active site of heme proteins, caused by ligand binding/release, are coupled to the biological function of the macromolecule. Because of this relationship, the determination of the magnitude and time scales of structural changes linked to ligand binding is crucial to the understanding of overall protein function. Studies involving photothermal techniques [including photoacoustic calorimetry (PAC) and phase grating spectroscopy] have revealed the time scale and magnitude of protein relaxation subsequent to CO photolysis on a picosecond to nanosecond time scale.[26,28,37] For example, PAC data demonstrate that CO photolysis from ferrous horse heart myoglobin (Mb) results in two relaxation processes with lifetimes of 30 and 700 ns.[26,28] The fast process gives rise to a volume decrease of 1.7 ml/mol with an enthalpy of 7.4 kcal/mol (relative to COMb). The evolution of the fast intermediate results in a second intermediate exhibiting a volume increase of 13.8 ml/mol and an enthalpy change of 14.3 kcal/mol (relative to COMb). The progressive structural changes that occur subsequent to CO photolysis have been rationalized in terms of CO migration through a ligand diffusion channel that requires breaking of a salt bridge between Arg-45 and the heme propionate. These changes also result in electrostriction of the solvent, which contributes to the initial volume changes in the protein.

In a previous study from our laboratory PAC and temperature/pressure-dependent transient absorption techniques were used to construct a complete volume and thermodynamic profile of CO binding to Fe(II) protoporphyrin IX [Fe(II)PPIX] encapsulated in cetylmethylammonium bromide (CTAB) micelles[38] (Fig. 6). This system is particularly attractive because CTAB micelles have dimensions similar to those of Mb and, in the presence of Fe(II)PPIX, contain a

[36] G. B. Jameson and J. A. Ibers, *in* "Bioinorganic Chemistry" (I. Bertini, H. B. Gray, S. J. Lippard, and J. S. Valentine, eds.). University Science Books, Sausalito, CA, 1994.

[37] D. C. Lamb, G. C. Lin, and A. G. Doukas, *Appl. Optics* **36,** 1660 (1997).

[38] R. W. Larsen, *Inorg. Chim. Acta* **288,** 74 (1999).

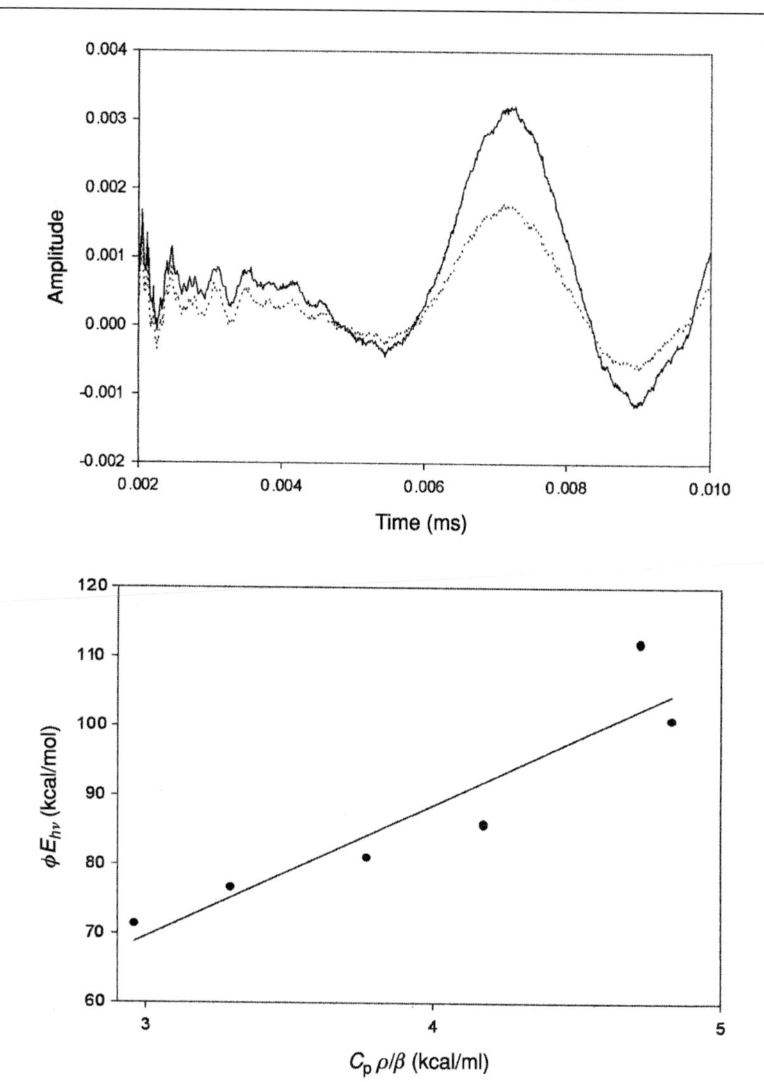

FIG. 6. Summary of PAC data for the photolysis and recombination of CO from ferrous heme model compounds. *Top:* Photoacoustic traces for the photolysis of CO from ferrous Fe(II) protoporphyrin IX encapsulated in CTAB micelles. The solid trace is for the sample, and the dotted line is the reference trace. *Bottom:* Plot of $\phi E_{h\nu}$ vs $C_p\rho/\beta$ for the Fe(II) protoporphyrin IX/CTAB data. See Ref. 38 for more details.

single heme center oriented such that the propionic acid side chains of the porphyrin make electrostatic contacts with the head groups of the detergent while the hydrophobic side of the porphyrin is buried in the interior of the micelle.[39] The results of the acoustic analysis reveal an enthalpy associated with ligand dissociation of 10 kcal/mol and a volume increase of 3 ml/mol that occurs on a time scale of less than 100 ns. The data are consistent with a model in which a heme conformational change occurs on formation of a four-coordinate transient species that distorts the micelle structure, resulting in micelle expansion. The PAC data further indicate that the volume changes associated with heme–CO bond cleavage and corresponding spin-state changes occur much faster than the observed intermediate in Mb. In addition, the magnitude and sign of the corresponding volume change are distinct (+3.2 ml/mol for CO–Fe(II)PPIX/CTAB bond cleavage vs. −1.7 ml/mol for the initial Mb intermediate). If similar heme dynamics occur in both CO–Fe(II)PPIX/CTAB and COMb, then the corresponding protein volume change is actually about −5 ml/mol. This initial volume change may be a response of the protein to the rapid volume change of the heme on ligand release and heme spin-state change and could be the "trigger" for the opening of the protein gate observed in the PAC COMb data.

We have also examined the thermodynamics and volume changes associated with CO photolysis from the water-soluble Fe(II) mesotetra-(4-sulfonatophenyl) porphyrin [Fe(II)4SP], using PBD (Fig. 7) and variable temperature/pressure transient absorption spectroscopy.[40] Previous studies of CO photolysis and recombination of a variety of Fe(II) porphyrin/base systems in water have provided a general mechanism for this type of reaction in which photolysis of the (B)(CO)Fe(II) porphyrin complex (where B is a proximal base) produces a four-coordinate Fe(II) porphyrin complex within the first few hundred femtoseconds subsequent to photolysis. In water, this is followed by the rapid binding of a water molecule to the fifth coordination site ($<1\ \mu s$) and subsequent rebinding of the CO.

The volume/enthalpy profile of ligand dissociation/binding obtained from the PBD and variable pressure/temperature transient absorption data is shown in Fig. 8. Typical values for the enthalpy of CO recombination to chelated hemes, which contain a nitrogenous base occupying the fifth coordination site, are on the order of −17 kcal/mol.[40] This is significantly lower than the −7 kcal/mol found for CO binding to (H_2O)Fe(II)4SP. Assuming that the presence of the sulfonatophenyl substituents at the methine positions of the porphyrin ring do not significantly alter the Fe(II)–CO bond strength, the roughly 10-kcal/mol difference in Fe(II)–CO bond energy can be attributed to the difference in the ligand at the fifth axial position of the heme iron. Early studies by Rougee and Brault[41] have

[39] J. Simplicio, *Biochemistry* **11,** 2525 (1973).

[40] T. G. Traylor, *Acc. Chem. Res.* **14,** 109 (1981).

[41] M. Rougee and D. Brault, *Biochemistry* **14,** 4100 (1975).

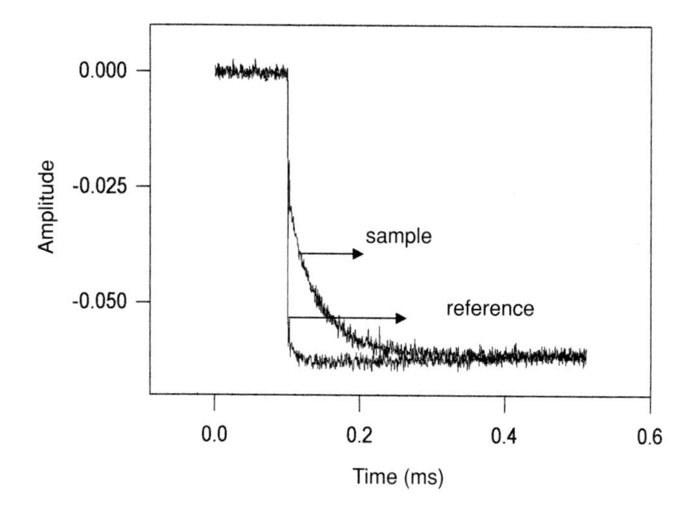

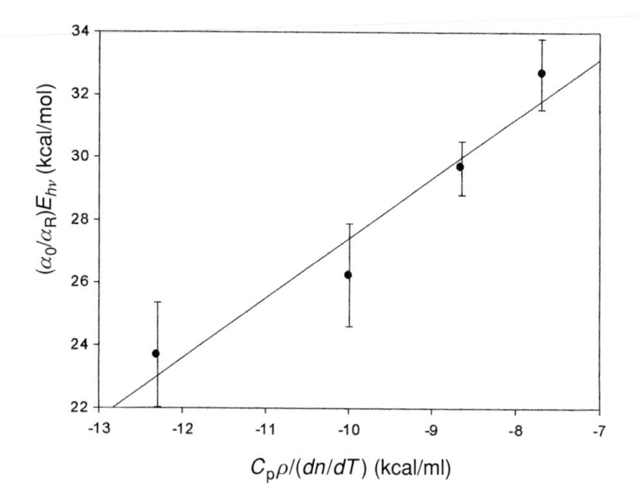

FIG. 7. PBD data for the photolysis and recombination of CO from ferrous heme model compounds. *Top:* PBD data for CO photolysis and recombination from COFe(II) [(tetrakis-4-sulfonatophenyl)porphyrin]. *Bottom:* Plot of $(\alpha_0/\alpha_R)E_{h\nu}$ vs $C_p\rho/(dn/dT)$ for the COFe(II)[(tetrakis-4-sulfonatophenyl)porphyrin] PBD data. See Ref. 44 for more details.

shown that strong field ligands trans to CO in deuteroheme complexes significantly increase the binding constant for CO relative to weak field ligands. For example, the binding constant for CO to (imidazole)Fe(II) deuteroheme is $4 \times 10^7\ M^{-1}$ ($\Delta G^\circ = -10.4$ kcal/mol) while the binding constant for CO to (H_2O)Fe(II) deuteroheme is roughly 0.1 M^{-1} ($\Delta G^\circ = -1.4$ kcal/mol). Assuming the entropy

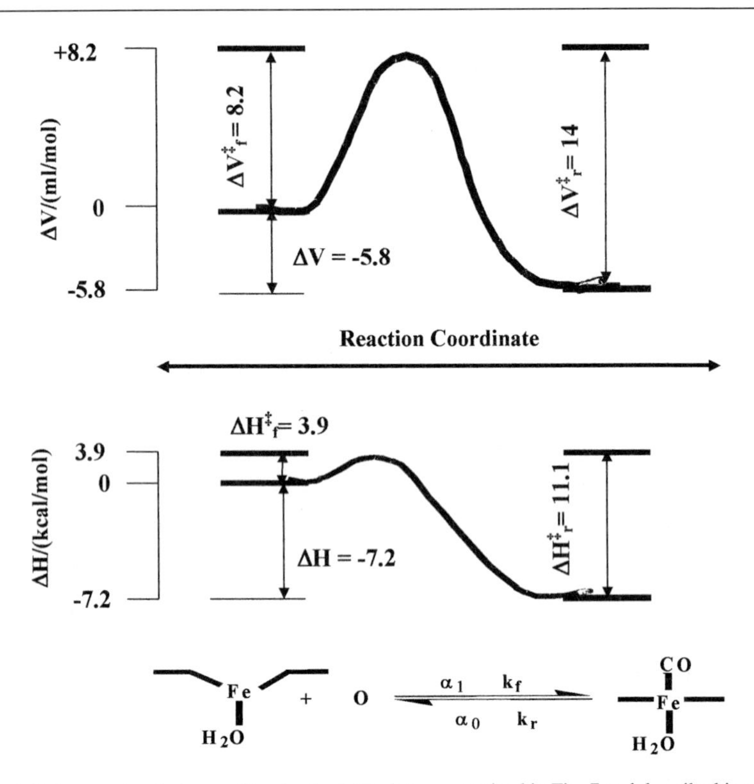

FIG. 8. Volume and enthalpy profiles for the PBD data summarized in Fig. 7 and described in text.

term (i.e., $T\Delta S$) to be similar in both cases, then the differences in the ΔG° values should be due only to differences in ΔH° of binding. This difference in ΔG° values for the imidazole–deuteroporphyrin/water–deuteroporphyrin system is roughly 9 kcal/mol, which is in good agreement with the differences in ΔH values between the chelated hemes and the $(H_2O)Fe(II)4SP$ system (10 kcal/mol). It has been suggested that the stronger affinity for CO by strong field trans ligands arises from the fact that σ-donating ligands strengthen π bonding with the carbonyl due to increased π backbonding with the Fe(II).[40]

Changes in partial molar volume (i.e., conformational changes) associated with CO binding to the heme can be obtained from temperature-dependent PBD (total volume change) and pressure-dependent transient absorption data (activation volume). The total volume change calculated for the dissociation reaction is $+5.8 \pm 0.4$ ml/mol, whereas the PBD data give a value of -5.8 ml/mol for the corresponding CO-rebinding reaction. The activation volume for CO recombination (determined from the pressure dependence of the rebinding rate constant) is $+8.2$ ml/mol. Thus, the activation volume for CO dissociation can be calculated using $\Delta V = (\Delta V^{\ddagger})_f - (\Delta V^{\ddagger})_r$, where $\Delta V = +5.8$ ml/mol and $(\Delta V^{\ddagger})_r = +8.2$ ml/mol gives $(\Delta V^{\ddagger})_f \approx +14$ ml/mol.

The observed volume changes provide insight into the mechanism of ligand binding to the heme. Bond formation and subsequent spin-state transition (high spin to low spin) are expected to contribute a significant negative $\Delta V^{\ddagger}$ (on the order of -10 ml/mol) to the overall activation volume.[42,43] Considering the dissociation reaction, the activation volume would have contributions from Fe–CO and Fe–H_2O bond cleavage, migration of both ligands out of the heme solvation shell, and a low-spin to high-spin transition. The calculated activation volume for ligand dissociation $(\Delta V^{\ddagger})_f \approx +14$ ml/mol is similar to $\Delta V^{\ddagger}$ expected for the spin-state transition (i.e., $\Delta V^{\ddagger}_{HS-LS} = -\Delta V^{\ddagger}_{LS-HS} \approx +10$ ml/mol). Thus, the activated complex resembles a high-spin heme within a ligand contact pair. Subsequent diffusion of the ligands from the heme reduces the volume (from the activated complex) by ~8.2 ml/mol, forming the equilibrium unliganded heme with an overall volume increase of +5.8 ml/mol. Because the equilibrium unliganded heme is in a high-spin conformation the volume increase (to form the activated complex) results from the formation of the heme–ligand contact pair. Previous studies have suggested that positive activation volumes for CO rebinding to heme in ethyleneglycol solutions arise from CO displacement of another ligand during the rebinding process.[42] If this were the case for the (CO)(H_2O)Fe(II)4SP system the binding of a water molecule to the five coordinate heme would occur subsequent to photolysis followed by displacement by CO. However, the transient difference spectrum suggests that the heme remains five-coordinate subsequent to CO photolysis.

The volume and thermodynamic profiles of CO binding to the water-soluble Fe(II)4SP are distinct from those previously determined for the Fe(II) protoporphyrin IX/CTAB system.[44] Carbon monoxide recombination to Fe(II)PPIX encapsulated in CTAB micelles has an associated ΔH of +10.4 kcal/mol and ΔV of -3.1 ml/mol relative to -7 kcal/mol and -5.8 ml/mol for CO recombination to Fe(II)4SP. The differences in the enthalpy values may be due to the difference in rebinding mechanism. In the case of the Fe(II)PPIX/CTAB system photolysis of the (H_2O)(CO)Fe(II)PPIX complex results in the formation of a four-coordinate Fe(II)PPIX within the micelle (as judged from the transient difference spectrum). Thus, formation of the equilibrium (H_2O)(CO)Fe(II)PPIX complex requires a concerted (on a nanosecond time scale) CO/H_2O-binding step that apparently increases the enthalpy of the overall rebinding reaction. The volume differences are not as pronounced as the differences in the enthalpies. In both cases volume contractions are observed on CO binding, consistent with the conversion of the heme to a

[42] D. J. Taube, H.-D. Projahn, R. van Eldik, D. Magde, and T. G. Traylor, *J. Am. Chem. Soc.* **112,** 6880 (1990).

[43] R. van Eldik, T. Asano, and W. J. Noble, *Chem. Rev.* **89,** 549 (1989).

[44] B. D. Barker and R. W. Larsen, *Inorg. Biochem.* **85,** 107 (2001).

low-spin electron configuration. Again, the differences in the magnitudes of these contractions may be due to the concerted two-ligand binding (H_2O/CO) for the Fe(II)PPIX/CTAB system versus the single-CO binding in the Fe(II)4SP system.

Applications to Heme Proteins: Study of Conformational Dynamics in Heme/Copper Oxidase

Heme/copper oxidases form a diverse class of respiratory proteins found in nearly all aerobic organisms.[45,46] Although these enzymes range widely in molecular weight and subunit composition, several common features are found throughout the class. The majority of heme/copper oxidases contain at least three subunits (SU I, SU II, and SU III), with SU I containing the majority of the redox-active metal centers (Fig. 9). In addition, these enzymes contain two heme chromophores (heme *a*, heme *b*, and/or heme *o*) and at least one copper ion. One of the two hemes contains a heme iron that is six-coordinate and low spin, which functions as a catalyst for electron transfer to the binuclear center. The binuclear center consists of the remaining heme (designated heme a_3, heme o_3, or heme b_3 depending on the organism), which contains a five-coordinate high-spin heme iron and a copper ion (designated Cu_B). In addition, heme/copper oxidases from higher organisms contain an additional binuclear copper cluster (designated Cu_A) that accepts electrons from cytochrome *c*. All members of this class catalyze the four-electron reduction of dioxygen to water and it is widely believed that most of these enzymes are energy transducing, that is, they couple redox energy to the active transport of protons across a membrane.

Although considerable progress has been made to characterize the structural and functional properties of terminal heme/copper oxidases, the mechanism by which electron transfer events are coupled to active proton translocation remains unknown. To complete the catalytic cycle, the enzyme complex musts undergo a series of conformational changes. Previously it has been shown that oxygen binding to the binuclear center induces conformational changes that may activate the proton-pumping cycle.[47,48] Furthermore, results of steady state high-pressure optical spectroscopy indicate large volume changes occurring during the transition between the intermediate states of bovine heart cytochrome *c* oxidase.[49–51] As

[45] S. M. Musser, M. H. B. Stowell, and S. I. Chan, *in* "Advances in Enzymology and Related Areas of Molecular Biology" (A. Meister, ed.), Vol. 71. John Wiley & Sons, New York, 1995.

[46] G. Buse, G. C. Steffens, R. Biewald, B. Bruch, and S. Hensel, *in* "Cytochrome Systems: Molecular Biology and Bioenergetics" (S. Papa, B. Chance, and L. Ernster, eds.). Plenum Press, New York, 1987.

[47] R. W. Larsen, *FEBS Lett.* **352,** 365 (1994).

[48] S. N. Niu and R. W. Larsen, *J. Biochem. Mol. Biol. Biophys.* **1,** 287 (1998).

[49] J. A. Kornblatt, G. H. B. Hoa, and K. Heremans, *Biochemistry* **27,** 5122 (1988).

[50] J. A. Kornblatt and G. H. B. Hoa, *Biochemistry* **29,** 9370 (1990).

[51] J. A. Kornblatt and M. J. Kornblatt, *Biochim. Biophys. Acta* **1099,** 182 (1992).

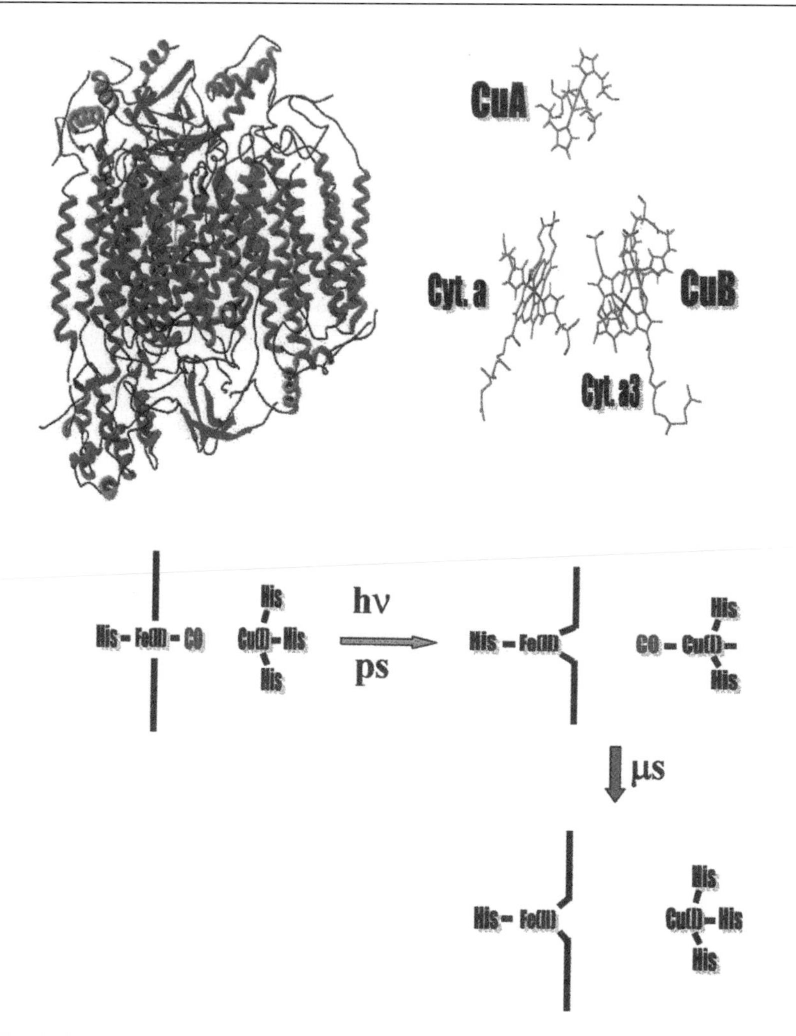

FIG. 9. Overview of bovine heart cytochrome *c* oxidase. *Top:* Crystal structure and metal site orientation. *Bottom:* Mechanism for CO dissociation from the binuclear center of the enzyme.

a first step in determining the role of conformational dynamics in modulating key reactions in the catalytic cycle of heme/copper oxidases we have examined conformational dynamics and thermodynamics of carbon monoxide binding to fully reduced forms of bovine heart cytochrome *c* oxidase (CcO) and the quinol oxidase cytochrome bo_3 from *Escherichia coli* (CbO).

Carbon monoxide binding to the fully reduced CcO and CbO occurs on a millisecond time scale at room temperature (pseudo first-order rate constant of ~100 ms^{-1} at 1 atm).[14,52] The corresponding activation enthalpy ($\Delta H^{\ddagger}$) for CO binding to binuclear center is 2.39 and 3.13 kcal/mol for CcO and CbO, respectively. The measurement of the pressure dependence of CO-binding reaction reveals corresponding activation volumes. Interestingly, the activation volume values for this reaction are significantly different for CbO and CcO (Larsen[53]; and our laboratory, unpublished results, 2001). We observe a negative activation volume of −9.0 ml/mol for CO rebinding to the binuclear center of CcO and a positive activation volume of 13.3 ml/mol for CbO. These results indicate that conformational dynamics are associated with ligand binding to heme/copper oxidases and that these structural changes are distinct between heme/copper oxidases from different species.

To better understand processes accompanying the ligand binding to heme/copper oxidases we have applied PAC to measure enthalpy and volume changes for ligand dissociation from heme/copper oxidases.[54,55] Figure 10 shows the acoustic wave for CO photodissociation from fully reduced CcO. The phase shift of the acoustic wave of the sample, compared with the reference, indicates volume and/or enthalpy changes occurring within ~100 ns to 5 μs. Deconvolution of the acoustic waves reveals two exponential decays. The first occurs with a lifetime shorter than the detection limit of our instrument (<100 ns) while the slower decay displays a lifetime ~3 μs (at 24°). The plots of $\phi_1 E_{h\nu}$ and $\phi_2 E_{h\nu}$ versus $\beta/C_p\rho$ give a large positive volume change, $\Delta V = 13.7$ ml/mol for the fast process and a smaller positive volume change, $\Delta V = 6.91$ ml/mol for the 3-μs decay. The value of ΔH could not be resolved because the absorption contribution of heme a and heme a_3 to the total absorption at the excitation wavelength is not known.

The corresponding acoustic wave for the CO photodissociation from the fully reduced CbO is shown in Fig. 11. The sample acoustic wave overlaps in time with the reference acoustic wave, indicating no volume changes occurring within ~100 ns to ~5 μs. A plot of the ratio of the signal amplitude of the sample to the signal amplitude of the reference scaled to $E_{h\nu}$ versus $\beta/C_p\rho$ reveals a negative volume change $\Delta V = -5.1$ ml/mol on the dissociation of CO (Fig. 11, bottom). To calculate the corresponding ΔH it must be considered that heme b and heme o_3 also contribute to the final acoustic wave, as do CO-bound cleavage and protein conformational changes. The acoustic signal for the CO photodissociation can be written as

$$(S/S_{\text{ref}})E_{h\nu} = [Q^{\text{Heme}\,b}(A_{\text{Heme}\,b}/A_{\text{TOT}}) + Q^{\text{Heme}\,o_3}(A_{\text{Heme}\,o_3}/A_{\text{TOT}})] + [\Delta V_{\text{con}}/(\beta/C_p\rho)] \tag{13}$$

[52] S. Brown, J. N. Rumbley, A. J. Moody, J. W. Thomas, R. B. Gennis, and P. R. Rich, *Biochim. Biophys. Acta* **1183,** 521 (1994).
[53] R. W. Larsen, *FEBS Lett.* **462,** 75 (1999).
[54] R. W. Larsen, J. Osborne, T. Langley, and R. B. Gennis, *J. Am. Chem. Soc.* **120,** 8887 (1998).
[55] R. W. Larsen and T. Langley, *J. Am. Chem. Soc.* **121,** 4495 (1999).

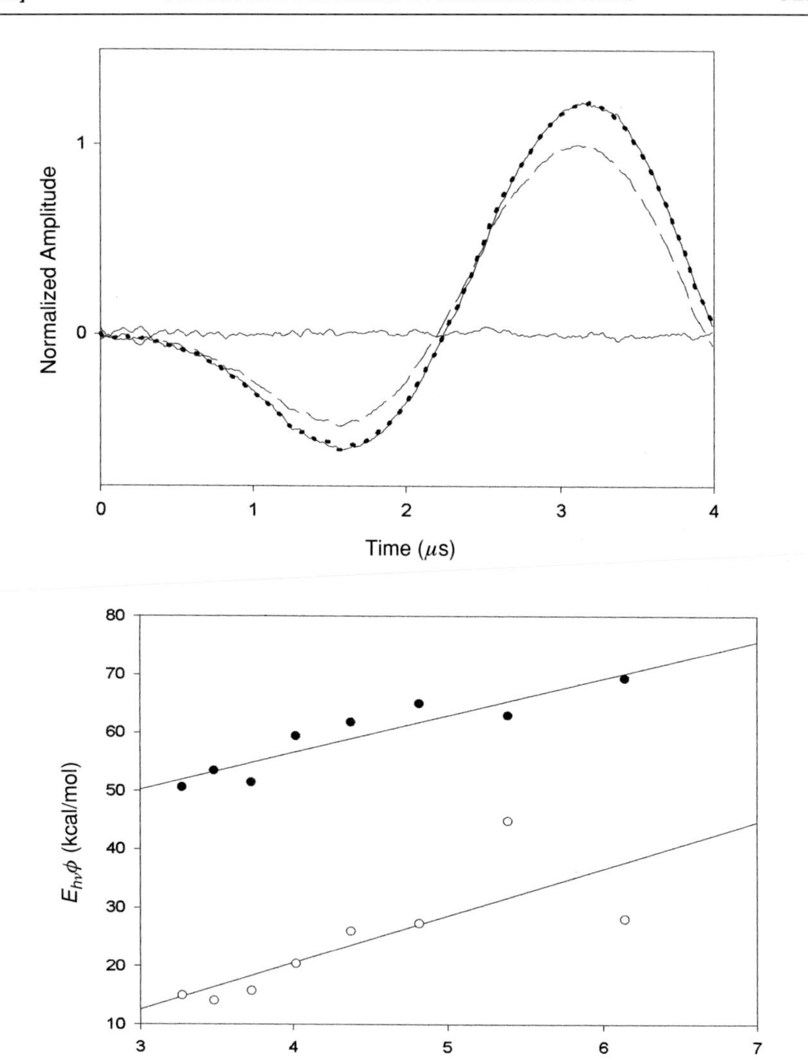

FIG. 10. Summary of photoacoustic results for CO photolysis from bovine heart cytochrome *c* oxidase. *Top:* Acoustic waveform of BCP (reference compound, dashed line), CO–CcO (solid line), CO–CcO fitted to two exponential decays (dotted line), and residuals of the fit. *Bottom:* Plot of $E_{h\nu}\phi_1$ (closed circles) and $E_{h\nu}\phi_2$ (open circles) (kcal/mol) vs. ($C_p\rho/\beta$) (kcal/ml) for the photolysis of CO from fully reduced CcO. The values for ϕ_1 and ϕ_2 where obtained for the two exponential decays derived from the Simplex deconvolution. The lifetimes for the two phases are as follows: 50 ns < (fast phase, ϕ_1), and ~3 μs (slow phase, ϕ_2). Conditions: 100 μM protein and 1 mM BCP (25 mM Tris–0.08% laurylmaltoside, pH 8). The excitation wavelength is 532 nm with an average power of ~75 μJ/pulse.

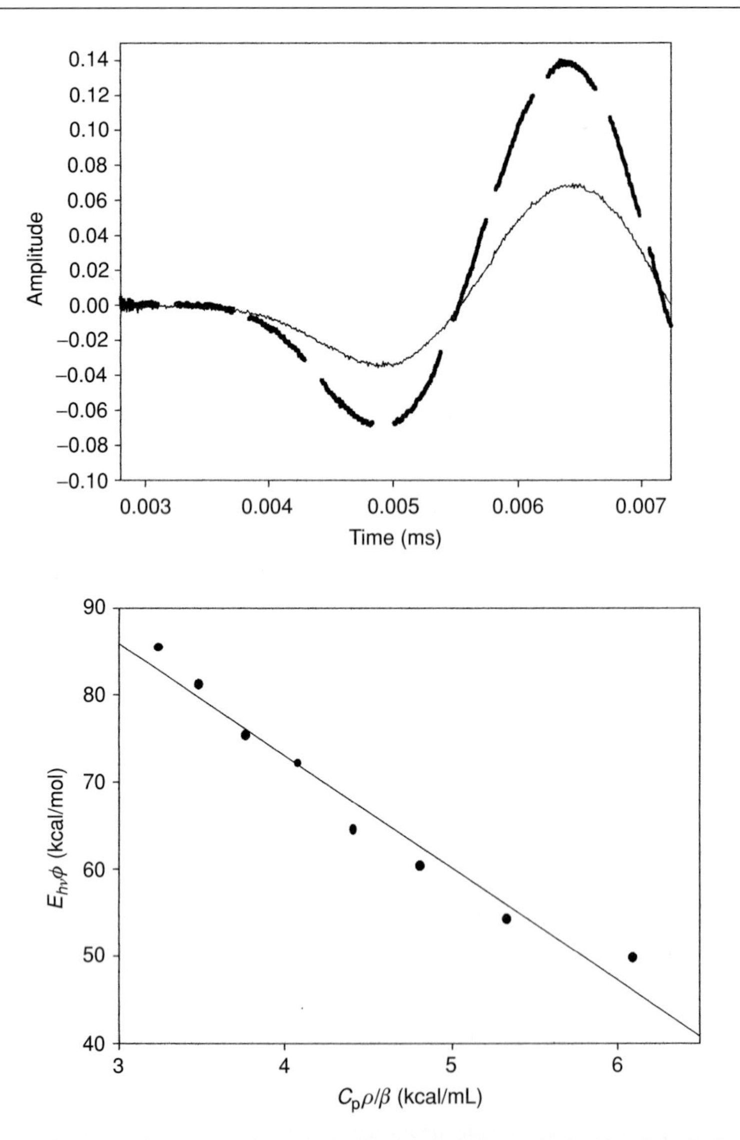

FIG. 11. Summary of photoacoustic results for CO photolysis from quinol oxidase from *Escherichia coli*. *Top:* Overlay of the acoustic waveforms of CO–cytochrome bo_3 (solid line) and BCP (dashed line). *Bottom:* Plot of $E_{h\nu}\phi$ (kcal/mol) vs. $C_p\rho/\beta$ (kcal/ml) for the photolysis of CO from fully reduced cytochrome bo_3. See Fig. 10 for conditions.

where A_{TOT} is the absorption at the excitation wavelength, $A_{Heme\,b/o_3}$ is the absorption of heme b or o_3. Because at 532 nm both hemes contribute equally to the total absorption,[54] the value of Q for each heme is one-half the value Q obtained from the plot in Fig. 11. ΔH for the CO photodissociation from CbO is then equal to $E_{h\nu} - Q^{\text{Heme}\,o}$ and $Q^{\text{Heme}\,o_3}$ is equal to half the value of Q. Using $E_{h\nu} = 53.74$ kcal/mol at 532 nm, a ΔH of 22.7 kcal/mol is obtained. This value corresponds to the enthalpy values obtained for CO dissociation from other heme protein.[40] However, for CbO this value represent not only CO dissociation from heme o_3 but also the ligand binding to Cu_B. This suggests that the actual ΔH value for CO binding to Cu_B is larger than the ΔH value for the CO dissociation from heme o_3. This can be explained by another endothermic process occurring during CO binding to Cu_B, such as the displacement of an endogenous ligand. The volume and enthalpy profiles for CO binding to CcO and CbO are summarized in Fig. 12.

The results of time-resolved optical spectroscopy and vibration spectroscopy methods have shown that after the photodissociation of CO from heme a_3/o_3 the

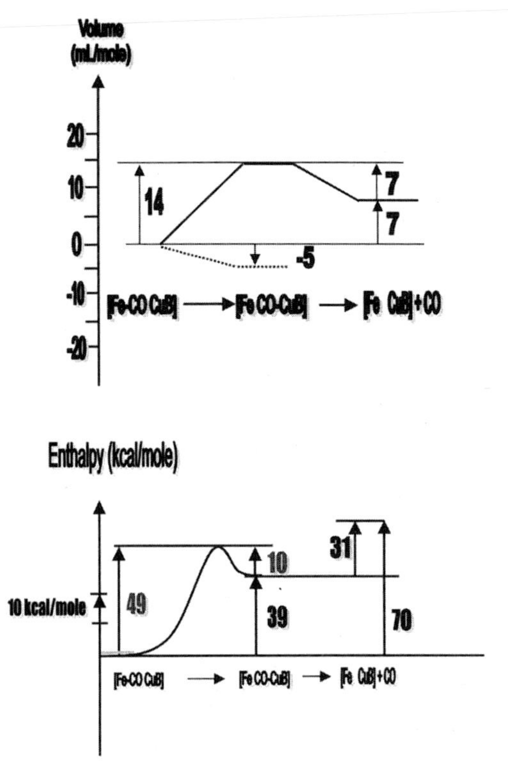

FIG. 12. Partial volume (*bottom*) and enthalpy profiles (*top*) for CO binding to heme/copper oxidases obtained from combined variable temperature/pressure TA and PAC. Solid lines are profiles for cytochrome c oxidase, and dotted lines are profiles for cytochrome bo_3.

ligand binds transiently to Cu_B ($t_{1/2} < 1$ ps).[14] Subsequently, CO dissociates from Cu_B and diffuses to the bulk solvent ($t_{1/2} = 1$ μs). The corresponding reaction is significantly slower for CbO. The CO–Cu_B complex is formed about 100 times slower than expected for a diffusional process, suggesting that the protein can limit access to the heme/copper active site. Because the time scale of the observed volume/enthalpy changes associated with the CO photodissociation from CbO and with the fast phase of the CO dissociation from the CcO is <50 ns the observed volume and enthalpy changes must arise from the binding of CO to Cu_B. Volume changes associated with this reaction include Fe–CO bond breaking, low-spin to high-spin transition of the heme a_3/o_3, CO–CuB bond formation, and possibly ligand dissociation from Cu_B and conformational changes of protein. The bond formation contributes a large negative ΔV and the bond breaking contributes a positive ΔV of a similar amplitude, and thus the overall contribution to the volume changes is zero.[42,43] The low-spin to high-spin transition of heme a_3/o_3 has a positive volume change ($\Delta V = 10$ ml/mol).[43] In the case of CcO we associate the fast volume expansion ($\Delta V = 13.7$ ml/mol) with the photoinduced ligand transfer form heme a_3 to Cu_B. The longer time scale positive volume change ($\Delta V =$ 6.9 ml/mol) may represent the relaxation of protein after the dissociation of CO from Cu_B because the time scale is similar to the Fe–His relaxation observed in resonance Raman and transient spectroscopy.[56] The photoacoustic data reveal different conformational changes accompanying ligand dissociation from heme o_3 in the CbO enzyme. The negative volume changes ($\Delta V = -5.1$ ml/mol) occurring after the photodissociation indicate than another process must contribute to the overall volume changes. It is possible that local protein structural changes such as displacement of a ligand from Cu_B accompany the ligand transfer reaction. This process may occur also in the CcO enzyme but with minimal contribution to the volume changes. Such a protein "gate" could account for the non-diffusion-limited ligand binding to Cu_B and would modulate access of the exogenous ligand to the active site. Studies utilizing Fourier transform infrared (FTIR) spectroscopy clearly demonstrated that CO binding to the Cu_B site in CcO perturbs the position of Glu-286.[57] This residue is located at the end of the D channel and site-directed mutagenesis of this residue in *Escherichia coli* or *Rhodobacter sphaeroides* results in diminished activity and no proton pumping.[58] Glu-286 is not directly linked to Cu_B or to any ligand of Cu_B, but several water molecules may facilitate the connection between Glu-286 and Cu_B. Iwata *et al.*[59] suggested the function of Glu-286 as a key residue in the proton translocation reaction. In bovine heart

[56] R. W. Larsen, M. R. Ondrias, R. A. Copeland, P. M. Li, and S. I. Chan, *Biochemistry* **28,** 6418 (1989).

[57] M. Tsubaki, H. Hori, and T. Mogi, *FEBS Lett.* **416,** 247 (1997).

[58] N. J. Watmough, A. Katsonouri, R. H. Little, J. P. Osborne, E. Furlong-Nickels, R. B. Gennis, T. Brittain, and C. Greenwood, *Biochemistry* **36,** 13736 (1997).

[59] S. Iwata, C. Ostermeier, B. Ludwig, and H. Michel, *Nature* (*London*) **376,** 660 (1995).

CcO the corresponding residue is Glu-242. The structure of fully oxidized enzyme shows the connectivity between Cu_B and Glu-242 via Pro-241 and His-240, with His-240 being one of the ligands of Cu_B. Interestingly, we have found different conformational responses to CO dissociation from heme a_3/o_3, indicating different coupling between the binuclear center and Glu-248/242.

Summary and Future Directions

The applications of PAC and PBD to iron porphyrins and heme/copper oxidases described in above are but a few examples of the utility of these methods in probing structure–function relationships in metalloproteins. The ability of both PAC and PBD to measure heat and molar volume changes complements transient optical techniques, which measure absorption changes of a localized chromophore. Combining photothermal methods with site-directed mutagenesis provides a means to determine regions of the protein giving rise to thermal/volume changes, thus providing a structural basis for the observed thermodynamic changes. Perhaps the greatest utility of PAC and PBD is in the study of time-resolved conformational dynamics and enthalpy changes in proteins that do not contain readily accessible optical signals. For example, PAC has now been applied to protein-folding problems involving apomyoglobin and small synthetic peptides.[60,61] In addition, we have applied both PAC and PBD to the study of small peptides containing photocleavable linkers that spontaneously fold subsequent to cleavage of the photolabile linker.[62] The possibility of utilizing PAC and PBD in the study of bacterial signal transduction, using caged chemoaffectors, is also currently being explored in our laboratory.

In summary, the relative ease with which any transient absorption instrument can be adapted to perform photothermal measurements as well as the ability of these methods to probe kinetics and thermodynamics of nonoptical transitions in proteins will lead to greater use of these methods in the biophysics community and greatly expand our fundamental understanding of protein structure–function relationships.

Acknowledgments

Much of the work described in this article was supported by the National Science Foundation and the American Heart Association, and their support is gratefully acknowledged.

[60] S. Abbruzzetti, E. Crema, L. Masino, A. Vecli, C. Viappiani, J. R. Small, L. J. Libertini, and E. W. Small, *Biophys. J.* **78,** 405 (2000).

[61] S. Abbruzzetti, C. Viappiani, J. R. Small, L. J. Libertini, and E. W. Small, *Biophys. J.* **79,** 2714 (2000).

[62] K. C. Hansen, R. S. Rock, R. W. Larsen, and S. I. Chan, *J. Am. Chem. Soc.* **122,** 11567 (2000).

[14] Spectroscopy and Microscopy of Cells and Cell Membrane Systems

By MOSHE LEVI, HUBERT ZAJICEK, and TIZIANA PARASASSI

Introduction

Spectroscopy and microscopy studies of cells and cell membranes yield important information about lipid and protein dynamics as well as lipid–protein interactions that play an important role in the regulation of cell function.

One of the challenges of performing spectroscopy and/or microscopy studies with biological cells is that the lipid and protein composition of the cell membrane and the intracellular organelles vary significantly from each other and therefore it is critical to perform subcellular fractionation of the cell and intracellular organelle membranes before performing the spectroscopy measurements. Furthermore, in polarized epithelial cells such as intestinal and renal cells there is also a marked difference between the apical and basolateral membrane transport proteins, hormone receptors, electrical resistance, lipid composition, and lipid fluidity.[1–8] Therefore it is critical to achieve highly purified apical and basolateral membrane isolation before spectroscopy and/or microscopy studies.

Spectroscopy Studies with Isolated Membranes

This section concentrates on apical and basolateral membranes isolated from the kidney to illustrate how studies with these plasma membrane fractions yield important information about differential lipid dynamics in apical versus basolateral membranes, mainly due to differential lipid composition.

Isolation of Brush Border Membranes and Basolateral Membranes from Kidney

Renal cortical brush border membranes (BBMs) and basolateral membranes (BLMs) can be isolated simultaneously from the same renal cortical homogenate.[7,8]

[1] M. Barac-Nieto, H. Murer, and R. Kinne, *Pflugers Arch.* **392,** 366 (1981).
[2] E. Frömer, *J. Physiol.* (*London*) **288,** 1 (1979).
[3] H. E. Ives, J. Yee, and D. G. Warnock, *J. Biol. Chem.* **258,** 13513 (1983).
[4] C. LeGrimellec, S. Carriere, J. Cardinal, and M.-C. Giocondi, *Am. J. Physiol.* **14,** F227 (1983).
[5] C. LeGrimellec, M.-C. Giocondi, B. Carriere, S. Carriere, and J. Cardinal, *Am. J. Physiol.* **242,** F246 (1982).
[6] Z. Taylor, D. S. Emmanouel, and A. I. Katz, *J. Clin. Invest.* **69,** 1136 (1982).
[7] B. A. Molitoris and F. R. Simon, *J. Membr. Biol.* **83,** 207 (1985).
[8] M. Levi, B. A. Molitoris, T. J. Burke, R. W. Schrier, and F. R. Simon, *Am. J. Physiol.* **252,** F267 (1987).

METHODS IN ENZYMOLOGY, VOL. 360
0076-6879/03 $35.00

For this purpose kidneys from adult Sprague-Dawley rats are rapidly removed and placed in a chilled isolation buffer consisting of 300 m*M* mannitol, 5 m*M* EGTA, 0.1 m*M* phenylmethylsulfonyl fluoride (PMSF), 16 m*M* HEPES (pH 7.40) with Tris. All steps are carried out at 4°. Thin slices from the superficial cortex of the kidneys are cut and are placed in 15 ml of the above-described chilled buffer. The slices are homogenized with a Polytron homogenizer (Brinkmann, Westbury, NY). An aliquot of homogenate is saved for subsequent (1) protein concentration and (2) enzyme activity analysis. Mg^{2+} precipitation is carried out by adding 21 ml of distilled water and 0.54 ml of a 1 *M* Mg^{2+} solution. The solution is shaken every 5 min for 20 min on ice and then centrifuged at 2790*g* for 15 min at 4° in a high-speed Beckman (Fullerton, CA) centrifuge, using a J-17 fixed-angle rotor.

The pellet (P_1) is saved for basolateral membrane isolation, whereas the supernatant (S_1) is processed for brush border membrane isolation. The supernatant is taken through the Mg^{2+} precipitation step again (Fig. 1). The resulting supernatant (S_2) is then centrifuged at 39,800*g* for 36 min at 4°. The pellet (P_3) representing brush border membranes (BBMs) is resuspended in a buffer consisting of 300 m*M* mannitol, 5 m*M* EGTA, 0.1 m*M* PMSF, 16 m*M* HEPES (pH 7.40) with Tris as described above and is kept on ice or frozen in liquid nitrogen for subsequent analysis of (1) protein concentration, (2) enzyme activity, (3) lipid composition, and (4) lipid fluidity.

The basolateral membrane fraction is isolated from the P_1 pellet. This pellet is resuspended in 15 ml of buffer, using a loose Dounce glass homogenizer, and taken through the Mg^{2+} precipitation step again. The resulting pellet (P_2) is resuspended in 15 ml of buffer, using a loose Dounce glass homogenizer, and centrifuged at 755*g* for 15 min at 4°. The resulting supernatant (S_3) is then centrifuged at 39,800*g* for 36 min at 4°. The pellet (P_4) is resuspended, using a Potter–Elvehjem Teflon–glass homogenizer, in 19 ml of 50% (w/v) sucrose and overlaid with a discontinuous sucrose gradient with 5 ml of 41% (w/v) sucrose and 12 ml of 38% (w/v) sucrose (Fig. 1). The tubes are centrifuged at 88,000*g* for 3 hr at 4° in a Beckman ultracentrifuge, using a swinging bucket SW-27 rotor. The top layer of the discontinuous gradient (38%) is harvested and washed in buffer and centrifuged at 39,800*g* for 36 min at 4°. The pellet representing basolateral membranes (BLMs) is resuspended in a buffer consisting of 300 m*M* mannitol, 5 m*M* EGTA, 0.1 m*M* PMSF, 16 m*M* HEPES (pH 7.40) with Tris as described above, and is kept on ice or frozen in liquid nitrogen for subsequent analysis of (1) protein concentration, (2) enzyme activity, (3) lipid composition, and (4) lipid fluidity.

Enzyme activity analysis of the cortical homogenate and BBM and BLM fraction for Na^+,K^+-ATPase (BLM-specific enzyme marker) and maltase, leucine aminopeptidase, and γ-glutamyl transpeptidase (BBM-specific enzyme markers) indicates that compared with the starting cortical homogenate the BBM and BLM fractions are enriched at least 10-fold, with less than 1.5-fold contamination with each other.[7,8]

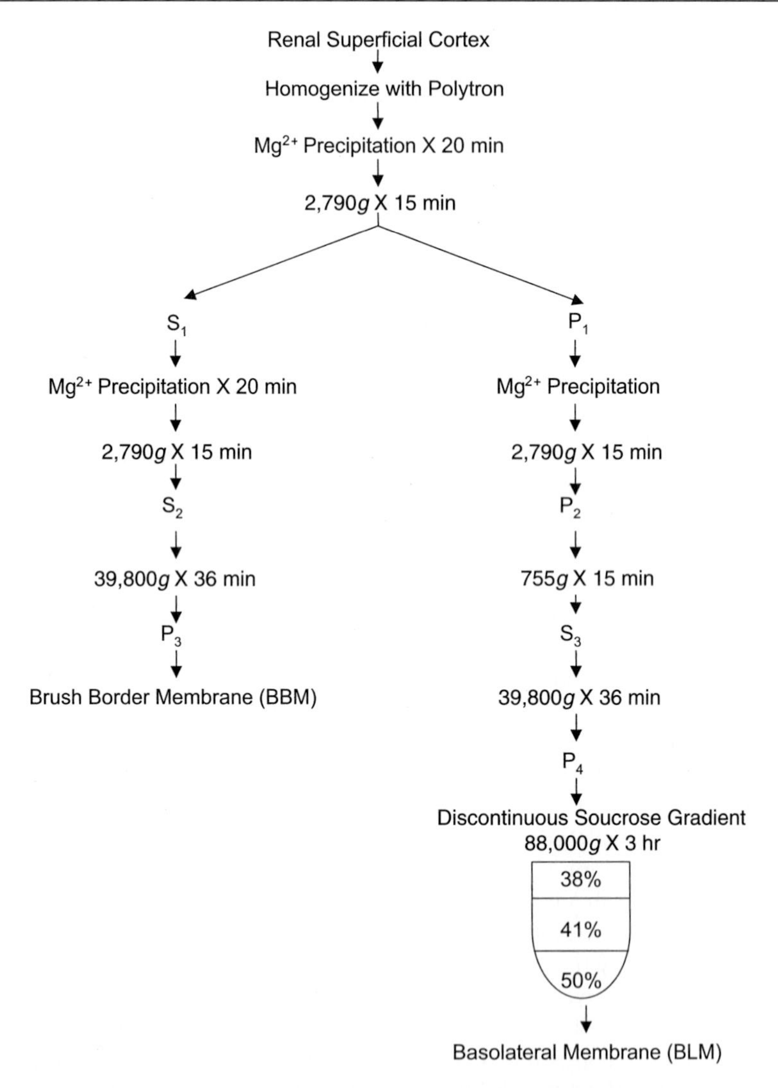

FIG. 1. Simultaneous isolation of brush border membranes and basolateral membranes from the same renal cortical homogenate.

Measurement of Lipid Fluidity of Brush Border and Basolateral Membranes

Brush border membrane and basolateral membrane fluidity is determined by (1) the steady state fluorescence polarization of 1,6-diphenyl-1,3,5-hexatriene (DPH), and by (2) the generalized polarization (GP) of 6-dodecanoyl-2-diethylaminonaphthalene (Laurdan; Molecular Probes, Eugene, OR).

TABLE I
BRUSH BORDER AND BASOLATERAL MEMBRANE LIPID FLUIDITY[a]

Source	DPH polarization	Laurdan generalized polarization
BBM	0.288	0.359
BLM	0.244	0.237

[a] As determined by DPH and Laurdan fluorescence spectroscopy.

The steady state polarization of DPH (P_{DPH}) is measured in a spectrofluorometer (PCI; ISS, Champaign-Urbana, IL) equipped with excitation and emission polarizers. The excitation wavelength is 360 nm and emission is viewed through a KV 399-nm filter.[9,10] P_{DPH} is determined by

$$P_{DPH} = \frac{I_{\|} - I_{\perp}}{I_{\|} + I_{\perp}} \quad (1)$$

where $I_{\|}$ and $I_{\perp}$ represent the intensities of the parallel and perpendicular components of the emission, respectively.[9,10]

P_{DPH} is markedly higher in BBMs compared with BLMs (Table I), which indicates that compared with BLMs the BBMs have decreased membrane fluidity.

The steady state emission spectrum of Laurdan is measured in a spectrofluorometer (PC1; ISS). The excitation wavelength is 340 nm, and emission is measured at 440 and 490 nm. In phospholipid vesicles and in BBMs and BLMs the emission maximum for Laurdan is 440 nm in the gel phase and 490 nm in the liquid–crystalline phase.[11–13] The emission spectrum of Laurdan is quantified by the generalized polarization (GP),

$$\mathrm{GP}_{\mathrm{Laurdan}} = \frac{I_{440} - I_{490}}{I_{440} + I_{490}} \quad (2)$$

I_{440} and I_{490} are the emission intensities at 440 and 490 nm, respectively.[11–13]

GP_{Laurdan} is markedly higher in BBMs compared with BLMs (Table I), which indicates that compared with BLMs the BBMs have decreased membrane fluidity.

Measurement of Lipid Composition of Brush Border and Basolateral Membranes

Total lipids are extracted by the method of Bligh and Dyer.[14] One milligram of BBM and BLM membrane protein is used for the extraction. Coprostanol (Supelco,

[9] M. Levi, D. M. Jameson, and B. W. Van Der Meer, *Am. J. Physiol.* **256,** F85 (1989).
[10] M. Levi, B. M. Baird, and P. V. Wilson, *J. Clin. Invest.* **85,** 231 (1990).
[11] T. Parasassi, G. DeStasio, G. Ravagnan, R. M. Rusch, and E. Gratton, *Biophys. J.* **60,** 179 (1991).
[12] M. Levi, P. V. Wilson, O. J. Cooper, and E. Gratton, *Photochem. Photobiol.* **57,** 420 (1993).
[13] T. Parasassi, M. Di Stefano, M. Loiero, G. Ravagnan, and E. Gratton, *Biophys. J.* **66,** 763 (1994).
[14] E. G. Bligh and W. J. Dyer, *Can. J. Biochem. Physiol.* **37,** 911 (1959).

Bellefonte, PA) is included as an internal standard for cholesterol determination. The original sample is extracted twice and the combined extract is evaporated to residue under nitrogen. The residue is resuspended in chloroform–methanol (1 : 1, v/v) and divided into aliquots for cholesterol, total, and individual phospholipid determinations.

The cholesterol sample is evaporated to a residue and resuspended in hexane and injected into a 530-μm 50% phenylmethyl silicone column in a Hewlett-Packard (Palo Alto, CA) model 5890 gas chromatograph with a flame ionization detector, run isothermally at 280°, with coprostanol serving as an internal standard. Area ratios are calculated with a Hewlett-Packard 3392A integrator, and cholesterol is expressed as nanomoles per milligram of protein.[15,16]

The phospholipid sample is dried to residue and resuspended in chloroform–methanol (1 : 1, v/v). Phospholipid content in the total lipid extract is determined by measuring the phosphorus content by the method of Ames and Dubin.[17] Individual phospholipid polar head group species are isolated by two-dimensional thin-layer chromatography (TLC).[18] The phospholipid sample is spotted onto Kieselgel silica gel 60-precoated TLC plates (Merck, Darmstadt, Germany). The plates are developed in two dimensions. The first-dimension solvent system consists of chloroform–methanol–acetic acid (65 : 25 : 10, v/v) and the second consists of chloroform–methanol–88% formic acid (65 : 25 : 10, v/v). This system results in an excellent separation of sphingomyelin (SM), phosphatidylcholine (PC), phosphatidylethanolamine (PE), phosphatidylserine (PS), and phosphatidylinositol (PI), as well as lysophosphatidylcholine (lysoPC) and lysophosphatidylethanolamine (lysoPE). After development, chromatograms are allowed to dry and then are briefly exposed to iodine crystals. Individual phospholipids are identified by comparison with cochromatography of authentic standards (Supelco). Areas of silica gel containing phospholipids are scraped into acid-washed test tubes. The phospholipids are reextracted from the gel by the Bligh and Dyer method, the silica gel is removed by centrifugation, and the phospholipids are separated into the chloroform phase. The sample is dried to a residue, and the phosphorus content is determined by the Ames and Dubin method. Samples are corrected for background by using an area of the TLC plate known to be free of phospholipids. Phospholipid content is expressed as nanomoles per milligram protein. The percent recovery of phospholipids from the TLC plate and reextraction from the gel is determined by dividing the sum amount of individual phospholipids by the total phospholipid applied to the TLC plate.

[15] M. Levi, B. Baird, and P. Wilson, *J. Clin. Invest.* **85,** 231 (1990).

[16] M. Levi, J. Shayman, A. Abe, R. H. McCluer, S. K. Gross, J. Biber, H. Murer, M. Lotscher, and R. E. Cronin, *J. Clin. Invest.* **96,** 207 (1995).

[17] B. N. Ames and D. T. Dubin, *J. Biol. Chem.* **235,** 769 (1960).

[18] J. D. Esko and C. R. H. Raetz, *J. Biol. Chem.* **255,** 474 (1980).

TABLE II
BRUSH BORDER AND BASOLATERAL MEMBRANE LIPID COMPOSITION

Lipid	Brush border membrane	Basolateral membrane
Cholesterol	409	336
Total phospholipid	482	742
Sphingomyelin	184 (38.4%)	128 (17.2%)
Phosphatidylcholine	102 (21.3%)	330 (44.2%)
Phosphatidylethanolamine	116 (24.3%)	228 (26.3%)
Phosphatidylserine	71 (14.7%)	38 (5.2%)
$\frac{\text{Cholesterol}}{\text{Total phospholipid}}$	0.848	0.453
$\frac{\text{Sphingomyelin}}{\text{Phosphatidylcholine}}$	1.80	0.39

Compared with BLMs the BBMs have a marked increase in cholesterol and sphingomyelin content and a decrease in total phospholipid and phosphatidylcholine content, resulting in marked increases in cholesterol to total phospholipid and of sphingomyelin to phosphatidylcholine mole ratios (Table II). The decrease in BBM fluidity is therefore explained by the significant increases in the mole ratios of cholesterol to total phospholipid and sphingomyelin to phosphatidylcholine.

Isolation of Lipid Fractions from Brush Border Membranes

To determine the effects of various lipids, especially cholesterol and sphingomyelin, on lipid fluidity, BBM lipid extracts are fractionated on silicic acid columns (Superclean LC-Si SPE tubes; Supelco).

The columns are first washed with 2 ml of chloroform–methanol (1 : 1, v/v). Lipid extracts in 50 μl of chloroform–methanol (1 : 1, v/v) are loaded on the column. Cholesterol is eluted from the column with 5 ml of chloroform, glycosphingolipids are eluted with 20 ml of acetone, and phospholipids are eluted with 5 ml of methanol. The glycosphingolipid and phospholipids fractions are combined to form the BBM lipids minus cholesterol fraction (BBM − cholesterol). Gas chromatographic analysis of this lipid fraction reveals that >95% of the original cholesterol is removed by this method.[19,20]

To further fractionate the phospholipid fraction and remove the sphingomyelin, the columns are first washed with 2 ml of chloroform–methanol (95 : 5, v/v). The phospholipid fraction in 50 μl of chloroform–methanol (1 : 1, v/v) is then applied to the column. Diphosphatidyl glycerol (DPG) and PA are eluted with 5 ml of

[19] C. Dietrich, L. Bagatolli, N. Thompson, M. Levi, K. Jacobson, and E. Gratton, *Biophys. J.* **80,** 1417 (2001).

[20] C. Dietrich, Z. N. Volovyk, M. Levi, N. L. Thompson, and K. Jacobson, *Proc. Natl. Acad. Sci. U.S.A.* **98,** 10642 (2001).

TABLE III
EFFECTS OF MEMBRANE PROTEINS, SPHINGOMYELIN, AND CHOLESTEROL ON BRUSH BORDER MEMBRANE LIPID FLUIDITY

Source	DPH polarization	Laurdan generalized polarization
BBM	0.284	0.393
BBM lipid extract	0.278	0.334
BBM lipids – sphingomyelin	0.268	0.303
BBM lipids – cholesterol	0.184	0.086

chloroform–methanol (95 : 5, v/v), PE and PS are eluted with 10 ml of chloroform–methanol (80 : 20, v/v), PI and PC are eluted with 10 ml of chloroform–methanol (50 : 50, v/v), and sphingomyelin and lysoPC are eluted with 10 ml of methanol. The phospholipid fractions minus the sphingomyelin are combined with the cholesterol and glycosphingolipid fractions described above to form the BBM lipids minus sphingomyelin fraction (BBM lipids − SM). Thin-layer chromatography of this lipid fraction shows that >95% of the original sphingomyelin is removed by this method.[19,20]

BBM lipid extraction and lipid fractionation studies indicate that although protein (BBM lipid extract vs. intact BBM) and sphingomyelin (BBM lipids − SM vs. BBM lipid extract) each contribute to the decrease in BBM lipid fluidity, cholesterol is by far the major mediator of the decrease in BBM lipid fluidity and BBM lipid dynamics (Table III).

Isolation of Lipid Rafts from Brush Border Membranes

To isolate detergent-resistant membrane fractions (lipid rafts), brush border membrane samples are subjected to detergent extraction at 4° for 30 min. Detergents utilized include Triton X-100 (Sigma-Aldrich, St. Louis, MO) and Lubrol WX (Serva, Heidelberg, Germany) at a concentration of 1% in a buffer consisting of 150 m*M* NaCl and 2 m*M* EGTA in 10 m*M* Tris-HCl (pH 7.5), supplemented with protease inhibitor cocktail.[21,22]

After detergent extraction at 4° the samples are centrifuged at 100,000*g* to obtain a detergent-resistant (DR) pellet and detergent-soluble (DS) supernatant. The pellet is resuspended in the same volume as the supernatant and the DR and DS fractions are analyzed for lipid composition (cholesterol and phospholipid) and lipid fluidity (Laurdan and/or DPH spectroscopy).

Compared with the intact BBM fraction the DR membrane fraction has a marked decrease in lipid fluidity (Table IV) which is mediated by increased levels

[21] D. A. Brown and J. K. Rose, *Cell* **68,** 533 (1992).
[22] K. Röper, D. Corbeil, and W. B. Huttner, *Nat. Cell Biol.* **2,** 582 (2000).

TABLE IV
LIPID DYNAMICS IN INTACT BRUSH BORDER MEMBRANE, DETERGENT-RESISTANT, AND DETERGENT-SENSITIVE MEMBRANE FRACTIONS

Source[a]	DPH polarization	Laurdan generalized polarization
BBM	0.230	0.298
BBM DR	0.267	0.390
BBM DS	0.119	0.213

[a] DR, detergent resistant; DS, detergent sensitive.

of sphingomyelin and cholesterol (results not shown). In contrast, the DS membrane fraction has a marked increase in lipid fluidity (Table IV), which is mediated by decreased levels of sphingomyelin and cholesterol (results not shown).

In addition, after detergent extraction at 4°, 0.5 ml of the membrane sample is mixed with 0.5 ml of 80% (w/v) sucrose, and then overlaid with 2 ml of 30% (w/v) and 2 ml of 5% (w/v) sucrose. The samples are then centrifuged at 100,000*g* for 18 hr.[21,22] At the end of centrifugation 0.5-ml fractions are recovered and are analyzed for lipid composition (cholesterol and phospholipid) and lipid fluidity (Laurdan and/or DPH spectroscopy). Alternatively, lipid rafts can also be fractionated, using the Optiprep gradient technique.[23–25]

Cell Culture System

The difficulties due to complex biochemical purification procedures can be overcome by directly imaging living cells in their growth medium. In this case we need a proper probe for the component or for the function we intend to visualize, and the appropriate experimental apparatus for the purpose we describe is the two-photon excitation fluorescence microscope.[26] For microscopy studies of cell membrane systems, our probe of choice is Laurdan, a lipophilic fluorophore whose spectral sensitivity to the polarity of membrane environments makes its generalized polarization (GP) function useful for the spatial resolution of membranes with different compositions and properties.

The choice of a two-photon excitation microscope is dictated by the virtual absence of damage with near-infrared excitation, which allows repeated scanning

[23] P. Keller and K. Simons, *J. Cell Biol.* **140,** 1357 (1998).

[24] F. Lafont, S. Lecat, P. Verkade, and K. Simons, *J. Cell Biol.* **142,** 1413 (1998).

[25] F. Lafont, P. Verkade, T. Galli, C. Wimmer, D. Louvard, and K. Simons, *Proc. Natl. Acad. Sci. U.S.A.* **96,** 3734 (1999).

[26] A. Diaspro, ed., "Confocal and Two-Photon Microscopy: Foundations, Applications and Advances." Wiley-Liss, New York, 2002.

of the same image frame without appreciable cell damage. Probes used for membrane studies often absorb in the ultraviolet region and rapidly photobleach, a condition mostly overcome by the use of near-infrared excitation.

This section concentrates on opossum kidney (OK) cells, an established cell line that has many similarities to the mammalian renal proximal tubule, and illustrates how alteration of cell membrane cholesterol content modulates lipid dynamics measured by Laurdan spectroscopy and/or imaging microscopy.

Cell Culture

OK cells[23] are grown in Dulbecco's modified Eagle's high-glucose medium (DMEM) supplemented with 10% (v/v) fetal calf serum (FCS), penicillin G (100 IU/ml), and streptomycin (100 μg/ml) in a humidified 5% CO_2–95% air atmosphere. The cells are grown to approximately 90% confluence and then are rendered quiescent for 24 hr by serum deprivation in Ham's F12–DMEM (1 : 1, v/v) supplemented with 4 m*M* L-glutamine, pH 7.3.

Cholesterol Modulation

Cells are seeded and grown in DMEM in the presence of 10% (v/v) FCS for 24 hr. Cells are then grown either in (1) regular DMEM in the presence of 5% (v/v) FCS, penicillin, and streptomycin, (2) DMEM plus 5% (v/v) lipoprotein-deficient serum (LPDS) for 2 days to deplete the cells of cholesterol, or (3) DMEM plus 5% (v/v) LPDS for 6 hr to upregulate their low-density lipoprotein (LDL) receptors and then grown on DMEM plus human LDL (400 μg/ml), for 2 days to enrich the cells with cholesterol. LDL is obtained from Calbiochem (La Jolla, CA) and LPDS from Intracel (Rockville, MD).

For fluorescence microscopy measurements, the cells are grown on coverslips. For fluorescence spectroscopy and lipid composition measurements the cells are grown in tissue culture dishes.

The effects of treatment with LPDS or LDL on cell cholesterol content are measured either by gas chromatography after lipid extraction of apical membranes (see above) or by fluorescence microscopy of intact cells after filipin staining. The effects of cholesterol enrichment and cholesterol depletion on cell membrane lipid dynamics are measured either by fluorescence spectroscopy of Laurdan in isolated apical membranes (see above) or by fluorescence microscopy of Laurdan GP images in intact cells.

Isolation of Apical Membrane-Enriched Fraction

OK cell monolayers are washed four times with 150 m*M* NaCl, 5 m*M* Tris-HCl (pH 7.2) and once with homogenization buffer, 5 m*M* HEPES–KOH (pH 7.2), pepstatin (1 μg/ml), and 0.1 m*M* phenylmethylsulfonyl fluoride. The cells are then scraped off the plate with a rubber policeman and homogenized by six passes through an 18-gauge needle. The lysate is centrifuged at 280*g* for 5 min (4°) to remove nuclear and cellular debris; the supernatant is centrifuged at 48,000*g* at 4° for 30 min. The pellet (crude membrane fraction) is resuspended in 1 ml of the homogenization buffer and applied to the top of a sucrose step gradient (40, 32.5, 29, and 10%, w/w) prepared with the homogenization buffer. The gradient is run for 90 min at 140,000*g*. Apical membranes are recovered from the 10/29% interphase.[27]

Cholesterol Imaging

Cells grown on coverslips and treated either with LDPS to achieve cholesterol depletion or with LDL to achieve cholesterol enrichment are rinsed with Tris-buffered saline (TBS) and fixed with 4% (w/v) paraformaldehyde in TBS, and then rinsed and incubated with filipin III (Sigma) in TBS, rinsed again, and mounted on a slide.[28] The cells are then viewed with two-photon excitation at 720 nm with a laser scanning confocal microscope (LSM 410; Zeiss, Jena, Germany) using a ×63 oil immersion objective.

Two-Photon Excitation Microscopy Laurdan Generalized Polarization Images

Laurdan labeling is performed directly on the coverslips in culture dishes, adding 1 μl of a 2 m*M* probe solution in dimethyl sulfoxide (DMSO) per milliliter of growth medium and incubating the mixture for 30 min in the dark. The cells are then gently washed with fresh medium and the coverslips are mounted on a microscope hanging drop slide, using fresh medium.[29] The data acquisition and image analysis for Laurdan GP microscopy measurements are performed as described in detail by Bagatolli *et al.* in [20] in this volume.[29a]

Treatment of OK cells with LPDS results in a significant decrease in cell cholesterol content. Laurdan GP imaging reveals heterogeneous distribution of GP values in the membranes of OK cells (Fig. 2). Compared with control cells the average GP value is decreased in LPDS-treated cells. The heterogeneous distribution of GP values is also seen in the apical membrane of OK cells. In contrast, treatment of OK cells with LDL results in a significant increase in cell cholesterol content. Laurdan GP imaging again reveals a heterogeneous distribution of GP values, and compared with control cells the average GP value is increased in LDL-treated cells (Fig. 3). The heterogeneous distribution of GP values is also seen in

[27] S. J. Reshkin, F. Wuarin, J. Biber, and H. Murer, *J. Biol. Chem.* **265,** 15261 (1990).

[28] M. I. Pörn and J. P. Slotte, *Biochem. J.* **308,** 269 (1995).

[29] T. Parasassi, E. Gratton, W. Yu, P. Wilson, and M. Levi, *Biophys. J.* **72,** 2413 (1997).

[29a] L. A. Bagatolli, S. A. Sanchez, T. Hazlett, and E. Gratton, *Methods Enzymol.* **360,** [20], 2002 (this volume).

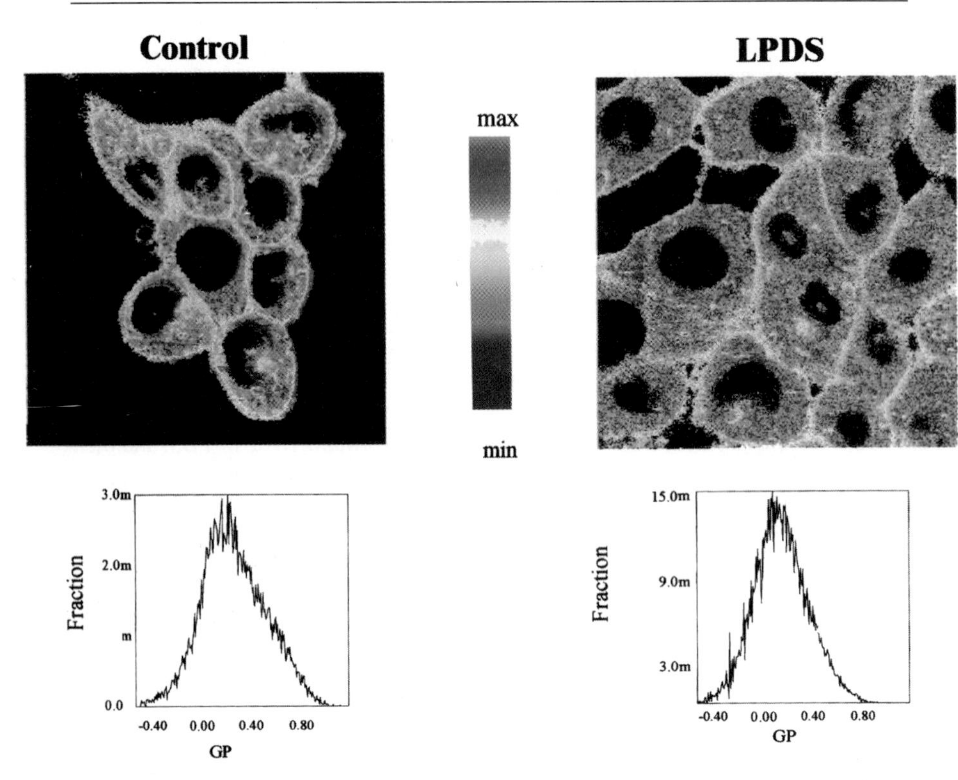

FIG. 2. Laurdan GP images of OK cells grown in the presence of 5% (v/v) FCS (control) and 5% (v/v) LPDS (cholesterol depletion).

the apical membrane of OK cells, where the GP values are higher than for other subcellular membranes and treatment with cholesterol results in a further increase in Laurdan GP, indicating a marked decrease in lipid dynamics.

Intact Tissues

Because of the deeper penetration of near-infrared with respect to shorter wavelengths, two-photon excitation microscopy is also the method of choice for whole tissue samples. Indeed, regardless of the thickness of the tissue sample, that is, regardless of slicing procedures, near-infrared excitation allows a penetration of at least 200 μm into the tissue. In addition, the tissue can be imaged under near-physiological conditions with no need for fixation and in a medium containing nutrients, so that the effect of eventual treatments can be directly imaged and characterized. When using the Laurdan probe, tissue labeling can be performed as efficiently as the labeling of cells in culture.

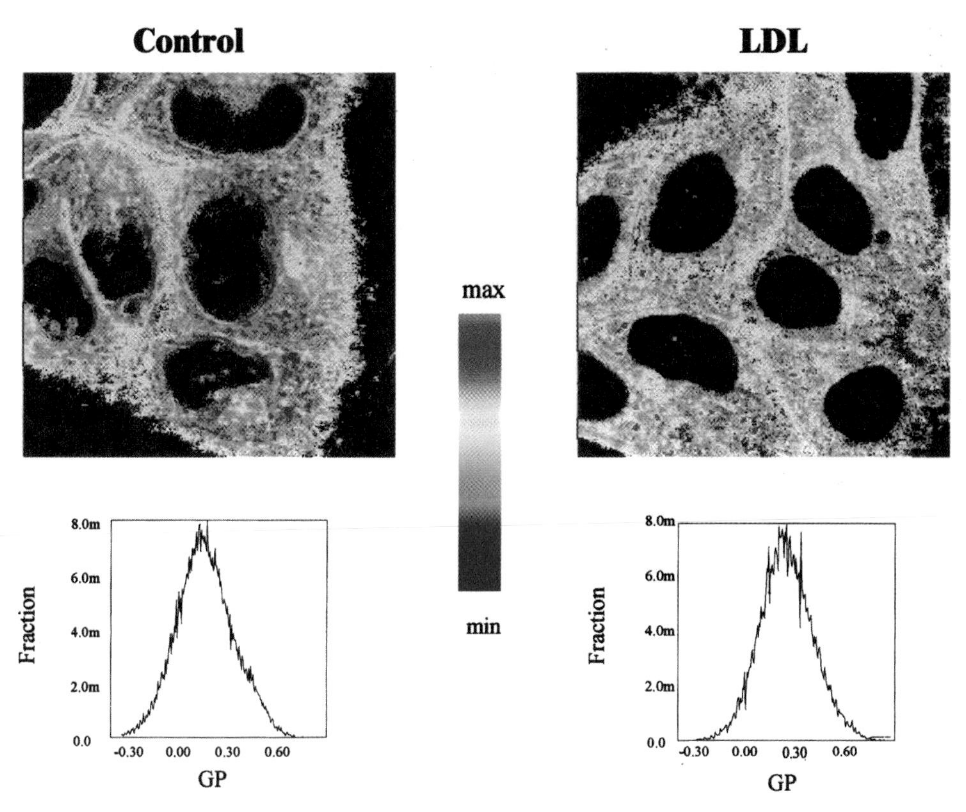

FIG. 3. Laurdan GP images of OK cells grown in the presence of 5% (v/v) FCS (control) and 5% (v/v) LPDS plus LDL at 400 μg/ml (cholesterol enrichment).

Imaging of Low-Density Lipoprotein Binding and Internalization into Fresh Rat Aorta

We label LDL with the lipophilic probe Laurdan.[30] Like other lipid probes currently used to label LDL, such as 1-1′-dioctadecyl-3,3,3′,3′-tetramethylindocarbocyanine perchlorate (DiI C_{18})[31] and 3-3′-dioctadecyloxacarbocyanine perchlorate (DiO C_{18}), Laurdan passively diffuses into all lipid components. Therefore, when labeled LDL binds to endothelial cells, all lipid components in the tissue, that is, cell membranes and cell organelles, are also labeled and imaged. Nevertheless, we can distinguish LDL from other tissue lipid components by imaging the Laurdan GP value instead of its fluorescence intensity.

[30] T. Parasassi, W. Yu, D. Durbin, L. Kuriashkina, E. Gratton, N. Maeda, and F. Ursini, *Free Radic. Biol. Med.* **28,** 1589 (2000).

[31] X. Gu, R. Lawrence, and M. Krieger, *J. Biol. Chem.* **275,** 9120 (2000).

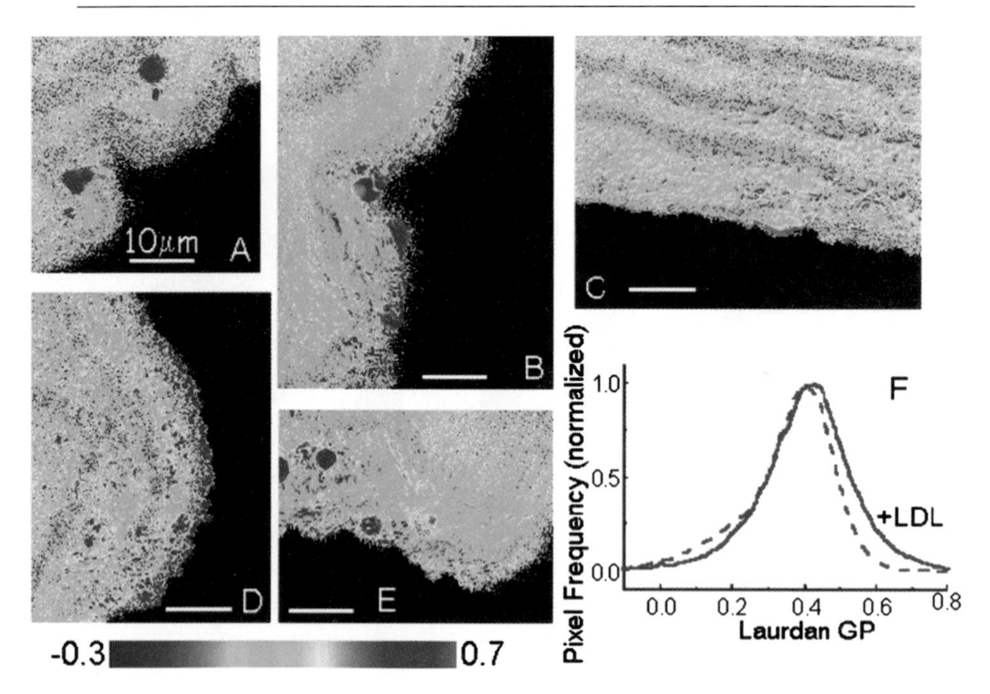

FIG. 4. Pseudo-color representation of Laurdan GP values in rat aorta rings incubated for 30 min at 37° with minimally oxidized LDL. (A, B, and D) The lumen is on the right; (C and E) the lumen is at the bottom. A GP color scale is at the bottom of the figure. A histogram of the normalized pixel frequency versus Laurdan GP values is reported in (F), averaged over four different images of control and LDL-incubated samples.

Because of the high concentration of cholesterol, the GP value in LDL is much higher than in cell membranes and organelles.[32] By collecting two images at two emission wavelengths, that is, by using bandpass filters at 440 and 490 nm, the GP image is simply calculated at each pixel according to Eq. (2) given above.

In Fig. 4 the GP image of an aorta ring section after incubation with LDL allows easy identification of LDL, bound to the endothelial cell layer facing the lumen (Fig. 4A and D), internalized and compartmentalized in large aggregates within the vessel tissue (Fig. 4A, B, and E). A histogram of pixel frequency versus Laurdan GP (Fig. 4F) shows a shift to higher GP values in the samples that have been incubated with LDL. Because of the deep penetration of two-photon excitation, the presented image is actually relative to the tissue interior. Details of the method used to obtain this image follow.

Rat aorta rings, about 0.5 mm thick, are sectioned from the ascending aorta of 6- to 8-week-old Long-Evans male rats, using a tissue chopper, immediately after

[32] R. Brunelli, G. Mei, E. K. Krasnowska, F. Pierucci, L. Zichella, F. Ursini, and T. Parasassi, *Biochemistry* **39,** 13897 (2000).

the animals are killed. Samples are maintained in Earle's balanced salt solution (EBSS) containing 24.6 m*M* glucose and 26.3 m*M* $NaHCO_3$, pH 7.4, saturated with 95% O_2 and 5% CO_2. Samples are imaged within a few hours by mounting them in hanging drop slides in fresh EBSS buffer.

Minimally oxidized human LDL is obtained by storage at 4° for 1–2 weeks of LDL isolated from healthy fasting volunteers by routine ultracentrifugation,[33] at a protein concentration of 3 mg/ml. LDL is then added to aorta rings in the EBSS and incubated for 30 min at 37° in the dark. Laurdan labeling is performed by adding 3 μl of a 2 m*M* solution of the probe in DMSO in the incubation medium. Tissue labeling is extremely efficient and serves for visualization during LDL binding and internalization.

Microscopy Measurements

The 770-nm excitation beam is from a titanium–sapphire laser (Mira 900; Coherent, Palo Alto, CA) pumped by an argon ion laser (Innova 310; Coherent). The microscope configuration and scanning routine have been previously described.[34] A frame scanning rate of 20 sec is used. The average laser power on the sample is on the order of 3 mW. A quarter-wave plate is used to change the laser light polarization from linear to circular. Autofluorescence is collected through a broad-band BG39 barrier filter. Images of Laurdan-labeled samples are acquired with two different optical bandpass filters (Ealing Electro Optics, Holliston, MA) centered in the blue (440-nm) and in the red (490-nm) regions of the Laurdan emission spectrum. The two filters are exchanged each time a full frame is scanned. From the images at the two emissions, the generalized polarization (GP) image is calculated as described.[34]

Imaging Autofluorescence of Extracellular Matrix Proteins in Rat Aorta

The images reported in Fig. 4 also show areas of low GP values, in blue, which are never observed in cell membranes.[34] Imaging the aorta tissue without any labeling reveals that these structures have a bright autofluorescence, which has been attributed to extracellular matrix protein.[30] The blue color of the GP images originates from the calculation of the GP, which for the spectral properties of this autofluorescence yield a low GP value. The attribution of aorta autofluorescence to extracellular protein is based on the morphology of the observed structures, clearly showing the tunica intima and the fibers in the tunica media, and on images of purified elastin. Indeed, blood vessel extracellular matrix is composed mainly of elastin and collagen, complex protein structures formed of several filaments bound

[33] M. J. Chapman, P. M. Laplaud, G. Luc, P. Forgez, E. Bruckert, S. Goulinet, and D. Lagrange, *J. Lipid Res.* **29,** 442 (1988).

[34] W. M. Yu, P. T. So, T. French, and E. Gratton, Fluorescence generalized polarization of cell membranes: a two-photon scanning microscopy approach. *Biophys. J.* **70,** 626 (1996).

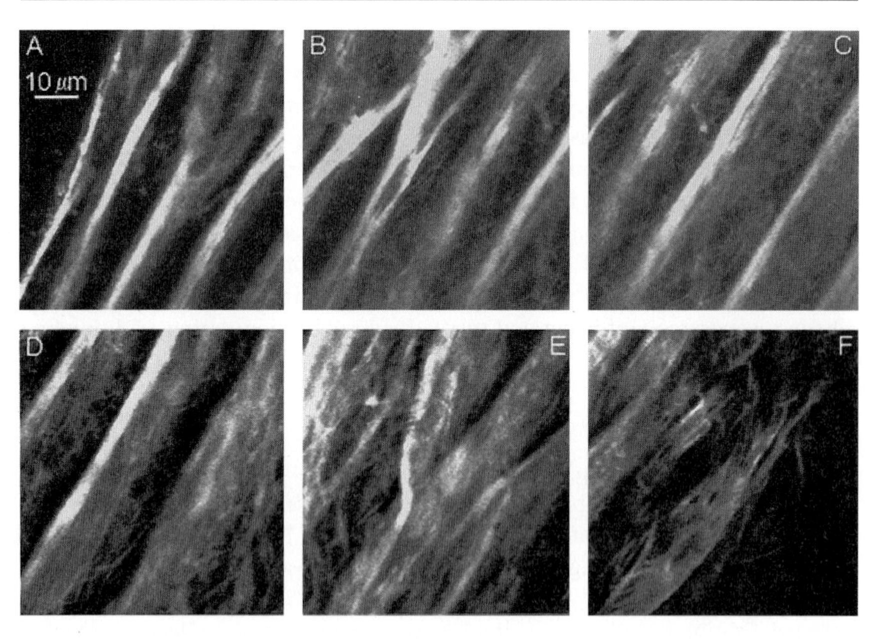

FIG. 5. Images of several adjacent frames of deep optical sections of a rat aorta cross-section ring showing the autofluorescence intensity of the matrix fibers from the lumen (A) to the ring exterior (F).

together by cross-links produced by modified amino acid residues. Desmosine forms cross-links in collagen and hydroxylysyl- and lysylpiridinoline form cross-links in elastin.[35] Both of them possess intense fluorescence with excitation and emission spectra in the range of our experimental setup. We use excitation at 770 nm, which corresponds to one-photon excitation at 385 nm, and we collect the fluorescence emission after passage through a broad-band BG39 filter.

The bright fluorescence is an interesting property of aorta matrix proteins. Their morphology and their physiological remodeling or pathological modification, such as bending and rupture, have a fundamental significance in the life of the blood vessel. Indeed, in our previous study we showed the occurrence of matrix proteolysis in rat aorta on brief incubation with oxidized LDL.[30]

When imaging vessel matrix proteins, an important reason suggests the use of two-photon excitation instead of the conventional one-photon excitation: the tissue can be observed under physiological conditions, with no need for fixation, labeling, or embedding in matrices and thin sectioning, procedures that may damage the tissue and alter the protein morphology. Actually, conventional fluorescence

[35] J. J. Baraga, R. P. Rava, M. Fitzmaurice, L. L. Tong, P. Taroni, C. Kittrell, and M. S. Feld, *Atherosclerosis* **88,** 1 (1991).

microscopy observation of matrix proteins in vessels shows a morphology different from what we observed when using fresh tissue.[36] Fixed samples show smaller, curly matrix fibers whereas in fresh tissue thick, elongated, straight fibers can be observed. This morphological observation has important consequences when considering that, starting only from the physiological ordered structure, we have been able to detect the modification induced by minimally modified LDL, resulting in curly and broken fibers.[30]

As an example of obtainable images, in Fig. 5 we report a series of images taken by scanning adjacent frames, from the lumen (top left) to the exterior (bottom right) of a portion of an aorta cross-section ring.

[36] A. A. Young, I. J. Legrice, M. A. Young, and B. H. Smaill, *J. Microsc.* **192,** 139 (1998).

[15] Photonics for Biologists

By IAN PARKER

Volumes 360 and 361 of *Methods in Enzymology* illustrate the wide array of powerful biophotonic approaches that have been developed to address biological problems. However, most biologists have little background in fields such as optics and laser technology, and may be daunted by the task of trying to understand how these techniques work or how to develop and construct photonic instrumentation for their own applications. This article is directed at such an audience and is divided into two main sections, which aim to provide (1) a primer in basic optics and functioning of the light microscope and (2) practical guidance on the functioning of photonic components and construction of complete biophotonic instrumentation.

Introduction to Basics of Optics

Classic Optics

This section covers only the basics of optics and microscopy; but it is surprising how far a rudimentary knowledge will take you. For further reading, see Refs. 1–4, and the Web resources listed in the Appendix. The catalogs of many of the

[1] M. Abramowitz, "Optics: A Primer." Olympus America, New York, 1984.
[2] E. Hecht and A. Zajac, "Optics." Addison-Wesley, Reading, MA, 1974.
[3] M. C. Gupta, "Handbook of Photonics." CRC Press, Boca Raton, FL, 1997.
[4] F. A. Jenkins and H. E. White, "Fundamentals of Optics." McGraw-Hill, New York, 1976.

0076-6879/03 $35.00

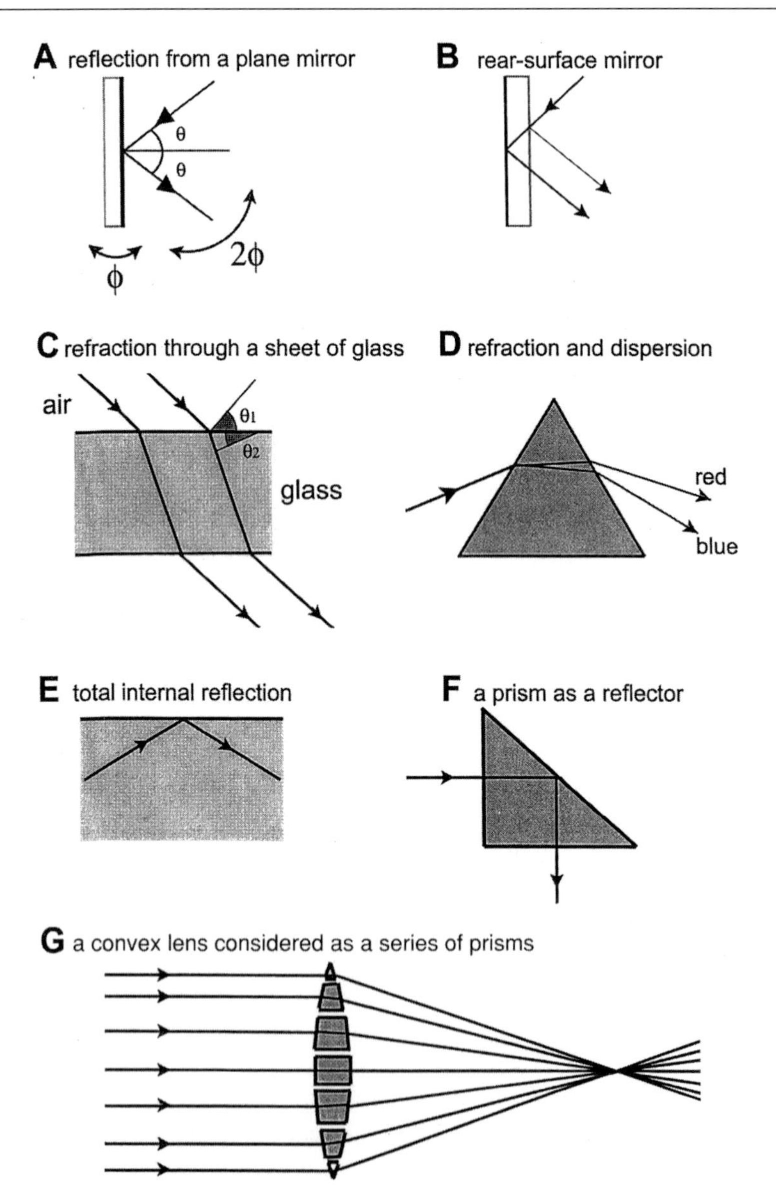

FIG. 1. Reflection and refraction at plane surfaces. (A) Reflection from a front-surface mirror. The incident and reflected rays subtend the same angle with regard to the mirror. If the mirror is rotated, light will move through an angle twice as great. (B) Rear-surface mirrors result in reflection both from the mirrored surface and, more faintly, from the front of the glass. (C) Light rays striking an air–glass interface at an angle are bent because of the change difference in refractive index (n): $n_{\text{air}} \sin\theta_1 = n_{\text{glass}} \sin\theta_2$. An inverse change of angle results when the light exits the far side of a sheet of glass

major optical vendors are also a valuable source of practical information on basic optics, as well as on the specifics of particular products.

Nature of Light. Light is a tiny part of the electromagnetic spectrum, extending from wavelengths of about 200 to 2000 nm. Within this range the human eye can see only from about 400 nm (violet) to 700 nm (deep red); a mere 2-fold span in wavelengths as compared, for example, with the $> 10^7$-fold span of radio waves. The speed of light in vacuum or air is about 3×10^8 m s^{-1}, which seems inconceivably fast until you consider that lasers can generate pulses of light as brief as 10 fs, so that each pulse has a length of only a few microns. Light has a dual nature, behaving as both a wave and as particles (photons). The energy of a photon increases as the reciprocal of the wavelength, so violet photons at 400 nm have twice the energy of infrared photons at 800 nm. A general rule seems to be that light will behave in whichever way presents the most obstacles for you. For example, when trying to image very small objects, the wavelength of light sets a limit to resolution, whereas when trying to detect very dim signals the discrete nature of individual photons introduces noise. In the latter regard, the statistical variance of the number of photons arriving per unit time increases as the square root of the mean photon flux, so that the signal-to-noise ratio improves proportional to the square root of light intensity.

Reflection at Plane Surfaces. Light incident on a plane mirror is reflected at an angle equal to the angle of incidence (Snell's law; Fig. 1A). If the mirror is rotated, the reflected light will rotate through an angle twice that of the mirror (an important point to remember when constructing laser scan systems). Optical mirrors are commonly made by depositing a reflective metal coating (silver or aluminum) onto the front of a flat glass substrate. However, reflection will occur at any interface between materials of different refractive index, to an extent proportional to the difference in indices (typically 4% for a glass–air interface). Thus, rear-surface mirrors should never be used, as reflection at the front glass surface causes a faint "ghost" reflection offset from the bright reflection from the rear-mirrored surface (Fig. 1B).

Refraction. The speed of light is fastest in a vacuum, and slows to varying extents in different materials. The ratio of the velocity of light in a vacuum to the velocity in a material is referred to as the refractive index (n) of that material.

---

with parallel sides, resulting in a lateral translation of the light beam without any net change in angle. (D) Glass components with nonparallel sides, such as prisms, cause deviation in the angle of light rays. Because the refractive index of glass varies with wavelength, shorter wavelengths are deviated more than longer wavelengths (dispersion). (E) If light rays within glass strike an air interface at a sufficiently shallow angle, they are reflected back into the glass without loss. (F) Use of total internal reflection within a right-angle prism to reflect a light beam. (G) A convex lens considered as a series of prisms.

Some examples include the following:

Medium	Refractive index
Vacuum	1
Air	1.0003
Water	1.333
Living cells	1.36
Optical glass	1.52
Diamond	2.42

When a beam of light passes an interface between materials of differing refractive index it bends (Fig. 1C), traveling at a steeper angle to the normal in medium of higher refractive index. (Imagine a line of people running at an angle along a beach and into the sea. They cannot run as fast when in the sea, so the angle of the line as a whole steepens.) For a sheet of glass with parallel sides the reverse process takes place at the glass–air interface on the opposite side. Thus, light traveling through a sheet of glass is not changed in direction, but is translated sideways, to an extent that depends on the angle of incidence and the thickness of glass (Fig. 1C).

In optical components such as prisms, the sides of which are not parallel, a net change in direction of light rays does result (Fig. 1D). Furthermore, because the refractive index of glass varies with wavelength, short wavelengths are bent more than longer wavelengths (dispersion).

Total Internal Reflection. Light rays traveling in a material of high refractive index (e.g., glass) are bent to a shallower angle when passing through an interface to a material of lower refractive index (e.g., air). If the angle of incidence is brought to a critical angle the refracted rays follow the surface of the interface, and at yet shallower angles of incidence the light is reflected back into the glass. This process is without loss (total internal reflection; Fig. 1E). The refractive index of glass is sufficiently high that a right-angle prism can thus be used as an efficient reflector (Fig. 1F), although there will still be losses at the air–glass interfaces and in transmission through the prism.

Simple Lenses. Imagine a series of prisms of progressively increasing steepness, arranged such that parallel rays of light striking each prism are bent so that they all pass through the same point (Fig. 1G). A convex lens with spherical surfaces provides a close approximation to this situation, and will focus parallel rays of light to a spot at a distance of one focal length behind the lens (Fig. 2A). The focal length becomes shorter for lenses with a smaller radius of curvature and higher refractive index. A lens with concave surfaces has a negative focal length, and diverges parallel rays such that they appear to have arisen from a point one focal length in front of the lens (Fig. 2B).

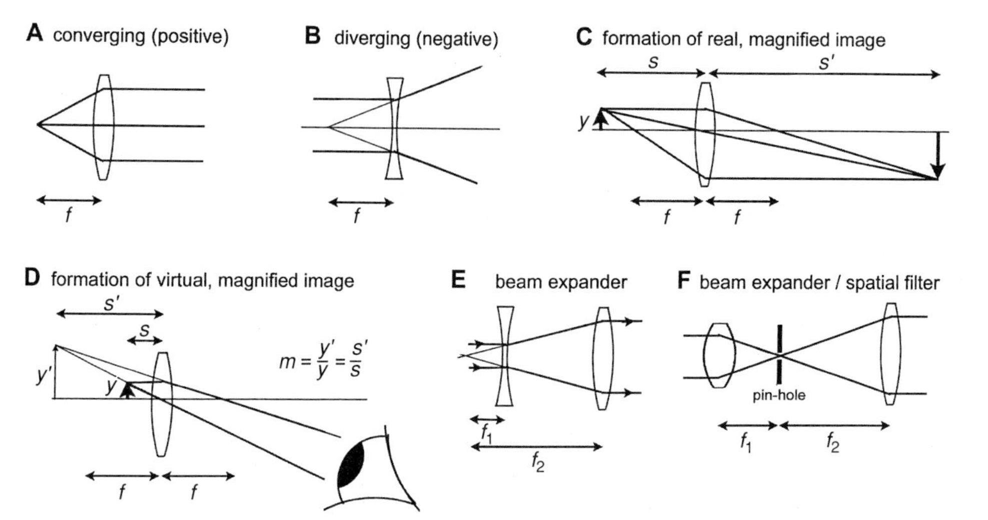

FIG. 2. Converging and diverging lenses. (A) A converging (positive) lens focuses a parallel beam of light to a spot at a distance one focal length (f) from the lens. (B) A diverging (negative) lens diverges a parallel beam of light so that its rays appear to have originated from a point one focal length behind the lens (as shown by thin lines). (C) An object placed further than one focal length from a converging lens will form a real image on the other side of the lens. The magnification m (image height y'/object height y) is given by s'/s. (D) An object placed closer than the focal length will not form a real image, but a virtual, magnified image can be viewed by the eye. This appears as right way up, behind the lens at the position marked by the thin arrow. Magnification is again given by s'/s. (E) A diverging and converging lens arranged as a beam expander. An incident beam of parallel light is expanded to form a parallel beam of greater diameter. The degree of expansion is given by f_2/f_1 (f_2 = focal length of converging lens, f_1 = focal length of diverging lens). (F) A beam expander made from two converging lenses of different focal length. The first (short focal length) focuses an incident laser beam to a spot, which is then recollimated by the second lens. A pinhole aperture can be placed at the focal point to act as a spatial filter and "clean up" the exiting beam.

An object placed anywhere in front of a convex lens at distances between one focal length and infinity will form a real image (i.e., one that can be projected onto a piece of card) on the opposite side of the lens. The degree of magnification depends on the respective object and image distances, as shown in Fig. 2C. If the object is placed within one focal length of the lens a real image is not formed (because light rays emerging from the lens are still diverging). However, a virtual image can be viewed by the eye (which focuses the light rays to form a real image on the retina), and appears upright with a magnification as shown in Fig. 2D.

A diverging lens and converging lens can be arranged as in Fig. 2E to form a beam expander, useful for expanding a laser beam before introducing it into a microscope so as to fill the aperture of the objective lens. Both the input and output beams are parallel, but the width of the output beam is expanded by a factor given by the focal length of the converging lens divided by that of the diverging lens. A beam expander can also be constructed from two converging lenses (Fig. 2F). This arrangement has the advantage that the beam is brought to a focus between the two lenses. A pinhole aperture of appropriate size placed here acts as a spatial filter, and will "clean up" the output beam so that it appears to have originated at a point source, even if the quality of the input beam is degraded.

Lens Aberrations. Most lenses are ground with spherical surfaces, because this can be done easily and inexpensively while maintaining high precision. Unfortunately, a spherical surface differs slightly from the parabolic profile required so that all parallel rays entering a lens will be brought to a focus at the same point. Instead, rays near the periphery are brought to a focus closer than central rays, so that spherical aberration causes the image of a specimen to be blurred. This problem can be minimized by utilizing only the central part of a lens.

A second major optical defect is chromatic aberration, which arises because light of different wavelengths is brought to a focus at different positions: for example, blue light is refracted to a greater extent than red, so that blue light will be focused to a spot closer to the lens than red light. For applications involving monochromatic laser beams chromatic aberration is not an issue, because only a single wavelength is present.

Both spherical and chromatic aberrations can be effectively corrected by use of more complex lenses employing multiple elements with materials of different refractive index and dispersion, or aspheric glass elements. The design of such lenses is complex. If aberrations from a simple lens are too great to be tolerated, the easiest solution is to purchase a microscope objective or other well-corrected compound lens.

Optical Microscope

Most techniques in optical biology involve interfacing an external optical system to a microscope. A good place to begin, therefore, is to describe the optical functioning of a "generic" modern commercial microscope, and consider how this may be adapted to uses beyond those envisaged by the manufacturers. For further readings an excellent starting point is the review by M. W. Davidson and M. Abramowitz (http://micro.magnet.fsu.edu/primer/opticalmicroscopy.html); and see also Refs. 5–7.

[5] S. Bradbury and B. Bracegirdle, "Introduction to Light Microscopy." BIOS Scientific Publishers, Oxford, 1998.

[6] M. Abramowitz, "Fluorescence Microscopy: The Essentials." Olympus America, New York, 1993.

[7] B. Herman and J. J. Ledmasters, eds., "Optical Microscopy: Emerging Methods and Applications." Academic Press, New York, 1993.

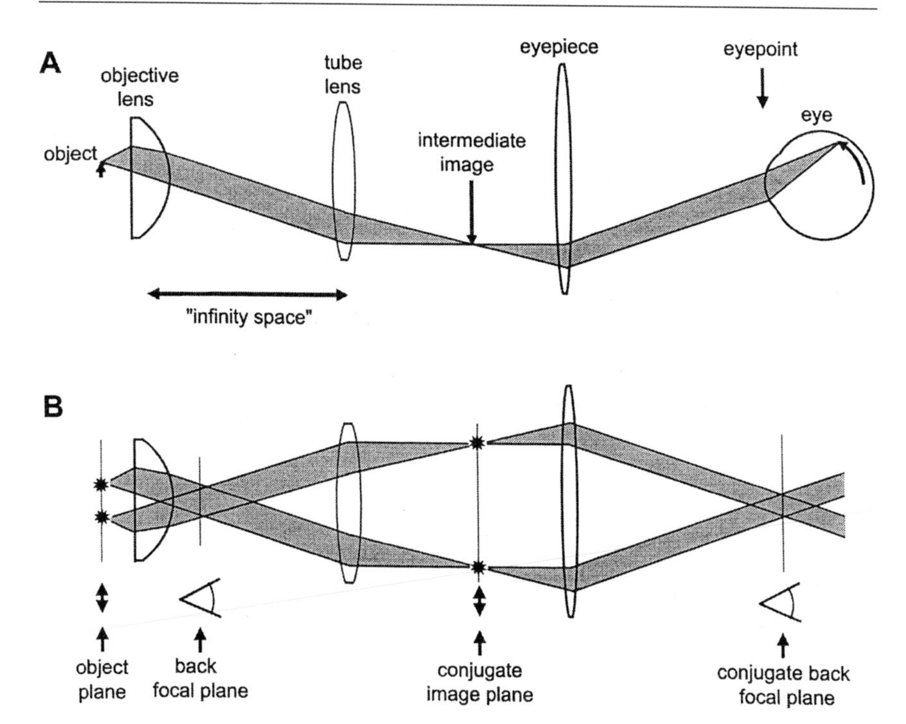

FIG. 3. Image ray paths within an infinity-corrected microscope. (A) Ray path illustrating the formation of a magnified virtual image. (B) Ray paths from two, laterally separated points in the specimen. Note that the lateral position at the object plane corresponds to the angle of parallel light rays at the back focal plane of the objective, and that these planes are reimaged at further points in the microscope.

Fundamentals of Image Formation by Microscope

RAY PATH IN INFINITY-CORRECTED MICROSCOPE. All the major microscope manufacturers have now adopted infinity-corrected optics, in which the optical train within the microscope follows the scheme diagrammed in Fig. 3A. When correctly focused for optimal correction of aberrations, light from a point source in the specimen, such as a tiny fluorescent bead, is focused by the objective lens to form parallel beams of light (i.e., an image is formed only infinitely far behind the lens). Thus, unlike earlier designs in which objectives formed a real image at a fixed distance within the microscope tube (usually 160 mm), infinity-corrected lenses do not directly form an image. Instead, a further "tube" lens, located in the microscope body or ocular head, is used to form an intermediate image. This lies at a conjugate image plane, at the position of the eyepiece graticule. Note that this is a real image. It can be projected on a piece of card or captured by a camera body placed at the conjugate image plane. The image is then further magnified by the microscope eyepiece, which, together with the lens of the eye, produces an erect, real image of the object on the retina.

A major advantage of using infinity-corrected objectives is the creation of a so-called infinity space between the objective and tube lens. Because light rays from any given point in the object plane are parallel, the separation between objective and tube lens is not critical, and accessories such as fluorescence illuminators and DIC (differential interference contrast) prisms can be introduced without degrading image quality or changing magnification or focus. It should not be thought, however, that the infinity space can be made infinitely long. Although light rays originating from any given point in the object plane are parallel to one another, these rays diverge from the optical axis at an angle that steepens depending on the distance of the point from the optical center (Fig. 3B). The width of the overall bundle of rays thus diverges, so that the infinity space can typically be extended only a few tens of centimeters before the image becomes vignetted because the tube lens no longer captures all rays.

CONJUGATE PLANES. Figure 3A also illustrates an intuitive point that is key to understanding how to interface external optics with microscopes. This is, the lateral position of a point in the object plane corresponds to the angle of parallel rays at the back focal plane of the objective. Thus, light shone through a pinhole aperture placed at the conjugate image plane will be focused to form a small spot at the object plane in the specimen; and this spot will move laterally if the pinhole is moved laterally. Conversely, the same result can be achieved by shining a parallel beam of light into the eyepiece (after removing the pinhole). This will again result in a focused spot at the object plane, but the lateral position of the spot can now be changed by varying the angle of the rays at the conjugate back focal plane (corresponding to the eyepoint, or position where the pupil of the eye would normally be located).

This concept is further illustrated in Fig. 4, showing the paths of illuminating light rays (A) and imaging light rays (B) within a microscope set up for bright-field illumination. A point in sharp focus in the specimen (i.e., at the object plane) is reimaged at several conjugate image planes, which lie at the positions of the retina of the observer, the eyepiece graticule, and the field diaphragm of the condenser (Fig. 4B). Thus, the object, the field diaphragm, and the graticule will all appear simultaneously in focus to the observer. On the other hand, the image field of the microscope should appear uniformly illuminated, without any visible structure from the lamp filament. This arrangement is known as Kohler illumination, and can be readily illustrated by looking at a flashlamp through a magnifying glass while moving the magnifying glass back and forth until the lens appears uniformly illuminated. At this point, light from the lamp is focused to form an image of the filament at the pupil of the eye.

In the microscope, light from the lamp is imaged and reimaged successively by the collector lens in the lamphousing, the condenser, the objective and other optics so that an image of the filament is formed at several conjugate back focal

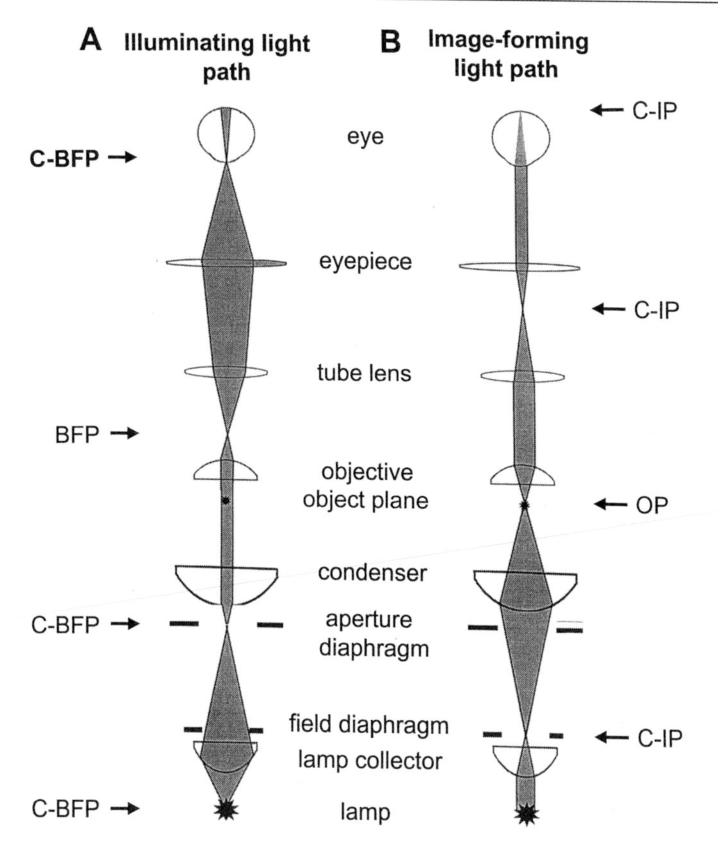

FIG. 4. Diagrams illustrating the paths of illuminating light and image-forming light within an infinity-corrected microscope set up for bright-field Kohler illumination. Note that a point in focus in the object plane (OP) is reimaged at conjugate image planes (C-IP) within the microscope (at the positions of the eyepiece graticule and the field diaphragm in the condenser), as well as at the retina of the eye. Correspondingly, in the illumination light path, the back focal plane of the objective (BFP) is reimaged at conjugate planes (C-BFP) at the exit pupil of the eyepiece and at the aperture diaphragm of the condenser.

planes, corresponding to positions of the aperture diaphragm in the condenser, the back focal plane of the objective, and the eyepoint of the eyepiece. Once again, the angle of light rays at each of these conjugate planes corresponds to the position of an object in the object plane. Conjugate back focal planes are, therefore, good places to put optics such as filters, because any dirt or scratches will not be in focus.

Fluorescence Microscopy. Fluorescence techniques form a large part of the biophotonic repertoire. The basic principle of conventional (one-photon) fluorescence is that a molecule (fluorophore) absorbs energy from an incident photon of relatively short wavelength (e.g., blue light), and then almost immediately (within nanoseconds) reemits a photon of lower energy (longer wavelength; e.g., green). Because of this shift to longer wavelengths (Stokes shift), filters can be used to block out the excitation light and visualize only the fluorescence. A regular bright-field microscope can, in principle, be modified for fluorescence use simply by placing a filter under the condenser to transmit only short wavelengths (excitation filter), and a second filter (barrier filter) beyond the objective to block the excitation light while transmitting emitted fluorescent light. However, this does not work well, because the excitation light is much more intense than the dim fluorescence, and is difficult to block completely.

Instead, modern fluorescence microscopes utilize epiillumination, in which the excitation light is introduced into the infinity space of the microscope and illuminates the specimen through the same objective lens that is used to image the fluorescence emission (Fig. 5). The key to this arrangement is the use of a dichroic mirror, which reflects the short-wavelength excitation light, while allowing the longer wavelength fluorescence emission to pass through toward the eyepieces. Because dichroic mirrors do not provide a perfect separation between excitation

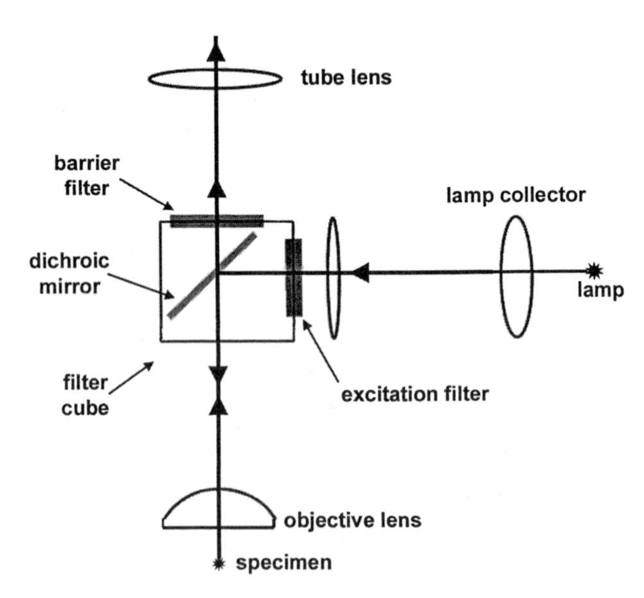

FIG. 5. Diagram of an epifluorescence microscope. A filter cube housing a dichroic mirror, excitation filter, and barrier (emission) filter is mounted in the infinity space behind the objective. Excitation light (usually from an arc lamp) is focused through the excitation filter and reflected by the dichroic mirror so that an image of the arc is formed at the back focal plane of the objective, resulting in even (Kohler) illumination of the specimen. Flourescent light emitted by the specimen is collected back through the objective, passes through the dichroic, and is imaged by the usual arrangement of tube lens and eyepiece. The barrier filter serves to block any excitation light that was transmitted through the dichroic mirror.

and emission wavelengths, additional excitation and barrier filters are incorporated. All three components are usually integrated within a filter "cube," and several cubes may be mounted in a turret or slider to allow ready interchange for work with different fluorophores.

As with bright-field imaging, epifluorescent illuminators are adjusted to provide Kohler illumination—the difference is only that the objective lens serves double duty as the condenser. In this regard, the numerical aperture of the objective (see below) assumes great importance for obtaining a bright image, because it determines the transmission of both excitation and emitted light. Regular fluorescence microscopes employ an arc lamp as an intense light source, which is focused to provide uniform (Kohler) illumination throught the field of view. In laser scan microscopes (e.g., confocal and multiphoton) the parallel laser beam is instead focused to a spot in the specimen, and the emitted fluorescence is detected point by point as the spot is scanned over the specimen.

Microscope Objective Lens. The image quality of a microscope is determined at the outset by the objective lens: all the rest of the microscope is really just a matter of ergonomics and added features. It is, therefore, useful to consider in more detail how microscope objectives can correct for many of the defects seen with simple lenses, and how objectives can be designed with optimal performance for particular applications.

Objective lenses are characterized by several parameters. The most obvious is the magnification, which is simply a function of the focal length of the lens. The magnification per se is less important than might be expected because, other things being equal, further magnification or reduction can be applied afterward. (For example, Olympus makes a "universal" ×20 objective, designed to be used with a magnification changer providing final magnifications equivalent to seperate objectives with magnifications between ×7 and ×80.) Instead, the numerical aperture (NA) of the objective (defined as shown in Fig. 6A) is the key parameter characterizing an objective.

NUMERICAL APERTURE. The numerical aperture is important in two major respects. First, the NA determines the resolving power of the microscope—finer details can be seen with objectives of high NA. An explanation for why this is so was first propounded by Ernst Abbe in the 19th century. In brief, light passing through a specimen emerges such that some remains undeviated (zeroth order), while the remainder is diffracted to form a fan at increasing angles (first order, second order, etc.) from the undeviated light (Fig. 6B). The greater the number of diffracted orders that are captured by the objective, the more accurately the image will represent the original object. However, because of the wave nature of light, a perfect representation can never be achieved. Thus, an infinitely small object appears as an Airy disk: a central spot of finite size surrounded by concentric rings of progressively reducing contrast. The size of this Airy disk decreases linearly with increasing NA of the objective lens and with decreasing wavelength of light (Fig. 6C). Resolution is commonly (although somewhat arbitrarily) expressed in

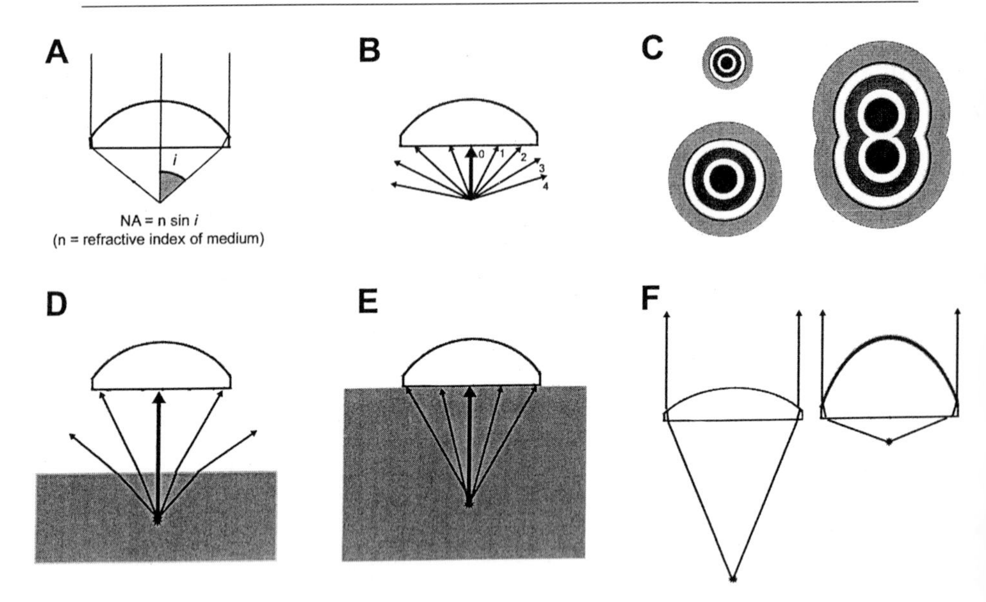

FIG. 6. Numerical aperture and resolution. (A) The numerical aperture (NA) of a lens is given by the sine of the angle i formed by the most peripheral rays that are collected by the lens, multiplied by the refractive index of the medium (e.g., 1 for a dry objective, 1.3 for a water-immersion objective). (B) Light from a specimen includes both undeviated rays (thick line; zeroth order) and diffracted rays (thin lines; first, second, etc., orders). The greater the number of diffracted orders collected by the objective, the more closely the image will represent the original object. (C) Airy disks and resolution. Objects smaller than the limit of resolution (e.g., 100 nm beads) form images of small, circular diffraction discs (Airy disks). High-numerical aperture lenses (which capture more of the higher orders of diffracted light) produce smaller Airy disks (upper left) than lenses of low numerical aperture (lower left). When two small objects approach close together their Airy disks begin to overlap (right) and, at the limit of resolution, the central spots can no longer be discriminated. (D and E) Higher resolution is achieved by water- or oil-immersion lenses. Illustrated is the imaging of an object in water (shaded) by a dry objective (D) and a water-immersion objective (E). A dry objective fails to capture higher order diffracted light because of refraction at the water–air interface. (F) High-numerical aperture objectives (right) usually have a shorter working distance (distance from the front surface of the lens to the object plane) than do low-aperture objectives (left).

terms of the minimum spacing between two infinitely small objects (conveniently approximated in practice by 50-nm-diameter latex beads or by the fine structure in diatoms) that still allows them to be discriminated (Fig. 6C), leading to the classic Rayleigh equation:

$$D = 1.22/(\lambda/2\text{NA})$$

where D is the minimum spacing and λ is the wavelength of illuminating light.

Second, the NA determines the light-gathering power of an objective. This is of particular importance for fluorescence microscopy, in which the fluorescence emitted by a specimen is usually dim, but cannot be increased merely by raising the intensity of the excitation light because of problems of photobleaching and phototoxicity to live cells. Each fluorescent molecule emits photons randomly in all directions. It is not possible in practice to collect more than a fraction of these, but the proportion that can be collected through an objective increases in proportion to the square of its NA.

The use of objectives with high NA is, therefore, almost always a good thing; but this does not come without cost. One problem is that the nature of the medium between the specimen and objective sets a fundamental limit to the maximum NA that can be achieved. For example, when an object in water is imaged by a "dry" objective, light rays at increasing angles from the optical axis (containing the higher orders of diffracted light) are refracted outward at the water–air interface, so that they escape the objective (Fig. 6D). Higher apertures can be achieved with an immersion medium (water or oil) between the specimen and objective (Fig. 6E), although this can get messy, particularly with inverted microscopes, with which there is a danger of oil running down into the interior of the lens. (Tricks to avoid this include use of a low-viscosity oil, and placement of a rubber O ring around the barrel of the lens as an oil trap.) Maximum practicable numerical apertures are about 0.9 for dry objectives, 1.2 for water immersion, 1.45 for oil immersion, and 1.65 when utilizing special oil and a coverglass with high refractive index. A second disadvantage is that a high aperture is usually achieved, in part, by minimizing the working distance between the front element of the objective and the object plane (Fig. 6F), which may restrict use with thick specimens and interfere with placement of microelectrodes under the lens. Finally, as noted earlier, problems of spherical and chromatic aberration increase with increasing lens aperture. These can be corrected, but involve more complex designs, so that high-NA lenses are generally more expensive.

Correction of Lens Aberrations. Any reputable objective lens should achieve diffraction-limited performance in the center of the field at a single wavelength, but outside of this limited specification there are several aberrations that may be corrected to different extents. The cost of objectives rises as a steeply nonlinear function of the degree and number of corrections that are applied. These are usually marked on the barrel by abbreviations (which, confusingly, differ slightly between manufacturers). Major corrections may include the following:

Chromatic aberration: Achromatic (Achro), fluorite (Fluor), and apochromatic (Apo) lenses, respectively, provide increasing levels of correction for chromatic aberration.

Field curvature: All three of the above-described objectives project a curved image. For some purposes this may not matter, but for others (e.g., imaging cultured cells on a flat coverslip) it does. Plan objectives provide a flat field.

Spectral transmission: Specialized lenses provide enhanced transmission in the UV (e.g., for Fura-2 imaging) or infrared (IR) (e.g., for IR DIC imaging of brain slices).

Interfacing to Microscopes

Utilizing a regular microscope for biophotonic applications involves getting light into or out of the microscope, often both at the same time. This is facilitated because the manufacturers generally provide several "ports" into the optical path, which, although intended for other purposes, can be readily adapted.

A first consideration is whether a given port provides access to the infinity space of the microscope (e.g., epifluorescence port) or whether it incorporates a tube lens or other focusing optics such that a conjugate image plane is formed (e.g., camera ports). Second, it is best to use a port located close to the objective, because the optical path then includes fewer elements that may absorb or scatter light, and problems of possible misalignment of sliders used to direct light to different ports are obviated. Finally, the use of an inverted microscope is greatly preferable to an upright microscope. Inverted microscopes are equipped with a greater number of ports, which are located near the base of the microscope body at a height convenient for post-mount components fitted to an optical bench.

Several ports may be used at the same time—for example, to introduce UV light for photolyzing caged compounds while simultaneously employing a confocal scanner to monitor signals from visible wavelength dyes. The trick is to arrange so that a dichroic mirror in a port close to the objective will transmit light introduced from more distant ports. For example, a UV-reflecting dichroic placed in the epifluorescence port immediately next to the objective will appear transparent to visible wavelengths passing toward the objective as well as from the specimen into the microscope body.

The epifluorescent port is, in optical terms, the best for providing external optical access to a microscope. It is located in the infinity space, immediately behind the objective, and the availability of sliders or turrets containing several different filter cubes allows the functioning of the port to be readily switched. The use of a dichroic mirror in the port allows, for example, introduction of short-wavelength light for fluorescence excitation while fluorescence emission is viewed through the regular microscope path. Alternatively, a front-surface mirror can be placed in a filter cube so that all light to and from the objective is diverted to the port. A practical disadvantage of the epifluorescence port in inverted microscopes is that it points backward, through the frame of the instrument, so that the infinity space is considerably extended before access is available at the rear of the microscope.

Use of modified side-pointing filter cubes (e.g., the Olympus "laser port") obviates this problem.

The side (video) port of an inverted microscope is perhaps the next best choice. It is located low down on the side of the microscope, and provides convenient access to a conjugate image plane (formed several centimeters away from the microscope body). Light is diverted to the port via a beam splitter cube, which directs all, or a large fraction of, the light to a port. The advantage of a partial split is that the specimen can still be viewed (dimly) through the eyepieces while using the port (e.g., to facilitate alignment procedures), but obviously necessitates some loss of light. Microscopes from some manufacturers incorporate both 100/0% and 85/15% beam splitters, but with other models it is necessary to specify a single type of beam splitter when ordering. Also, some models make provision for a further port, which is accessed from the underside of the microscope body. The 35-mm camera port of an inverted microscope can also be used, but its physical location on the front of the microscope is inconvenient for locating more than small components, and its light path includes intermediate optics to project a magnified real image.

Finally, the phototube of a trinocular eyepiece head again provides access to a conjugate image plane, but its elevated location close to the eyepieces makes interface with external optics inconvenient. However, this may be the only readily available option on an upright microscope, short of machining a custom port that can be sandwiched in the infinity space between the eyepiece head and the microscope body.

Constructing Optical Systems

Although commercial instruments are now available to serve many biophotonic applications, powerful incentives remain for an investigator to construct his or her own optical system. These include the use of new techniques for which no commercial systems yet exist; the ability to adapt an instrument to suit particular needs; the considerable (typically >5-fold) cost saving that can be achieved; and the intimate familiarity gained of both basic principles and practical implementation. The purpose of the following sections is to demonstrate that such a task is easier than might be expected, and to provide practical methods and tips for the biologist without prior optical experience. This is written on the basis of the author's experience in building confocal and multiphoton imaging and photolysis systems,[8–11] but the general principles are applicable to most biophotonic techniques.

[8] N. Callamaras and I. Parker, *Methods Enzymol.* **307,** 152 (1999).

[9] N. Callamaras and I. Parker, *Methods Enzymol.* **291,** 497 (1998).

[10] N. Callamaras and I. Parker, *Cell Calcium* **26,** 271 (1999).

[11] M. J. Sanderson and I. Parker, *Methods Enzymol.* **360,** [19], 2002 (this volume).

Although optical design is a highly complex business, several factors facilitate the task of a biologist attempting to construct an optical instrument. First, most systems involve interfacing external optics to a microscope, where the hard work in terms of optical design (corrections for spherical and chromatic aberrations, flatness of field, etc.) has already been incorporated into the objective lens (which is why lenses can cost many thousands of dollars). What emerges from the microscope port are parallel beams of light that can be imaged without introducing appreciable distortions by simple lenses of low numerical aperture, and without knowledge of more than the elementary lens formulas. Second, a large industry has developed to support the laser and optical telecommunications sectors, so that a vast range of optical (lenses, prisms, mirrors, etc.) and optomechanical (tables, posts, mirror mounts, etc.) components are available "off-the-shelf." Finally, many applications involve the use of lasers, which serve as an almost perfect light source, emitting a narrow, parallel beam of (usually) monochromatic light. Laser beams can be directed at will by steering mirrors, allowing the layout of optical components to be arranged for physical convenience, and can be focused by simple lenses without problems of spherical or chromatic aberrations. Assembling an optical system, therefore, usually involves little more than bolting together standardized parts (a grown-up Lego set), and can be accomplished with only a fairly rudimentary background in optics.

The following sections deal with (1) photonic components for generating, manipulating and detecting light (lasers, lenses, photomultipliers, etc.), (2) optomechanical components (mounts, tables, etc.) for assembling the optical components together in the correct alignment, and (3) an example illustrating the design and construction of a simple biophotonic instrument.

Photonic Components

Light Sources

CONVENTIONAL SOURCES. Commonly used nonlaser light sources include tungsten–halogen bulbs and arc lamps. In all of these, light is emitted in all directions from a source of finite size (e.g., a filament array covering a few square millimeters in a halogen bulb, or a <1-mm sphere in a small arc lamp). Unlike with a laser (which behaves as a perfect point source), it is impossible to effectively collect all of the emitted light or to focus the source down to a diffraction-limited spot. Furthermore, light from conventional sources is polychromatic (spanning a wide range of wavelengths), and is incoherent (light waves are not in phase). Nevertheless, conventional sources are generally preferable to lasers for applications requiring uniform illumination of extended areas rather than the formation of an intense, focused spot.

Halogen lamps (Fig. 7F) emit a smooth, continuous spectrum, with most energy in the infrared and progressively less at shorter wavelengths. They provide little

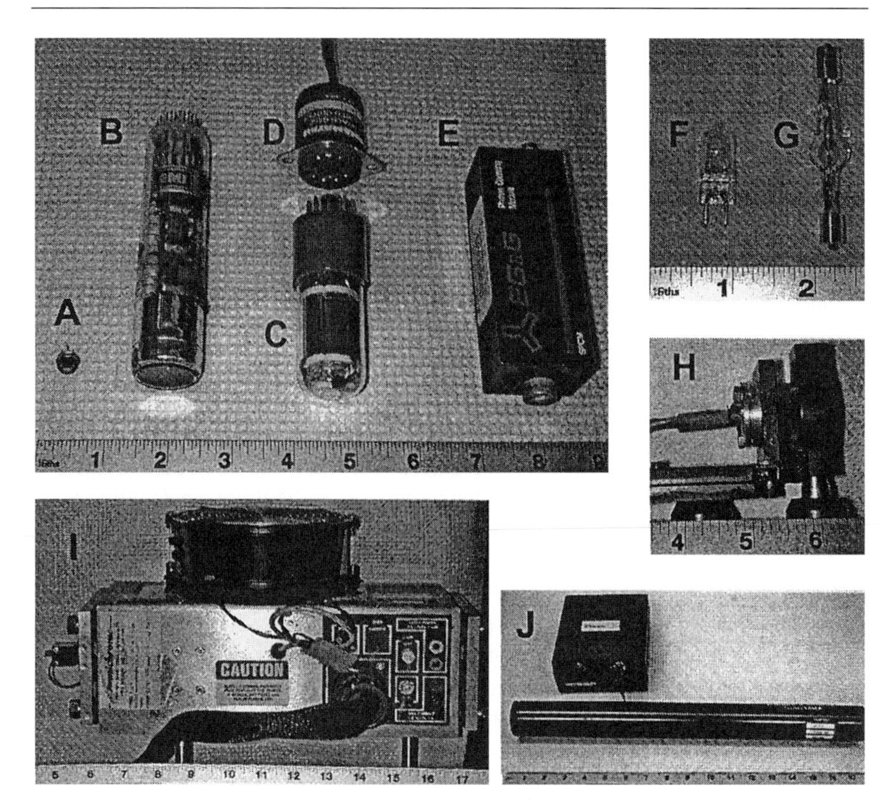

FIG. 7. Detectors (A–E) and light sources (F–J). Scales are in inches. (A) A PIN photodiode. (B and C) Head-on and side-on photomultiplier tubes. (D) Photomultiplier socket with built-in high-voltage power supply. (E) Avalanche photodiode photon-counting module. (F) Tungsten–halogen light bulb. (G) Xenon arc lamp bulb. (H) Laser diode with collimating lens. (I) Air-cooled argon ion laser. (J) Helium–neon laser and power supply.

useful UV output. Advantages include simplicity, low cost, long life, and the ability to obtain a highly stable output when operated from a well-regulated DC power supply.

Arc lamps work by establishing an electrical discharge in a pressurized gas enclosed in a quartz bulb (Fig. 7G). The arc is smaller and more intense than the filament of a halogen lamp, so that light can be collected more efficiently and focused much more intensely on a small specimen. Also, arc lamps have a strong output in the UV (down to $<$250 nm), and provide a more "white" illumination in the visible. Xenon arc lamps show a relatively uniform spectrum throughout the UV and visible spectra, whereas the output of mercury arc lamps is concentrated in a few sharp peaks (365, 46, 546, and 577 nm). Disadvantages shared by all arc lamps include a less stable output (the arc position tends to wander, and bulbs should always be mounted vertically to minimize this); the short life and high cost of bulbs; and the possibility that bulbs will explode, destroying not only themselves but also the lens and mirror within the lamp-housing! Furthermore, xenon lamps require a high-voltage pulse to "ignite" them, which can damage nearby electronic equipment (amplifiers, computers) if turned on before the lamp.

LASERS. The laser is the key invention that has made many biophotonic techniques possible. In most instances (e.g., confocal microscopy, laser trapping) the important property of the laser beam is that light rays are almost perfectly parallel, allowing the entire output to be focused to an intense, diffraction-limited spot. The fact that laser light is (usually) monochromatic is an incidental advantage; its coherence is sometimes a nuisance, leading to interference effects and laser speckle. Whereas lasers serve as an almost perfect light source from an optical perspective, they may have practical drawbacks including high cost, short lifetime, high power consumption and associated heat generation, and need for cooling fans or water supply. However, continuing developments in laser technology (most notably the availability of diode lasers and diode-pumped systems) now mean that almost all lasers can be operated as "turn-key" devices without requiring any special expertise from the user. Primary considerations in selecting a laser include the desired operating wavelength, required power, whether the beam is continuous or pulsed, and the beam quality. The best quality is known as TEM_{00}, meaning that the light intensity across the width of the beam is distributed as a Gaussian profile. The following list describes some of the types of laser commonly used for biophotonic applications, but is by no means comprehensive.

Argon ion lasers: The argon ion laser has been the workhorse for laser-scan fluorescence imaging for many years. Argon ion lasers produce a bright blue line at 488 nm, which is well matched to many fluorophores [e.g., fluorescein isothiocyanate (FITC), green fluorescent protein (GFP) variants, and numerous green Ca^{2+} indicators). Small air-cooled units (Fig. 7I) are available with powers up to 100 mW, which is more than sufficient for fluorescence imaging. Multiline argon ion lasers produce outputs also at 457 and 514 nm, which can be selected by interference filters. Argon–krypton lasers further add lines at 647 and 676 nm; but they have a poor reputation for reliability. A drawback of ion lasers is their remarkable inefficiency (1 kW of electricity gives $<$100 mW of light), resulting in the need for a cooling fan. This is best located remotely (the space above the ceiling tiles is a good place) to reduce noise and heating, and coupled to the laser through flexible tubing. The laser tube has a finite lifetime (about 3 years under typical operating conditions), which shortens steeply as the laser is operated at higher powers. A solid-state 488-nm laser has been introduced (Sapphire; Coherent, Aubum, CA) as an argon ion laser replacement, and offers much higher efficiency (no cooling required) and a small laser head.

Helium–neon lasers: Helium–neon lasers are simple and inexpensive, with models available giving lines at 543 nm (green), 594 nm (yellow), 612 nm (orange), and 633 nm (red). They are usually packaged as a long (about 1-m) cylindrical laser head, and a small, separate high-voltage power supply (Fig. 7J). Their power output is only a few milliwatts at the shorter wavelengths, but can be a few tens of milliwatts for red He–Ne lasers. Beam quality is excellent.

Laser diodes: Semiconductor lasers producing outputs at several wavelengths in the red and infrared have been available for several years, with powers ranging from <1 mW to >1 W. These long wavelengths are not generally useful for fluorescence applications, but are applicable for techniques such as laser trapping. However, novel diode lasers operating at 400 nm (violet) and 450 nm (deep blue) have become available, and are likely to find numerous biophotonic applications. The light output from all laser diodes is diverging and astigmatic, so special optics (often packaged together with the diode in a module) are needed to generate a well-collimated beam. Advantages of diode lasers include their high efficiency, relatively low cost, and the small size of the laser head (Fig. 7H).

Nd:YAG lasers: Crystals of Nd:YAG (neodymium-doped yttrium aluminum garnet) form a highly efficient lasing medium with an infrared output at 1064 nm, which can be frequency doubled or tripled to provide lasers with outputs at 532 nm (green) and 355 nm (UV). Models using different mechanisms to pump the crystal provide either a continuous beam or a pulsed output (e.g., pulses of about 5 ns in duration at repetition rates up to 20 Hz). The power during each pulse can be high, and can be utilized, for example, for cell ablation or localized photolysis of caged compounds.

LASER SAFETY. Laser systems should be constructed with covers that entirely enclose the beam during normal operation. Use a low optical table so that the laser beam is below eye level, even when sitting. Post warning signs, and restrict access to the room. If a laser beam is exposed, wear appropriate protective eyewear unless you need to see the beam in order to align it. In that case, attenuate the laser power as much as possible. Beams that you cannot see (UV and IR) are more dangerous than those you can, and pulsed lasers (e.g, Nd:YAG) cause greater damage than continuous lasers of equivalent power. Accidents happen most often when the user is tired or frustrated. Years of experience in working with lasers does not confer immunity to accidents; rather, it induces complacence. Finally, although the main danger from the relatively low-power laser beams used in biophotonic techniques is limited to eye damage, attempts to repair high-voltage power supplies (e.g., in argon ion and flash lamp-pumped lasers) can result in fatal electrocution.

The back of a white business card is the standard tool for visualizing laser beams. If the beam is in the UV, it can be seen by rubbing the card with a yellow fluorescent highlighter pen. Infrared detector cards allow visualization of IR beams—but these do not work well and IR viewers are a better solution.

Detectors. Light detectors can work in one of two modes: analog or photon counting. In analog mode the electrical output varies in a continuously graded fashion with the photon flux incident on the detector. In photon-counting mode, a discrete electrical pulse is generated for each photon that is detected. Photon counting is not inherently more sensitive than analog detection, but provides lower noise (analog detection introduces "multiplicative" noise, as the electrical signals produced by successive photons of the same energy vary in amplitude), and is preferable for low light levels. Other important characteristics to consider when choosing a detector include quantum efficiency (the percentage of incident photons that actually generate a signal), dark current or noise (the analog current or false counts generated in complete darkness), relative sensitivity at different wavelengths, and speed of response. Available, highly sensitive detectors are based on both vacuum tube technology (photomultipliers; Fig. 7B and C) and solid state technology (photodiodes; Fig. 7A and E).

PHOTOMULTIPLIERS. A photon incident on the cathode of a photomultiplier dislodges a photoelectron, which is then amplified by acceleration through a cascade of electrodes (dynodes) held at increasing high voltages. The result is that individual photons produce current pulses at the anode that are sufficiently large that essentially no further noise is introduced by subsequent amplification or electrical processing. The photomultiplier output can be used in photon-counting mode by employing electrical circuitry that discriminates signals above a certain threshold, and converts them to stereotyped pulses of fixed amplitude and duration. Alternatively, the anode current can be simply filtered and converted to an analog voltage signal by using a current-to-voltage converter. Advantages of photomultipliers include their large detector area (25 mm in diameter or greater), low dark count (<10 counts s^{-1} on selected tubes; even lower if cooled), wide dynamic range, and ability to cope with high photon fluxes ($>10^8$ counts s^{-1}). Drawbacks include a relatively poor quantum efficiency (typically $<20\%$ and decreasing at longer wavelengths), large physical size, and susceptibility to magnetic fields. Photomultipliers require a high-voltage supply, a resistor chain to provide dynode voltages, and a sensitive current amplifier. However, these are available as modules from manufacturers, and can even be obtained built into the photomultiplier socket (Fig. 7D), so the use of photomultipliers is simpler than might be anticipated. The photomultiplier gain can be readily controlled simply by varying the high-voltage supply; this is a commonly used approach, but not theoretically ideal because an optimal signal-to-noise ratio is achieved only throughout a narrow voltage range.

PHOTODIODES. Light incident on a semiconductor junction generates an electrical current. In regular (PIN) photodiodes (Fig. 7A) each photon produces only a single electron charge, which becomes swamped by noise originating in the diode itself and in the subsequent amplifier. Such photodiodes are, therefore, not suitable for photon counting or for low light level detection. However, they have a high quantum efficiency (60% at 550 nm, increasing toward the infrared), are small and inexpensive, and respond rapidly. Although wide-area photodiodes are available, their noise decreases and speed of response increases as the detector area is decreased, so photodiodes are best suited to applications in which light of moderate or high intensity can be focused down to a small spot.

Avalanche photodiodes combine many of the advantages of photomultipliers and solid state detectors. They work by applying a high reverse bias (voltage) across a specialized photodiode. An incident photon causes a transient breakdown, leading to a large flood of electrons so that the diode provides inherent gain (many electrons from a single photon) similar to a photomultiplier. Avalanche diodes provide the high quantum efficiency of a photodiode together with a low dark count similar to a photomultiplier. They are generally used in photon-counting mode, and offer unmatched sensitivity at low light levels. Complete modules (Fig. 7E) are available that include the diode and associated circuitry: only a low-voltage power supply is required, and the output gives a TTL pulse for each detected photon. Disadvantages of avalanche photodiodes include their small detector area (a few hundred microns for lowest dark noise), and relatively low maximum count rate (about 10^7 s^{-1}).

Optical Components

Mirrors. Use of front-surface mirrors is an absolute requirement to avoid secondary reflections from the front glass surface of a back-coated mirror. However, despite the use of protective coatings, the front coating is delicate; do not touch it with your fingers, and try to minimize the need for cleaning. Silver-coated mirrors are a good general purpose solution, and have reflectivity better than 95% from about 450 nm well into the infrared. For use at shorter wavelengths, UV-enhanced aluminum mirrors reflect down to <200 nm, but their reflectivity is only about 85%. Dielectric mirrors provide the ultimate in reflectivity (>99%) when every photon counts, but are more expensive than metal mirrors, and work only over specified wavelength bands (i.e., separate mirrors are needed for UV, visible, and IR), which depend somewhat on the angle of incidence and polarization of the light.

Spectral Filters. Two types of filters are available that selectively transmit particular wavelengths of light: colored glass filters, (the absorption of which varies with wavelength), and interference filters made by combining many thin layers of dielectric materials to produce constructive and destructive interference in transmitted light. Factors to consider when choosing a filter include the desired wavelength range for transmission, how steeply transmission cuts off outside this range, the peak transmission, and how well the filter blocks wavelengths well outside its pass band.

Colored glass filters work superbly well as long-pass filters. Their transmission at long wavelengths is high (>99%), blocking at short wavelengths is excellent ($<10^{-5}$), and the cut-on (increase in transmission with wavelength) is sharp. They are simple, inexpensive, and available for a wide range of wavelengths and in several sizes. For short- and band-pass applications, however, colored glass filters serve less well. For these applications, interference filters are generally preferable. They allow a precise control center wavelength, bandwidth, and steepness of cut-on and cut-off, and are available with characteristics tailored to match the spectra of many fluorophors. One characteristic to note is that, unlike colored glass filters, interference filters are sensitive to the angle of the incident light, and are thus best used with fairly well-collimated (parallel) light beams. In some instances this characteristic can be utilized for "tilt tuning": shifting the transmission to shorter wavelengths by deliberately tilting the filter in the light path. A further characteristic is that they may display unexpected behavior outside their specified spectral range—for example, a filter designed to transmit at 400 nm may also transmit at 800 nm. To mitigate this latter problem, interference filters are often cemented together with a colored glass filter. Such composite filters should be mounted with the "shiny" side facing a strong light source.

Dichroic mirrors are a form of interference filter specifically designed to reflect a range of wavelengths while transmitting the remainder. They are mostly used to reflect short wavelengths and transmit longer wavelengths, but dichroic mirrors with band-pass and long-wavelength reflection are also possible. Like interference filters, the characteristics of dichroic mirrors depend on angle, and they are usually specified for incidence at 45°. They should be mounted with the correct orientation (reflective surface facing the incident light) to minimize spurious reflections from the back surface of the glass substrate.

Neutral Density Filters. Neutral density (ND) filters made with partially reflecting metal coatings on a glass substrate provide a convenient means of attenuating laser and other light beams with fairly uniform attenuation over a wide (400- to 2000-nm) spectral range. They are usually specified in terms of optical density [OD = −log(transmission): e.g., an OD of 1 will transmit 10% of the incident light, an OD of 2 transmits 1%, etc.]. Filters should be inclined at a small angle to a laser beam, as light reflected back into the laser cavity can cause unstable fluctuations in power output. ND filters are available as individual components (Fig. 8D); but these tend to become covered by fingerprints and are easy to lose. A better solution is to use ready-assembled wheels containing sets of ND filters (Fig. 8F). For example, a unit with two six-place wheels containing filters graduated in steps of 0.1 and 0.5 OD allows selection of a wide range of densities between 0 and 3.0 (100 to 0.1% transmission) in increments of 0.1. Alternatively, continuously varying ND filters are available, constructed both as glass disks, with density varying with rotational angle (Fig. 8E), and as rectangular filters, with density varying linearly with length. Such continuous filters are most useful when a smooth control of intensity is needed, without requiring precise knowledge of the degree of attenuation.

Lenses. Simple (singlet) lenses are available with both surfaces curved (e.g., equiconvex) or with one flat surface (e.g., planoconvex). The latter is preferable for focusing parallel and near-parallel beams, and should be oriented so that the "work" done in bending the rays is shared between both surfaces (i.e., with the flat surface adjacent to the focal point). Equiconvex lenses are preferred for imaging at magnifications close to unity ($<\times 5$), as their symmetry causes aberrations to cancel. Meniscus lenses (with surfaces of opposite curvature) are used to change the focal length of another lens.

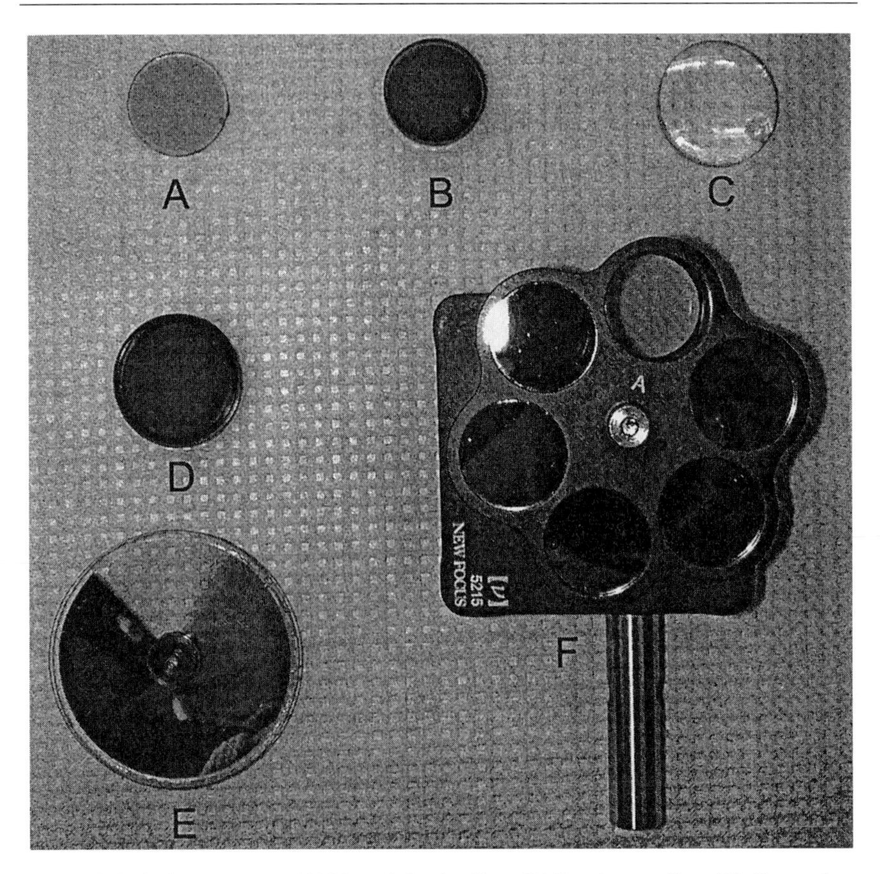

FIG. 8. Optical components. (A) Neutral density filter. (B) Interference filter. (C) Convex lens. (D) Polarizing filter. (E) Circular continuously variable neutral density filter. (F) Six-position neutral density filter wheel.

Aspheric lenses can be made with short focal lengths while minimizing spherical aberration, and are good for high throughput applications, such as concentrating light onto the sensitive area of a detector or collecting the maximum possible light from a lamp. However, their optical quality is generally poor, and aspheric lenses are not well suited to form sharp images.

Cylindrical lenses focus light in only one dimension, and can thus be used to focus a laser beam as a line, rather than a spot.

Achromatic doublet lenses comprise two closely spaced, often cemented, combinations of converging and diverging elements with differing refractive index. Their focal length remains virtually constant across the visible spectrum (i.e., almost no chromatic aberration), and spherical aberration is also much reduced, so that they are superior to singlet lenses even for monochromatic applications.

Types of Glass and Optical Coatings. The extent to which light is transmitted through an optical component (lens, prism, etc.) depends both on how much light is reflected at each air–glass interface, and on how much is absorbed in passing through the glass. In typical use, reflection from an untreated glass surface is about 4%, amounting to an 8% loss in light passing through a single lens. This rapidly compounds in multielement systems and, in addition to the reduction in intensity, multiple reflections increase stray light and reduce image contrast. The use of antireflective coatings substantially mitigates this problem, and most optics can be ordered with broad-band antireflective coatings that reduce reflection to <0.5% per surface over a specified wavelength range (e.g., 375–650 nm).

Absorption of light by the glass substrate is usually not an issue. Regular optical glass (BK7) transmits well at wavelengths from about 350 to 2000 nm, and fused silica can be used for wavelengths deeper into the UV (down to 200 nm).

Cleaning Optics. If an optic needs cleaning, first blow off the dust, which otherwise acts like sandpaper. Then clean it gently with a solvent and lens tissue. Always use lens tissue (not a Kimwipe or clothing), and never use dry tissue, which will scratch. A good solvent is a mix of 60% acetone and 40% methanol—both of high purity. Gentle cleaning is best accomplished by the "drop and drag" technique. Place the dusted optic face up on a clean surface. Lay an unfolded lens tissue on top of it, apply a few drops of solvent, and slowly drag the soaked tissue across the face of the optic. For heavier cleaning, fold a tissue several times, soak it with solvent, and slowly wipe the folded edge across the optic, using fingers or plastic forceps. Do not touch the part of the tissue used for cleaning, and use each tissue only once.

Optomechanical Hardware

The method of construction described and recommended here is based on laying out optical components, using standard mounts affixed to an optical breadboard or table (Fig. 9). This approach offers great flexibility. Off-the-shelf parts are available to cover most eventualities, and components can be readily adjusted, added, or recycled to build new projects. A basic stock of parts is a good lifetime investment, as they will never wear out and are unlikely to become obsolete.

Breadboards and Tables. Optical systems are most conveniently built on some form of optical "breadboard": a flat, rigid metal surface, usually drilled and tapped with a regular grid of mounting holes on 1-inch (or 25-mm) centers, and used to support components on pillar or postmounts (see below). These are available in a wide range of sizes (from a few inches to room-filling) from most optical suppliers. For small instruments (up to about 2 ft^2) a thick sheet of aluminum suffices. However, for larger sizes a thicker construction is used, comprising a honeycomb core (providing high rigidity for low weight) sandwiched between upper and lower

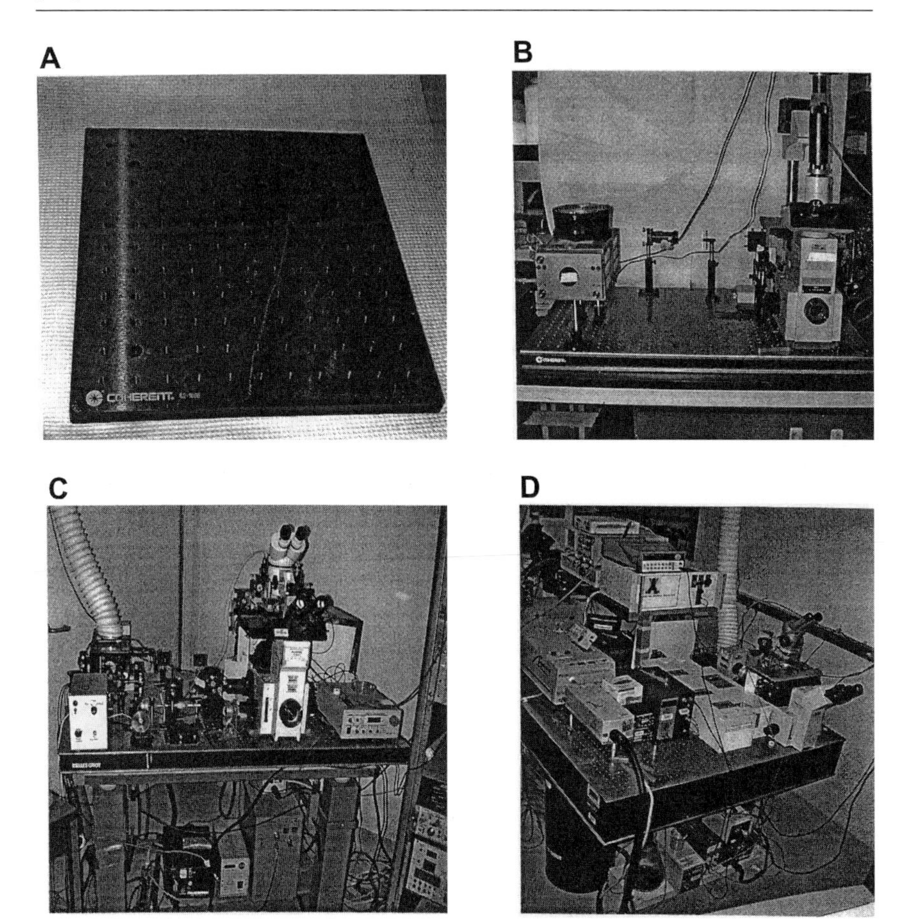

FIG. 9. Optical base plates, breadboards, and tables. (A) Small (1-ft^2) base plate, made from aluminum plate with tapped holes at 1-inch centers. (B) Optical breadboard, used to construct a total internal reflection microscope. Little vibration isolation was required, so the breadboard is simply laid on the laboratory bench. (C) Larger breadboard used as a free-standing table to construct a confocal microscope. (D) Large (8 × 3 ft) optical table with air-suspended feet, used to construct a multiphoton microscope.

steel plates. These are available as relatively thin and light breadboards that can be laid on top of a regular laboratory bench, and as thicker, heavier units that function as independent tabletops. Optical tables with sizes up to about 4 × 3 ft can be put into place without assistance, but larger sizes require professional installation.

The method used to support the breadboard or table depends on the degree of vibration isolation that is needed. In a rigid building without any major vibration source (e.g., nearby traffic, elevator machinery) it may suffice simply to place a breadboard on the laboratory bench, using squash balls as a primitive antivibration mount. The ultimate vibration isolation is achieved by self-leveling pneumatic isolators, which work best with heavier tables. However, these are not without their problems, including a tendency to instability if the load on the table changes (a carelessly rested elbow), or if the load is asymmetrically distributed or has a high center of gravity.

Posts and Pillars. The holders for various optical components (mirrors, lenses, etc.) are typically mounted at a height of a few inches above the optical table. Two main systems can be used to achieve this. The first employs steel pillars: 1-inch-diameter rods, available in a range of preset lengths, which are clamped down with metal forks (Fig. 10A and B). Pillar mounts provide high rigidity, but have the disadvantage that the height of components cannot be readily and continuously adjusted (e.g., to match the height of a microscope port). Post mounts (Fig. 10C, D, and G) are often preferable; the optical mount bolts onto a stainless steel post (0.5-inch diameter rod, available in lengths of 1.5–6 inches) that, in turn, fits into a post holder (1-inch-diameter tube, available in various heights). The post holder can either be screwed directly to a tapped hole in the breadboard or, to allow flexibility in lateral positioning, mounted via a base plate (Fig. 10E and F). Most importantly, a spring-loaded clamp allows the height of the post to be easily adjusted, and then

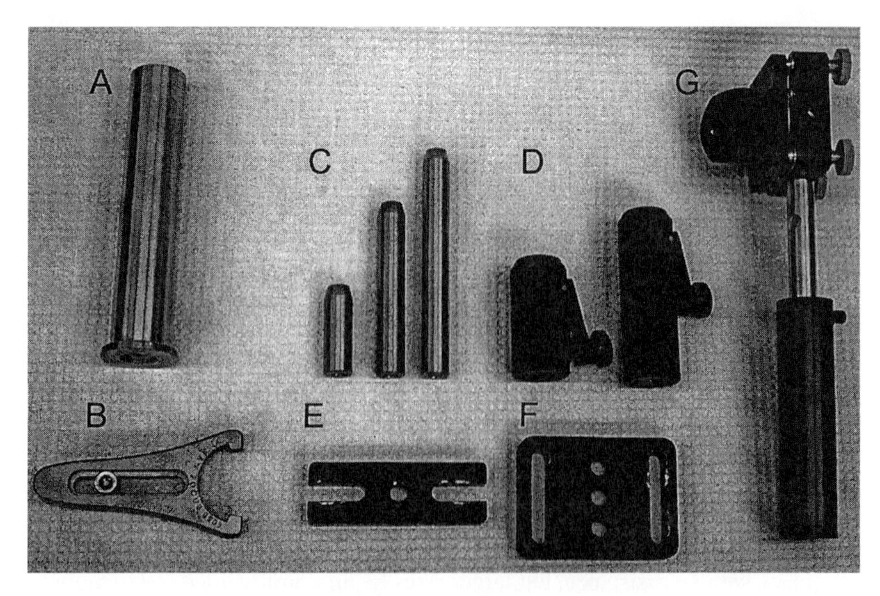

FIG. 10. Pillar and post mount components. (A) Pillar mount. (B) Clamping fork for the pillar. (C) Mounting posts of varying length. (D) Post holders of varying length. Note the spring-loaded clamping devices (lever and screw on right) which secure the post. (E and F) Narrow and wide base plates. A post holder is bolted onto the central hole, and the assembly is then attached to an optical table using the slotted holes. (G) Assembled unit, with a mirror mount attached to a post and post holder.

locked firmly in place. Thus, the height of the optical axis can be set anywhere between about 2.5 and 10 inches above the table top—although it is better kept fairly low if possible, as longer posts are less stable and tend to vibrate.

Imperial and Metric Sizing. Optomechanical components are available in both imperial (e.g., 1-inch hole spacing, 0.5-inch-diameter posts) and metric (25-mm hole spacing, 12-mm posts) sizes. There is no functional reason to prefer one over the other, and when first purchasing components it is simply a matter of deciding on one system and sticking to it. In some cases the sizes are sufficiently close that components are interchangeable; e.g., 1-inch (25.4-mm) base plates will work on tables with 25-mm mounting hole arrays. However, there are many incompatibilities (mounting screw threads are different, and inch posts will not fit metric post holders), so care is needed not to mix up metric and inch catalog numbers when ordering.

Mirror Mounts. Mirror mounts are kinematic mounts that allow the angle of mirrors and other optical components to be precisely adjusted. The most common design (Fig. 11A) comprises a mounting plate supported at three points, two of

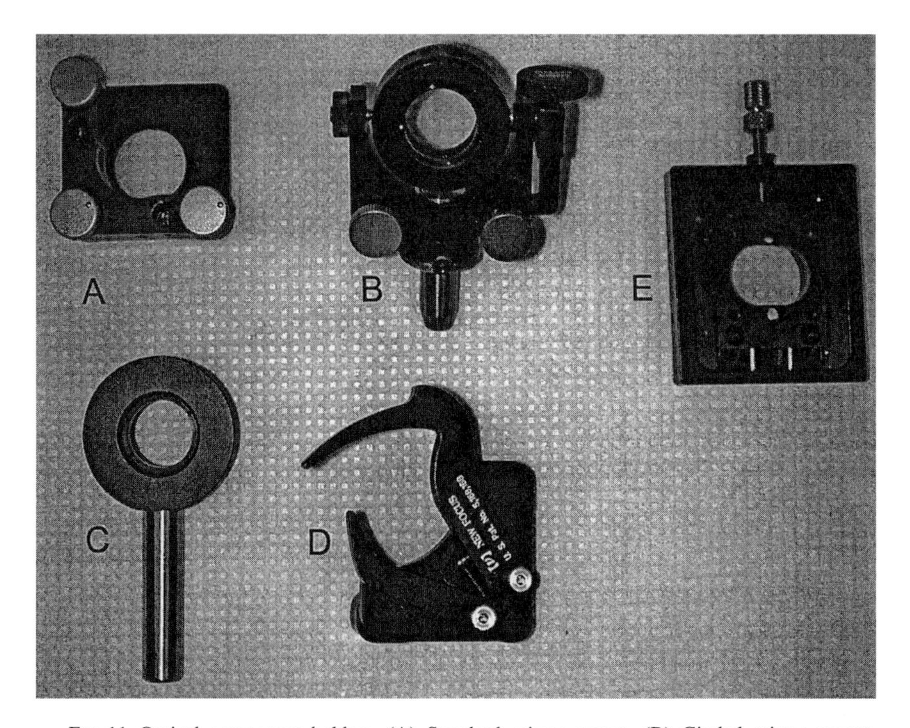

FIG. 11. Optical component holders. (A) Standard mirror mount. (B) Gimbal mirror mount. (C) Mount for a 1-inch-diameter lens or filter. (D) Adjustable, self-centering mount (Opticlaw; New Focus, San Jose, CA), capable of holding components of varying diameters up to 2 inches. (E) Translation stage, providing precise lateral positioning.

which have adjusting screws that independently control vertical and horizontal tilt. The expense of top-quality units is well justified, as these employ fine-pitch screw threads and sapphire seats that promote smooth positioning and long-term stability. Versions with the adjusting screws on top (rather than on the back) allow alignment while a laser beam remains fully enclosed beneath covers, but have more backlash than regular models and are thus best avoided unless they offer a particular ergonomic advantage. Regular mirror mounts introduce a slight translation of the laser beam in addition to rotation (because the mounting plate moves forward as it is tilted). This is not usually a problem, but if it is, gimbal mounts (Fig. 11B) allow pure rotation around an axis in the center of the front plane of the mirror.

Component Holders. Simple component holders (e.g., Fig. 11C) allow the mounting of optics such as lenses and filters with standard diameters (e.g., 1 inch/25 mm). For optics of nonstandard size, the Opticlaw mount (New Focus, San Jose, CA; Fig. 11D) provides an ingenious solution. Other available holders are threaded to accept microscope objectives, and specialized rotary holders can be used with polarizers and prisms. Linear translation slides (Fig. 11E) allow for fine lateral movement of components, and are available with manual and motorized actuators.

In most instances optics can be held in place by the set-screw or threaded ring supplied with a holder, but improvization is sometimes needed. Double-sided adhesive tape works well for components with flat surfaces (e.g., prisms), and Blu-tack (Bostik, Leicester, UK) is a semipermanent, but easily removable, material used to fix lenses and other small components. UV-curing adhesive is the method of choice to cement optical components together. Quick-setting epoxy is good for permanent mounting of optical and mechanical components, but cyanoacrylate adhesive ("Super Glue") is best avoided because its vapor deposits a film on optics.

Scanners. As described above, it is possible to scan a focused spot of laser light across a specimen by varying the angle of a laser beam introduced into a microscope at a conjugate back focal plane. This is most conveniently done with a front-surface mirror that is rotated by a galvanometer. Complete units are available from two major manufacturers [Cambridge Instruments (Cambridge, MA) and General Scanning/Lumonics (Bedford, MA)] and comprise a motor (moving coil or moving magnet) with an integral rotational position sensor, an attached mirror, and a separate driver/servo printed circuit card. Use of a negative feedback servo ensures that a given input voltage results in a precise angle of rotation of the mirror. The speed of response depends on the inertia of the rotating mass, so that higher speeds can be attained with smaller mirrors. Current systems with mirrors of about 3 mm (sufficiently large for most purposes) respond to small step changes within 200 μs, and can be used for linear (sawtooth waveform) scanning at repetition rates of 1 kHz or higher. However, scanning at video rate requires even faster rates (15 kHz), which cannot be attained by linear galvanometer systems. One solution is to use a resonant mirror, which is designed to rotate sinusoidally at a fixed frequency determined by its mechanical construction (a rotational equivalent of a tuning fork). Another approach (although one that appears to have seen little practical implementation) is to use a multifaceted polygonal mirror rotating at high speed.

Scanning in two dimensions (x–y) presents a problem in that two mirrors, rotating around orthogonal axes, are generally required, and both cannot be placed together at a conjugate back focal plane. Various solutions are possible. The simplest is to locate both mirrors as close to each another as possible, with the conjugate back focal plane located midway between them so as to minimize scanning errors. A second is to use relay optics to reimage the conjugate plane from one mirror to the second. A third is to adopt a single-mirror system, in which a small galvanometer providing the fast x scan is mounted on a larger galvanometer so that rotation of the entire x-galvanometer/mirror assembly provides the slower y scan.

Shutters. Uniblitz shutters (Vincent Associates, Rochester, NY) are simple, robust electromechanical devices, available with apertures from 2 to 35 mm. Units with small apertures are suitable for switching laser beams and allow open times as short as 1 ms, whereas shutters with larger apertures operate more slowly but can be used in the illumination or imaging light paths of a microscope. Special driver units are needed to operate the shutter, and can be either "dumb" (with the open time determined by the duration of a TTL pulse) or include built-in timers. Shutter housings carry a threaded hole for mounting on standard optical posts, and can be ordered with adaptors to fit microscopes from all major manufacturers. Shutters blades that face powerful arc lamp or laser sources should be uncoated (shiny steel), but black coatings are available to reduce reflections for low light level applications.

Pockels cells and acousto-optic devices provide means for shuttering and modulating laser beams at rates (microseconds) much faster than is possible with mechanical shutters.

Example: Building a Total Internal Reflection Microscope

This section gives a practical example of constructing a relatively simple biophotonic instrument—a microscope for evanescent wave imaging—to illustrate the processes involved in going from theory to practical implementation.

Theory. When light undergoes total internal reflection at an interface from high to low refractive index (Fig. 1E) it travels a small distance into a medium of low refractive index. This evanescent wave can thus be used to excite fluorescence that is confined to a thin plane immediately next to the interface, thereby providing an optical sectioning effect.[12] The great advantage is that the section is

[12] D. Axelrod, *Methods Enzymol.* **361,** in press (2003).

much thinner (about 30–100 nm) than can be achieved by confocal microscopy. A major disadvantage is that only objects immediately adjacent to the interface can be viewed and, unlike confocal microscopy, the section cannot be focused throughout a specimen. Total internal reflection microscopy (TIRFM) is readily applicable to live cells, because the refractive indices of water and biological tissue are appreciably lower than that of glass.

The simplest way to implement TIRFM is to place a fluorescent specimen on a glass block, and introduce excitation light through the side at an appropriate angle to provide total internal reflection. However, the specimen must then be viewed from above, which both limits physical access and degrades optical quality, because light must pass through the thickness of the specimen before reaching the microscope objective. A more elegant approach is to utilize excitation by an off-axis beam introduced through the objective lens of an inverted microscope. This is possible using oil-immersion objectives with high NA, so that the light rays from the periphery of the lens make an angle sufficiently shallow to undergo total internal reflection at the interface between a cover glass and aqueous medium. Fluorescence excited within the evanescent layer can then be viewed through the objective (as with conventional epifluorescence microscopy; Fig. 5) without passing through intervening tissue, and the space above the specimen is open for manipulation (positioning of pipettes, addition of drugs, etc.).

Optical Design. For through-the-lens TIRFM laser light must be introduced at the periphery of a high-aperture objective lens so as to produce Kohler illumination of the evanescent wave at the specimen. Figure 12A shows an optical layout to achieve this result, with the light path of illuminating rays shown shaded. A parallel laser beam is focused by lens L1, and reflected by a dichroic mirror in the microscope epifluorescence port to form a spot at the back of the objective lens. The beam is directed onto the dichroic by mirror M3, which is mounted on a linear translation stage so that the position of the spot can be moved across the back focal plane and thereby change the angle of incidence of light emerging from the objective to achieve optimal conditions for TIRFM. Fluorescence emitted from the specimen is collected by the objective and imaged through the microscope eyepieces after passing through the dichroic mirror and a long-pass barrier filter to block laser light. Lens L1 acts as a tube lens, so that a conjugate image plane is formed behind the lens. An aperture placed at this point thus appears in sharp focus through the eyepieces, and can be used to restrict the illuminated area in the specimen. The aperture needs to be uniformly filled by laser light to give even illumination, but the beam emerging from the laser is too narrow to provide a sufficient field. Thus, a beam expander is used to fill the aperture and provide a usefully large illuminated area within the microscope field of view.

Practical Implementation. The key component in implementing this design is a newly available lens from Olympus (Plan Apo ×60, oil immersion, TIRFM)

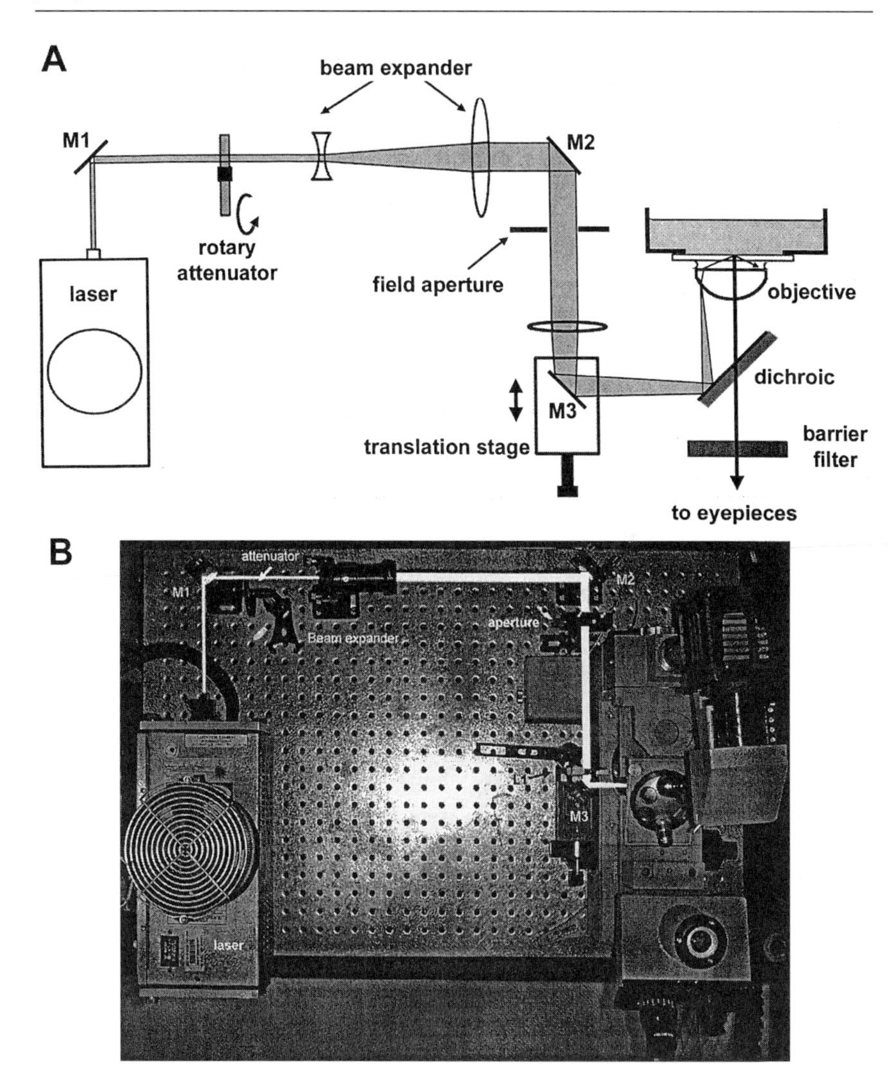

FIG. 12. Schematic layout (A) and final design (B) of a through-the-lens total internal reflection microscope. See text for further details.

that provides an extremely high numerical aperture (1.45) while still working with cover glasses and immersion oil of regular refractive index. This is mounted on an inverted microscope, and a dichroic mirror reflecting $\lambda < 500$ nm is fitted in a custom-modified cube mounted in the epifluorescence port so that laser light can be introduced from the side, rather than the rear of the microscope. The cube also incorporates a barrier filter (color glass, 510-nm long pass) to block reflected or scattered laser light from entering the microscope light path. An old Olympus IX50 microscope was used for this project. This was designed for 160-mm tube-length objectives and, in order to allow the modern infinity-corrected TIRFM objective to be used, the Telan (diverging) lens normally fitted in the microscope nosepiece to create an infinity space was removed.

Figure 12B shows a top view of the completed instrument. A 3 × 4 ft optical breadboard is sufficiently large to accommodate the microscope, laser, and other components. To provide working space around the microscope, the laser is mounted to the side with the beam directed by steering mirrors M1 and M2. All optical components, including the laser head, are raised on post mounts to bring the height of the laser beam up to the microscope epifluorescence port. A 30-mW argon ion laser provides more than sufficient power, and a continuously variable neutral density filter in the laser beam allows the intensity at the specimen to be readily controlled. The particular model of laser used provides only a single 488-nm wavelength. If a multiline laser were used a laser line filter would be used to select the desired wavelength. A "ready-made" beam expander provides a fivefold expansion of the laser beam, although this could equally well be constructed (for much lower cost) from two separate lenses. Lens L1 is mounted just before the adjustable mirror (M3) that directs light onto the dichroic mirror, and has a focal length such that the laser beam is focused to a spot at the back of the objective. To achieve correct TIRFM conditions this spot must be positioned near the periphery of the objective back aperture, so the mirror mount holding M3 is mounted, in turn, on a translation stage allowing it to be moved back and forth along the laser beam by a micrometer screw. This motion changes the position at which the reflected beam is incident on the dichroic mirror, and hence the lateral position of the spot at the objective lens.

Initial alignment involves ensuring that the laser beam is centered and parallel to the optical axis of the objective lens. It is easiest to begin with all lenses (beam expander, L1, and the objective) removed. Position the translation stage in the middle of its travel so that it lines up with the dichroic, and place a piece of translucent paper over the empty nosepiece position. Adjust M1 and M2 to center the laser beam on M3, and then adjust the angular position of M3 to center the beam in the nosepiece aperture. Note that rotation of M3 alone changes the angle of the beam going into the microscope, whereas coordinated adjustment of both M2 and M3 can be used to translate the beam (move it from side to side or up and down) without changing its angle. Iterative changes in both these adjustments can then be applied to align the beam so it is both centered and parallel to the optical axis of the microscope, by alternately viewing the position of the beam at the nosepiece and projected onto the ceiling. Once achieved (and this is easier to do in practice than describe!) the beam expander and then L1 can be introduced, with each centered so that the laser beam is not deflected, and the objective lens replaced.

Finally, comes the satisfaction of viewing your first evanescent wave image. A good test specimen is to put a drop of water containing a dilute suspension of 0.1-μm fluorescent beads on a coverslip. Without moving the translation stage, focus on the bottom of the droplet. You should see a central fluorescent area with stationary beads that have adhered to the coverslip in sharp focus, and other beads that are out of focus to varying extents and are dancing randomly in the water by Brownian motion. Now move the translation stage to bring the laser beam to the periphery of the objective. When total internal reflection is achieved the adherent beads will remain bright, but the motile beads disappear except for brief flashes as they encroach into the evanescent layer near the coverslip.

Appendix

Photonics Publications and Web Sites

The following journals offer free subscriptions, and provide an excellent guide to recent developments in techniques and equipment.

Photonics Spectra/Biophotonics International
Laurin Publications
P.O. Box 4949
Pittsfield, MA 01202-4949
Phone: (413) 499-0514
FAX: (413) 442-3180
Website: www.photonics.com

Laser Focus World
Penwell Publishing
P.O. Box 3293
Northbrook, IL 60065-3293
FAX: (847) 291-4816
Website: http://lfw.pennnet.com/home.cfm/

The following provides a wide-ranging introduction to all aspects of microscopy, with helpful interactive Java tutorials and an extensive reference list. See especially the review chapter on Optical Microscopy by Davidson and Abramowitz.

Molecular Expressions Microscopy Primer
Website: http://micro.magnet.fsu.edu/primer/index.html

For a clear introduction, starting from the very basics, look for the following:

Patterns in Nature: Light and Optics
Website: http://acept.la.asu.edu/PiN/rdg/readings.shtml

Vendors

The following lists of vendors are not intended to be comprehensive, but primarily include companies from whom I have purchased products. Full listings are available in the Buyers Guides published by Photonics Spectra and Laser Focus World (see above for addresses), and through the Web sites of these publishers.

General Optical Suppliers

The companies listed below all sell an extensive range of optical and optomechanical components, and their catalogs are a great resource for information on optical design as well as on specific products.

Coherent Auburn Group
Catalog Division, A91
2303 Lindbergh Street
Auburn, CA 95602-9976
Toll Free: (800) 343-4912
Phone: (530) 889-5365
FAX: (530) 889-5366
E-mail: info service@cohr.com
Website: www.coherentinc.com

Edmund Industrial Optics
101 East Gloucester Pike
Barrington, NJ 08007-1380
Toll Free: (800) 363-1992
Phone: (856) 573-6250
FAX: (856) 573-6295
E-mail: sales@edmundoptics.com
Website: www.edmundoptics.com

Melles Griot
Photonics Components Div.
16542 Millikan Avenue
Irvine, CA 92606
Toll Free: (800) 835-2626
Phone: (949) 261-5600
FAX: (949) 261-7790
E-mail: sales@irvine.mellesgriot.com
Website: www.mellesgriot.com

New Focus, Inc.
5215 Hellyer Avenue, Suite 100
San Jose, CA 95138-1001
Phone: (866) 683-6287
Fax: (408) 284-4824
E-mail: contact@newfocus.com
Website: www.newfocus.com

Newport Corp.
1791 Deere Avenue
Irvine, CA 92606-4814
Toll Free: (800) 222-6440
Phone: (949) 863-3144
FAX: (949) 253-1680
E-mail: info@newport.com
Website: www.newport.com

Spindler & Hoyer Inc.
459 Fortune Boulevard
Milford, MA 01757-1745
Phone: (508) 478-6200
FAX: (508) 478-5980
E-mail: info@linos-photonics.com
Website: www.spindlerhoyer.com

Thorlabs Inc.
435 Route 206 North
Newton NJ 07860
Phone: (973) 579-7227
FAX: (973) 300-3600
Website: www.thorlabs.com

Optical Filters

These vendors offer specialized filters and dichroic mirrors for fluorescence microscopy. A wide range of colored glass and interference filters is also available from many of the general optical suppliers.

Chroma Technology Corp.
72 Cotton Mill Hill
Unit A-9
Brattleboro, VT 05301
Toll Free: (800) 824-7662
Phone: (802) 257-1800
FAX: (802) 257-9400
E-mail: sales@chroma.com
Website: www.chroma.com

Omega Optical Inc.
210 Main Street
Brattleboro, VT 05301
Toll Free: (866) 488-1064
Phone: (802) 254-2690
FAX: (802) 254-3937
E-mail: info@omegafilters.com
Website: www.omegafilters.com

Scanners

Both of these companies manufacture a wide range of galvanometer-based scanners. GSI also produce resonant scanners.

Cambridge Technology Inc.
109 Smith Pl.
Cambridge, MA 02138
Phone: (617) 441-0600
FAX: (617) 497-8800
E-mail: scanners@camtech.com
Website: www.camtech.com

GSI Lumonics
OPTICAL Scanning Products Group Div.
4E Crosby Drive
Bedford, MA 01730
Phone: (781) 275-1300
FAX: (781) 275-3844
E-mail: scanning@gsilumonics.com
Website: www.gsilumonics.com

Lasers

This list includes only a small fraction of the worldwide laser manufacturers. Some of the general optical suppliers (Melles Griot, Coherent Auburn) sell diode, argon ion, and He–Ne lasers. Evergreen Laser provides efficient, low-cost service and refurbishment for ion lasers.

Coherent Photonics Group
5100 Patrick Henry Drive
Santa Clara, CA 95054
Phone: (408) 764-4983
FAX: (408) 988-6838
E-mail: Tech-sales@cohr.com
Website: www.coherentinc.com

Continuum
3150 Central Expressway
Santa Clara, CA 95051
Phone: (800) 956-7757
FAX: (408) 727-3550
E-mail: continuum@ceoi.com
Website: www.continuumlasers.com

Melles Griot
Laser Division
2051 Palomar Airport Rd., #200
Carlsbad, CA 92009
Toll Free: (800) 645-2737
Phone: (760) 438-2131
FAX: (760) 438-5208
E-mail: sales@carlsbad.mellesgriot.com
Website: www.mellesgriot.com

Spectra-Physics
335 Terra Bella Avenue
Mountain View, CA 94043
Phone: (650) 961-2550
FAX: (650) 968-5215
E-mail: sales@splasers.com
Website: www.spectra-physics.com

Evergreen Laser Corp.
9G Commerce Circle
Durham, CT 06422
Phone: (860) 349-1797
FAX: (860) 349-3873
E-mail: elc@connix.com
Website: www.evergreenlaser.com

Detectors

Hamamatsu and Electron Tubes produce a wide range of photomultipliers and accessories. PerkinElmer (previously EG & G, Inc.) supply avalanche photodiode photon-counting modules.

Electron Tubes, Ltd.
Bury St.
Ruislip, Middlesex
HA4 7TA United Kingdom
Phone: 44-1895-630771
FAX: 44-1895-635953
E-mail: info@electron-tubes.co.uk
Website: www.electron-tubes.co.uk
Sales offices:
Electron Tubes Inc.
100 Forge Way, Unit F
Rockaway, NJ 07866
Phone: (800) 521-8382
FAX: (973) 586-9771
E-mail: sales@electrontubes.com

Hamamatsu Corp.
360 Foothill Rd.
Bridgewater, NJ 08807-0910
Phone: (908) 231-0960
FAX: (908) 231-0405
E-mail: usa@hamamatsu.com
Website: www.hamamatsu.com
E-mail: opto@perkinelmer.com

PerkinElmer Optoelectronics Headquarters
2175 Mission College Boulevard
Santa Clara, CA 95054
Toll Free: (800) 775-6786
Phone: (408) 565-0830
FAX: (408) 565-0703
E-mail: opto@perkinelmer.com
Website: www.perkinelmer.com

Flash Lamps and Photolysis Systems

Chadwick-Helmuth Co.
4601 North Arden Drive
El Monte, CA 91731
Phone: (626) 575-6161
FAX: (626) 350-4236
E-mail: chadwick@chadwick-helmuth.com
Website: www.chadwick-helmuth.com

Cairn Research, Ltd.
Unit 3G
Brents Shipyard Industrial Estate
Faversham, Kent ME13 7DZ
United Kingdom
Tel: +44 0 1 79 559 0140
FAX: +44 0 1 79 559 0150
E-mail: sales@cairnweb.com
Website: www.cairnweb.com

Rapp Optoelektronik
Gehlenkamp 9a
D-22559 Hamburg
Germany
Phone: +49 40 811330
FAX: +49 40 814906
E-mail: info@rapp-opto.com
Website. www.rapp-opto.com

Shutters

Vincent Associates
1255 University Ave.
Rochester, NY 14607
Toll Free: (800) 828-6972
Phone: (716) 473-2232
FAX: (716) 244-6787
E-mail: vincent@frontier.net
Website: www.uniblitz.com

Electronic Components

Digi-Key Corporation
701 Brooks Ave. South
Thief River Falls, MN 56701-0677
Phone: (800) 344-4539
FAX: (218) 681-3380
Website: www.digikey.com

Allied Electronics
7410 Pebble Drive
Fort Worth, TX 76118
Phone: (817) 595-3500
FAX: (817) 595-6444
Website: http://www.alliedelec.com

Imaging Software

Research Systems Inc.
4990 Pearl East Circle
Boulder, CO 80301
Phone: (303) 786-9900
FAX: (303) 786-9909
E-mail: info@researchsystems.com
Website: www.researchsystems.com

Universal Imaging Corp.
One Ridgewood Place
Downington, PA 19335
Phone: (610) 873-5610
FAX: (610) 873-5499
E-mail: sales@universal-imaging.com
Website. www.universal-imaging.com

Machine Tools and Parts

Rutland and Enco sell tools and accessories, from hand tools to industrial size lathes. Small Parts is a wonderful source for metal and plastic stock, tubing, gears, etc.

Airgas-Rutland Tool
2225 Workman Mill Rd.
Whittier, CA 90601-1437
Phone: (800) 727-9787
FAX: (800) 444-4787
Website: www.rutlandtool.com
www.airgas.com

Enco Manufacturing Co.
400 Nevada Pacific Highway
Fernley, NV 89408
Phone: (800) 873-3626
Website: www.use-enco.com

Small Parts Inc.
13980 N.W. 58th Court
P.O. Box 4650
Miami Lakes, FL 33014-0650
Phone: (800) 220-4242
FAX: (800) 423-9009
Website: www.smallparts.com

Miscellaneous

Microscope Immersion Oils

Cargill Laboratories
55 Commerce Rd.
Cedar Grove, NJ 07009
Phone: (973) 239-6633
FAX: (973) 239-6096
Website: www.cargille.com

Laser Safety Products

Kentek Corp.
19 Depot St.
Pittsfield, NH 03263
Phone: (800) 432-2323
FAX: (603) 435-7441
Website: www.kentek-lasers.com

Acknowledgments

I thank Drs. Isabel Ivorra, Yong Yao, Nick Callamaras, Gil Wier, and Jonathan Marchant for help in building many generations of the laser microscope. This work was supported by a grant (GM 48071) from the National Institutes of Health.

[16] Imaging at Low Light Levels with Cooled and Intensified Charge-Coupled Device Cameras

By PINA COLARUSSO and KENNETH R. SPRING

Overview

The charge-coupled device (CCD) camera has become an essential component of the light microscope, particularly in fluorescence microscopy. The CCD is composed of a large matrix of photosensitive elements (often referred to as "pixels," shorthand for picture elements) that simultaneously capture an image over the entire detector surface. Although fluorescence microscopy is considered a low-light level technique, the emitted light flux may vary over several orders of magnitude. In the past, "low light level" referred to conditions in which a video-rate camera is incapable of capturing an image. The weakest light flux that can be detected by a tube or solid-state video-rate camera is about 10^{-2} lx (or 10^{-3} ft-c), equivalent to the nighttime incident light of a full moon or to the levels observed in polarized light microscopy near extinction. The most demanding applications in fluorescence microscopy may require as much as four orders of magnitude greater sensitivity.

Besides the physical limits of fluorescence, the need for sensitive detectors is driven by the low collection efficiency of conventional wide-field microscopes. The geometric constraints imposed by the numerical aperture of the objective lens as well as the inevitable losses in the optical train of the microscope lead to a loss of 80% or more of the original fluorescence signal. Depending on the properties of the detector, the resultant electronic signal may represent as little as 3% or as much as 16% of the fluorescence emission.

Two types of low-light level devices—cooled and intensified CCD cameras—are widely used for imaging fluorescence emission. It is our intent to inform the reader about the relative strengths and weaknesses of cooled and intensified CCD cameras as well as about some potential future developments in the field of detectors. Nonimaging detectors such as photomultiplier tubes or photodiodes are not considered in this article.

Basic Principles

Electronic imaging with a CCD involves three stages: (1) interaction of a photon with the photosensitive surface, (2) storage of the liberated charge, (3) readout or measurement of the stored charge. The CCDs in modern cooled and intensified cameras contain large arrays of silicon photoconductors (typically 1200×1000) incorporating channels for transfer of the signal to the output amplifier. An individual pixel absorbs a photon in a specialized region known as the depletion layer,

0076-6879/03 $35.00

resulting in the liberation of an electron and a corresponding positively charged hole in the silicon crystal lattice. External voltages are applied to the individual pixels to control the storage and movement of the generated charges. Initially, the pixels act as "wells," storing the charge for a given time interval; the amount of stored charge is proportional to the light flux on the pixel. After the charges are collected, a series of voltage steps is applied in order to shift the charge along the transfer channels to the readout amplifier, where the signal is converted to a voltage.

The charge storage capacity of a potential well in a CCD is largely a function of the physical size of the individual pixel. Most pixels are square in present-day CCDs, and the charge storage capacity of the potential well may be approximated by the area of the pixel in microns multiplied by 1000. Thus, a 4-μm-square pixel will have a charge storage capacity of about 16,000 electrons or holes. This full-well capacity (FWC) determines the maximum signal that can be sensed in the pixel.

Sources of Noise

The minimum detectable signal is determined both by the photon statistical noise and the electronic noise of the CCD. Conservatively, a signal can be discriminated from noise only when it exceeds the noise by a factor of ~2.7, that is, a signal-to-noise ratio (S/N) of 2.7. How low a signal will yield an S/N of 2.7? Even with an ideal noiseless detector there will be inherent noise associated with the signal arising from the random variation of the photon flux. This photon statistical noise is equal to the square root of the number of photons. Therefore, the maximum achievable S/N for a perfect detector is given by $S/S^{1/2}$, equal to $S^{1/2}$; thus, a light flux of eight photons will be required to achieve a S/N of 2.7.

In practice, a detector adds its own noise to the photon statistical noise. The electronic noise in a CCD primarily originates from two sources: dark noise and readout amplifier noise. Dark noise is the random fluctuation in the amount of electronic charge that is accumulated in a potential well when the detector is in total darkness. Such dark charge accumulation is a consequence of thermal excitations that liberate electrons or holes. Dark noise simply cannot be subtracted as it represents the uncertainty in the magnitude of the dark charge accumulation in a relevant time interval. Dark noise has been drastically reduced in modern CCDs by the storage of holes rather than electrons, as there is a far greater likelihood of thermal generation of electrons than holes. In addition, cooling the detector reduces the accumulation of dark charge by an order of magnitude for every 20° of temperature reduction. Cooling to −30°, or even to 0°, reduces dark noise to a negligible quantity in most fluorescence microscopy applications.

Readout amplifier noise is the other major electronic noise source in a CCD and the major concern in most cooled CCD cameras. Readout noise may be thought of as a "toll" that must be paid for converting the accumulated charge into a voltage. Because it is noise, its magnitude cannot be precisely determined but only approximated by an average value. Readout noise increases with the speed of measurement of the accumulated charge in each pixel. Higher speeds require higher amplifier bandwidths and, inevitably, higher amplifier bandwidth is associated with greater noise. Cooling the CCD also helps to reduce readout amplifier noise, but there is a bottom limit below which overall performance is reduced. Equipping the CCD with multiple output taps, so that more than one readout amplifier is used to extract the image information, is an increasingly common strategy for achieving higher speeds with reduced amplifier bandwidth and, hence, noise. The image information is then obtained in the form of blocks that are later stitched together by processing software.

An additional benefit of cooling the CCD is an improvement in the charge transfer efficiency (CTE). Each time a packet of charge is shifted along the transfer channels there is a possibility of leaving some behind. When this occurs, the image is less crisp because the charge from multiple pixels is inadvertently admixed, leading to blurring. The large format of modern CCDs necessitates a high CTE to prevent the loss of a significant magnitude of the charge from pixels that are far from the readout amplifier. The farthest pixel undergoes thousands more transfer steps than the pixel closest to the amplifier. Typical CTE values in a cooled CCD are 0.9999 or greater and charge loss is negligible. When the CTE is lower than 0.999, such loss occurs and the region farthest from the output amplifier appears dimmer than that near it and the camera is said to exhibit "shading."

Quantum Efficiency

In addition to readout noise, quantum efficiency (QE) is an important determinant of the minimum detectable signal in a CCD camera. QE is a measure of the likelihood that a photon of a particular wavelength (and therefore energy) will be captured in the detector. If the photon is not detected because it never reaches the depletion layer or passes completely through it, no signal is generated. CCDs typically used in fluorescence microscopy can detect photons with wavelengths between 400 and 1100 nm.

The latest generation CCD sensors have a QE of 70% in the green, so 7 of every 10 incident photons are detected in this range of the visible spectrum. If the readout noise is relatively high, the charge generated by a detected photon cannot be discriminated from the amplifier noise and the signal goes undetected. Assuming a detectable signal must be about 2.7 times larger than the noise, a CCD camera that has a readout noise of 10 electrons per pixel plus signal-dependent noise given by $S^{1/2}$ requires a minimum detectable signal of 35 electrons. Because present-day CCD cameras utilize unity gain readout amplifiers, each photon that is detected results in liberation of a single electron or hole. If the QE is 70%, 50 photons must impinge on a single pixel before a signal can be identified with certainty. If a video-rate CCD camera is used, the readout noise is much higher, ~150 electrons per pixel, and the minimum detectable signal is then about 450 electrons per pixel or 643 photons per pixel.

Additional strategies are employed to improve QE in both the visible and ultraviolet wavelength ranges. The charge transfer gates that lie on the surface of the CCD sensor are not perfectly transparent and absorb or reflect blue or UV light. In the back-illuminated CCD sensor, the detector is flipped over so that light is incident on the back surface, which has been substantially thinned by etching. A QE as high as 90% can be achieved in such back-thinned detectors, but these devices are delicate and relatively costly. Performance of back-thinned CCDs in the ultraviolet wavelengths may be further improved by the use of special antireflection coatings. The addition of wavelength conversion phosphors to the front face of conventionally oriented sensors also has been used to enhance UV sensitivity.

Dynamic Range

Dynamic range is the extent of signal variation that can be quantified by a detector. It is typically expressed in units of 2^N, where N is the dynamic range in bits; a detector that can support $2^8 = 256$ discrimination levels (i.e., gray levels) has a dynamic range of 8 bits. Most CCD camera manufacturers specify the dynamic range of the camera as the ratio of the FWC to the readout noise. Thus a camera with a 16,000 electron FWC and a readout noise of 10 electrons would have a dynamic range of 1600, or between 10- and 11-bit resolution. This specification represents the maximum dynamic range within a scene in which some regions are just at saturation and others are lost in the noise. This is not equivalent to the maximum achievable S/N, a parameter that is also a function of FWC. Because the FWC represents the maximum signal that can be accumulated, the associated photon statistical noise is the square root of the FWC, or 126.5 electrons for an FWC of 16,000 electrons. The maximum S/N is given by the signal, 16,000 electrons, divided by the noise, 126.5 electrons, and is equal to 126.5, the square root of the signal itself. Electronic noise, as well as stray light, will decrease the maximum achievable S/N to values below the value of 126.5 in our example as they both diminish the effective FWC by filling the wells with charge that is not signal.

Image Integration

When the light flux incident on the CCD is low, as is the case in fluorescence microscopy, the period of image integration before readout may be lengthened to accumulate more charges in each well. Image integration on the sensor (on-chip integration) is essential for most applications of CCD cameras, cooled or room temperature, under low light-level conditions. The maximum integration period is a function both of the sensor FWC and of the rate of dark charge accumulation. Ideally, the integration period should be sufficiently long to nearly fill the wells of the pixels in the brightest regions. Most microscopy applications do not require integration periods longer than 30 sec, and dark charge is still a small fraction of FWC for such a time interval.

Careful examination of the image obtained after a long integration period in total darkness often reveals a small number of white spots that represent "hot" pixels. These are individual sensors with abnormally high rates of dark charge accumulation that saturate long before the neighboring pixels exhibit significant signal. The usual practice is to discard the information from the hot pixels by subtraction of a "spot mask" or background image.

Accumulation of charge beyond the full-well capacity may result in "blooming," a situation in which the excess charge spills out into adjacent pixels and contaminates their wells. Most CCDs are equipped with antiblooming capabilities, in which the excess charge is siphoned off and discarded instead of being allowed to spread to nearby pixels. In either case the remaining charge fails to faithfully represent the incident light flux and the upper limit of the dynamic range of the detector is exceeded. The pixel is saturated and the stored intensity information is compromised.

Binning

When the incident light flux is low, the process of binning may improve the *S/N* of a CCD camera. Binning is an operation by which the signals from several adjacent pixels are pooled and treated as if they came from one large sensor (superpixel). Binning involves shifting the charge from a number of adjacent pixels into the output node and delaying the readout until the signal from all of the selected pixels has accumulated in the node. Binning trades off spatial resolution for sensitivity. The improvement in *S/N* comes from the fact that the pooled signal from several pixels is readout only once, so that the relative contribution of the readout amplifier noise is reduced in proportion to the number of pixels that are binned. When 3×3 binning is employed, the signal is increased 9-fold while the noise is one-ninth that of sampling each of the nine pixels individually. The only physical limitation to the number of pixels that can be combined is the charge storage capacity of the serial register and output node of the CCD; both are typically two to three times larger than the FWC of a single pixel. This is not usually a significant concern because binning is employed when the signal is weak and the amount of stored charge does not approach the FWC. The associated reduction in spatial resolution is often a more important limitation to the extent of binning.

Image Intensification

Image intensification is a widely used alternative to prolonged on-chip integration and binning when the image must be obtained in a short time period. Image intensifiers are an integral part of many low light-level cameras, particularly video-rate devices. When the intensifier is coupled to a CCD camera, the device is designated an ICCD (intensified CCD). Image intensifiers are three-stage devices: the first stage is photon detection by a photocathode and the liberation of photoelectrons, the second stage is an electron accelerator and/or multiplier, and the third stage is conversion of the electron beam to visible light or a detectable electronic signal.

Photons impinge on the front surface of a photocathode that absorbs their energy and releases an electron from its back surface. The released electron is accelerated by a substantial voltage and strikes the wall of a microscopic capillary in a microchannel plate (MCP) electron multiplier. The impact of the electron on the capillary surface results in the release of a cloud of electrons that cascade through the capillary driven by a voltage gradient along its length. The capillary is slightly tilted so that subsequent additional impacts with the capillary wall occur and result in a large increase in the number of electrons that emerge from the end of the channel in the MCP. A high voltage accelerates the electron cloud across a small gap to impact on a screen coated with a phosphor that converts the electron energy to light. The process results in an overall light gain of as high as 10^6, although 10^5 is more commonly achieved. Thus, a single detected photon at the input of the intensifier may result in an emission at the output of tens or hundreds of thousands of photons. This output light flux is usually relayed to the CCD sensor by a tapered fiber optic coupler that is designed to collect as much of the emitted light as possible.

The gain of image intensifiers is controlled by variation in the voltage across the MCP and can usually be adjusted over about a 1000-fold range. MCP gain can be adjusted rapidly (within a few milliseconds) and reproducibly so that frequent gain alteration is an integral part of many quantitative imaging systems using an ICCD. When the gain of the MCP is fixed, image intensifiers have a limited dynamic range within a scene compared with a CCD camera. The dynamic range is largely determined by the maximum current that can flow through a region of the MCP. Well-designed ICCD cameras exhibit a 1000-fold dynamic range within a scene and, therefore, just achieve 10-bit resolution. When the gain of the intensifier is adjusted, a much larger intensity range can be accommodated but the limitations still apply for the dynamic range within a scene. Because each detected photon results in the generation of thousands of photons at the output window of the image intensifier, the potential wells in the CCD sensor to which the intensifier is coupled may quickly fill to overflowing with charge. Therefore, unless the CCD employed has large pixels, dynamic range limitations will also arise in the CCD itself. Thus, just a few frames of integration on the CCD sensor may saturate regions of interest and preclude the accurate measurement of their intensity. Often, when integration is performed with an ICCD, the gain of the intensifier is substantially reduced to prevent such regional saturation.

Spatial resolution of an ICCD is always less than that of the CCD alone. The intensification process tends to blur the signal from individual photons and obscure fine image details. The extent of the resolution loss is a function of the size of the MCP capillaries, the geometry of the fiber optic coupling, and the pixel size and number of the CCD sensor. The latest designs utilize 6-μm-diameter MCP capillaries, tapered fiber optic couplers, and CCDs with array sizes of 1000×1000 or greater. The spatial resolution achieved is about 75% of that of the CCD alone.

Gating

The photocathode of the image intensifier can be rapidly turned off or on (gated) by altering its voltage from negative to positive. Gating effectively shutters the intensifier, eliminating the need for a mechanical shutter and allowing the intensified camera to tolerate much higher input light intensities. When the input light intensity is overly bright, gating can be used to reduce the "on" time to as little as a few nanoseconds. Gated intensified CCD cameras are used in spectroscopy and high-speed imaging for sampling the photon fluxes that are obtained during such brief exposures.

Electron-Bombarded Charge-Coupled Device

The first intensified video camera, the silicon intensifier target or SIT, utilized a photocathode and large accelerating voltage to generate energetic photoelectrons that impacted directly on the target of a silicon diode vidicon tube. Because of the high energy of the electrons, each electron impact generated many charge pairs in the silicon diode target. The SIT differed from the image intensified cameras that followed because the electrons were not used to generate photons by impacting on an output phosphor but instead acted directly on the detector surface. The electron-bombarded charge-coupled device (EBCCD) is the modern day, solid-state embodiment of the SIT. The photosensitive surface is a photocathode as in the SIT and image intensifiers, and the photoelectrons are accelerated by a large voltage, but they impact onto a back-thinned CCD, where they liberate a large number of charge carriers.

The EBCCD has several advantages over the old SIT, including faster response, lower geometric distortion, and better uniformity of sensitivity across the detector surface. The spatial resolution of the EBCCD is generally higher than that of a comparable ICCD, but the EBCCD has a substantially smaller dynamic range than a cooled CCD with the same size pixels. Each detected photon generates hundreds of charges, so the potential wells of the CCD fill quickly. As with the ICCD, this limitation can be partially overcome by using a CCD with larger pixels.

Application of Cooled and Intensified Cameras in Fluorescence Microscopy

When a veteran video camera company representative was once asked about the basis for deciding between a cooled or intensified CCD camera for fluorescence microscopy, he replied "It's all a matter of time." Indeed, given sufficient time for integration, a cooled CCD camera will always outperform an ICCD with a comparable detector. However, some biological events occur on such fast time scales and yield so few photons that a useful image cannot be obtained with a cooled CCD camera. The emphasis on the application of fluorescence microscopy to study the dynamics of living cells results in many such situations. The high illumination intensities or prolonged integration periods needed to capture an image with a cooled CCD camera may produce such serious photodynamic damage. How does one decide which detector to use?

The ICCD has the strengths of high speed and sensitivity but the weaknesses of limited dynamic range and resolution, both bit and spatial depth. The cooled CCD has the virtues of high quantum efficiency, large dynamic range, virtually perfect linearity, and excellent spatial resolution. These strengths come at the price of limited speed, both in the period of integration required to capture an image with acceptable S/N and in the time taken to read that image out. The break-even point is determined by the experimental requirements, particularly the speed/sensitivity concern.

A rule of thumb for minimization of photodynamic damage during live cell imaging by reducing the excitation light intensity can be summarized as follows: "If you can see the fluorescence through the eyepieces, the excitation light is too bright." The limiting sensitivity of the dark-adapted human eye is about 10^{-4} lx or 10^{-5} ft-c, a light level that requires an integration time of about 200 ms on a cooled CCD camera to produce an acceptable S/N. If the biological event of interest occurs in less than this time, an ICCD is the appropriate detector. In many fluorescence microscopy applications involving living cells, the concentration of fluorophore in the cell or organelle may be limited either by loading limitations or by interaction of the probe with the process of interest, such as the undesired calcium buffering by the calcium indicator dye. Under these conditions the signal may be so weak that a longer integration time is required for a cooled CCD to capture a useful image, even when binning is employed. Again, the ICCD may prove more appropriate under these circumstances because the gain may be increased to compensate for the weak signal while maintaining the requisite sampling speed.

To illustrate their relative merits, images of the same specimen were recorded with cooled and intensified CCD cameras. Figure 1 shows images of the actin cytoskeleton of a bovine aortic endothelial cell labeled with fluorescent phalloidin. The images were obtained with a spinning disk confocal attachment (Ultraview; PerkinElmer Wallac, Gaithersburg, MD) coupled to an inverted microscope equipped with a $\times 63/1.25$ numerical aperture water immersion objective lens (Nikon, Melville, NY). The left-hand column in Fig. 1 shows the images obtained with an intensified CCD camera with a 1024×1024 pixel sensor producing a 10-bit digital output (ICCD-4000 DF; Video Scope, Dulles, VA). The right-hand column in Fig. 1 shows the images obtained with a cooled CCD camera with a similar sensor size (1280×1024) and a 12-bit digital output (Orca-ER; Hamamatsu,

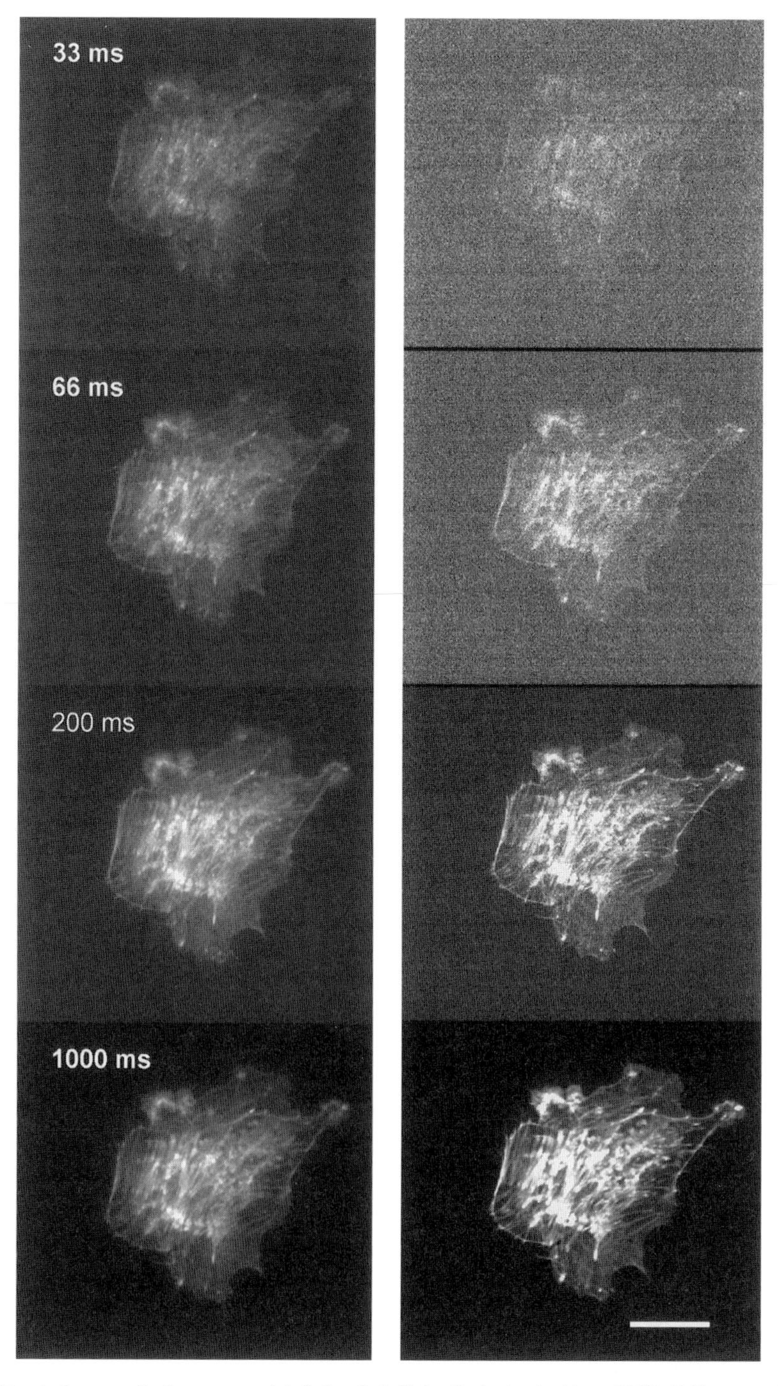

FIG. 1. Images of a fluorescence-labeled endothelial cell obtained with an ICCD (*left*) or a cooled CCD (*right*) at different integration times. Bars 20 μm. See text for details.

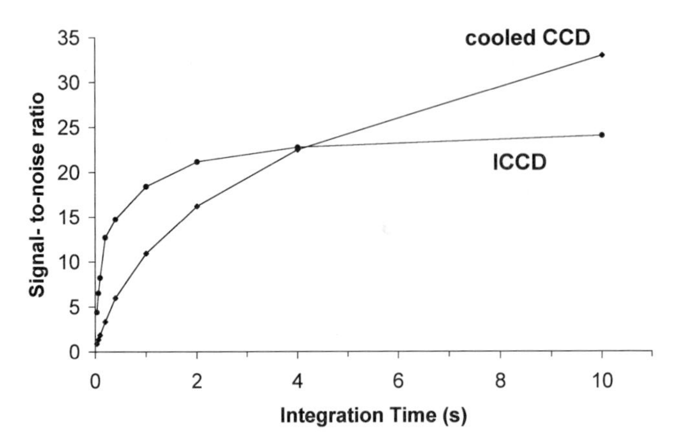

FIG. 2. *S/N* on the ordinate is plotted against integration time on the abscissa for ICCD (filled circles) and cooled CCD (filled diamonds) images of a uniform, featureless field of view. See text for details.

Bridgewater, NJ). The excitation light intensity was reduced to a level comparable to that employed for live cell imaging (approximately 50 $\mu W/cm^2$ at the back aperture of the objective lens). Both cameras were operated at maximum gain.

When the on-chip integration time was 33 ms, the ICCD produced a usable image whereas the cooled CCD did not. The images from the cooled CCD became visually acceptable only at integration times of 200 ms or longer. Although such qualitative inspection of images gives the investigator a sense of whether the information content of an image is sufficient, a far more meaningful measure comes from signal-to-noise measurements.

The graph in Fig. 2 shows the *S/N* for both detectors obtained with dim and monochromatic transmitted light illumination of a uniform field of view. *S/N* can be properly determined only from an absolutely uniform region that is illuminated at a constant intensity, conditions that are readily achieved with transmitted light illumination of a blank field of view. The graph shows that the *S/N* of the ICCD exceeds that of the cooled CCD at all integration times up to 4 sec. The theoretical maximum *S/N* for a 10-bit camera is 32 (the square root of 1024) whereas that of the 12-bit camera is 64 (the square root of 4096). The ICCD achieves a maximum *S/N* of 23, about 72% of theoretical, whereas the cooled CCD has a maximum *S/N* of 33, about 52% of theoretical. In agreement with the visual observations from Fig. 1, the *S/N* of the cooled CCD is unacceptably low (below 3) when the integration period is less than 200 ms whereas the ICCD produces an acceptable *S/N* at times as short as 33 ms.

It is important to note that the dynamic range limitations of the ICCD can make it difficult to capture events that exhibit large variations in brightness, as in specimens containing both thick and thin regions. When imaging calcium in neurons, for example, the cell body may be 40 times thicker than the dendrites, which requires a minimum of 12-bit resolution to simultaneously monitor the fluorescence from both sites. Because ICCDs generally have at most a 10-bit dynamic range, one would have to capture several images at different gain settings with an ICCD. High gain could be used to visualize the small structures and the low gain could be used to capture the image of the thick cell body. With a cooled CCD, however, it would be possible to effectively image both sites at once.

There is a cross-over point in the performance characteristics of both detectors as illustrated in Fig. 2. When light levels are high enough, the *S*/*N* will be determined primarily by the photon statistical noise and not by the detector, and therefore other variables (e.g., resolution and dynamic range) become the primary factors in determining the quality of the image. When light levels are extremely low, as in photon-limited imaging of single molecules or weak fluorescence, camera noise and quantum efficiency are the most important factors.

Future Developments in Imaging Detectors

Two promising developments in sensor technology may prove useful to light microscopists. As discussed in Basic Principles (above), the readout amplifier in a conventional CCD has unity gain and each detected photon liberates a single electronic charge. Many conventional CCD cameras incorporate subsequent amplification stages that allow multiplication of the output of the readout amplifier. This mode of amplification has the disadvantage of increasing both the signal and the readout noise to the same extent. Ideally, the readout amplifier itself should have variable gain, so that the signal can be increased without any amplification of the readout amplifier noise. A new design has been announced for a low-noise charge multiplication stage interposed between the serial register and the readout amplifier of a CCD. Charges generated in the photosensitive portion of the CCD are transferred as usual to the serial register; from there they travel through a specialized region in which a variable number of impact ionization steps may occur. Each of these steps leads to charge carrier multiplication without a significant increase in noise. Thus, the signal that reaches the readout amplifier may be increased up to 50-fold, depending on the gain that is selected, while the readout amplifier noise is virtually unaffected. A single electronic charge resulting from the detection of one photon could result in as many as 50 charges at the output node. For a CCD with a readout noise of 10 electrons, the *S*/*N* would be 5 for a single photon event.

Low noise amplification of the stored charge before readout is preferable to the amplification of the incident light flux that occurs in ICCD cameras or to the electron multiplication that occurs in the EBCCD. In both of these devices amplification of the signal substantially reduces the dynamic range because the charge storage wells are filled at a rate that is directly related to the amplification factor.

Another promising development is the improved performance of the CMOS (complementary metal oxide semiconductor) sensors. These devices have large charge storage capacity, fast readout of regions of interest, and extremely large dynamic range. Several (three to six) amplifiers are required for each pixel and, to date, it has been problematic to obtain uniform images with a homogeneous background because of the inherent difficulties of balancing the gain in all of the amplifiers. CMOS sensors also exhibit relatively high noise associated with the requisite high-speed switching. Both of these deficiencies are being addressed and sensor performance is nearing that required for scientific imaging.

[17] Filters and Mirrors for Applications in Fluorescence Microscopy

By C. MICHAEL STANLEY

Introduction

There have been considerable changes in the world of microscopy since the days of Hooke and van Leeuwenhoek. As researchers constantly search for more details in both the physical and biological world, we have been witness to amazing advances in the technology to observe the very small. One of the most prominent changes in microscopy has been the fairly recent explosion of applications associated with fluorescence microscopy. This article serves as a short summary of fluorescence; applications in fluorescence; and, specifically, the filters and mirrors required for some of the current methods of fluorescence microscopy.

Fluorescence is a phenomenon in which a substance (atom or molecule) absorbs light of a specific wavelength, and rapidly radiates light of another (usually longer) wavelength.[1,2] The process of fluorescence was first observed and documented by Brewster as early as 1838. However, it was Stokes who did much of the early work, and who coined the term "fluorescence." When done properly, and viewed through a microscope cofigured for fluorescence, the result is that the fluorescent substance will be a bright color against a nearly black background.[3] This provided a totally different view through the microscope when compared with transmitted wide-field illumination.

[1] S. C. Watkins, coordinator, "Quantitative Fluorescence Microscopy" (handbook for course), Mount Desert Island Biological Laboratory, Salisbury Cove, ME (2001).

[2] W. T. Mason, ed., "Fluorescent and Luminescent Probes for Biological Activity: A Practical Guide to Technology for Quantitative Real-Time Analysis." Academic Press, New York, 1993.

[3] M. Abramowitz, "Fluorescence Microscopy: The Essentials." Olympus America, New York, 1993.

METHODS IN ENZYMOLOGY, VOL. 360
0076-6879/03 $35.00

Students are routinely taught the basic techniques of microscopy in beginning biology and histology classes. In addition to simple transmitted light morphometry, many of these techniques involve the visualization of specimens that are stained with colorimetric dyes. These dyes can be used to distinguish one type of tissue from another, and even selected/specific regions within a single cell. These dyes work by absorbing specific colors of the white light source used, resulting in a different color being transmitted to the eye/detector. These stains/dyes are used extensively in pathology and histology, as well as a variety of other disciplines as general markers and for identification of gross specimens. Unfortunately, these dyes are not specific or selective.

The advent of fluorescence techniques has provided an avenue for microscopists to improve resolution, sensitivity, and specificity. The current state of the art allows extreme specificity, thanks in large part to Coons (1941)[3a] and his development of antibody labeling with fluorescent probes.[4] The resolution limit of present fluorescence techniques allow for visualization to single molecules, well beyond the diffraction limits of the microscope optics alone. Sensitivity levels allow researchers to observe reactions involving extremely low levels of reagents, as well as their immediate environment (i.e., calcium levels and pH). With developments in optical components, fluorescence also provides a much better signal-to-noise ratio compared with previous techniques. All of these combine to make fluorescence an invaluable tool for the modern laboratory.[5]

The substances that exhibit the properties of fluorescence are called fluorescent molecules, fluorochromes, or fluorophores. The absorption of the initial photon is termed "excitation," and functions to move the fluorophore from the ground state to an excited state. This process occurs rapidly, typically in about 10^{-15} sec. Whether a photon will be absorbed is a function of the specific arrangement of electrons in the fluorochrome. These electron levels represent their quantum states of electronic, vibrational, and rotational energy. The absorption of a photon is an all-or-nothing event. The transition to the excited state is a probabilistic event, and therefore will result in a band of possible wavelengths, not one exact/specific wavelength.[6]

Immediately after absorption of the photon, there is an initial loss of energy to heat, which lowers the electron to an intermediate level. This process requires about 10^{-11} sec. From this intermediate, excited singlet state, there are three principal options available: (1) intersystem crossing, which can lead to a prolonged emission of light called phosphorescence; (2) nonradiative conversion; or (3) emission of a photon as the electron falls back to the ground state (in fact, a variety of probabilistic

[3a] A. H. Coons, H. J. Creech, and R. Jones, *Proc. Soc. Exp. Biol. Med.* **47,** 200 (1941).

[4] X. F. Wang and B. Herman, eds., "Fluorescence Imaging Spectroscopy and Microscopy." John Wiley & Sons, New York, 1996.

[5] B. Herman, "Fluorescence Microscopy," 2nd Ed. Springer-Verlag, New York, 1998.

[6] F. W. D. Rost, "Fluorescence Microscopy," Vol. I. Cambridge University Press, Cambridge, 1992.

ground states are available). We are primarily interested in the third option, the emission of a photon.

This emission, according to Stokes, is "almost always" of a longer wavelength than the absorbed photon. This is due to the energy losses in the conversion, such that the absorbed photon of higher energy (shorter wavelength) is converted to a less energetic form (longer wavelength). This shift/change in the emission wavelength relative to the excitation photon is called the Stokes shift, and may be represented by the mean absorption and emission wavelengths. For example, fluorescein isothiocyanate (FITC) is usually listed with an absorption maximum of 485 nm and an emission maximum of 520 nm. Because of the probabilistic nature of the different available quantum states the absorption is actually a fairly wide spectrum, and the emission is typically a mirror image shifted to longer wavelengths.

In reality, the absorption and emission are both fairly large curves, and the best way to adequately identify these characteristics is with absorption/emission spectra (see Reichman[7]).

Early fluorescence microscopes used transmitted, diascopic light for fluorescence.[7] These scopes were built with the illumination source, and all other optics, in line with the specimen and the eye. Typically there was a light source, lens, a filter to reject all light except that required for fluorescence, condenser, specimen, lens, and a barrier/emission filter for blocking out the excitation light while transmitting the emission of the fluorochrome to the eye. All these elements were in a straight line relative to each other.

Even after the development of the dark-field condenser, placed before the specimen to supply oblique illumination, these designs were difficult to use and produced unwanted signal (i.e., noise). Part of this noise resulted from the fact that even the early illumination sources were tens of thousands of times brighter than the emission signal. The fact that the illumination source was shining straight into the emission optic compounded this difficulty.

Current fluorescence microscopes are typically based on an episcopic design, with the excitation source at a 90° angle to the emission path. Episcopic designs first appeared in the late 1920s. However, it was Blumberg (in the late 1940s), and then Ploem (in 1967), who truly set the stage. Ploem's commercial design required that a special mirror be used at 45°, to reflect the excitation/illumination light to the sample, while at the same time allowing the longer emission light to pass through to the detector (eye). The first mirrors were 50/50 beam splitters, meaning that they reflected 50% of the light that hit them (regardless of wavelength), and transmitted 50% of the light hitting them (also regardless of wavelength). The great advantage of this design was that the excitation light (thousands of times brighter than the emission light) was significantly reduced

[7] J. Reichman, "Handbook of Optical Filters for Fluorescence Microscopy." Chroma Technology, Brattleboro, VT, 2000.

in the emission/detection path. This meant that the signal-to-noise (S/N) ratio was significantly increased compared with diascopic illumination, and the blocking characteristics of the emission filter were enhanced. The resulting image was much brighter, against a darker background.

Later designs of episcopic fluorescence exploited the fact that mirrors could be made (see spectra of dichroics, below) that reflected the excitation light efficiently (+95%) while transmitting more than 90% of the emission wavelengths. This was a tremendous improvement over the early 50/50 designs, which were at best 25% efficient. This new, special optic is referred to as a dichromatic beam splitter (dichroic or simply mirror). This optic revolutionized fluorescence microscopy.

The background for the current popularity in fluorescence techniques is multidimensional. The microscopes, and all other equipment used, had to evolve to a point of highest possible throughput and resolution. The biology and chemistry of fluorescence had to advance considerably and there had to be an accompanying advance in filter technology, well beyond the days of using liquid filters and simple absorption glass.

The current art of interference filters is a process of depositing thin layers of materials onto a substrate. The layers are of alternating high and low refractive index, and in terms of thickness are usually a quarter of a wavelength of light each.[8] The substrates to which these thin layers are applied are either float glass or fused silica/quartz. The individual layers are basically colorless, but the intersection of each layer creates a reflection pattern. These reflections combine through wave interference to reflect specific wavelengths while transmitting different wavelengths. For explicit details refer to Reichman.[7]

Basic Filter Sets for Epifluorescence

The correct choice of filters depends on the fluorochrome used, the light source, and the detector employed. The "typical" filter set is made up of three optics, mounted into a holder/mount. These three optics are (1) the exciter filter, (2) the dichromatic beam splitter/mirror, and (3) the emission/barrier filter.

The primary function of the exciter filter is to block all light from the source, except for a band of light that corresponds to the absorption spectra of the fluorochrome. The light sources are typically mercury or xenon, and both produce a wide band of light from ultraviolet to near infrared, most of which must be prevented from reaching the specimen. Using FITC as an example, the excitation light should be passed through a bandpass filter, which will pass only light from 460 to 500 nm. This optic would be named "hq480/40×" and is shown in Fig. 1, which has 480 nm as the center wavelength (cwl), with a full-width half-maximum

[8] J. D. Rancourt, "Optical Thin Films: User Handbook." SPIE Optical Engineering Press, Bellingham, WA, 1996.

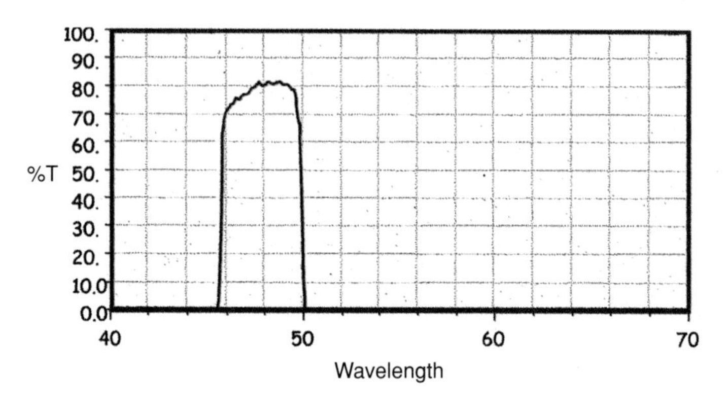

FIG. 1. hq480/40× optic or exciter filter. T, Transmission.

(fwhm) of 40 nm. This optic is composed of layers of dielectrics deposited onto float glass; they are fragile and must be sealed by an epoxy lamination process. The most harmful substance to these materials is moisture, which must be excluded from the hygroscopic layers of the thin film. This optic is used at 0° of incidence, perpendicular to the beam path. There are new developments in cube design that tilt this optic very slightly, 3 or 4°, in order to reduce the chance of internal reflections. This tilt is especially important if using a laser light source (see Laser Scanning Confocal Microscopy, below). Because this optic is not in the emission beam path, at least in episcopic illumination, it is typically not ground and polished, and the only tolerance is that it should have less than 6 arc-min of wedge. However, great care should be taken to ensure that the substrate is pinhole free. There should be no measurable defects, inclusions/omissions, or bubbles in the coating or the lamination epoxy. A tiny hole that allows "other" wavelengths of light to pass will significantly reduce the signal-to-noise ratio within the system.

The second optic in the beam path is the dichromatic beam splitter, also called the mirror or simply dichroic. Because of its complicated function, even veteran microscopists understand this optic only poorly. In a standard epidesign the dichroic has two primary functions: it must reflect the excitation light to the sample, and it must transmit the emission light to the detector. This optic is inserted into the beam path at a 45° angle (angle of incidence, aoi). It is made of quartz or fused silica to minimize autofluorescence because it is in both the excitation path and the emission path. The dichroic is also dielectrically coated with special/specific materials to provide optimal performance, and to reduce any autofluorescence of the materials being used for the thin film. The coating on the mirror is much more durable, and typically does not need to be laminated or sealed from the environment.

To produce an optic that reflects a specific range of wavelengths, and then transmits another range of wavelengths, is difficult. Using the FITC example, this mirror must reflect 460- to 500-nm light (from the source via the excitation filter)

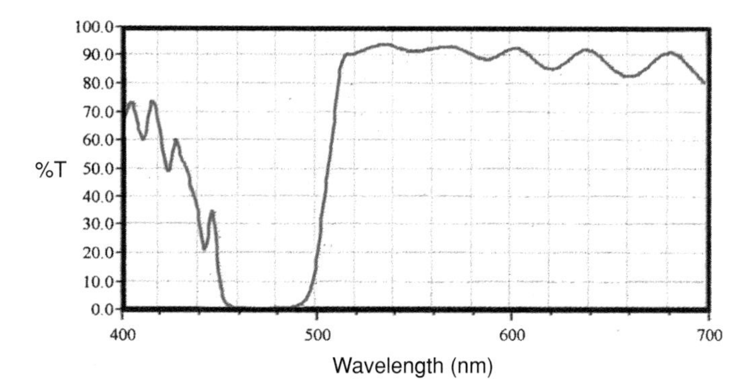

FIG. 2. q505lp mirror.

to the sample. This typically means that the light is reflected either down (upright microscope) or up (inverted design), and therefore changes the excitation light path by 90°. The emission from FITC will be of a longer wavelength, around 525 nm, and it must be allowed to pass straight through the mirror and on to the emission filter.

The dichroic mirror must be made to exact physical specifications, as well as spectral characteristics, because it is in both the excitation and the emission beam paths. This optic should have a surface flatness of less than 10 waves/inch, transmitted wavefront of one wave/inch, and a wedge of less than 1 arc-min. Much more exact standards are not unusual, as noted in the following sections.

It should be noted that one of the confusing issues surrounding the dichroic/mirror is the terminology of naming. Mirrors are named by their 50% cut. A mirror that is called a "505dclp" refers to a dielectric coating that transmits 50% at 505 nm. This means that the mirror is neither fully reflective nor fully transmissive at that point. The most confusing issue is that this mirror does not reflect 505 nm and less, and it does not transmit 505 nm and longer. See Fig. 2 for a typical q505lp design. Notice that this optic is not fully (+95%) reflective until 500 nm and is not fully (+85%) transmissive until 510 nm. Because of certain laws of physics this transition cannot be perfectly vertical. Also note that this mirror does not reflect "everything" below the transition point, as many people believe. A standard, or typical dichroic, design reflects only 30–60 nm, and then it begins to transmit again on the lower wavelength side, as in Fig. 3. The transmission is also not continuous; the mirror will stop transmitting at some point on the long wavelength side as shown in Fig. 4. There are designs for mirrors that do reflect much wider bandpass, called extended reflection, as seen in Figs. 4 and 5. Designs are also available that have extended transmission regions. The naming/terminology of the mirror does add to this confusion somewhat, but the important issue is to know that there are distinct/definitive limits on both reflection and transmission. The design of the mirror can be changed for specific requirements, but only if those requirements are known. Mirrors that are part of a set with specific excitation and

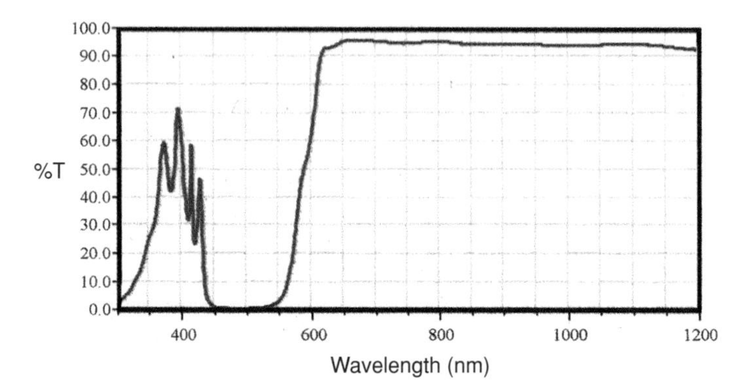

FIG. 3. 585dclp mirror.

emission filters will be self-explanatory. All other mirrors should be specified with needed reflection band and transmission band, and the manufacturers should be allowed to choose the correct design and name.

The third optic is the emission filter, also called the barrier filter. The primary function of this optic is to block out (reject) the light from the excitation filter. The excitation light in a fluorescence microscope can easily be 100,000 times brighter than the fluorescence emission from the fluorochrome. This means that if even a small amount of this excitation light leaks into the detector (either the eye or the camera) the fluorescence signal will be either partially or completely masked. For this reason the manufacturers of the emission optic must know precisely which excitation filter/optic is being used with the emission filter. The emission optic must be made to block the excitation wavelengths to an optical density (OD) of 5.5 or greater.

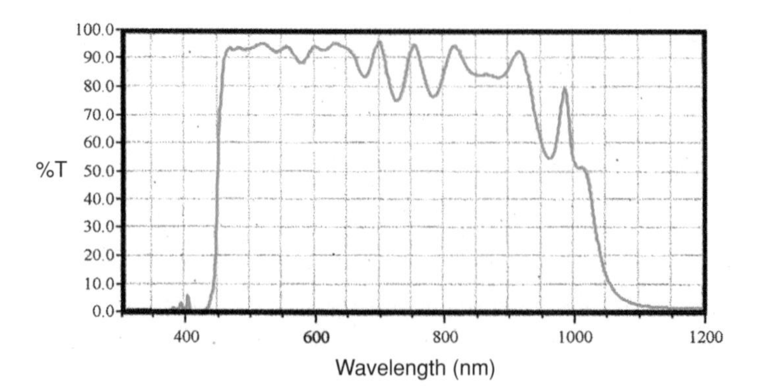

FIG. 4. 450dcxr mirror.

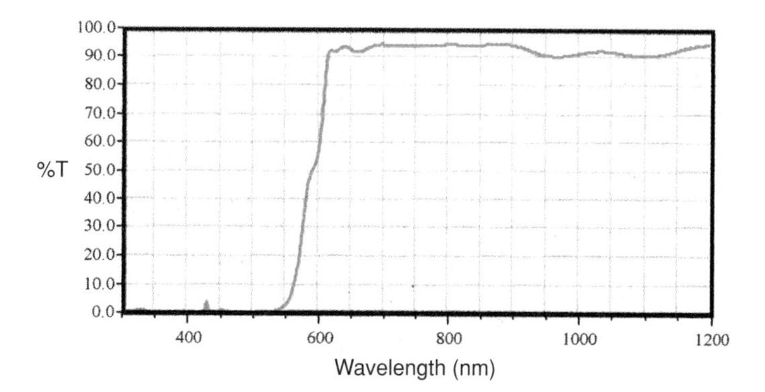

FIG. 5. 585dcxr mirror.

The secondary job of the emission filter is to pass only that light/wavelengths corresponding to the emission spectra of the fluorochrome of interest. From the FITC example, this optic would be a bandpass from 510 to 560 nm, shown in Fig. 6 as 535/50m. There are special applications in which a long-pass design might be required. A long-pass filter, 510lp, would transmit from 515 nm (the 510 refers to the 50% transmission point) to over 750 nm. This does allow for added collection of photons, but it also permits light to pass from either other fluorochromes in the sample and/or from endogenous molecules that may also fluoresce. Typically, high-resolution biological protocols are done with a bandpass emission filter. Emission filters are typically ground and polished with a transmitted wavefront of less than one wave/inch and a wedge of less than 1 arc-min. This optic is now routinely tilted by 2–3° to minimize any internal reflections within the system.

Failure to match the exciter with the mirror, and with the correct emission filter, will result in less than optimal images. In the case of ultraviolet illumination, mismatching the optics may also be harmful to the human eye.

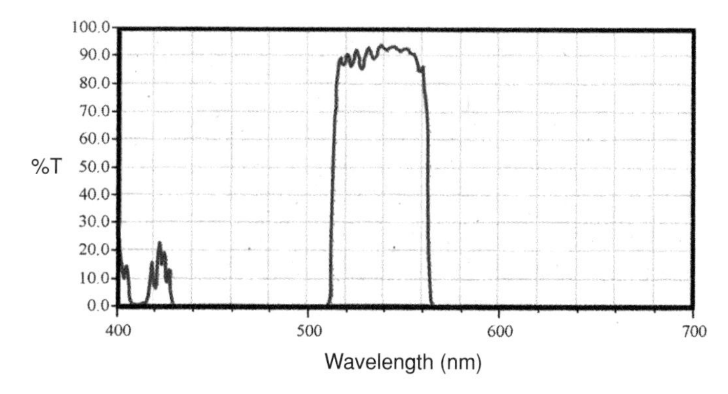

FIG. 6. hq535/50m optic or emission filter.

Green Fluorescent Protein

Green fluorescent protein (GFP) has rapidly become an integral part of biological research. In fact, few applications have ever had the overall effect that has been witnessed by the development and subsequent use of GFP.[9] The green bioluminescence from *Aequorea victoria* is actually the result of energy transfer from the aequorin protein to the green fluorescent protein *in vivo*. Once this acceptor protein was purified and cloned, it became available to researchers as a tag for a large variety of cellular tasks. The protein has 238 amino acids, and is a 27-kDa barrel with the chromophore in the center. The wild-type green protein has a maximum absorption at 390 nm with maximum emission at 509 nm. There is a secondary absorption at about 470 nm.

Genetic mutations have been developed to increase the efficiency of the protein as a fluorophore, and to shift the excitation and emission spectra to different colors/wavelengths. There are four mutants that are commonly available for fluorescence work. These forms have been termed "enhanced" because of their increased quantum yields and stabilities as compared with the wild-type protein.

The primary proteins now available, with absorption/emission maxima, are as follows: (1) blue (eBFP, enhanced blue fluorescent protein): 385 nm, 450 nm; (2) cyan (eCFP, enhanced cyan fluorescent protein): 434 nm, 480 nm; (3) green (eGFP, enhanced green fluorescent protein): 489 nm, 508 nm; and (4) yellow (eYFP, enhanced yellow fluorescent protein): 511 nm, 527 nm.

There is also a fifth mutant protein from the jellyfish, although less commonly used: (5) Sapphire (and also referred to as uvGFP): 390 nm, 511 nm. There is also now a red protein from a sea anemone, *Discosoma striata*. (6) DsRed (RFP, red fluorescent protein): 555 nm, 595 nm.

It should be pointed out that eBFP is difficult to work with in wide-field epifluorescence because of its poor quantum yield and rapid bleaching characteristics. Also, the "yellow" GFP is not yellow to the naked eye. It is yellow shifted from the green form, but with a maximum emission of about 530 nm it appears green to most observers. With fairly wide emission spectra it is difficult to separate CFP from GFP, and impossible to separate GFP from YFP, for imaging. The most commonly used pairs for dual labeling/expression are the CFP/YFP, CFP/dsRed, and GFP/RFP pairs. Some investigators have tried BFP with GFP.

From the standpoint of filters, the GFP mutants are treated like any other fluorophores. The excitation/emission set is designed to best fit with the absorption/emission spectra for the protein, coupled with the appropriate mirror. It should be noted that the proteins can show some variation that may be attributed to pH,

[9] K. F. Sullivan and S. A. Kay, eds., "Methods in Cell Biology," Vol. 58: "Green Fluorescent Proteins." Academic Press, New York, 1999.

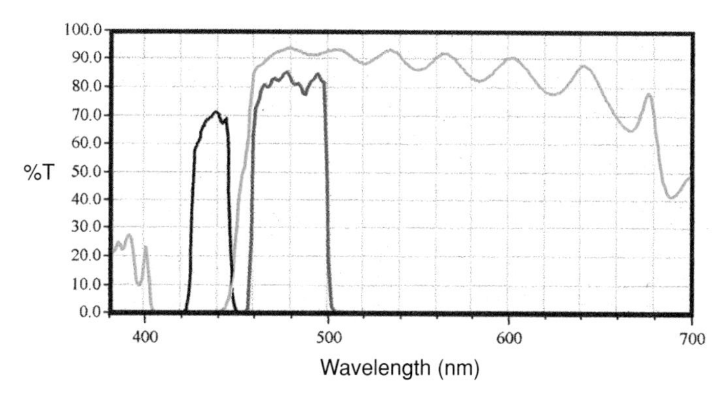

FIG. 7. CFP filter set: d436/20× (black), 455dclp (green), d480/40m (blue).

or to slight mutational effects that may show as emission shifts. Two examples of filter sets designed specifically for the GFP proteins are shown in Fig. 7 (CFP set) and Fig. 8 (YFP set). The enhanced GFP mutant is typically viewed with an FITC filter set, but if expression is low or if photons are limited there is a custom set for GFP (Fig. 9).

Laser Scanning Confocal Microscopy

In 1957 Marvin Minsky, a young postdoctoral fellow at Harvard University, applied for the first patent for a point scanning confocal microscope. He had some interesting ideas that would eventually revolutionize microscopy.[10,11] In summary, a confocal microscope uses a second objective lens in place of the condenser, or uses one objective lens for both condenser and objective. The field of illumination in the confocal is restricted by an aperture, typically called the pinhole. The field of view is also restricted by an aperture/pinhole, which is placed at the image plane conjugate to the illumination point and the first aperture. This "confocal" arrangement of apertures results in an image, at the detector, that is an optical slice/section of the specimen. Therefore a sample that might be several millimeters thick can be reduced to a focus field of microns in the z axis. This configuration results in less out-of-focus light reaching the detector, thereby increasing the signal-to-noise ratio. The original design used a stage that was moved within the beam of illumination light. The modified/modulated light from the specimen was sent to a photoelectric cell for viewing with an oscilloscope. The illumination, and subsequent detection, would occur point by point by moving the stage.

[10] M. Minsky, *Scanning* **10,** 128 (1988).

[11] M. Minsky, United States Patent 3013467. Microscopy apparatus (1957).

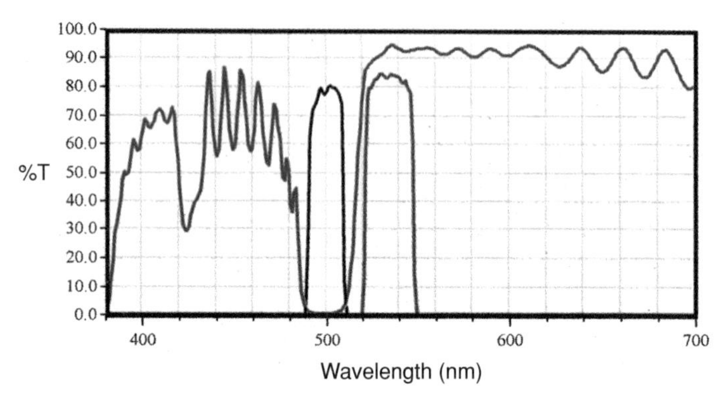

FIG. 8. YFP set: hq500/20× (black), q515lp (blue), hq535/30m (red).

Although the Minsky patent was approved, there were several technical difficulties that had to be overcome. Because of the restricted apertures this system required a bright/intense light source, which was not readily available. Although Minsky used a long-persistence oscilloscope for viewing, there was at the time no way to faithfully record the image. It took other inventors, and further developments, for this special microscope to be useable by the average investigator.

The development of intense monochromatic light sources, lasers, solved the illumination problem. Computers, with digitizer boards, made recording and displaying the image trivial. Therefore, the modern scanning confocal microscope uses a laser, or combination of lasers, for the illumination source. The scan is now done by carefully controlled galvanometer mirrors that move the diffraction-limited spot of illumination across the field in a raster motion, much like a modern

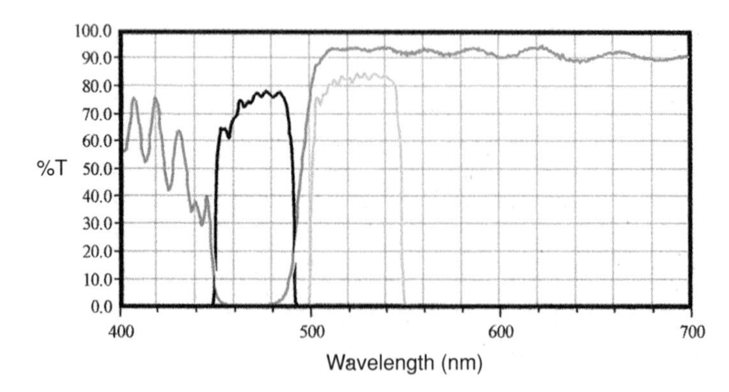

FIG. 9. eGFP filter set: hq470/40× (black), q495lp (dark gray), hq525/50m (light gray).

television, controlled by a computer.[12] The scattered, reflected, fluorescence light from the sample is sent to a photomultiplier tube (PMT), while the image is formed onto a computer screen one point at a time. Please see the reference list for more details of this incredible story, especially Pawley.[13]

Although Minsky listed many advantages of the scanning confocal microscope, perhaps the most important feature is its ability to optically section the specimen.[13–15] Historically, the finest details of a cell, or tissue, were achieved by fixing the tissue and then carefully cutting it into thin sections (1 μm or less) for viewing/imaging. This process required killing the cell/tissue, and the techniques for cutting (microtomy) could take years to perfect. Confocal optical sectioning allows users to image thick tissue without microtomy. It also allows users to image living cells/tissues/organisms with high resolution. Live cell imaging has therefore become an important subset of confocal microscopy.

This system does place some unusual requirements on the optical elements used for fluorescence imaging. A major misconception is that the laser light output from a specific laser produces only one wavelength. In reality, almost all lasers will produce harmonics and/or scatter light. Although these secondary lines may be weak compared with the primary line, they can, nonetheless, greatly reduce the signal-to-noise ratio. If the fluorescence emission happens to be in the same wavelength region as the harmonic (or noise from the laser), the fluorescent signal may be completely masked. Therefore, the first element in the beam path is a laser clean-up filter, which is a modified excitation filter. Because of the collimated; coherent nature of laser light, and the relatively small size of the beam, this optic should be ground and polished. Note from the preceding discussion that a wide-field microscope does not require polishing and grinding. This optic should also have good transmission characteristics, including wavefront distortion of less than one wave/inch. The wedge should be minimized, less than 1 arc-min, so that different clean-up filters may be used in the same beam without realignment. This clean-up filter is typically 10 nm wide (fwhm), and blocks all other light from the laser source (the maximum range would be from UV to 1200 nm) and is coated with antireflective (AR) material for maximum transmission. With the new, more powerful lasers now available, the AR coating may not be necessary. These optics are made by using interference layers with maximum reflection characteristics to avoid thermal damage. For the 488-nm line of an argon laser, this would be a

[12] B. R. Masters, ed., "SPIE Milestone Series," Vol. MS 131: "Selected Papers on Confocal Microscopy." SPIE Optical Engineering Press, Bellingham, WA, 1996.

[13] J. B. Pawley, ed., "Handbook of Biological Confocal Microscopy," 2nd Ed. Plenum Press, New York, 1990.

[14] B. Matsumoto, ed., "Methods in Cell Biology," Vol. 38: "Cell Biological Applications of Confocal Microscopy." Academic Press, New York, 1993.

[15] S. W. Paddock, ed., "Methods in Molecular Biology," Vol. 122: "Confocal Microscopy: Methods and Protocols." Humana Press, Totowa, NJ, 1999.

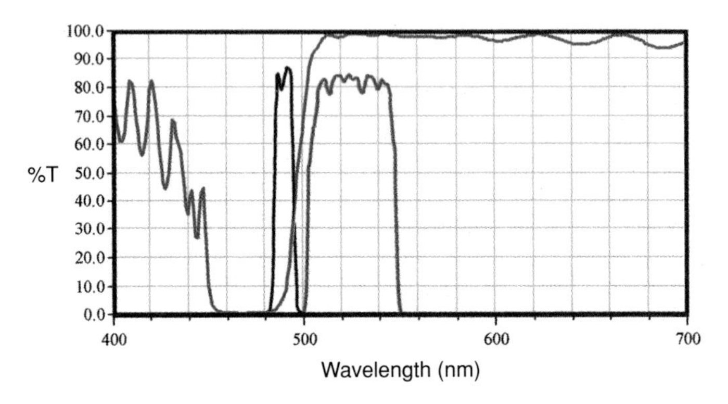

FIG. 10. 488 laser set: d488/10× (black), q497lp (blue), hq525/50m (red).

488/10× as shown in the typical laser set in Fig. 10. The laser clean-up filter should be mounted such that it is at 2–4° from zero. This will prevent reflected laser light from going back into the laser cavity, where it can reduce the life of the laser.

The dichroic in a confocal microscope is also selected for higher flatness specifications, and better wavefront distortion, when compared with wide-field illumination systems. Both are typically specified at better than one wave/inch. As the lasers have increased in size considerably, there are also considerations for the power load applied to the primary mirror. The mirrors are all coated with AR material. Another consideration for all mirrors is polarization effects. All optics in the beam path at an angle can act as polarizers, and this becomes particularly important because most laser sources are polarized. Figure 11 shows a mirror that will reflect more than 96% of a 488-nm laser line if the laser is s-polarized. If, however, the laser is p-polarized, the same mirror would transmit more than 97% of the laser line.

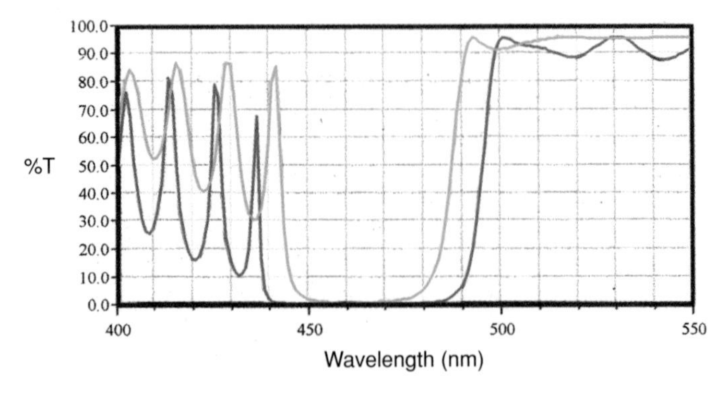

FIG. 11. Polarization effect of mirror: reflects 488-nm laser over 96% in s-polarization (purple, right), transmits 488-nm laser over 97% if p-polarized (green, left).

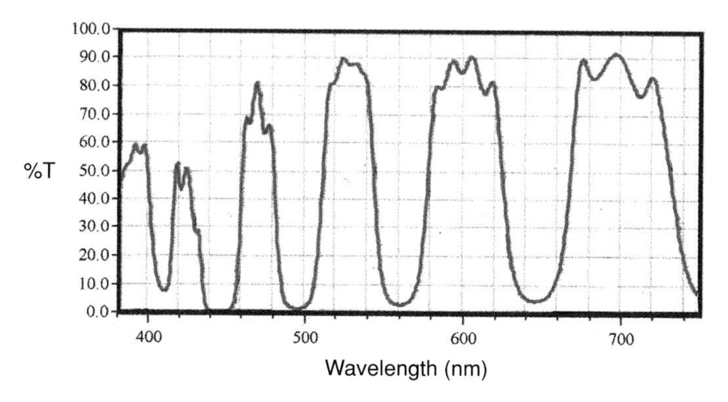

FIG. 12. 488/568/647 polychroic mirror.

As in all fluorescnce, the primary duty of the emission filter is to block the excitation source. In confocal systems this blocking usually does not involve as wide a region of the spectrum, when compared with wide-field sources. However, because of the power output of the laser lines, the blocking may have to be specifically designed to be greater at that particular wavelength. The actual design of the emission filter for a confocal is still somewhat in debate. Because the detector is typically a PMT, in which the image is formed one pixel (picture element) at a time, these optics should not necessarily be ground and polished. However, because of the increased resolution standards placed on the confocal image, most investigators believe that grinding and polishing, as a normal emission filter, is not wasted effort.

Filter sets for laser systems are built around the laser emission lines, and not necessarily for the maximum absorption of a particular dye. This does present several questions as to which fluorochrome(s) may be appropriate for specific laser systems. Fortunately, there are many more choices now available from the manufacturers of dyes. This also has been addressed by the laser manufacturers, as they are constantly developing new lasers for use with the variety of fluorochromes.

There is also the option of using a polychroic primary mirror, for registration or speed, that will reflect more than one laser line. See Fig. 12 for a mirror designed to reflect the 488-, 568-, and 647-nm laser lines simultaneously. It must be noted that any emission filters used with multiple laser lines must be maximally blocked at all laser lines, unless the emitters are used sequentially with only one laser.

Multiphoton Microscopy

The development of practical multiphoton microscopy has dramatically changed optical filter design. Although most microscopy still utilizes one-photon excitation, more and more emphasis is being placed on multiphoton optics. This has created completely new opportunities for filter designers, but it has also created incredible challenges.

Two-photon excitation is a phenomenon whereby two photons, with half the energy and therefore double the wavelength, arriving at a fluorochrome simultaneously may elicit an emission response.[16] If a fluorochrome is maximally excited with light at 500 nm, then two photons at 1000 nm will cause the same effect. This theory has been extant since the 1930s, but the tools to commercialize it are more recent developments.

Traditional one-photon optical systems are built with an excitation filter and a long-pass dichroic, coupled with an emission filter designed primarily to block the shorter wavelength of the one-photon excitation. Single-photon confocal designs use the same set of optics, as shown in Fig. 10.

The advent of two-photon (multiphoton) systems turned this scheme completely around. Multiphoton microscopy utilizes an excitation wavelength that is in the long red or near-infrared (650–1100 nm, usually), which in turn requires a short-pass dichroic design to reflect the longer wavelength excitation while transmitting the shorter/visible fluorescence emissions. These mirrors are not new in optics investigations, but were traditionally used as emission beam splitters only. This new technology also demands that the emission filter block the longer wavelengths to a much greater extinction/optical density, while still blocking the shorter wavelengths.

When laser manufacturers developed tunable lasers, a new and greater demand was placed on the primary mirrors. Now, the short-pass mirror was required to reflect over a broad range of wavelengths (i.e., 715–1100 nm) while transmitting as little as possible. Short-pass mirrors are much more difficult to design and manufacture in any design, but to extend the reflective region requires extreme control. A typical extended reflection short-pass dichroic is shown in Fig. 13. Because the dielectric components used in these mirror designs, the mirror will transmit maximally only down to about 400–420 nm. At that point, the materials will begin to absorb.

Depending on the particular application and wavelength needed for excitation, the mirror must be made with different transitions from reflection to transmission. Compare Fig. 13 (675dcspxr) with Fig. 14 (700dcspxr). The 700dcspxr short-pass design transmits more of the red to the detector, but does not fully (>90%) reflect until a longer wavelength, compared with the 675dcspxr design.

There is also a strong polarization component to be considered. Because most of the primary mirrors/dichroics are used at 45° as the angle of incidence, this optic will be polarizing. The laser itself is also usually polarized, and therefore the

[16] A. Periasamy, ed., "Methods in Cellular Imaging." Published for the American Physiological Society by Oxford University Press, New York, 2001.

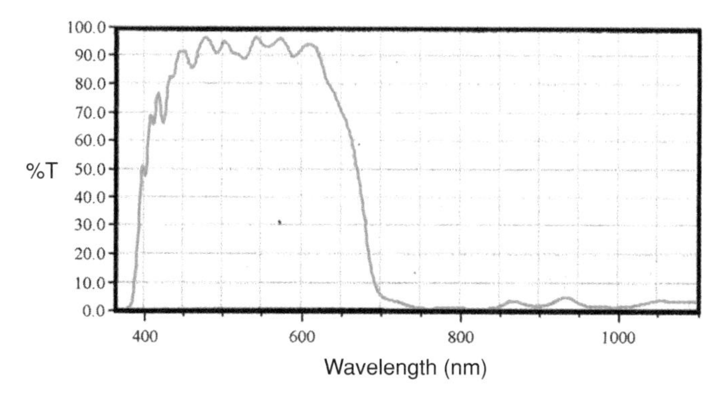

FIG. 13. 675dcspxr dichroic mirror.

designer of the mirror must take this into consideration. Figure 15 shows a typical extended reflection short-pass mirror with s-, random, and p-polarization.

Note that with this design a p-polarized laser will be reflected at 740 nm, but not at 710 nm. This is an extremely important consideration because most mirrors are made and identified using random polarization. This number is calculated with random polarization, unless otherwise specified. Therefore the mirror in Fig. 15 would be called a 700dcspxr, but it will not reflect a laser at 700 nm unless it is s-polarized. Manufacturers can, and do, design mirrors that have specific s- and p-polarization reflections/transmissions. These optics must be carefully designed and manufactured, as a slight change in the dielectric can make dramatic changes in the mirror.

As mentioned above, the primary job of the emission filter is to block the excitation. In single-photon design this puts the emphasis of the blocking range at the shorter wavelength. In multiphoton microscopy this is reversed, with the additional blocking requirements placed on the longer wavelength side. This may

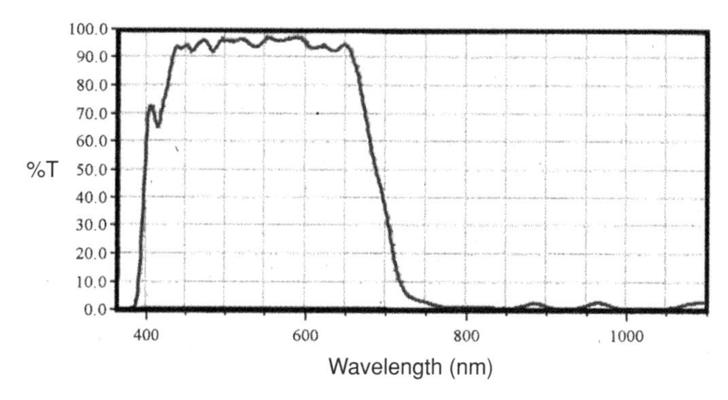

FIG. 14. 700dcspxr dichroic mirror.

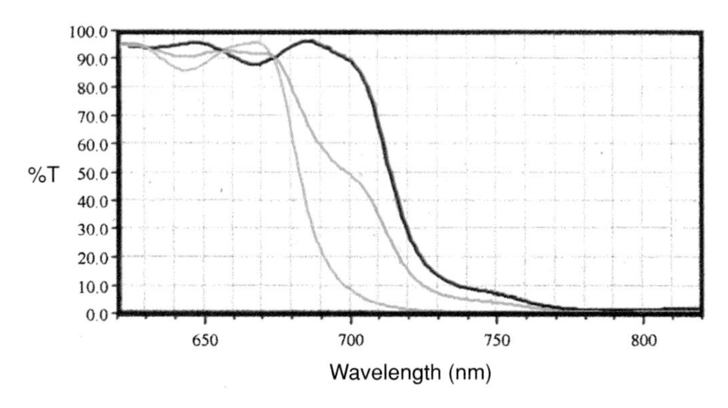

FIG. 15. 700dcspxr mirror with s-, random, and p-polarization (left to right).

sound like a simple difference, but it is not. There is also the problem associated with the large/strong lasers that are typically used now in multiphoton applications.

In a standard epifluorescence microscope, with a mercury or xenon burner, a blocking range of OD 4.5 or greater is adequate. For most single-photon laser systems, with lasers less than 80 mW, a blocking of OD 5.5 is usually fine. However, with some of the new multiphoton lasers the blocking may need to exceed OD 8. For some applications, and with different designs of filters, this can be difficult to accomplish. It is also nearly impossible to actually measure with the current technology. The latest generation of spectrophotometers will measure accurately to only OD 6.5.

Antireflection (AR) coatings are another area of coating design that has changed because of multiphoton microscopy. The AR coatings are applied to increase transmission characteristics, by reducing reflections at the interface. These coatings are wavelength dependent. An AR coating designed to reduce reflections in the UV will not work well in the near infrared (NIR), and vice versa. Therefore, care must be taken in specifying and manufacturing the AR coatings.

There are also applications that require single and multiphoton excitation simultaneously, or at least without changing the primary mirror. Figure 16 shows a custom mirror for reflecting the 488-nm laser (single-photon microscopy), and the NIR laser (for multiphoton microscopy).

Some researchers have reported that the current sapphire lasers used in multiphoton applications may produce light in the red region, along with their NIR emissions. The red light may be in the range of 620–650 nm. This red emission can contaminate the fluorescence emission, and appear as "noise" to the detectors. It was originally thought that this noise was the NIR light leaking into the detectors, even with theoretical OD 8+ blocking with the emission filters. One possible solution is to use rg715 absorption glass in the illumination path of the laser. Other researchers choose to use a laser clean-up filter in the excitation path, such as a d825/25×.

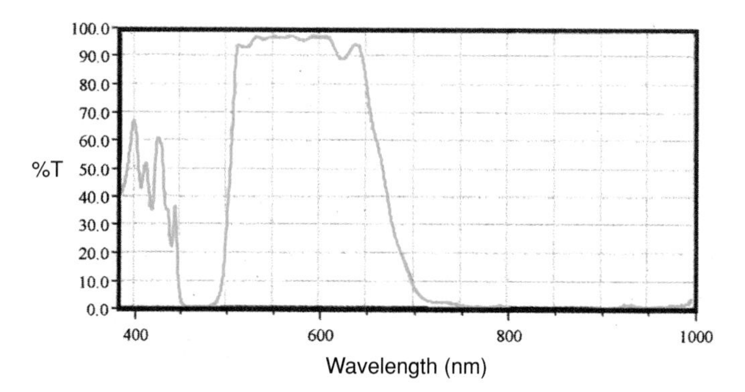

FIG. 16. 488/nir polychroic mirror.

Fluorescence Resonance Energy Transfer

Fluorescence resonance energy transfer (FRET) is a process involving two fluorochromes with overlapping spectra. One fluorochrome must have an emission spectrum that coincides with the absorption spectrum of the second fluorochrome. If the two molecules are close together in physical space, the energy from the first (donor) molecule may be transferred to the second (acceptor) molecule. This is a nonradiative process whereby excitation of the shorter wavelength fluorochrome is transferred to the longer wavelength fluorochrome. If this occurs, the fluorescence of the donor (shorter wavelength emission) will be quenched/reduced, and there will be emission from the acceptor (longer wavelength emission). The emission from the acceptor is via energy transfer, and not by direct stimulation of light energy. This can occur only if the two molecules are within 100 angstroms (Å) of each other. In fact, some believe that they must be less than 60 Å apart.[5] FRET can therefore be used to determine both intramolecular and intermolecular reactions, in both spatial and temporal terms. It can be used to follow and record the interactions of proteins, enzymes, DNA and RNA, all at well below the resolution limits of the optical microscope. Several excellent sources describing the details of FRET are listed in Refs. 5, 15, 16, and 18.

In theory, this sounds like a fairly simple optical arrangement: an excitation filter and dichroic to match the donor molecule, and an emission filter to match the emission of the acceptor, as shown in Fig. 17. The microscopy would involve exciting the donor only, and imaging/recording only for acceptor emission. The presence of acceptor emission, without acceptor illumination/excitation, should indicate FRET. However, it is not quite that simple.

Unfortunately there are many problems with the above-described scenario. One of the first problems to be discovered was that the small signals involved could easily become lost in the noise of the microscope. The next logical step was to ratio the two emissions, instead of trying to measure only the FRET emission.

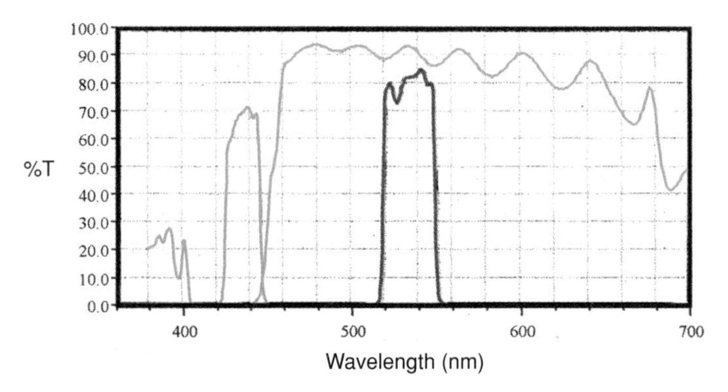

FIG. 17. FRET donor excitation filter (green) and dichroic (blue) with emission filter to match acceptor emission (red) (see text).

The optical filter set for this process is shown in Fig. 18. It should also be noted that any emission filter used in this type of design must be made to block the appropriate excitation band. This division process makes it possible to divide out much of the noise, and enhance the true FRET emission. Although division of two images is considerably more difficult than addition or subtraction algorithms, most morphometry software packages now have a module that allow for ratiometric analysis.

A common pair of fluorochromes serves as an example: two of the mutants of green fluorescent protein, cyan and yellow. The cyan fluorescent protein (CFP) has an absorption maximum of 434 nm, with emission at 480 nm. The yellow fluorescent protein (YFP) has a maximum absorption of 511 nm with emission at 527 nm. The emission curve for the cyan protein overlaps the absorption curve for the yellow protein, which is one of the prerequisites for FRET. This pair has become popular for FRET imaging, and is therefore important in this discussion.

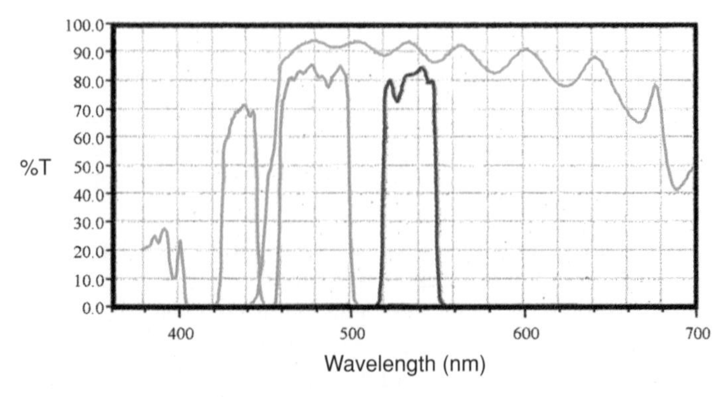

FIG. 18. FRET set: Donor exciter and mirror. Donor emitter and acceptor emitter for ratioing the two emissions.

The simplified version, which some researchers still try to use, is to excite the CFP with light of about 436 nm, and record/image any emission around 530 nm (see Fig. 17). Some believe that the presence of emission at 530 nm indicates an FRET relationship because this is well above the emission maximum of CFP at 480 nm. Unfortunately, it has been shown repeatedly that cyan protein has an extensive emission tail that extends well beyond 550 nm. Although the cyan emission is considerably weaker than the 480-nm emission it can nonetheless obscure the possible FRET signal, which itself is extremely weak in most cases. As mentioned above, the first solution was to ratio the two images/intensities, CFP emission divided by YFP emission (see Fig. 18). This will indeed help, but again there is a complication. Before the division can occur, the contamination of the yellow (acceptor) signal by the cyan emission (true emission) must be subtracted. Because we know that the cyan alone will emit some signal in the yellow emission band (520–550 nm) it must be removed/subtracted. This signal from cyan would not occur if 100% of the cyan were coupled (within FRET distances) of an appropriate yellow molecule. However, this never occurs; there will always be some CFP that is not within this distance and therefore it will react as a normal fluorochrome. It will be excited at 435 nm and it will emit at 480 nm. It will also have that tail, which will show up in the yellow emission window of 520–550 nm. Subtraction of this cyan emission signal from the total FRET-plus-cyan signal seems simple and straightforward, until it is discovered that this contamination can be considerably stronger than the FRET signal.

Another major issue in wide-field microscopy is the inherent noise within the system. Mercury light sources are notoriously noisy, and are considered by many as unacceptable for ratio imaging. Yet this is the most common light source currently being used. Xenon light sources are considered more stable and less noisy, but do not offer the advantages of energy spikes that are present in the mercury source. There is also the rest of the noise produced by the microscope itself. Even when properly aligned and cleaned there will be several percentage points of noise in most microscopes, before the signal reaches the camera/detector. If the measured ratio for a common FRET relationship is about 15%, it would be easy to lose much of that in the noise of the microscope, making reliable statistics unlikely.

Another issue is motion. Because the two images must be ratioed, which is typically done on a pixel-by-pixel basis with a software package, there cannot be any movement in the interval between acquisition of the two images. Of course, the best way to avoid this is to acquire both images at once with a dual detection system. This is not always practical. The next best case is to use a fairly rapid emission filter wheel. Acquire the image/data from the donor (CFP) while exciting the CFP, and then move the wheel to acquire the image from the acceptor channel (YFP emission) as quickly as possible. It must be kept in mind that much of this work is done on live cells, in aqueous solution, which itself presents other issues involving motion. Also, the acquisition time of the camera/detector must be short enough to minimize any inherent or brownian movement. This means that the camera must be sensitive enough to collect the photons, and that enough photons are present at the detector to form the image. Meanwhile, the microscope along with all attachments must be kept stable. Vibration isolation tables are typically used.

Many of the above-described problems can be minimized if FRET is done with a laser scanning confocal microscope. The diffraction-limited spot, coupled with the PMT detection of only that one spot in time, greatly reduces the overall noise in the system and makes it much easier to dissect out the FRET relationship. It is also common for confocals to have multiple detectors so that the two images may be acquired simultaneously. See the following section on confocal applications, and Refs. 5, 7, 10, and 13 for more details.

Of course, even with all the problems associated with FRET it has been shown to work successfully in many cases. The optical arrangement of filters and mirror is fairly straightforward. The exciter filter should match the absorption maximum of the donor molecule, whereas the dichroic must reflect the exciter and have transmission to match the emission maximum of the acceptor molecule. A common mistake is to use the dichroic mirror from the acceptor set instead of the dichroic for the donor. For our CFP/YFP example, the optics would include 436/20×, 455dclp, 480/40m, and 535/30m. See Fig. 18 for the spectra.

There are researchers who do not have emission filter wheels, or dual detectors for either of the techniques listed above. Some try to use two separate cubes to measure FRET. One cube would be a typical donor (CFP) cube: 440/20×, 455dclp, and 480/40m. The other cube would have the same exciter and dichroic, but with the acceptor (YFP) emission (such as 440/20×, 455dclp, and 535/30m; see Fig. 17). There are several problems with this approach. There is the constant threat of motion as the two cubes are moved into/out of position. This is lessened with a turret system, but can still be a problem. There is the added time required to make this movement, which may result in motion of the cells. The microscope and the two cubes must be nearly perfectly machined and aligned, otherwise the two images will not align at the camera/detector. And there is the possibility that the two cubes are not image registered. In other words, the image from one cube may go to a slightly different location on the detector, due to beam deflection and wedge associated with either the mirror and/or the emission filter. Many software packages now provide an algorithm that will realign two images relative to each other, but it is an added step.

Live Cell Imaging

Advances in microscopy and staining have made it routine to image live cell fluorescent preparations.[17,18] There are several important factors that affect filter

[17] S. Inoue and K. R. Spring, "Video Microscopy: The Fundamentals," 2nd Ed. Plenum Press, New York, 1997.

[18] G. Sluder and D. E. Wolf, eds., "Methods in Cell Biology," Vol. 56: "Video Microscopy." Academic Press, New York, 1998.

selection in this technique. The most important issue is to reduce the excitation/illumination light to the sample as much as possible. The exposure time to the camera/detector must always be considered in the context of illumination power, as a sort of balancing act: there must be enough light to acquire reasonable images without damaging the cells.

Many years ago it was thought that only UV light would actually damage, or kill cells. Many studies have shown that almost any light, regardless of the wavelength, if strong enough, can damage cellular structure. The illumination power can be adjusted by simple neutral density (ND) filters. Most ND filters are measured by using a log scale of optical density, such that an ND of 0.1 has a transmission of 10%, and an ND of 1.0 has a transmission of 0.1%. Another way to reduce light load at the sample is to use narrower excitation filters, such as a 20- or 30-nm fwhm filter instead of a 40- or 50-nm fwhm filter.

The mirror in this configuration is fairly standard, reflecting the excitation light and transmitting the emission wavelengths. However, it should always be AR coated for better transmission.

The emission filter for live cell studies must first fully block the excitation light as usual, but it must then be as transmissive as possible to increase the efficiency of collection of photons at the detector. This usually means wider emission filters, but that depends on the autofluorescence characteristics of the cells. It certainly means using an AR coating on the emission filters, for that extra few percent of transmission. It also means that the emission filter should be blocked for the excitation, but minimally blocked anywhere else, so as to increase the transmission characteristics.

As with all the applications described above, the exact filters/mirrors may require some experimentation and testing. For a better understanding of basic light and the physics of light, everyone should consider texts by Ditchburn[19] and Sir Isaac Newton.[20]

Acknowledgment

This work was made possible by the employees/owners of Chroma Technology Corp. All spectra are from lot numbers preceding 33000.

[19] R. W. Ditchburn, "Light." Dover Publications, New York, 1991.

[20] Sir Isaac Newton, "Opticks, or a Treatise on the Reflections, Refractions, Inflections and Colours of Light," 4th Ed., London, 1730. Reprinted by Dover Publications, New York, 1979.

[18] Resolution in Optical Microscopy

By JAMES E. N. JONKMAN, JIM SWOGER, HOLGER KRESS, ALEXANDER ROHRBACH, and ERNST H. K. STELZER

I. Introduction

Optical microscopes are fundamentally limited in the resolution they can achieve. The resolution depends on the wavelength of the light (both incident and detected), on the numerical aperture (NA) of the optical arrangement, and on the specimen to be observed or the experiment to be performed. In fact, "real" specimens cannot be imaged at the theoretically achievable resolution of the instrument. Living cells and tissues, for example, are optically thick and inhomogeneous, and hence a focused beam of light cannot maintain an ideal profile while penetrating the sample. Live specimens are also dynamic and sensitive to photobleaching and thermal damage, which imposes a limit on the duration for which they can be observed and on the power of the incident light. In short, the inherent properties of a specimen restrict the way in which it can be imaged and, consequently, limit the achievable resolution.

Two types of microscope are referred to often in this discussion of high-resolution optical microscopy, and hence they deserve brief introductions. In the confocal fluorescence microscope depicted in Fig. 1, a beam splitter deflects a laser beam toward an objective lens, which focuses the beam to a diffraction-limited spot in the specimen. Fluorescence is excited throughout its illumination cone, but only fluorescence emitted from the focal point is imaged through the confocal pinhole to the detector. The fluorescence emitted from above or below the plane of focus is strongly suppressed. An image is generated point-by-point by scanning the focused spot of laser light laterally across the specimen, and by moving the specimen axially through the focal plane. A series of such images forms a three-dimensional data set of the specimen. Confocal microscopes have a slightly better resolution than conventional microscopes; but their main advantage is that they perform optical sectioning. Another microscope that performs optical sectioning is the two-photon microscope. In a two-photon microscope, the excitation light is delivered in short, high-power pulses at approximately twice the single-photon excitation wavelength. Under these conditions, there is a reasonable probability that two photons will interact with the same fluorescent molecule at the same time in the vicinity of the focus. Because the combined energy of both photons is required to excite the fluorophore, fluorescence is generated only close to the focal point. Both confocal and two-photon techniques perform optical sectioning by instrumental means.

0076-6879/03 $35.00

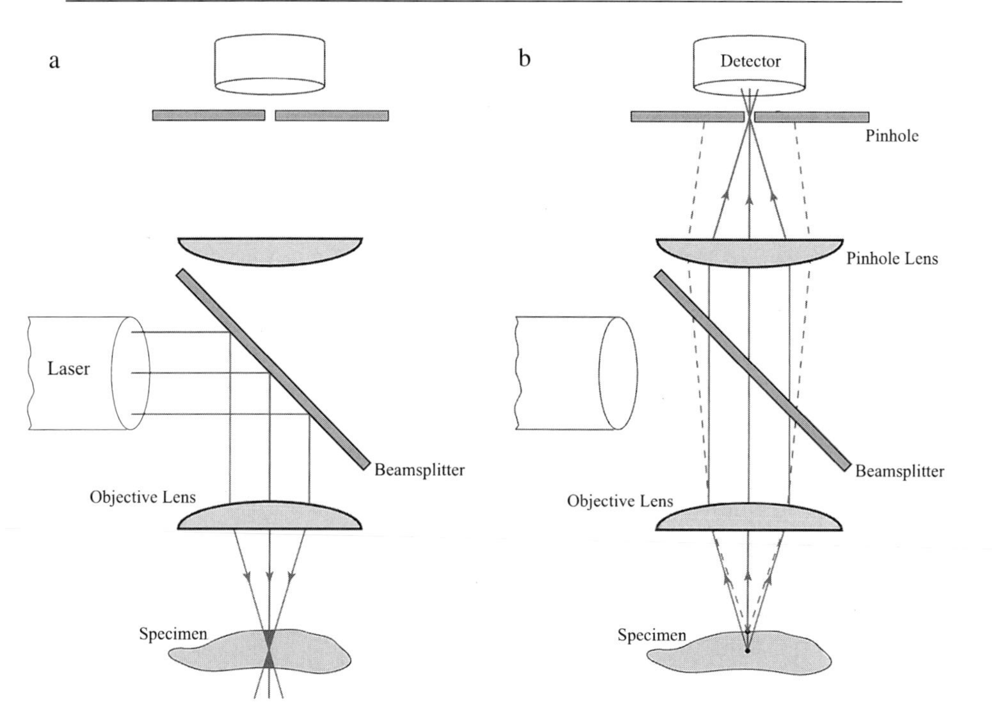

FIG. 1. Confocal fluorescence microscopy schematic. (a) The plane wave of light emitted by a laser is focused to a diffraction-limited spot in a specimen, exciting fluorophores throughout its illumination cone. (b) Fluorescence generated at the focal point is imaged through a pinhole onto a large-area detector (solid line), while fluorescence generated above the focal point (dashed line) is blocked by the pinhole. Similarly, fluorescence generated below and to either side of the focus (not shown) is blocked.

We begin this article by introducing the concept of point-spread functions, which allow us to compare the three-dimensional resolution of various light microscope arrangements. Next, we define resolution in a quantitative way, and explain how resolution is related to precision and optical sectioning. We then detail several factors affecting resolution, including noise, pixelation, and aberrations. We conclude by presenting and comparing several optical and computational techniques for improving resolution in optical microscopy.

II. Point-Spread Function as Resolution Quantifier

A. *Point-Spread Function Definition*

A useful tool for comparing the performance of optical microscopes is the point-spread function (PSF). The PSF can be defined in two complementary ways: the amplitude PSF is the electromagnetic field distribution in the focal region when a plane wave is focused; and it is the amplitude and phase distribution of the electromagnetic field in the image plane when observing a point-like light source (e.g., Born and Wolf,[1] pp. 484–499). We cannot observe amplitude fluctuations directly in the visible regime; but we can observe the modulus squared of the amplitude PSF, namely the intensity PSF. The intensity PSF (hereafter referred to simply as the PSF) can be measured by taking images of, for example, a subresolution bead as it is scanned through the focus of a microscope. Real specimens are, of course, rarely point sources. However, real specimens can be regarded as a superposition of many pointlike, subresolution objects. The extents of the PSF of a microscope in the two lateral directions and the axial direction are related to its resolution. Therefore, comparing the PSFs of various microscope arrangements provides a useful evaluation of their lateral and axial performances.

B. Point-Spread Functions for Illumination and Detection

Following the principles of Huygens, according to McCutchen[2] and Goodman,[3] the three-dimensional field distribution $h(x, y, z)$ in the focal region can be obtained by taking the Fourier transform of the generalized lens aperture $A(\mathbf{k})$:

$$h(x, y, z) = \int A(\mathbf{k}) \cdot e^{ikr} d^3\mathbf{k} \tag{1}$$

where $\mathbf{k} = (k_x, k_y, k_z)$ is the wave vector with the wave number $|\mathbf{k}| = n2\pi/\lambda$. The vacuum wavelength is λ and the index of refraction of the medium is n. For a detailed treatment of how the generalized lens aperture $A(\mathbf{k})$ is obtained, see Rohrbach and Stelzer.[4] The aperture function $A(\mathbf{k})$ includes information about the incident field strength at the lens, the transmission behavior of the lens, and the polarization of the incident field. Throughout this article, all calculations are performed by assuming that a circular objective lens aperture with a radius R is illuminated by a Gaussian beam propagating in the z direction, with waist $w = 2R$; that the incident light is linearly polarized in the x direction; and that the objective lens obeys the sine condition (see, e.g., Born and Wolf[1]). In addition, for the detection of fluorescent light we multiply the illumination aperture function by a factor that narrows the PSF slightly in the lateral direction, for reasons described in Visser *et al.*[5]

[1] M. Born and E. Wolf, "Principles of Optics." Cambridge University Press, Cambridge, 1999.

[2] C. W. McCutchen, *J. Opt. Soc. Am.* **54,** 240 (1964).

[3] J. W. Goodman, "Introduction to Fourier Optics." McGraw-Hill, San Francisco, 1996.

[4] A. Rohrbach and E. H. K. Stelzer, *J. Opt. Soc. Am. A* **18,** 839 (2001).

[5] T. D. Visser, G. J. Brakenhoff, and F. C. A. Groen, *Optik* **87,** 39 (1991).

The intensity PSF $|h(x, y, z)|^2$ is calculated as the product of the amplitude PSF and its complex conjugate:

$$|h(x, y, z)|^2 = h(x, y, z) \cdot h^*(x, y, z) \qquad (2)$$

It describes the three-dimensional intensity distribution of a beam of light focused by a lens, or equivalently the spatial absorption pattern of a uniform fluorophore solution in the vicinity of the focus. It also describes the probability that a photon, emitted from a certain location in the object, will hit a point detector in an image plane of the microscope. In these two senses, the PSFs are referred to as the illumination PSF and detection PSF, respectively.

Figures 2a and b show calculated illumination PSFs for microscopes that use pointwise single-photon illumination, such as the confocal microscope, in the XZ and YZ planes, respectively. The calculations were performed with an illumination wavelength (λ_{ill}) of 488 nm and a numerical aperture (NA) of 1.2, and with the assumption that the refractive index of the medium was that of water, $n = 1.33$. These parameters, summarized in Table I, were chosen for their relevance in imaging living specimens expressing the widely used enhanced green fluorescent protein (EGFP).[6,7] Figure 2e and f shows calculated illumination PSFs for microscopes that use pointwise two-photon illumination (i.e., for a two-photon microscope) in the XZ and YZ planes, respectively. The two-photon illumination wavelength ($\lambda_{\text{2p-ill}}$) was 900 nm, and the numerical aperture and refractive index were again NA = 1.2 and $n = 1.33$. It is interesting to observe that the illumination PSFs are broader in the direction of the beam's original polarization (i.e., in the XZ plane) than in the orthogonal direction. Figure 2c and Fig. 2g, which are identical, show the detection PSF for fluorescence microscopes using a detection wavelength (λ_{det}) of 520 nm. The fluorescence detection PSF is the same for conventional, confocal, and two-photon fluorescence microscopes.

C. Confocal and Two-Photon Point-Spread Functions

The system PSF of a microscope can be defined as the product of its illumination and detection PSFs. For example, in a wide-field fluorescence microscope the specimen is uniformly illuminated, which can be described by a uniform PSF (not shown). The fluorescence is detected in a spatially resolved manner by imaging details of the specimen with a camera or with the eye of an observer, and can be described by the detection PSF shown in Fig. 2c. Its system PSF is, therefore, the product of a constant and the detection PSF: in other words, the detection PSF shown is also a description of the system PSF of a wide-field fluorescence microscope.

[6] C. Xu, W. R. Zipfel, J. B. Shear, R. M. Williams, and W. W. Webb, *Proc. Natl. Acad. Sci. U.S.A.* **93,** 10763 (1996).

[7] S. Jakobs, V. Subramaniam, A. Schönle, T. M. Jovin, and S. W. Hell, *FEBS Lett.* **479,** 131 (2000).

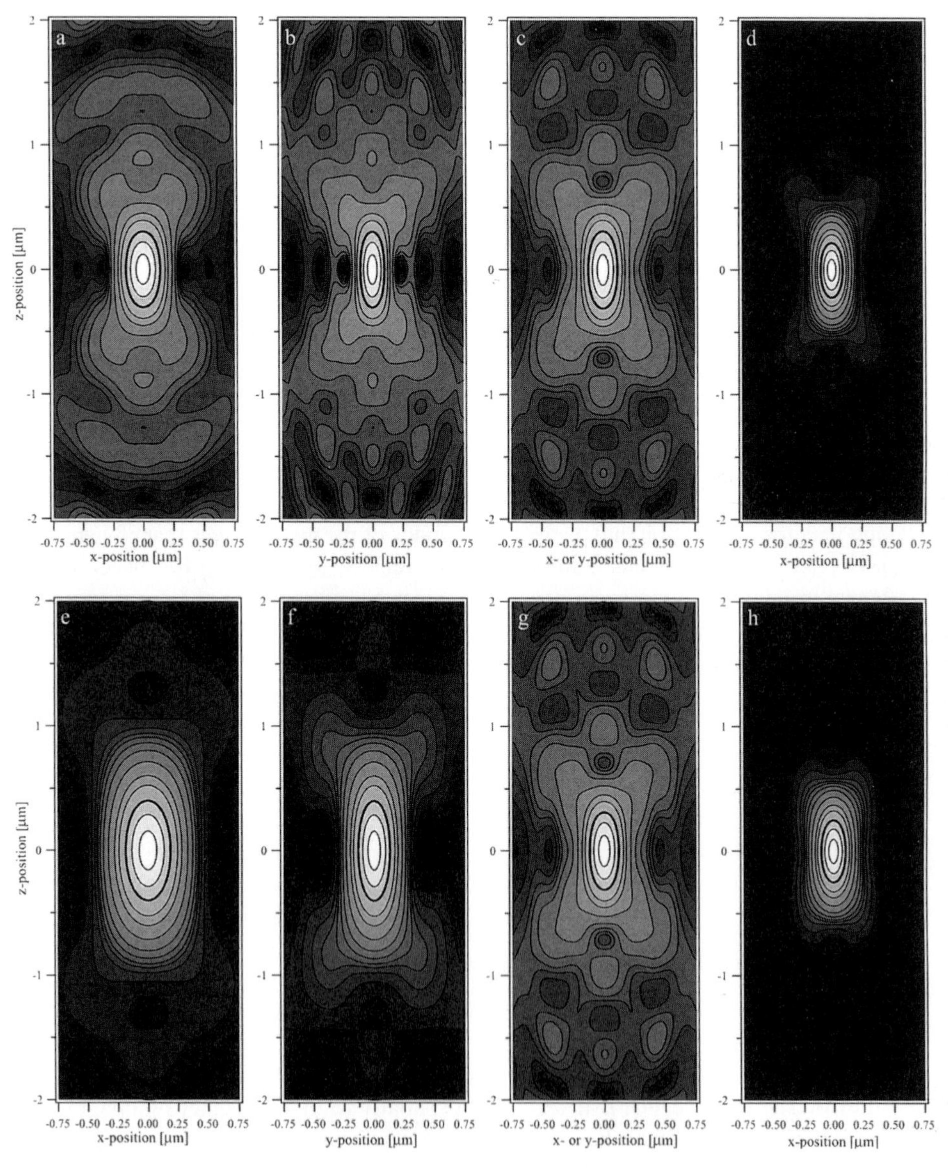

FIG. 2. Calculated point-spread functions for fluorescence microscope arrangements, in which the beam propagates in the z direction and is polarized in the x direction. Illumination PSFs in the (a) XZ and (b) YZ planes ($\lambda_{\text{ill}} = 488$ nm). (c) Fluorescence detection PSF ($\lambda_{\text{det}} = 520$ nm). (d) Confocal fluorescence PSF in the XZ plane, which is the product of the XZ illumination PSF and the detection PSF. Two-photon illumination PSFs in the (e) XZ and (f) YZ planes ($\lambda_{\text{2p-ill}} = 900$ nm). (g) Fluorescence detection PSF [identical to (c)]. (h) 2p-confocal PSF in the XZ plane. (d) and (e) should be compared to understand the performance differences between confocal and two-photon microscopy. All PSFs are calculated with NA $= 1.2$, $n = 1.33$. Contour lines are drawn for intensities of 0.9, 0.7, 0.5 (boldface), 0.3, 0.2, 0.1, 0.05, 0.03, 0.02, 0.015, 0.01, 0.005, 0.002, and 0.001. This spacing of the contour lines follows that found in M. Born and E. Wolf, "Principles of Optics." Cambridge University Press, Cambridge, 1999.

TABLE I
PARAMETERS USED FOR POINT-SPREAD FUNCTION CALCULATIONS

Parameter	Value
Illumination wavelength, single photon (λ_{ill})	488 nm
Illumination wavelength, two photon ($\lambda_{2p\text{-}ill}$)	900 nm
Detection wavelength (λ_{det})	520 nm
Numerical aperture (NA)	1.2
Refractive index of medium (n)	1.33

The concept of the system PSF is more obvious when applied to the confocal fluorescence microscope, in which the objective lens forms an image of both the illumination pinhole and the detection pinhole in the object plane. Only fluorophores in the volume shared by the illumination and detection PSFs are both excited and detected. Hence, the confocal fluorescence PSF is the product of the illumination and detection intensity PSFs, and is given by

$$|h_{\text{confocal}}(x, y, z)|^2 = |h_{\text{det}}(x, y, z)|^2 |h_{\text{ill}}(x, y, z)|^2 \tag{3}$$

The confocal PSF shown in Fig. 2d is the product of the illumination PSF in the XZ plane (Fig. 2a) and the detection PSF (Fig. 2c). The confocal system PSF in the YZ plane (not shown) is slightly narrower. However, in practice it is difficult to achieve the ideal illumination and detection assumed by the calculated PSFs, and therefore the broader PSF is usually the more realistic one.

In a two-photon microscope, fluorescence excitation is proportional to the probability of absorbing two longer wavelength (lower energy) photons. The illumination wavelength is approximately twice as long as in the single-photon case.[6,8] The PSF of a two-photon microscope is, therefore, the square of the illumination intensity PSF[9]:

$$|h_{\text{two-photon}}(x, y, z)|^2 = \left(|h_{\text{ill}}(x, y, z)|^2\right)^2 \tag{4}$$

The two-photon PSF was calculated using $\lambda_{2p\text{-ill}} = 900$ nm, because it is appropriate for EGFP, and is shown in Fig. 2e and f for the XZ and YZ planes, respectively.

The two-photon PSFs were calculated by assuming that a large-area (i.e., nonconfocal) detector is used. Although it is rarely done, two photon-excited fluorescence can also be detected in a confocal manner by placing a pinhole in front of the detector.[10] The two-photon confocal (2p-confocal) PSF is the product of the

[8] A. Fischer, C. Cremer, and E. H. K. Stelzer, *Appl. Optics* **34,** 1989 (1995).

[9] C. J. R. Sheppard and M. Gu, *Optik* **86,** 104 (1990).

[10] E. H. K. Stelzer, S. Hell, S. Lindek, R. Stricker, R. Pick, C. Storz, G. Ritter, and N. Salmon, *Optics Commun.* **104,** 223 (1994).

two-photon PSF (Fig. 2e) and the detection PSF (Fig. 2g):

$$|h_{\text{2p-confocal}}(x, y, z)|^2 = \left(|h_{\text{ill}}(x, y, z)|^2\right)^2 |h_{\text{det}}(x, y, z)|^2 \tag{5}$$

The 2p-confocal PSF, shown in the XZ plane in Fig. 2h, is still broader in both the axial and lateral directions than that of a comparable single-photon confocal microscope.

Comparing the widths of the system PSFs for wide-field fluorescence, confocal, and two-photon microscopes, in the two orthogonal planes, gives a comparison of their respective resolutions. To help evaluate the extents of the PSFs presented in Fig. 2, line scans through the PSFs in the lateral and axial directions are presented in Fig. 3, and the full width at half-maximum (FWHM) extents of the central peak are given in Table II. First, we note that the polarization of the beam has a substantial effect on the width of the PSFs in the lateral directions: the lateral illumination curves are considerably broader in the direction of the beam's original polarization (the x direction) than in the orthogonal direction. In fact, in the XZ plane the illumination PSF using $\lambda_{\text{ill}} = 488$ nm is even broader than the fluorescence detection PSF using $\lambda_{\text{det}} = 520$ nm. The average width of the illumination PSF in the two lateral directions is, however, still almost 10% narrower than the fluorescence detection PSF.

By comparing the confocal fluorescence microscope with one that uses conventional fluorescence detection, we observe that in both the lateral and axial directions the confocal PSF is narrower by about 30%. This means that a confocal microscope can provide somewhat improved resolution, both laterally and axially, over an equivalent conventional fluorescence microscope. In practice, however, the lateral resolution enhancement is difficult to observe: practical considerations such as the size of the confocal pinhole, and the influence of noise, pixelation, and aberrations (Section IV), broaden the ideal confocal PSF. In the axial direction, the main advantage of the confocal microscope over the conventional fluorescence microscope is not the somewhat reduced width of its PSF, but rather its optical sectioning property (discussed below).

Because the two-photon microscope uses an illumination wavelength nearly twice that of the confocal microscope, the PSF of the two-photon microscope is somewhat broader than the PSF of a conventional fluorescence microscope. Moreover, the PSF of the two-photon microscope is substantially broader (more than 80%) than the confocal PSF, both axially and laterally. Obviously, the advantage of a two-photon microscope over a conventional microscope is not related to the width of its PSF. Rather, the two-photon microscope also performs optical sectioning, which a conventional fluorescence microscope cannot perform. Proponents of two-photon microscopy observe an advantage over (single-photon) confocal microscopes for imaging deep into biological specimens,[11,12] especially in

[11] V. E. Centonze and J. G. White, *Biophys. J.* **75,** 2015 (1998).

[12] A. Periasamy, P. Skoglund, C. Noakes, and R. Keller, *Microsc. Res. Tech.* **47,** 172 (1999).

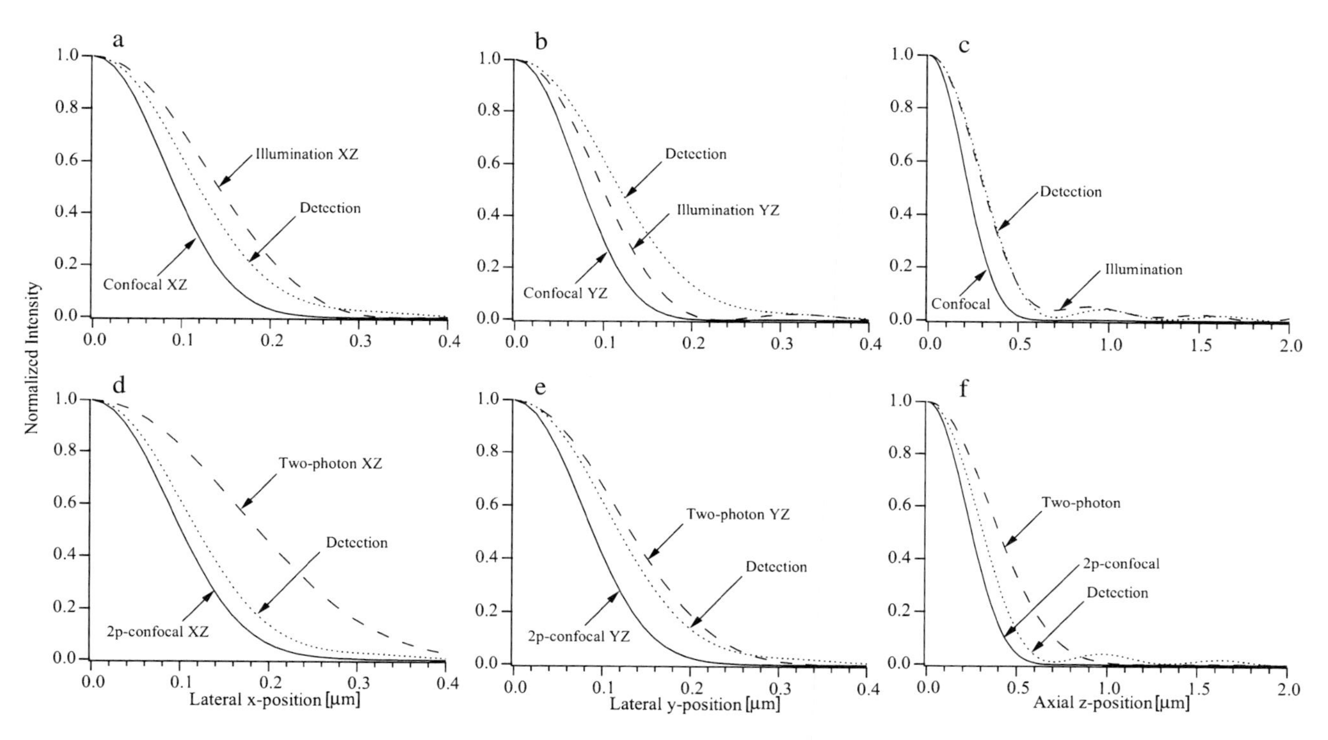

FIG. 3. Line scans showing two orthogonal extents of each of the PSFs of Fig. 2. (a) Lateral extents of the illumination, detection, and confocal PSFs in the XZ and (b) YZ planes. (c) Axial extents of the illumination, detection, and confocal PSFs. (d) Lateral extents of the two-photon, detection, and 2p-confocal PSFs in the XZ and (e) YZ planes. (f) Axial extents of the two-photon, detection, and 2p-confocal PSFs.

TABLE II
FULL WIDTH AT HALF-MAXIMUM EXTENTS OF CALCULATED POINT-SPREAD FUNCTIONS[a]

PSF	FWHM extent (μm)		
	x-lateral	y-lateral	Axial
Illumination (ill)	0.28	0.20	0.61
Detection (det)	0.24	0.24	0.62
Confocal = ill × det	0.18	0.15	0.44
Two-photon = $(\text{ill})^2$	0.38	0.26	0.80
Two-photon confocal = $(\text{ill})^2 \times \text{det}$	0.20	0.18	0.50

[a] As shown in Fig. 2. The parameters used for the calculations are summarized in Table I.

conjunction with a low-magnification, high-numerical aperture objective lens.[13,14] Such an advantage is realized only when the two-photon microscope is operated with a large-area detector; and therefore the properties of a two-photon microscope with confocal detection (2p-confocal) is not generally relevant.

The confocal and two-photon confocal PSFs were calculated by assuming that the diameter of the confocal pinhole was equal to the FWHM extent of the detection PSF. Changing the pinhole diameter for confocal detection has a dramatic effect on resolution, as detailed elsewhere.[10,15–18]

III. Quantifying Resolution

A. Lateral and Axial Resolution

In a general sense, resolution is defined as the minimum separation between two point objects, such that they can still be distinguished. In any practical optical microscope, the image of even an ideal pointlike object (i.e., the point-spread function) will have a finite extent that is determined by both the measurement system and the sample environment. Thus, if the two point objects of interest are insufficiently separated, they will appear merged into one object in the image.

[13] E. Beaurepaire, M. Oheim, and J. Mertz, *Optics Commun.* **188,** 25 (2001).

[14] M. Oheim, E. Beaurepaire, E. Chaigneau, J. Mertz, and S. Charpak, *J. Neurosci. Methods* **111,** 29 (2001).

[15] M. Gu and C. J. R. Sheppard, *J. Mod. Optics* **40,** 2009 (1993).

[16] T. Wilson, *in* "Handbook of Biological Confocal Microscopy" (J. B. Pawley, ed.), p. 167. Plenum Press, New York, 1995.

[17] R. Gauderon, P. B. Lukins, and C. J. R. Sheppard, *Microsc. Res. Tech.* **47,** 210 (1999).

[18] M. Gu and X. S. Gan, *Bioimaging* **4,** 129 (1996).

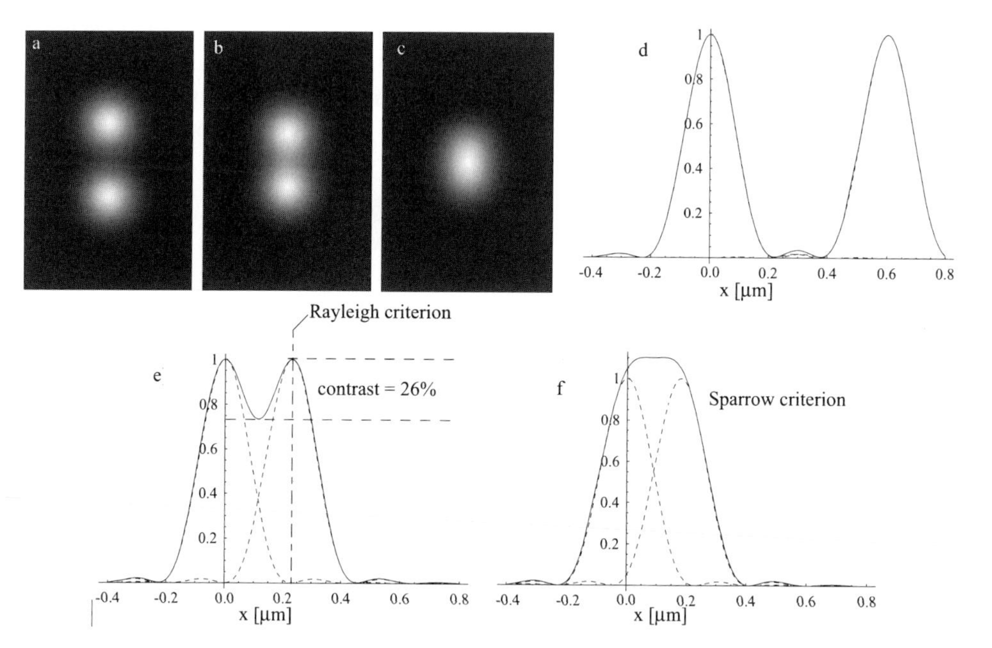

FIG. 4. Two point objects are resolved if their separation is sufficient to produce a contrast between them. (a and d) Well-separated objects are clearly resolved. (b and e) Point objects separated by the distance specified by the Rayleigh criterion yield a 26% contrast. (c and f) Point objects separated by the distance specified by the Sparrow criterion yield "just" no contrast.

This idea is illustrated in Fig. 4, in which we simulate the images of two pointlike objects, separated by various distances. In Fig. 4a, the point centers are separated by a relatively large distance, and the distinction between the two points is clear (i.e., they are well resolved). In Fig. 4b the points are closer together, and although they are not completely isolated, it is still clear that this is the image of two distinct points. Finally, Fig. 4c shows two unresolved points: when simply looking at an image such as this, it is not possible to determine whether it contains two distinct points.

The above-described qualitative description of resolution is sufficient to distinguish it from a more vague term such as "image quality," which is usually a subjective term based on the viewer's sense of aesthetics. For instance, a high-resolution image might still be relatively noisy, or have lower contrast, than a lower resolution image that is visually pleasing.

Resolution is, then, the minimum separation between two point objects, such that they can still be distinguished. However, there is some ambiguity as to what constitutes this minimum separation. The curves shown in Fig. 4d are the lateral components of the PSFs for two well-separated pointlike objects of equal intensity, imaged by conventional fluorescence microscopy. These objects are certainly resolved. The well-known Rayleigh criterion (e.g., Hecht,[19] p. 465) suggests that two point objects are still resolved when the first minimum of one of the curves is aligned with the central maximum of the second curve, as depicted in Fig. 4e. This distance is usually given as

$$r_{xy} = \frac{1.22\lambda_{\rm em}}{2n \sin\alpha} \approx \frac{0.61\lambda_{\rm em}}{\rm NA} \quad (6)$$

where n is the refractive index of the medium, α is the opening angle of the lens aperture, and the numerical aperture is $\mathrm{NA} = n \sin\alpha$. Using another criterion, called the Sparrow criterion (Hecht,[19] p. 465), the two objects are no longer resolved when the dip between their combined curves just disappears, as depicted in Fig. 4f.

To quantify resolution we define the term "contrast" as the difference between the maximum intensity of the two objects and the minimum intensity found between them. Using this definition, the Rayleigh criterion corresponds to the separation at which a 26.4% contrast is achieved. The Sparrow criterion defines the resolution as the cutoff distance: that is, the separation at and below which zero contrast is achieved. Essentially, any contrast between 0 and 100% can be used to define a resolution; but the 26% contrast provided by the Rayleigh criterion means that Eq. (6) is reasonable for most cases. A thorough treatment of the relationship between contrast and resolution in fluorescence microscopy has been presented elsewhere.[20]

For pointlike objects, the same approach can be applied to determine the axial resolution of a microscope. As in the lateral direction, a reasonable contrast between pointlike objects is achieved when they are separated by the distance between the central maximum and first minimum of the axial component of the PSF. In a conventional fluorescence microscope, this distance is (Born and Wolf,[1] p. 491):

$$r_z = \frac{\lambda_{\rm em}}{n(1 - \cos\alpha)} \quad (7)$$

where n is the index of refraction of the specimen. Another approach, which derives the lateral and axial resolutions on the basis of the uncertainty principle,[21] yields, in addition to Eq. (7), a description of the lateral resolution that is valid over all opening angles α from zero to π:

$$r_{xy} = \frac{\lambda}{n(3 - 2\cos\alpha - \cos 2\alpha)^{1/2}} \quad (8)$$

[19] E. Hecht, "Optics." Addison-Wesley, Reading, MA, 1998.
[20] E. H. K. Stelzer, *J. Microsc.* **189,** 15 (1998).
[21] S. Grill and E. H. K. Stelzer, *J. Opt. Soc. Am.* **16,** 2658 (1999).

For $\alpha = \pi$ Eqs. (7) and (8) generate the same number. To determine the resolution of an optical microscope experimentally, it is common to measure the FWHM extents of the PSF of the microscope (as we did in Section II) rather than attempting to measure the position of the intensity minimum of the PSF.

B. Precision

Care must be taken not to confuse precision with resolution. Resolution has been defined as the distance between point objects in an image, such that they are discernable as separate objects with a given contrast (say, 26%). However, the location of a single point object can be measured to a precision much greater than the resolution of the microscope by calculating the intensity-weighted "center of mass equivalent" of the object. Many applications take advantage of precision to make measurements in the region of tens of nanometers. For example, the photonic force microscope[22] tracks the position of a single bead to within 10 nm, in order to determine the three-dimensional structure of molecules. Single-particle tracking in video sequences[23] yields precisions as fine as 20 nm. Finally, surface topologies can be determined to within a fraction of a micron, using confocal microscopes.[24]

C. Optical Sectioning

One of the features of wide-field microscopy of thick samples is that the image contains information about both the portion of the sample that lies in the focal plane, and the out-of-focus regions. Although the in-focus portion may be clear and sharp (to within the resolution of the microscope, as discussed above), the out-of-focus light is naturally blurry, and can contribute a background that masks the in-focus information. This effect is apparent when the two images in Fig. 5 are compared. In thin regions of the sample, such as the region indicated by the arrow in Fig. 5a, details are not obscured by out-of-focus light: here the optical sectioning provided by the confocal microscope has no obvious advantage. However, in thicker regions, such as in the center of the image, rejection of out-of-focus light by confocal microscopy (Fig. 5b) clearly enhances the contrast of the in-focus features.

As we noted in Section II, the axial extent of the confocal fluorescence point-spread function is better, but not much better, than that of the conventional fluorescence microscope. Therefore, for pointlike objects the axial resolution in a confocal microscope is only marginally better than in a conventional instrument. However, because resolution depends on contrast (Section III,A), optical sectioning can improve the image resolution even though it may not improve the FWHM of the PSF.

[22] E.-L. Florin, A. Pralle, J. K. H. Hörber, and E. H. K. Stelzer, *J. Struct. Biol.* **119,** 202 (1997).

[23] M. J. Saxton and K. Jacobson, *Annu. Rev. Biophys. Biomol. Struct.* **26,** 373 (1997).

[24] R. W. Wijnaendts van Resandt, H. J. B. Marsman, R. Kaplan, J. Davoust, E. H. K. Stelzer, and R. Stricker, *J. Microsc.* **138,** 29 (1985).

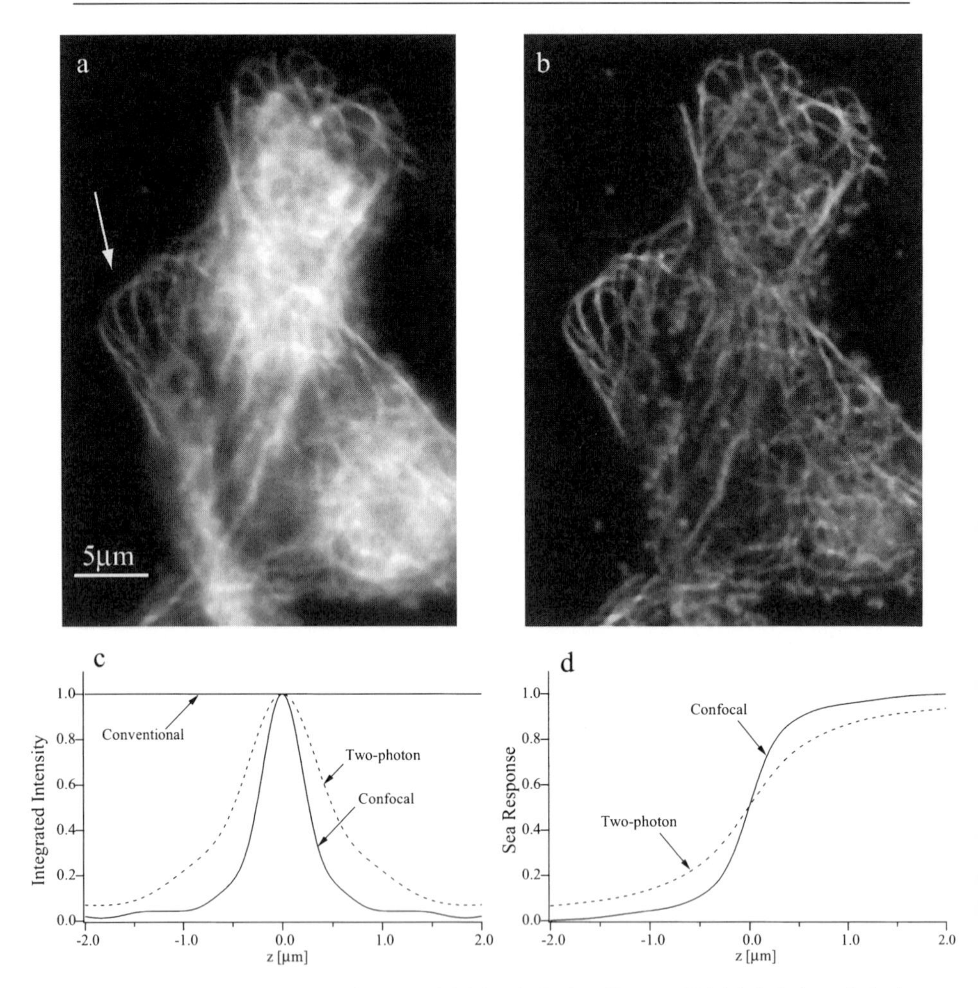

FIG. 5. (a) Wide-field image of Alexa 488 (Molecular Probes, Eugene, OR)-labeled microtubules in a fibroblast cell. Thin regions of the sample (such as the region indicated by the arrow) are well resolved, but in thick regions details are obscured by fluorescence emitted above and below the focal plane. (b) In the corresponding confocal image, optical sectioning distinguishes details in the focal plane. (c) The integrated intensity of the detection PSF for conventional fluorescence microscopy is constant as a function of depth; but the integrated intensities of the confocal and two-photon PSFs peak at the focus. In contrast to conventional fluorescence microscopes, both confocal and two-photon techniques observe only fluorophores in the neighborhood of the focus; hence they perform optical sectioning. (d) The sea response of the confocal microscope has a greater slope than the two-photon microscope sea response, resulting in a better resolution along the optical axis in a confocal microscope.

To understand how optical sectioning works, consider the intensity PSF of a microscope, integrated over its lateral extent. For the illumination PSF, this integral is

$$E_{\mathrm{ill,int}}(z) \propto \int_{r=0}^{r=\infty} |h_{\mathrm{ill}}(r, z)|^2 2\pi r \, dr = \text{constant} \tag{9}$$

which is independent of the axial position. At each depth in the specimen, the distribution of the illumination intensity may differ but the total optical power remains the same. The laterally integrated intensity of the detection intensity PSF is similarly constant, which means that a conventional fluorescence microscope has no optical sectioning capability. In fact, we are only rephrasing the conservation of energy, using optical terminology.

In contrast, the integrated intensity PSF for the confocal microscope is

$$E_{\mathrm{confocal,int}}(z) \propto \int_{r=0}^{r=\infty} \left(|h_{\mathrm{ill}}(r, z)|^2\right)\left(|h_{\mathrm{det}}(r, z)|^2\right) 2\pi r \, dr \tag{10}$$

which has a maximum in the focal plane (Fig. 5c). This is the explanation for the depth discrimination capability of a confocal microscope.[24,25] A two-photon microscope also exhibits optical sectioning because its integrated PSF peaks at the focus. However, because the illumination wavelength for a two-photon microscope is longer than that of its single-photon counterpart, the peak is not as sharp, and consequently the optical sectioning performance is poorer than that of a comparable confocal microscope.

The best practical illustration for optical sectioning is to record the "sea response," that is, the change in intensity while focusing through the coverslip into a thick layer of fluorophore dissolved in the immersion medium of the lens. The confocal sea response is the integral over the extent of the fluorescent sea, which occupies the range $-\infty < z < z_0$:

$$E_{\mathrm{confocal,sea}}(z_0) \propto \int_{z=-\infty}^{z=z_0} \int_{r=0}^{r=\infty} \left(|h_{\mathrm{ill}}(r, z)|^2\right)\left(|h_{\mathrm{det}}(r, z)|^2\right) 2\pi r \, dr \tag{11}$$

This, and the equivalent two-photon sea response, are plotted in Fig. 5d. The derivative of the sea response curve, which is simply the integrated intensity, can be used to characterize the axial resolutions of the confocal, two-photon, and other microscope arrangements for thick specimens.

A method for performing optical sectioning using a wide-field fluorescence microscope has been reported.[26–29] A grid pattern is projected onto the object,

[25] I. J. Cox, C. J. R. Sheppard, and T. Wilson, *Appl. Optics* **21,** 778 (1982).

[26] R. Juskaitis, T. Wilson, M. A. A. Neil, and M. Kozubek, *Nature* (*London*) **383,** 804 (1996).

[27] M. A. A. Neil, R. Juskaitis, and T. Wilson, *Optics Lett.* **22,** 1905 (1997).

[28] M. A. A. Neil, T. Wilson, and R. Juskaitis, *J. Microsc.* **189,** 114 (1998).

[29] M. A. A. Neil, A. Squire, R. Juskaitis, P. I. H. Bastiaens, and T. Wilson, *J. Microsc.* **197,** 1 (2000).

and images taken at three spatial positions of the grid are processed in real time to produce optically sectioned images that can be similar to those obtained with confocal microscopes. This method works reasonably well for thin specimens; but for thick specimens the grid pattern is probably distorted as it penetrates the specimen, and the optically sectioned image cannot be properly reconstructed.

D. Extending Depth of Focus

Changing the diameter of the detection pinhole in confocal microscopy alters the effectiveness of optical sectioning and can be used to increase the thickness of the optical image slices. A series of sharply focused XY slices can be put together to form a three-dimensional (3-D) image of the object, which can be, for example, rotated computationally to be viewed from any direction. This 3-D image is often reduced to an extended-focus two-dimensional image by projecting the data onto the XY plane, as if the 3-D data set were viewed from the top. The resulting extended-focus image has the contrast and resolution of a confocal image, together with what is effectively an arbitrarily large depth of focus.

Unfortunately, laser scanning microscopes (such as the confocal microscope) are generally slow, requiring as much as minutes to record a series of, say, 20 images. The observation of dynamic processes in live specimens is typically restricted by this scanning speed to the imaging of only one focal plane for each time point in the experiment. Some specialized confocal and two-photon microscopes speed up the imaging process by scanning many points across the specimen in parallel[30,31]; but these instruments are less flexible [e.g., they cannot be used for photobleaching experiments such as fluorescence recovery after photobleaching (FRAP)] and optically less efficient than conventional confocal microscopes.

Another way to produce an extended depth-of-focus image is to introduce a transmission or phase filter in the back focal plane (BFP) of the objective lens of a microscope. The simplest way to generate an axially extended point-spread function is to use an annular transmission filter, which blocks the small-angle rays. More light is transmitted by choosing phase filters in the BFP. This technique is also called wave front coding.[32] A typical phase distribution is the phase axicon $\phi(x, y) = 1 - c[(x^2 + y^2)^{1/2}]$, which can produce the same elongated focus as an annular transmission filter. Figure 6a shows the extended depth of such a modified PSF for an NA $= 1.2$ water immersion lens, where the half-maximum (thick black contour line) is compared with the half-maximum of a conventional PSF (indicated by a thick gray line). Such an elongated PSF can be used to reduce the number of slices in a thick biological specimen, when the filter is used for confocal imaging.

[30] G. S. Kino, *in* "Handbook of Biological Confocal Microscopy" (J. B. Pawley, ed.), p. 155. Plenum Press, New York, 1995.

[31] M. Straub, P. Lodemann, P. Holroyd, R. Jahn, and S. W. Hell, *Eur. J. Cell Biol.* **79,** 726 (2000).

[32] E. R. Dowski and W. T. Cathey, *Appl. Optics* **34,** 1859 (1995).

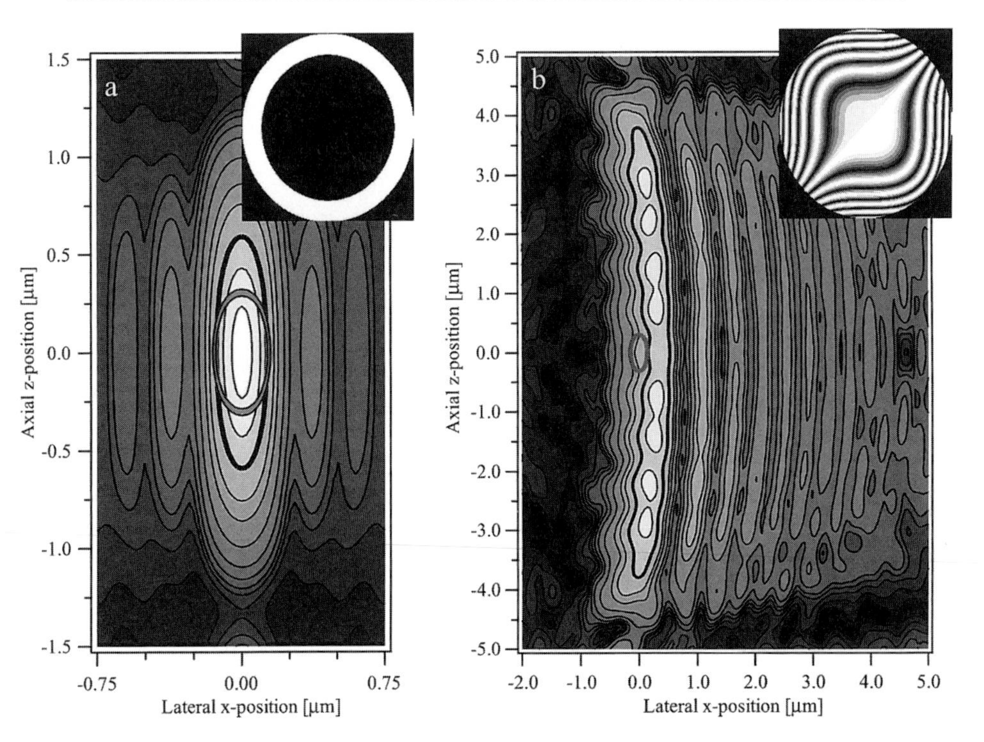

FIG. 6. (a) PSF with a 2-fold extended depth of focus as a result of annular filtering in the back focal plane (BFP) (*inset*). (b) Asymmetric PSF with a 14-fold extended depth of focus as a result of the phase BFP filter (*inset*). The half-maximum contour line of the PSFs (boldface line) is compared with the half-maximum contour line of a conventional PSF (thick gray line). Notice the different axis scaling in (a) and (b).

A dramatic improvement in image acquisition speed can be achieved by combining wave front coding with wide-field microscopy, which acquires complete XY image planes without scanning. This idea is especially useful when the specimen is thinner than the axial width of the modified PSF. For example, the phase distribution $\phi(x, y) = \phi_0(x^3 + y^3)$ produces an asymmetric PSF with a 14-fold extended depth of focus when $\phi_0 = 20\pi$.[33] This asymmetric PSF is shown in Fig. 6b, together with the field behind the BFP phase filter $\propto \cos[\phi(x, y)]$. It is remarkable that for slices that have about the same thickness as the axial PSF, nearly the same lateral resolution is obtained with only a modest reduction in signal-to-noise ratio in comparison to conventional fluorescence microscopy.

[33] S. Bradburn, W. T. Cathey, and E. R. Dowski, *Appl. Optics* **36,** 9157 (1997).

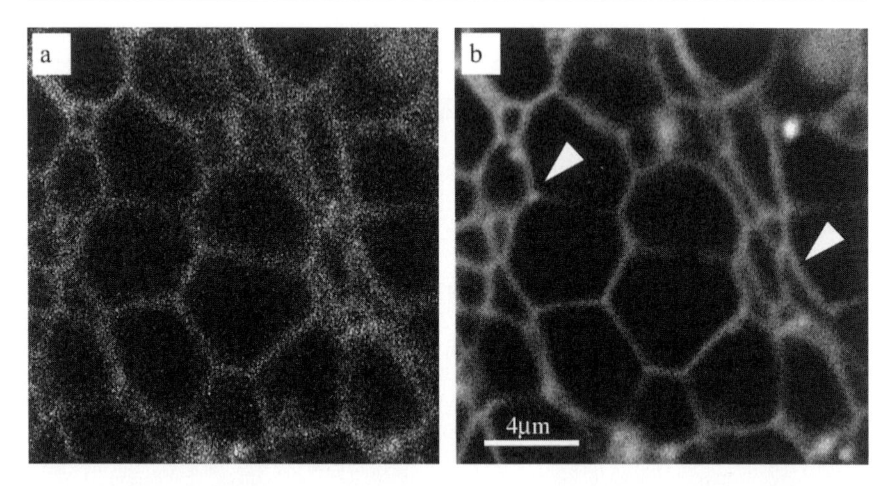

FIG. 7. Confocal images of an Alexa 488-labeled *Drosophila* embryo, with line averaging set to (a) 1 and (b) 16. Some features, such as those indicated by the arrowheads, become visible only when the noise is reduced. Noise reduces contrast, and therefore reduces the resolution of an image. Averaging independently recorded images reduces noise and hence improves the contrast, which means that the theoretical resolution of the optical system can be more fully utilized.

IV. Factors Affecting Resolution

So far, we have assumed that images are acquired under ideal conditions. Here, we outline three factors that contribute to a reduction in contrast and, therefore, in the resolution of microscope images.

A. *Noise*

Noise affects the resolution of a recorded microscope image. Thin, fixed, immunofluorescently labeled specimens usually emit enough fluorescence photons to provide a high signal-to-noise ratio (SNR). However, many confocal and two-photon microscopy applications (including live-cell imaging with GFPs) involve imaging low concentrations of fluorophores, deep within specimens that readily absorb and scatter photons. In these situations, noise can strongly contribute to or even dominate the image. Noise creates an uncertainty in the quantification of the fluorescence intensity for a given pixel.[34] This results in an uncertainty in the measured contrast. Because we must accept an error in our estimation of the intensity we must accept that the contrast is, in general, underestimated in a noisy image. Of course, a reduction in contrast due to noise also means a reduction in resolution, as demonstrated in Fig. 7. Particularly in the regions of the image marked with arrowheads (Fig. 7b), features become visible with 16-frame averages that were

[34] K. Carlsson, *J. Microsc.* **163,** 167 (1991).

not distinguished in a single image frame. For each type of specimen, it is necessary to find an appropriate balance between the scanning speed, the illumination intensity, and the resolution improvement provided by a reduced pinhole diameter (e.g., Gauderon and Sheppard[35]).

B. Pixelation

So far, point-spread functions and their components have been presented as smooth and continuous. However, real microscope images are recorded with a finite number of pixels in all three dimensions. The pixels will not generally fall exactly on the peak intensity of one point object, or on the intensity minimum that separates it from a neighboring point object. The result is that maxima tend to be underestimated, whereas minima are overestimated. The result is that pixelation reduces the contrast and, consequently, the resolution of images recorded on the microscope. The often-quoted Nyquist sampling criterion says that it is necessary to sample at least twice for each resolution element.[36] However, the Nyquist criterion (like the Sparrow criterion) is a cutoff criterion: sampling at the Nyquist frequency does not guarantee a minimum contrast between features. Although with finite-sized pixels it is never possible to achieve the ideal contrast, sampling at a rate of 8 pixels per central maximum of the PSF guarantees a contrast of at least 17%, for objects that would have a contrast of 26% in an unpixelated image.[20]

C. Aberrations

In our discussion thus far, it has generally been assumed that imaging conditions were ideal, that is, that the theoretical PSF provides a good description of the actual imaging process. However, aberrations induced by the sample or by elements of the microscope itself result in a distortion of the PSF and a decrease in resolution. For thin specimens imaged close to the coverslip, a high numerical aperture objective (e.g., an oil immersion lens, with NA = 1.4) provides the highest resolution. However, for biological specimens in an aqueous medium, using an oil immersion lens results in aberrations that severely degrade the imaging performance further into the specimen.[34,37–41] PSFs for a confocal microscope arrangement that take into account a change in refractive index (immersion oil refractive index $n = 1.518$; sample refractive index $n = 1.33$) have been calculated

[35] R. Gauderon and C. J. R. Sheppard, *Appl. Optics* **38,** 3562 (1999).

[36] R. H. Webb and C. K. Dorey, *in* "Handbook of Biological Confocal Microscopy" (J. B. Pawley, ed.), p. 55. Plenum Press, New York, 1995.

[37] S. Hell, G. Reiner, C. Cremer, and E. H. K. Stelzer, *J. Microsc.* **169,** 391 (1993).

[38] T. Wilson and A. R. Carlini, *J. Microsc.* **154,** 243 (1989).

[39] D. B. Allred and J. P. Mills, *Appl. Optics* **28,** 673 (1989).

[40] P. Török, P. Varga, Z. Laczik, and G. R. Booker, *J. Opt. Soc. Am. A* **12,** 325 (1995).

[41] A. Rohrbach and E. H. K. Stelzer, *Appl. Optics* **41,** 2494 (2002).

and verified experimentally.[37,42] They show that aberrations due to refractive index mismatch become severe at a depth of 15 μm. These aberrations result in a factor-of-two increase in the axial FWHM extent of the PSF, as well as an axial focal shift of 2.5 μm from the nominal focal position and a 60% reduction in detected intensity (relative to the intensity detected at the coverslip). Similarly, in a two-photon microscope arrangement, experiments using oil immersion objectives to image into biological specimens show a strong reduction in fluorescence intensity as a function of depth, as well as a decrease in axial resolution.[43,44] A water immersion lens with a lower NA performs better in biological specimens than the high-NA oil immersion objective.

Another class of distortions is illustrated in Fig. 8, in which fluorescent beads were imaged through an ~70-μm starfish oocyte (*Asterina miniata*). In Fig. 8a, it is clear that the bead that is not overshadowed by the egg (lower right) is well resolved, whereas that lying beneath the egg (upper left) is substantially distorted by the presence of the material between the egg and the microscope objective. (The overshadowed bead is also substantially dimmer: its brightness has been enhanced here by a factor of 5 for visualization purposes.) These spatially varying distortions make quantitative microscopy challenging, particularly for thick, three-dimensional samples.

Sample-induced spatially varying distortions are difficult to correct: because they change from sample to sample (or even with time, in a live sample), it is not possible to characterize them in advance. In biological microscopy, such effects are most often due to variations in the refractive index of the sample itself, so that the thicker the material through which imaging is attempted, the more distorted the image becomes. This effect is illustrated schematically in Fig. 8c and d. In Fig. 8c, the ideal case of a nonabsorbing, nonrefracting, nonscattering sample is shown. In the sample, a given point, p, emits a spherical wave front that, undistorted by the sample or the microscope optics, can be imaged to a diffraction-limited spot, $I(p)$, which is simply the PSF of the microscope. In contrast, Fig. 8b shows the case of a realistic, thick sample, the structure of which significantly affects the light passing through it. In this case, the wave front emitted by p that emerges from the sample is no longer a perfect spherical wave, but has accumulated distortions as it passes through the sample. Its image is, therefore, not the ideal PSF, but rather some distorted version thereof. Although not explicitly shown in Fig. 8d, these distortions vary with the position of p within the sample. This makes it impossible to make corrections with conventional deconvolution techniques, because these require a "spatially invariant PSF."

[42] S. W. Hell and E. H. K. Stelzer, *in* "Handbook of Biological Confocal Microscopy" (J. B. Pawley, ed.), p. 347. Plenum Press, New York, 1995.

[43] H. Jacobsen, P. E. Hänninen, E. Soini, and S. W. Hell, *J. Microsc.* **176,** 226 (1994).

[44] H. C. Gerritsen and C. J. De Grauw, *Microsc. Res. Tech.* **47,** 206 (1999).

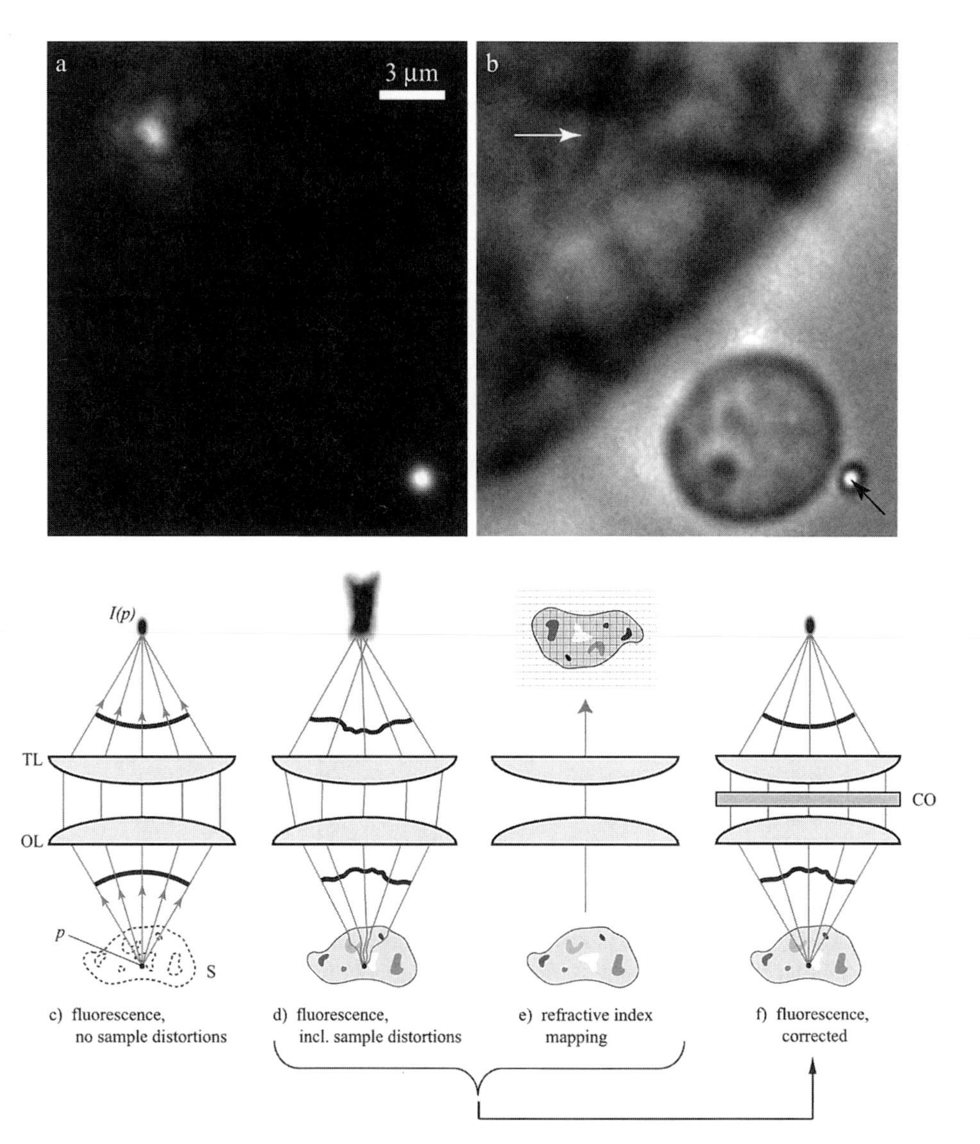

FIG. 8. Image distortions by sample-induced aberrations, and strategies for correction. (a) Fluorescence image of two 1-μm-diameter latex beads, using a water immersion, NA $= 0.75$ objective lens. The uppermost bead lies beneath a 70-μm-diameter starfish (*Asterina miniata*) oocyte, which induces substantial aberrations and required the image brightness beneath the egg to be enhanced 5-fold. (b) Corresponding transmission image, showing the location of the egg (*upper left*) as well as debris on the coverslip (*lower right*). An arrow indicates the location of the uppermost bead. (c) In an ideal sample, the image of a point, p, is the diffraction-limited PSF, $I(p)$. (d) In a realistic (refracting) sample, the wavefront emitted by p is distorted, so that a blurred image $I(p)$ is formed by the microscope. (e) To correct for these sample-induced aberrations, a 3-D refractive index map of the sample could be generated. (f) The refractive index map allows correction of the distortions, using corrective optics or software, resulting in an image that more closely resembles the ideal case. TL, Tube lens; OL, objective lens; S, sample; CO, corrective optics or software.

Because of modern advances in both imaging hardware and software, it is now becoming feasible to attempt to correct both of these types of spatially varying distortions, at least in some instances. A general scheme, adapted from Kam *et al.*,[45] is presented in Fig. 8d–f. The idea is straightforward: if the cause of the aberrations is known, an attempt can be made to correct them. In the case of distortions caused by refractive index variations in the sample, if the variations are known their effects on the wave fronts emitted by the sample can be calculated and, on the basis of this information, corrected. So, in addition to the distorted image of the sample (Fig. 8d), a "map" of the refractive index of the sample is also measured (Fig. 8e). Given this map, it is possible to determine the aberrations that the wave front from point p underwent while passing through the sample, and therefore the correction needed to "deaberrate" the wave front and return it to spherical form. This is illustrated schematically in Fig. 8f, where the correction optics or software removes the distortions, and the result is, ideally, a diffraction-limited image as in Fig. 8c.

Various methods have been proposed to achieve the first step in this process, the generation of the refractive index map. Kam *et al.* measured a stack of differential interference contrast (DIF) images of the sample, which can be integrated to yield a 3-D map of the sample refractive index (or, more precisely, a map of its optical density). This technique is reasonably straightforward, although care must be taken to ensure that the DIC signal is truly linear in the differential phase. Holographic techniques can also be used to determine the refractive index of the sample (or, more precisely, the optical phase shift that it induces) by recording the fringes generated by the interference of a beam passing through the sample with a reference beam.[46] Although large refractive index variations can be measured with this technique, the issue of "phase wrapping" can complicate the analysis when the refractive index changes rapidly. This issue is resolved by measurements such as the transport-of-intensity technique proposed by Barty *et al.*,[47] although the calculations involved can be intensive. See Paganin and Nugent[48] for a review of noninterferometric phase measurement techniques.

The final step in the correction process, depicted in Fig. 8f, is also the most challenging. One approach is based on a hardware, adaptive optical element for the "correction optics" indicated in Fig. 8f. An adaptive optical element, such as a micromirror array or a liquid–crystal spatial light modulator, can be programmed to induce predetermined variations (most often phase) in a wave front. Because the wave front aberrations can be calculated from the refractive index map or, in some cases, directly measured,[49] the adaptive optical element can be used to preaberrate

[45] Z. Kam, B. Hanser, M. G. L. Gustafsson, D. A. Agard, and J. W. Sedat, *Proc. Natl. Acad. Sci. U.S.A.* **98,** 3790 (2001).

[46] S. Schedin, G. Pedrini, and H. J. Tiziani, *Appl. Optics* **39,** 2853 (2000).

[47] A. Barty, K. A. Nugent, D. Paganin, and A. Roberts, *Optics Lett.* **23,** 817 (1998).

[48] D. Paganin and K. A. Nugent, *Adv. Imaging Electron Phys.* **118,** 85 (2001).

[49] M. A. A. Neil, M. J. Booth, and T. Wilson, *Optics Lett.* **25,** 1083 (2000).

the beam so that the net effects (adaptive optics plus sample distortions) cancel each other out. The use of adaptive optics has become relatively common in astronomy for the correction of aberrations caused by atmospheric turbulence in earth-based telescopes. However, such techniques have only relatively recently been applied in microscopy.

Another approach to the aberration correction is a software, rather than a hardware, solution.[45] In this case, instead of actively adjusting the optical wave fronts to remove distortions, the 3-D refractive index map is used to calculate the aberrated PSF for each point in the sample volume. This information, together with the distorted fluorescence image, can be combined through the use of a space-variant deconvolution to yield an approximation to the ideal image of the sample. Although this technique does require extensive computer postprocessing of the recorded data, it can be implemented with standard microscopy hardware (fluorescence and DIC imaging systems) and avoids the need for special adaptive optics.

The effects of sample distortions in optical microscopy are familiar to anyone with even limited experience in biological imaging. Traditionally, they have either been avoided or ignored, but techniques are now emerging that hold promise for removing or correcting these effects, which will be an inestimable benefit for high-resolution microscopy of thick biological samples.

V. Techniques for Improving Resolution

A. General

One obvious method for improving the resolution of an optical microscope is to decrease the wavelength of the illumination. Both lateral and axial resolutions are proportional to the wavelength; hence the resolution can be improved in both directions simultaneously by using a fluorophore that has shorter absorption and emission wavelengths. For single-photon (conventional or confocal) fluorescence, excitation wavelengths generally range from 350 to about 650 nm.[50] For two-photon microscopy, the wavelengths range from about 690 nm (for UV-excitable dyes) to 1100 nm.[6,8] Of course, the specimen usually limits the choice of fluorescence probe. For example, live specimens are sensitive to ultraviolet (below 350-nm) light exposure.[51,52] For live-cell imaging of protein dynamics, there are several fluorescent proteins available whose absorption spectra range from 430 to 570 nm.[7] For a single fluorescent label, it is usual to select a GFP mutant that

[50] R. Y. Tsien and A. Waggoner, *in* "Handbook of Biological Confocal Microscopy" (J. B. Pawley, ed.), p. 267. Plenum Press, New York, 1995.

[51] K. Carlsson, K. Mossberg, P. J. Helm, and J. Philip, *Micron Microsc. Acta* **23,** 413 (1992).

[52] M. Montag, J. Kukulies, R. Jörgens, H. Gundlach, M. F. Trendelenburg, and H. Spring, *J. Microsc.* **163,** 201 (1991).

performs best in terms of photostability, quantum efficiency, size, and so on, as opposed to one that offers the best resolution (EGFP is currently a good choice in many cases). For double labeling, choice is limited to combinations of spectrally separated GFP mutants, such as cyan and yellow fluorescent proteins (CFP and YFP), or EGFP and DsRed. In essence, the specimen and the nature of the experiment usually determine which wavelengths can be used.

Another way to improve the resolution (especially axially) is to use an objective lens with a higher numerical aperture. The lateral resolution of a microscope is inversely proportional to the NA of the objective lens, and the axial resolution is inversely proportional to the square of the NA. Again, there are limitations. One factor that limits the NA of the objective lens is the working distance of the lens: an NA = 1.4 oil immersion objective lens with a 63 times magnification (63×) generally has a working distance of about 250 μm; whereas an NA = 0.9, 63×, water immersion lens can have a working distance as great as 2.0 mm. A relatively long working distance is required for imaging thick specimens below the specimen surface. Also, spherical aberrations induced by refractive-index mismatch prevent the use of an oil immersion objective to focus into live (water-based) specimens, as discussed in Section IV,C.

B. Deconvolution

A recorded microscope image is the convolution of the actual structure under observation with the point-spread function of the imaging system. Provided that detailed knowledge of the shape of the system PSF is available, the PSF can be computationally deconvolved from the image, and an image can be recovered that more closely resembles the actual structures. A resolution enhancement in both the lateral and axial directions can be achieved for conventional fluorescence images, as well as for confocal and two-photon fluorescence images. Unfortunately, although the system PSF can be measured for ideal specimens, specimen-induced aberrations affect the PSF locally so that for real specimens the PSF can never be fully characterized (see Section IV,C). Also, deconvolution does not provide true optical sectioning for thick specimens. In many software packages deconvolution is combined with the inclusion of *a priori* knowledge. For example, it might be assumed that the objects are spherical, and if the lateral extent can be determined then the axial extent is also known and identical. For a detailed treatment and comparison of deconvolution techniques, see, for example, Refs. 53–55.

[53] T. J. Holmes, S. Bhattacharyya, J. A. Cooper, D. Hanzel, V. Krishnamurthi, W. Lin, B. Roysam, D. H. Szarowski, and J. N. Turner, *in* "Handbook of Biological Confocal Microscopy" (J. B. Pawley, ed.), p. 389. Plenum Press, New York, 1995.

[54] P. J. Verveer, M. J. Gemkow, and T. M. Jovin, *J. Microsc.* **193,** 50 (1999).

[55] M. M. Falk and U. Lauf, *Microsc. Res. Tech.* **52,** 251 (2001).

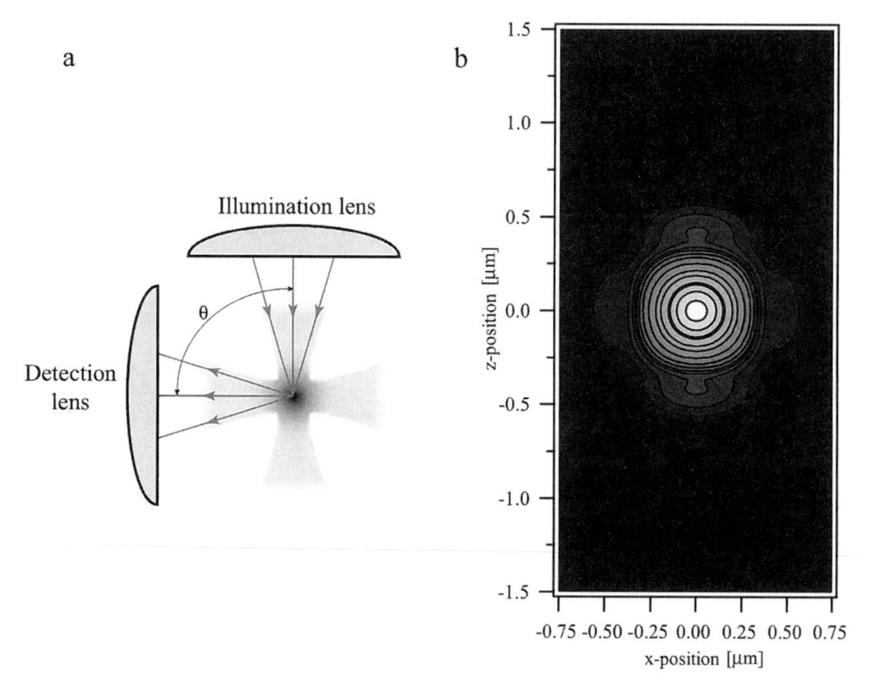

FIG. 9. (a) In theta microscopy, the axes of illumination and detection are tilted relative to each other by an angle θ, close to 90°. (b) The PSF of the theta microscope in the *XZ* plane, calculated using $\theta = 90°$ and NA $= 0.9$, is nearly isotropic. The FWHM extents are 0.30 and 0.28 μm in the lateral and axial directions, respectively.

C. *Confocal Theta Microscopy*

Looking back at Fig. 2, it can be seen that all the PSFs presented there have much greater axial extents compared with their lateral extents. Confocal theta microscopy overcomes this elongation of the observation volume to achieve an almost isotropic resolution.[56] Theta microscopy works by changing the spatial configuration of the illumination and detection volumes by using different axes for illumination and detection, as depicted in Fig. 9a. One objective lens is used to illuminate the sample. A second objective lens, whose optical axis is at an angle θ to the illumination axis, is used to collect the emitted fluorescence. The system PSF, given by

$$\begin{aligned} |h_{\text{Confocal Theta},90°}(x, y, z)|^2 &= |h_{\text{ill}}(x, y, z)|^2 |h_{\text{det},90°}(x, y, z)|^2 \\ &= |h_{\text{ill}}(x, y, z)|^2 |h_{\text{det}}(z, y, x)|^2 \end{aligned} \tag{12}$$

[56] E. H. K. Stelzer and S. Lindek, *Optics Commun.* **111,** 536 (1994).

is almost isotropic for the angle $\theta = 90°$ [57]; but angles between 70° and 110° still result in acceptably small volumes.[58] The confocal theta PSF shown in Fig. 9b was calculated using the parameters given in Table I, except that the numerical aperture was NA = 0.9. Both the lateral and axial extents of the confocal theta PSF are comparable to the lateral extent of the confocal fluorescence PSF of similar numerical aperture.

There are several ways to realize such an optical arrangement. Figure 9a illustrates the most obvious arrangement, in which two long-working-distance microscope objectives are positioned in such a way that an angle θ of nearly 90° between illumination and detection axis is achieved.[59] In practice, this arrangement is difficult to realize on a standard microscope base. A more elegant solution, which can be adapted to any standard confocal microscope, is the single-lens theta microscope arrangement.[60,61] In both arrangements, the specimen must be mounted in a special way (e.g., on a glass capillary) and scanned through the focus of the objective. Theta microscopy can also be realized in a nonconfocal arrangement using two-photon excitation.[62]

Theta microscopy is not intended for systems that require the resolution provided by a high-NA objective. Theta microscopy is, however, excellently suited for optical systems that use low-NA, long-working-distance objective lenses,[63] such as those used for imaging whole embryos. A detailed discussion of the resolution, efficiency, and working distance achieved in a single-lens theta microscope has been published elsewhere.[64]

D. 4Pi Confocal Microscopy

In 4Pi microscopy, the sample is illuminated and/or detected coherently through two opposing objective lenses having a common focus.[59,65,66] The use of two objective lenses essentially doubles the angular aperture (approaching the maximum 4π solid angle—hence the name), and therefore provides an improvement in resolution, most notably in the axial direction. An axial FWHM extent (for the central peak of the PSF) of 75 nm has been confirmed experimentally for a 4Pi confocal

[57] S. Lindek, N. Salmon, C. Cremer, and E. H. K. Stelzer, *Optik* **98,** 15 (1994).
[58] S. Lindek, R. Pick, and E. H. K. Stelzer, *Rev. Sci. Instrum.* **65,** 3367 (1994).
[59] S. Lindek, E. H. K. Stelzer, and S. W. Hell, *in* "Handbook of Biological Confocal Microscopy" (J. B. Pawley, ed.), p. 417. Plenum Press, New York, 1995.
[60] S. Lindek, T. Stefany, and E. H. K. Stelzer, *J. Microsc.* **188,** 280 (1997).
[61] J. Swoger, S. Lindek, T. Stefany, F.-M. Haar, and E. H. K. Stelzer, *Rev. Sci. Instrum.* **69,** 2956 (1998).
[62] S. Lindek and E. H. K. Stelzer, *Optics Lett.* **24,** 1505 (1999).
[63] E. H. K. Stelzer, S. Lindek, S. Albrecht, R. Pick, G. Ritter, N. J. Salmon, and R. Stricker, *J. Microsc.* **179,** 1 (1995).
[64] S. Lindek, J. Swoger, and E. H. K. Stelzer, *J. Mod. Optics* **46,** 843 (1999).
[65] S. Hell and E. H. K. Stelzer, *J. Opt. Soc. Am. A* **9,** 1259 (1992).
[66] S. Hell and E. H. K. Stelzer, *Optics Commun.* **93,** 277 (1992).

arrangement.[67,68] Here, we discuss the type of 4Pi microscopy called 4Pi(A), in which coherent illumination through both objective lenses is used together with incoherent detection through only one lens.[65,66]

In 4Pi(A) microscopy, coherent illumination wave fronts from the two objective lenses interfere in the focal volume. The 4Pi-illumination PSF is calculated by adding two counterpropagating amplitude PSFs with a controllable phase difference between them:

$$|h_{4\text{Pi,ill}}(x, y, z)|^2 = |h_{\text{ill}}(x, y, z) + e^{i\phi} h_{\text{ill}}(x, y, -z)|^2 \tag{13}$$

The 4Pi illumination PSF (not shown) is strongly modulated along the z axis. The 4Pi confocal PSF is calculated, as usual, by multiplying the 4Pi illumination PSF with the standard detection PSF (Fig. 2c):

$$|h_{4\text{Pi(A)}}(x, y, z)|^2 = |h_{4\text{Pi,ill}}(x, y, z)|^2 |h_{\text{det}}(x, y, z)|^2 \tag{14}$$

The resulting PSF is shown in Fig. 10a (for the XZ plane), together with line scans along the z axis in Fig. 10c. Line scans for the confocal and 2p-confocal microscopes are shown for comparison. The 4Pi confocal PSF has a narrow central peak in the axial direction (FWHM $= 0.12$ μm), but experiences severe ringing: almost 60% of the total illumination intensity occurs within the axial side lobes. A significant improvement in 4Pi confocal microscopy has been realized by the use of two-photon excitation. Combining 4Pi illumination with two-photon excitation and confocal detection results in the system PSF shown in Fig. 10b, whose FWHM extent is marginally broader than the 4Pi confocal microscope, but whose axial sidelobes are substantially reduced. Such a microscope arrangement has been extensively investigated.[65,66,69,70] Further improvements have been the use of point deconvolution to computationally remove the axial side lobes, and active compensation of relative phase changes between the interfering wave fronts.[71–73] A two-photon 4Pi confocal microscope has achieved an axial resolution of 145 nm for observing fluorescently labeled actin fiber bundles in a fixed fibroblast cell, at depths of up to 5 μm in a fixed cell,[73] that is, under ideal circumstances.

E. Structured Illumination

The use of structured illumination (also called harmonic excitation, standing wave illumination, or modulated excitation) to improve the resolution in

[67] S. W. Hell, S. Lindek, and E. H. K. Stelzer, *J. Mod. Optics* **41,** 675 (1994).
[68] S. W. Hell, S. Lindek, C. Cremer, and E. H. K. Stelzer, *Appl. Phys. Lett.* **64,** 1335 (1994).
[69] M. Gu and C. J. R. Sheppard, *Optics Commun.* **114,** 45 (1995).
[70] P. E. Hänninen, S. W. Hell, J. Salo, E. Soini, and C. Cremer, *Appl. Phys. Lett.* **66,** 1698 (1995).
[71] S. W. Hell, M. Schrader, P. E. Hänninen, and E. Soini, *Optics Commun.* **128,** 394 (1996).
[72] S. W. Hell, M. Schrader, and H. T. M. van der Voort, *J. Microsc.* **187,** 1 (1997).
[73] M. Schrader, K. Bahlmann, G. Giese, and S. W. Hell, *Biophys. J.* **75,** 1659 (1998).

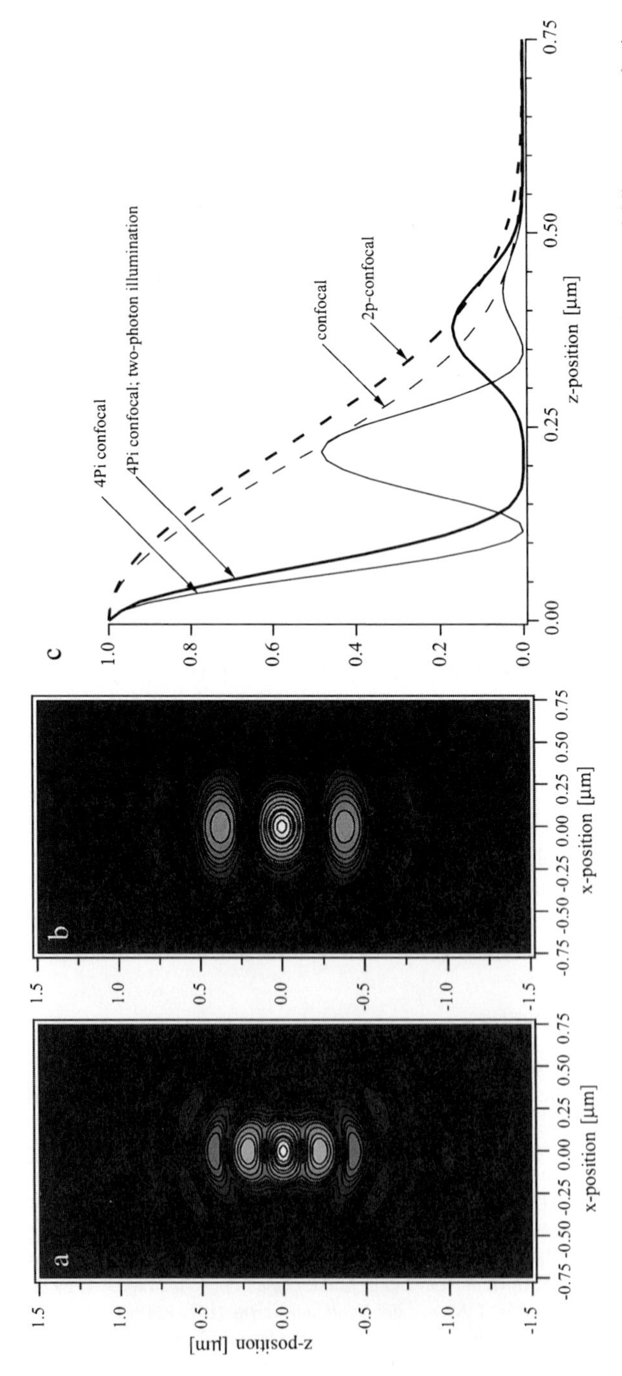

FIG. 10. Point-spread functions for 4Pi(A) confocal microscopy using (a) single-photon and (b) two-photon illumination. (c) Axial line scans emphasize the severe ringing from which the 4Pi techniques suffer. Two-photon illumination helps damp the side lobes somewhat, but also reduces the resolution. Axial line scans for the confocal and 2p-confocal PSFs are shown as references.

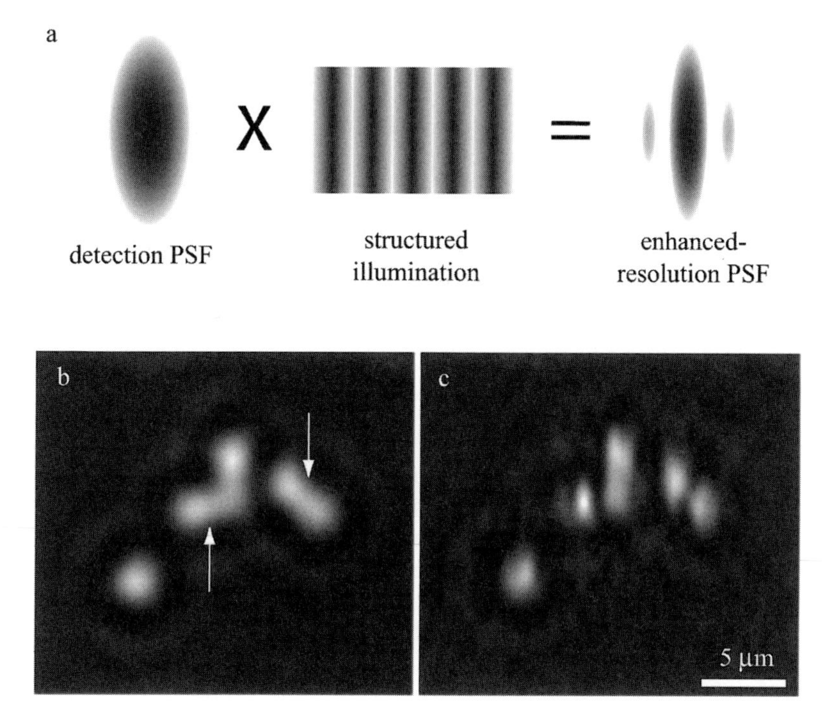

FIG. 11. (a) Schematic representation of structured illumination microscopy. (b) Wide-field image of a group of fluorescent latex beads. (c) Corresponding structured illumination image using vertical interference fringes. The images were acquired with: NA = 0.25 (in air); illumination = 488 nm; detection above 550 nm; illumination fringe spacing = 0.84 μm; bead diameter = 1.0 μm. The same CCD camera, filters, etc., were used for both images, and the same deconvolution algorithm was applied.

fluorescence microscopy is a technique that has seen increased popularity.[74–77] In this technique, instead of uniform illumination (i.e., wide field) or point illumination (i.e., confocal), the sample is illuminated by a spatially varying intensity pattern generated by the interference of two or more beams. For illumination through a given objective lens, structured illumination allows a theoretical doubling of the lateral resolution of a fluorescence microscope, and, in more complex configurations, substantial improvement of the axial resolution as well.[78]

The mechanism behind the resolution improvement in structured illumination is illustrated in Fig. 11a. As discussed in Section II,C, the system PSF is the product of the illumination and detection PSFs: in this case, this amounts to the

[74] J. T. Frohn, H. F. Knapp, and A. Stemmer, *Proc. Natl. Acad. Sci. U.S.A.* **97,** 7232 (2000).
[75] M. G. L. Gustafsson, *J. Microsc.* **198,** 82 (2000).
[76] B. Albrecht, A. V. Failla, R. Heintzmann, and C. Cremer, *J. Biomed. Optics* **6,** 292 (2001).
[77] J. T. Frohn, H. F. Knapp, and A. Stemmer, *Optics Lett.* **26,** 828 (2001).
[78] M. G. L. Gustafsson, D. A. Agard, and J. W. Sedat, *J. Microsc.* **195,** 10 (1999).

detection PSF, which we have seen is equivalent to the PSF of a traditional wide-field microscope, modulated by the illumination pattern, as indicated in Fig. 11. In fact, the imaging process is somewhat more complicated than this, involving the measurement of multiple images with phase shifts in the illumination pattern and a substantial amount of postprocessing. Note that the resolution improvement is primarily in the direction perpendicular to the interference fringes (i.e., in the horizontal direction in Fig. 11a). To achieve an isotropic enhancement of the resolution, images that contain fringe patterns in multiple directions must be combined.

In Fig. 11b and c, a structured illumination image of a group of 1-μm fluorescent latex beads is compared with a similar wide-field image. In the wide-field image (Fig. 11b), for which the illumination was uniform, it can be seen that while the isolated bead appears as a circular spot, beads that are closer together are not quite resolved. In contrast, beads that are slightly displaced from each other in the horizontal direction in the structured illumination image (indicated by arrows in Fig. 11b) are well separated. However, beads separated by a primarily vertical offset are not substantially better resolved in the structured illumination image than the wide-field one. The reason for this anisotropic resolution is that the illumination interference fringes (not shown) were oriented vertically. By using multiple fringe orientations, a more isotropic resolution improvement could be achieved, as has been shown in the literature (e.g., Gustafsson[75]).

F. Stimulated Emission Depletion

Stimulated emission depletion (STED) microscopy is a promising technique for achieving enhanced three-dimensional resolution.[79–81] The technique relies on quenching the fluorescence from the outer part of the focal spot by stimulated emission depletion.[82] The process is indicated schematically in Fig. 12. A visible illumination pulse excites the fluorophores in a standard PSF volume. A second, depletion pulse of slightly longer wavelength follows a few picoseconds after the excitation pulse. Using an optical phase plate, the depletion pulse is shaped so that the PSF is intense in a ring around the focal point, and dark at its center. The depletion pulse stimulates quasi-instantaneous, coherent emission in the focal ring immediately after excitation; the longer lived fluorescence in the (nonstimulated) center of the focal volume is subsequently detected. By proper tailoring of the depletion PSF, both axial and lateral resolution improvements can be achieved. An axial resolution of about 100 nm was achieved using STED, corresponding to a 5-fold improvement over a similar confocal microscope arrangement. For further comments on the potentials and limitations of this technique, see Weiss.[83]

[79] T. A. Klar and S. Hell, *Optics Lett.* **24,** 954 (1999).

[80] T. A. Klar, S. Jakobs, M. Dyba, A. Egner, and S. W. Hell, *Proc. Natl. Acad. Sci. U.S.A.* **97,** 8206 (2000).

[81] T. A. Klar, M. Dyba and S. W. Hell, *Appl. Phys. Lett.* **78,** 393 (2001).

[82] S. W. Hell and J. Wichmann, *Optics Lett.* **19,** 780 (1994).

[83] S. Weiss, *Proc. Natl. Acad. Sci. U.S.A.* **97,** 8747 (2000).

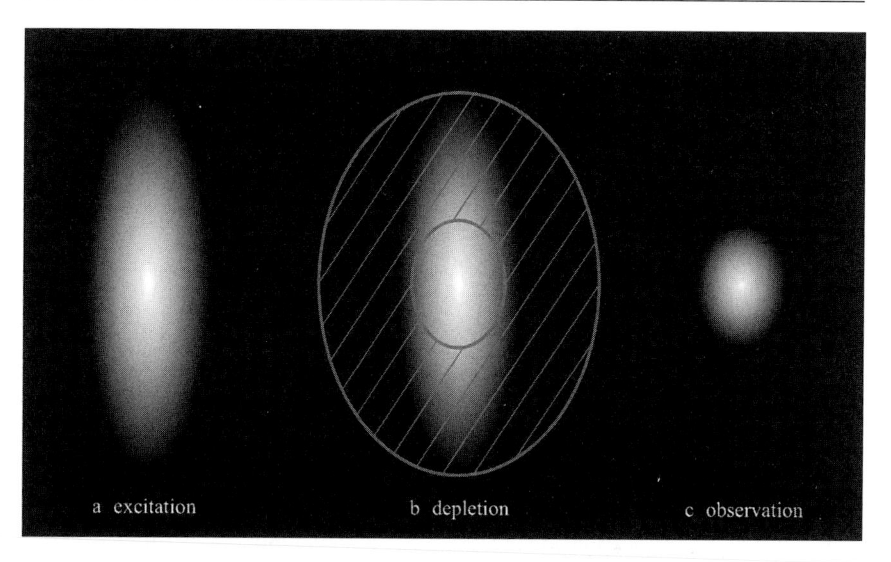

FIG. 12. Mechanism for resolution improvement using stimulated emission depletion. (a) The traditional illumination PSF excites the fluorophores within its volume. (b) The depletion pulse, the spatial extent of which is indicated by diagonal lines, depletes the excitation in the outer regions of the PSF by stimulated emission. (c) The resulting observation PSF, which determines the image resolution, is substantially reduced in size, compared with (a).

VI. Conclusions

Optical microscopy remains an important tool for the probing of biological processes and biophysical parameters, especially in living cells and tissues. Driving the development of new optical microscopes are demands for higher resolution, faster image acquisition, greater flexibility (i.e., allowing a greater range of imaging-based experiments), and, of course, reduced cost. Each new microscope must find a niche in one or more of these four areas that makes it applicable.

For relatively thin specimens, wide-field fluorescence microscopy provides high-resolution images quickly, with little effort, and at little expense. Transmission contrasts, such as phase contrast or differential interference contrast (DIC), can easily be used to supplement the fluorescence images in order to visualize unlabeled structures in the specimen. Structured illumination techniques (Section V,E) can be used to improve both the lateral and axial resolution of wide-field images for thin specimens, although these techniques are computationally demanding and not yet commercially available. Projecting a grid into a relatively thin specimen provides an inexpensive method for obtaining optically sectioned images with a wide-field microscope.

For thick specimens, confocal microscopes provide optical sectioning to achieve high-resolution, three-dimensional images. A laser-scanning confocal microscope is extremely flexible, offering the ability to scan or magnify arbitrary regions. Multiple excitation wavelengths provide optimal spectral separation of multiple fluorophore signals (e.g., for the combination CFP–YFP; Ellenberg *et al.*[84]). Water immersion objective lenses provide excellent resolution when imaging deep inside biological specimens. Confocal microscopes provide an excellent base for the study of protein dynamics in living cells,[85] including live-specimen imaging experiments such as fluorescence recovery after photobleaching (FRAP),[86] and fluorescence resonance energy transfer (FRET).[87] Spinning disk confocal microscopes are much less flexible than their laser scanning counterparts; but they find a niche in the observation of fast processes in living cells.

Two-photon microscopes also provide optical sectioning for observing thick specimens. However, the resolution of a two-photon microscope and its optical sectioning performance are worse than those of a comparable confocal microscope. Considerably higher optical powers are needed for two-photon excitation of fluorophores; and the excitation efficiency (in terms of collected photons per bleached fluorophore) is worse than in the single-photon case.[88] The broad two-photon absorption cross-sections of many dyes do not permit selective excitation of fluorophores in multiply labeled specimens. Because illumination is confined to the focal plane, two-photon microscopy results in reduced photobleaching outside of the focal volume; however, photobleaching in the focal plane is higher than in the confocal case.[89] Despite these well-known drawbacks, the two-photon microscope seems to find a niche in imaging deep within scattering specimens,[11,90] especially when a low-magnification, high-NA lens is employed.[13,14] Such applications benefit from two aspects of the two-photon microscope: first, there is reduced scattering in biological specimens at infrared excitation wavelengths, allowing for deeper penetration of the excitation beam. Second, the fact that all emitted fluorescent photons must stem from the confined illumination volume permits the use of large-area detectors, which collect photons more efficiently than point detectors. Two-photon microscopes operated with confocal detection [e.g., the multiphoton multifocal microscope (MMM) proposed by Straub and Hell][91] cannot take advantage of this enhanced detection efficiency, and are hence not useful for imaging deep inside scattering specimens.

Acknowledgments

The authors thank Luisa Pieroni, Valentina Greco, and Eric Karsenti for the specimens.

[84] J. Ellenberg, J. Lippincott-Schwartz, and J. F. Presley, *Trends Cell Biol.* **9,** 52 (1999).

[85] J. Lippincott-Schwartz, E. Snapp, and A. Kenworthy, *Nat. Rev. Mol. Cell Biol.* **2,** 444 (2001).

[86] J. White and E. Stelzer, *Trends Cell Biol.* **9,** 61 (1999).

[87] P. van Roessel and A. H. Brand, *Nat. Cell Biol.* **4,** E15 (2001).

[88] L. Brand, C. Eggeling, C. Zander, K. H. Drexhage, and C. A. M. Seidel, *J. Phys. Chem. A* **101,** 4313 (1997).

[89] G. H. Patterson and D. W. Piston, *Biophys. J.* **78,** 2159 (2000).

[90] D. W. Piston, *Trends Cell Biol.* **9,** 66 (1999).

[91] M. Straub and S. W. Hell, *Bioimaging* **6,** 177 (1998).

[19] Video-Rate Confocal Microscopy

By MICHAEL J. SANDERSON and IAN PARKER

Introduction

The ability to observe cell function through the light microscope is a simple and direct method for obtaining a wealth of experimental information. The versatility of this approach has been greatly extended by the development of a diversity of fluorescent reporter dyes and molecular probes. Cell activities involving changes in ion concentrations (e.g., Ca^{2+} or pH) or the expression of proteins can be readily observed. However, a constant challenge is the requirement for high resolution both in the spatial and temporal domains. Typically, fluorescent molecules that are outside the plane of focus reduce image contrast and resolution. Thus, specimens with substantial tissue thickness or depth (i.e., almost everything except cultured or thin layers of cells) are difficult study with wide-field microscopy. Fortunately, confocal laser scanning microscopy (CLSM) provides one solution to this problem. With CLSM, a small aperture rejects the out-of-focus light and the final image represents a thin slice through the specimen with greatly improved axial resolution. However, a disadvantage of CLSM is that it requires that the specimen is, in general, illuminated with a scanning point of light. This has the consequence that the time needed to create an image is governed by the time to scan the specimen and the time or sensitivity of the instrumentation to detect fluorescent light from a small point source. Although most modern confocal microscopes employ a laser to produce a high-intensity point source for the excitation illumination (and thereby increase the amount of fluorescence emitted), the time taken to acquire an image can be relatively long. A slow acquisition time is unimportant when studying fixed or slowly changing specimens but precludes the study of many rapid biological processes.

Fast [30 frames per second (fps) or video-rate] confocal microscopes have been previously manufactured, but unfortunately, the cost of these microscopes was usually beyond the budget of individual investigators. These microscopes included the Nikon (Melville, NY) RCM 8000, a system that reflects a design described in detail by Tsien and Bacski[1]; the Noran (Middleton, WI) Odyssey; and the Bio-Rad (Hercules, CA) RTS2000. A limited number of sales of these high-speed confocal microscopes may explain the withdrawal from the market of one of the leading manufacturers (Noran). A consequence of this withdrawal will be the lack of support for some users—a situation to be avoided. One way to

[1] R. Tsien and B. J. Bacski, *in* "Handbook of Biological Confocal Microscopy" (J. B. Pawley, ed.), p. 459. Plenum Press, New York, 1995.

0076-6879/03 $35.00

circumvent all these problems and to acquire a high-speed confocal microscope, at an affordable price and with the assurance of continued maintenance, is to build it yourself.

Contrary to what might be expected, this requires relatively little experience in optics, electronics, or programming. The CLSM featured here is based on the specifications described by Callamaras and Parker[2] and represents the first microscope built (independently, but with considerable help) to that design by another investigator. The details of the original design are reiterated here, together with detailed practical instructions, diagrams for construction, and extensive appendices listing parts and supplies. In addition, several improvements are described, including the ability to correct, in real time, the image distortion arising from the variation in scanner speed, and to record (and play back) large numbers of images directly to (and from) a computer hard drive at both 30 (400 lines ×420 pixels) and 60 (200 lines ×420 pixels) frames per second.

Design Considerations and Methods

Although the history and details of confocal microscopy design cannot be addressed in this article, these are extensively reviewed in the text *Handbook of Biological Confocal Microscopy*.[3] This book is recommended for studying the practical issues of confocal microscopy.

Microscope Selection

Basic Microscope. An inverted microscope is the preferred choice for the construction of the CLSM because this type of microscope usually has an optical or camera port near bench level, which facilitates the integration of optical components with the microscope (Figs. 1 and 2). The choice between a 160-mm tube-length microscope and an infinity-corrected optical microscope is less critical. Both the Nikon Diaphot 200 (160-mm tube length) and the Olympus (Melville, NY) IX70 (infinity-corrected optics) have been successfully used as the base for a CLSM.

Image Redirection. The projection of an image to the optical side port is usually achieved by redirecting the light from the objective with an internal prism. The options available include redirecting 100% of the light so that the specimen can no longer be seen via the eyepieces or redirecting 80% of the light so that the specimen can be observed simultaneously via the eyepieces. Because the excitation light follows the same pathway as the emitted light in the CLSM, the properties of the redirecting prism will also influence the excitation illumination. As a result,

[2] N. Callamaras and I. Parker, *Cell Calcium* **26,** 271 (1999).

[3] J. B. Pawley, ed., "Handbook of Biological Confocal Microscopy." Plenum Press, New York, 1995.

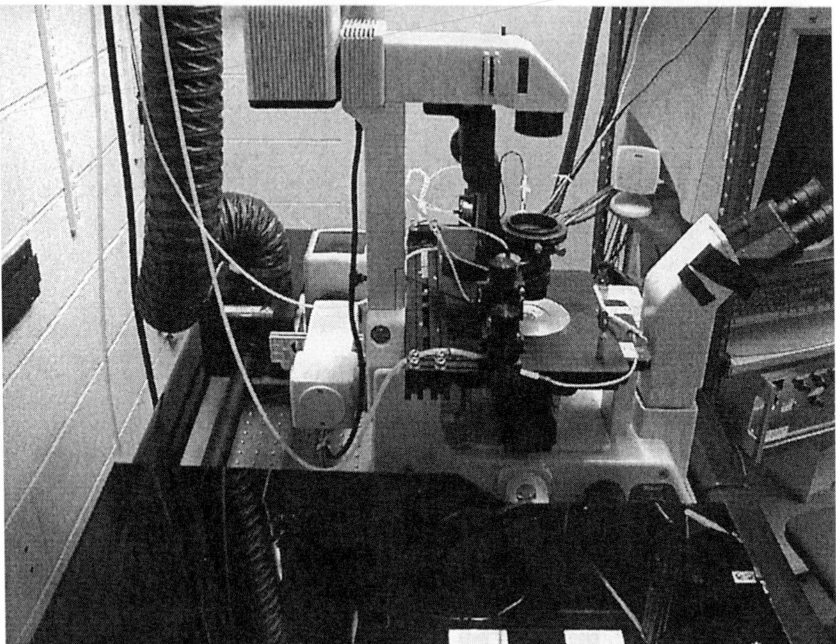

FIG. 1. A front view (*top*) and side view (*bottom*) of the confocal scanning laser microscope, illustrating the basic layout of the instrument on an air table. The black box contains the optical components. The flexible hose serves to exhaust cooling air to a ceiling air-conditioning vent.

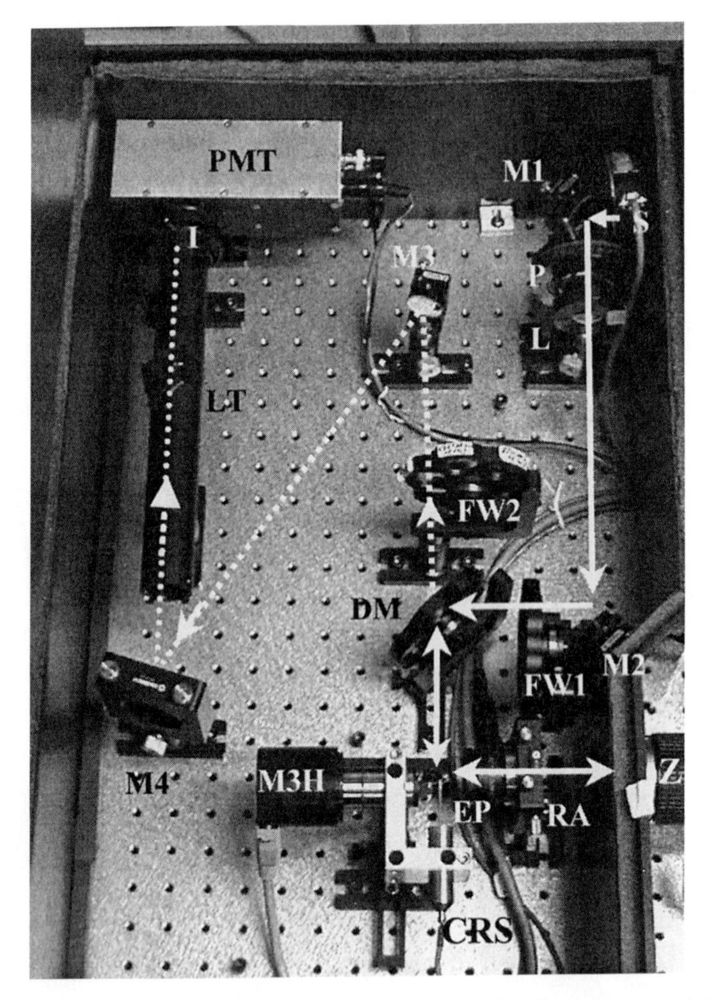

FIG. 2. A top view of the layout of the optical components and pathway for the excitation and emitted light in the CLSM. The solid white line indicates the path of the excitation argon laser light (488 nm). The laser enters the system through the shutter (S) and is reflected by mirror 1 (M1) to pass through the polarizer (P) and concave lens (L). The light is reflected by mirror 2 (M2), through filter wheel 1 (FW1) to the dichroic mirror (DM). The mirrors of the vertical scanner (M3H) and horizontal scanner (CRS) reflect the light through the eyepiece (EP) and rectangular aperture (RA) into the zoom adapter (Z) of the microscope. The emitted light (dotted line, >510 nm) initially follows the path of the excitation light until it reaches the DM. The emitted light passes through the DM and filter wheel 2 (FW2) to mirror 3 (M3) for reflection to the photomultiplier (PMT) via mirror 4 (M4), the light tube (LT), and confocal aperture (I). The optics are surrounded by a black Plexiglas box.

the choice of prism must be considered in view of a number of advantages and disadvantages. By redirecting all the light, the instrument will be most sensitive because all the excitation light reaches the specimen and all the emitted light can be monitored. In addition, stray light from the eyepieces is never a problem. However, with this approach, the alignment and location of the area or line being scanned by the system are more difficult to determine. By contrast, the simultaneous viewing of the specimen and scan area greatly facilitates alignment and specimen selection. A simple solution would be to have both options of 80 or 100% redirection available. Although we have not implemented this solution, the Nikon Diaphot 300 is equipped with this option. It is likely that some fine adjustments to the alignment would be required when switching between prisms.

SAFETY NOTE. If any laser light can reach the eyepieces, a safety barrier filter must be installed in the binocular microscope head. A long-pass 520-nm filter will block the 488-nm laser line but will transmit the fluorescent light.

Conventional Fluorescence Microscopy. The ability to quickly find the region of interest in the specimen is critical for successful observations and an initial search is facilitated by a wide field of view. Unfortunately, a wide field of view is not compatible with the small scanning area of the CLSM, but a wide field of view can be obtained by viewing the specimen by conventional epifluorescence illumination, using a mercury (Hg) or xenon (Xn) bulb. The excitation wavelength of light is directed to the specimen by inserting a standard filter cube carrying a band-pass excitation filter (485 nm) and dichroic mirror (505 nm) together with an appropriate emission filter (520 nm). Once the area of interest is found and centered in the field, the optics are easily changed for CLSM by sliding the filter block out of the way.

Phase-Contrast Microscopy. When examining fluorescent specimens with CLSM, it is often essential to obtain a nonfluorescent or transmitted light image of the specimen. This allows the spatial orientation and organization of the specimen to be interpreted. From our experience, the study of different tissues emphasizes the variability in this requirement. The study of calcium signaling in relatively large oocytes, which have a homogeneous cytosol, has little need for structural images. By contrast, the study of intercellular calcium signaling in multicellular systems cannot be achieved without structural information. Although the current CLSM design does not incorporate the acquisition of transmitted confocal images, this would be possible by adding a transmitted light detector after the condenser lens. An alternative approach is to make use of the conventional optics of the microscope. The Diaphot 300, like many inverted microscopes, has an additional camera port that can be selected. With a long working distance phase-contrast condenser and phase-contrast objective, phase-contrast images can be obtained with a separate 35-mm or charge-coupled device (CCD) camera. Use of a phase-contrast objective does not appear to cause any obvious degradation of the confocal image.

Microscope and Optical Mountings. It is advisable to establish the system on an air-suspended optical table, not only to isolate the system from external

TABLE I
MAJOR WAVELENGTHS OF EXCITATION LIGHT PRODUCED BY DIFFERENT LASERS AND CORRESPONDING FLUORESCENT DYES

Excitation wavelength (nm)	Dye	Laser
364	Indo-1	UV argon
488	Fluo-3, Fluo-4	Argon, air cooled (argon–krypton)
	FITC	
	Oregon Green	
	Fura Red	
	GFP	
514	YFP	
543	Cy-3	Green He–Ne
	TRITC	
	DiI	
568	TRITC	Krypton (argon–krypton)
	DsRed	
594	Texas Red	Yellow He–Ne
647	Cy5	Krypton (argon–krypton)

vibration, but for ease of aligning and integrating optical components (Figs. 1 and 2). A common selection is a table with a stainless steel top with threaded holes (0.25 inch, 20 thread) centered on a 1-inch grid. The alignment of optical components is achieved with a range of adjustable mounting posts that are available from all the major optical companies. Because the exact height requirements for each component vary, it is advisable to have a selection of posts and post holders ranging from 1 to 4 inches available. The posts screw directly into the air table or can be mounted on an adjustable base when their alignment does not coincide with the tabletop hole pattern.

Laser Selection

A major component of the CLSM is the excitation laser. Consequently, the choice of laser or lasers should be considered carefully in view of future experimental goals. The development of a wide range of fluorescent dyes, which can be used to detect a variety of biological molecules, provides the means for a versatile instrument.[4] Although it is not feasible to design an instrument for all possibilities, a wide range of options can be accessible if the major laser line excitation wavelengths are available. These major wavelengths include 364, 488, 514, 543, 568, 594, and 647 nm. Some of the lasers that produce the appropriate excitation lines for various dyes are summarized in Table I.

[4] R. Y. Tsien and A. Waggoner, *in* "Handbook of Biological Confocal Microscopy" (J. B. Pawley, ed.), p. 267. Plenum Press, New York, 1995.

For the basic system, a single air-cooled argon laser provides laser lines of 488 and 514 nm, which can be used to excite a variety of dyes. However, if multiple dye experiments are required, the choice of an argon–krypton laser would add lines at 568 and 647 nm. Alternatively, a second green helium–neon laser can be incorporated to provide a line at 543 nm. The choice of lasers depends mainly on the application and resources available. It should also be noted that certain laser lines fade faster than others in mixed gas lasers. As a result, separate lasers for each line may be more cost effective.

The argon laser (maximum output, 100 mW) is frequently operated at about 25 mW (in stand-by mode) to conserve tube life. Therefore, a less powerful laser can be purchased to save costs. However, the argon laser model used is compact and has the advantage of having some reserve capacity for weak signals. The laser is purchased with a cooling fan unit. Because the laser generates a considerable amount of heat, it is advantageous to plumb the exhaust vent into the air conditioning system. To lessen noise, a mounting location above the ceiling tiling is also advantageous. A switched 20-A, 120-V power supply is required, and is preferably dedicated to the laser alone.

Laser Scanning Mechanism

The scanning of the laser beam across the specimen is achieved with two scanning mirrors mounted at right angles. For alignment, each scanner is mounted in a simple bracket so that the mirror can be moved along its axis as well as rotated before being clamped in place (Figs. 3 and 4). The complete assembly is mounted on a rail for alignment. The horizontal scan is generated by a mirror mounted on a resonant scanner that oscillates at a fixed frequency of about 8 kHz (Counter Rotating scanner, CRS; GSI Lumonics, Bedford, MA). This mirror rotates through an angle (up to 26°) with an angular velocity that reflects a cosine function. The fact that the mirror moves in a predictable manner means that a software algorithm can be performed to correct the image distortion that results from the change in mirror velocity (see Software for Correction of Image Distortion, below). The pair of controller boards supplied with the scanner provide all the necessary electronics for the operation of the scanning mirror. The user needs only to supply power and a control voltage to regulate the magnitude of the scan angle. The magnitude of the scan angle determines the horizontal magnification. These boards also provide the horizontal synchronization signal.

The vertical scan is generated by a second mirror mounted on an M3H scanner (GSI Lumonics) that rotates in response to a sawtooth waveform at 30 Hz. The mirror attached to the M3H is mounted "off axis" at the end of a short extension arm or paddle. This arrangement minimizes the amount of beam rotation induced by scanning mirrors.[2] The M3H scanner is also controlled by an independent driver board (OATS driver) and requires only a power supply and two input voltages (a sawtooth waveform and an offset) for position control. GSI Lumonics

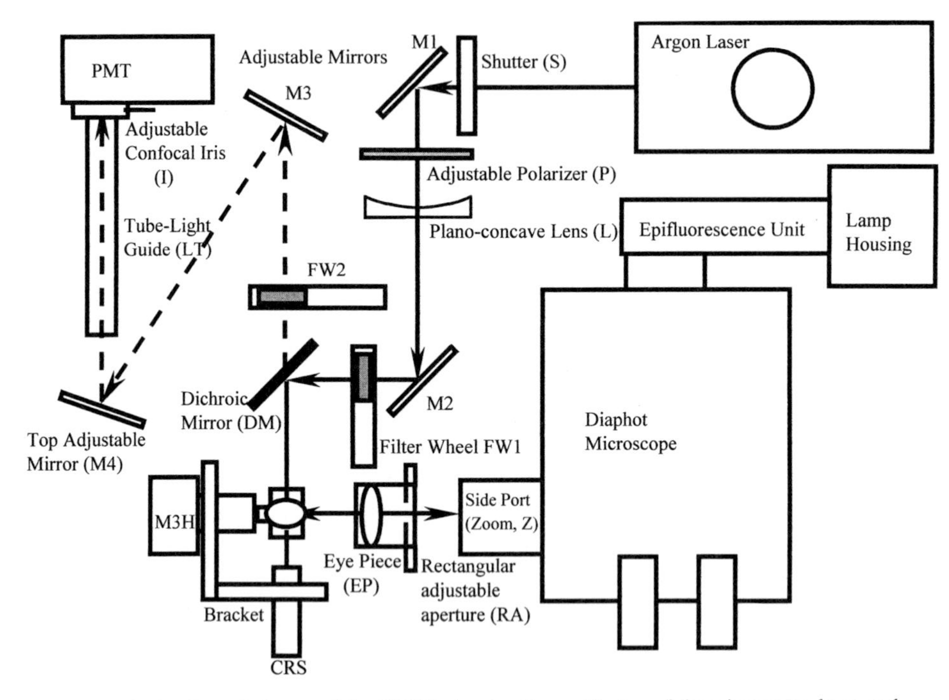

FIG. 3. A schematic layout of the CLSM, showing the positioning of the microscope, laser, and components shown in Fig. 2. Abbreviations are the same as for Fig. 2.

recommends the replacement of the M3H scanner with an improved M3S scanner and "MiniSax" driver. These components have similar characteristics and should be easily substituted for the M3H scanner in the design. The major difference between the M3H and M3S scanner is that the scan angle amplitude control is reduced from ± 5 to ± 3 V.

When purchasing the M3H scanner (or M3S), it is important to request that the scanner be set up or "tuned" with its driver board and with the paddle mirror attached for a 1- to 3-V, 30- to 60-Hz sawtooth waveform. This minimizes resonance frequencies from distorting the scan. It is also important to note that the control cable from the OATS board to the M3H scanner is not a 15-pin straight-through cable (wiring is not pin to pin). However, a 15-pin cable can be used to extend this cable. Because the documentation provided with these boards is sparse, we indicate pin locations and connections in Fig. 8.

Photomultiplier Tube

The detector used for the CLSM is a photomultiplier tube (PMT), which produces an analog voltage that is proportional to the light intensity. Although the

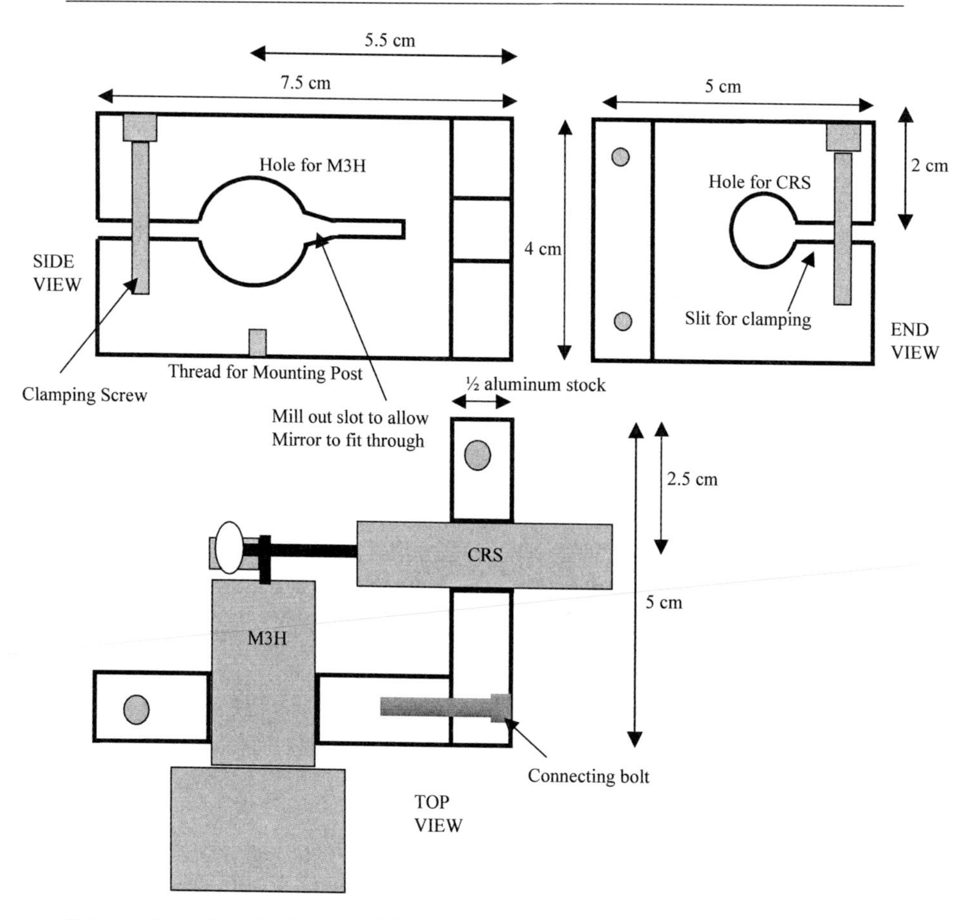

FIG. 4. A detailed construction plan of the mounting bracket that carries the CRS and M3H scanners.

quantum efficiency of photomultipliers is lower than that of solid-state detectors (e.g., avalanche photodiodes), they are still the best detector for use at the high photon count rates ($>10^8$ photons s^{-1}) required for video-rate imaging. Because the wavelength of the light emitted from the specimen is between 500 and 650 nm, the photomultiplier tube was selected to have maximal sensitivity at these wavelengths (R3896; Hamamatsu, Bridgewater, NJ). This photomultiplier replaces the tube used in earlier designs.[2] The photomultiplier is mounted in a specialized socket (C7247-01) that also contains the amplification circuit to provide a positive DC voltage. The amplifier has a bandwidth of DC-5 MHz and matches the required filtration. The high-voltage power supply (C4900-00) is a circuit board component that is incorporated into the system control box. The socket and power supply replace components previously described[2] because the old components are now out of production. The photomultiplier and socket are enclosed in a custom-built housing (Figs. 2, 3, and 6).

Frame Capture Board

The final image is constructed by a video frame capture board in a PC that sequentially digitizes the signal from the photomultiplier tube. Digitization is synchronized with the position of the scanning laser beam by control signals provided by the CRS and custom circuitry. We have used the MV-1000-20 digitizer board (Mu-Tech, Billerica, MA). The configuration of the frame grabber must be set up at the time of installation. For the MV-1000-20, this requires the setting of a variety of jumpers and the modification of a driver file (available on request). Other frame grabber boards may be used, but they must have the following capabilities: the ability to accept externally applied vertical and horizontal synchronization signals, the ability to provide a pixel clock at 10–15 MHz, the ability to DC-couple the incoming data signal, and the ability to match the AD converter to the parameters of the PMT signal. The Raven (BitFlow, Woburn, MA) and the Meteor II Multichannel (Matrox Imaging, Dorval, PC, Canada) may also be suitable. An AC-coupled signal can be used if a reference voltage (for clamping the black level) can be incorporated into the PMT signal. The knife edge of the rectangle aperture can be used for this purpose. It is important to ensure that the frame grabber is set up correctly and that the intensity of the image reflects the intensity of the PMT signal. A simple check is the observation of a uniform gray intensity across the whole image in response to a constant voltage in place of the PMT input. Varying this voltage should vary the gray intensity in a linear fashion.

Image Capture, Storage, and Retrieval

A major consideration of all imaging systems is the method of storing and viewing the acquired images. The reason for the current CLSM design was the need to collect images with good temporal resolution (30–60 fps). However, the consequence of this design is that large numbers of images are generated in short periods of time. Recording images to a video cassette recorder (VCR) is the simplest option. Although analog [video home system (VHS)] recorders have many disadvantages, the new generation of digital video (DV) recorders allows high-quality frame-by-frame replay, and can be digitally interfaced to a PC through an IEEE 1394 (FireWire) adapter. A similar approach is to use an optical memory disk recorder to record images. This has the advantage of random access to individual frames, but the costs and availability of recording disks are becoming prohibitive.

Software for Real-Time Recording to Hard Drive. A better solution to the problem of image storage and management has become possible with the availability of computer hard drives capable of writing uncompressed video data in real time. We have implemented this feature using a real-time recording system called "Video Savant" developed by IO Industries (London, ON, Canada). This software package runs under the operating system Windows NT or 2000. The system requires one or two high-speed small computer systems interface (SCSI) disks configured as a volume set and a compatible peripheral component interface (PCI)-based frame grabber. The Mu-Tech frame grabber board was chosen because it is both supported by Video Savant and interfaces with external synchronization signals. High-speed (10,000 rpm) SCSI drives are currently available in capacities of 30 GB or more and, when combined, two SCSI disks can record at rates up to 45 MB/sec. The software requires a fully compatible PC mother board and we have assembled our own PC platform rather than buying a standard package.

Software for Imaging at Video Rates. A fundamental concept for the design of a video-rate CLSM is the ability to scan the specimen at 30 fps. Because the CRS mirror oscillates at ~8 kHz (line period, 125 μs) it nominally produces 265 lines in 1/30 sec. Each line consists of a forward sweep of the specimen (62.5 μs) followed by a reverse sweep of the specimen as the mirror oscillates from side to side. In a previous design,[2] information resulting from the reverse sweep of the mirror could not be captured and a video frame was built from 512 lines every 1/15th of a second. The reverse scan would still contribute to specimen bleaching. Therefore we have exploited the reverse scan of the mirror to collect data so that the frame rate can be doubled. This is achieved by initially collecting frames that are twice the normal width (forward and reverse sweep) and half the height (200 lines $\times$1200 pixels) (Fig. 5). At the end of both the forward and reverse sweep of the CRS mirror, the vertical sawtooth is incremented one line so that the vertical scan progresses smoothly across the specimen. A custom-written algorithm of Video Savant software processes this image (before the next image is acquired) by reversing the order of the pixels of the reverse scan line and interlacing them with the lines of the forward scan. Because it difficult to determine the exact phase relationship between the horizontal synchronization signal (H sync) and the position of the CRS mirror (adjustable with potentiometer R48 CRS controller board), the algorithm operates around the central plane of symmetry, where the CRS mirror position is precisely known. The frame grabber is configured to acquire an even number of pixels from a specified starting pixel that can be adjusted to center the pixel array with the reverse point of the CRS mirror. The correct configuration is readily identified by the registration of the images from the forward and reverse scans. This process produces a normal image of 400 lines $\times$600 pixels at 30 fps. The right-hand side of the image corresponds to the CRS mirror at its reversal point. The left-hand side of the image corresponds to the CRS mirror position about 24° into its scan period (Fig. 5). By reducing the number of vertical lines collected (determined by the output selected from the binary counter; Fig. 8), image rates of 60 fps (200 lines) or 120 fps (100 lines) can be achieved. In the extreme case,

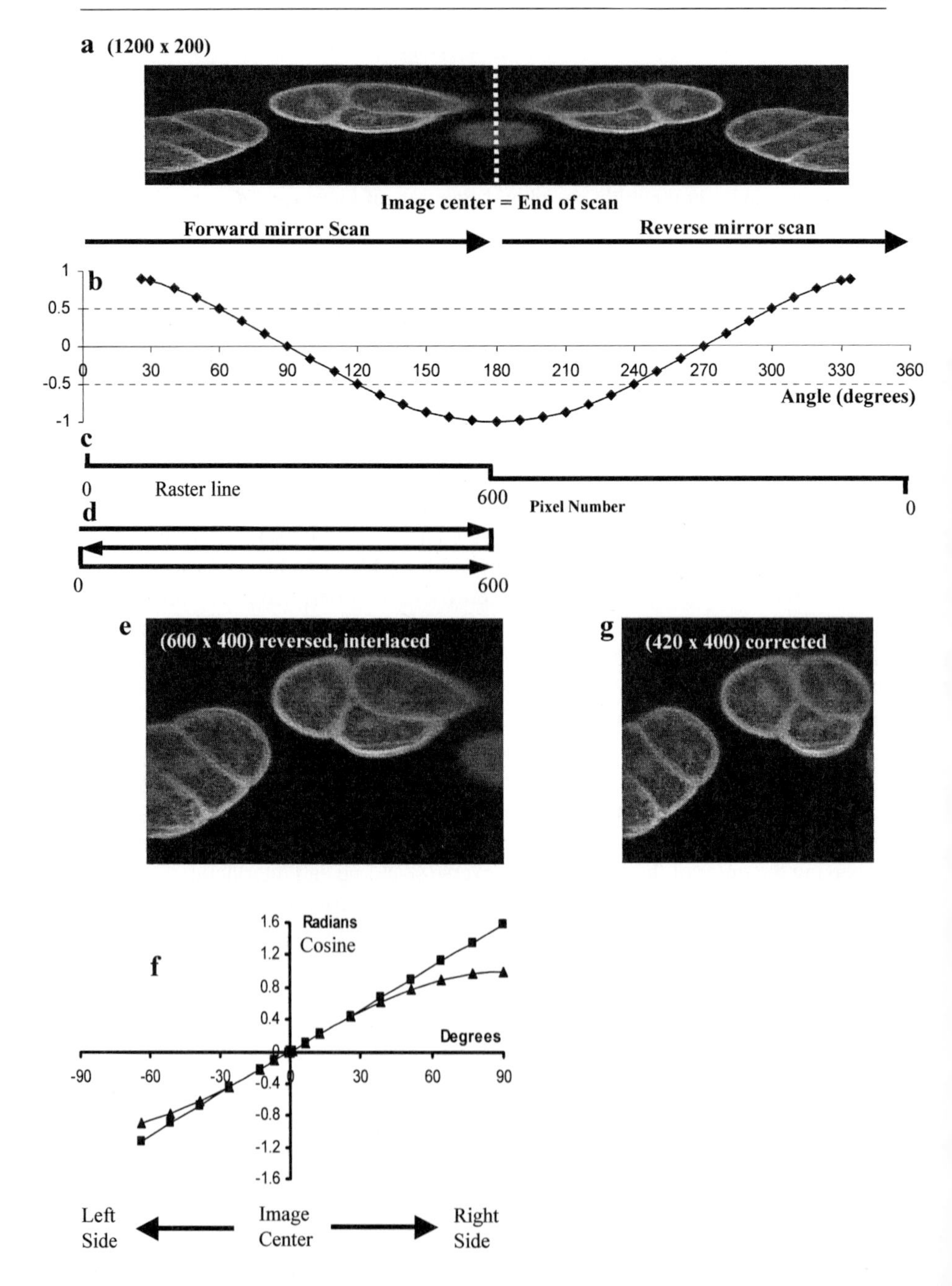
a (1200 x 200)
Image center = End of scan
Forward mirror Scan
Reverse mirror scan
b
Angle (degrees)
c
0 Raster line
600 Pixel Number
0
d
0
600
e
(600 x 400) reversed, interlaced
g
(420 x 400) corrected
f
Radians
Cosine
Degrees
Left Side
Image Center
Right Side

only a single horizontal line is collected (line-scanning mode); the vertical scan mirror is stationary and used only to select the position of horizontal line.[5]

Software for Correction of Image Distortion. Because rapid scanning is achieved with a resonant scanner, the angular velocity of the scan rate is not constant and varies as a cosine function. At the extreme ends of the scan, the mirror is instantaneously stationary. As the mirror moves through the scan it reaches a maximum velocity at the central position before slowing to zero at the opposite end of the scan. To digitize the image, frame grabbers commonly employ a pixel or sampling clock that has a constant interval and this assumes a linear translocation of the beam. As a result, the image appears stretched at the edges because the actual sampling occurs at almost the same position while the pixel position is assumed to be progressing linearly. A simple solution to this problem is to record or utilize data from the central 66% of the mirror scan, where the velocity of the mirror is almost linear.[2] In this case, the blades of the rectangular aperture are used to mask the area of interest in order to protect the specimen exposed to the extremes of the scan line from bleaching. Toward each end of the scan, the dwell time of the laser for each pixel is rapidly increasing. This translates into a longer illumination time and thereby increases the risk of photobleaching.

An alternative way to use a greater extent of the scan is to perform a spatial correction for each pixel. We have developed an algorithm (Fig. 5), which is now included in Video Savant, and works in real time for distortion-free viewing and image recording. This algorithm is not hardware dependent and will work for

[5] I. Parker, N. Callamaras, and W. G. Weir, *Cell Calcium* **21,** 441 (1997).

FIG. 5. Procedures for obtaining a corrected final image, using a nonlinear sinusoidal resonant scan mirror. (a) The initial unprocessed image acquired by the MV-1000-20 frame grabber consists of 1200 pixels ×200 lines. Each line is constructed from the forward and reverse scan of the CRS mirror to generate a mirror-image pair of the specimen. (b) The velocity profile of the CRS mirror as it moves forward and in reverse across the specimen to form the image seen in (a). The collection of image pixels is initiated and terminated when the mirror is 24 or 336° into its oscillatory period. As a result, the distortion of the image is most noticeable toward the center of the image shown in (a). (c) To avoid scanning the same section of the specimen, the vertical positions of the forward scans and the reverse scans are incremented by a single linewidth each time the mirror changes direction. (d) The effective raster pattern used to form the real image is obtained by reversing the order of the pixels in the reverse scan and interlacing them beneath the pixels in the forward scan. Numbers represent the real pixel location. (e) The resulting image after pixel reversal and line interlacing. The image is 600 pixels wide and 400 pixels high. (f) The relationship of the actual velocity profile (triangles) and the assumed velocity profile (resulting from a constant pixel clock frequency, squares) of the CRS mirror as it scans the specimen to form the image shown in (e). Image distortion is corrected by transforming the assumed pixel location to the real pixel location. (g) The resulting distortion-free image of that shown in (e). The corrected image has fewer pixels per line because most of the pixels toward the ends of the scan of the distorted image are duplicates.

images of any pixel width. The algorithm is applied after image reversal and uses the right-hand side of the image (reversal point of the CRS mirror) as a reference point (Fig. 5f) in order to determine the center pixel of the image. From the center of the image (in both directions),

$$P_{\text{new}} = P_{\text{old}}/\text{correction factor}$$

where the correction factor is the velocity factor of the pixel position or the angular position (from the center, in radians) [ranges from 1 at the center of the image to 1.57 (90°) at the end of the sweep].

Software for Image Acquisition, Analysis, and Presentation. The major impetus for incorporating commercial software for image recording was to avoid software development in the laboratory. While this can have many advantages for customization, software development is difficult and time consuming. Video Savant has solved most of the problems associated with rapid image acquisition and has provided a simple user interface. In addition, this company has been willing to write custom software to perform image reversal and distortion correction. To record images, all that is necessary is to define a file of the appropriate size and highlight the number of frames to be recorded. Clicking a control panel that mimicks a VCR initiates image acquisition. Image acquisition can be either continuous at 30 fps or discontinuous with single frames (time-lapse), sequences of frames, or averaged frames being recorded in response to a synchronization signal that also controls the laser shutter. The synchronization pulses can be generated by Video Savant and the Mu-Tech board or by an external stimulator (Fig. 9). Playback follows the same procedure with the options of looped play back and different playback speeds. Playback can be performed at speeds faster than the acquisition speed, which greatly facilitates the recognition of changes in the image intensity. Video Savant has a number of image analysis tools including averaging and image division and subtraction.

Sequences of complete images or regions of interest can be exported in a variety of file formats. We have found that the most convenient file format is a tagged image file (TIF) stack. This format is fully compatible with NIH Image (Scion, Frederick, MD; free software for PC) and other analysis packages. Images can be pseudo-colored during acquisition or playback and can be saved as MPEG (moving picture experts group) movie files. Image sequences can be archived to CD-ROM (compact disk-read only memory).

Computer System

For Video Savant to work reliably, a compatible mother board is required. Use of the very latest chip sets may cause incompatibilities and it is advisable to install a tested, even if older, mother board. The major characteristics required are an Intel (Santa Clara, CA) BX support chip set for a Pentium III processor, and an advanced graphics port (AGP) slot. A mother board with the ability to hold up to

1 GB of memory is an advantage. We have chosen the ASUS mother board P3BF with an 800-MHz processor. This processor comes only in the FCPGA format and an adapter card is required to install it in the mother board (slot 1 type). The second major requirement is a high-speed SCSI controller and two high-capacity, high-speed SCSI drives. The fastest version for the SCSI is currently the Ultra 160. It is possible, with the increasing speed of SCSI drives, that only one drive will be required, but two will always give the advantage of increased capacity. A video graphics card by ATI is recommended because these cards have worked consistently. A 21-inch monitor is recommended because the microscope images will be viewed on this screen at a resolution of 1024×796. The operating system (OS) of choice is Windows 2000, to keep pace with software development and networking. This OS installs easily on the large-capacity drives. A CD-ROM or DVD-ROM (digital video disk-read only memory) or DVD-RAM (digital video disk-random access memory) drive is recommended to archive and back up images files. Most of the other computer components are not critical. In view of these special requirements, we recommend purchasing components separately and assembling the computer in the laboratory. The frame grabber board is installed in the computer in a PCI slot and its software drivers are installed according to the manufacturer. Alternatively, IO Industries will custom build, on request, a computer system with the frame grabber and Video Savant software installed.

Image Acquisition along Axial Plane

It is frequently required, especially in thick specimens, to examine the tissue at several levels. With thin optical sections produced at video rate by the CLSM, this simply requires a change in focus. However, the rapid reproducible excursions along the z axis, which are necessary to compare activity at different planes within a tissue, are unattainable by manual focus adjustment. It is therefore recommended that a piezoelectric focusing device (Polytec PI, Auburn, MA) be incorporated into the design. The sections at precise depths through the pollen grains shown in Fig. 11 were obtained with this equipment. The unit consists of a piezoelectric linear translator that attaches the objective to the microscope turret. The adapter raises the objective position by about 5/8 inch and this requires that the microscope stage be similarly raised. The incorporation of the adapter will slightly lengthen the optical path, and should be, if possible, completed before aligning other optics.

Microscope Construction

Optical Arrangement

The construction of the microscope is similar to that described by Callamaras and Parker.[2] Little fabrication is required because most components are mounted with optical supports or studs that screw into the air-table top. In this design, only a single laser and single photomultiplier are considered. If a multiwavelength system is required this should be designed into the system at an early stage. The exact location of most components is not critical, but rather more of a convenience, and the use of multiple mirrors is an easy way of containing a long optical path on a small bench. Of course, alignment always remains critical. For laser safety and the exclusion of room light, the optical elements are surrounded by a black Plexiglas case with an opening lid. For sound control (the resonant mirror emits a noticeable 8-kHz whine), the enclosure is lined with sound-absorbing foam.

The overall layout of the optical system is described with reference to Figs. 1, 2, and 3. As noted, most components are mounted on standard post mounts. An exception is the bracket to hold the scan galvanometers. This is no longer available from General Scanning (Watertown, MA), and Fig. 4 shows mechanical drawings to construct a custom bracket.

The microscope body is placed off center on the optical table to allow adequate space on the left side of the microscope adjacent to the camera port. A zoom projection lens is inserted into the camera port that forms a conjugate image plane approximately 45 mm from the end of the lens. This lens is optional and, if not used, it may be necessary to remove the casing of the camera port to determine the position of the conjugate image plane.

The laser is positioned behind the microscope and its height is adjusted so that the emitted laser beam is just below the center of the side port. A shutter (S; Figs. 2 and 3) controls the access of the beam to the optics. The laser beam is steered through 90° by mirror 1 (M1) to a rotatable polarizer (P) that provides a variable attenuation of the intensity of the linearly polarized laser beam. A planoconcave lens (L; 500-mm focal length) serves as a simple means to expand the laser beam so that it overfills the back aperture of the objective lens. The laser light thus appears to originate from a point source 500 mm behind the lens, and the lens should be positioned so that the total length between this point and the dichroic mirror (DM) is roughly the same as the distance from the DM to the confocal aperture. After reflection by mirror 2 (M2), the laser beam passes through a laser excitation line filter housed in a rotating filter wheel (FW1) for easy exchange of excitation filters. The filtered laser beam is reflected toward the scanning mirrors by the DM, which reflects the excitation light and transmits the emitted light. The laser beam reflects off the two scanning mirrors and is focused to a point source by the eyepiece lens on the conjugate plane of the microscope. A rectangular, adjustable aperture is used to define the area scanned. Both the eyepiece and the bracket carrying the scanning mirrors are mounted on sliding rails to facilitate alignment. The optics of the microscope relays the scanning spot to the specimen. The emitted fluorescence returns along the same pathway and is descanned by the mirrors before passing through the DM. The emitted fluorescence is filtered through an emission filter in a second filter wheel (FW2) and reflected to the photomultiplier (PMT) by two mirrors (M3 and M4). To reduce stray light from reaching the PMT, the aperture to the PMT is shielded with a long tube (LT). The confocal aperture, an adjustable iris (I), is placed at the end of the tube, just in front of the PMT. By extending the path length of the emitted beam before it is detected by the PMT, the beam is expanded, thus eliminating the need for a small pinhole aperture. Instead, an adjustable iris provides the ability to regulate the depth of the confocal plane and to match this with the light availability.

Optical Alignment

Whereas the placement of the optical components is relatively straightforward, their alignment is more difficult, especially the first time. However, once alignment has been achieved and experience has been gained, it is easy to realign the system from scratch or just keep it in alignment. All the control electronics should be completed before alignment. For eye safety, the laser power should be reduced to a minimum.

Step 1. Focus the microscope on a grid reticule and then look through the side port eyepiece and adjust its position until the reticule is in focus and centrally placed. This process is greatly facilitated by mounting the eyepiece on a sliding rail.

Step 2. Adjust the height of the bracket carrying the CRS mirror so that the mirror is approximately level with the central axis of the eyepiece.

Step 3. Swing the objective lens out of position, and place a piece of translucent paper or ground glass over the open aperture. Remove the planoconcave lens (L; Fig. 3) and emission filter (FW2). Turn on the laser and center the laser beam (using mirrors M1 and M2) onto the scanning mirrors. With the scanning mirrors set to their central positions (default for CRS, positioning required for the M3H), rotate the bodies of the scanners so that the beam is reflected into the microscope. Watch for maximum illumination at the open objective position. Clamp the scanners and fine-adjust the beam location with adjustments of mirrors M1 and M2.

Step 4. Activate the CRS and ensure the illumination is centered. Manually, slowly vary the position of the MH3 mirror and watch the illumination. When the beam is filling the aperture correctly the illumination will initially not change much with the movement of the M3H mirror. The illumination will abruptly disappear when the mirror deflection is too great. Replace the objective, activate the M3H scanner and maximize the illumination with the mirror controls (M1). Replace the planoconcave lens and align its vertical and horizontal positions so that the illumination remains centered. Fine-tune, again, with mirror controls (on M1).

Step 5. Place a reflective object on the stage and focus on it (a piece of tin foil pressed against a slide; focus on an edge). This will reflect light back through the system that can be detected on the transmission side of the dichroic mirror. (The back of a business card is the traditional tool for tracking laser beams.) Adjust the focus until a spot of light is formed. Using the card, follow the beam along its path and ensure that it strikes mirror M3 centrally. Similarly, locate the reflected beam and adjust its path, with adjustments to M3, to be directed at the second mirror (M4). Repeat the process in order to align the beam on the photomultiplier aperture. Start with the aperture fully open and monitor the signal intensity with an oscilloscope. Once the beam is roughly aligned the signal can be maximized with the adjustment screws of the mirror (M4). Continue to maintain a maximal signal by moving the beam with the adjustment screws while the aperture is reduced in size, ensuring, throughout, that the reflective specimen remains in focus.

Step 6. Replace the foil slide with a pollen grain slide and focus on the grains. Connect the signal to the imaging board to see an image. Small adjustments to the illumination beam and emitted beam may be required to increase the brightness of the image and the sharpness of the image. A uniform layer of dye (e.g., 10 μM fluorescein, sandwiched between a slide and cover slip) should be examined to confirm even illumination.

Control Electronics

A single control box is recommended to house a variety of circuits and centralize the control functions. The circuits in the control box are as follows:

Photomultiplier power and amplification (Fig. 6)
Laser control and power regulation (Fig. 7)
Generation of timing signals for the M3H from the CRS (Fig. 8)
General power and shutter control (open, time lapse) (Fig. 9)

The construction of the electronics is relatively straightforward, requiring some machining of the box faceplates to accept switches, dials, and receptacles (Fig. 10, Appendices B and C) and the soldering of electronic components to PC boards. Several power supplies are required (±5, ±15, and ±24 V) (Fig. 9). Independent power switches and LED (light-emitting diode) indicators are useful so that various parts of system can be operated in isolation.

Photomultiplier Circuit. The high voltage for the PMT is provided by module C4900-00 (Fig. 6). The module requires power and a 50-kΩ potentiometer to control the gain of the PMT by varying the applied high voltage. The PMT current signal is converted into a voltage by an amplification circuit built in the socket. A front panel switch controls the power to the PMT.

Laser Control. The argon laser is supplied with a 25-pin control card that plugs into the power supply (Fig. 7). For convenience of monitoring and adjusting the board, the laser control board is incorporated into the central control box. A 25-pin parallel cable connects the power supply of the laser to the control box. The controls on the laser board are simply extended to the front panel of the control box. A panel voltmeter is added to provide a digital readout of the operating power of the laser. The voltage applied to the meter is scaled with a resistor (voltage divider) so that 1 mV = 1 mW.

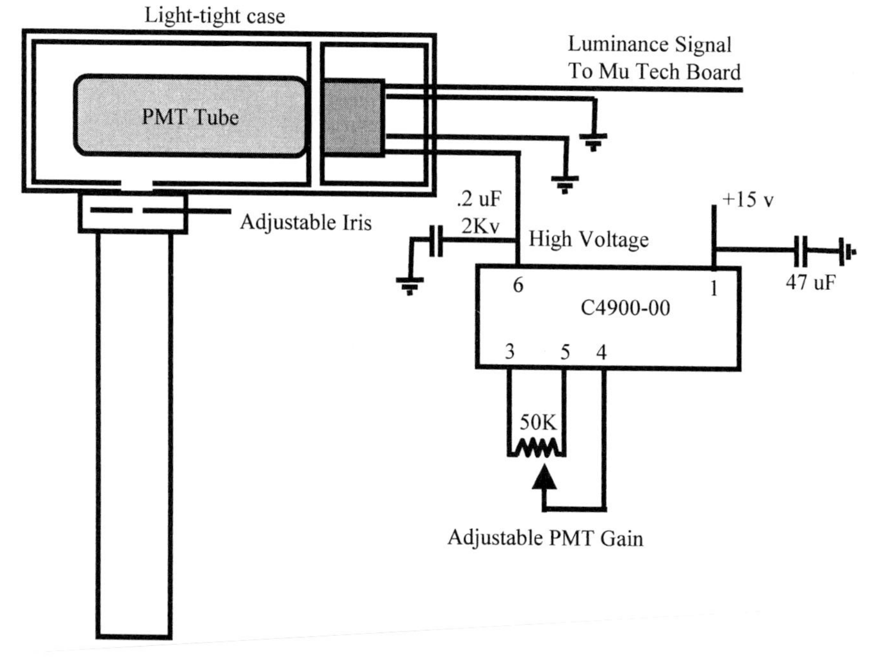

FIG. 6. Circuits associated with the photomultiplier, and layout of the photomultiplier/confocal aperture housing.

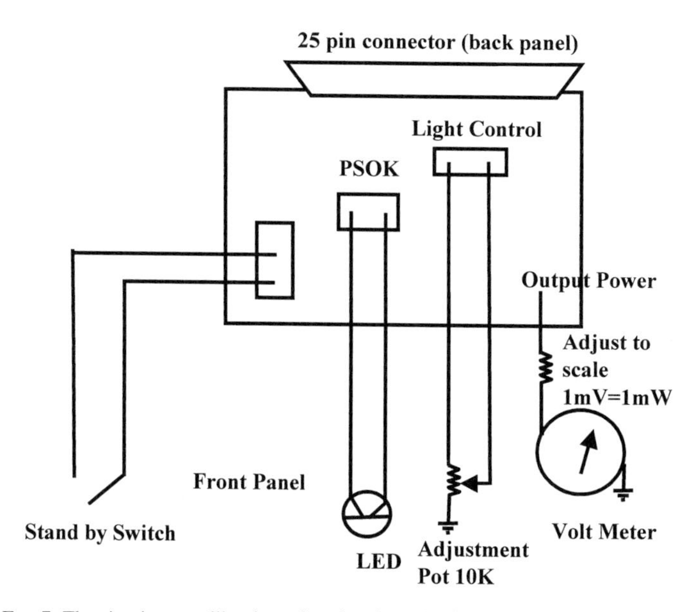

FIG. 7. The circuit controlling laser function from the front panel of the control box.

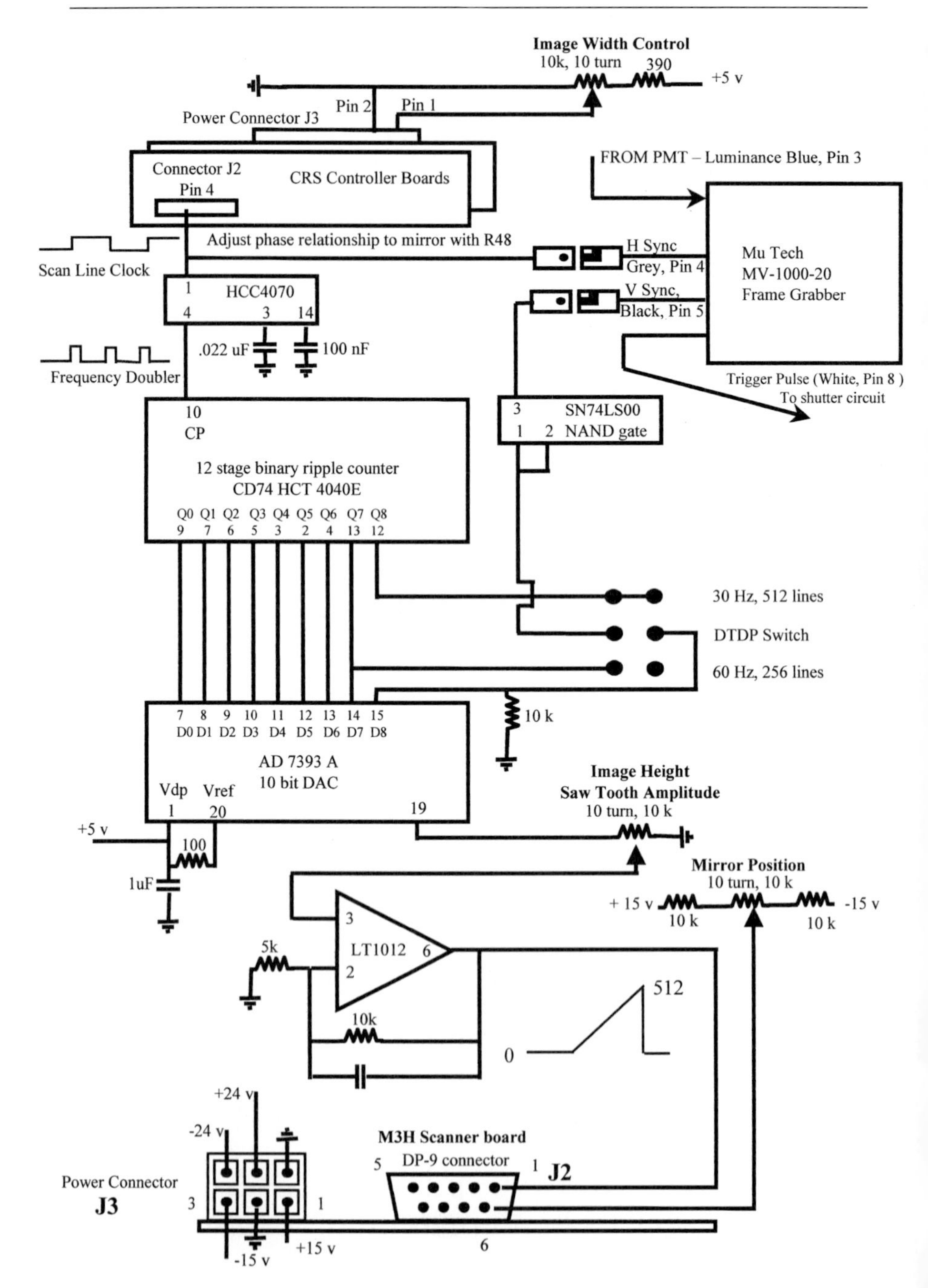

FIG. 8. Circuits generating the sawtooth ramp drive signal to the M3H scanner and horizontal (line) and vertical (frame) sync pulses to the frame grabber board.

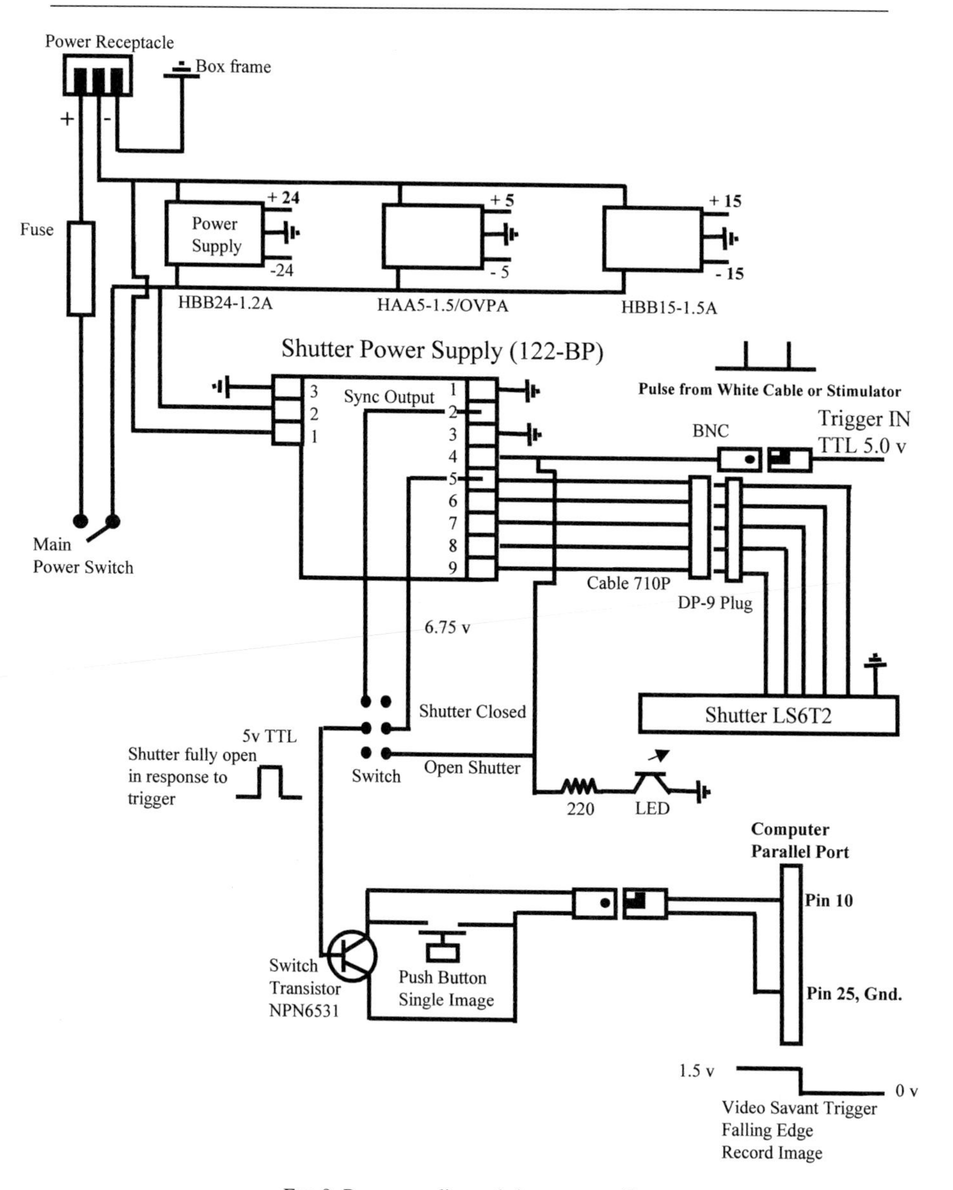

FIG. 9. Power supplies and shutter controller.

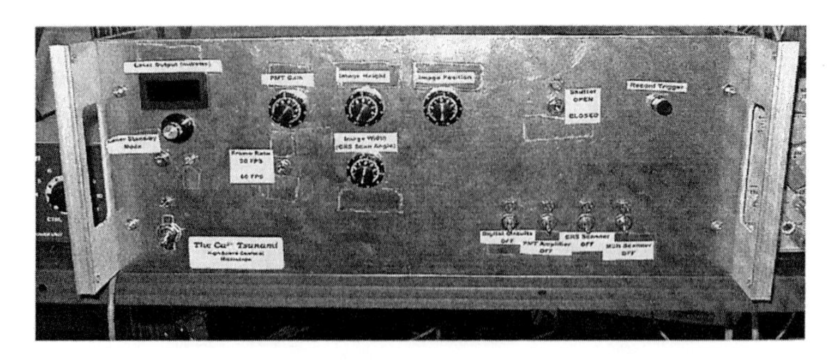

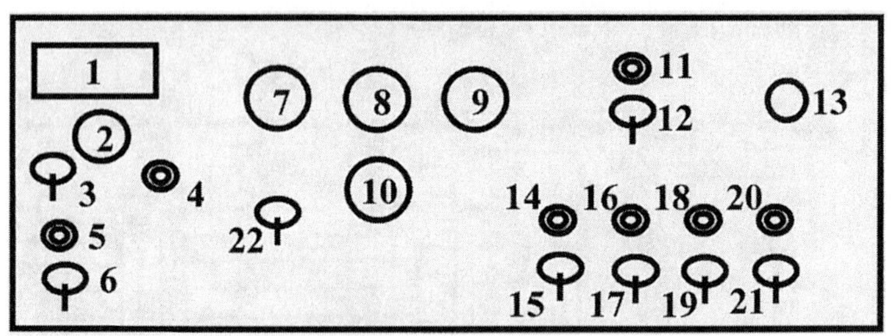

Front Panel of Control Box

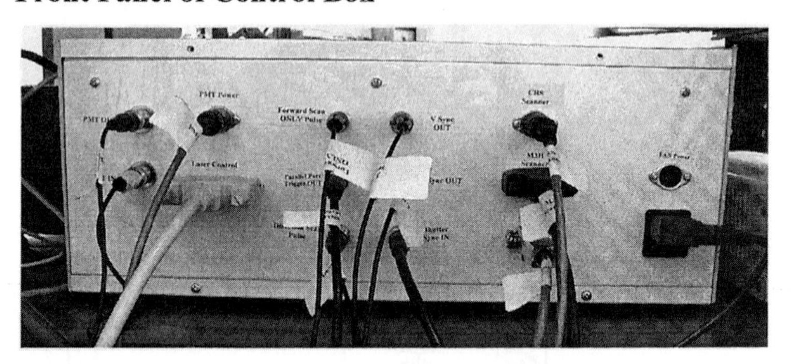

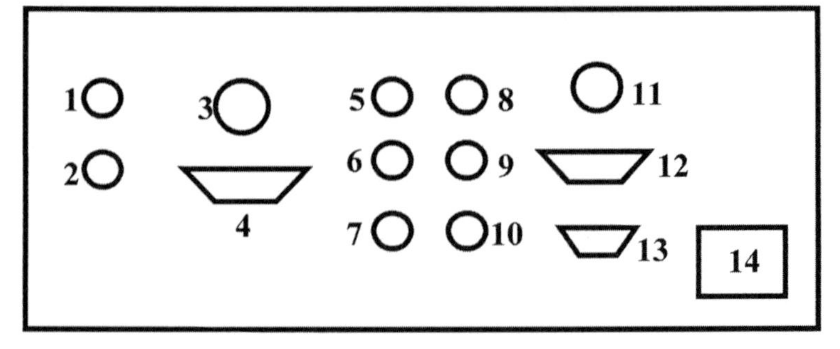

Back panel of control box

FIG. 10. The layout of the front and back panels of the control box. Refer to Appendices B and C for identification of each component.

Generation of Timing Signals. The key circuit of the CLSM generates, the horizontal synchronization signal (H sync) of the CRS (Fig. 8), the vertical synchronization signal (V sync), and sawtooth waveform that drives the vertical scanning mirror. A separate front panel switch controls power for these circuits. The CRS control boards generate a square-wave timing signal from pin 4 of connector J2 that indicates the direction of the CRS mirror rotation. This signal can also be obtained at pin 8, connector J3 if jumper W1 is installed. The exact phase relationship of this signal to the mirror position can be fine-tuned with potentiometer R48. When the signal changes from high to low or vice versa, the mirror reverses direction. The falling edge of this signal serves as the H sync signal. As a result a forward scan and a reverse scan constitute one horizontal line.

To avoid scanning the same part of the tissue twice, the scan line must be advanced at the end of each forward or reverse scan and this is achieved by the slower movement of the M3H scanner mirror. The H sync signal is fed to a frequency-doubling integrated chip (HCC4070) and, as a result, a pulse is generated at the end of each scan line. Each pulse is applied to the input of a 12-stage binary counter. The count is converted to an analog voltage by the 10-bit digital-to-analog convertor (DAC) that is further manipulated by the operational amplifier LT1012. As the count progresses from 0 to 512, a sawtooth waveform is generated that is applied to and displaces the M3H scanner (J2, pin 1). In essence, each pulse advances the scanning position of the laser. The amplitude of the M3H displacement is controlled by a 10-kΩ potentiometer (front panel). An offset voltage to control the vertical or central position of the M3H mirror is provided (J2, pin 6) by a second 10-kΩ potentiometer (front panel) (Fig. 8).

Every 512 counts, line Q8 of the binary counter goes high and indicates the end of the current image. The signal is feed through an NAND gate to invert it and form the V sync signal. The selection of 512 lines (30 Hz) or 256 lines (60 Hz) per image is possible with a switch. Line Q7 serves as the control line and Q8 is disconnected from the DAC. A reset pulse to the binary counter is not required because lines above Q8 are not connected to the DAC.

The H and V sync signals are fed to the frame grabber to synchronize the digitization of the PMT luminance signal. The pixel clock is set by the frame grabber board and determines the horizontal pixel resolution (e.g., 12 MHz = 1500 pixels per full line in 125 μs). Each horizontal line contains a double scan (forward and reverse) by the CRS mirror and each frame contains 256 horizontal lines (at 30 Hz) (Fig. 8). Software is used to reconstruct the viewable image in real time (Fig. 5).

M3H Scanner Control Board. The M3H scanner control board requires a heat sink and should be bolted to the control case. A 15-pin flat ribbon cable is used to connect the M3H board (J1) to the back panel of the control box. The M3H board is connected to the M3H scanner with the custom cable (provided) and a 15-pin extension cable. Power to the M3H board is applied via the 6-pin plug J3. Control signals are applied to J2 as described above. A front panel switch controls power to the M3H.

CRS Control Boards. CRS control boards [pixel clock board (PCB) and driver board (DB)] require heat sinks and are bolted to the control box. A flat ribbon cable (provided) connects the two boards together (PCB, J2–DB, J2). The board is connected to the scanner head via the back panel of the box and a 5-pin connector and cable attached to DB, J1. Additional flat ribbon cables are required for pins PCB, J1 (24 pin, data from this connector not used in the current configuration) and DB, J3 (16 pin). Power is provided through J3. The amplitude of the scan of the CRS is controlled by a variable voltage provided by a 10-kΩ potentiometer (front panel) applied to pin 2, J3 (Fig. 8). An additional resistor of 390 Ω is added, in series, for voltage protection (maximum, <5 V). A separate front panel switch controls power to the CRS.

Shutter Control. An independent, fully integrated power supply and controller (122-BP) is used for shutter (LS6T2) control (Fig. 9). For manual control of the shutter, a front panel switch is used. For automatic control, the shutter is placed in the closed position with the manual switch and a shutter trigger pulse (TTL, 5.0 V) to pin 4 of the controller results in the shutter opening. The generation of this trigger pulse is controlled from within Video Savant. When the shutter is 80% open, a synchronization pulse is generated at pin 5, and this is applied to a switch transistor. The activation of the transistor grounds the parallel port trigger of the Video Savant software. As a result, image acquisition is initiated when the shutter is fully open. However, to prevent the shutter closing prematurely, the duration of the trigger pulse must be slightly longer (one or two frame periods) than the time needed to record the required number of images. Closure of the shutter resets the system. Time-lapse recording is achieved by the trigger pulse frequency.

Results and Discussion

This version (MJS1.0) of the CLSM took approximately 1 year to assemble. However, much of that time was taken up in waiting for parts to be delivered, with the realization that, once they had arrived, other parts were still required. Following this experience, we have attempted to provide a full parts list that will allow investigators to order all the parts at once. Furthermore, additional time was required to implement a number of design improvements over the original model[2]; specifically the incorporation of software for forward and reverse scanning, the correction of image distortion, and the ability to save large numbers of images to hard disk. We consider the design to be now beyond the beta-testing stage and, if the detailed instructions provided here are followed, envisage that an investigator should be able to construct the CLSM within 4 months. Despite the modest cost of the components required for construction, the performance of the CLSM is comparable to that of commercial instruments costing hundreds of thousands of dollars.

Figure 11 demonstrates the ability of the CLSM to acquire a series of thin confocal sections at different planes. The specimens illustrated are "spiky" and lobed pollen grains that provide a convenient test for checking the operation of the CLSM and comparing its performance with other microscopes. The shape of each pollen grain is easily determined and the details of its inner structure are clear. The iris aperture easily regulates the extent of rejection of out-of-focus light so that image brightness and slice thickness can be optimized in the final image. A full comparison of images of pollen grains has been presented by Callamaras and Parker.[2]

The video-rate CLSM has been extensively used for imaging elementary intracellular calcium dynamics.[6,7] In conjuction with the photolytic release of inositol trisphosphate, the rapid localized release of calcium by clusters of IP_3 receptors or "Ca^{2+} puffs" was correlated with the generation and propagation of intracellular Ca^{2+} waves in oocytes.

Figure 12 also shows the versatility of the CLSM for imaging dynamic processes in thick tissue. Our research addresses asthma, a common lung disease that is mediated by the hyperactivity of small airway smooth muscle cells (SMCs). Most studies have employed cultured SMCs, but a major criticism of this approach is that the isolation of SMCs radically alters their phenotype and cellular physiology. The loss of SMC contractility is the most common form of function loss. In addition, the SMCs isolated are rarely those of the small airways but more often from the trachea.

We, therefore, developed procedures to cut slices ($\sim$75 μm thick) of mouse lung in order to study, *in situ,* the Ca^{2+} signaling that occurs within airway epithelial and smooth muscle cells (SMCs). In a lung slice, the structural relationships between the epithelial cells, the SMCs, the airways, and surrounding alveolar tissue are extremely well maintained. Furthermore, the airway SMCs retain the ability to perform repetitive contractions for at least 1 week (in organ culture). These slices can be loaded with fluorescent dyes for monitoring Ca^{2+} (using AM esters). However, many other cells besides the SMCs take up the dye and, as a result of the slice being several cells thick, out-of-focus fluorescence makes conventional wide-field observation impossible. Using the CLSM, it has been possible to locate and study SMCs. The addition of a variety of drugs [e.g., acetylcholine (ACH)] induces rapid increases in Ca^{2+} followed by Ca^{2+} oscillations. These oscillations can persist for some time but are generally inhibited by the removal of ACH with esterase. The cessation of the oscillation is accompanied by a substantial relaxation of the muscle (Fig. 12). We believe this is the first time Ca^{2+} signaling in airway SMCs *in situ* has been studied.[8]

[6] J. S. Marchant and I. Parker, *EMBO J.* **20,** 65 (2001).
[7] J. S. Marchant and I. Parker, *Br. J. Pharmacol.* **132,** 1396 (2001).
[8] A. Bergner and M. J. Sanderson, *J. Gen. Physiol.* **119,** 187 (2002).

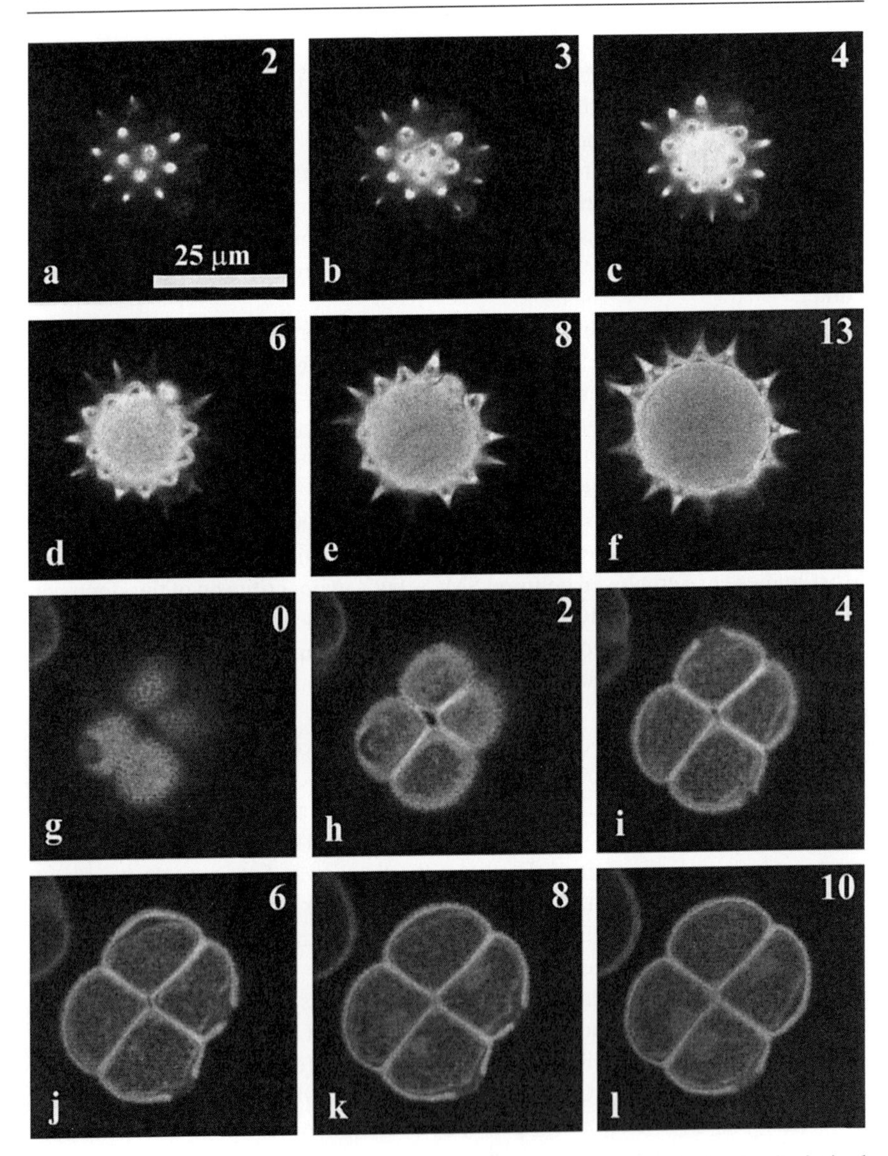

FIG. 11. A series of optical sections through two different pollen grains (a–f and g–l) obtained with the CLSM and the piezoelectrical focusing adapter. The relative depth position of each section (in micrometers) is indicated at the top right of each image. Each image is an average of eight video frames (acquired at 30 fps). Image width and height: 50 μm.

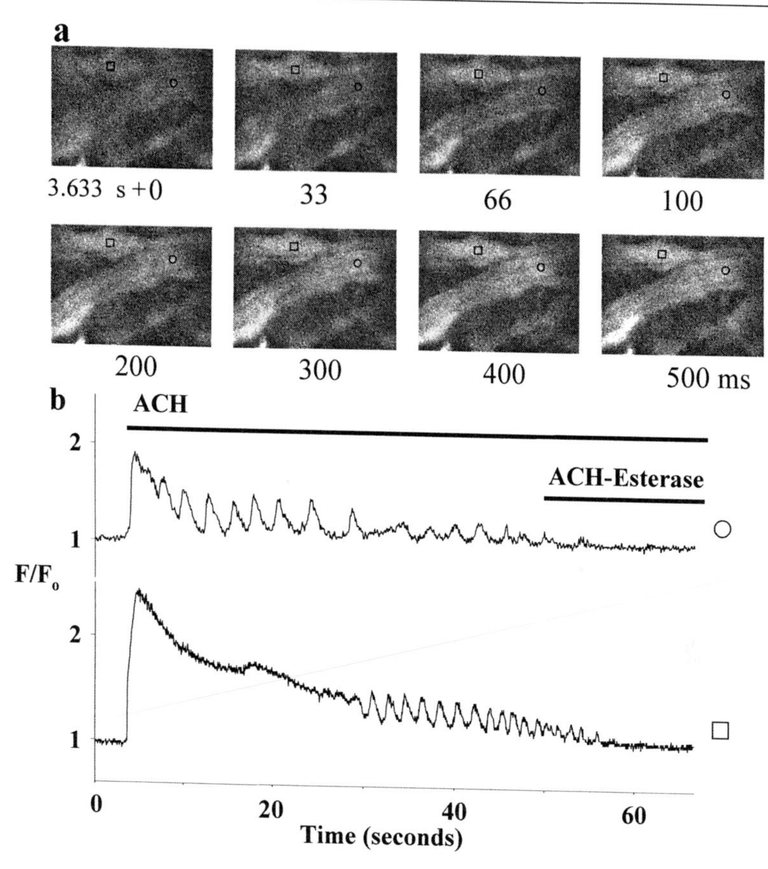

FIG. 12. Confocal imaging of dynamic intracellular Ca^{2+} signals within a slice of lung tissue. (a) A series of selected images recorded from airway smooth muscle cells (SMCs) *in situ* within a lung slice at 30 fps. The acquisition time from the beginning of the sequence is indicated (in milliseconds) below each image. A single frame at each time point is shown. (b) An extended pixel point analysis of the images recorded in (a). The change in fluorescence ratio (F/F_0) was determined from two different cells as indicated by the square and circle. More than 1800 frames were used. These SMCs lie parallel to an airway that is orientated from bottom left to top right. The SMCs were loaded with Oregon Green. Acetylcholine (1 m*M*) was added to the slice over the time range indicated by the bar, inducing a rapid increase in intracellular Ca^{2+} in many SMCs, accompanied by a large, slow contraction. The Ca^{2+} increase occurred within 100 ms and is documented by the first four images in (a). Because the cells are contacting, after ACH application, it was necessary to track the location of the cells so that the same pixel area could be monitored from each individual SMC. Following the initial Ca^{2+} increase, the intracellular Ca^{2+} began to decline and intracellular Ca^{2+} oscillations occurred with a declining baseline and amplitude. The presence of these oscillations could continue for several minutes but the application acetylcholine esterase (85 units/ml; indicated by a bar) inhibited the Ca^{2+} oscillations. This reduction in Ca^{2+} was accompanied by a large relaxation of the SMCs.

An other advantage arising from the use of the video-rate CLSM is that it provides rapid feedback when changing focus or moving the objective. This makes the use of the microscope simple and interactive. The ease of recording thousands of images on reusable media removes any inhibition about performing experiments because of wastage of media. In addition, the instant replay of images, either forward or backward at any speed, or at one image at a time, allows for an initial interpretation of the data during the experimental period.

Summary

The CLSM described here can be built with relatively little electronic or optical experience and with a budget of approximately \$20,000–\$30,000 (excluding microscope and table). This cost is substantially less than that of commercial counterparts. However, this CLSM has excellent spatial and temporal resolution and the convenience of digital recording and playback. By building the CLSM, the investigator ensures long-term support and reliability of the instrument as well as the potential for future modifications and improvements. Finally, the sense of accomplishment of building your own instrument should not to be underestimated.

Appendices

APPENDIX A
MANUFACTURER AND COMPONENT LIST

Item/vendor	Part/catalog number	Description	Notes
Air-cooled argon laser			
Melles Griot Laser Division	532-A-A04	Ion-laser Head 100-mW argon laser 5956	
2051 Palomar Airport Drive	05224	Cooling fan assembly AC2-10B	
Carslbad, CA	176B-120B	Ion-power supply lab 120 V 4852	
(760)-438-2131			
Antivibration table			
Newport Corp.	VW-3046-opt-021022	Air table, stainless steel top (30 × 46 inches), 1-inch holes centered holes, 0.25, inch, 20 thread	
1791 Deere Ave.			
Irvine, CA 92714			
(714)-863-3144			
Scanning mirror assemblies			
GSI Lumonics	000-30150B	CRS 8-kHz scanner including driver boards, cables, pixel clock, and mirror	
4E Crosby Drive			
Bedford, MA	000-3008001	M3H scanner	
(781)-275-1300	E11-132095	OATS driver board for M3H	
	310-146611	Paddle mirror assembly	
	312-153261	72-inch extension cable	
		Tune OATS driver for 30- to 60-Hz sawtooth waveform	

APPENDIX A (*continued*)

Item/vendor	Part/catalog number	Description	Notes
Photomultiplier			
Hamamatsu Corporation	R3986	Photomultiplier tube (side on)	
360 Foothill Road, Box 6910	C4900-00	Power supply	
Bridgewater, NJ 08807	C7247-01	Socket with amplifier (side on)	
1-(800)-524-0504			
Imaging boards			
Mu-Tech	90-10002-E00	MV-1000-20MHz	
85 Rangeway Road	91-00VC7-001	MVC-7 cable	
Billerica, MA 01862			
(978)-663-2400			
Software			
IO Industries		Video Savant with hard disk recording (standard edition)	1
102-252 Pall Mall Street			
London, ON N6A 5P6, Canada	Available on request	A custom-built compatible computer system with imaging board installed	
519-663-9570			
Contact: Andrew Sharpe			
Basic computer			
Treasure Chest Computers	P3BF	ASUS mother board, 440 BX, APG slot 1, 5 PCI	2
tccomputers.com			
	EN7237	Case	
	IT277356	Pentium III 800-MHz processor FCPGA	
	ASUA370	Slot 1 to socket 370 card	
	XPERT 2000	ATI video card, APG slot	
	G810	Large-monitor 21-inch ViewSonic	3
		Yamaha CD-RW, 16 × 10 × 40	4
		Windows 2000 operating system	5
		Floppy drive, sound card, mouse, keyboard, network card	6
	32 × 64100S	Recommend 2 × 256 MB of memory (can add up to 1 GB)	7
Hard drives			
DC Drives	A19160	Adaptec PCI Ultra 160 SCSI controller	8
3716 Timber Drive	TN318200LW	Quantum hard drive, 18.2 GB, 10,000 rpm, 160 SCSI (two required)	9
Dickinson, Texas 77539			
1-(800)-786-1160	ST320420A	Seagate, 20 GB/7200 rpm, ATA/66	10
Shutter			
Vincent Associates	LS6T2	Laser shutter	11
1255 University Ave.	122-BP	Open frame shutter driver	12
Rochester, NY	710P	7-Pin cable	
1-(800)-828-6972			
Optical components			
Nikon: See local dealer	84220	CFW ×10 eyepiece (additional to microscope)	13
Omega Optical	XL06	488NB3 argon laser line filter	
P.O. Box 753	XF2037	Dichroic filter 500 DRLP	14
Brattleboro, VT 05302	XC100	Filter cube for Nikon to hold filters	15

continued

APPENDIX A (*continued*)

Item/vendor	Part/catalog number	Description	Notes
Optical components			
www.omegafilters.com	XF3006	18-mm long-pass OG15	
(802)-254-2690	XF22	18-mm excitation filter 485DF22	
	XF22	18-mm band pass 530DF30	
	XF22	Dichroic filter 505DRLP	
Opto-mechanics			16
Coherent Auburn Group	61-1137	Rectangular aperture	
2303 Lindbergh Street	61-3497	12-inch rail (1)	17
Auburn, CA 95602	61-3513	Rail carrier, 1 inch (2)	
1-(800)-343-4912	53-9775	Filter wheel (2)	
	53-2432	Top adjustable mirror mount	
	OG-515	Long-pass glass filter	
Edmund Scientific	K54863	Iris mount, 38 mm	
101 East Gloucester Pike	K53907	Iris, 37-mm OD	
Barrington, NJ 08007	K52557	Polarizer, 42 mm	
(856)-573-6250	K52572	Rotary optic holder	
	K55177	0.25-Inch, 20 thread screws, 0.75 inch long	
OptoSigma	034-2230	Broad brand mirror, 1 inch diameter (4)	
2001 Deere Ave.	112-0250	Mirror mount (3)	
Santa Ana, CA 92705	112-0264	Mirror holder, 1 diameter (3)	
(949)-851-5881	199-0161	Ball driver/4 20 (1)	
	148-0210	Post, 1 inch (~5)	18
	148-0220	Post 1.5 inch (~10)	19
	148-0230	Post 2 inch (~10)	20
	148-0240	Post 3 inch (~5)	21
	148-1310	Post holder 1 inch (~10)	
	148-1320	Post holder 2 inch (~10)	
	147-0443	Post mounting bases (~20)	
	111-0080	Lens holder, fixed	
Microscope			
Nikon: See local dealer	90101	Inverted, Nikon Diaphot 300, with 80/20 and 100/100 light-directing cubes	22
	85006	Objective, ×40 or ×60 oil, NA ~1.3	23
	84220	Objective low power, ×10, aid for aligning laser beam	24
	90131	Epifluorescence system for scanning the specimen	25
	See Omega Optical	Filter cube/holder	26
	See Omega Optical	Filters set for Fluo-3, Ex 485, Dichroic 505, Barrier 510LP	26
	90145	3-Position cube changer cassette	27
	87530/87531	Hg lamp housing and socket	28
	78589	Hg bulb	29
	87505	Condenser lens	30
	79444	Zoom lens	31

APPENDIX A (*continued*)

Item/vendor	Part/catalog number	Description	Notes
Chiu Technical Corp. 252 Indian Head Rd. Kings Park, NY 11754	MX-75/100R	100-W power supply for Hg or Xn	
Polytec PI 23 Midstate Drive	P-721.10	Piezoelectric focus with LVDT	
Suite 212 Auburn, MA 01501 (508)-832-3456	E-662.LR	LVPZT amplifier controller	
Electronics components			
Newark Electronics 217 Wilcox Ave. Gaffney, SC 29340 1-(800)-463-9275	99F1014	Enclosure for electronics (control box) Valuline HC 14104	
FLW	HAA5-1.5/OVPA	Power supply ±5 V	
350 Cadillac Ave.	HBB15-1.5A	Power supply ±15 V	
Costa Mesa, CA 92626 www.flw.com	HBB24-1.2A	Power supply ±24 V	
Digi-Key Corp.		Front panel	32
701 Brooks Ave. South	73JA103-ND	10 kΩ, ten-turn potentiometer (3)	
Thief River Falls, MN 56701	73JA503-ND	50 kΩ, ten-turn potentiometer (1)	
1-(800)-344-4539	412KL-ND	Potentiometer knob, black (4)	
www.digikey.com	CKN1038-ND	4PDT toggle switches (4)	
	67-1147-ND	3-mm red LED (5)	
	67-1148-ND	3-mm green LED (2)	
	RLC010-ND	Mini volt meter (1)	
	381N103-ND	10 kΩ, 1W, 1 turn plastic potentiometer (1)	
	8558K-ND	Knob for potentiometer (1)	
	CKN1021-ND	SPDT switch (1)	
	CKN1035-ND	DPDT switch (2)	
	CKN1121-ND	Push-button switch (1)	
		Back panel	32
	ARFX1064-ND	Panel mount BNC receptacle (8)	
	CP-1250-ND	Panel mount 5-pin receptacle (2)	
	MFR15-ND	Sub D 15-pin connector (1)	
	A2047-ND	Sub D 9-pin connector (1)	
	Q204-ND	Plug receptacle	
		Other	
	VFP-KIT-ND	Shrink tubing kit (various sizes)	
	1602-KIT-ND	Nuts and bolts kit (various sizes for mounting)	
	J216-ND	Stand-off threaded spacer for circuit boards	
	J212-ND	Stand-off threaded spacer for circuit boards	
	3122K-ND	Isolating nylon washers, no. 6 screw	
	F-KIT-ND	Various capacitors	
	PHD1-KIT-ND	Various ceramic capacitors	
		Hook-up wire	

continued

APPENDIX A (*continued*)

Item/vendor	Part/catalog number	Description	Notes
		Cables	
	CP1050-ND	In-line plug, 5 pin (4) (CRS and PMT cable)	
	MMR09-ND	In-line D plug, 9 pin (shutter cable)	
	A120-100-ND	Multiconductor cable to make cables	
	A3213-100-ND	4-Conductor cable	
		Chips	
	296-1626-5-ND	Quad, 2 input, NAND gate, SN74LS00 (1)	
	296-2118-5-ND	12-Bit binary counter, CD74HCT4040E (1)	
	CD4070BCN-ND	Quad, 2-input exclusive OR gate (1) Replacement part for HCC4070	
	LT1012-CN8-ND	Op amp LT1012 (1)	
	AE9808-ND	IC socket, 8 pin, gold (for op amp)	
	AE9814-ND	IC socket 14 pin, gold (for NAND gate)	
	AE9816-ND	IC socket 16 pin, gold (for binary counter)	
	AE9820-ND	IC socket 20 pin, gold (for AD converter)	
	Analog Devices	10-Bit ADC	
		Connectors for M3H boards	
	109M-ND	DB-9 connector, male for ramp and position	
	WM3702-ND	6-Pin power connector housing—Mini Fit	
	WM2501-ND	Pins for above	
	M7PXK-1506R-ND	D subcable, single end male, pin	
		Connector for CRS board	
	M1AXK-1636R-ND	Single-ended socket connector/cable for J3 (power)	
	M1AXA-2436R-ND	Single end socket/cable for J1 (data)	
Radio Shack: local stores in United States	2760168	PC boards	
		Various resistors	
	2750602	SPST power switch (1)	
Analog Devices 3 Technology Way P. O. Box 9106 Norwood, MA 02062 (781)-329-4700	AD7393AN	10-Bit ADC	
L-Com 45 Beechwood Drive North Andover, MA 01845 (978)-682-6936	CC174-10	Coaxial cable, 10 ft, BNC-BNC (8)	
	CSM15MF-10	MF15 pin parallel cable, pin to pin (1)	
	CSM25MF-10	MF25 pin parallel cable (1)	
Miscellaneous			
Carolina Biological Supply P.O. Box 6010 Burlington, NC 27216 1-(800)-334-5551	30-4264	Slide of mixed pollen grains	

APPENDIX A (*continued*)

1. When purchasing, request the drivers for Mu-Tech board that have the ability to perform (a) pixel reversal and (b) image distortion correction, (c) live display during recording, and (d) advanced recording as developed for Dr. Sanderson. An evaluation version of the software is available at their Web site.
2. This mother board works with the Mu-Tech card and Video Savant. Other mother boards may not. Contact IO Industries for other compatible mother boards. It is not worth buying the latest thing.
3–7. The specifications of these computer components frequently change with technology advances. A review of each component is recommended at the time of purchase for the current model.
4. A read–write CD provides the option of archiving and back-up of data. The RW-CD-ROM is rapidly being replaced by RW-DVD (ROM or RAM).
6. No specific requirements for the basic components of the computer.
8. A PCI controller card for the SCSI drives is preferable to a controller built into the mother board, as this allows for upgrades in disk access speed without changing the whole system.
9. These SCSI drives have probably been superseded by newer drives that have a larger capacity, faster access speed, and are cheaper.
10. The Seagate hard disk is for the operating system and can be anything. A large capacity and fast access time are recommended.
11. The LS6T2 was used because it has a wider aperture, making alignment less critical. However, this shutter cannot open and close as fast as the 3-mm-diameter shutter. The Teflon-coated blades are not usually used for high-powered lasers, but these seem to be no problem in our systems.
12. The power supply/controller for the shutter is a simple way to integrate the shutter into the control box.
13. This eyepiece needs to be machined and attached to the rectangular aperture such that the knife edge shutters are at the focal plane of the eye piece.
14. The size and shape of filter will depend on mounting: Square or circular.
15. This can be custom made by a machine shop or substituted with a filter wheel.
16. All optical mounts and components can be bought from any optics company.
17. Rail is cut into two parts to use for sliding mount of eyepiece and scanner mirrors.
18–21. The exact number of posts at set heights that are required is difficult to predict because this will depend on the alignment of the system. Having enough on hand to swap components avoids considerable frustration.
22–31. Original Nikon Diaphot 200/300 parts are no longer in production. A substitute microscope may have to be purchased.
26. A filter cube cassette allows the filter set to be easily and quickly inserted into or removed from the illumination pathway.
32. All electrical components can be purchased from any vendor. It is recommended that the catalog number is checked before purchasing.

APPENDIX B
LAYOUT OF FRONT PANEL CONTROLS[a]

Item	Indicator/control	Function	Part number
1	Mini volt meter	Indicator of laser output power (mW)	RLC010-ND
2	10 kΩ potentiometer knob	Control of laser output power	381N103-ND/8558K-ND
3	Switch	Placement of laser in stand-by mode	CKN1021-ND
4	Green LED	Indicator of laser status	67-1148-ND
5	Red LED	Indicator of main power	67-1147-ND
6	Switch	Main power switch	RS 2750602
7	10-Turn, 50 kΩ potentiometer	PMT gain control	73JA503-ND/412KL-ND
8	10-Turn, 10 kΩ potentiometer	Image width, CRS scan control	73JA103-ND/412KL-ND
9	10-Turn, 10 kΩ potentiometer	Image height, M3H scan control	73JA103-ND/412KL-ND
10	10-Turn, 10 kΩ potentiometer	Vertical image position, M3H scan control	73JA103-ND/412KL-ND
11	Green LED	Shutter open indicator	67-1148-ND
12	Switch	Shutter open/closed, trigger control	CKN 1035-ND
13	Push button	Trigger to record single frame	CKN1121-ND
14	Red LED	Indicator of power to circuits	67-1147-ND
15	Switch	Switch for power to circuits	CKN1038-ND
16	Red LED	Indicator of power to PMT	67-1147-ND
17	Switch	Power switch for PMT	CKN1038-ND
18	Red LED	Indicator of power to CRS	67-1147-ND
19	Switch	Power switch for CRS	CKN1038-ND
20	Red LED	Indicator for power to M3H	67-1147-ND
21	Switch	Power switch for M3H	CKN1038-ND
22	Switch	Select 30/60 Hz	CKN1035-ND

[a] Part numbers are included in Appendix A. See Fig. 10 for design.

APPENDIX C
LAYOUT OF BACK PANEL CONNECTIONS[a]

Item	Connector	Function	Part number
1	Female BNC	In PMT luminance	ARFX1064-ND
2	Female BNC	Out PMT luminance	ARFX1064-ND
3	Female 5 pin	High voltage and power to PMT	CP-1250-ND
4	Female Sub D 25 pin	Built-in to argon laser control card	Laser part
5	Female BNC	Not used	ARFX1064-ND
6	Female BNC	Parallel port trigger out	ARFX1064-ND
7	Female BNC	Not used	ARFX1064-ND
8	Female BNC	V sync out	ARFX1064-ND
9	Female BNC	H sync out	ARFX1064-ND
10	Female BNC	Shutter trigger in	ARFX1064-ND
11	Female 5 pin	Power and control connector to CRS	CP-1250-ND
12	Female sub D 15 pin	Power and control to M3H scanner	MFR15-ND
13	Female sub D 9 pin	Shutter control	A2047-ND
14	Plug receptacle	Main power	Q204-ND

[a] Part numbers are included in Appendix A. See Fig. 10 for design.

Acknowledgments

Supported by NIH Grant 49288 to M.J.S. and by NIH Grant 48071 to I.P. We thank Dr. Albrecht Bergner for collecting data from lung slices.

[20] Giant Vesicles, Laurdan, and Two-Photon Fluorescence Microscopy: Evidence of Lipid Lateral Separation in Bilayers

By LUIS A. BAGATOLLI, SUSANA A. SANCHEZ, THEODORE HAZLETT, and ENRICO GRATTON

Introduction

The critical issues in membrane biophysics today are centered on the molecular dynamics of the bilayer structure. The interplay between lipids that results in the formation of domains on a bilayer surface and the interactions among these domains and relevant membrane-associated biomolecules is of particular interest. There has been extensive research in model systems to identify the forces involved in lipid domain formation and lipid diffusion in simple phospholipid mixtures, using nuclear magnetic resonance (NMR), electron spin resonance (ESR), infrared spectroscopy, calorimetry, and fluorescence spectroscopy. Much has been learned about lipid behavior, but even binary lipid systems are still only partially understood. Work on multicomponent bilayers and on cell membranes is difficult because of the multifaceted nature of lipid interactions. In a review article, Maggio fittingly describes the complexity of membrane dynamics as "... a complex and fascinating ecological problem at the molecular level that involves the concerted modulation of environmental, supramolecular, topological and temporal events."[1] From a biological viewpoint, we are at the beginning of our work. It will be essential to develop new techniques and new analysis strategies to effectively identify relevant local forces that give rise to the global membrane effects.

Studies of physical and chemical events in simple artificial lipid systems, in which the lipid composition and the environmental conditions (such as temperature, ionic strength, and pH) can be systematically varied, provide information that can help us understand the behavior of biological membranes. As models of cellular membranes, liposomes have had significant impact on our understanding of the molecular aspects of lipid–lipid and lipid–protein interactions since their introduction by A. D. Bangham *et al.* in 1965.[2] Liposome studies generally

[1] B. Maggio, *Prog. Biophys. Mol. Biol.* **62,** 55 (1994).
[2] A. D. Bangham, M. M. Standish, and J. C. Watjins, *J. Mol. Biol.* **13,** 238 (1965).

0076-6879/03 $35.00

involve aqueous suspensions consisting of small unilamellar vesicles (SUVs; mean diameter, 100 nm), large unilamellar vesicles (LUVs; mean diameter, 400 nm), and multilamellar vesicles (MLVs), these being the most popular model systems. An array of different experimental techniques (differential scanning calorimetry, fluorescence spectroscopy, NMR, low-angle X-ray scattering, and electron spin resonance, to mention a few[3–12]), including theoretical treatments using computer simulations,[13,14] have been used to determine parameters regarding the thermodynamic and kinetic aspects of liposome bilayers. However, these experimental approaches produce mean parameters on the basis of data collected from bulk solutions of many vesicles and lack information about lipid organization at the level of a single vesicle, a quality that can be provided by microscopy. In this article, we describe how microscopy techniques, which provide information about membrane microorganization by direct visualization, and spectroscopy techniques, which provide information about molecular interaction, order, and microenvironment, can be combined to give a powerful new tool to study lipid–lipid and protein–lipid interactions.

Our aim in this article is to elaborate on the use of fluorescence spectroscopy tools in fluorescence microscopy. Advances in optical methods, optical components, acquisition electronics, and detectors have greatly enhanced our ability to collect increasingly detailed spectroscopic data with the imaging optics of a light microscope. The continued development of confocal microscopy (both one-photon and two-photon approaches), which has greatly increased the information available through imaging, has allowed for rapid advances in fluorescence correlation spectroscopy and three-dimensional particle-tracking methods, both of which can now be performed in the microscope environment. At the present time, there are a number of laboratories actively advancing this concept, that is, performing spectroscopy in a microscope, for a variety of protocols and generating exciting results in studies ranging from cell physiology to the mechanics of polymer motion on surfaces.

The advantages of using a microscope as the optical arrangement are clear. The light collection efficiency of a well-designed microscope is greatly enhanced

[3] A. G. Lee, *Biochim. Biophys. Acta* **413,** 11 (1975).

[4] B. R. Lentz, Y. Barenholtz, and T. E. Thompson, *Biochemistry* **15,** 4529 (1976).

[5] S. Mabrey and J. M. Sturtevant, *Proc. Natl. Acad. Sci. U.S.A.* **73,** 3862 (1976).

[6] K. Arnold, A. Lösche, and K. Gawrisch, *Biochim. Biophys. Acta* **645,** 143 (1981).

[7] M. Caffrey and F. S. Hing, *Biophys. J.* **51,** 37 (1987).

[8] E. J. Shimshick and H. M. McConnell, *Biochemistry* **12,** 2351 (1973).

[9] W. L. C. Vaz, *Mol. Membr. Biol.* **12,** 39 (1995).

[10] B. Maggio, G. D. Fidelio, F. A. Cumar, and R. K. Yu, *Chem. Phys. Lipids* **42,** 49 (1986).

[11] L. A. Bagatolli, B. Maggio, F. Aguilar, C. P. Sotomayor, and G. D. Fidelio, *Biochim. Biophys. Acta* **1325,** 80 (1997).

[12] B. Piknova, D. Marsh, and T. E. Thompson, *Biophys. J.* **72,** 2660 (1997).

[13] K. Jørgensen and O. G. Mouritsen, *Biophys. J.* **69,** 942 (1995).

[14] T. Gil, J. H. Ipsen, O. G. Mouritsen, M. C. Sabra, M. M. Sperotto, and M. J. Zuckermann, *Biochim. Biophys. Acta* **1376,** 245 (1998).

over other optical arrangements. Indeed, collecting every possible photon is essential for any protocol that requires single, or near-single, molecule sensitivity. In addition, the flexibility of fluorescence microscopes creates for the spectroscopist a malleable optical compartment that can be designed and readily redesigned as needed. Of equal importance as the sensitivity and flexibility of a microscope is the addition of spectroscopy to the ability to collect spatially resolved information. With a properly designed system, we can perform quantitative spectroscopic studies at each pixel and build an information image related to the sample at hand, be it a living cell, an extended surface polymer, or a giant unilamellar vesicle. Researchers have been interested in this marriage of technologies and examples can be found in the early ratio imaging studies, fluorescence lifetime imaging techniques, and fluorescence polarization imaging.

We present here examples of the use of the confocal ability of a two-photon scanning microscope with a multicolor detection attachment to study phospholipid phase behavior in unsupported lipid bilayers, giant unilamellar vesicles. We take advantage of the sectioning ability of the two-photon excitation effect to independently examine the surface and cross section of these spherical bilayers as they are perturbed by temperature and lipid makeup. The fluorescent membrane probe Laurdan [6-lauroyl-2-(*N*,*N*-dimethylamino)naphthalene; Molecular Probes, Eugene, OR] is used to monitor the membrane and changes in the membrane as perturbations are introduced. Although other probes and methodologies can be applied, we limit our discussion primarily to Laurdan and to techniques that can be used with this unique fluorophore.

Microscopy Techniques and Giant Unilamellar Vesicles

A distinct advantage of giant unilamellar vesicles (GUVs; mean diameter, 20–30 μm) over SUVs and LUVs is that GUVs can be observed under the microscope. That fact allows a variety of experiments to be performed at the level of a single vesicle.[15,16] As Menger and Keiper mentioned in their review article, GUVs are being examined by multiple disciplines with multiple approaches and objectives.[15] Admittedly, the use of GUVs in membrane biophysics is still in an early developmental stage, with only a few laboratories active in this area. Examples include elegant studies using GUVs and transmission microscope techniques to investigate changes in the physical properties of membranes through the calculation of elementary deformation parameters.[17–23] The mechanical properties of

[15] F. M. Menger and J. S. Keiper, *Curr. Opin. Chem. Biol.* **2,** 726 (1998).

[16] P. L. Luisi and P. Walde, eds.,"Giant Vesicles." John Wiley & Sons, London, 2000.

[17] E. Evans and R. Kwok, *Biochemistry* **21,** 4874 (1982).

[18] D. Needham and E. Evans, *Biochemistry* **27,** 8261 (1988).

[19] O. Sandre, L. Moreaux, and F. Brochard-Wyart, *Proc. Natl. Acad. Sci. U.S.A.* **96,** 10591 (1999).

[20] P. Meléard, C. Gerbeaud, T. Pott, L. Fernandez-Puente, I. Bivas, M. D. Mitov, J. Dufourcq, and P. Bothorel, *Biophys. J.* **72,** 2616 (1997).

the lipid bilayer explored by these authors are critical for our interpretation of the morphological changes that have been observed in cell membranes.

Investigations of lipid–lipid interactions, in particular the study of phase coexistence in lipid bilayers, single components, and natural and artificial lipid mixtures, have been performed with GUVs. Although the GUV membrane model is an attractive system with which to study lipid-phase equilibria in single vesicles, using fluorescence microscopy, only a few studies have been reported in the literature.[24–32] Among these studies, we recall the seminal contribution by Haverstick and Glaser. These authors reported the first visualization of lipid domain coexistence (Ca^{2+}-induced lipid domains) in GUVs composed of artificial and natural lipid mixtures, using fluorescence microscopy and digital image processing.[24] In addition, Glaser and co-workers also studied lipid domain formation induced by addition of proteins and peptides in GUV membranes.[33–35]

In addition to their use in membrane research, GUVs have been used as membrane models in a number of studies of lipid–protein and lipid–DNA interactions.[36–40] In these reports, a novel approach was taken, consisting of depositing femtoliter amounts of DNA or protein solutions onto GUVs and then monitoring the subsequent binding events and vesicle morphology changes by optical microscopy (phase contrast or epifluorescence).[36,38–40] Membrane protein incorporation on GUVs has also been reported in the literature and offers new opportunities to address issues related to lipid–membrane protein interactions and microscopic organization at the level of single vesicles.[41]

[21] P. Meléard, C. Gerbeaud, P. Bardusco, N. Jeandine, M. D. Mitov, and L. Fernandez-Puente, *Biochimie* **80,** 401 (1998).

[22] E. Sackmann, *FEBS Lett.* **346,** 3 (1994).

[23] H.-G. Döbereiner, E. Evans, M. Kraus, U. Seifert, and M. Wortis, *Phys. Rev. E* **55,** 4458 (1997).

[24] D. M. Haverstick and M. Glaser, *Proc. Natl. Acad. Sci. U.S.A.* **84,** 4475 (1987).

[25] L. A. Bagatolli and E. Gratton, *Biophys. J.* **77,** 2090 (1999).

[26] J. Korlach, P. Schwille, W. W. Webb, and G. W. Feigenson, *Proc. Natl. Acad. Sci. U.S.A.* **96,** 8461 (1999).

[27] L. A. Bagatolli and E. Gratton, *Biophys. J.* **78,** 290 (2000).

[28] L. A. Bagatolli and E. Gratton, *Biophys J.* **79,** 434 (2000).

[29] L. A. Bagatolli and E. Gratton, *J. Fluoresc.* **11,** 141 (2001).

[30] L. A. Bagatolli, E. Gratton, T. K. Khan, and P. L. G. Chong, *Biophys J.* **79,** 416 (2000).

[31] C. Dietrich, L. A. Bagatolli, Z. Volovyk, N. L. Thompson, M. Levi, K. Jacobson, and E. Gratton, *Biophys. J.* **80,** 1417 (2001).

[32] G. W. Feigenson and J. T. Buboltz, *Biophys. J.* **80,** 2775 (2001).

[33] D. M. Haverstick and M. Glaser, *Biophys. J.* **55,** 677 (1989).

[34] M. Glaser, *Comments Mol. Cell. Biophys. J.* **8,** 37 (1992).

[35] L. Yang and M. Glaser, *Biochemistry* **34,** 1500 (1995).

[36] R. Wick, M. I. Angelova, P. Walde, and P. L. Luisi, *Chem. Biol.* **3,** 105 (1996).

[37] M. L. Longo, A. J. Waring, L. M. Gordon, and D. A. Hammer, *Langmuir* **14,** 2385 (1998).

[38] P. Bucher, A. Fischer, P. L. Luisi, T. Oberholzer, and P. Walde, *Langmuir* **14,** 2712 (1998).

[39] M. I. Angelova, N. Hristova, and I. Tsoneva, *Eur. Biophys. J.* **28,** 142 (1999).

[40] J. M. Holopainen, M. I. Angelova, and P. K. J. Kinnunen, *Biophys. J.* **78,** 830 (2000).

[41] N. Kahya, E.-I. Pècheur, W. de Boeij, D. A. Wiersma, and D. Hoekstra, *Biophys. J.* **81,** 1464 (2001).

Fluorescence Microscopy and Lipid-Phase Coexistence

Fluorescence spectroscopy has been and still is extensively used to study the physical properties of lipid bilayers. One example is the study of the thermotropic behavior of lipid bilayers, in particular the presence of lipid-phase coexistence in model systems.[4,11,12,42] In general, experiments designed to detect lipid-phase coexistence in bilayers by fluorescence techniques (cuvette studies) are performed with liposome suspensions and, as mentioned above, these experiments do not provide direct information about the size and shape of the lipid domains and their time evolution. It is this information that allows correlation of the microscopic organization at the surface of single vesicles with the physical parameters, determined at the molecular level, of the lipid bilayer (lipid mobility, lipid hydration, etc).

Epifluorescence microscopy was applied to artificial membranes to observe phase coexistence in monolayers at the air–water interface.[43,44] In such experiments, fluorophores that display preferential partitioning into one of the coexisting phases are used to obtain information about lipid domain shape directly from the fluorescence images. Information about the lipid domain phase state is then deduced from the known partitioning of the particular fluorescent molecule. However, care must be taken when interpreting these effects. We have demonstrated that at the phase coexistence temperature region N-Rh-DPPE (Lissamine rhodamine B 1,2-dihexadecanoyl-*sn*-glycero-3-phosphoethanolamine, triethylammonium salt), a commonly used lipid fluorophore, preferentially partitions to the fluid phase in bilayers composed of dilauroylphosphatidylcholine–distearoylphosphatidylcholine (DLPC–DSPC) mixtures whereas in those composed of DLPC–DPPC (dipalmitoylphosphatidylcholine) mixtures N-Rh-DPPE shows a high affinity for the more ordered lipid domains.[27–29] This finding clearly shows that it is difficult to directly establish, from the fluorescence image alone, the nature of the lipid-phase state, and that it is not possible to generalize the affinity of the fluorescent molecule for the different lipid phases without a careful probe characterization.[29]

Consider the two membrane fluorophores DiI C_{20} (1,1′-dieicosanyl-3,3,3′,3′-tetramethylindocarbocyanine perchlorate) and Bodipy-PC [2-(4,4-difluoro-5,7-dimethyl-4-bora-3a,4a-diaza-*s*-indacene-3-pentanoyl)-1-hexadecanoyl-*sn*-glycero-3-phosphocholine], which preferentially partition into the ordered and disordered lipid phases, respectively. These probes were utilized in DLPC–DPPC–POPS (palmitoyloleoylphosphatidylserine) GUVs to directly visualize gel- and fluid-phase coexistence by confocal fluorescence microscopy.[26] Even though two different regions of the lipid membrane are clearly defined from the fluorescence images showing gel–fluid phase coexistence, the phase states of the coexisting lipid domains were determined by measuring probe translational diffusion, using

[42] L. M. S. Loura, A. Fedorov, and M. Prieto, *Biochim. Biophys. Acta* **1467,** 101 (2000).

[43] R. M. Weis and H. M. McConnell, *Nature (London)* **310,** 47 (1984).

[44] H. Möhwald, A. Dietrich, C. Böhm, G. Brezesindki, and M. Thoma, *Mol. Membr. Biol.* **12,** 29 (1995).

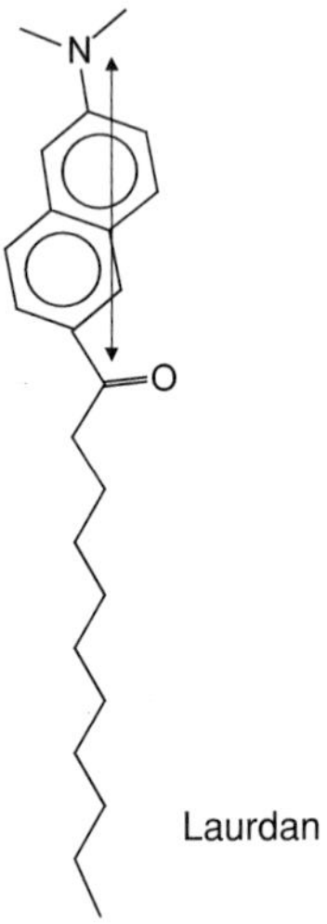

FIG. 1. Laurdan molecule. The double-headed arrow indicates the orientation of the excited state dipole.

fluorescence correlation spectroscopy (FCS).[26] In this case, additional spectroscopic methods were necessary to characterize the physical state of the membrane and to verify the impression given by the fluorescence images.

The fluorescence spectroscopic parameters measured in traditional experiments involving liposome solutions can also be measured at the level of single vesicles by using fluorescence microscopy. Fluorescence polarization, fluorescence lifetime, or phase-dependent fluorescence emission shifts can be obtained to directly determine the lipid domain phase state of the different membrane structures seen in the fluorescence images. These methods require a well-characterized fluorophore, one of the many commonly used in cuvette experiments, GUVs, a fluorescence microscope, and some way to obtain spectroscopic images. In the following sections, we focus on a particularly interesting and useful membrane probe, Laurdan, which has unique spectroscopic properties for monitoring local lipid dynamics. We emphasize how this probe can be used to simultaneously visualize regions of the membrane with different properties and characterize the properties of these regions by a relatively simple spectroscopic measurement.

Laurdan and Possibility of Performing Fluorescence Spectroscopy and Imaging Simultaneously

Laurdan (Fig. 1) belongs to the family of polarity-sensitive fluorescent probes first designed and synthesized by G. Weber for the study of the phenomenon of solvent dipolar relaxation.[45–48] This family of probes also includes 6-propionyl-2-(*N,N*-dimethylamino)naphthalene (Prodan) and 2′-(*N,N*-dimethylamino)-6-naphthoyl-4-*trans*-cyclohexanoic acid (Danca). The advantages of Laurdan in the study of lipid–lipid interactions have been amply demonstrated by the more than 100 articles published in the 1990s. The unique characteristics of Laurdan to determine lipid lateral organization in bilayers are particularly useful in studying lipid-phase coexistence.[49,50] These characteristics can be divided into four fundamental properties[29,49,50]: (1) the electronic transition dipole of Laurdan is aligned parallel to the hydrophobic lipid chains; (2) Laurdan shows a phase-dependent emission spectral shift, that is blue in the ordered lipid phase and blue-green in the disordered lipid phase (this effect is attributed to the reorientation of water molecules present at the lipid interface near the fluorescent moiety of Laurdan); (3) Laurdan distributes equally into the solid and liquid lipid phases; and (4) Laurdan is negligibly soluble in water.

The characteristics of Laurdan encourage its use in fluorescence microscopy, in particular to resolve different domains by simple analysis of Laurdan fluorescence images at the level of single vesicles. Unfortunately, the extent of Laurdan photobleaching under the epifluorescence microscope is severe, making it almost impossible to collect images of Laurdan-labeled specimens for more than few seconds.[29] However, the use of two-photon excitation fluorescence microscopy[51,52,53] helps circumvent this problem. The fact that two-photon absorption in the microscope is confined to the focal volume without excitation in areas above and below the focal plane (because of insufficient photon flux) dramatically reduces the extent of probe photobleaching during image collection. In addition, Laurdan has a good two-photon absorption cross section, which permits collection of Laurdan fluorescence. Furthermore, the sectioning (confocal) effect of two-photon excitation allows the researcher to collect images from different focal planes; a useful feature that was an essential component in a number of studies aimed at visualizing lipid-phase coexistence in GUVs.[25,27–29,53,54]

[45] G. Weber and F. J. Farris, *Biochemistry* **18,** 3075 (1979).

[46] R. B. Macgregor and G. Weber, *Nature (London)* **319,** 70 (1986).

[47] T. Parasassi, F. Conti, and E. Gratton, *Cell. Mol. Biol.* **32,** 103 (1986).

[48] M. Lasagna, V. Vargas, D. M. Jameson, and J. E. Brunet, *Biochemistry* **35,** 973 (1996).

[49] T. Parasassi and E. Gratton, *J. Fluoresc.* **5,** 59 (1995).

[50] T. Parasassi, E. Krasnowska, L. A. Bagatolli, and E. Gratton, *J. Fluoresc.* **8,** 365 (1998).

[51] W. Denk, J. H. Strickler, and W. W. Webb, *Science* **248,** 73 (1990).

[52] B. R. Master, P. T. C. So, and E. Gratton, *in* "Fluorescent and Luminescent Probes," 2nd Ed., p. 414. Academic Press, New York, 1999.

[53] L. A. Bagatolli, T. Parasassi, and E. Gratton, *Chem. Phys. Lipids* **105,** 135 (2000).

[54] T. Parasassi, E. Gratton, W. Yu, P. Wilson, and M. Levi, *Biophys. J.* **72,** 2413 (1997).

Using Laurdan Imaging to Define Lipid Domain Shape and Phase State

In addition to the phase-dependent emission spectral shift of Laurdan, the location of the Laurdan excited state dipole, which is parallel to the lipids in the bilayer, offers an additional advantage to ascertain lipid-phase coexistence from the fluorescence images, using the photoselection effect.[25,27–29,31,54] Because of the photoselection effect, only those fluorophores that are aligned parallel, or nearly so, to the plane of polarization of the excitation light are excited. For this reason, when observing the top or bottom surface of a spherical lipid vesicle displaying gel- and fluid-phase coexistence, only fluorescence coming from the fluid part of the bilayer can be observed (Fig. 2). In the fluid phase, a component of the transition dipole of Laurdan is always parallel to the excitation polarization because of the relatively low lipid order, that is, the wobbling movement of the Laurdan molecule. In contrast, Laurdan molecules present in the lipid gel phase have a greatly restricted wobbling motion so that few molecules are excited on the top or bottom GUV surface and little fluorescence intensity is observed. This photoselection effect exists only at the top and bottom surface of the vesicle, where the Laurdan molecules are oriented along the z axis (the excitation light polarization plane is defined as the x–y plane and light propagates in the z direction).

GUVs displaying fluid-ordered/fluid-disordered and gel/fluid lipid-phase coexistence are shown in Fig. 3. In Fig. 3, we exploit the photoselection effect to recognize the tightly and loosely packed bilayer regions occurring at the phase coexistence temperatures.[27–29,31] In the case of gel- and fluid-phase coexistence (Fig. 3B) Laurdan fluorescence originates from the fluid region even though Laurdan is homogeneously distributed in the bilayer (see above). In this case, the fluid region is associated with a red-shifted fluorescence emission (high extent of water dipolar relaxation). In the GUV displaying fluid-ordered/fluid-disordered phase coexistence (Fig. 3A), the less ordered region displays enhanced intensity compared with the more ordered region because of the photoselection effect. In addition, the fluorescence emission of Laurdan at the fluid-ordered bilayer regions is blue shifted, when compared with the emission from the fluid-disordered bilayer regions, allowing identification of the phase of the lipid domain directly from the fluorescence image.[29,31]

Phase State and Shape of Lipid Domains

When fluid domains are embedded in a fluid environment, circular domains will form because both phases are isotropic and the line energy (tension), which is associated with the rim of two demixing phases, is minimized by reducing the area-to-perimeter ratio. This last situation is observed in the DOPC–cholesterol–sphingomyelin mixture (Fig. 3A) and is consistent with considerations about the lipid domain phase state obtained from fluorescence images using Laurdan.[29,31]

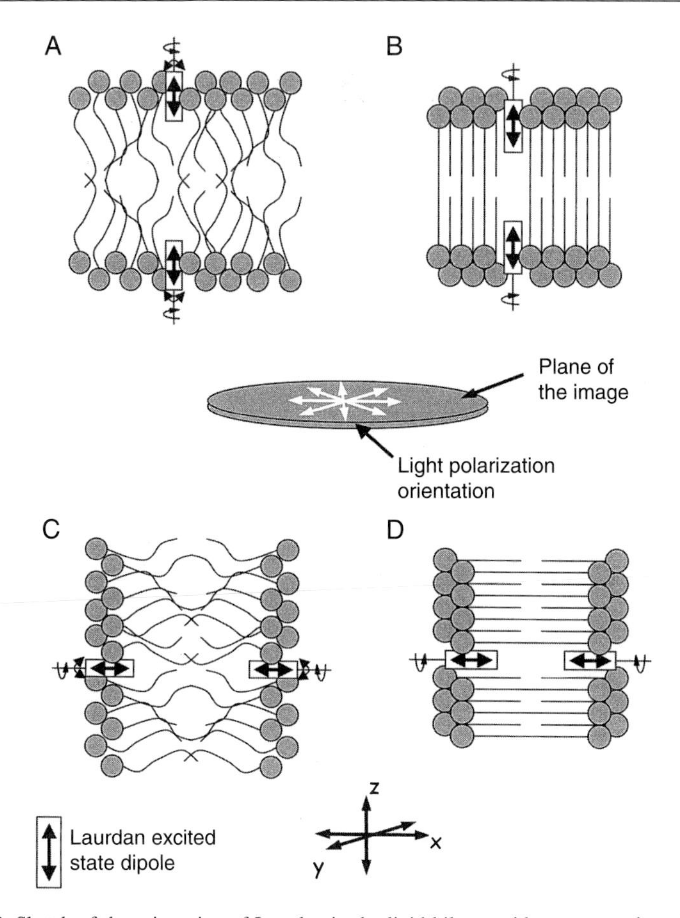

FIG. 2. Sketch of the orientation of Laurdan in the lipid bilayer with respect to the polarization plane of the excitation light (circular polarized light). (A and C) Fluid-phase (loosely packed) bilayer observed at the polar region (A) and in the equatorial region of the GUV (C). In these last two situations green emission is observed in both cases—the emission intensity is lower in (A) with respect to (C) because of the photoselection effect. (B and D) Gel-phase bilayer observed at the polar region (B) and in the equatorial region of the GUV (D). In this case only blue emission is observed in (D) [being zero in (B)] because of the photoselection effect (see text).

This contrasts starkly with that found in the gel–fluid coexistence where round domains are not observed (Fig. 3B).[27,29]

Laurdan Generalized Polarization Function

A way to quantify the extent of water dipolar relaxation, which in turn is related to the phase state of the lipid interfaces, is based on a useful relationship between the emission intensities obtained at the blue and red sides of the Laurdan emission

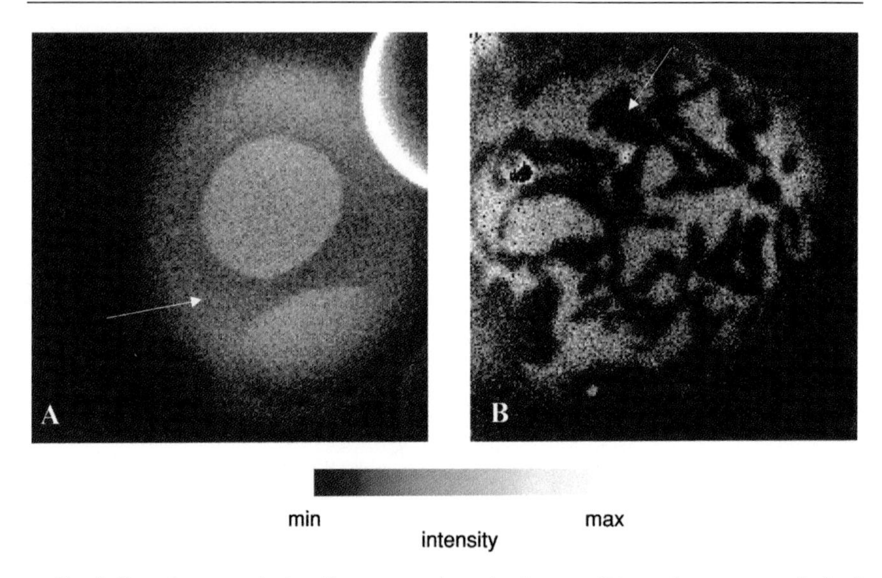

FIG. 3. Two-photon excitation fluorescence intensity images (false color representation) of GUVs formed of DOPC–cholesterol–sphingomyelin (1 : 1 : 1, mol/mol) (A) and DMPC–DSPC (1 : 1, mol/mol) (B) labeled with Laurdan. The images were taken at the top part of the GUV at temperatures corresponding to the phase coexistence region (24° for DOPC–cholesterol–sphingomyelin and 45° for DMPC–DSPC). The arrows indicate the fluid-ordered (DOPC–cholesterol–sphingomyelin) and gel (DMPC–DSPC) domains. GUVs diameter ~30 μm.

spectrum. This relationship, called generalized polarization (GP),[49,50] was defined in analogy with the fluorescence polarization function as

$$\mathrm{GP} = \frac{I_\mathrm{B} - I_\mathrm{R}}{I_\mathrm{B} + I_\mathrm{R}} \tag{1}$$

In this function, the relative parallel and perpendicular orientations in the classic polarization function were substituted by the intensities at the blue and red edges of the emission spectrum [I_B and I_R, respectively, in Eq. (1)], using a given excitation wavelength.[49,50] The generalized polarization parameter contains information about solvent dipolar relaxation processes that occur during the time that Laurdan is in the excited state, and it is related to water penetration in the phospholipid interfaces (see above). Therefore, GP images can be constructed from the intensity images obtained with blue and green bandpass filters on the microscope, allowing further characterization of the phase state of the coexisting lipid domains.[25,27–31] It is important that the GP images be obtained at the equatorial region of the GUVs to avoid artifacts due to the photoselection effect because of the particular locations of the Laurdan probe in the lipid bilayer (see above).[29]

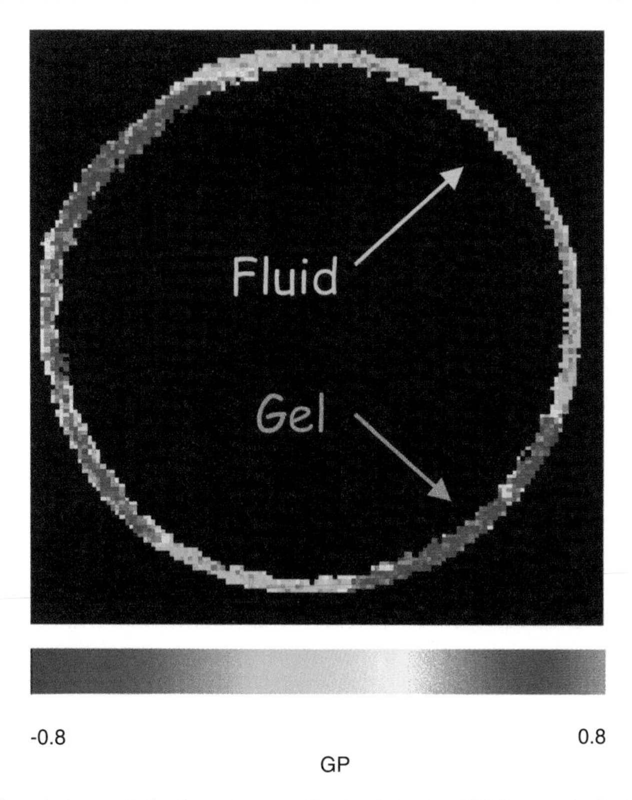

FIG. 4. Two-photon excitation Laurdan generalized polarization (GP) image of a single GUV composed of DLPC and DAPC (1 : 1, mol/mol), obtained with circular polarized light. The temperature was 55°, corresponding to the gel- and fluid-phase coexistence temperature regime. The image corresponds to the equatorial region of a 30-μm-diameter vesicle. The arrows show the fluid and gel lipid domains spanning the bilayer.

As an example, the GP image of DLPC–DAPC (diarachydoylphosphatidylcholine) is shown in Fig. 4. In this phospholipid binary mixture the GP measured in the tight- and loose-packed regions of the lipid bilayer corresponds to that observed at temperatures below and above the main phase transition temperature in GUVs composed of a single phospholipid component. The fact that in the DLPC–DAPC mixture the physical characteristics of the coexisting phases resemble those observed in the "pure" gel- and fluid-phase states suggests highly energetic and compositional differences between the coexisting lipid domains (low miscibility between the components of the binary mixture).[28,29] Interestingly enough, the observation that lipid domains span the lipid bilayer is obtained directly from the GP images (see Fig. 4). This finding is in agreement with the observations of GUVs composed of DLPC–DPPC–POPS mixtures at room temperature, using standard one-photon confocal microscopy.[26] Lipid domain spanning the lipid bilayer is a general phenomenon in samples displaying fluid/fluid and gel/fluid phase coexistence (GUVs composed of artificial and natural lipid mixtures).[27–29,31] This last feature suggests a mechanism for how differentiated domains in the outer monolayer of biological membranes are coupled to cytoplasmic signal transduction pathways: a question that has eluded simple answers.[31]

Laurdan Generalized Polarization Images Providing Information about Domains Smaller than Microscope Resolution

When the size of the lipid domain is smaller than the size of the image pixel, it is not possible to determine the lipid domain shape by the above-mentioned experimental approach. However, it is possible to determine lipid domain coexistence by obtaining Laurdan GP images at the equatorial region of the GUV, using linear polarized light.[25,54] If the lipid domain size is smaller than the microscope resolution, each pixel of the image (we assume a pixel has a size comparable to the microscope resolution) will present an average GP value in the equatorial section of the GUV. We can discriminate pixels with high and low GP values because the polarized light, which photoselects appropriately oriented Laurdan molecules, also selects Laurdan molecules associated with high GP values. Essentially, if the image contains separate domains (pixels) of different GP values, because of lipid-phase coexistence, the higher GP value domains appear parallel to the orientation of the polarized excitation light and not in the perpendicular direction. This effect was used to ascertain lipid domain coexistence in model and natural membranes[54–56] and in GUVs composed of pure phospholipids.[25]

To further illustrate this effect, let us analyze the fluorescence intensity and the GP function of Laurdan in the GUV equatorial plane (Fig. 5). If we excite with polarized light (the plane of polarization corresponds to the direction 0° axis in Fig. 5), the intensity will vary as we go around the equatorial section and the extent of change in the intensity will depend on the local orientation of the probe. Let us first consider the case in which the bilayer is homogeneous. In the gel phase, the intensity will greatly vary as we go around the section, but in the liquid–crystalline phase the intensity will vary much less. If the probe is randomly oriented, the intensity will not change as we go around the section. If we now look at the GP changes (Fig. 5), again assuming that the sample is homogeneous, the GP should be independent of the intensity. However, if the sample displays microheterogeneity, the GP value will also change as we go around the GUV

[55] W. Yu, P. T. So, T. French, and E. Gratton, *Biophys. J.* **70,** 626 (1996).

[56] T. Parasassi, W. Yu, D. Durbin, L. Kuriashkina, E. Gratton, N. Maeda, and F. Ursini, *Free Radic. Biol. Med.* **28,** 1589 (2000).

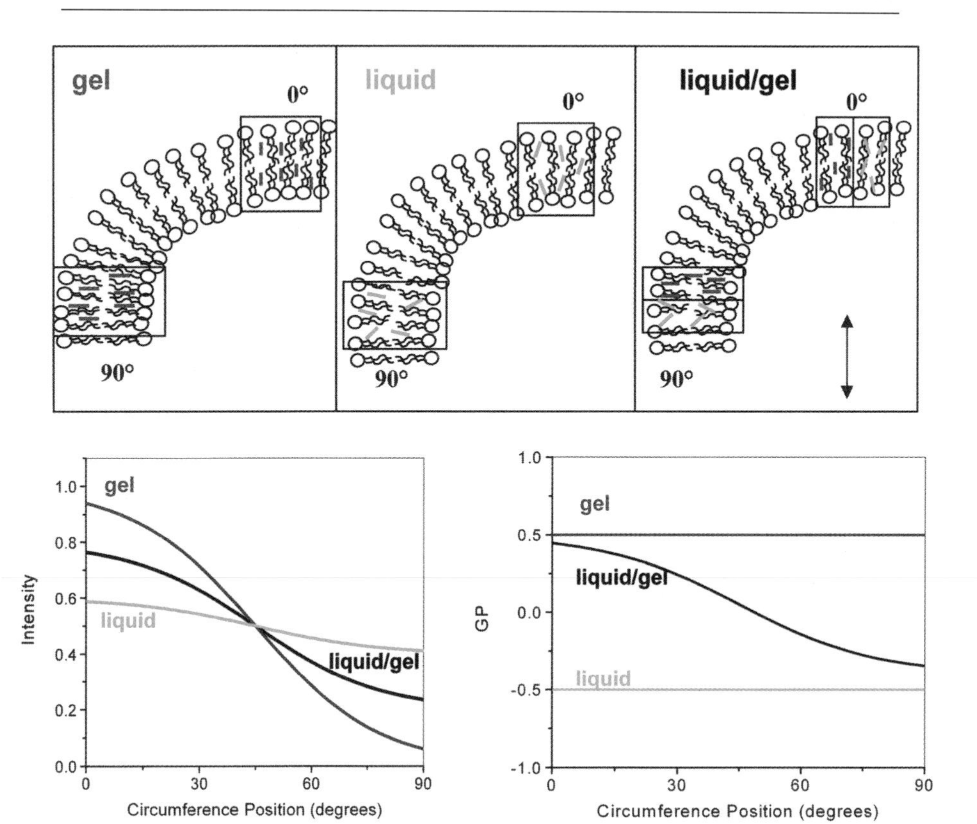

FIG. 5. *Top:* Schematic drawing of the orientation and mobility of Laurdan within a phospholipid bilayer under gel, liquid, and phase coexistence conditions. The excitation light is assumed to be polarized along the 0° axis and the excited state Laurdan molecules are indicated in red (gel) and green (fluid). *Bottom:* Laurdan intensity and Laurdan GP are plotted as a function of the perimeter of a circular vesicle. The curves are labeled to match the lipid states illustrated (*top*).

perimeter because, along the polarization axis, we are preferentially exciting the gel phase with high GP, but in the perpendicular direction we are exciting only the liquid–crystal microdomains.

Measurements of a GUV composed of a simple lipid above, at, and below the phase transition is shown in Fig. 6A and B. At low temperature, the GUV is in the gel phase. There is little microheterogeneity because the intensity varies greatly but the GP is essentially constant around the equatorial section (red line in Fig. 6D). Instead, at the phase transition and above, both the intensity and the GP vary (black and green lines respectively in Fig. 6C and D). As we further increase the temperature, the membrane heterogeneity decreases gradually and even at a temperature well above the main transition temperature, we can still observe the membrane microheterogeneity. This unexpected result suggests that there is an

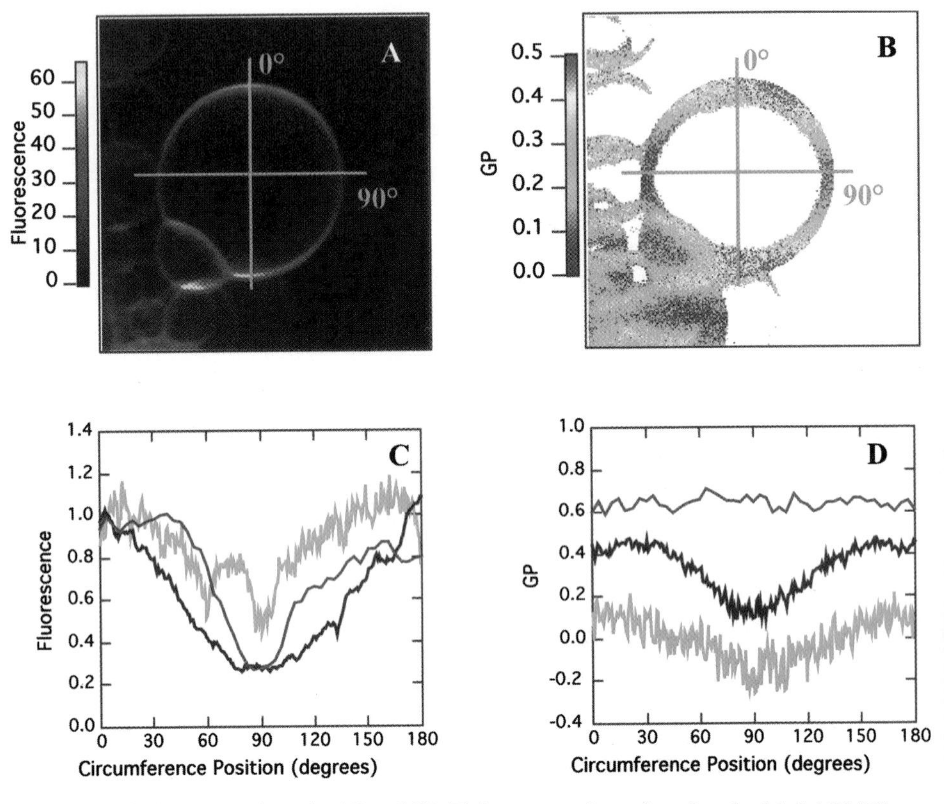

FIG. 6. Fluorescence intensity (A) and GP (B) images are shown for a Laurdan-labeled DMPC giant unilamellar vesicle at 25°, near the main phase transition temperature of 24.5°. The excitation light was polarized along the 0–180° axis. Fluorescence intensity (C) and GP (D) as a function of GUV circumference position are shown for Laurdan-labeled vesicles in liquid phase, DMPC at 41° (green line), in the phase transition region, DMPC at 25° [black line, from the images in (A) and (B)], and in gel phase, DPPC at 32° (red line).

intrinsic microheterogeneity for phospholipids in the fluid phase[28] that does not appear to be true for the gel-phase phospholipid domains.

Laurdan Generalized Polarization Providing Information about Phase State of Lipid Membrane during Action of Interfacial Hydrolysis: Case of Phospholipase A_2

In this section, we show the use of the microscopy techniques described above to study a particular problem of an enzymatic reaction in lipid vesicles. Laurdan was used to characterize the changes in the phase state of vesicles during enzymatic hydrolysis of membrane phospholipids.

Secreted phospholipase A_2 ($sPLA_2$) has been shown to directly hydrolyze phospholipids of GUVs made by the electroformation process.[36] It is not surprising that this enzyme was chosen for examination of the interactions between protein and GUVs because this class of enzyme is particularly active against aggregated phospholipids, hydrolyzing the *sn*-2-acyl chain of phospholipids to produce free fatty acid and lysophospholipid. As the substrate is hydrolyzed the reaction products become new membrane components, giving rise to hydrolysis-associated changes in the bilayer that can be explored through the use of membrane-reporting fluorophores. Although seemingly simple, the hydrolysis of a phospholipid bilayer is a complex, nonequilibrium process. The rapid production of fatty acid and lysolipid by the enzyme creates a complex membrane dynamic that, in turn, will be self-perturbing and affect continued $sPLA_2$ actions. Indeed, $sPLA_2$ activity is sensitive to the membrane character and is modified by a host of membrane factors that include the phospholipid headgroup size, headgroup charge, phospholipid phase state, phospholipid dynamics, and membrane curvature.[57–63] This interaction between organized phospholipid substrate and $sPLA_2$ has long intrigued researchers and has made these enzymes ideal model systems for examining protein and membrane interactions.

Direct visualization of $sPLA_2$ action on unsupported bilayer membranes has primarily come from images of lipid vesicles demonstrating morphological changes, using electron microscopy techniques.[63,64] Atomic force microscopy has also been used to visualize the effect of $sPLA_2$ hydrolysis on supported bilayers.[65] Images of the time-dependent changes in palmitoyloleoylphosphatidylcholine (POPC) GUVs due to the presence of $sPLA_2$ have been reported by Wick and co-workers.[36] The approach taken by these authors consisted of injecting femtoliter volumes of enzyme solution into, or onto, GUVs by use of a microinjector and then monitoring the influence of $sPLA_2$ on single vesicles, using conventional microscope techniques (phase contrast). This elegant work presents a simple, yet powerful experimental arrangement to study the interactions between proteins and lipid interfaces. In describing the $sPLA_2$ attack on GUVs, Wick and co-workers reported two kinds of effects: gradual shrinking of the GUV and immediate destruction of the GUV.[36] In our hands, using this same general approach, we have found

[57] W. R. Burack, Q. Yaun, and R. L. Biltonen, *Biochemistry* **32,** 583 (1993).

[58] J. D. Bell, M. Burnside, J. A. Owen, M. L. Royall, and M. L. Baker, *Biochemistry* **35,** 4945 (1996).

[59] W. R. Burack, M. E. Gadd, and R. L. Biltonen, *Biochemistry* **34,** 14819 (1995).

[60] C. R. Kensil and E. A. Dennis, *Biochemistry* **254,** 5843 (1979).

[61] J. A. F. Op den Kamp, M. T. Kauerz, and L. L. M. van Deenen, *Biochim. Biophys. Acta* **406,** 169 (1975).

[62] J. C. Wilschut, J. Regts, H. Westenberg, and G. Scherphof, *Biochim. Biophys. Acta* **508,** 185 (1978).

[63] W. R. Burack, A. R. G. Dibble, M. M. Allietta, and R. L. Biltonen, *Biochemistry* **36,** 10551 (1997).

[64] T. H. Callisen, *Biochemistry* **37,** 10987 (1998).

[65] M. Grandbois, H. Clausen-Schaumann, and H. Gaub, *Biophys. J.* **74,** 2398 (1998).

A

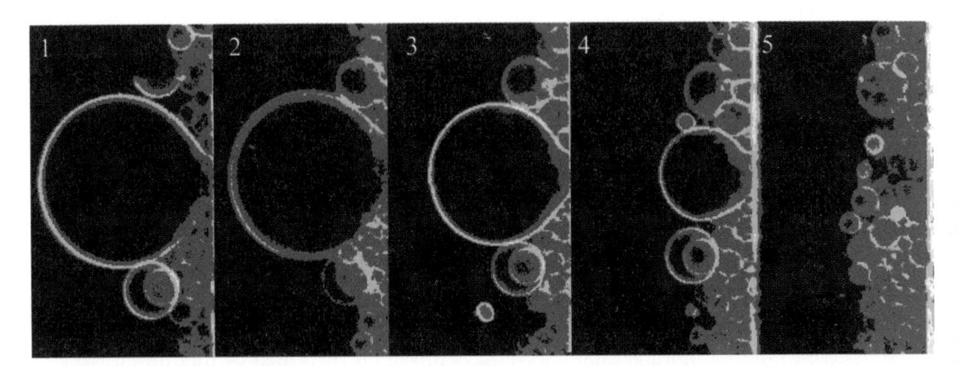

B

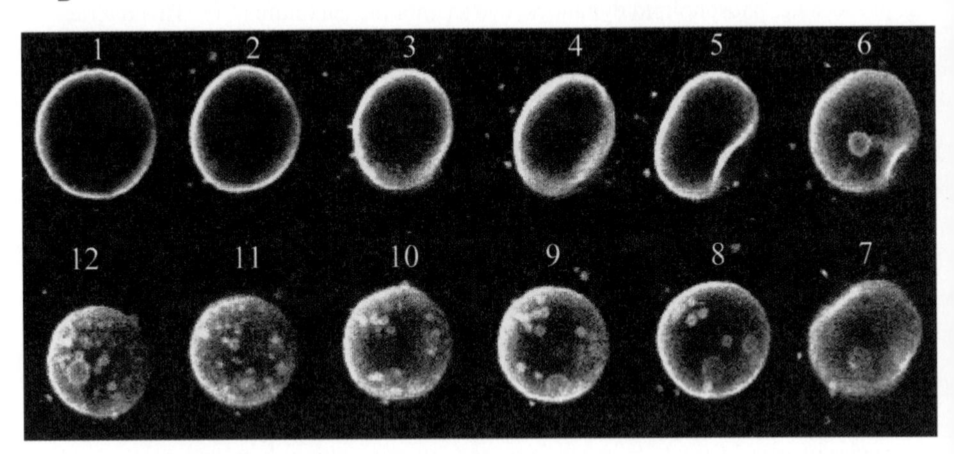

FIG. 7. sPLA$_2$-induced morphology changes in GUVs labeled with Laurdan. sPLA$_2$, isolated from *Crotalus atrox* venom, was added to the bulk solution as indicated. The images shown are as follows: (A) POPC vesicles at 20° and (B) DMPC vesicles at 31°.

that the most common effect is the steady reduction in size of the GUV, presumably following the time course of hydrolysis of the membrane phospholipids (Fig. 7A). The shrinking is surprising for a number of reasons. Within the phospholipase research community, the assumption has long been made that in the hydrolysis of vesicles only the outer layer of the vesicle is hydrolyzed and the vesicle remains intact.[66–68] The evidence to support these claims appears to be solid, although

[66] S. Gul and A. D. Smith, *Biochim. Biophys. Acta* **367,** 271 (1974).

[67] M. K. Jain, C. J. A. van Echteld, F. Ramirez, J. de Gier, G. H. De Haas, and L. L. M. van Deenen, *Nature* (*London*) **284,** 486 (1980).

[68] M. K. Jain, B.-Z. Yu, J. Rogers, G. N. Ranadive, and O. G. Berg, *Biochemistry* **30,** 7306 (1991).

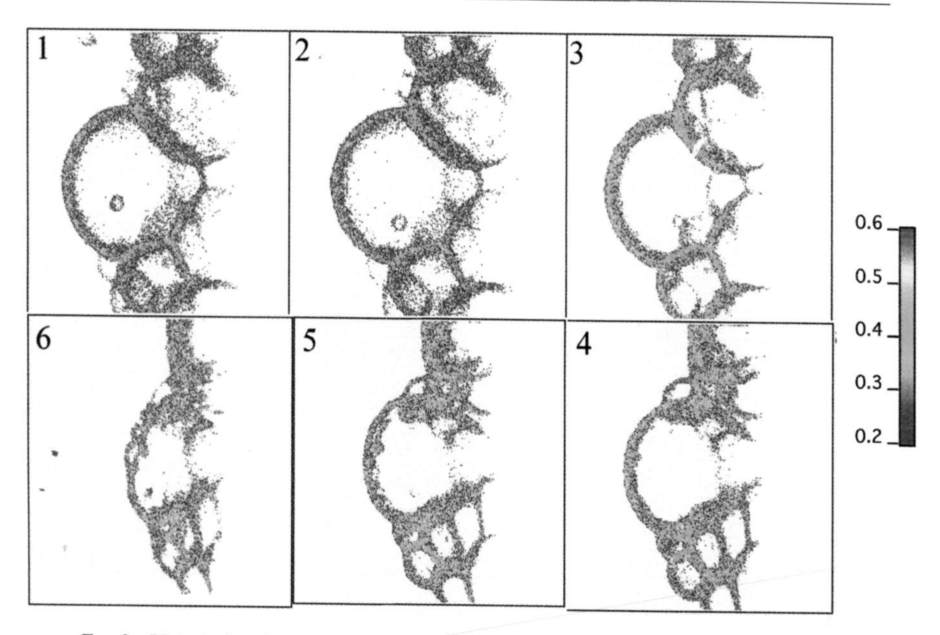

FIG. 8. $sPLA_2$-induced Laurdan GP changes in DMPC GUVs equilibrated at 26°. *Crotalus atrox* venom $sPLA_2$ was added to the bulk solution immediately before the initial image. The buffer contained 0.5 m*M* Tris (pH 8.0) and 0.2 m*M* $CaCl_2$.

there have been suggestions that vesicle morphology may be significantly affected by $sPLA_2$ lipid hydrolysis.[59,69] What are the forces involved and why should the GUV system, which is merely another model membrane bilayer, show different characteristics than any other bilayer system?

Shrinking is not the only effect that can be observed with $sPLA_2$ hydrolysis of the GUV membrane surface. On occasion, a gross distortion of GUV shape after $sPLA_2$ addition can be seen. Although not a common observation, this distortion underscores the fact that significant structural forces are induced through phospholipid hydrolysis and the creation of products in the membrane. A clear example of this effect is shown in Fig. 7B. The formation of vesicles inside the primary GUV during the hydrolysis is apparent in this sequence (Fig. 7B) and suggests that membrane stress can be relieved through expulsion of smaller vesicles. How can the nature of these changes be examined? Membrane probes, such as Laurdan, offer the researcher one method to dissect time-dependent fluctuations in membrane water penetration, which, in turn, are related to the packing and kinetics of the membrane phospholipids (see above). A GP image series monitoring the time-dependent changes in a GUV membrane after $sPLA_2$ addition is shown in Fig. 8. Laurdan had been added to the GUVs after their formation and allowed

[69] W. R. Burack, A. R. G. Dibble, and R. L. Biltonen, *Chem. Phys. Lipids* **90,** 87 (1997).

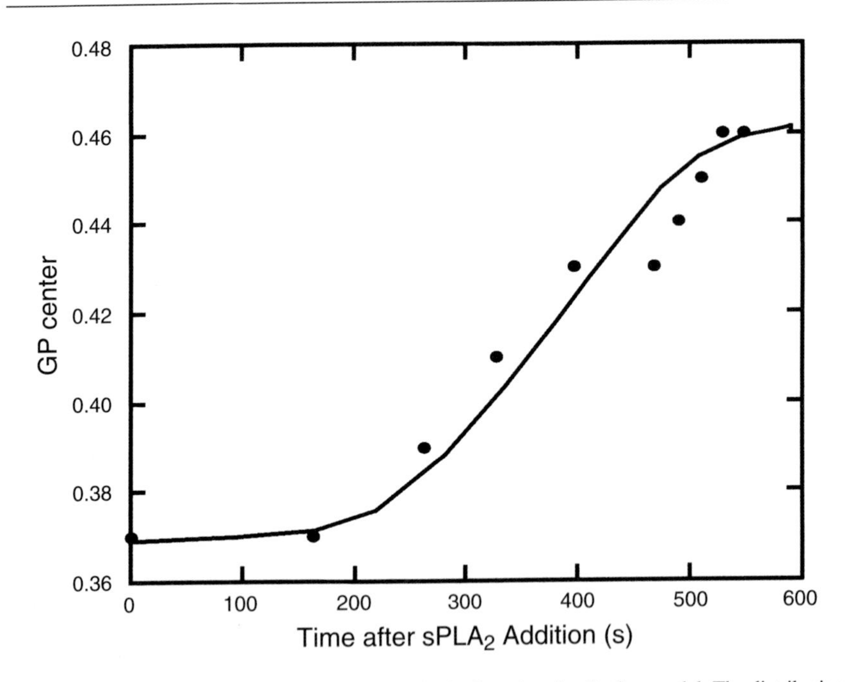

FIG. 9. The membrane GPs were well fit to a single Gaussian distribution model. The distribution center is given as a function of time after *Crotalus atrox* venom $sPLA_2$ addition. These data are derived from the image series that is selectively represented in Fig. 8.

to disperse throughout the sample chamber for approximately 15 min. GP images of the vesicles were then calculated from the simultaneously collected red and blue images, as described earlier in this article. The shrinking is readily apparent, although not as extreme as in other cases that we have observed. As hydrolysis proceeds, the average GP shifts from approximately 0.37 to 0.46, indicating an overall decrease in membrane polarity. The results are shown in the plot of the average GP as a function of time in Fig. 9. These data are consistent with results from bulk solution studies of large unilamellar vesicles.[58,70,71] In our case, however, we have the additional ability to examine the membranes of individual vesicles and observe the ongoing events at the local microscopic level. The GP images in Fig. 8 show the progressive formation of small solid domains (red-yellow), showing us that the shift in average GP values (Fig. 9) is the consequence of small solid domains formed as hydrolysis proceeds. The melting temperature of

[70] J. B. Henshaw, C. A. Olsen, A. R. Farnbach, K. H. Nielson, and J. D. Bell, *Biochemistry* **37,** 10709 (1998).

[71] H. A. Wilson, J. B. Waldrip, K. H. Nielson, A. M. Judd, S. K. Han, W. W. Cho, P. J. Sims, and J. D. Bell, *J. Biol. Chem.* **274,** 11494 (1999).

myristic acid is significantly higher than that of the root phospholipid and would be expected to give rise to higher GP values as its concentration builds in the membrane. The increase in average GP can be due to interdigitation of the fatty acid in the membrane, reducing water penetration, or fatty acid domain formation. The fact that small multipixel domains with higher GP are observed suggests that the fatty acid is not homogeneously distributed and small, fatty acid-rich regions are the source of the GP change. The other hydrolysis product, lysophospholipid, is relatively water soluble and would largely partition into the bulk phase (see below).[72–74] Phospholipid monolayer studies by Grainger and co-workers show a similar phenomenon through monitoring of fluorophore-labeled fatty acid and, interestingly, the authors also observed that the fatty acid coalesced with $sPLA_2$ to form enzyme–product domains.[75–77] Our hydrolysis-induced higher GP microdomains are consistent with this hypothesis, but we have not as yet been able to identify preferential binding of the *Crotalus atrox* $sPLA_2$ to these regions.

The shape changes shown in Fig. 7 are complex and can be linked to what we have already inferred from the Laurdan GP images. As already stated, the $sPLA_2$ hydrolysis of bilayer phospholipids generates two products: fatty acid and lysophospholipid. At the pH used in these studies, pH 8.0, both products have highly favorable partition coefficients into the lipid phase. However, under the conditions of the GUV system the concentration of lipid is extremely low, on the order of 0.3 μM. This creates a large volume for the soluble phase in proportion to the lipid, hydrophobic phase. The equilibration of fatty acid and lysolipid into the soluble phase creates structural stress on the outer leaflet of the membrane, where $sPLA_2$ initially has access and begins active lipid hydrolysis. Given the partition coefficients for myristoyl-lysophosphatidylcholine (0.5×10^6)[76] and myristic acid (2.6×10^6),[78] the products would be expected to partition into the aqueous phase. In contrast, bulk solution measurements with large/medium/small unilamellar vesicles are usually performed at lipid concentrations of $>100\ \mu M$, where a significant fraction of the products would be expected to remain in the bilayer. We would not expect to observe as great a degree of vesicle distortion in these systems. However, the fact that we see a GP change in the GUVs as hydrolysis proceeds suggests that fatty acid concentration is, in fact, building up in the membrane, despite its partition coefficient. It would be reasonable to assume that kinetic effects and organizational forces of the membrane help to

[72] A. Anel, G. V. Richieri, and A. M. Kleinfeld, *Biochemistry* **32,** 530 (1993).
[73] S. D. Brown, B. Baker, and J. D. Bell, *Biochim. Biophys. Acta* **1168,** 13 (1993).
[74] J. D. Bell, M. L. Baker, E. D. Bent, R. W. Ashton, D. J. B. Hemming, and L. D. Hansen, *Biochemistry* **34,** 11551 (1995).
[75] A. Reichert, H. Ringsdorf, and A. Wagenknecht, *Biochim. Biophys. Acta* **1106,** 178 (1992).
[76] D. W. Grainger, A. Reichert, H. Ringsdorf, and C. Salesse, *Biochim. Biophys. Acta* **1023,** 365 (1990).
[77] D. W. Grainger, A. Reichert, H. Ringsdorf, and C. Salesse, *FEBS Lett.* **252,** 73 (1989).
[78] E. D. Bent and J. D. Bell, *Biochim. Biophys. Acta* **1254,** 349 (1995).

retain product, at least on the time scale of these experiments (minutes). Clearly, the interaction of $sPLA_2$ and a lipid bilayer is complex and there is much more to be done to understand the microscopic scenario that is now unfolding. The Laurdan GP images have been instrumental in our interpretation of the hydrolysis events and have provided clues as to the fate of the individual hydrolysis products. Further work with other fluorophores and under varying experimental conditions should prove productive in our studies of the intricate lipid dynamics involved with this and similar systems. Readers are encouraged to explore Ref. 79 that compares the action of $sPLA_2$ on GUVs composed of single-component phospholipids and phospholid binary mixtures displaying phase coexistence.

Conclusions

In this article we have pointed out the importance of generating new experimental tools to explore the concerted phenomena occurring at different mesoscopic levels of the lipid membrane. The combination of Laurdan, GUVs, and two-photon fluorescence microscopy has been extremely useful in producing a microscopic picture of lipid-phase coexistence in the GUV bilayer model system. Laurdan is a unique probe, giving simultaneous information about morphology and phase state of lipid domains from fluorescence images. After 30 years of study of the physical and chemical aspects of lipid–lipid interactions (in particular the lateral lipid separation) in model systems, the importance of the relationship between lipid composition and membrane organization is now well recognized.[80] We have shown with several examples that it is possible to visualize and characterize regions of different properties in artificial membranes. It is important to realize that the same kind of studies can be done in cells. The characterization of Laurdan parameters in different cell membranes is a necessary step that will help to understand the role of lipid composition and organization in biological membranes.

Acknowledgments

This work is supported by a grant from the National Institutes of Health (RR03155 to S.A.S., E.G., and T.L.H.) and the Fundacion Antorchas (to L.A.B.). MEMPHYS is supported by the Danish National Research Foundation.

[79] S. Sanchez, L. A. Bagatolli, E. Gratton, and T. Hazlett, *Biophys. J.* **82,** 2232 (2002).
[80] D. Brown, *Proc. Natl. Acad. Sci. U.S.A.* **98,** 10517 (2001).

[21] Biological Near-Field Microscopy

By PHILIP G. HAYDON

Introduction

Optical microscopy is an invaluable tool in the life sciences. However, a significant constraint with this approach is that resolution is limited by diffraction to about one-half the wavelength of the light. In biological imaging this results in an ability to resolve structures to about 250 nm. Because proteins such as ion channels are on the order of 10 nm in diameter, lens-based microscopy is unable to resolve the distribution of these proteins at even relatively low densities. Thus, there is considerable interest in developing microscopy modalities that have higher resolution than does classic lens-based microscopy. There has been considerable interest in the application of various forms of scanning probe microscopy in attempting to reach this goal. We initially needed to determine whether the resolution afforded by atomic force microscopy (AFM) could be used to study protein distribution. Although individual atoms can be resolved with AFM, it is rarely possible to detect proteins in cell membranes by this approach because of the compliance of cells. However, gap junction structures have been resolved[1] and it has proved possible to disclose the spatial distribution of ion channels that had been tagged with gold particles in chemically fixed samples.[2] However, tip–sample convolutions led to difficulties in the identification of channel-attached particles, which has made the application of this high-resolution technique to biological imaging problematic. A further difficulty with this method is that it has been difficult to perform double-label experiments to determine spatial colocalization of different biomolecules.

The development of an alternative form of scanning probe microscopy, near-field scanning optical microscopy (NSOM), in which subdiffraction optical resolution (~50 nm) can be achieved,[3–5] offers an alternative approach to the study of biological samples. The principle of near-field microscopy was first described in 1928 by Synge,[6] who suggested the possibility of extending the resolution of the light microscope by illuminating samples through a minute aperture that was significantly smaller than the wavelength of illuminating light. Ash and Nicholls

[1] J. H. Hoh, G. E. Sosinsky, J. P. Revel, and P. K. Hansma, *Biophys. J.* **65,** 149 (1993).
[2] P. G. Haydon, E. Henderson, and E. F. Stanley, *Neuron* **13,** 1275 (1994).
[3] E. Betzig and R. J. Chichester, *Science* **257,** 189 (1992).
[4] J. Hwang, L. K. Tamm, C. Bohm, T. S. Ramalingam, E. Betzig, and M. Edinin, *Science* **270,** 610 (1995).
[5] J. K. Trautman, J. J. Macklin, L. E. Brus, and E. Betzig, *Nature (London)* **369,** 40 (1997).
[6] E. H. Synge, *Phil. Mag.* **6,** 356 (1928).

0076-6879/03 $35.00

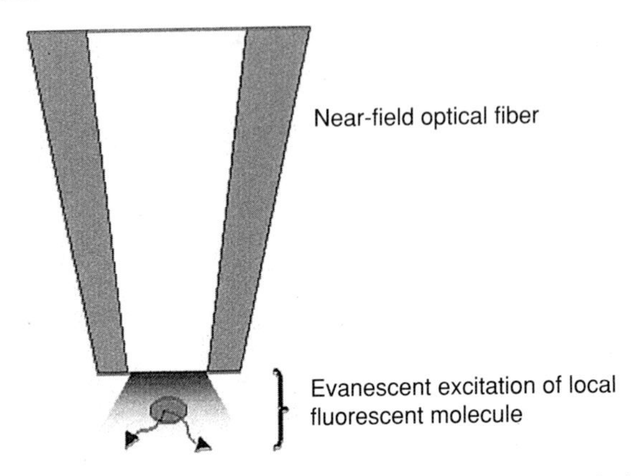

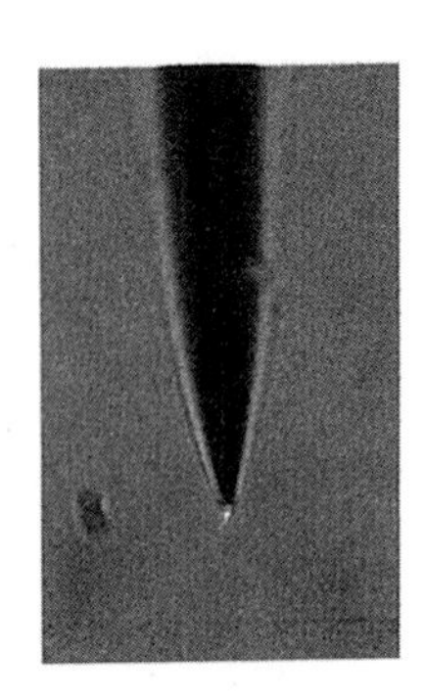

FIG. 1. Evanescent wave excitation of local fluorophore. *Left:* An evanescent wave at the tip of a near-field optical fiber can be used to locally probe for fluorescent molecules. *Right:* Photomicrograph of a near-field optical fiber that is connected to a laser source. A point of excitation energy can be seen at the tip of the optical fiber.

(1972)[7] put this method into practice by using 3-cm radiation and a restricted aperture and demonstrated resolution equivalent to $\lambda/60$. When the sample is located within the "near field" of the aperture (~50 nm), it is illuminated by a point of energy having dimensions that are regulated by the aperture size. Monitoring the transmitted, reflected, or fluorescence signal while raster scanning the aperture (or the sample) generates a subdiffraction resolution image. The development of NSOM with visible wavelengths of light was limited because of the difficulty of making subdiffraction apertures. The breakthrough that allowed NSOM to be broadly applied to studies was the development of an optical fiber as a subwavelength illumination source.

Near-field optical fibers are produced by heat pulling (or chemically etching) an optical fiber to a fine tip. Aluminum is then directionally evaporated onto the fiber to leave a small (20- to 50-nm-diameter) aperture at its tip. By coupling a laser to the unpulled end of the fiber, it is possible to supply sufficient energy to near-field fibers to generate an evanescent wave at the working end of the fiber (Fig. 1). Using near-field probes for point illumination of a specimen in combination with sample scanning methods, NSOM offers a subdiffraction resolution imaging method that can be applied to the spectroscopic studies of individual molecules.[5,8,9]

The successful operation of a near-field microscope requires (1) a subdiffraction illumination source (the near-field optical fiber), (2) a piezo-controlled

[7] E. A. Ash and G. Nicholls, *Nature (London)* **237,** 510 (1972).
[8] E. Betzig and R. J. Chichester, *Science* **262,** 1422 (1993).
[9] X. S. Xie and R. C. Dunn, *Science* **265,** 361 (1994).

feedback method to control the position of the optical fiber in relation to the sample, (3) an optical detector (photomultiplier tube or avalanche photodiode), and (4) a raster-scanning system to scan the sample in relation to the optical fiber.

Commercially available NSOMs [suppliers include Nanonics Imaging (Jerusalem, Israel; www.nanonics.co.il) and TM Microscopes (Sunnyvale, CA; www.thermomicro.com] were initially developed to attach to standard inverted optical microscopes for use in the physical sciences. The most difficult technical hurdle to overcome with NSOM has been the control of the vertical position of the near-field optical fiber in relation to the sample. In initial studies in the physical sciences this was of particular importance because monolayers of molecules attached to glass coverslips were studied and the aperture of the optical fiber had to be maintained within about 10 nm of the substrate while raster scanning the sample (or the optical fiber). Contact between the tip of the optical fiber and the glass substrate causes the tip to break, which increases the diameter of the aperture and reduces the resolution of imaging. To achieve fine control over tip–sample separation, a shear force, noncontact feedback method was developed. In this feedback method the optical fiber is vibrated at its resonant frequency while a z piezo is used to advance the tip to the sample (or vice versa). As the tip sample separation closes to within tens of nanometers, shear forces between the tip and the sample lead to a phase shift in the response of the fiber, which can then be locked onto to maintain a constant separation. This feedback method has proved reliable and permitted dramatic studies of individual molecules.

Despite the potential for high-resolution imaging provided by NSOM, there have been few examples in which this technique has been used effectively with biological specimens.[4,10–12] However, those that have been successful have demonstrated that NSOM has the potential to revolutionize the biosciences.[4,13–15] The paucity of successful biological applications for NSOM has probably arisen because the original development of the technique was performed in the physical sciences. Samples in the physical sciences have characteristics different from those in the life sciences. Unlike the planar samples on a coverslip used in the physical sciences, cells have significant thickness, must be imaged in solution, and, finally, are highly compliant samples. Repeated attempts have been made to use shear force feedback with cells in solution, with little success.

Two alternative feedback methods have been used for near-field microscopy applied to biological samples. In the first, the optical fiber also acts as a cantilever, and the optical lever approach of AFM has been used to detect contact between the

[10] V. Subramaniam, A. K. Kirsch, and T. M. Jovin, *Cell Mol. Biol. (Noisy-le-Grand)* **44,** 689 (1998).

[11] A. Lewis, A. Radko, N. Ben Ami, D. Palanker, and K. Lieberman, *Trends Cell Biol.* **9,** 70 (1999).

[12] E. J. Sanchez, L. Novotny, G. R. Holtom, and X. S. Xie, *J. Phys. Chem.* **101A,** 7019 (1997).

[13] J. Hwang, L. A. Gheber, L. Margolis, and M. Edidin, *Biophys. J.* **74,** 2184 (1998).

[14] C. W. Hollars and R. C. Dunn, *Biophys. J.* **75,** 342 (1998).

[15] S. A. Vickery and R. C. Dunn, *Biophys. J.* **76,** 1812 (1999).

tip and the sample.[11] Excellent images have been obtained by this approach. When using this feedback method, contact between the tip of the optical fiber and the sample does not damage the tip (at least within a working range of applied forces) because the fiber does not approach the hard substrate from a purely vertical direction. In the second method the optical fiber is repeatedly translated in the vertical axis to the surface of the cell while slowly raster scanning in the x and y axes. Even without a strict feedback method, subdiffraction images can be achieved by this approach when using relatively thick hydrated samples. In the remainder of this article, this open loop approach for collecting near-field images is discussed.

Instrumentation Setup of Biological Near-Field Microscope

To perform our near-field studies we use an integrated near-field illumination and positioning source with a confocal detection pathway (NeD_{NF}; Prairie Technologies LLC, Middleton, WI; www.prairie-technologies.com) that is attached to a Nikon (Melville, NY) Diaphot 300 inverted microscope (for further description see Fig. 2). Near-field optical fibers (50-nm aperture diameter; Nanonics Imaging) are mounted on a triple axis motorized manipulator (25-mm movement in each axis), which also contains three axes of piezo-controlled movement (20 μm in each axis; 2-nm precision). Optical fibers are positioned above cells by visualizing the relative position of cells and the optical fiber with an ORCA 100 digital camera (Hamamatsu Photonics, Hamamatsu, Japan; www.hamamatsu.com) that is connected to one arm of our optical detection pathway. Using this system we are able to control the position of the near-field fiber with respect to the sample, using calibrated software in which we can define the x and y image coordinates, and then command the motors to drive the optical fiber to the appropriate location. An aperture is then positioned at the secondary image plane, around the image of the tip of the near-field fiber. Positioning is performed by two motors that are calibrated such that aperture positioning is controlled from the camera image. After confirming the appropriate positioning of the aperture, a mirror is moved out of the optical pathway so that all photons are directed to a photomultiplier tube (PMT). The PMT output is fed through a preamplifer to an integrator circuit. To commence near-field studies we control the positioning of the optical fibers with piezoelectric devices controlled by voltage ramp software and data acquisition cards supplied by Prairie Technologies LLC.

Extraction of Near- from Far-Field Fluorescence

The fluorescence signal that is detected from cells has been shown to be a composite signal containing both near- and far-field components. Although the evanescent wave of a near-field optical fiber locally excites fluorescent molecules within about 100 nm of the aperture, some energy excitation frequently escapes

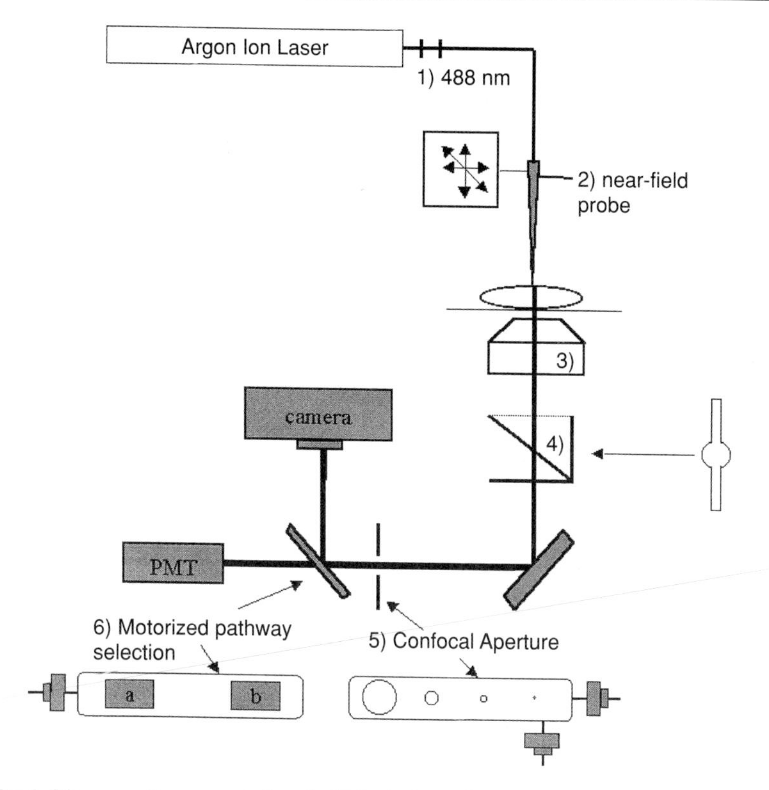

FIG. 2. Diagrammatic representation of a biological near-field microscope. The output of an argon ion laser (1) is coupled to a single-mode fiber that in turn connects to the near-field optical fiber. The near-field fiber is positioned above biological samples, which are mounted on an inverted microscope, using a three-axis motor (25-mm movement range) and piezo (15-μm movement range) manipulator (2). Fluorescent emission from the sample is collected with (3) an oil immersion $\times 60$ objective (1.4 NA). A triple filter set (4) beneath the objective permits wide-field fluorescence imaging of the sample as well as selection of fluorescent emission when the sample is excited by 488-nm energy provided by the near-field fiber. The positioning of the optical fiber in relation to the biological sample is controlled through a camera interface. Once the optical fiber is positioned, a motorized pinhole plate (5) containing a series of apertures is used to position a confocal aperture in the detection pathway to reduce the contribution of far-field fluorescence. Once the aperture is positioned, photons are directed to a photomultiplier tube (PMT), using a motorized selector (6), which directs light from the microscope to a camera (position a), or to PMTs (position b). The output of the PMT is connected to an integrator circuit and data acquisition devices.

from this local region and excites fluorescence in the far field throughout the thickness of the sample. As a consequence, if an image is generated only by measuring the peak fluorescence intensity at each pixel, the highest resolution information that is contained within the near field can be masked by far-field fluorescence.

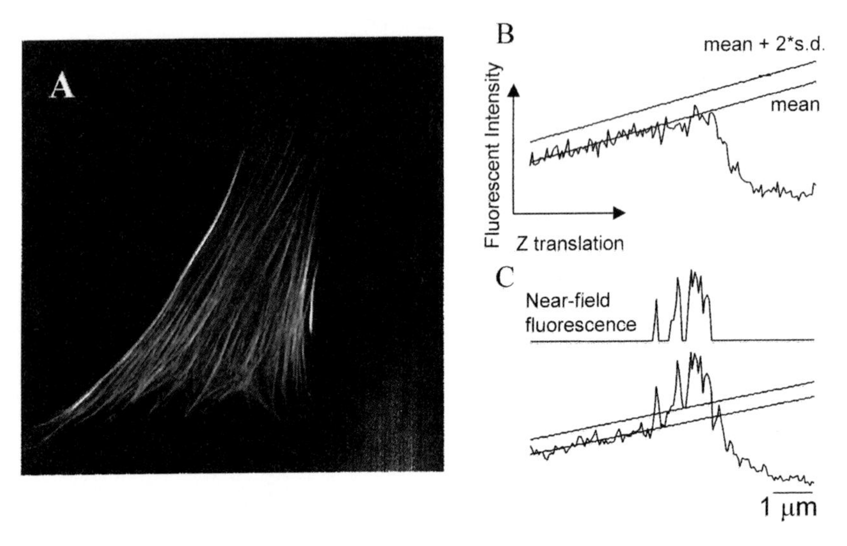

FIG. 3. Oregon Green 488 phalloidin-labeled actin filaments can be detected with near-field optical fibers. (A) Confocal image of an astrocyte labeled with Oregon Green 488 phalloidin. (B and C) Relation between the fluorescence intensity of Oregon Green 488 phalloidin and the z position of a near-field optical fiber. In (B) the fluorescent signal is shown when a near-field fiber approaches a region of an astrocyte that does not contain a labeled actin filament beneath the tip of the near-field optical fiber. Only a linear increase in fluorescence intensity due to far-field excitation of distant filaments is detected. By contrast, in (C), a step increase in fluorescence is detected when the evanescent wave of the near-field optical fiber illuminates a labeled actin filament. Superimposed on the data are two regressions corresponding to the linear regression of a baseline fit, as well as a linear regression that is offset by 2 standard deviations from the baseline. Note that as the tip interacts with the sample, the nonlinear increase in fluorescence exceeds these regressions. The upper trace represents the near-field fluorescence that was extracted from the composite signal (lower trace). Points that exceed the linear regression corresponding to the mean plus 2 standard deviations for three consecutive points were identified as caused by evanescent excitation of the sample. [Reprinted from R. T. Doyle, M. J. Szulzcewski, and P. G. Haydon, *Biophys. J.* **80,** 2477 (2001).]

We have developed a simple method to extract the near-field fluorescence that arises from local evanescent wave excitation.[16] Figure 3 shows the principle of the approach that we use. In this example, an Oregon Green 488 phalloidin (Molecular Probes, Eugene, OR)-stained astrocyte (Fig. 3A) was probed with a near-field optical fiber (Fig. 3B, C). Because phalloidin labels actin filaments, this sample provides regions that are locally devoid and enriched in fluorescently labeled filaments. Figure 3B shows a profile of the fluorescence intensity as the optical fiber is translated in the z axis to the sample and falls to a region devoid of local actin filaments. As the source of excitation approaches the sample the

[16] R. T. Doyle, M. J. Szulzcewski, and P. G. Haydon, *Biophys. J.* **80,** 2477 (2001).

fluorescence intensity increases because of far-field excitation of distant actin filaments. Once contact is made between the sample and the optical fiber, further extension to dimple the cell results in a reduction in fluorescence intensity. In contrast to this typical far-field fluorescence curve, when the same optical fiber locally excites a phalloidin-stained filament, a step in fluorescence intensity is reached as the evanescent wave locally excites the sample (Fig. 3C). Because the high-resolution fluorescence information is contained within this step change in intensity we subtract the background far-field fluorescence to reveal the local near-field signal. When using this subtraction protocol at each x, y coordinate we have found that it is possible to obtain subdiffraction resolution images from samples with resolution down to about 50 nm; for example, Figure 4A shows a confocal image of an actin-labeled astrocyte with a 2×5 μm region of filaments scanned with a biological near-field microscope (Fig. 4B and C). Figure 4B shows the composite image that results from extracting the peak fluorescence in each z

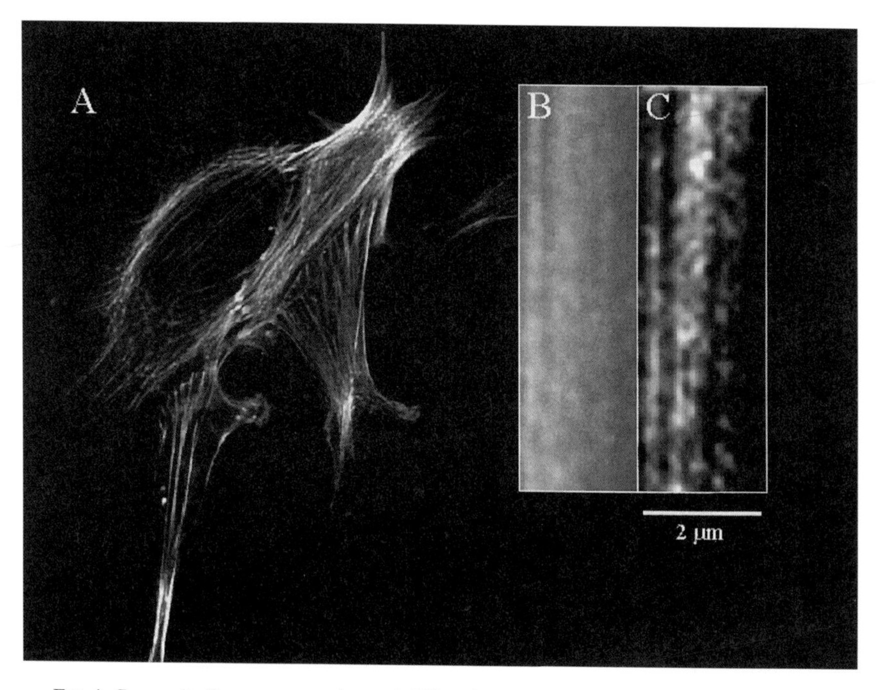

FIG. 4. Composite fluorescence and near-field imaging of actin bundles. (A) Confocal image of an astrocyte labeled with phalloidin. (B and C) Images that were generated over one actin bundle, using a near-field optical fiber as the excitation source [different cell than in (A)]. In (B) the peak intensity of fluorescence is shown (composite of near and far-field fluorescence) whereas in (C) the extracted near-field fluorescence image is presented. In these images a 3×3 Gaussian filter was used for display purposes. [Modified from R. T. Doyle, M. J. Szulzcewski, and P. G. Haydon, *Biophys. J.* **80,** 2477 (2001).]

translation, and in Fig. 4C the extracted near-field fluorescence signal from the same scan is shown.

Future Opportunities for Biological Near-Field Microscopy

We have demonstrated that biological near-field microscopy can be applied to hydrated, chemically fixed cells, but could it also be used to study living cells? Although little time has been spent using this technique to probe cells, initial investigations indicate that biological near-field microscopy has the potential to be an invaluable tool to address certain questions. However, the local nature of the evanescent wave limits these studies to cell surface measurements. Because image capture is a slow serial process that can take minutes to generate an image, it is unlikely that it will provide high-resolution images of molecular distribution in living cells. Instead, this approach could be exploited for point measurements or for very local scans, such as line scans. We anticipate that it will be possible to use biological near-field microscopy to study local calcium dynamics and report the accumulation of calcium beneath the plasma membrane during and immediately after the opening of calcium channels. Indeed, we have determined that the z translation of the optical fiber does not damage living hippocampal neurons and have demonstrated that we can locally excite calcium indicators with near-field optical fibers. A second potential live-cell application of this technique will be in the study of vesicle dynamics. Total internal reflection microscopy has revealed the dynamics of individual labeled vesicles in nerve terminals of bipolar cells.[17] Perhaps the enhanced x and y resolution of biological near-field microscopy will further improve the resolution of the study of vesicle dynamics in nerve terminals. Perhaps it will become possible to simultaneously resolve the immediate accumulation of calcium beneath an ion channel and the subsequent fusion of the adjacent vesicle with the nerve terminal. Irrespective of the specific applications, it is clear that biological near-field microscopy has advanced to such a stage that it can now be applied to living cells to address biological problems beneath the diffraction limit.

Acknowledgments

The author thanks Yolande Haydon for expert assistance in editing this manuscript, and the National Institutes of Health for financial support. The author has equity interest in Prairie Technologies LLC.

[17] D. Zenisek, J. A. Steyer, and W. Almers, *Nature (London)* **406,** 849 (2001).

[22] Fluorescence Lifetime-Resolved Imaging: Measuring Lifetimes in an Image

By ROBERT M. CLEGG, OLIVER HOLUB, and CHRISTOPHER GOHLKE

Introduction

Fluorescence lifetime-resolved imaging (FLI) refers to the technique of recording a fluorescence image whereby the fluorescence lifetimes of the fluorophores in a sample are resolved at every location of the image. This can result in either the actual determination of the fluorescence lifetimes (or more often the apparent fluorescence lifetime), or the determination of a spectroscopic parameter that is possible only when the fluorescence signal is lifetime resolved in the measurement. The spectroscopic information related to the lifetime of the fluorescence decay can then be displayed to the experimenter in an image format, where every pixel of the image contains the lifetime-resolved information. FLI measurements are analogous to normal intensity fluorescence imaging measurements and are acquired on the same samples, except that the information related to the fluorescence lifetime is directly recorded in addition to the normal measurement of the fluorescence intensity.

Standard nonimaging fluorescence measurements are easily carried out in a cuvette in either a steady state or lifetime-resolved fashion. An extensive literature on this subject exists and lifetime-resolved fluorescence measurements have provided a wealth of valuable information about biological systems. Of course, there is no directly measured spatial information available in a cuvette-type spectroscopic measurement. In contrast to this single-channel experimental format, fluorescence measurements made in an imaging environment, such as in a fluorescence microscope, are usually limited to spectrally resolved intensity measurements (where wavelengths are selected with a simple optical filter). On the other hand, although normal fluorescence microscope instrumentation does not provide temporal resolution of the fluorescence signal, advanced techniques and instrumentation are available that reveal detailed morphological information with high spatial resolution. For this reason, fluorescence imaging has provided an enormous wealth of information in cellular biology. Fluorescence microscopy has become a familiar measurement in essentially every field of cellular biology.

The FLI measurement combines the spatial resolution of a fluorescence image of a spatially extended fluorescing sample with the temporal dimension of a lifetime-resolved measurement. As the FLI technique emerges and becomes available to more researchers, the enhanced and more refined information content of time-resolved fluorescence measurements will extend significantly the capability of the investigator to reveal physical details on the molecular scale in fluorescence

0076-6879/03 $35.00

images of biological samples. This is not a review of the many excellent publications and outstanding contributions that have been made on different FLI instruments.* The major aim of this article is to acquaint the biologist with the information offered by time-resolved fluorescence spectroscopy and to describe how the experimental measurements are related to the fundamental mechanism of fluorescence. An understanding of the different pathways of deexcitation available to a molecule in an excited state leads naturally to an appreciation of the knowledge that can be gained by temporally resolving the emission of a fluorescence signal. Awareness of the fundamental time-dependent mechanisms that play a role in fluorescence and of the interdependence of the different deexcitation pathways available to an excited fluorophore will hopefully lead to new, innovative experiments on specific biological systems. In addition, as with all techniques, awareness on a fundamental level helps to avoid pitfalls, problems, and false interpretations. We also survey briefly the present methods of FLI instrumentation and discuss their comparative advantages, although this is not the main focus of the article. We stress at the beginning of this article that lifetime-resolved imaging is not confined to the optical microscope. All fluorescence signals inherently contain lifetime information, whether they are made on a macroscopic object, in a microscope, or in an endoscope.

Measuring Fluorescence Lifetimes and Lifetime-Resolved Images

Because fluorescence lifetime measurements are not usually made in the microscope, it is helpful to review briefly why such measurements are desirable. The reason is simple: fluorescence is inherently a time-resolved phenomenon. The kinetic rate of leaving an excited state depends on how many, and which, pathways are available to the excited molecule for relaxing to the ground state. The probability (rate) of passing out of the excited state through any particular kinetic pathway depends on the molecular environment of the fluorophore. It is this dependence on the environment, the kinetic competition between different pathways, and the sensitivity of the measured fluorescence signal to physical events on the scale of molecular dimensions, that makes fluorescence a highly informative method of measurement. Fluorescence imaging is of course useful simply to report on microscopic morphology, or to report whether a particular molecular component is in a particular location. These simple imaging modes are the most common

*The authors wish to express their gratitude and appreciation to the community of scientists who have been instrumental in the development of FLI. It is a definite pleasure to work within the "FLI community." There are many aspects of FLI that are not covered in this article, and no details are given of any particular study or instrument. However, the essence of this article is not a literature review. Therefore, we could not do justice in this article to the many innovative contributions of many research groups. The reader can find this information either directly in the original references that have been given, or in the extended references.

applications of fluorescence in microscopy. However, much more detail is available from the fluorescence emission when the rates of decay can be determined. The experimental parameters and molecular interpretations of fluorescence are most informative when we consider directly the time-resolved decay from the excited state. In addition to the mechanistic and molecular information available by fluorescence lifetime measurements, separating the fluorescence signal into its elementary lifetime components provides a way to distinguish quantitatively different fluorophores. It can be used to increase the contrast between different components in an image. Lifetime resolution also permits the detection of the same fluorophore in different environments. Reliable measurement can be made of the fraction of fluorophores in selected isomeric states (such as protonated and deprotonated forms in pH measurements).[1–3] A consideration of the "life of an excited fluorophore" will reveal clearly the experimental possibilities afforded by fluorescence imaging instrumentation with lifetime resolution. We discuss this in an expository fashion because these basic considerations lie at the heart of lifetime-resolved measurements. An appreciation of the fundamental mechanisms contributing to the decay rate leads inevitably to a better awareness of the experimental techniques and interpretations.

In general, in biophysical measurements we are not concerned with the same details that interest a spectroscopist, who is often interested in the mechanism of radiative emission from the excited state. We are usually interested in using fluorophores as probes. The fluorophores report to us the makeup and characteristics of their physical and chemical environment. We can gain this information by recording the dynamics, spectral dispersion, polarization, and intensity of the emitted photons. The physical properties of the environment of a fluorophore are manifested through the effect of the molecular surroundings on the rates of deexcitation pathways of the excited molecules.

Pathways and Rates of Deexcitation

To exemplify what we have just said, we discuss briefly the different mechanisms and pathways by which a molecule can become deactivated from the excited state. Consider a fluorophore that is in an excited state. It is not important how the molecule got there—it could have become excited by absorbing a photon, by receiving energy from a nearby excited molecule, or by chemical means. In FLI we excite the molecules with light. The measured rate by which an excited molecule leaves the excited state is the sum of all the rates of the different kinetic pathways available to the molecule. The most common pathways are

[1] T. J. Rink, R. Y. Tsien, and T. Pozzan, *J. Cell Biol.* **95,** 189 (1982).

[2] H. Szmacinski and J. R. Lakowicz, *Anal. Chem.* **65,** 1668 (1993).

[3] K. Carlsson, A. Liljeborg, R. M. Andersson, and H. Brismar, *J. Microsc.* **199,** 106 (2000).

(1) the intrinsic rate of fluorescence (this is calculated from quantum mechanics and depends on the details of the excited and ground state electric configurations), (2) internal conversion (which is dependent on the interaction of the molecule with its environment through thermal, vibrational interactions), (3) static quenching (by forming a complex in the excited or ground state that lowers the probability of emission or absorption of a photon), (4) dynamic quenching that is due to collisions with surrounding "quencher" molecules), (5) energy transfer, FRET (which is due to the transfer of the excited energy to a neighboring molecule or a nearby boundary—this is a powerful method for determining molecular proximity), (6) excited state reactions other than quenching, such as charge transfer, molecular isomerizations, and bimolecular reactions, and (7) intersystem crossing (the mechanism by which the molecule passes from an excited singlet state to an excited triplet state, or vice versa). As one can see, there are many different mechanisms available and they all provide us with diagnostic information. In general, the total rate of leaving the excited state, which is the reciprocal of the measured excited state lifetime, is the summation of the rates of all the possible pathways. This is usually depicted in a "Jablonski" diagram (or perhaps this should also be termed a Perrin–Jablonski diagram[4,5]; see Fig. 1). All these pathways compete with each other. It is important to realize that when measuring fluorescence lifetime we are not measuring the rate of proceeding only through the fluorescence pathway; on the contrary, we are measuring the total rate of leaving the excited state, which is the sum of all the rates for leaving the excited state. This simple fact exemplifies the power of fluorescence in general, especially the lifetime-resolved measurements. The fluorescence lifetime carries the signature of all the different pathways of deexcitation. By carrying out the appropriate measurements we can obtain information on all the different mechanisms of deexcitation listed above. This is a wealth of information concerning the molecular environment of the fluorophores. We discuss some particular cases below. Discussing fluorescence from the natural point of view of competing kinetic rates emphasizes unequivocally that fluorescence is a convenient and sensitive method for measuring kinetic mechanisms and molecular configurations on a molecular scale. To be able to carry out lifetime-resolved fluorescence measurements in an image in contrast to simply measuring the intensity (which is the integration of the fluorescence signal over a proscribed time) increases appreciably the molecular and diagnostic information available to the experimenter.

Interdependence of Pathways of Deexcitation

We demonstrate the interdependence of the different pathways and the determination of intrinsic rates of any particular pathway by the following simple general considerations. Assume that we measure the rate of deexcitation when all

[4] B. Nickel, *EPA Newslett.* **58,** 9 (1996).
[5] B. Nickel, *EPA Newslett.* **61,** 27 (1997).

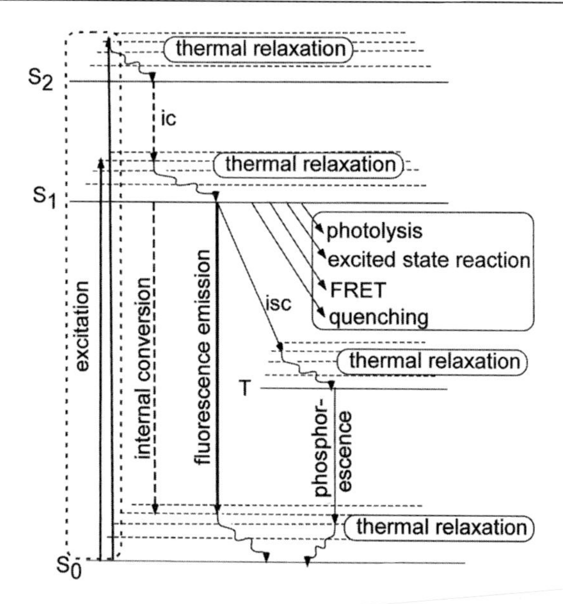

FIG. 1. Jablonski energy level diagram. Energy increases vertically up. The arrows represent the transition of a quantum change. The transitions begin at the lowest vibrational levels of each electronic state (S_0, S_1, and S_2) for both the absorption (upward arrows) and the deexcitation process (downward arrows). The transition from the excited state to the ground state always happens from the lowest level of S_1 (first singlet excited state). The deexcitation transitions other than the emissive, internal conversion, and intersystem crossing transitions are gathered together in a box at the right of the diagram. Via intersystem crossing (isc) the excited molecule passes from the singlet (S_1) to the triplet state (T). Emission from the triplet state is phosphorescence. The other deexcitation transitions from the triplet state, which are not shown, are similar to those from the singlet state. The electronic transitions usually leave the molecule in an excited vibrational level of the end electronic state. This vibrational excitation energy relaxes thermally very rapidly, within picoseconds, to the lower vibrational levels of the corresponding electronic states. Two possible absorption (excitation) transitions are shown to two different singlet excited states (these are both contained within the dotted box). If the S_2 state becomes excited, the molecule immediately relaxes by internal conversion to the S_1 state.

the pathways of deexcitation are operative. The total rate of deexcitation from the excited state D^* to the ground state D can be depicted as a chemical transformation,

$$D^* \xrightarrow{\sum_i k_i} D$$

That is, the average time that the molecule stays in the excited state (the average lifetime of the excited state τ_{D^*}) is inversely related to the sum of all the different pathways of deexcitation, $1/\tau_{D^*} = \sum_i k_i$. The sum extends over all possible modes of deexcitation for any particular molecule. Now, say we make another measurement of the excited state lifetime under conditions where pathway j cannot take place. Then the measured lifetime will be $(1/\tau_{D^*})_{i \neq j} = \sum_{i \neq j} k_i$. Obviously, the rate of deactivation in the presence of pathway j is greater than in its absence;

that is, $(1/\tau_{D^*})_{i\neq j} \leq 1/\tau_{D^*}$. Whenever we allow an additional pathway of excitation, the rate of decay becomes faster. How can we measure the rate of decay of the excited state? We can determine the lifetime of D^* by measuring some experimental variable along any of the chosen pathways. Usually we choose to measure the rate of fluorescence decay (but it could be any other measurable parameter that has the required time resolution). Define the measured parameter to correspond to pathway f (where we have chosen the letter f because we usually measure fluorescence). The lifetimes measured along pathway f in the presence and absence of pathway j will be $1/\tau_{f,\text{meas}} = \sum_i k_i$ and $(1/\tau_{f,\text{meas}})_{i\neq j} = \sum_{i\neq j} k_i$. Note that k_f is one member of both sums (since we are measuring fluorescence, f cannot be the same as j). Also, note that the $1/\tau_{f,\text{meas}}$ is the rate of decay of the measured fluorescence signal, which is the same as the overall rate at which the excited state is depleted. $1/\tau_{f,\text{meas}}$ is not the intrinsic rate of exiting the excited state by way of the fluorescence pathway (this rate is k_f). That is, the probability of fluorescence per unit time (k_f) is the same in both sums; however, the total pool of excited molecules becomes depleted faster than k_f because there are other pathways for deexcitation that are simultaneously actively depleting the excited state. The fluorescence decay signal mirrors the total decay of the excited state population. We have belabored this point, because although this is obvious, it is often misunderstood. However, once this is understood, all the different ways of measuring the different decay processes are easily understood. For instance, in order to determine k_j we simply subtract the two measured inverse decay times, $1/\tau_{f,\text{meas}} - (1/\tau_{f,\text{meas}})_{i\neq j} = \sum_i k_i - \sum_{i\neq j} k_i = k_j$. Note that we have measured fluorescence in order to determine the rate of a pathway (j), but pathway j has nothing to do with fluorescence. We use this reasoning below to demonstrate how to use fluorescence lifetimes to investigate all physical processes related to the decay of the excited state other than fluorescence. The obvious reason for choosing fluorescence to investigate all the other nonemissive pathways is because it is easy to detect photons on the nanosecond time scale. By using the sensitivity and temporal capabilities provided by fluorescence we can derive quantitative estimates of all the other dynamic processes happening in the immediate environment of the excited fluorophore. The lifetime of decay of fluorescence is a direct fingerprint, a probe, of the overall molecular dynamic situation, and (unless new molecular complexes are formed) is unaffected by the concentration of emitting molecules. Through developments of FLI instrumentation, the significant advantages of time-resolved fluorescence measurements are now available for imaging experiments. This is the major motivation for making FLI measurements.

Specific Features of Different Deexcitation Pathways

In the following we discuss specific features that elucidate the rationale for measuring the dynamic decay from the excited state, rather than just measuring the steady state intensity. Spectroscopists have long taken advantage of the unique molecular information that is available when the dynamic decay of the emission from a molecule in an excited state is resolved by measuring it directly in the time domain.[6] During the time that a molecule is in an excited state, it interacts intimately with its molecular environment. Although a molecule spends only a short time in the excited state—in general, picoseconds to nanoseconds—the eventual emission of a photon bears the historical imprint of this sojourn in the excited state. During this short time many events can happen. We will discuss several major pathways of deexcitation available to an excited chromophore in some detail, because these are the basis for interpreting lifetime measurements.

Intrinsic Rate of Emission: Fluorescence

Every molecule has a particular rate, called the intrinsic fluorescence rate that defines the longest average time the molecule can stay in the excited state. This intrinsic spontaneous rate of emission is defined wholly by the quantum mechanical nature of the excited and ground state electric configurations. It can be thought of as the rate (that is, the probability per unit time) at which an isolated molecule (isolated from all other molecules) that is in the lowest (first) singlet excited state will leave this excited state by emitting a photon, thereby returning to its electric ground state. The intrinsic rate of emission is the lower limit of the rate for leaving the excited state; that is, the average fluorescence emission cannot happen slower than this. However, as mentioned above, the intrinsic fluorescence rate is not the rate that is measured. The measured rate of fluorescence is the sum of all the competing processes leading away from the excited state. The intrinsic emission rate is the same as the Einstein spontaneous emission rate calculated in most textbooks on spectroscopy and quantum mechanics.[7–17] It is also related to the

[6] R. B. Cundall and R. E. Dale, eds., *in* "Time-Resolved Fluorescence Spectroscopy in Biochemistry and Biology," Vol. 69, p. 785. NATO ASI Series, Series A, Life Sciences. Plenum, New York, 1983.

[7] J. R. Lakowicz, "Principles of Fluorescence Spectroscopy," 2nd Ed. Kluwer Academic/Plenum Publishers, New York, 1999.

[8] T. Förster, "Fluoreszenz organischer Verbindungen." Vandenhoeck and Ruprecht, Göttingen, 1951.

[9] R. S. Becker, "Theory and Interpretation of Fluorescence and Phosphorescence." Wiley Interscience, New York, 1969.

[10] C. A. Parker, "Photoluminescence of Solutions." Elsevier, Amsterdam, 1968.

[11] S.-H. Chen and M. Kotlarchyk, "Interactions of Photons and Neutrons with Matter—an Introduction," p. 400. World Scientific, River Edge, NJ, 1997.

[12] W. T. Silfvast, *in* "Laser Fundamentals," p. 521. Cambridge University Press, Cambridge, 1996.

[13] P. W. Atkins and R. S. Friedman, *in* "Molecular Quantum Mechanics," 3rd Ed., p. 544. Oxford University Press, Oxford, 1997.

[14] E. Lippert and J. D. Macomber, *in* "Dynamics of Spectroscopic Transitions: Basic Concepts," p. 580. Springer-Verlag, Berlin, 1995.

[15] D. P. Craig and T. Thirunamachandran, *in* "Molecular Quantum Electrodynamics: An Introduction to Radiation Molecule Interactions," p. 324. Dover, Mineola, NY, 1984.

[16] W. Kauzmann, *in* "Quantum Chemistry. An Introduction," p. 744. Academic Press, New York, 1957.

[17] L. I. Schiff, *in* "Quantum Mechanics," p. 544. McGraw-Hill, New York, 1968.

uncertainty in the energy of a system that has a finite lifetime through the relation $\Delta E \Delta t = \Delta E \tau_f \sim \hbar$. The energy and time are not noncommuting operators (there is no quantum mechanical operator for the time, and the energy is well defined by the Schroedinger equation). Thus, this uncertainty between the value of an energy and the time that the system is in the state is not a true Heisenberg uncertainty relation.[13,17,18] Nevertheless, the measurement time is limited by the lifetime of the decay, and this leads to a broadening of the energy. The important aspect of this relation is that there is a lower limit to the spectral width of a spectroscopic transition and this lower limit is related to the lifetime of the excited state (lifetime broadening). That is, there is a complementary relation between the possible energy distribution of an excited molecule (the natural line width) and the lifetime of the excited state. For this presentation it is important only to recognize that the intrinsic rate of emission is a basic quantum mechanical property of an isolated molecule, and is a property calculable in principle from its energy levels and wave functions of its quantum states. The fundamental physical reason that the excited molecule undergoes spontaneous emission (i.e., in the absence of an external field) is because of the coupling of the excited and ground states through the interaction with the zero-point level of the radiation field. This is a consequence of the uncertainty principle that requires a nonzero lowest quantum state of the radiation field. This is the correct quantum mechanical explanation for spontaneous emission; there is no fully valid classic description. However, for our purposes, it is necessary only to know that this rate of spontaneous emission does not change unless the electronic structure of the molecule changes.

In general, the molecule leaves the excited state faster than this intrinsic spontaneous rate because it interacts with its environment, and this leads to a faster deactivation. The intrinsic spontaneous emission radiative rate is usually assumed to be a constant, independent of the environment, and also temperature independent. However, in the presence of a high electric field, as when the fluorophore is close to a metal surface with a sharp curvature, the intrinsic rate can be affected. We do not discuss this interesting phenomenon, but refer the reader to the literature.[19–23] This phenomenon has until now not been of much use for biological fluorescence measurements, but it is interesting for the future, especially from the point of view of lifetime measurements (it is a way to shorten the intrinsic lifetime).

[18] L. D. Landau, "*Quantum Mechanics,*" 3rd Ed., Vol. 3. Butterworth-Heinemann, New York, 1997.
[19] E. J. Sánchez, L. Novotny, and X. S. Xie, *Phys. Rev. Lett.* **82,** 4014 (1999).
[20] H. F. Hamann, M. Kuno, A. Gallagher, and D. J. Nesbitt, *J. Chem. Phys.* **114,** 8596 (2001).
[21] H. F. Hamann, A. Gallagher, and D. J. Nesbitt, *Appl. Phys. Lett.* **76,** 1953 (2000).
[22] J. Fiurásek, B. Chernobrod, Y. Prior, and I. Sh. Averbukh, *Phys. Rev. B* **63,** 45420 (2001).
[23] J. Enderlein, *Appl. Phys. Lett.* **80,** 315 (2002).

Thermal Relaxation

The dynamics of the thermal interactions are manifested in the rate of fluorescence decay. Thermal interactions with the solvent surrounding the fluorophore, or with the immediate surrounding molecular matrix, will reduce the time a molecule spends in the excited state. This process is known as an internal conversion, and leads to a nonradiative transition from the excited to the ground state. Internal conversion shortens the time spent in the excited state by providing another pathway (other than the intrinsic emission) for leaving the excited state. The rates of passage out of the excited state are additive, so that the intrinsic radiative emission rate and the nonradiative thermal internal conversion add together leading to a faster exit out of the excited state than by the intrinsic emission pathway alone. The thermal relaxation will happen in the absence of all other additional deexcitation processes. Because these interactions with the neighboring molecules are thermally controlled and are coupled through vibrations and collisions, they will be temperature dependent, and dependent on the composition (remember that the intrinsic rate of fluorescence is essentially independent of temperature). Through this vibronic coupling, leading to a nonradiative pathway from the excited state, information about the environment can be derived. The vibronic coupling is directly observable through the fluorescence lifetime. The coupling of the thermal environment does not only lead to a separate pathway to the ground state. These intramolecular interactions between the vibrational states of the excited molecule, and the interaction between the molecule and its thermal environment also elicit and facilitate rapid relaxations from higher vibrational levels of excited molecules to the lowest vibrational states of the first excited state (resulting in a Boltzman distribution among the lowest vibrational levels). This is one reason why the energy of emission is always less than the energy of excitation (the Stokes shift). Thus, the measured fluorescence emission will always decay faster than just due to the intrinsic emission radiative rate that one might calculate. The thermal relaxation (to the lowest vibrational levels of the first excited state and from this state to the ground state) requires overlap of the vibrational energies of the molecule with vibrational states of the solvent. Except for a few odd cases, the thermal relaxation to the lowest vibrational states of the first excited electronic state always takes place much faster than the eventual decay of the excited electronic state. Biological samples are almost always in condensed media and usually measured at ambient temperatures, or not far from it. Thus the intrinsic rate of fluorescence plus thermal nonradiative deactivations will set the apparent longest decay time normally observed for fluorophores in biophysical measurements.

Molecular Relaxation of Solvent or Molecular Matrix Environment

In biophysics we always measure fluorescence in a complex condensed matter environment, such as an aqueous environment or in an apolar surrounding such as in lipid membranes. These condensed molecular environments often consist of concentrated biological components and couple strongly to the excited molecule. In a polar environment it often occurs that the solvent molecules (or the neighboring molecular matrix) reorient around the excited molecule (this happens especially in an aqueous solvent). This solvent relaxation will take place if the dipole moment of the excited molecule differs from the dipole moment of the ground state. This is observed in a polar aqueous environment (water has a large dipole moment of 1.85 debye) and when internal charge transfer takes place after the molecule is excited into the excited state. These local dipole relaxations of the molecular environment around the dipolar excited state lead to a further decrease (following the much faster thermal relaxations to the lowest vibrational states) in energy of the excited molecule before emission. This phenomenon further increases the red shift—that is, a shift of the emission to lower energies. The dipolar interactions between the excited molecule and polar solvent molecules cause the solvent dipoles to reorient, forming more favorable electrostatic interactions. Obviously, solvent relaxation is dependent on the temperature and the viscosity of the molecular environment. Solvent relaxation contributes significantly to the Stokes shift in a highly polar solvent environment; that is, the energy (wavelength) of the emission is less than (longer than) the energy (wavelength) of excitation. This shift, caused by the polar solvent relaxation, makes it easier to select the emission photons from photons due to scattered excitation light with normal wavelength filters (larger shift of the fluorescence wavelength maximum from the excitation wavelength). Often the solvent relaxation occurs before the molecule exits the excited state. Interestingly, if the "solvent" or "matrix" relaxation occurs on the same time scale of the fluorescence emission we can gain valuable information about the relaxation of the solvent (or molecular matrix) dipoles from fluorescence lifetime measurements. There is usually no effect of solvent relaxation on the real lifetime of fluorescence emission (unless dynamic quenching is involved). However, if the relaxation takes place on the same time scale as fluorescence, the measured dynamic signal will be affected because the measured fluorescence intensity will be affected by the shift of the fluorescence emission out of (or into) the bandwidth of the optical filter. This would usually be classified as an artifact; however, it is clear that the time-dependent wavelength shift of the fluorescence emission contains valuable information on the molecular environment. Strong solvent interactions and charge transfer processes can also change the fluorescence lifetime by changing the overlap of the excited and ground state wave functions. This would be a true effect on the lifetime of the excited state.

In general, the rate of solvent relaxation is temperature dependent; however, because we measure fluorescence in microscopes close to room temperature, the temperature dependence is often not pronounced. If the temperature is changed, an effect of the temperature on the relaxation of the solvent must be expected. If solvent relaxation takes place, it is usually interesting to take the temperature dependence into account. These measurements can also give us interesting information concerning the rigidity of the molecular surroundings and the effective viscosity affecting the rotational diffusion of the surrounding "solvent" molecules. In addition, it may help to identify the dipolar molecule interacting with the fluorophor. Admittedly, this is difficult to do in an environment of a biological cell. Usually a shift in the emission spectrum is done by recording the spectral (wavelength) dispersion to detect the red shift directly. Even in cuvette-type measurements spectrum shifts are usually recorded by a steady state fluorescence spectrum; however, the emission spectrum can be recorded in a time-resolved mode, giving the shift of the spectrum as a function of the time. Such a time-resolved experiment records directly the relaxation of the polar "solvent." This is also possible in imaging mode, but has not yet been reported. It would be necessary to record the spectral shift as a function of time following the excitation for every pixel of the image. It should be clear that vibrational relaxation processes affect the lifetime of the fluorescing molecule directly by contributing a competing pathway for deexcitation. The solvent relaxation can lead to an apparent decay process in the signal (as discussed above) but it does not, in general, change the lifetime of the excited state. No new pathway for deexcitation is produced. Both processes provide valuable information about the molecular environment of the fluorescence probe. Detailed measurements utilizing either of these relaxation processes have not yet been employed in FLI; however, considering the heterogeneous environment in a living cell, it is clear that this dynamic information would be valuable. Spectral shifts of fluorophores have been used extensively in fluorescence imaging without temporal resolution. This has been done especially by the group of Gratton, to gather detailed information of the polarity changes and micro-organization in the molecular environment of certain fluorophores, for example, Laurdan (which has the sensitive polarity-sensitive chromophore, Prodan) in biological membranes.[24,25]

Quenchers: Dynamic

Dynamic quenchers must effectively collide with an excited molecule in order to compete with other pathways of deexcitation. Because the quenchers move by random diffusion and the excited state lifetime is on the order of nanoseconds, only those quencher molecules that are close enough to the excited molecule to collide with it before the molecule deexcites by another pathway will be effective. Singlet oxygen is the main perpetrator of dynamic quenching in normal biological milieu. Oxygen is often undesirable from a spectroscopic point of view, but usually unavoidable in biological samples. However, many other molecules can

[24] C. Dietrich, L. Bagatolli, Z. Volovyk, N. Thompson, M. Levi, K. Jacobson, and E. Gratton, *Biophys. J.* **80,** 1417 (2001).

[25] L. A. Bagatolli and E. Gratton, *Biophys. J.* **77,** 2090 (1999).

function as effective quenchers, such as Br^-, I^-, and acrylamide (nonpolymerized). Many effective "collisional quenchers" interact with the spin of the excited electron, effectively instigating a transfer to the triplet state of the excited molecule (intersystem crossing, see below), which removes the singlet excited state, and thereby decreases the prompt fluorescence signal. This perturbation is usually accomplished through spin-orbit coupling, and it is especially effective if the collisional quencher has an unpaired weakly held electron (such as I^- and Br^-). In general, only smaller charged ions are effective quenchers due to their rapid diffusion; however, charged, or highly polarizable, groups on macromolecules can also effectively quench fluorophores that are attached, either covalently or simply bound, to the macromolecules. For FLI, the principal effect of quenchers is that the fluorescence lifetime is shortened. Often in cellular imaging it is difficult to correlate the intensity of fluorescence with the concentration of the probe, because a decrease in intensity could come from a smaller number of fluorescing molecules or from a decrease in intensity due to quenching. FLI can easily distinguish these two possibilities and this is important when quantifying concentrations. If the lifetimes have been shortened due to dynamic quenching, the correct concentrations can be calculated if the lifetime is measured. This is a powerful application of FLI, and is possible only if the fluorescence lifetime can be determined at every pixel of the image.

Dynamic quenching involves diffusion of the quencher (and sometimes the diffusion of the fluorophore). According to the discussion above concerning the contribution of all the competitive kinetic pathways contribution to the fluorescence decay rate, we can measure the rate of dynamic quenching by simply measuring the rate of fluorescence decay in the presence and absence of quenching:

$$1/\tau_{f,\mathrm{meas}}^{+\mathrm{quencher}} - 1/\tau_{f,\mathrm{meas}}^{-\mathrm{quencher}} = \sum_i k_i - \sum_{i \neq \mathrm{quencher}} k_i = k_{\mathrm{quencher}}$$

If the quencher is known, and if the movement of the quencher is controlled by diffusion (which it usually is), then we can determine the effective viscosity of the environment. This we can do because the rate of dynamic quenching is inversely dependent on the viscosity of the environment and representative dimensions of the quencher molecule and the fluorophore. For instance, if the quencher is spherical (with radius a), and only the quencher diffuses, then $k_{\mathrm{quencher}} \propto 1/6\pi\eta a$, where a is the radius of the quencher, and η is the viscosity. Thus, the lifetime can provide not only corrections to dynamic quenching, but can also furnish indications of the rigidity (effective viscosity) of the molecular environment of a fluorescence probe through the dynamics of quenching (actually it provides information about the relative mobility of the fluorescence probe and the quencher molecule). This could be deduced only with great difficulties from steady state experiments in an image; but it is a simple experiment for FLI.

Excited State Reactions

It is clear that if excited state reactions occur, this will decrease the lifetime of the excited state because it is a competing pathway. If the fluorophore in the excited state reacts with a reaction partner selectively, then the lifetime-resolved imaging can be used to map the location of the reactive component. All that is required is to record the fluorescence lifetime, and apply the same analysis as discussed above for dynamic quenching (quenching is essentially an excited state reaction that does not destroy the fluorophore). A useful application of this measurement involves the formation of eximers, or exiplexes. For instance, pyrene forms eximers—dimers of pyrene where one of the reaction partners is a molecule in the excited state. This is often observed by the appearance of a red-shifted emission of the fluorescence due to the eximer emission. However, the eximer is an independent chemical species and as such it is the product of an excited state reaction, and will shorten the emission lifetime (and lower the intensity) of the original independent excited pyrene molecules.

Förster Fluorescence Resonance Energy Transfer

Fluorescence resonance energy transfer (FRET) is a process by which the extra energy of a molecule in an excited electronic state (called the donor, D) is transferred to a molecular chromophore called the acceptor (A). This physical transfer of energy usually takes place over a D–A separation of 0.5 to 10 nm. The energy is not emitted by D as a photon, and no photon is absorbed by A. There is a Coulomb charge–charge interaction between the D and A molecules; this mutual perturbation between the excited D molecule and the A molecule takes place electrodynamically through space as a dipole–dipole interaction. If the conditions are right–for instance, if the spectral overlap of the emission and absorption spectra of D and A is sufficient, and if the quantum yield of the D and the absorption coefficient of A are great enough—there is a significant probability that the excitation energy of D will be transferred to A. The rate of energy transfer, which is a strong function of the separation distance between D and A, leaves D in its electronic ground state and A in an electronic excited state. If energy transfer is appreciably favorable, the probability of photon emission of D is diminished, and the lifetime of D in the excited state is measurably shortened. As an observer we can learn a great deal by studying the rate or the efficiency of the transfer. From these measurements we can learn whether D and A are close together. FRET is sensitive to distance changes between 0.5 and 10 nm. The rate of energy transfer between single donor and acceptor molecules is proportional to $1/R^6$, where R is the distance between the centers of the two chromophores. From the efficiency (or equivalently the rate) of energy transfer, we can gain quantitative information about the distance, and we can sometimes learn about the relative orientation between D and A (the effectiveness of FRET depends on the orientation of the transition dipoles of D and A). These physical measurements can be coupled to a wide variety of biological assays that yield specific information about the environments of the chromophores. FRET is probably the major reason why many people want to make FLI measurements. Förster showed[8,26,27] that the rate of energy transfer could be expressed as

$$k_{\mathrm{ET}} = \frac{1}{\tau_{F_{\mathrm{D}}^{-\mathrm{A}}}} \left(\frac{R_0}{R} \right)^6 \tag{1}$$

$R_0$is the value of R where the rate of energy transfer, k_{ET}, equals the rate of deexcitation from the excited state in the absence of the acceptor (which is the rate of fluorescence, $1/\tau_{F_{\mathrm{D}}^{-\mathrm{A}}}$). R_0 can be calculated from knowledge of the relative orientations between the transition dipoles and the spectral overlap of the emission spectrum of the donor and the absorption spectrum of the acceptor.[8] Using the general expressions of competitive rates derived above, it is simple to determine the efficiency of energy transfer using fluorescence lifetimes. One has only to measure (or know) the fluorescence lifetime of the donor in the absence of the acceptor, and then the efficiency can be measured by simply forming the following ratio:

$$\begin{aligned} E &= \frac{\text{rate of energy transfer}}{\text{total deexcitation rate}} = \frac{\left(1/\tau_{F_{\mathrm{D}}^{+\mathrm{A}}}\right) - \left(1/\tau_{F_{\mathrm{D}}^{-\mathrm{A}}}\right)}{\left(1/\tau_{F_{\mathrm{D}}^{+\mathrm{A}}}\right)} = \frac{k_{\mathrm{ET}}}{\tau_{F_{\mathrm{D}}^{+\mathrm{A}}}^{-1}} \\ &= \frac{k_{\mathrm{ET}}}{\tau_{F_{\mathrm{D}}^{-\mathrm{A}}}^{-1} + k_{\mathrm{ET}}} = \frac{1}{\left(k_{\mathrm{ET}}\tau_{F_{\mathrm{D}}^{+\mathrm{A}}}\right)^{-1} + 1} = \frac{1}{1 + (R/R_0)^6} \end{aligned} \tag{2}$$

The efficiency of energy transfer can also be called the quantum yield of energy transfer, because it is the fraction of times that excited molecules follow the ET pathway of deexcitation. It is difficult to quantify FRET measurements in imaging experiments (fluorescence microscope) using steady state fluorescence because standards must be used to calibrate the fluorescence signals; that is, we must compare the fluorescence intensity in the presence and absence of acceptor. These calibrations are relatively simple and accurate to make in a cuvette environment, but in imaging environments it is almost impossible to know what the concentrations are. Without knowing the concentration, or at least relative concentration, of a fluorophore (the donor) it is impossible to interpret the fluorescence intensity in terms of energy transfer efficiency. Usually the variability of concentrations between different biological cells, or the distribution within a biological cell, is unknown. Several methods have been developed to circumvent this difficulty and determine E when measuring the steady state fluorescence in images. One popular method involves forming a ratio of differences and sums of two measured fluorescence intensities. One measures either the ratio using two excitation wavelengths (exciting mainly the

[26] T. Förster, *Naturwissenschaften* **6,** 166 (1946).
[27] T. Förster, *Ann. Phys.* **2,** 55 (1948).

donor or acceptor), or the ratio of the fluorescence intensities using two emission wavelengths (measuring mainly the fluorescence of the acceptor and donor).[28–32] Usually the actual efficiency is not calculated, but numbers proportional to the efficiency. These ratios are difficult to make accurately unless one can record spectra, because the donor and acceptor fluorescence usually both contribute to the measured signals and it is difficult reliably to sort out their individual contributions. Another steady state method uses photodecomposition of the acceptor in order to measure the change in donor fluorescence in the presence and then (after photodestruction) absence of the acceptor. The success of this method rests on the assumption that the acceptor when photolyzed does not still absorb in the spectral region of donor emission. The latter technique has the great disadvantage that it cannot be used for following real-time events, or following events over an extended time period. Once the acceptor has been photolyzed, it is usually not replaced, so it is a one-shot experiment. In addition, the assumption must be made that the photolyzed acceptor molecules do not interfere with the measurement or the viability of the cell.

FLI overcomes these difficulties in making reliable FRET measurements in an image. As can be seen from the equations above, there is no reason to calibrate intensities; one has only to be able to measure the lifetimes accurately. This is the great advantage of lifetime measurements. Given that nowadays it does not take too much time to measure a lifetime-resolved image, FLI is the method of choice if one has the instrumentation. We defer further discussion of the advantages of FLI for FRET measurements until we have considered the methods.

Intersystem Crossing

The first time-resolved imaging experiments were carried out on samples with delayed emission: phosphorescence and delayed fluorescence.[33] The great advantage of measuring the lifetime-resolved emission of delayed luminescence is that the time range of microseconds to seconds is easier to handle. The ground states of most fluorophores are singlet states, and the first excited state is also a singlet state (this means that the electrons in the highest occupied electronic levels, in the ground and the excited electron configurations, are paired with opposite spins). This has the consequence that the transition from the excited state to the ground state is "spin-allowed" and therefore takes place in the nanosecond time scale. Due to spin-orbit coupling involving the electron that has been elevated to the excited state, there is a probability that the spin of the excited electron will

[28] G. Y. Fan, H. Fujisaki, A. Miyawaki, R.-K. Tsay, R. Y. Tsien, and M. H. Ellisman, *Biophys. J.* **76,** 2412 (1999).

[29] G. Bright, G. Fisher, J. Rogowska, and D. Taylor, *Methods Cell Biol.* **30,** 157 (1989).

[30] K. Dunn and F. Maxfield, *Methods Cell Biol.* **56,** 217 (1998).

[31] R. Silver, *Methods Cell Biol.* **56,** 237 (1998).

[32] N. Optiz, *Biomed. Tech. (Berl.)* **43** (suppl.), 452 (1998).

[33] G. Marriott, R. M. Clegg, D. J. Arndt-Jovin, and T. M. Jovin, *Biophys. J.* **60,** 1374 (1991).

flip, creating a triplet state (where two electrons have parallel spins). The energy of the triplet state is lower than the corresponding singlet state. The two unpaired electrons in the triplet state are farther apart than when the electrons have the same spin, lowering their electrostatic repulsion energy (a consequence of the Pauli exclusion principle). Without spin-orbit coupling such a pure spin transition is not allowed, and therefore the transition from a singlet to triplet state is usually not highly probable. This means that the rate of the singlet-triplet transition of the excited state (intersystem crossing) is usually much slower than the normal intrinsic rate of fluorescence or the deactivation by internal conversion. Coupling of orbital magnetic moments with the spins perturbs this symmetry, and the transition becomes partially allowed. Molecular groups or atoms with electrons having large orbital magnetic moments can bring about this transition. These molecular components can be covalently attached to the fluorophore (such as in erythrosin or eosin, which are essentially fluorescein molecules with four covalent iodines or bromines) or diffusing freely in solution (such as iodide molecules at high enough concentration). If the probability of a singlet–triplet transition is high enough, then the formation of a triplet state will become a viable competitor with the other deexcitation processes. This will then become a competing kinetic pathway, decreasing significantly the measured lifetime of the fluorescence emission from the singlet state, and if the triplet can emit a photon a long-lived emission decay is produced. Triplet states usually have much longer intrinsic lifetimes than singlet states because a triplet-to-singlet transition is required for its decay to the ground state. This has several important consequences, especially in a complex environment of a biological cell. Because of the unpaired available electron of the excited triplet molecule, the triplet is often highly reactive, and it can react with triplet scavengers such as singlet oxygen (unless removed, oxygen is usually in solution at approximately 5 m*M* concentration), often leading to excited state decomposition (photolysis) reactions of the fluorophore. Photodecomposition usually takes place via the triplet state. The triplet state is only short-lived unless oxygen is removed from solution; therefore, phosphorescence emission is usually not observed. However, if the oxygen is rigorously removed, then some fluorophores, such as acridine orange and tryptophan, exhibit a high triplet emission quantum yield (phosphorescence). Some protein molecules emit tryptophan phosphorescence with a long lifetime even in oxygenated solution.[34,35] This is because the excited tryptophan is well protected from oxygen or from other deactivating processes that would quench the triplet state. Phosphorescence emission takes place in the microsecond-to-second time range—much longer than fluorescence, and this extends the time range for probing dynamic processes with lifetime-resolved measurements. DLIM (delayed luminescence imaging microscopy) makes use of this long delay to completely avoid background fluorescence that is much faster than

[34] M. Gonnelli and G. B. Strambini, *Biochemistry* **34,** 13847 (1995).
[35] G. B. Strambini and M. Gonnelli, *Biochemistry* **29,** 196 (1990).

the phosphorescence.[33,36] Examples of lifetime-resolved delayed luminescence imaging can be found in these publications. A disadvantage of long-lived delayed luminescence is that the emission rate of photons is slow, lowering the sensitivity.

Slow Luminescence without Intersystem Crossing

Lanthanides exhibit long lifetime decays that are not due to triplet states.[37,38] The spectroscopic emission and absorption of lanthanides involve transitions between internal lower electronic orbitals that are well protected from the solvent environment, and this leads to long lifetimes of the excited states. The long lifetimes are accompanied by narrow emission peaks (see discussion above concerning the energy-time uncertainty, $\Delta E \tau_f \sim \hbar$) that can effectively be separated from other broad emissions with narrow band filters. The use of lanthanides has the great advantage that the delayed emission can be observed after the excitation has completely subsided, and this is easy to carry out (the lifetimes can be as long as a few milliseconds). Time-resolved spectroscopy using lanthanides has found application in sensitive assay systems[39–40a], and have also been employed in imaging microscopy.[33,36,41,42] The absorption coefficients of lanthanides are extremely low by themselves (on the order of 1–10 M^{-1} cm^{-1}). However, it is possible to choose organic chelate structures with high absorption coefficients that efficiently capture the ions (and hold them for long times) and transfer essentially 100% of their excitation energy to the lanthanide ions.[38] This is an effective way to excite the ions. The narrow emission energy bandwidth of the ions can be separated easily from other broadband fluorescence, and the long lifetime emission can be resolved temporally from prompt fluorescence. A review of measuring delayed emission from lanthanides has appeared in this series.[43] Imaging using the lanthanides is also an active development.[36,44]

Photolysis: Process and Interpretation of Measurements

The pathway of photolysis competes directly with all the other pathways of deexcitation. The difference is that the product of photolysis is a chemical species

[36] G. Vereb, E. Jares-Erijman, P. R. Selvin, and T. M. Jovin, *Biophys. J.* **74,** 2210 (1998).

[37] J. Chen and P. R. Selvin, *Bioconjug. Chem.* **10,** 311 (1999).

[38] M. Xiao and P. R. Selvin, *J. Am. Chem. Soc.* **123,** 7067 (2001).

[39] L. Seveus, M. Vaisala, I. Hemmila, H. Kojola, G. M. Roomans, and E. Soini, *Microsc. Res. Tech.* **28,** 149 (1994).

[40] L. Seveus, M. Vaisala, S. Syrjanen, M. Sandberg, A. Kuusisto, R. Harju, J. Salo, I. Hemmila, H. Kojola, and E. Soini, *Cytometry* **13,** 329 (1992).

[40a] A. Periasamy, M. Siadat-Pajouh, P. Wodnicki, X. F. Wang, and B. Herman, *Microsc. Anal.* **11,** 33 (1995).

[41] E. J. Hennink, R. De Haas, N. P. Verwoerd, and H. J. Tanke, *Cytometry* **24,** 312 (1996).

[42] G. Marriott, M. Heidecker, E. P. Diamandis, and Y. Yan-Marriott, *Biophys. J.* **67,** 957 (1994).

[43] P. R. Selvin, *Methods Enzymol.* **246,** 300 (1995).

[44] R. R. De Haas, N. P. Verwoerd, M. P. Van der Corput, R. P. Van Gijlswijk, H. Sitari, and H. J. Yanke, *J. Histochem. Cytochem.* **44,** 1091 (1996).

that is not fluorescent (in the case of photobleaching) or at least the reaction product fluoresces differently than the original fluorophore. That is, the product of this pathway is not returned to the pool of fluorescence-competent ground state molecules. Obviously, the photodestruction pathway limits the number of photons on the average that a single fluorescent molecule can emit. This rate is usually orders of magnitude slower than the other competing rates, but since it is a dead-end terminal pathway, it will always eventually win. Of course, there will be a statistical distribution of the number of photons emitted by any molecule (before irreversible destruction) that is repeatedly excited since the choice of any pathway is a stochastic probabilistic event.

By observing the rate of photobleaching we can deduce the rate of other kinetic processes in the excited state by the same competitive reaction methods already discussed. This method originally suggested and shown by Hirshfeld[45] has been championed by Jovin and others[46–51] in biological fluorescence imaging. Because the overall ensemble-measured rate of photolysis is slow (by many orders of magnitude) compared with the other processes that take place in nanosecond times in the excited state it is easy to carry out in most laboratories. It is not used to measure the rate of fluorescence directly, but has found most use in determining the efficiency of FRET in image samples. There is a certain probability per unit time that an excited molecule will undergo a photochemical degradation. In the presence of other effective deexcitation pathways (such as FRET) the overall ensemble rate of photolysis will be diminished; therefore, measuring the rate of photolysis of a donor in the presence and absence of an acceptor provides a method for determining the efficiency of energy transfer using equations similar to those describing the use of the fluorescence lifetime for determining the efficiency of FRET. The reasoning is analogous, except that we are directly measuring the rate of photon destruction, by using fluorescence to measure the number of molecules that have not undergone photolysis.[52]

$$E_{\mathrm{ET}} = \frac{k_{\mathrm{PB}}^{-\mathrm{A}} - k_{\mathrm{PB}}^{+\mathrm{A}}}{k_{\mathrm{PB}}^{-\mathrm{A}}} = \frac{(1/\tau_{\mathrm{PB^{-A}}}) - (1/\tau_{\mathrm{PB^{+A}}})}{(1/\tau_{\mathrm{PB^{-A}}})} = 1 - \frac{\tau_{\mathrm{PB^{-A}}}}{\tau_{\mathrm{PB^{+A}}}} = \frac{1}{1 + (R/R_0)^6} \qquad (3)$$

Again we see how simple it is to measure E_{ET} when measuring directly the kinetics

[45] T. Hirschfeld, *Appl. Opt.* **15,** 3135 (1976).

[46] T. Jovin and D. Arndt-Jovin, *in* "Cell Structure and Function by Microspectrofluorometry," p. 99. Academic Press, San Diego, CA, 1989.

[47] T. Jovin and D. Arndt-Jovin, *Annu. Rev. Biophys. Chem.* **18,** 271 (1989).

[48] U. Kubitscheck, R. Schweitzer-Stenner, D. Ardnt-Jovin, T. Jovin, and I. Pecht, *Biophys. J.* **64,** 110 (1993).

[49] G. Szabo, P. Pine, J. Weaver, M. Kasari, and A. Aszalos, *Biophys. J.* **61,** 661 (1992).

[50] R. Young, J. Arnette, D. Roess, and B. Barisas, *Biophys. J.* **67,** 881 (1994).

[51] A. K. Kenworthy, N. Petranova, and M. Edidin, *Mol. Biol. Cell.* **11,** 1645 (2000).

[52] R. M. Clegg, *in* "Fluorescence Imaging: Spectroscopy and Microscopy" (X. F. Wang and B. Herman, eds.), p. 179. John Wiley & Sons, New York, 1996.

of a dynamic process. We mention that another simple way to measure FRET using photobleaching is to measure the fluorescence intensity of the donor in the presence of the acceptor and after photobleaching the acceptor. After photobleaching of the acceptor the donor fluorescence increases.

Unifying Feature of Extracting Information from Excited State Pathways

The unifying feature of the above mechanisms of deexcitation is that they all compete with each other, and each pathway of deexcitation contributes separately to the overall rate of excited state deactivation. Fluorescence is only one of several possible pathways out of the excited state. The separate rates of any particular individual pathway can be determined by measuring the dynamic aspects of the fluorescence signal in the presence and absence of this pathway. In essence, any pathway could be used to make the measurement of any other pathway; however, we choose fluorescence because it is the easiest and most versatile to handle. In addition, fluorescence is extremely sensitive (down to the level of single-molecule detection) and the measurements can be done with picosecond accuracy and high reproducibility.

Other Parameters That Are Related to Lifetime-Resolved Fluorescence: Dynamic and Steady State Measurements

There are kinetic processes that affect the time course of the time-resolved acquired fluorescence signal and can be measured by fluorescence experiments, but are not direct competing kinetic pathways of deactivation out of the excited state. There are also processes that produce effects on both time-averaged steady state signals as well as time-resolved signals, just as is the case for the other direct competing processes. We discuss a few of these that are common.

Spectral Relaxation

We have already considered the effect of relaxation of the dipoles of the surrounding solvent matrix on the spectral dispersion of the fluorescence emission. If this relaxation happens on the same time scale as the fluorescence, the peak of the fluorescence spectrum will move as the excited state is decaying. This does not affect the rate of diminution of the excited state (so it is not a competing kinetic process for the diminution of the excited state). However, depending on the light filter used to select the observed emission bandwidth, the dynamics of the recorded fluorescence signal could be affected by this solvent relaxation. In addition, the relaxation of the spectral dispersion could be observed on this fast time scale. Of course, the recorded steady state spectrum also contains the integrated dynamic information about the solvent relaxation. If the solvent relaxes rapidly due to the fluorescence emission, the red shift will be greater. If the solvent relaxation is slow compared with the fluorescence emission, the red shift is less.

Anisotropy Decay

The decay of the extent of polarization (due to molecular rotation) does not change the excited state lifetime. But if the excitation light is polarized, the measured rate of fluorescence decay will show signs of polarization (indeed polarization effects are always present). This will be seen in the steady state fluorescence as well as in the dynamic decay properties. It is important to remember that the incoming excitation light always polarizes the sample, even if the excitation light is unpolarized in the plane perpendicular to the direction of the light beam (it will be polarized in the third direction). Because all optical instrumentation tends to partially polarize light passing through, the fluorescence emission will always be partially polarized. This is true unless the effect of anisotropy on the measured fluorescence lifetime is avoided by using the "magic angle,"[7,53] which involves using polarized excitation and emission. The angle between the excitation and emission polarizer axes is adjusted such that the polarization of a random sample is zero (this angle is 54.5°, and is called the magic angle). But this is difficult to achieve (almost impossible) in imaging experiments. In this respect, as a word of caution, one must realize in addition that the usual sample in a fluorescence microscope is not randomly oriented. Of course, the fluorescence signal integrated over all angles (which can be done using an integrating sphere) also shows the expected fluorescence decay (with the lifetime of the excited state) without the possible effects of polarization decay. The angle between the direction of the excitation light beam and the direction of emission detection also affects the degree of polarization. These problems have been analyzed and methods for correction given.[54] General references to polarization can be found in any textbook on fluorescence.[7]

For a steady state anisotropy measurement the fluorescence signal is integrated over time to achieve the steady state signal. This steady state fluorescence is polarized. If the rotational correlation times are known, it is possible to calculate fluorescence lifetime from steady state anisotropy measurements. The extent of molecular rotation that takes place before fluorescence is emitted will depend on the statistical probability that the molecule remains in the excited state—the shorter the lifetime, the higher the (steady state) anisotropy. The excited state lifetime presents a window to observe competing molecular relaxation processes. If this observation window is short compared with the molecular rotation of the fluorophore, then the fluorescence emission will be highly polarized. If the decay of the emission is long compared with the molecular rotation, then the emission will be fully unpolarized light. The expression linking the fluorescence lifetime to the degree of polarization for the simple case of a freely rotating molecular sphere

[53] R. D. Spencer and G. Weber, *J. Chem. Phys.* **52,** 1654 (1970).

[54] M. Koshioka, K. Sasaki, and H. Masuhara, *Appl. Spectrosc.* **49,** 224 (1995).

(where the fluorophore transition dipole is attached rigidly to the sphere) is

$$r_{\text{steady state}} = r_0 \frac{1}{1 + 6D_{\text{rot}}\tau_F} \tag{4}$$

r_0 is the maximum anisotropy (before rotation at zero time), which is 0.4 if the absorption and emission transition dipoles are exactly parallel. D_{rot} is the rotational diffusion constant of the molecular sphere, and τ_F is the measured fluorescence lifetime of the chromophore. For a simple sphere (with diameter R_s) there is only one rotational diffusion constant, $D_{\text{rot}} = k_B T / 8\pi\eta R_s^3$. η is the viscosity and k_B is Boltzmann's constant. Such a simple relationship is valid only for a sphere, but this is often a good first approximation. Obviously, if we know, or measure, τ_F then we can calculate D_{rot} by measuring $r_{\text{steady state}}$. On the other hand, as mentioned above, if we know D_{rot} we can calculate τ_F. The only required relationship between the fluorescence lifetime and the dynamic processes responsible for polarization decay is that the fluorescence signal lasts long enough so that molecular rotations can be observed. For small molecules, and macromolecules similar in size to soluble proteins, rotational correlation times of picoseconds to hundreds of nanoseconds are expected, and this is in the range that can be observed by fluorescence with nanosecond lifetimes. If we know τ_F and D_{rot} then we can determine the effective viscosity of the solvent surrounding the rotating sphere. We do not have room to discuss the time-resolved expressions of rotation and anisotropy decay. However, if we can measure the time dependence of the anisotropy (as with an FLI instrument) we can measure even more accurate and reliable molecular information concerning molecular rotations.

In general, if there is only one component of fluorescence, the anisotropy measurement can be carried out such that only the anisotropy decay is observed without the interference of the decay of the fluorescence intensity. This is done by analyzing the polarization extent with the anisotropy function, which parses the fluorescence decay out of the fluorescence signal, leaving only the decay of the rotational depolarization. Alternatively, if the fluorescence lifetime is known by measuring under "magic angle" conditions, the total dynamic signal involving the sum of the intensity and polarization decay, can then be uniquely analyzed. For this article the important point is that the polarization of a fluorescence sample is due to orientation of the molecules. In general, following excitation, these molecules will change direction by rotational diffusion. This is a dynamic process that does not compete with pathways of deexcitation, but does convolute the rotational dynamics into the measured fluorescence decay. This can be an advantage if these rotations are to be studied. On the other hand, these rotational effects can interfere with fluorescence lifetime measurements unless precautions are taken, and this is difficult to do in an imaging experiment. The dynamic rotational movements tell us a great deal about the molecular environment of the macromolecules to which fluorescence probes are attached, and dynamic fluorescence measurements are necessary to extract this information. There have been many reviews of rotational diffusion, but only more recently have FLI enthusiasts turned their attention to this measurement. It is more difficult that just measuring the fluorescence lifetime-resolved image, but this will be an active development in the future.

Steady State Quenching Measurement

Another example where steady state measurements can be used to gather lifetime information is the measurement of the extent of dynamic quenching. Dynamic quenching will be more effective the longer the excited state of the fluorophore. If a known concentration of quencher is calibrated for its effectiveness with a certain fluorophore, then the intensity of fluorescence can be used as a measure of the fluorescence lifetime, provided the concentration (and therefore the fluorescence intensity in the absence of the quencher) of the fluorophore is known. However, in contrast to the polarization measurements, dynamic quenching does compete directly with the basic deactivation pathways. Static quenching (forming a ground state nonfluorescent species) does not affect the rate of deactivation of the activated state, and therefore does not affect the fluorescence decay rate.

Experimental Realization(s)

In general, there are two optical configurations for gathering image data and two approaches to acquire the dynamic information. As mentioned above, there are other parameters that mirror the fluorescence lifetimes. But we consider only the common direct methods.

Two Modes of Image Acquisition

Optical imaging measurements can be carried out in full-field (wide-field) or scanning modes. In full-field mode all regions of an imaged object are illuminated at the same time and the images are captured as a whole picture; all pixels are measured simultaneously. This is a highly parallel method of imaging, and usually involves capturing the final images using CCD cameras, or some other parallel array detector, as well as full field-modulated illumination. In scanning mode, a (usually) diffraction-limited focused beam of modulated excitation light is scanned over the object (as in a scanning confocal microscope) and the data for each pixel are gathered serially with a single channel photo detector (PM or diode). Each of these methods has been reviewed.[55–67] Advantages of each method are given in Table I. The list given in Table I is to be considered only as a guideline; due to technology improvements the capabilities of the instruments are changing.

[55] R. M. Clegg, P. C. Schneider, and T. M. Jovin, *in* "Biomedical Optical Instrumentation and Laser-Assisted Biotechnology" (A. M. Verga Scheggi, ed.) p. 143. Kluwer Academic Publishers, Dordrecht, 1996.

General Experimental Requirements. In general, any direct method of fluorescence lifetime measurement (by this we mean detecting the temporal decay of fluorescence directly) involves modulating the excitation light in some characteristic way and following the time-delayed response of the fluorescence emission. In order to acquire fluorescence decays in the nanosecond time region, the modulation of the excitation light—or the participation of some molecular kinetic process in the excited state—must take place at least on the same time scale as the lifetimes of the excited states. Usually the detection techniques are subdivided into time and frequency domain methods, and they are often discussed separately. Actually both methods are similar; both involve experimentally deconvoluting the time course of the fluorescence emission from the time dependence of the excitation pulse. Usually the excitation consists of a train of pulses. The shape of the pulses vary from short repetitive pulses—pulse duration down to 100 fs—to continuous sinusoidal modulation. The pulses (modulation) are repeated as often as necessary to achieve a good signal-to-noise ratio. In the time domain the signal is recorded and analyzed directly as a function of time. In the frequency domain this time dependence is expressed in terms of a phase lag and degree of demodulation.

Time Domain Measurements. If the fluorescence decay is acquired and analyzed as a time-relaxing signal, then the acquisition and analysis is usually said to take place in the time domain.[68] Time domain measurements require that the

[56] R. M. Clegg and P. C. Schneider, *in* "Fluorescence Microscopy and Fluorescent Probe" (J. Slavik, ed.), p. 15. Plenum Press, New York, 1996.

[57] M. vande Ven and E. Gratton, *in* "Optical Microscopy: Emerging Methods and Applications" (B. Herman and J. J. Lemasters, eds.), p. 373. Academic Press, New York, 1992.

[58] T. W. J. J. Gadella, "Fluorescent and Luminescent Probes," 2nd Ed., p. 467. Academic Press, New York, 1999.

[59] K. Dowling, M. J. Dayel, S. C. W. Hyde, C. Dainty, P. M. W. French, P. Vourdas, M. J. Lever, A. K. L. Dymoke-Bradshaw, J. D. Hares, and P. A. Kellet, *IEEE J. Select. Top. Quantum Electronics* **4,** 370 (1998).

[60] T. French, P. T. So, C. Y. Dong, K. M. Berland, and E. Gratton, *Methods Cell Biol.* **56,** 277 (1998).

[61] C. Y. Dong, P. T. C. So, T. French, and E. Gratton, *Biophys. J.* **69,** 2234 (1995).

[62] J. R. Lakowicz and H. Szmacinski, *in* "Fluorescence Imaging Spectroscopy and Microscopy" (X. F. Wang and B. Herman, eds.), p. 273. John Wiley & Sons, New York, 1996.

[63] X. F. Wang, A. Periasamy, P. Wodnicki, G. W. Gordon, and B. Herman, *in* "Fluorescence Imaging Spectroscopy and Microscopy" (X. F. Wang and B. Herman, eds.), p. 313. John Wiley & Sons, New York, 1996.

[64] P. T. C. So, T. French, W. M. Yu, K. M. Berland, C. Y. Dong, and E. Gratton, *in* "Fluorescence Imaging Spectroscopy and Microscopy" (X. F. Wang and B. Herman, eds.), p. 351. John Wiley & Sons, New York, 1996.

[65] P. I. Bastiaens and A. Squire, *Trends Cell Biol.* **9,** 48 (1999).

[66] D. Phillips, *Analysit* **119,** 543 (1994).

[67] K. Koenig and H. Schneckenburger, *J. Fluoresc.* **4,** 17 (1994).

[68] D. V. O'Conner and D. Phillips, "Time-Correlated Single Photon Counting." Academic Press, London, 1984.

TABLE I
ADVANTAGES OF SCANNING TWO-PHOTON AND FULL-FIELD FLUORESCENCE LIFETIME-RESOLVED IMAGING

Scanning 2-$h\nu$ FLI	Full-field FLI
Spatial confinement of excitation–diffraction limited focusing	Simultaneous collection of all pixels
Confocal effect: 0.3 m × 1 m ($h\nu_{ex}$ = 700 nm, NA = 1.3)	Even x–y illumination (low iris effect)
Little or no photo damage outside of 2-$h\nu$ region	Simplicity of optical construction and operation; attach to any microscope
Depth of penetration	FLIE (endoscopy)
3-D images possible without deconvolution	3-D possible with image deconvolution
UV excitation (localized) because of two-photon excitation	CCD data acquisition (long integration times possible without unreasonable total measurement time)
PM detection, multifrequencies, Fourier spectrum	Phosphorescence (DLIM)
Detection straightforward, same as single-channel cuvette measurement	Real-time (video rate) lifetime-resolved image acquisition
Localized photoactivation of caged compounds; localized rapid kinetic acquisition	Time resolution for image kinetics in millisecond range

data acquisition be fast enough to record the fluorescence intensity for several (sometimes hundreds) consecutive time periods that are a fraction of the total decay time. The time progression of the fluorescence decay is recorded at specific times by delaying the period of observation following the start of each excitation pulse. The length of the delay before starting the measuring period is varied, and in this way the time course of the fluorescence decay is captured. In a scanning mode this is naturally accomplished for every recorded pixel separately. The scanning method is similar in many respects to the techniques used for cuvette based lifetime measurements. In the full-field mode all pixels of the image are recorded simultaneously at every time period. This requires that there be a means of time-gating the image; this is done using a fast gating image intensifier that is placed before the final image recorder that is usually a CCD. This method is easy to understand because the measurement takes place directly in the time domain. One simply excites the molecules with a short pulse and records the time decay directly in the time following the pulse. The reader is referred to the literature for details of the instrumentation and analysis.[63,69–78] Synchrotron radiation has also been used for lifetime measurements in microscopes, having the advantage

[69] K. Carlsson, A. Liljeborg, R. M. Andersson, and H. Brismar, *J. Microsc.* **199,** 106 (2000).

[70] M. J. Cole, J. Siegel, S. E. Webb, R. Jones, K. Dowling, M. J. Dayel, D. Parsons-Karavassilis, P. M. French, M. J. Lever, L. O. Sucharov, M. A. Neil, R. Juskaitis, and T. Wilson, *J. Microsc.* **203,** 246 (2001).

[71] H. C. Gerritsen, J. M. Vroom, and C. J. de Grauw, *IEEE Eng. Med. Biol. Mag.* **18,** 31 (1999).

that the wavelength range of 250–700 nm is available for excitation.[79] Time domain dynamic fluorescence measurements using near-field optical microscopy have also been reported.[80] A technique for recording the lifetime information in the time domain, using single photon-counting techniques and a quadrant detection scheme to identify the location of fluorescence emission from an object is being developed based on new time and space-correlated single photon counting MCP-PMT instrumentation.[81] This technique has a 10-ps resolution and has can accurately determine multiple lifetimes as is customary in normal single-channel time-correlated single-photon counting using time-to-amplitude converters.

Frequency Domain Measurements. In the frequency domain a continuous high-frequency (HF) repetitive train of pulses that can be of any shape excites the fluorophores.[82] Because this method is not as familiar as the direct time domain measurements, we describe this in more detail. If the HF modulation is not purely sinusoidal (i.e., a sinusoid signal of one frequency) the repetitive pulse train will contain the frequency components required to define the repetitive pulse shape according to a Fourier analysis. If the excitation light is modulated with a pure sinusoid, then only the fundamental frequency component of the repetitive pulse train is present. The fluorescence signal will have the same frequency components as the excitation light but the fluorescence frequency components will be delayed (phase shifted) and demodulated (amplitude reduced) relative to the excitation light frequency components in characteristic ways depending on the fluorescence lifetimes. It is the relation between the phase delay and fractional diminution (the demodulation) of the fluorescence signal relative to the phase and modulation of the excitation light that allows the determination of the fluorescence decay time (s). As the frequency is increased, the phase of the fluorescence is increasingly delayed (from 0 to 90° and the demodulation (the ratio of the AC to DC component of the fluorescence to the same AC/DC ratio of the excitation light) is decreased from

[72] V. Barzda, C. J. de Grauw, J. Vroom, F. J. Kleima, R. van Grondelle, H. van Amerongen, and H. C. Gerritson, *Biophys. J.* **81,** 538 (2001).

[73] R. Sanders, A. Draaijer, H. C. Gerritsen, P. M. Houpt, and Y. K. Levine, *Anal. Biochem.* **227,** 302 (1995).

[74] M. Kohl, J. Neukammer, U. Sukowski, H. Rinneberg, D. Wohrle, H. J. Sinn, and E. A. Friedrich, *Appl. Phys. B Photophys. Laser Chem.* **B56,** 131 (1993).

[75] T. Oida, Y. Sako, and A. Kusumi, *Biophys. J.* **64,** 676 (1993).

[76] X. F. Wang, T. Uchida, M. Maeshima, and S. Minami, *Appl. Spectrosc.* **45,** 560 (1991).

[77] R. Cubeddu, A. Pifferi, P. Taroni, A. Torricelli, G. Valentini, F. Rinaldi, and E. Sorbellini, *IEEE J. Select. Top. Quantum Electronics* **5,** 923 (1999).

[78] T. Minami and S. Hirayama, *J. Photochem. Photobiol. A,* **53,** 11 (1990).

[79] C. J. R. Van Der Oord, H. C. Gerritsen, F. F. G. Rommerts, D. A. Shaw, I. H. Munro, and Y. K. Levine, *Appl. Spectrosc.* **49,** 1469 (1995).

[80] D. A. M. Smith, S. A. Williams, R. D. Miller, and R. M. Hochstrasser, *J. Fluoresc.* **4,** 137 (1994).

[81] K. Kemnitz, *in* "New Trends in Fluorescence Spectroscopy: Applications to Chemical and Life Sciences" (B. Valeur and J.-C. Brochon, eds.), p. 381. Springer-Verlag, Berlin, 2001.

[82] D. M. Jameson, E. Gratton, and R. D. Hall, *Appl. Spectrosc. Rev.* **20,** 55 (1984).

one to zero. The range of high frequency over which the phase and demodulation changes are measured is determined by the inverse of the time decay of the fluorescence (it is usually between 20 and 200 MHz). The data are usually not recorded directly in the time frame of the high frequency modulation (i.e., the 100-MHz modulation signal is not recorded directly). The detector amplification is modulated also at a high frequency in order to mix the modulation frequency of the fluorescence with the modulation frequency of the amplification factor of the detector. These hetero- and homodyne techniques (respectively, where either the two frequencies are different but close, or where the frequencies are exactly the same) are similar to the way FM radios work. If the modulation frequency of the detector is close to, but not exactly equal to, the modulation frequency of the excitation light, the two frequencies mix to form the difference frequency that is much lower than the primary modulation frequency. This is called heterodyning, and all the phase and modulation information that determines the fluorescence lifetime decay at the high frequencies is preserved in the lower heterodyne (the difference) frequency signal. But it is much easier and usually less expensive to handle the lower frequency signals. Alternatively, if the modulation frequencies of the excitation light and the detector are exactly the same, then the dynamic phase and modulation information is transferred to a DC level. This constitutes a homodyne measurement, and the phase and demodulation of the fluorescence relative to the excitation light (which is set by the fluorescence decay kinetics) is determined by varying the phase between the high frequency modulation of the excitation light and the detector. The fluorescence signal (in the case of FLI this is the whole image) is then recorded at several phase delay settings. These methods have been described in detail for imaging platforms.[55,56,58,83–88] Although the description of the frequency domain experiment sounds different from the time domain, in essence they are identical. In both cases—time and frequency domains—we must describe the time-dependent shape of the fluorescence signal relative to the excitation waveform. In the time domain the fluorescence signal is described as a sum of exponentials, and the exponential rate constants define the fluorescence lifetimes (and other artifacts, as discussed above). In the frequency domain the fluorescence signal is described as a sum of sinusoids (which are the Fourier components of the excitation waveform) with characteristic phase and modulation amplitudes that are related to the fluorescence lifetimes. The fluorescence lifetimes are then extracted from the phase and demodulation values, which themselves are functions of the modulation frequency. The time and frequency domains are Fourier transforms of each other. The advantages of each technique are related more to the

[83] J. R. Lakowicz and K. W. Berndt, *Rev. Sci. Instrum.* **62,** 1727 (1991).
[84] T. W. J. Gadella, Jr., T. M. Jovin, and R. M. Clegg, *Biophys. Chem.* **48,** 221 (1993).
[85] A. Squire, P. J. Verveer, and P. I. Bastiaens, *J. Microsc.* **197,** 136 (2000).
[86] C. G. Morgen, A. C. Mitchell, and J. G. Murray, *J. Microsc.* **165,** 49 (1992).
[87] C. G. Morgan, A. C. Mitchell, and J. G. Murray, *Trans. R. Microsc. Soc.* **1,** 463 (1990).
[88] O. Holub, M. Seufferheld, C. Gohlke, Govindjee, and R. M. Clegg, *Photosynthetica* **38,** 581 (2000).

availability and cost of instrumentation, rather than anything fundamental about the measurement process.

Performance Goals and Comparisons. There are several performance goals to be considered in designing and constructing the lifetime-resolved imaging instruments. Some of the major considerations are (1) the accuracy of the lifetime determination, (2) the sensitivity of the measurement, (3) the time of data acquisition, data analysis, and display, (4) the ease of making the measurement, (5) the type and complexity of data analysis required, and (6) the cost. As usual, tradeoffs must be made depending on the most relevant requirements for a particular application. Figure 2 shows the basic components of the different measurement modes in schematic form. There are many variations for the equipment and the reader is referred to the original literature. This is a rapidly changing field, and some of the later publications contain reference to earlier literature.[64,71,85,88–90]

Scanning Modes. In essence the scanning modes of data acquisition are similar to the classic one-channel nonimaging techniques.[91] The only difference is that the excitation light is scanned over the object and the lifetime is measured at different locations of the image (this is not meant to imply that the instrumentation is trivial). The scanning technique acquires data serially—each point of the image is measured separately. For this reason in order to acquire the images in reasonable times, the measurement at each pixel must be carried out rapidly. The critical limitation on the speed of point-by-point acquisition is the number of photons required to make a statistically sound lifetime measurement. Scanning imaging methods can acquire well-resolved three-dimensional images by using confocal methods, and the lifetime-resolved scanning imaging techniques have the same advantage. Another advantage of scanning techniques is that two-photon excitation can be employed. The excitation beam is highly focused, producing locally (in approximately a 1-μm^3 volume) at a high intensity, and the fluorophore can then be excited with two photons simultaneously. The modern pulsed Ti-sapphire lasers are ideal for the two-photon excitation, and the short light pulses of this laser and the repetition frequency of the laser pulses are suitable for lifetime measurements.[57,92,93] This laser is often used in both frequency and time domain measurements. The pulse is short, which is needed for the time domain, and the repetitive frequency is in the right range for frequency domain measurements. These scanning methods do not need to use spatial deconvolution techniques to

[89] P. C. Schneider and R. M. Clegg, *Rev. Sci. Instrum.* **68,** 4107 (1997).

[90] R. Cubeddu, G. Canti, A. Pifferi, P. Taroni, and G. Valentini, *Proc. SPIE. Int. Soc. Opt. Eng.* **2976,** 98 (1997).

[91] T. Wilson and C. Sheppard, "Theory and Practice of Scanning Optical Microscopy." Academic Press, London, 1984.

[92] E. Gratton, N. P. Barry, K. Beretta, and A. Celli, *Methods Cell Biol.* **25,** 103 (2001).

[93] P. So, K. Konig, K. Berland, C. Y. Dong, T. French, C. Buhler, T. Ragan, and E. Gratton, *Cell. Mol. Biol. (Noisy-le-Grand)* **44,** 771 (1998).

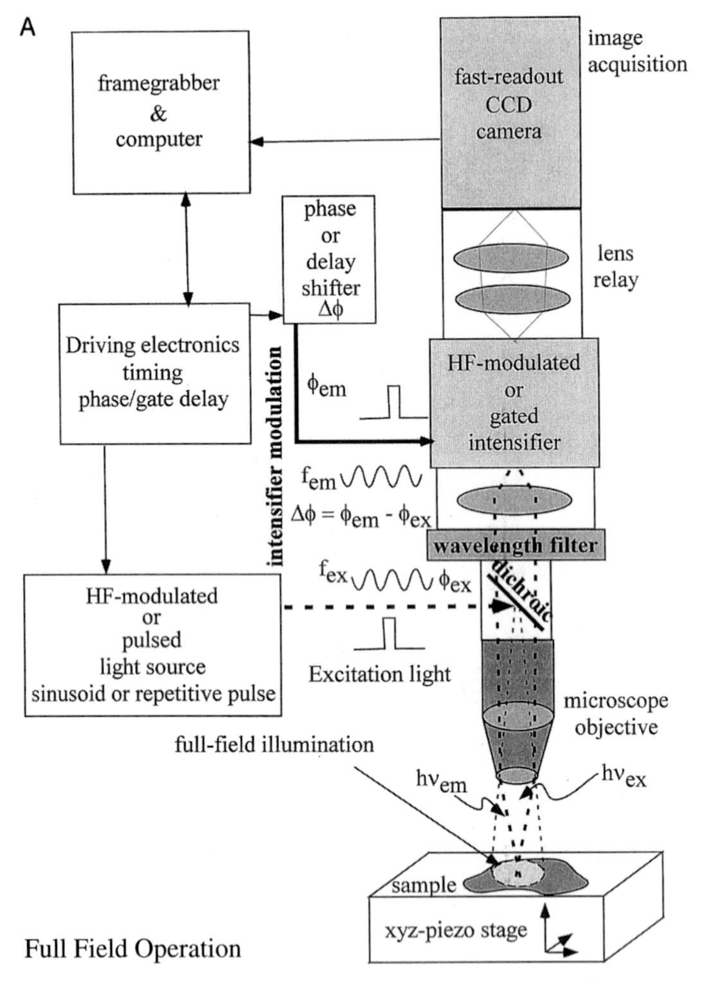

FIG. 2. A schematic of the instrumentation for FLI measurements. This is a much abbreviated schematic of the instrumentation, incorporating only the interconnections between the essential parts. The full-field (A) and scanning (B) modes are depicted separately. Both frequency- and time-domain implementations have been indicated in each separate diagram. The text should be consulted for more detailed explanations. The microscope is depicted only as a microscope objective. An epi-illuminated upright configuration is depicted. The excitation light is modulated either as a repetitive pulse or as a sinusoidal (or some other repetitive continuous pulse train). The fluorescence signal varies in time as a pulse train with the same basic repetitive frequency as the excitation light. In both modes, the time decay of the fluorescence (at every location of the illuminated sample) decays as a sum of exponentials that is convoluted with the shape of the excitation pulse. In the frequency domain, measurements are made of the phase and modulation of the Fourier components of the fluorescence. These phase and modulation parameters are compared with the corresponding phase and modulation components of the excitation light pulse. In the frequency domain measurement this is accomplished by modulating a detector [in the full field mode (A), the intensifier amplification is modulated; in the scanning mode (B), the photomultiplier is modulated] at the same frequencies as the excitation modulation frequency components (this constitutes homodyne detection) or at a frequency that is different, but close, to the excitation frequencies (this constitutes heterodyne detection).

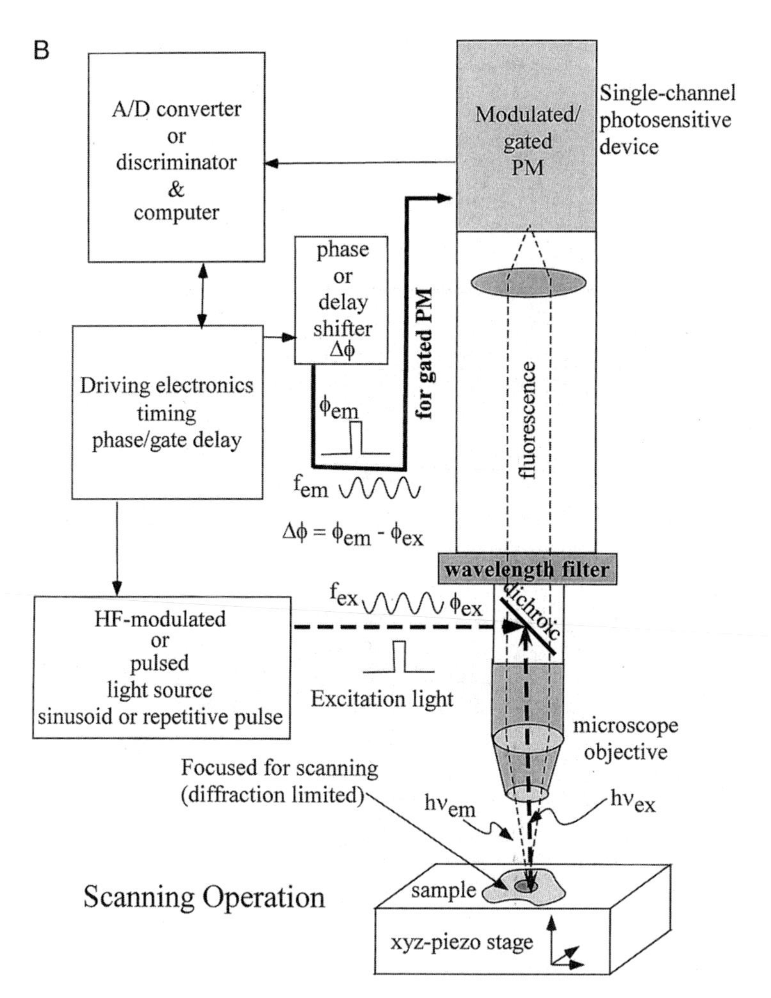

In homodyne detection the high frequency components are transferred to DC. In heterodyne detection the time-varying signal is transferred down to the heterodyne frequency (the difference frequency; this is usually between 10 and 10^4 Hz). In both cases the homodyne and heterodyne signals retain all the phase and modulation information (i.e., the lifetime information) of the high-frequency signals. In the time domain, the signal is chopped with a time delay (relative to the excitation pulse) that is varied so that the time response of the fluorescence decay signal is recorded in sequential measurements. This is accomplished either with a fast gated intensifier in the full-field mode, or directly on the photomultiplier. For many applications involving photomultiplier detection, the experiment can be carried out with photon counting, where the time of arrival of photons is determined following the synchronized excitation pulse. For most modes of operation the light source is a laser. More recently, a common excitation source is a Ti:Sapphire laser with an 80-MHz repetitive pulse of a hundred or a few hundred femtoseconds. If long lifetimes are present, the pulses can be selected (resulting in a slower repetitive frequency) with a pulse-picker. The experiments are computer controlled, and the data are collected through a frame grabber (for full-field applications with a CCD camera), DAC, or counting interface (through a discriminator) for photon counting. Specialized driving electronics are usually required for operating the timing circuitry and the modulation high-power signals for the modulated detection circuitry (intensifier or photomultiplier).

remove out-of-focus fluorescence because the volume that is imaged through a pinhole in a confocal arrangement or the volume excited in a two-photon excitation arrangement is naturally limited in the z direction. These confocal or two-photon scanning methods have also been extended to multifocus arrangements (using multilens objectives), allowing parallel lifetime-resolved data acquisition of many point-focused spots.[94,95]

Full-Field Modes. Full-field imaging has the great advantage that the measurement of every pixel in the image (and this can be up to 1 million pixels) is made simultaneously, saving considerable time.[60,84,85,96] This makes full-field lifetime-resolved imaging the method of choice for real-time imaging applications. All full-field instruments require a high-frequency modulatable image detector, and this is usually accomplished by using a microchannel plate high-frequency modulated intensifier. The drawback of these devices is the cost and the noise of the cathode (which has the same noise characteristics as a photomultiplier). But they are commercially available and can be operated at high frequencies with almost no deleterious iris effects (for the frequency domain) or gated into the subnanosecond range (for the time domain). The output of the intensifier (which is a phosphor screen onto which the accelerated electrons from the microchannel plate are focused, pixel by pixel) is focused onto a CCD camera and, after averaging, the image is read into a computer for analysis. This whole process (data acquisition, image analysis, and image display) can now be made at video rates when the frequency domain is used.[88] There is a large selection of CCD cameras with excellent noise characteristics and fast frame grabbers to transfer the data to the computer. PCs are now fast enough to control the experiment, gather the data, analyze the data to extract the lifetime information, and display the resultant images.[88]

Data Analysis

Analysis of the lifetime-resolved data at every pixel of the recorded image is similar (actually identical) to single-channel measurements. The time domain measurements are analyzed as exponentials (convoluted with the form of the excitation pulse) and the frequency domain measurements are analyzed by determining the phase and demodulation of the fluorescence relative to the excitation light modulation for all the frequency components (see Clegg and Schneider[56] for a detailed account). There are different ways to do these analyses; this is an extensive episode of data analysis with a long history, and we do not go into details that can be found in the article by Clegg and Schneider.[56] If one is careful to pay attention to statistics, most techniques of analysis will yield identical results. Because of the large number of recorded picture elements, the challenge is to carry out the data analysis

[94] A. Schonle, M. Glatz, and S. W. Hell, *Appl. Optics* **39,** 6306 (2000).

[95] J. R. Lakowicz, I. Gryczynski, H. Malak, M. Schrader, P. Engelhardt, H. Kano, and S. W. Hell, *Biophys. J.* **72,** 567 (1997).

[96] A. G. Harpur, F. S. Wouters, and P. I. Bastiaens, *Nat. Biotechnol.* **19,** 167 (2001).

on hundreds of thousands of pixels in a reasonably short time. The choice of methods depends more on the experience of the research group and on the quality of the data, than on any theoretical motivation. Just as in single-channel measurements, averaging must often be employed to reduce the statistical noise. However, there are characteristics of the image measurement that provide new opportunities for analysis (besides the obvious key advantage of having acquired the spatial distribution of fluorescence lifetimes in an extended sample). This has been treated in detail.[97–100] Major improvements in data analysis can be achieved by global analysis; if one can assume that the lifetimes of the different components (or some of them) are the same everywhere in the image then a global analysis will improve the quality of the fit considerably. There are several different ways to carry out the data analysis. In addition to the above references we refer the reader to an article that deals with some aspects of the data analysis.[88]

Display of Lifetime-Resolved Images

The way in which the final images are displayed is important because FLI produces data sets that are more complex than the normal intensity images. This complexity is growing as we add spectral and polarization capabilities in our repertoire for FLI imaging. It is clear from the number of different physical processes that affect the lifetime of the excited state that a fluorescence signal can encompass a large assortment of effects. The important critical aspects of a statistical assessment of the data are comparable to the methods of analysis used in single-channel work. The statistical requirements for a good fit are the same. But displaying the information for 10,000 to 100,000 pixels conveniently and informatively is demanding. This is especially challenging when the data are being updated actively and displayed in real time. The extent of detail and the type of display will depend on the time available. In general, except for the usual image analysis and display tasks, commercial software is often not particularly adequate for many needs of FLI because it is either not fast enough, or the peculiar analysis that is employed for analyzing FLI data is not available. Of course, given enough time, many of the commercial image manipulation and improvement algorithms available can be used, especially for display. We have been concentrating on rapid data acquisition and display that is aimed at medical imaging applications. These methods are also useful for cellular biological applications where many samples must be identified, selected, and analyzed. The judicious use of shading, lighting, color coding, and displaying contours can expediently and instructively accentuate critical elements of the data analysis to the user. This is demonstrated in Fig. 3, which explains the display encoding. The software is written so that the display can be

[97] R. Pepperkok, A. Squire, S. Geley, and P. I. Bastiaens, *Curr. Biol.* **9,** 269 (1999).
[98] A. Squire and P. I. Bastiaens, *J. Microsc.* **193,** 36 (1999).
[99] P. J. Verveer, A. Squire, and P. I. Bastiaens, *Biophys. J.* **78,** 2127 (2000).
[100] P. J. Verveer, A. Squire, and P. I. Bastiaens, *J. Microsc.* **202,** 451 (2001).

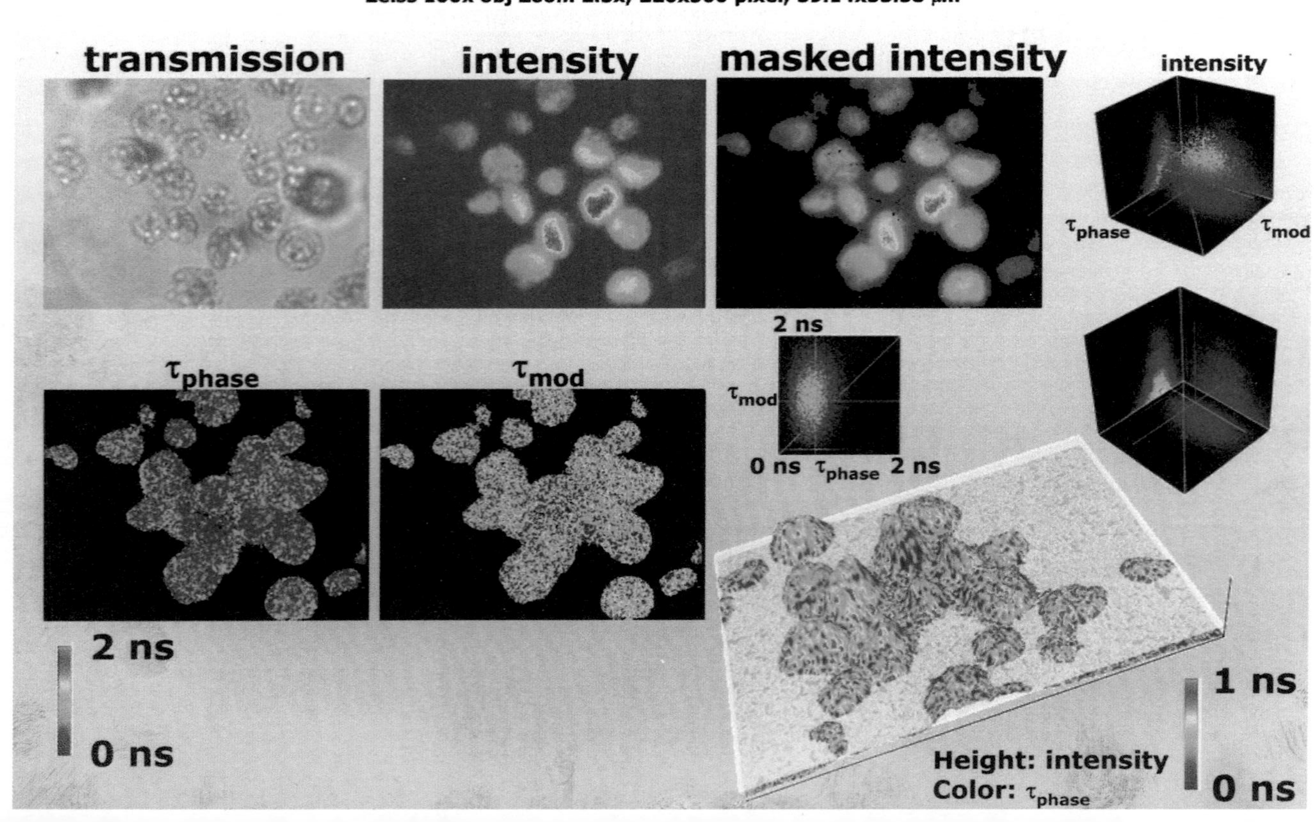

Lifetime measurements of Chl a fluorescence of single WT cells (C. reinhardtii)
in microcapillary in dark for 30 h (not swimming); 1300 μmol photons/(m² s)
Zeiss 100x obj Zoom 2.5x; 220x300 pixel; 39.14x53.38 μm
transmission
intensity
masked intensity
intensity
τphase
τmod
τphase
τmod
2 ns
τmod
0 ns τphase 2 ns
2 ns
0 ns
1 ns
0 ns
Height: intensity
Color: τphase

modified and adapted actively by the user. The image display can be fashioned by the user actively as well as changing the mode of data acquisition and analysis. Close attention has been paid to ensure that the display modes are rapid. Some information is available[88] and details are given elsewhere. Figure 3 demonstrates the considerations.

Summary

We have given an overview of what one can gain by lifetime-resolved imaging and reviewed the major issues concerning lifetime-resolved measurements and FLI instrumentation. Instead of giving diverse selected examples, we have discussed the underlying basic pathways of deexcitation available to the molecules in the excited state. It is by traversing these pathways that compete kinetically with the fluorescence pathway of deactivation—and therefore affect the measured fluorescence lifetime—that we gain the information that lifetime-resolved fluorescence provides. It is hoped that being aware of the diversity of pathways available to an excited fluorophore will facilitate potential users to recognize the value of FLI measurements and inspire innovative experiments using lifetime-resolved imaging. FLI gives us the ability within a fluorescence image of measuring and quantifying dynamic events taking place in the immediate surroundings of fluorophores as well

FIG. 3. Example of FLI data and presentation modes. This composite presentation of FLI data represents the breadth of information that is acquired in an FLI measurement. The data have been taken with a full-field frequency-domain FLI microscope, using a single frequency of modulation (see Ref. 88). The information and results of analysis of FLI data must be presented such that they can be understood and recognized reasonably well in a comprehensive fashion. The data must be acquired, analyzed, and displayed as rapidly as possible. The data acquisition described in Ref. 88 has been written so that the user can adjust several options as the experiment is running. The analysis has been streamlined to take a minimum of time; the major portion of time per experiment (image capture) is required for the data transfer. A level of fluorescence intensity can be set (masked) so that all pixels with lower intensity will not be analyzed or displayed (top three images). It is useful to be able to compare the transmission image at the same time that the fluorescence image is acquired and analyzed. Three-dimensional plots give the user the ability to oversee the interplay between the two analyses of lifetimes (from the phase and demodulation) and the intensity of fluorescence. A quick plot of the two lifetime determinations is useful, because in the frequency mode the modulation lifetime is always longer that the phase lifetime if there is more than one lifetime component. If the lifetimes are distributed along the diagonal of a $\tau_{\text{phase-versus}}$-$\tau_{\text{modulation}}$ plot, this is a sign that there is only a single lifetime. From this plot the time distribution of the lifetimes is easily discernable. The color coding of the two lifetime images (lower left) shows the spatial distribution of lifetimes and it can be seen from the color coding that the modulation lifetimes are longer that the phase lifetimes. These images can be collected, analyzed, and displayed rapidly, and a subset of them can be acquired and displayed in real time (video rate). The display in the lower right corner is useful. It is a three-dimensional projection with color coding. The lifetimes are color coded, and the intensity of the pixels corresponds to the height. It is easy to recognize spatial correlations between lifetimes and intensities with this display. By employing lighting, shadowing, and texturing this display is informative.

as locating the fluorescent components within the image. Just as measurements in cuvettes, lifetime-resolved imaging extends considerably the potential information that can be derived from a fluorescence experiment. Our purpose has been to arouse an appreciation for the broad application of fluorescence lifetime-resolved measurements in imaging. We have given only general design characteristics of the instrumentation and discussed the characteristics that distinguish imaging from the single channel lifetime-resolved measurements. We have not provided details of the instrumentation or the presented many examples. These are available in the literature, and given in the references, and they are continually and rapidly growing.

[23] Fluorescence Resonance Energy Transfer Imaging Microscopy

By VICTORIA E. CENTONZE, MAO SUN, ATSUSHI MASUDA, HANS GERRITSEN, and BRIAN HERMAN

Introduction

For centuries the light microscope has been used to examine cells and tissues to determine the proximity of structures. Such information has been useful to determine the juxtaposition of cellular structures. Under conditions free from aberration, the limit of resolution limit (r_{Airy}) of the light microscope as defined by Rayleigh's criterion,

$$r_{Airy} = 0.61\lambda_0/NA_{obj} \quad (1)$$

where NA, the numerical aperture of the objective, is considered to be approximately 0.2 μm. Therefore two objects must be approximately 0.2 μm apart from one another to be visualized as being separate. If two objects cannot be seen as distinct structures (i.e., they are closer than 0.2 μm), then they may be considered coincident in space. Coincidence may represent a proximity close enough that a molecular association could be possible, although the light microscope does not have sufficient resolution to determine whether coincidence is equivalent to molecular interaction. However, using the technique of fluorescence resonance energy transfer (FRET) as a "molecular ruler," it is now possible to determine whether two molecules are within a distance of 10–100 Å of one another, thus being close enough for molecular interaction to occur.[1,2]

[1] C. G. Dos Remedios and P. D. J. Moens, *J. Struct. Biol.* **115,** 175 (1995).

[2] B. Herman, *in* "Fluorescence Microscopy" (B. Herman, ed.). BIOS Scientific Publishers, Oxford, 1998.

0076-6879/03 $35.00

Principles of Fluorescence Resonance Energy Transfer

Resonance energy transfer (RET) is a process by which a donor fluorophore (D) in its excited state may transfer its excitation energy to a nearby chromophore (acceptor, A). The process is nonradiative (not mediated by a photon) and is achieved through dipole–dipole interactions. In principle, for such a transfer to occur several major criteria must be satisfied. First, the donor molecule must have an emission spectrum that overlaps the absorption spectrum of the acceptor molecule (Fig. 1). When energy transfer occurs the fluorescence of the donor molecule is quenched in the presence of the acceptor and if the acceptor is a fluorescent molecule itself, it demonstrates increased, or sensitized, emission.

The second criterion that needs to be satisfied is that the donor and acceptor molecules must be within 10–100 Å of each other. The efficiency of energy transfer between donor and acceptor molecules falls off as the sixth power of the distance separating the donor and acceptor molecules. As such, the ability of the donor molecule to transfer its excitation energy nonradiatively to the acceptor molecule drops off precipitously with increasing distance between the molecules. The distance dependence of the RET process provides that basis for its usefulness in the study of molecular interactions. In the case of biological molecules containing donor and acceptor fluorophores, transfer of energy will occur only between molecules that are also close enough to interact biologically with one another.

Another criterion that must be met is that the fluorescence lifetime of the donor molecule must be of sufficient duration to allow FRET to occur. The rate of energy transfer (K_T) and the efficiency of energy transfer (E_T) are directly related to the lifetime of the donor molecule in the presence and the absence of the acceptor.

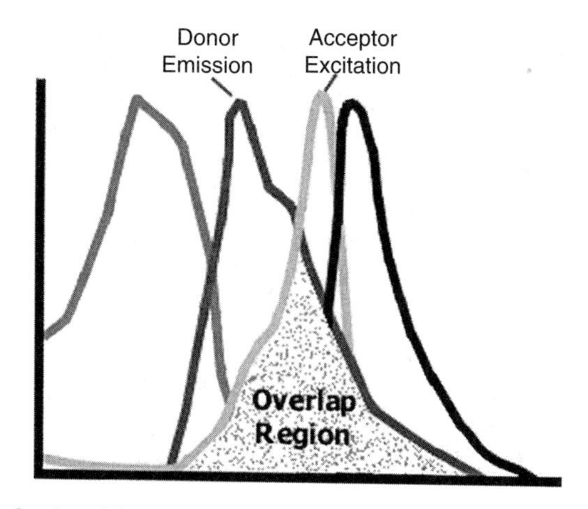

FIG. 1. Overlap of donor emission spectrum with acceptor excitation spectrum.

The rate of energy transfer is

$$K_T = (1/\tau_D)[R_0/R]^6 \tag{2}$$

where R_0 is the Förster critical distance, τ_D is the lifetime of the donor in the absence of the acceptor, and R is the distance separating the donor and acceptor molecules. The Förster critical distance, R_0, is the distance between donor and acceptor molecules at which the transfer rate equals the donor decay rate in the absence of acceptor. Typically this value is between 20 and 60 Å, which is on the order of protein dimensions.[3] R_0 is expressed as

$$R_0 = [\kappa^2 \times J(\lambda) \times \eta^{-4} \times Q_D]^{1/6} \times 9.7 \times 10^2 \tag{3}$$

illustrating that the relative orientation of the transition dipoles of the fluorophore (κ^2), the integral of the region of overlap between the donor emission and acceptor absorbance spectra [$J(\lambda)$], the refractive index of the surrounding medium (η), and the quantum yield of the donor (Q) will all affect R_0. The efficiency of energy transfer, E_T, is related to R by

$$R = R_0[(1/E_T) - 1]^{1/6} \tag{4}$$

where

$$E_T = 1 - (\tau_{DA}/\tau_D) \tag{5}$$

where τ_{DA} is the lifetime of the donor in the presence of the acceptor and τ_D is the lifetime of the donor in the absence of acceptor. Thus, by a simple measurement of donor fluorescence lifetime in the presence and absence of acceptor it is possible to determine the distance between the donor and acceptor molecules.

Molecule Labeling for Fluorescence Resonance Energy Transfer Analysis

Specimen preparation and imaging parameters need to be optimized to maximize the detection of FRET. Appropriate donor and acceptor probes need to be selected on the basis of characteristics of their absorption and emission spectra. The emission spectrum of the donor must have substantial overlap with the absorption spectrum of the acceptor and at the same time should not have overlap with its own absorption spectrum to minimize donor–donor self-transfer. Once excited, the donor requires a sufficiently long fluorescence lifetime to enable energy transfer to occur. In addition, the acceptor should not be excitable at the excitation maxima of the donor. Ideally, specimens should be prepared in such a way that the concentrations of donor and acceptor molecules are tightly controlled, as the efficiency of FRET is related to the relative concentration of donor and acceptor (the

[3] P. I. H. Bastiaens and R. Pepperkock, *Trends Biochem. Sci.* **25,** 631 (2000).

efficiency of FRET is maximized when many acceptor molecules surround a single donor). When conjugating the donor and acceptor reagents to biologically active molecules it is important to consider the location of the reagents with respect to the tertiary structure of the interacting molecules. The donor and acceptor reagent must be placed as close to the site of likely molecular interaction without their presence resulting in disruption of the biological activity. In some cases it might be necessary to truncate the interacting molecule in order to place the donor or acceptor reagent close enough to a site of molecular interaction for FRET to occur.[4] In most biological systems, however, it is unlikely that all of these parameters can be controlled.[2]

In Vitro Derivatization of Molecules

In spite of these caveats several pairs of fluorophores have proved successful as reporters of FRET. Secondary antibodies conjugated with fluorescein isothiocyanate (FITC) and rhodamine have been used to colocalize the tumor suppressor protein p53 and human papillomavirus E6 protein in human cervical carcinoma cell lines. Analysis of the double-labeled cells for FRET showed that p53 protein binds HPV-16/18 E6 protein in the cytoplasm, thus preventing p53 from acting as a tumor suppressor in the nucleus.[5] The interaction of the antiapoptotic protein Bcl-2 with Beclin, a protein encoded by tumor suppressor gene, has also been evaluated. Previous studies using yeast two-hybrid systems indicated a high degree of interaction between these two proteins. FRET imaging microscopy demonstrated significant interaction between Bcl-2 and Beclin but not between Bcl-2 and Beclin mutants or between Bcl-2 and the SERCA (sarcoplasmic or endoplasmic reticulum Ca^{2+}) ATPase.[6] Other investigators have been successful in measuring FRET between antibody-labeled molecules. Cy3 and Cy5 conjugated directly to proteins, antibodies, or other ligands have been used as reporters of FRET.[7–9] Because Cy3 and Cy5 fluorophores have well-separated emission spectra they are the FRET pair of choice for single-molecule FRET.[10]

Although tagging the proteins of interest with antibodies will work in some FRET systems there is a risk of false negative results. Successful use of FRET in this system depends on being able to label sites close enough to the point of molecular interaction. In addition, it is possible that even though the proteins

[4] M. G. Erickson, B. A. Alseikhan, B. Z. Peterson, and D. T. Yue, *Neuron* **31,** 973 (2001).
[5] X. H. Liang, M. Volkmann, R. Klein, B. Herman, and S. J. Lockett, *Oncogene* **8,** 2645 (1993).
[6] G. W. Gordon, G. Berry, X. H. Liang, B. Levine, and B. Herman, *Biophys. J.* **74,** 2702 (1998).
[7] P. I. H. Bastiaens and T. M. Jovin, *Proc. Natl. Acad. Sci. U.S.A.* **93,** 8407 (1996).
[8] A. K. Kenworthy and M. Edidin, *J. Cell Biol.* **142,** 69 (1998).
[9] A. K. Kenworthy, N. Petranova, and M. Edidin, *Mol. Biol. Cell* **11,** 1645 (2000).
[10] T. Ha, T. Enderle, D. F. Ogletree, D. S. Chemla, P. R. Selvin, and S. Weiss, *Proc. Natl. Acad. Sci. U.S.A.* **93,** 6264 (1996).

of interest may interact, the size of either a fluorophore-conjugated primary or secondary antibody bound to the proteins of interest could result in the donor and acceptor fluorophores being separated by a length >100 Å, thus being too far apart to undergo energy transfer.

Direct conjugation of fluorophores to proteins is the preferred method of labeling proteins for FRET. The "cysteine-light" proteins have aided conventional methods for *in vitro* derivatization. Through genetic engineering, reactive cysteine residues can be placed at desired positions in the protein to be labeled, thus allowing the placement of conventional SH-reactive fluorescent dyes in a site-specific manner during derivitization. Fluorescently labeled cysteine-light proteins have been used in FRET studies to investigate kinesin binding states,[11] voltage- and ligand-gated channels, Na^+,K^+-ATPase pumps,[12] and several DNA-binding proteins.[13] Rare earth elements, or lanthanide atoms, are also being developed as FRET donor molecules. The unique luminescence properties of lanthanide ions (mainly Tb^{3+} and Eu^{3+}) may be used to great advantage as donor reagents because of their extremely long luminescence lifetime (on the order of milliseconds) and large Stokes shift (>150 nm). These unusual spectroscopic properties make it possible to measure sensitized emission with essentially no contaminating background. The extremely long lifetimes make them particularly useful for diffusion-enhanced FRET detection. Lanthanide donors have been used to measure distances between protein–DNA complexes[14] and long distances within myosin,[15] and to detect conformational changes in voltage-gated ion channels.[16]

Genetically Encoded Tagged Molecules

In vitro derivitization of proteins is flexible in that a variety of FRET pair combinations may be generated. However, because the labeling must be done *in vitro,* there arise concerns regarding the biological activity of the derivatized proteins and the need to introduce the protein into cells without perturbation of cellular function. This problem is exacerbated as the amount of exogenously labeled protein that can be introduced into cells is limited ($\leq$10% of total concentration of that protein in the cell), leading to low fluorescent signal. An alternative method for introducing labeled proteins into cells is to transfect the cells with DNA constructs containing the sequence of the protein of interest fused in frame to the sequence of donor

[11] S. Rice, A. W. Lin, D. Safer, C. L. Hart, N. Naber, B. O. Carragher, S. M. Cain, E. Pechatnikova, E. M. Wilson-Kubalek, M. Whittaker, E. Pate, R. Cooke, E. W. Taylor, R. A. Milligan, and R. D. Vale, *Nature* (*London*) **402,** 778 (1999).

[12] Y. K. Hu and J. H. Kaplan, *J. Biol. Chem.* **275,** 19185 (2000).

[13] D. A. Leonard and T. K. Kerppola, *Nature Struct. Biol.* **5,** 877 (1998).

[14] E. Heyduk, T. Heyduk, P. Claus, and J. R. Wisniewski, *J. Biol. Chem.* **272,** 19763 (1997).

[15] M. Xiao, H. Li, G. E. Snyder, R. Cooke, R. G. Yount, and P. R. Selvin, *Proc. Natl. Acad. Sci. U.S.A.* **95,** 15309 (1998).

[16] A. Cha, G. E. Snyder, P. R. Selvin, and F. Bezanilla, *Nature* (*London*) **402,** 809 (1999).

and acceptor fluorescent proteins. The family of green fluorescent protein (GFP) mutants has made this approach feasible as the chimeric protein is expressed in a normal cellular background and can be used to observe directly the fate and function of the protein. The constructs may contain a genetically encoded targeting sequence that will direct the expressed protein to specific subcellular compartments, thus enabling the observation of biochemical processes.[17] For example, genetically engineered proteins have been used to determine molecular interactions between putative apoptotic regulatory proteins.[18] GFP–FRET constructs have also been used in other studies of apoptosis that involve directly detecting the activity of caspases, the enzymes responsible for proteolysis of many proteins during apoptosis.[19]

Cells prepared for FRET measurements may be observed either in the living state or after fixation. Dynamic FRET measurements may be made on living cells; however, special care needs to be taken to prevent undue stress on the cells that might affect the processing being observed. To mimic normal growth conditions in the microscope, it should be outfitted with a heated stage to maintain the cells at 37°. Evaporation can be minimized by keeping coverslip chambers covered or by overlaying the medium with a thin layer of mineral oil (pharmacy grade). To maintain the cells in a 5% (v/v) CO_2 environment a small chamber, or isolette, can be constructed to mount over the cell. Heated, humidified 5% CO_2 gas is flowed through the chamber for the duration of the experiment. When a CO_2 source is not available the cells should be changed into a HEPES-buffered medium during the period of observation. Numerous designs of such cell viability chambers have been described and many are available commercially.

Steady state FRET measurements may be more easily made on fixed material that is mounted for microscope observation. Ideally, fixation should be rapid, giving little opportunity for perturbation to the physiological state at the time of fixation. Freshly made paraformaldehyde in a buffer that maintains osmotic balance and pH in the cells gives good preservation while introducing little background. After fixation the cells should be mounted on slides, using a mounting medium that contains an antioxidant to protect against photobleaching during microscopic observation (for protocol, see Table I).

Measurement of Fluorescence Resonance Energy Transfer

Steady State Fluorescence Resonance Energy Transfer Imaging

A range of microscope techniques may be used to detect FRET. Fluorescence intensity-based detection of FRET is achieved by measuring changes in the relative

[17] R. Rizzuto, M. Brini, P. Pizzo, M. Murgio, and T. Pozzan, *Curr. Biol.* **5,** 635 (1995).

[18] N. P. Mahajan, K. Linder, G. Berry, G. W. Gordon, R. Tsien, R. Heim, and B. Herman, *Nat. Biotechnol.* **16,** 547 (1998).

[19] N. P. Mahajan, D. C. Harrison-Shostak, J. Michaux, and B. Herman, *Chem. Biol.* **6,** 401 (1999).

TABLE I
PROTOCOL FOR FIXATION AND MOUNTING OF CELLS FOR FRET MEASUREMENTS

1. Medium is rinsed from transfected cells, using PHEM[a] or other appropriate buffer.
2. Cells are fixed for 20 min at room temperature with freshly prepared 4% (w/v) paraformaldehyde, pH 7.2–7.4, in PHEM buffer
3. After fixation, cells are rinsed twice with buffer
4. Cells are incubated with $NaBH_4$ (1 mg/ml), three changes for 5 min each
5. Cells are rinsed with buffer and mounted in Vectashield (Vector Laboratories, Burlingame, CA) or Mowiol (Polysciences, Warrington, PA) containing and *p*-phenylenediamine (1 mg/ml)
6. Excess mounting medium is removed and the coverslips are sealed with epoxy or nail hardener[b]

[a] PHEM: 60 m*M* PIPES, 25 m*M* HEPES, 10 m*M* EGTA, 2 m*M* $MgCl_2$.
[b] Sealant should not contain acetone or other denaturants that might denature the fluorescent protein, thus destroying its ability to fluoresce.

amounts of emission intensity of the donor molecule and the acceptor molecule. When FRET occurs, there is an increase in acceptor emission (I_A) and a concomitant decrease in donor emission (I_D). Following a condition that induces a change in the relative distance between two molecules the ratio of the donor and acceptor emissions changes (I_A/I_D). Thus, microscopic observation of FRET may be achieved by preferential excitation of a donor fluorophore and detection of the increase in sensitized emission of an acceptor fluorophore along with the reduction in donor fluorescence from quenching due to energy transfer.

One of the most common problems with intensity-based measurements of FRET is the difficulty of selecting appropriate donor and acceptor fluorophores that meet the criterion outlined above. In most cases overlap of donor emission and acceptor absorbance is not ideal. Some direct excitation of the acceptor usually occurs at the wavelengths employed to maximize donor absorbance and some emission of the donor occurs at the maximal emission of the acceptor. In addition, routine excitation, dichroic, and emission filters used for fluorescence detection are not 100% efficient. As a result, it is necessary to correct for direct excitation of the acceptor at the donor excitation wavelengths, emission of the donor at the acceptor emission maximum, and inappropriate signal and leakage through filter sets. Although numerous approaches have been developed whereby FRET may be detected with a single filter set, two filter sets, or three filter sets,[6,20,21] none of these methods fully separates FRET from non-FRET signals. In addition, they do not correct for leakage or cross-talk in the filter sets. As an example, Fig. 2 shows a montage of images collected for analysis by the three-filter set method described by Gordon *et al.*[6] These images demonstrate the need for extensive corrections to account for cross-talk.

[20] J. W. Erikson and R. A. Cerione, *Biochemistry* **30,** 7112 (1991).
[21] D. C. Youvan, W. J. Coleman, C. M. Silva, J. Petersen, E. J. Bylina, and M. M. Yang, *Biotechnology* **1,** 1 (1997).

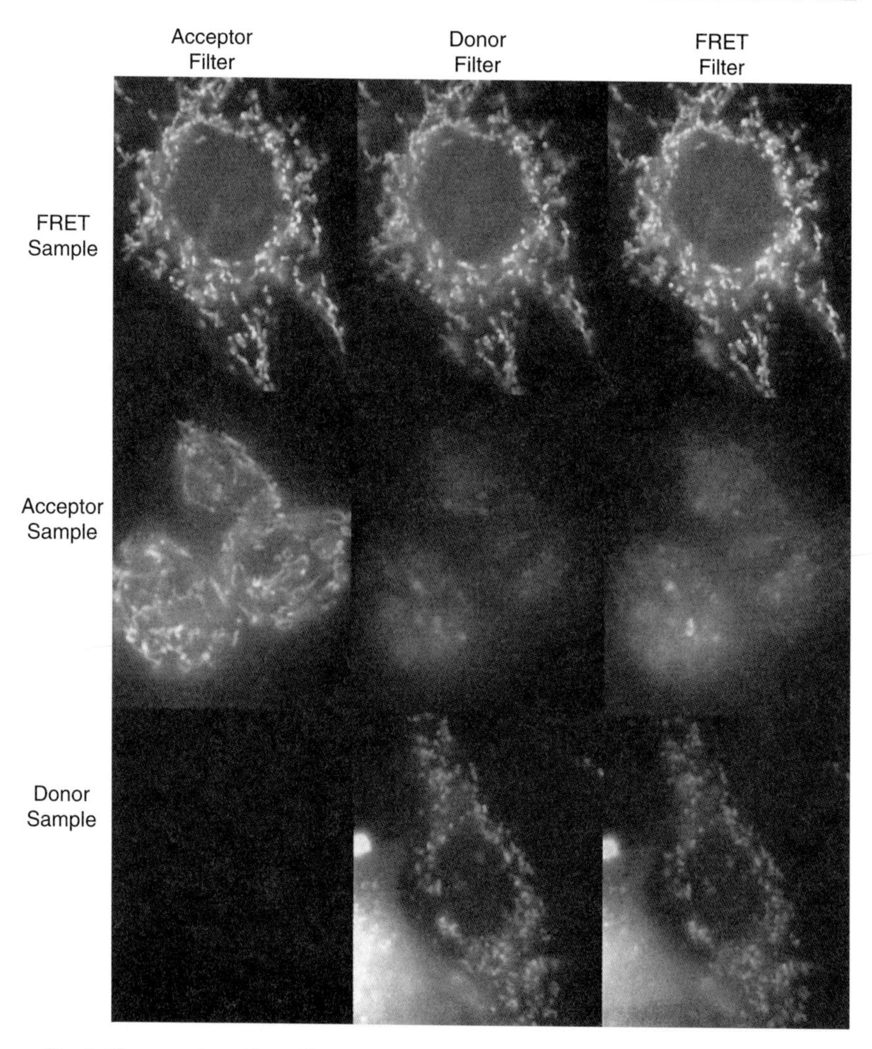

FIG. 2. Three-specimen/three-filter set data. These images depict the type of data acquired for analysis by the three-specimen/three-filter set method to measure FRET. *Top row:* A BHK-21 cell expressing a mitochondrially targeted construct that links CFP with YFP via an amino acid sequence encoding the caspase 2 recognition sequence. From left to right, the panels in the top row represent the signal observed through the acceptor filter set, the donor filter set, and the FRET filter set. *Middle row:* Cells expressing only mitochondrially targeted YFP (acceptor), viewed with the same filter sets and acquired under the same conditions for each filter set. *Bottom row:* A cell expressing only mitochondrially targeted CFP (donor) as viewed through the same filter sets and acquired under the same conditions for each filter set. Signal intensity values from images like those in the middle and bottom rows are used to correct for signal cross-talk between filters.

To address these deficiencies, a three-filter set/three-specimen method for calculation of normalized (corrected) FRET (FRETN) has been developed. This method corrects for filter set cross-talk and accounts for the effect of varying donor and acceptor concentrations on FRET efficiency.[6]

This three-filter set method is the most conservative and the most general approach to the calculation of FRET and can be implemented on any microscope or microfluorimeter. For each measurement of FRET, three samples must be prepared: one containing donor only (d), one containing acceptor only (a), and one that contains both donor and acceptor (f). Each of the samples is imaged with an acceptor filter set (A), a donor filter set (D), and a FRET filter set (F). The FRET filter set is configured to have an exciter and dichroic that are matched with the donor filter set and an emitter filter that is matched with the acceptor filter set. Filter sets used for the studies described here are from Chroma Technology (Brattleboro, VT) and are configured as follows:

CFP (donor set): exciter, 425/40×; dichroic, 460 nm; emitter, 495/20 nm
YFP (acceptor set): exciter, HQ525/10×; dichroic, 535 nm; emitter, HQ560/40
FRET: exciter, 425/40×; dichroic, 460 nm; emitter, HQ560/40

When preparing to acquire a set of data to be analyzed for FRETN it is essential to optimize the acquisition settings for each filter set. Acquisition settings for the filter sets are determined by imaging a representative area of the donor sample with the donor filter set. Exposure time, light intensity, and camera settings are adjusted to acquire images with good signal-to-noise ratio (S/N). It is reasonable to select settings that will utilize approximately 60–80% of the dynamic range to allow for sample variation. For example, using the Orca II 16-bit camera (Hamamatsu Photonics Systems, Bridgewater, NJ) acquisition parameters are adjusted to acquire images with maximum intensities of approximately 11,500 intensity units. This is repeated for the acceptor sample and acceptor filter set, and the FRET sample with the FRET filter set. Once the acquisition settings are determined for a filter set they must not be altered regardless of the sample being acquired. The acquisition is controlled by MetaMorph (Universal Imaging, Downingtown, PA) and the optimal parameters for each filter set are saved as individual digital camera settings. A journal has been written to capture, from a given sample, a stack of images, first using the A filter and configuration, followed by the D filter and configuration, and finally by the FRET filter and configuration. Although the fluorescence filters and neutral density filters may be changed manually, it would be far more time efficient to have motorized filter wheels that can be driven by the acquisition software such as is possible with MetaMorph.

For every specimen at least 10 fields are imaged with each of the three-filter sets. The stacks of images generated are then analyzed, using a custom-written journal for MetaMorph (courtesy of N. Glicksman, Universal Imaging). First, regions of local background in each image of a stack are measured to determine the average fluorescence contribution from this source. Next, regions of interest are marked. The analysis journal then measures the average intensity within each of the regions of interest in each of the planes of the stack. The value for local background in the corresponding plane is subtracted and the result is logged to a spreadsheet (Excel; Microsoft, Redmond, WA). These intensity measurements represent nine different categories: Aa (acceptor sample, acceptor filter), Ad (acceptor filter, donor sample), Af (acceptor filter, FRET sample), Da (donor filter, acceptor sample), Dd (donor filter, donor sample), Df (donor filter, FRET sample), Fa (FRET filter, acceptor sample), Fd (FRET filter, donor sample), and Ff (FRET filter, FRET sample). The relationship between these nine values is derived in Gordon *et al.*[6]

In the FRETN algorithm the donor, acceptor, and sensitized emission signals measured from a FRET sample are first expressed as the sum of signals contributed from the donor and acceptor fluorescence.

$$\mathrm{Df} = \mathrm{Dfd} + \mathrm{Dfa} \tag{6a}$$

$$\mathrm{Ff} = \mathrm{Ffd} + \mathrm{Ffa} \tag{6b}$$

$$\mathrm{Af} = \mathrm{Afd} + \mathrm{Afa} \tag{6c}$$

Each of the donor contributions (Dfd, Ffd, and Afd) is then expressed as the difference between the fluorescence that would have occurred if there were no FRET ($\overline{\mathrm{Dfd}}$) and the fluorescence loss due to FRET (FRET1). $\overline{\mathrm{Dfd}}$ is proportional to the total concentration of the donor-labeled species. FRET1 is proportional to the concentration of interacting (bound) donor-labeled and acceptor-labeled species. Each of the acceptor contributions (Dfa, Ffa, and Afa) is then expressed as the sum of the fluorescence that would have occurred if there were no FRET ($\overline{\mathrm{Afa}}$) and the fluorescence increase due to FRET. $\overline{\mathrm{Afa}}$ is proportional to the total concentration of the acceptor-labeled species. The next step is to express all three equations in terms of $\overline{\mathrm{Dfd}}$, $\overline{\mathrm{Afa}}$, and FRET1, with the result shown in Eqs. (6a)–(6c)

$$\mathrm{Df} = \overline{\mathrm{Dfd}} - \mathrm{FRET1} + \overline{\mathrm{Afa}}\frac{\mathrm{Da}}{\mathrm{Aa}} + G \cdot \mathrm{FRET1}\frac{\mathrm{Da}}{\mathrm{Fa}} \tag{7a}$$

$$\mathrm{Ff} = (\overline{\mathrm{Dfd}} - \mathrm{FRET1})\frac{\mathrm{Fd}}{\mathrm{Dd}} + \overline{\mathrm{Afa}}\frac{\mathrm{Fa}}{\mathrm{Aa}} + G \cdot \mathrm{FRET1} \tag{7b}$$

$$\mathrm{Af} = (\overline{\mathrm{Dfd}} - \mathrm{FRET1})\frac{\mathrm{Ad}}{\mathrm{Dd}} + \overline{\mathrm{Afa}} + G \cdot \mathrm{FRET1}\frac{\mathrm{Ad}}{\mathrm{Fd}} \tag{7c}$$

There are seven new quantities introduced in Eqs. (7a)–(7c). Six of the quantities represent the fluorescence of specimens containing either only donor or only acceptor as measured directly from the samples. The seventh is G, which is the factor relating the increase in acceptor signal due to FRET using the FRET filter set to the loss of donor signal due to FRET using the donor filter set. The ratios of these signals represent the amount of cross-talk measured through the fluorescence filter sets. Equations (7a)–(7c) have three unknowns: $\overline{\text{Dfd}}$, $\overline{\text{Afa}}$, and FRET1. Solving for the unknowns yields the values needed to calculate FRETN as defined by

$$\text{FRETN} = \frac{\text{FRET1}}{\overline{\text{Dfd}} \cdot \overline{\text{Afa}}} \propto \frac{[\text{bound}]}{[\text{total d}] \cdot [\text{total a}]} \tag{8}$$

where [total d] is the total concentration of donor-labeled molecules and [total a] is the total concentration of acceptor labeled molecules. These calculations have been incorporated into a simple Excel spreadsheet format that is available to interested parties. The analysis journal is designed to log average intensity values directly to the appropriate columns of this spreadsheet.

Earlier studies from this laboratory used sensitized emission to demonstrate the direct interaction between the antiapoptotic protein Bc1-2 and the proapoptotic protein Bax.[18] In the original constructs Bcl-2 was fused to blue fluorescent protein (BFP) and Bax was fused to GFP. Since that time improvements to the two-fusion FRET protocol have been implemented that include labeling with more photostable fluorescent proteins, CFP and YFP, generating constructs with commercially available vectors that contain multiple cloning sites, and using alternate transfection reagents resulting in more efficient transfection with less background fluorescence (Table II). Figure 3 summarizes the results of two-fusion FRET analysis, showing that by calculation of FRETN it can be determined that under normal conditions Bcl-2 can form homodimers with itself and can also form a heterodimer with Bax. Bax is also shown to form a loose homodimer that is minimally affected by the presence of Bcl-2. Each of these interactions shows a marked decrease on the

TABLE II

PROTOCOLS FOR GENERATION OF MUTANT GFP-TAGGED Bax AND Bcl-2[a]

1. Gene sequences for Bax and Bcl-2 are inserted in-frame into the multiple-cloning site (MCS) of the vectors pECFP-N1 and pEYFP-N1, respectively (Clontech, Palo Alto, CA)
2. Each vector is transformed into DH5α competent cells[b] to generate DNA for transfections
3. DNA is isolated from DH5α cells, using a Mini-Prep kit (Qiagen, Chatworth, CA)
4. DNA sequence for each vector is verified by an independent DNA sequencing laboratory. Glycerol stocks are made of transformed DH5α cells for future experiments
5. DNA constructs are preincubated with a transfection agent at room temperature [30 min in medium without serum LipofectAMINE (Boerhinger Manheim, Mannheim, Germany) or 15 min in normal medium FuGENE (Roche Applied Science, Penzberg, Germany)]. For two fusion FRET experiments, a mixture of DNA is prepared along with DNA from each of the individual constructs
6. Cells grown on coverslips or in coverslip chambers [MatTek (Ashland, MA) or Nunc (Roskilde, Denmark)] are transfected with DNA for 3 hr, after which they are rinsed and fresh medium is added to the cultures

[a] Fluorescent proteins may be observed in cells between 6 and 24 hr posttransfection, at which point the living cells are measured for FRET or the cells are fixed and mounted for future use.

[b] According to T. Maniatis, J. Sambrook, and E. F. Fritsch, "Molecular Cloning: A Laboratory Manual," 2nd Ed. Cold Spring Harbor Laboratory Press, Cold Spring Harbor, NY, 1989.

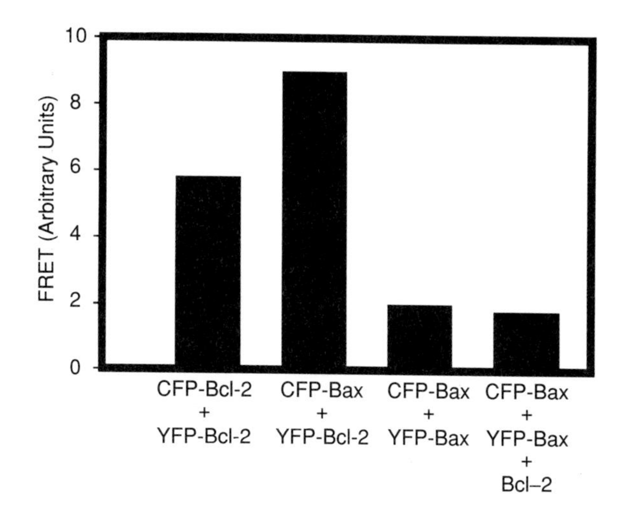

FIG. 3. Interaction between Bcl-2 and Bax, measured by the three-specimen/three-filter set (FRETN) method (see Ref. 6). FRET analysis shows that Bcl-2 can form homodimers as well as heterodimers with Bax. Bax shows some ability to form homodimers that are minimally disrupted by the presence of Bcl-2.

induction of apoptosis. The biological relevance of these findings is that regulation of the interactions between Bcl-2 and Bax family members with themselves and each other may serve as a mechanism for modulating the apoptotic process.

FRETN analysis has also been used to study the induction of active apoptotic enzymes, caspases, as measured by detecting changes in sensitized emission from a construct that contains a caspase-specific substrate site.[19] Fluorescent substrates for caspase activity are constructed by linking a donor and acceptor pair of fluorophores (CFP and YFP) via a four- to six-amino acid peptide caspase recognition sequence. These constructs may also be fused with a mitochondrial targeting sequence.[17] In the absence of caspase activity, the fluorophores remain linked and FRET is high. However, on induction of apoptosis caspase activity increases, the linker between the donor and acceptor is cleaved, the distance between the CFP and YFP increases and the FRETN decreases. Constructs containing the caspase-2 substrate, VDVAD, and the caspase-9 substrate, LEHD, have been used to detect enzymatic activity in mitochondria after induction of oxidative stress by exposure to 100 μM *tert*-butylhydroperoxide (*t*-BOOH) (Fig. 4). Time course studies indicate that caspase 2 is activated shortly after the induction of oxidative stress (within the first 2 hr of *t*-BOOH treatment) whereas caspase-9 activation lags behind capsase-2 activation and is maximal within 6 hr of treatment. The reduction in sensitized emission is not seen when cells are preincubated with zVAD.fmk, a broad-range caspase inhibitor (Fig. 4C). It is interesting to note that sensitized emission from the mCGY construct is significantly greater than for mC2Y. Because the linker in CGY is a noncleavable chain of glycines, it represents the maximum

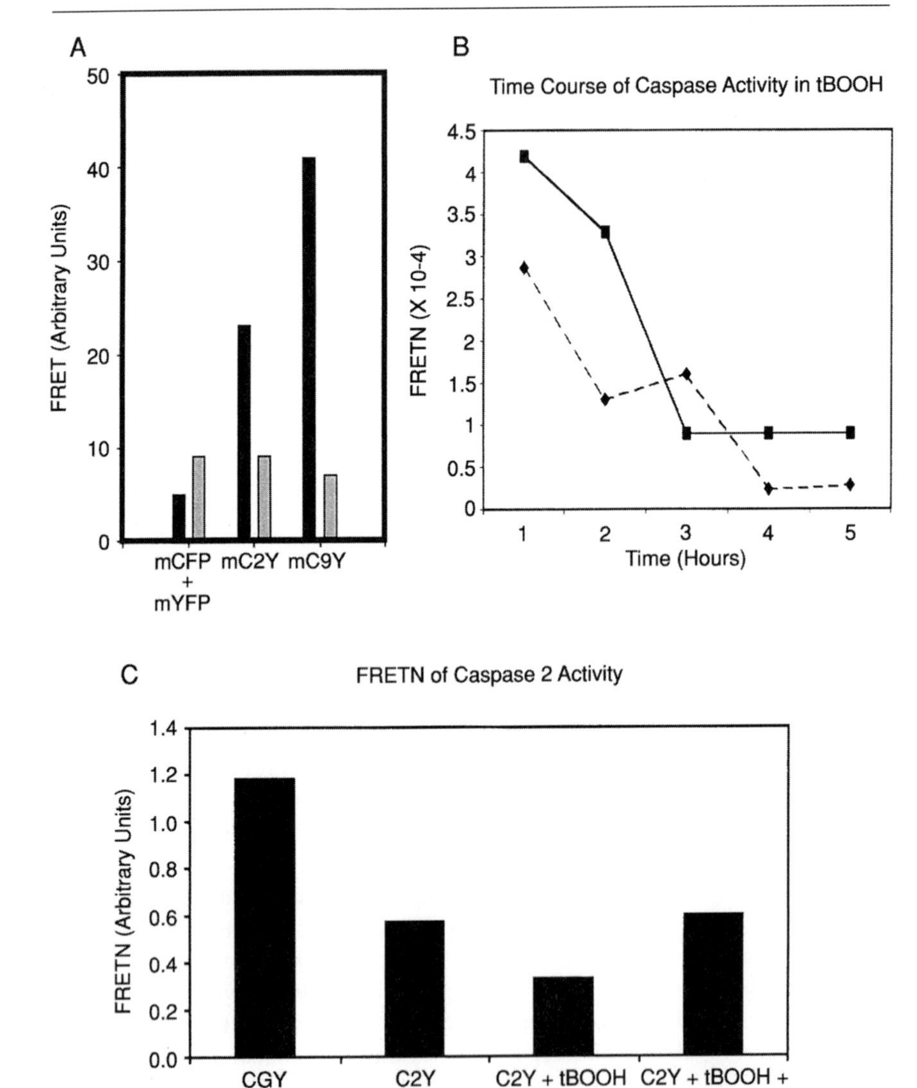

FIG. 4. Mitochondrial caspase activation following oxidative stress. (A) Mitochondrially targeted mC2Y and mC9Y show pronounced FRET signal (solid columns) above that seen with unlinked mCFP and mYFP cotransfected into mitochondria. After a 12-hr treatment with *t*-BOOH (100 μM) to induce oxidative stress, the amount of FRET is at control levels (shaded columns). (B) Time course data demonstrate that mC2Y (– – –) and mC9Y (—) FRET caspase substrates initially exhibit pronounced FRET signal. After 2 hr of exposure to 100 μM *t*-BOOH, mC2Y exhibits markedly reduced FRET signal, indicating that caspase 2 is activated and capable of cleaving the mitochondrially localized caspase substrate. At this same time point, mC9Y FRET is only slightly reduced, showing a significant

FRET signal possible in the biological system. The fact that C2Y shows lower FRETN may indicate an endogenous level of caspase activity in normal mitochondria.

Photobleaching Methods to Determine Fluorescence Resonance Energy Transfer

Donor Photobleaching. Selective photobleaching of the donor molecule (pbFRET) can also be used to measure FRET. The technique takes advantage of the sensitivity of a fluorophore to photodamage only when in its excited state.[22,23] Because of this, fluorophores having longer lifetimes are more susceptible to photobleaching. Because the process of resonance energy transfer decreases the lifetime of the excited state of the donor molecule it will result in a decreased rate of donor photobleaching relative to that seen for the donor in the absence of FRET. In some respects this method is less complicated than measurement of sensitized emission; however, it does require the fitting of time constants to photobleaching curves. These curves may contain multiple components,[24] thus adding a different form of complication to the measurement of FRET. The pbFRET method has been used to probe the structure and function of molecules in the plasma membrane.[25,26]

Acceptor Photobleaching. FRET can also be detected by observing the change in donor emission quenching both before and after selectively photobleaching the acceptor molecule.[7,27,28] The efficiency of energy transfer can be expressed as

$$E_{\mathrm{T}} = 1 - (I_{\mathrm{DA}}/I_{\mathrm{D}}) \tag{9}$$

where I_{DA} and I_{D} represent the steady state donor fluorescence intensity in the presence and absence of FRET. Using the same specimen donor fluorescence, intensity is measured in specific regions of interest. The acceptor molecule in those

[22] T. M. Jovin and D. J. Arndt-Jovin, *Annu. Rev. Biophys. Biophys. Chem.* **18,** 271 (1989).

[23] T. M. Jovin, D. J. Arndt-Jovin, G. Marriott, R. M. Clegg, M. Robert-Nicoud, and T. Schormann, *in* "Optical Microscopy for Biology" (B. Herman and K. Jacobson, eds.), p. 575. New York, 1990.

[24] L. Song, E. J. Hennink, I. T. Young, and H. J. Tanke, *Biophys. J.* **68,** 2588 (1995).

[25] S. Damjanovich, G. Vereb, A. Schaper, A. Jenei, J. Matko, J. P. Pascual Starink, G. Q. Fox, D. J. Arndt-Jovin, and T. M. Jovin, *Proc. Natl. Acad. Sci. U.S.A.* **92,** 1122 (1995).

[26] Z. Bacsó, L. Bene, A. Bodnár, J. Matkó, and S. Damjanovich, *Immunol. Lett.* **54,** 151 (1996).

[27] P. I. H. Bastiaens, I. V. Majoul, P. J. Verveer, H. D. Soling, and T. M. Jovin, *EMBO J.* **15,** 4246 (1996).

[28] F. S. Wouters, *EMBO J.* **17,** 7179 (1998).

decrease after 6 hr of exposure to *t*-BOOH, after which the FRET signal essentially levels off for the duration of the experiment. (C) Sensitized emission from CGY shows the maximum sensitized emission possible in mitochondria, as it consists of a noncleavable glycine linker between CFP and YFP. C2Y shows lower sensitized emission, presumably due to endogenous caspase activity. This level of FRETN is further lowered by treatment with *t*-BOOH to induce oxidative stress. The effects of *t*-BOOH on FRETN are counteracted by pretreatment of cells with zVAD.fmk, a broad-range caspase inhibitor.

regions of the specimen is then selectively photobleached with wavelengths of light at the absorption maximum of the acceptor. After photobleaching the emission intensity of the donor is measured in the same regions. Analysis of the change in donor intensity before and after photobleaching of the acceptor is performed on a pixel-by-pixel basis to determine the FRET efficiency. An advantage of this method is that it requires only a single sample and that the energy transfer efficiency can be directly related to both the donor fluorescence and the acceptor fluorescence.

We have used this approach to monitor mitochondrial caspase activity during oxidative stress. Samples prepared to study caspase-2 activity are analyzed by both the FRETN techniques previously described as well as by acceptor photobleaching to determine FRET efficiency. A 510 LSM confocal microscope (Zeiss, Thornwood, NY) used to perform acceptor photobleaching. Cells are imaged in the "multitracking" mode, to visualize fluorescent protein expression. To perform multitrack acquisition, a set of three "track" configurations must be preset. Each track consists of a specific LSM configuration that defines the excitation laser source and emission pathway to visualize a particular signal. To visualize emission from C2Y-containing cells the three tracks are analogous to acceptor filter sets [track 1—illumination with a 514-nm laser; emission passed through an NFT 515-nm dichroic and an HQ 550/50 bandpass filter (Chroma Technology, Brattleboro, VT) to photomultiplier (PMT) 1], a donor filter set [track 2—illumination with a 453-nm laser; emission reflected off of an NFT 515-nm dichroic and directed through an HQ 487/37 bandpass filter (Chroma Technology, Brattleboro, VT) to PMT 2], and a FRET filter set [track 3—illumination with a 453-nm laser; emission passed through an NFT 515-nm dichroic and an HQ 550/50 bandpass filter (Chroma Technology, Brattleboro, VT) to photomultiplier (PMT) 1]. Images from each track are displayed in separate image channels. Regions of interest (ROIs) are identified and bleaching parameters are defined. A "timed bleach" is then performed; one averaged image is acquired by multitracking, and then the acceptor probes in the predetermined ROIs are selectively photobleached, after which an averaged postbleaching image is acquired. The prebleach and postbleach images are contained in one image stack and fluorescence intensity can be measured in each ROI from each channel. Figure 5A shows the image of a cell expressing mC2Y as acquired by track 1. Figure 5B shows the same cell after photobleaching. The box indicates the region of laser irradiation. During the photobleaching process the acceptor signal is destroyed but the donor signal remains unaffected. A summary of FRET efficiency results using acceptor photobleaching is shown in Fig. 5C for cells expressing either mCGY or mC2Y under normal conditions and after treatment with *t*-BOOH to induce oxidative stress. Comparison of this graph with that in Fig. 4C shows that similar results are obtained by the FRETN approach as are obtained by acceptor photobleaching.

Acceptor photobleaching has been successfully used to determine FRET between a variety of biologically interactive molecules. This method of measuring FRET is simple and yields quantitative information. However, because it requires

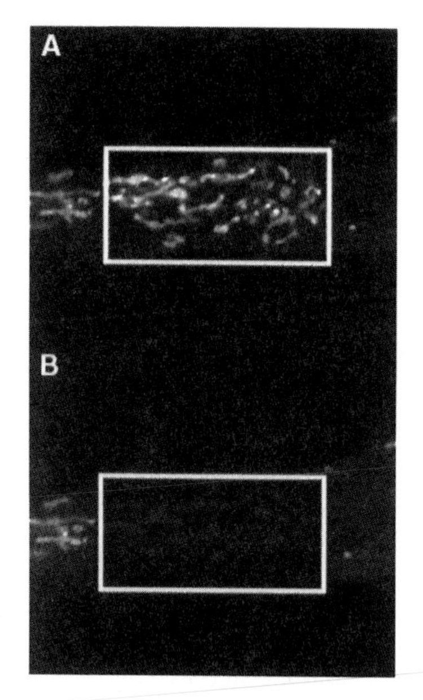

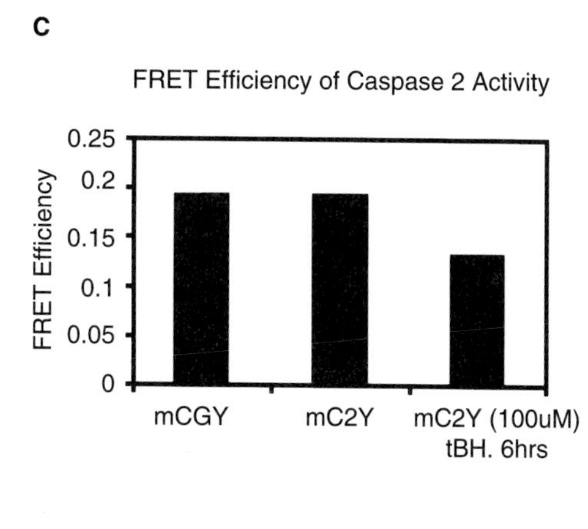

FIG. 5. Analysis of FRET efficiency by acceptor photobleaching. Cells expressing mC2Y were imaged in the multitracking mode with a Zeiss LSM 510 microscope (A). Regions of interest were identified and photobleached (see box). YFP signal, CFP signal, and FRET signal were recorded for each region of interest before, during, and after selectively photobleaching the acceptor probe. The ratio between CFP intensity before and after photobleaching was used to calculate FRET efficiency [see Eq. (9)]. (C) Summary of FRET efficiency for mCGY and mC2Y under normal conditions and during oxidative stress (100 μM t-BOOH treatment). These data are similar to the responses measured by FRETN analysis of the same samples (Fig. 4C).

the photodestruction of the acceptor molecule it is best applied to observations of fixed samples or for single time points in living cells. Other ratiometric methods are better suited to monitor FRET over time.

Lifetime

Time-resolved fluorescence spectroscopy is a well-established technique for studying the emission dynamics of a fluorescent molecule during its excited state, that is, the distribution of times between the electronic excitation of a fluorophore and the radiative decay of the electron from the excited stated producing an emitted photon. Lifetime measurements can yield information about the molecular microenvironment of a fluorescent molecule. Factors such as ionic strength, hydrophobicity, oxygen concentration, binding to macromolecules, and the proximity of molecules that can deplete the excited state by resonance energy transfer can all modify the lifetime of a fluorophore. Measurements of lifetime can therefore be used as indicators of these parameters. Furthermore, these measurements are generally absolute, being independent of the concentration of the fluorophore. This can have considerable practical advantages.

The development of fluorescence lifetime imaging microscopy (FLIM)[29–31] represents an exciting advancement. FLIM combines the advantages of lifetime spectroscopy with fluorescence microscopy by revealing the spatial distribution of a fluorescent molecule together with information about its microenvironment. Lifetime imaging systems have been developed using wide-field,[29] confocal,[32] and multiphoton[33,34] imaging modes.

There are two methods commonly employed to measure fluorescent lifetimes: frequency-domain or phase-resolved methods[31,35] and time-domain pulsed methods.[36] Frequency-domain (phase-resolved) lifetime measurements utilize sinusoidally modulated light as an excitation source and lifetimes are calculated from the phase shift and (de)modulation depth of the fluorescence emission signal. The extraction of complicated (multicomponent) lifetimes by phase modulation can require prolonged (i.e., excessive) exposure of the sample to damaging excitation energies, which may result in inadequate temporal resolution for biological processes. The development of global analysis approaches may obviate some of these concerns. Time-domain lifetime measurements employ pulsed excitation sources and the fluorescent lifetime is determined directly from the fluorescence signal or by photon-counting detection. This requires detection systems with sufficient temporal resolution to capture most of the emitted photons from each excitation pulse. The relative merits of each of these approaches will likely depend on the biological question and sample being examined.

Lifetime measurements have been shown to be a sensitive indicator of FRET.[37] Live-cell FRET measurements are more feasible using fluorescence lifetime imaging methods because lifetimes of fluorescent molecules are independent of concentration and light path length. Thus, a determination of FRET may be made using one measurement and may be undertaken without the need for photodestruction of either the donor or the acceptor molecule. FRET reduces the fluorescence lifetime of the donor molecule as energy is transferred to the acceptor from its excited

[29] J. R. Lakowicz, H. Szmacinski, K. Nowaczyk, K. W. Berndt, and M. Johnson, *Anal. Biochem.* **202,** 316 (1992).

[30] X. R. Wang, A. Periasamy, and B. Herman, *Crit. Rev. Anal. Chem.* **23,** 369 (1992).

[31] T. W. J. Gadella, T. M. Jovin, and R. M. Clegg, *Biophys. Chem.* **48,** 221 (1993).

[32] R. Sanders, A. Draaijer, H. C. Gerritsen, P. M. Houpt, and Y. K. Levine, *Anal. Biochem.* **227,** 302 (1995).

[33] D. W. Piston, B. R. Masters, and W. W. Webb, *J. Microsc.* **178,** 20 (1995).

[34] T. French, P. T. So, D. J. Weaver, Jr., T. Coelho-Sampaio, E. Gratton, E. W. Voss, Jr., and J. Carrero, *J. Microsc.* **185,** 339 (1997).

[35] J. R. Lakowicz and K. Berndt, *Rev. Sci. Instrun.* **62,** 1727 (1991).

[36] A. Periasamy, X. F. Wang, P. Wodnicki, G. W. Gordon, S. Kwon, P. A. Diliberto, and B. Herman, *J. Microsc. Soc. Am.* **1,** 13 (1995).

[37] T. W. Gadella, Jr. and T. M. Jovin, *J. Cell Biol.* **129,** 1543 (1995).

state.[38] Therefore, by comparing the donor fluorescence lifetime in the presence of the acceptor (τ_{DA}) with that of the donor in the absence of the acceptor (τ_D), it is possible to calculate the FRET efficiencies at each pixel,[39]

$$E_T = 1 - (\tau_{DA}/\tau_D) \tag{10}$$

Fluorescence lifetime measurements require the sample to be exposed to continued (sinusoidally modulated) excitation (phase modulation) of repetitive high-frequency excitation light pulses (time domain). For such measurements in living cells, this is appropriate so long as the act of imaging does not affect cell physiology. The reference measurement (donor alone) must be acquired under the same environmental and imaging conditions as the experimental measurement, although in the absence of the acceptor molecule. This measurement may be conveniently made after photobleaching the acceptor at the termination of an experiment.[40] One of the greatest advantages of measuring FRET by observing changes in fluorescence lifetime is that even FRET between pairs with similar emission spectra can be readily distinguished on the basis of lifetime measurement.[41]

Previously methods for measurement of fluorescence lifetimes have been limited to an estimate of a single lifetime at a given pixel of an image. This method cannot easily resolve within the same pixel a mixture of different molecular species or molecules in different biochemical states. Time-domain measurements fit a multiexponential function to the measurements in order to resolve multiple components. Frequency-domain methods must rely on multiple-frequency fluorescence lifetime imaging microscopy (mfFLIM[42]). Global analysis algorithms have been applied to FLIM and mfFLIM data[43] to improve the accuracy and precision of lifetime measurements. The use of global analysis is based on the assumption that some of the parameters are identical for each experimental manipulation, allowing for the analysis of multiple experiments simultaneously, using a biexponential model. Because all pixels are analyzed together it must also be assumed that the fluorescence lifetimes do not vary spatially. The result of such an analysis is a determination of fluorescence lifetimes and their fractional contributions to the steady state fluorescence. Whether these are valid assumptions is not clear.

Until more recently there have not been any commercially available instruments for the detection of fluorescence lifetimes. There are now at least two systems that can be purchased for use with scanning microscope systems. The TCSPC module from Becker & Heckle (Berlin, Germany) is a time-correlated

[38] R. M. Clegg, *in* "Fluorescence Imaging Spectroscopy and Microscopy" (X. F. Wang and B. Herman, eds.), p. 179. John Wiley & Sons, New York, 1996.

[39] T. Ng, A. Squire, G. Hansra, F. Bornancin, C. Prevostel, A. Hanby, W. Harris, D. Barnes, S. Schmidt, H. Mellor, P. I. Bastiaens, and P. J. Parker, *Science* **283,** 2085 (1999).

[40] F. S. Wouters and P. I. H. Bastiaens, *Curr. Biol.* **9,** 1127 (1999).

[41] A. Harpur, F. Wouters, and P. I. Bastiaens, *Nat. Biotechnol.* **19,** 167 (2001).

[42] A. Squire, P. J. Verveer, and P. I. H. Bastiaens, *J. Microsc.* **197,** 136 (2000).

[43] P. J. Verveer, A. Squire, and P. I. H. Bastiaens, *Biophys. J.* **78,** 2127 (2000).

single photon-counting module that receives single-photon pulses from the PMT of the microscope. By syncing to the scanning module (Frame Sync and Line Sync) to the pulsed excitation laser source, the module relates the time of detection of the emitted photon in relationship to the laser pulse and its point of origin within the scanned area. Over time, a distribution of the photon density as a function of x, y, and time during the fluorescence decay builds up in the memory of the board. The result can be interpreted as a sequence of fluorescence images obtained at different times after the excitation pulse.

Another commercially available FLIM module is the high-speed lifetime module LIMO (Nikon Europe, The Netherlands). This system, developed in the laboratory of H. Gerritsen (Utrecht University, Utrecht, The Netherlands), is based on time-gating electronics. A pulse of illumination causes the excitation of the fluorophores, and photons emitted from each excitation event are then counted by one of four to eight electronically time-gated accumulation windows. Photons are accumulated for a period of time at each pixel, up to 20 μs, to provide a representation of the exponential decay of the fluorophores in that pixel. Selection of an individual pixel or array of pixels will give the average lifetime. In our laboratories, preliminary experiments to measure fluorescence lifetime using the LIMO module have shown promising results. A scanning light microscope was used to scan a femtosecond-pulsed Ti:Saph laser across the specimen to excite CFP by two-photon absorption. The fluorescence emission was collected by a fast photomultiplier tube (R1894; Hamamatsu Photonics Systems) and the PMT signal was gated by the LIMO module. The average fluorescence lifetime detected in cells expressing mitochondrially targeted CFP was measured as 3.1 ns. However, in cells expressing mCGY the CFP emission shows a shortened lifetime ($\sim$2.6 ns), as is expected considering that these constructs exhibit significant FRET. Further experiments will be conducted to measure lifetimes in cells expressing FRET constructs under normal conditions and under oxidative stress. Because only a single measurement of donor lifetime is necessary to determine whether FRET is occurring, measurements performed in living cells can now be done more readily. In this way determination of molecular interactions can be made with high temporal resolution.

Of the various methods available for the measurement of FRET it appears that no one method is without disadvantages. Some methods are appropriate for live cell imaging whereas others are not. Some methods require significant corrective calculations or are based on assumptions of unknown validity whereas other methods require expensive equipment. It has yet to be determined which method is the most reliable to implement on living specimens to give the most accurate measurement. Continued research in this area is needed to determine optimal approaches for the measurement of FRET and the development of more refined algorithms for the analysis of data.

[24] Fluorescence Resonance Energy Transfer Imaging Microscopy and Fluorescence Polarization Imaging Microscopy

By YULING YAN and GERARD MARRIOTT

Introduction

Measurements of fluorescence resonance energy transfer (FRET) efficiency between a donor–acceptor pair can provide unique information about changes in molecular proximity and structural dynamics within a protein complex.[1,2] A particularly attractive feature of the FRET approach is that these proximity measurements can be conducted within large, functional molecular complexes under physiological conditions. For this reason FRET-based distance measurements are often used to validate molecular models of complex systems derived from X-ray diffraction studies.[2–5] In this article we show how the combination of FRET imaging microscopy (FRETIM; Heidecker *et al.*[2]) and fluorescence polarization imaging microscopy (FPIM; Kinosita *et al.*[6]) can be used to improve the precision of proximity measurements between specific loci within functional macromolecular assemblies. In particular, we describe how FPIM studies of single molecular complexes may be used to experimentally determine the value of the orientation factor (κ^2). Although we focus our attention on the actin filament, the methodologies described herein may be applied to any system that can be imaged at the level of a single macromolecular complex.

Considerations in Fluorescence Resonance Energy Transfer Measurements of Molecular Proximity

Labeling Donor and Acceptor Sites. FRET between a fluorescent donor (D) probe and a fluorescent or nonfluorescent acceptor (A) probe will occur if (1) the distance separating the donor and the acceptor is between 1 and 10 nm; (2) the emission spectrum of the donor overlaps with the absorption spectrum of the acceptor; and (3) the angle between the donor dipole moment and the

[1] J. R. Lakowicz, "Principles of Fluorescence Spectroscopy," 2nd Ed. Kluwer Academic Press, New York, 1999.

[2] M. Heidecker, Y. Yan-Marriott, and G. Marriott, *Biochemistry* **34,** 11017 (1995).

[3] K. C. Holmes, D. Popp, W. Gebhard, and W. Kabsch, *Nature* (*London*) **347,** 44 (1990).

[4] M. Miki, S. I. O'Donoghue, and C. G. Dos Remedios, *J. Muscle Res. Cell Motil.* **13,** 132 (1992).

[5] J. H. Gerson, E. Bobkova, E. Homsher, and E. Reisler, *J. Biol. Chem.* **274,** 17545 (1999).

[6] K. Kinosita, Jr., H. Itoh, S. Ishiwata, K. Hirano, T. Nishizaka, and T. Hayakawa, *J. Cell Biol.* **115,** 67 (1991).

0076-6879/03 $35.00

acceptor dipole moment is other than 90°.[7] These three conditions can easily be satisfied by using spectrally characterized donor and acceptor probe pairs and by defined labeling chemistry.[8–10] In practice, specific and nonperturbing labeling of proteins is achieved through the thiol group of a cysteine residue—in the case of actin the most reactive cysteine is Cys-374. Other loci on actin that have been labeled include Gln-42 (using fluorescent cadaverine reagents and transglutaminase; Gerson *et al.*[5]) and the nucleotide-binding site (using fluorescent analogs of ATP; Moens *et al.*[10]).

It is absolutely essential to ensure that every donor probe is within the Förster transfer distance of an acceptor probe if the FRET efficiency is measured under steady state conditions of excitation and emission. This is usually achieved through specific and stoichiometric labeling of the donor and acceptor sites. Multiple or substoichiometric labeling of donor and/or acceptor probes will result in either an overestimation or underestimation of FRET efficiency and molecular distance. In the case of actin, we usually find that Cys-374 is labeled to less than 70%. These conjugates are still useful in FRET-based distance determinations if the acceptor site can be stoichiometrically labeled, or if FRET efficiency is determined using fluorescence lifetime measurements.

κ^2 Problem. A key advantage of the FRET/FPIM approach to determine proximity is that we can measure absolute values of fluorescence polarization (FP) for probes on a single actin filament. The FP value yields information about the orientation of the donor or acceptor probe with respect to the filament axis and, together with the published structural relationships for the actin filament,[3,4,11] we can calculate the real (absolute) value of κ^2 for the donor–acceptor probe pair. This unique capability allows us to determine proximity relationships between donor–acceptor probes with higher precision compared with solution measurements that assume a random orientation of dipole moments for the donor and acceptor probes, that is, $\kappa^2 = 2/3$; the reader should refer to studies that seek to justify this assumption, for example, Dale and Eisinger,[12] Luo *et al.*,[13] and Dong *et al.*[14] However, it is important to recognize that this latter assumption can only place limits on the values that κ^2 may take around an average value of 2/3 and does not involve experimental determinations of the κ^2 value as is the case with our approach. Finally

[7] T. Förster, *Ann. Phys.* **90,** 21 (1948).

[8] D. L. Taylor, J. Riedler, J. A. Spudich, and L. Stryer, *J. Cell Biol.* **89,** 362 (1981).

[9] A. A. Kasprzrak, R. Takashi, and M. F. Morales, *Biochemistry* **27,** 4512 (1988).

[10] P. J. D. Moens, J. D. Yee, and C. G. Dos Remedios, *Biochemistry* **33,** 13102 (1994).

[11] A. Orlova, V. E. Galkin, M. S. VanLoock, E. Kim, A. Shvetsov, E. Reisler, and E. H. Egelman, *J. Mol. Biol.* **312,** 95 (2001).

[12] R. E. Dale and J. Eisinger, *Biopolymers* **13,** 1573 (1974).

[13] Y. Luo, J. L. Wu, J. Gergely, and T. Tao, *Biophys. J.* **74,** 3111 (1998).

[14] W.-J. Dong, J. Xing, M. Villain, M. Hellinger, J. M. Robinson, M. Chandra, J. D. Solaro, and H. C. Cheung, *J. Biol. Chem.* **274,** 31382 (1999).

microscope-based FP imaging, together with digital image processing, adds a new spatial dimension to analyze protein structure and dynamics. For example, changes in the FP value for a fixed and oriented probe on a single thin filament sliding on myosin may result from protein motions during motility[6,15] that may not be revealed from solution measurements.

Theory

Fluorescence Resonance Energy Transfer between Donor–Acceptor Pair

The FRET approach for determining molecular proximity has a long and validated history (reviewed by Clegg[16]). Here we present essential elements of the Förster theory[7] that describe the relationship between molecular distance and the rate of energy transfer or FRET efficiency for a fluorescent donor (D) and an acceptor (A) probe pair.[2,7,17]

Rate of Energy Transfer. The rate of energy transfer (k_T) between the donor–acceptor probe pair, derived by Förster,[7] is expressed as follows:

$$k_{\mathrm{T}} = \frac{\phi_{\mathrm{D}}\kappa^2}{\tau_{\mathrm{D}}R^6}\left[\frac{9000(\ln 10)}{128\pi^5 N n^4}\right]J(\lambda) \tag{1}$$

where ϕ_D is the quantum yield of the donor in the absence of acceptor; κ^2, the orientation factor, describes the relative orientation in space between the transition dipole moments of the donor and acceptor (Fig. 5); n is the refractive index of the medium, which is usually taken as that of water (1.33); N is Avogadro's number (6.023×10^{23}); τ_D is the lifetime of the donor in the absence of the acceptor; R is the distance between the donor probe and the acceptor probe; and $J(\lambda)$ is the degree of spectral overlap between the donor emission and the acceptor absorption, expressed by the overlap integral:

$$J(\lambda) = \int_0^{\infty} F_{\mathrm{D}}(\lambda)\varepsilon_{\mathrm{A}}(\lambda)\lambda^4 d\lambda \tag{2}$$

where $F_D(\lambda)$ is the corrected fluorescence intensity of the donor in the wavelength range λ to $\lambda + \Delta\lambda$, with the total intensity normalized to unity, and $\varepsilon_A(\lambda)$ is the molar extinction coefficient of the acceptor probe at the wavelength λ.

[15] K. Kinosita, Jr., N. Suzuki, S. Ishiwata, T. Nishizaka, H. Itoh, H. Hakozaki, G. Marriott, and H. Miyata, *Adv. Exp. Med. Biol.* **332,** 321 (1997).

[16] R. M. Clegg, *in* "Spectroscopy and Microscopy" (X. F. Wang and B. Herman, eds.), p. 179. John Wiley & Sons, New York, 1996.

[17] L. Stryer, *Annu. Rev. Biochem.* **4,** 819 (1978).

Because the rate of transfer k_T depends on the distance (R), we can rewrite Eq. (1) in terms of the Förster distance (R_0) as

$$k_T = \frac{1}{\tau_D}\left(\frac{R_0}{R}\right)^6 \tag{3}$$

Förster Distance. R_0 is the distance where the rate of energy transfer (k_T) equals the rate of the decay of the donor emission in the absence of acceptor ($1/\tau_D$), which can be written as [from Eq. (1)]

$$R_0^6 = \frac{9000(\ln\ 10)\phi_D\kappa^2}{128\pi^5 N n^4} J(\lambda) \tag{4}$$

and by assigning values to the physical constants and expressing wavelength in centimeters and the overlap integral $J(\lambda)$ in units of M^{-1} cm^3, a simplified form of R_0^6 (in units of cm^6) is obtained as follows:

$$R_0^6 = 8.79 \times 10^{-25}[\kappa^2 n^{-4}\phi_D J(\lambda)] \tag{5}$$

The value of R_0 can be determined once the values of $J(\lambda)$, ϕ_D, n, and κ^2 are entered. The κ^2 value is usually taken as 2/3 unless it can be measured directly by our FRETIM/FPIM approach as described below.[2]

Fluorescence Resonance Energy Transfer Efficiency. FRET efficiency (E_T) is related to the rate of transfer (k_T) by

$$E_T = \frac{k_T}{k_T + k_f + k_i} \tag{6}$$

where k_f is the rate of fluorescence emission of the donor, $k_f = 1/\tau_D$; and k_i includes the rates of other deactivating processes of the excited state of the donor. Because these processes are the same for FITC–ph (fluorescein–phalloidin; the donor probe in our studies)-labeled filaments and FITC–ph/TRITC–ph (tetramethylrhodamine–phalloidin; the acceptor probe)-colabeled filaments, they can be ignored in the calculation. Equation (6) is then simplified as

$$E_T = \frac{k_T}{\tau_D^{-1} + k_T} \tag{7}$$

In practice, the efficiency of FRET is determined by measuring the quantum yield of the donor, in the absence and presence of the acceptor, or by determining the fluorescence lifetime of the donor in the absence and presence of the acceptor (see the next section).

Methods

Proteins and Protein Labeling

Proteins. Actin and myosin are purified from rabbit leg muscle, using methods described in Heidecker *et al.*[2] The concentration of G-actin is calculated with an extinction coefficient (at 290 nm) of 0.63 mg/ml.[18]

Donor/Acceptor Probes. Fluorescent phalloidin derivatives serve as alternative donor and/or acceptor probes for FRET-based determinations of molecular proximity on actin.[2] FRET and FP measurements of fluorescent phalloidin on actin filaments offer several advantages over probes attached to specific amino acids on actin. These advantages include the following: (1) they have a single, high-affinity, saturable binding site; (2) they have no adverse effects on the structure of the actin monomer or actin filament on binding of the probe; (3) the phalloidin probes are closer to the helix axis compared with Cys-374-labeled probes[10]; and (4) phalloidin derivatives are used as functional probes for imaging actin filaments in living cells and in motility assays.[2,19]

Protein Labeling. Actin filaments are labeled with a stoichiometric amount of either FITC–ph or TRITC–ph by adding 1.5 equivalents of fluorescent phalloidin (from a 1-mg/ml methanol stock solution) over the actin monomer concentration (10 μM) in the filament. Filaments colabeled with the two fluorescent phalloidins are prepared by adding 0.75 equivalent of each phalloidin with respect to the G-actin concentration in the filament. Fluorescently labeled F-actin samples (10 μM) are left overnight (or longer) at 4° to ensure complete saturation of binding sites. Excess dye is removed by centrifuging the fluorescent filaments at 100,000*g* for 60 min at 4° and resuspending the pellet in AB buffer [25 m*M* imidazole, 25 m*M* KCl, 1.0 m*M* dithiothreitol (DTT), 4 m*M* $MgCl_2$, pH 7.4] or F buffer (5 m*M* Tris, 0.2 m*M* ATP, 0.2 m*M* $CaCl_2$, 1 m*M* DTT, pH 8.0). The resuspended pellet is analyzed the following day by absorption and fluorescence spectroscopy to determine the degree of labeling. Actin filaments prepared in this way and stored in the presence of 0.1% (w/v) sodium azide are stable for several weeks as monitored by their maximum sliding velocity (5 μm/sec) on a surface of heavy meromyosin (HMM).

Characterization of Fluorescently Labeled Actin Filaments

Absorption spectra of actin filaments labeled with stoichiometric amounts of FITC–ph or TRITC–ph and colabeled with FITC–ph/TRITC–ph are recorded with a standard absorption scanning spectrophotometer. FITC–ph and TRITC–ph

[18] D. J. Gordon, Y. Z. Yang, and E. D. Korn, *J. Biol. Chem.* **251,** 7474 (1976).
[19] S. J. Kron and J. A. Spudich, *Proc. Natl. Acad. Sci. U.S.A.* **83,** 6272 (1986).

(Fluka, Ronkonkoma, NY; Sigma, St. Louis, MD) have extinction coefficients of 72,000 and 85,000 $M^{-1}cm^{-1}$, respectively.[20] Fluorescence spectra and solution-based FP measurements on these filaments are conducted with an SLM-AB2 fluorescence spectrophotometer (SLM, Urbana, Il). It is important to correct steady state fluorescence excitation and emission spectra for the detection response of the instrument (as described in Marriott *et al.*[21]), using "magic angle" (54°/0°) excitation/emission. The steady state excitation polarization spectra are recorded in an L-format configuration that collects data in the four combinations of angles of excitation and emission polarizers (0°/0°, 0°/90°, 90°/0°, and 90°/90°) for 2 sec.

In Vitro Motility Assay

Single actin filaments decorated with FITC–ph or TRITC–ph are imaged in rigor or during ATP-dependent sliding on an HMM-covered surface in an *in vitro* motility assay,[22,23] using a hybrid protocol based on Kron *et al.*[22] and Kinosita *et al.*[6] In our version of this assay we use a flowthrough chamber consisting of a large (25 × 50 mm) and smaller (24 × 24 mm) coverslips separated by greased strips of Parafilm paper. The large coverslips are prepared by washing in 10% HCl and absolute ethanol and drying under an N_2 stream, and are stored or used after the following steps: sonifying in 0.1 *M* KOH for 30 min, washing in ethanol, and drying under nitrogen and then coating with a monolayer of nitrocellulose (Electron Microscopy Sciences, Fort Washington, PA) dissolved to 0.1% (v/v) isoamyl acetate. The chamber is flushed with protein solutions and buffers from one end and removed from the other end with a dry piece of filter paper as previously described by Kron *et al.*[22] We do not use bovine serum albumin (BSA) in our buffers because it may contain plasma gelsolin, which could shorten the average length of actin filaments. F-actin filaments are imaged within a few minutes of the final wash buffer to limit the dissociation of phalloidin molecules. The *in vitro* motility assay is performed according to Heidecker *et al.*[2] Basically, HMM at 50 μg/ml is laid on a coverslip coated with nitrocellulose for 90 sec in AB buffer. The chamber is washed with AB buffer (×2). Actin filaments labeled with TRITC–ph are added to the chamber at a concentration of 20 n*M*. After 90 sec the filaments are washed out with AB buffer containing glucose oxidase (GO)–catalase (CAT) and 5 m*M* DTT. Filament sliding is triggered by the addition of 1 m*M* ATP in AB/GO/CAT/DTT buffer.

[20] A. Waggoner, R. deBiasio, P. Conrad, G. R. Bright, L. Ernst, K. Ryan, M. Nederlof, and D. L. Taylor, *Methods Cell Biol.* **29,** 449 (1989).

[21] G. Marriott, K. Zechel, and T. M. Jovin, *Biochemistry* **27,** 6214 (1988).

[22] S. J. Kron, Y. Y. Toyoshima, T. Q. Uyeda, and J. A. Spudich, *Methods Enzymol.* **196,** 399 (1991).

[23] G. Marriott and M. Heidecker, *Biochemistry* **35,** 3170 (1996).

Experimental Determinations of Fluorescence Resonance Energy Transfer Efficiency

Solution-Based Measurements of Fluorescence Resonance Energy Transfer Efficiency between Donor and Acceptor on Actin Filaments

Considerations in Fluorescence Resonance Energy Transfer Measurements. Trivial causes of donor quenching that could affect the measurement of FRET efficiency are ruled out by ensuring that (1) the fluorescence intensity originating from FITC–ph-labeled actin is linear with the degree of saturation and so is independent of the degree of occupancy—this rules out fluorescein self-quenching; (2) trivial reabsorption of fluorescence (inner filter effect) does not occur—this can be ensured if the optical density of the solution in the 2-mm cuvette is below 0.2 unit at the maximum absorption wavelength of the probe. This can also be confirmed if the FP value of the labeled filaments, as a function of the optical density of the solution, is constant. In our system the steady state polarization spectra (at the emission wavelength of 580 nm) are found to be identical, for example, for solutions of TRITC–ph-labeled F-actin at 3.8 and 1.9 μM (Fig. 1A); and (3) FRET is restricted to fluorescent phalloidins within the same filament and is not a result of a non-Förster interfilament transfer—this is confirmed in our system by analyzing excitation polarization spectra of filaments labeled with either FITC–ph or TRITC–ph shortly after they have been gently mixed in the same cuvette. We find that the FP values of 4.0 μM FITC–ph-labeled F-actin and 3.8 μM TRITC–ph-labeled F-actin are equivalent before and shortly after being mixed (Fig. 1B, after 1 min of mixing). The FP of FITC–ph emission in this mixed actin filament preparation decreases with time because of an exchange of FITC–ph and TRITC–ph probes on the actin filaments (Fig. 1B, after 1 hr of mixing). This effect proves that hetero-FRET occurs between FITC–ph and TRITC–ph on the same filament.

Donor/Acceptor Labeling Ratio on Actin Filaments. The donor/acceptor (D/A) ratio is expressed in terms of the concentration ratio of actin monomer to fluorescent phalloidin and determined by absorption spectroscopy as described earlier. The labeling ratio for FITC–ph in colabeled filaments is determined after correcting for the contribution of TRITC–ph absorption at 492 nm.

Spectral Overlap Integral. The overlap integral between the FITC–ph (donor) emission spectrum and the TRITC–ph (acceptor) absorption spectrum, $J(\lambda)$, is $4.9 \times 10^{14}\ M^{-1}\ \text{cm}^{-1}\ \text{nm}^4$ at $\lambda = 552$ nm. This value is calculated from Eq. (2), using the corrected and normalized emission spectrum of FITC–ph-labeled filaments (Fig. 2, solid line). $\varepsilon(\lambda)$, the molar absorptivity of TRITC–ph-labeled filaments, as a function of wavelength, is based on a value of 85,000 $M^{-1}\ \text{cm}^{-1}$ at $\lambda = 552$ nm.[20] The κ^2 value is assumed to be 2/3 for a random dipole distribution in these solution studies and the refractive index (n) is taken as 1.4.[24]

[24] R. H. Fairclough and C. R. Cantor, *Methods Enzymol.* **XLVIII,** 347 (1978).

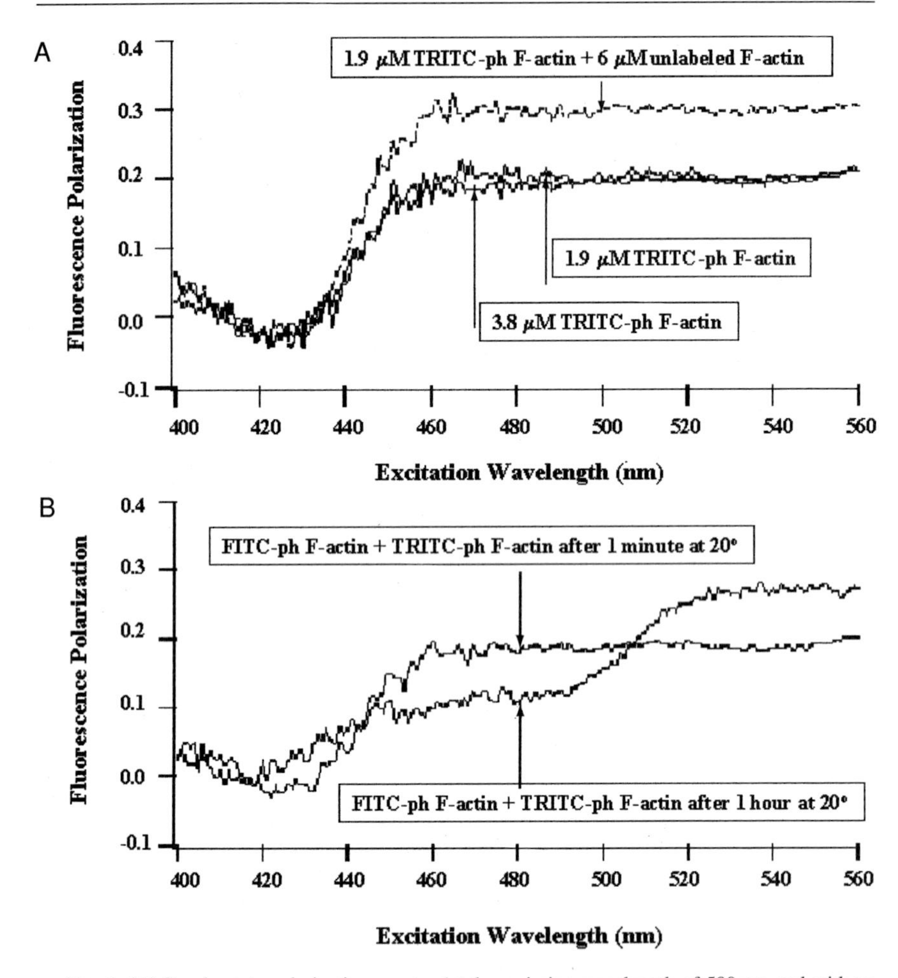

FIG. 1. (A) Steady state polarization spectra (at the emission wavelength of 580 nm and with an excitation bandpass of 2 nm) of solutions of TRITC–ph-labeled F-actin at 1.9 μM (solid line), 3.8 μM (dotted line), and 1.9 μM in the presence of 6 μM unlabeled F-actin (dash-dotted line). (B) Excitation polarization spectra of 4.0 μM FITC–ph-labeled F-actin and 3.8 μM TRITC–ph-labeled F-actin after 1 min of mixing and after 1 hr of mixing.

CALCULATING FLUORESCENCE RESONANCE ENERGY TRANSFER EFFICIENCY AND MOLECULAR PROXIMITY BASED ON QUANTUM YIELD MEASUREMENTS. One method to calculate the FRET efficiency is to measure the decrease in the relative quantum yield of the donor probe (FITC–ph) on filaments colabeled with acceptor (TRITC–ph) compared with filaments labeled with donor alone as described above. In practice, we measure the integrated intensity under the corrected

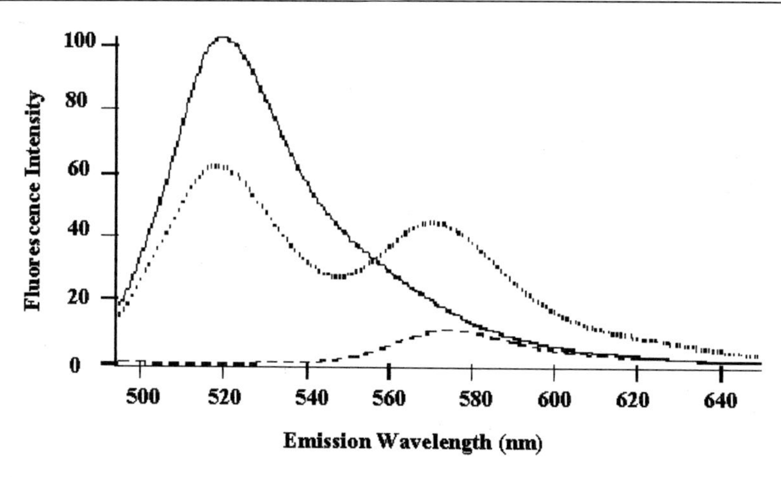

FIG. 2. Steady state, corrected fluorescence emission spectrum of FITC–ph-labeled F-actin, normalized to 4.4 μM (solid line) and steady state corrected fluorescence emission spectrum of 4.4 μM FITC–ph/4.5 μM TRITC–ph-labeled F-actin (dotted line) corrected for indirect excitation of TRITC–ph (dashed line). Excitation wavelength was at 488 nm with a 2-nm bandpass.

emission spectrum of FITC–ph on actin filaments labeled with FITC–ph alone (donor; Fig. 2, solid line) and colabeled (1 : 1) with FITC–ph/TRITC–ph (Fig. 2, dotted line; D–A). The FRET efficiency is computed as follows:

$$E_{\mathrm{T}} = 1 - \frac{\Phi_{\mathrm{DA}}}{\Phi_{\mathrm{D}}} \tag{8}$$

where Φ_{D} and Φ_{DA} are the quantum yields of the donor in the absence and presence of the acceptor, respectively. The absolute quantum yield of FITC–ph in donor-only filaments, $\Phi_{\mathrm{D}} = 0.37$, is determined through comparisons of the integrated emission spectra of a solution of FITC–ph-labeled filaments with the integrated emission spectrum of exactly the same concentration of disodium fluorescein in 0.01 M NaOH, which has a quantum yield of 0.91.[24]

The calculated value of $E_{\mathrm{T}} = 0.408$ between FITC–ph and TRITC–ph on actin filaments is based on the assumption that each FITC–ph in filaments is labeled with 1 : 1 ratio of FITC–ph. However, a statistical evaluation of the random labeling of phalloidin sites reveals that only three of four FITC–ph molecules satisfy this condition, which leads to a corrected value of 0.51 for E_{T}. When all of the constant and measured parameters are introduced into the Förster equation [Eq. (1)], we obtain a distance of 3.7 nm between adjacent fluorescent phalloidin probes on actin.

CALCULATING FLUORESCENCE RESONANCE ENERGY TRANSFER EFFICIENCY AND MOLECULAR PROXIMITY BASED ON TIME-RESOLVED MEASUREMENTS. A second method to calculate the FRET efficiency is to measure the decrease in lifetime of the donor probe in the presence of the acceptor. Time-resolved fluorescence intensity decay measurements of filament preparations labeled with donor only and donor–acceptor fluorescent phalloidins are made in solution, using magic angle excitation of FITC–ph with 450-nm light pulses with a full-width half-maximum (FWHM) of 2 ps (argon-ion/Tsunami laser; Spectra-Physics, Mountain View, CA) operating at 4 MHz. The fluorescence is collected with a monochromotor set at an emission wavelength of either 530 or 600 nm. The FRET efficiency is computed by using the relationship

$$E_{\mathrm{T}} = 1 - \frac{\tau_{\mathrm{DA}}}{\tau_{\mathrm{D}}} \tag{9}$$

where τ_{D} and τ_{DA} are the fluorescence lifetimes of the donor in the absence and presence of the acceptor, respectively.

Decay rate (or lifetime) data from donor only- and donor–acceptor-labeled filaments are analyzed by the IGOR data analysis program (Wavemetrics, Lake Oswego, OR). The FRET efficiency is calculated to be $E_{\mathrm{T}} = 0.52$ [Eq. (9)], using an average lifetime of FITC–ph, τ_{D}, of 1.44 ns (i.e., a decay rate of $k_{\mathrm{f}} = 6.94 \times 10^{8}\ \mathrm{s}^{-1}$) and an average lifetime of FITC–ph in the presence of TRITC–ph, τ_{DA}, of 0.69 ns; corresponding to a transfer rate of $k_{\mathrm{T}} = 7.5 \times 10^{8}\ \mathrm{s}^{-1}$ [Eq. (7)]. Workup of the Förster equation [Eq. (1)] gives a calculated distance of 3.7 nm between adjacent phalloidin probes on actin, which is in good agreement with that obtained by quantum yield measurement (3.7 nm).

Fluorescence Resonance Energy Transfer Imaging Microscopy

Microscope Workstation. The microscope workstation used to record FRET on single actin filaments is schematized in Fig. 3. Excitation light is provided by a mercury arc lamp and the excitation wavelength is selected with an optimized filter set (Keio filter set; Chromatech, Brattleboro, VT). This set consists of an interference filter (450–480 nm), a dichroic mirror that reflects the excitation at 450–480 nm, and an emission interference filter with transmittance between 515 and 565 nm (donor emission). A quartz fluar objective [×100, 1.3 NA (numerical aperture); Zeiss, Thornwood, NY] is employed to excite and collect fluorescence from single actin filaments.

Determining Fluorescence Resonance Energy Transfer Efficiency from Images of Single Actin Filaments. One often-cited drawback of the FRETIM approach is the difficulty in comparing quantum yields of the donor probe in the absence and presence of acceptor in the same image field. We have solved this problem by using a motility assay of actomyosin contraction in which hundreds of individual actin filaments labeled with either donor probes, acceptor probes, or 1 : 1 mixture of donor–acceptor probes are imaged in a single image field. Measuring FRET efficiency for each of these colabeled filaments allows us to calculate proximity relationships between phalloidin sites even when these filaments are sliding at 5 μm/sec in the presence of 1 m*M* ATP.

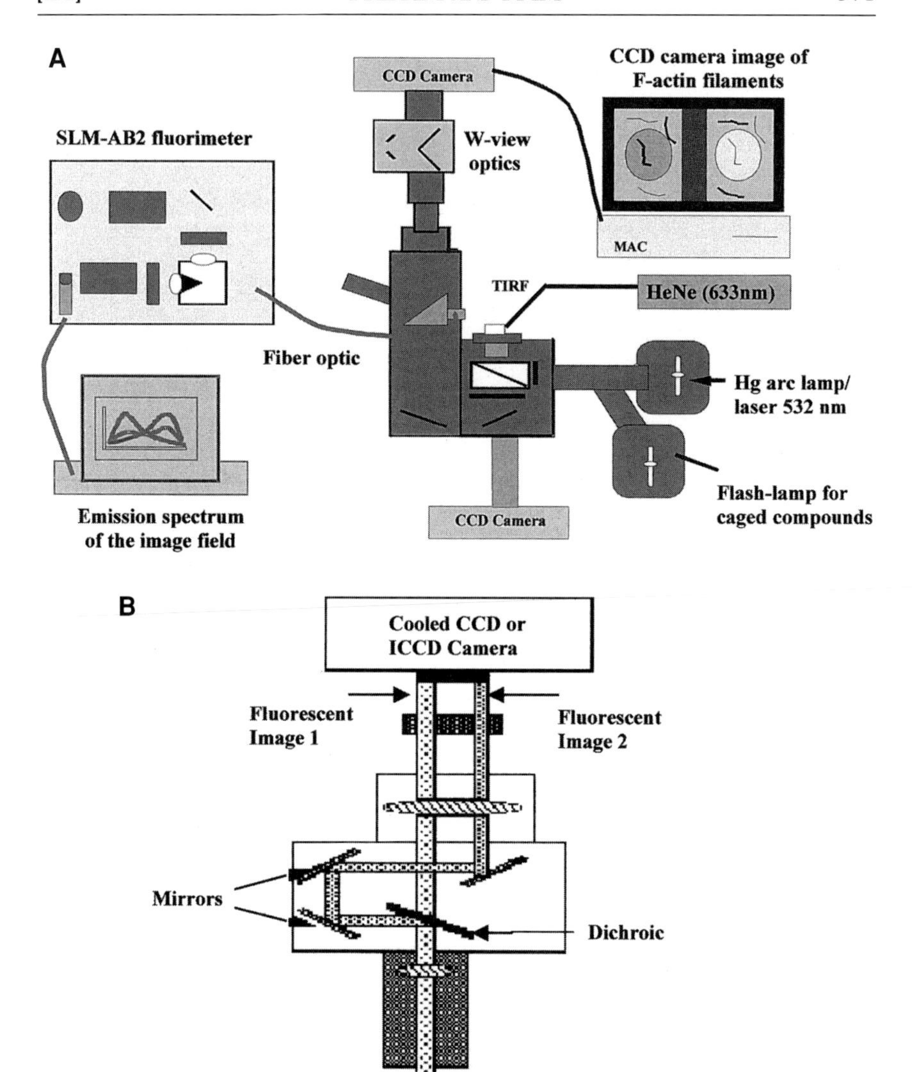

FIG. 3. (A) Schematic representation of the FRET and FP imaging microscope used to record accurate measurements of FRET and FP on single actin filaments. (B) Splitting off the fluorescence emission to a fluorometer via a fiber optic can be used to record the fluorescence spectra. W-view imaging is achieved by separating the TRITC and FITC emissions, using a pair of dichroic mirrors.

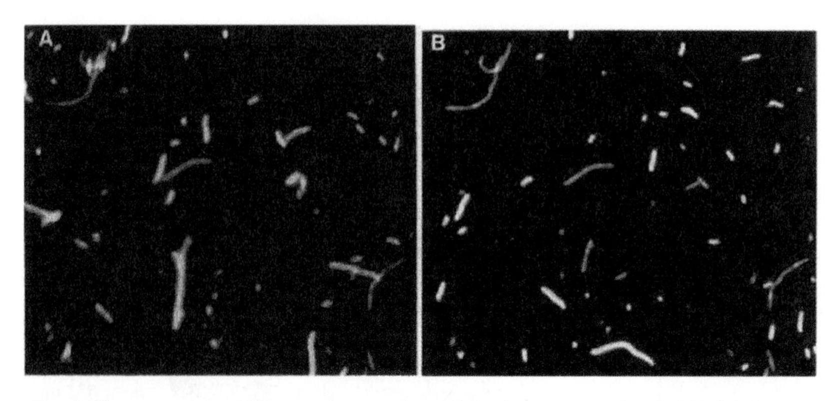

FIG. 4. Fluorescence intensities on donor only- and donor–acceptor-colabeled filaments are used to calculate FRET efficiency. (A) Donor only- and donor–acceptor-colabeled filaments are identified in the image field, using 450- to 480-nm excitation and collecting emission between 515 and 565 nm. (B) Donor–acceptor-colabeled and acceptor only-labeled actin filaments are identified in the image field, using 546-nm excitation and collecting the emission beyond 580 nm.

Here, we describe three different methods for measuring the quantum yield of donor-only and donor–acceptor probes on single actin filaments.

1. The first method quantifies the decrease in quantum yield of the donor (FITC–ph) emission in the presence of the acceptor (TRITC–ph) within single filaments in an image field by exciting the donor probes between 450 and 480 nm and collecting images of the donor emission between 515 and 565 nm. These images identify two types of filaments labeled with either donor only or donor and acceptor. A second image recorded with 546-nm excitation and a filter collecting the acceptor emission (>580 nm) enables us to unequivocally identify the colabeled filaments (Fig. 4B). The FRET efficiency is calculated for filaments in images obtained on the basis of donor excitation and donor emission (Fig. 4A), using the raw intensity value (12-bit camera using Image J software; Scion, Rockville, MD). In practice the fluorescence intensities from closely situated, single actin filaments that are labeled with either 100% donor probe or 50% donor–50% acceptor probes are measured along the entire length of each filament. This is converted into an average intensity per unit length (typically 1 μm). Similarly, a background signal is measured in the immediate vicinity of these filaments and subtracted from the fluorescence of the corresponding filament (Fig. 4A). This operation is repeated for multiple donor only- and donor–acceptor-labeled filaments within the image field.

The intensity of the D–A pair is doubled to account for the fact that donor-only filaments have twice as many donor probes as the colabeled filaments. A value of $E_T = 0.45$, obtained from the image in Fig. 4A, is corrected to account for a statistical factor of 1.25 for the population of FITC–ph on the filament not having an adjacent TRITC–ph acceptor—the average E_T value for six different filaments in Fig. 4A is 0.55 ± 0.02 (or $k_T = 8.55 \times 10^8$ s^{-1}).

2. In the second method the donor probes on single filaments are excited with 450- to 480-nm light and images of the donor and acceptor emissions in the same image field are recorded simultaneously through a double-view (W-view) device (Kinosita *et al.*[6]; Fig. 3B). The acceptor emission (sensitized emission) image provides qualitative proof that FRET is occurring in the system. The W-view adapter separates the contributions from the donor and acceptor emissions via a pair of dichroic beam splitters, which reflect light between 515 and 565 nm (donor emission) and transmit light beyond 580 nm (acceptor emission) (see Fig. 4). The adjustable mirrors redirect the donor emission onto a second dichroic that is reflected to the left-hand side of the charge-coupled device (CCD) camera [or a Hamamatsu Photonics (Bridgewater, NJ) intensified CCD (ICCD) camera]. The two fluorescent images are optically separated, using an adjustable slit device placed at an intermediate image plane in a Zeiss Axiovert 35 microscope (the slot for the 35-mm camera cross-hair adaptor). Quantitative analyses of FRET efficiency from the donor excitation/emission image are conducted as described above.

3. In the third method, the efficiency of FRET between donor and acceptor probes on single actin filaments is measured by microspectrofluorimetry. Here the emissions from an image field containing single filaments that are labeled with either donor probes only or donor–acceptor probes are directed via a quartz fiber optic cable onto the emission monochromotor of the SLM-AB2 spectrofluorimeter. The emission spectra from the two preparations are recorded and the FRET efficiency is calculated from these spectra as described for the solution measurement.

Quantitative Analyses of Molecular Proximity from Image Data. Assuming that the values of the quantum yield for FITC–ph on actin filaments and other parameters including refractive index, orientation factor, and fluorescent lifetime are unchanged in the motility assay (Fig. 4A), we calculated a distance of 3.6 nm between adjacent phalloidin-binding sites. The difference in this distance compared with that for filaments in solution (3.7 nm) could result from structural changes in the filament induced by myosin binding to actin.

Fluorescence Polarization Imaging Microscopy Based Experimental Determination of Orientation Factor κ^2 for Donor–Acceptor Probes on Single Thin Filaments

FPIM can be used to experimentally determine the κ^2 value, provided the fluorescent probes are aligned and relatively immobile on a single molecular assembly such as a single actin filament.

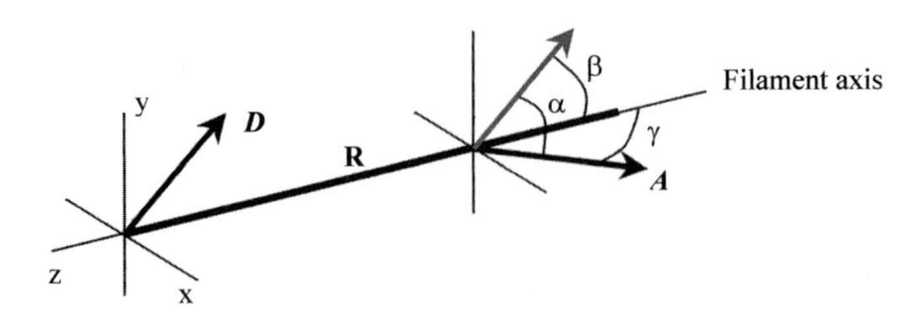

FIG. 5. Geometric relationship between donor and acceptor probes on an actin filament.

Orientation Factor κ^2. κ^2 is related to the orientations of the dipole moments of the donor and acceptor by Eq. (10):

$$\kappa^2 = (\cos\alpha - 3\cos\beta\cos\gamma)^2 \tag{10}$$

where β and γ are the angles the donor and acceptor dipole moments make with the filament axis, respectively; and α is the angle between the donor and the acceptor dipole moments (Fig. 5).

The value chosen for κ^2 can greatly affect the precision of the distance determination.[1] Unfortunately, the orientation of neither the donor nor the acceptor probe is known, although when the probes exhibit a large degree of rotational freedom during their excited state lifetime it might be justifiable to assume a random orientation factor of $\kappa^2 = 2/3$. Otherwise this assumption leads to a worst-case error of about 35%.[1] Dale and Eisinger[12] have shown how limits on the κ^2 value can be established if the steady state anisotropy values for the donor and acceptor probes are known.[13,25] The donor and acceptor probes on actin described in this article have high FP values (>0.35) and are aligned on the filament; so the assumption that κ^2 has a value of 2/3 may not be correct.

Dealing with κ^2 *Problem.* The angles β and γ (Fig. 5) can be determined experimentally from absolute measurements of FP on single filaments oriented in a parallel or perpendicular direction (Fig. 8, p. 578). The high steady state FP value of the conjugates suggests that the angle describing the distribution of the donor and acceptor dipoles on each probe is small. Consequently, the contribution of this angle can be ignored when calculating the inclination angles to the filament axis (β and γ). In the case of single filaments labeled with FITC–ph and TRITC–ph, the corrected FP values are similar and exceed the Perrin limit for randomly distributed fluorophores of 0.5 unit, indicating that these filaments are aligned along the filament. The fluorescent phalloidin D–A probes occupy equivalent sites on the filament and therefore the angles β and γ are identical (30°; Kinosita *et al.*[6]). The angle α can be calculated from the molecular model of the actin filament[3] as shown by Heidecker *et al.*[2] α, β, and γ are used to calculate the exact value of κ^2, using

[25] M. She, J. Xing, W. J. Dong, P. K. Umeda, and H. C. Cheung, *J. Mol. Biol.* **281,** 445 (1998).

Eq. (10) (see also Kinosita *et al.*[6] and Heidecker *et al.*[2]). For FITC–ph/TRITC–ph bound to single actin filaments, κ^2 is 1.42.[2] Because the FPIM technique, together with the FRETIM technique, yields absolute measurements of FP, κ^2, and FRET efficiency for thousands of spatially resolved filaments within a single preparation, the precision of the calculated distance can be superior to that of values measured in a fluorometer. However, the FP values determined with the microscope must be corrected for instrumentation artifacts.

Considerations in FPIM

Instrument Correction Factor. A correction factor is necessary in microscope-based FP measurements because transmission of the parallel and perpendicular components of the emission through the microscope are not identical. To eliminate this source of artifact, a reference fluorophore with a low polarization value is required. We use ethidium bromide dissolved in ethanol (10 μg/ml, with a decay time of 23 ns), having a steady state FP value of 0.008 ± 0.002, to determine the instrument correction factor. The corrected polarization value is calculated as follows:

$$p = \frac{I_z/I_y - K}{I_z/I_y + K} \quad (11)$$

where K is the correction factor.

High Numerical Aperture Effect. Because a high-powered numerical aperture objective (NA = 1.3) is used in these studies, the FP values are reduced because of the well-known numerical aperture effect.[26] We analyze this depolarization factor by comparing FP values for a solution of TRITC as a function of viscosity, using a high-NA microscope and a low-NA fluorometer. Figure 6 shows the high-NA effect: Fig. 6A presents threewise FP measurements for a 0.1 μM solution of TRITC in water–glycerol mixtures performed over a range of viscosity (1–1000 cP) under low-NA excitation/emission (<0.5 NA; SLM-AB2) in an FPIM workstation (shown in Fig. 3) using a 1.3-NA objective with polarized excitation and natural light excitation. Figure 6B presents FP measurements of a 0.1 μM solution of TRITC in 95% glycerol as a function of NA. The FP value drops off with higher NA. The solid line in Fig. 6B represents the expected (theoretical) FP value when considering only an emission cone effect. The experimental data show that the FP value is also diminished by the excitation cone effect. It can be seen from Fig. 6 that the high NA decreases the value of FP by up to 32%.

We have found some advantages in using natural light excitation for microscope-based FP measurements of TRITC–ph on single actin filaments. Now, as might be expected, the FP value for rhodamine solutions measured in a microscope using natural light excitation is zero (Fig. 6A). However, using natural light excitation TRITC–ph probes aligned along single actin filaments have

[26] D. Axelrod, *Biophys. J.* **26,** 557 (1979).

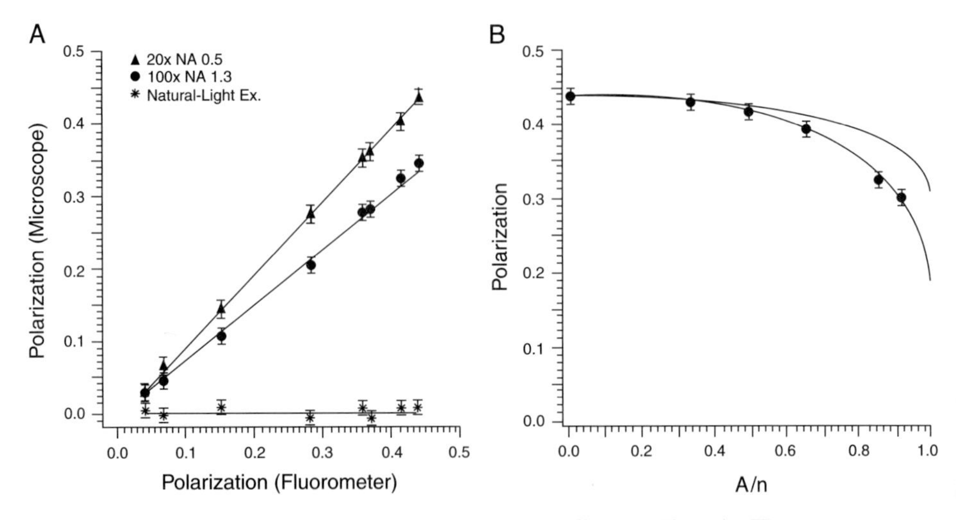

FIG. 6. Correcting for the high-numerical aperture (NA) effect. (A) Threewise FP measurements for a 0.1 μM solution of TRITC in water–glycerol mixtures were performed over a range of viscosity (1–1000 cP) under low-NA excitation/emission (<0.5 NA; SLM-AB2; ▲) in an FPIM workstation (shown in Fig. 3) using a 1.3-NA objective with polarized excitation (●) and natural light excitation (∗). (B) FP measurements of a 0.1 μM solution of TRITC in 95% glycerol as a function of NA (●). The solid line represents the theoretical FP value when considering only the emission cone effect.

nonzero values of FP that may exceed the Perrin limits for solution studies (+0.5 and −0.33; Fig. 8, p. 578). This result is expected because TRITC–ph binds to a unique site on the filament and because we image the fluorescence emission from a single filament. Actin filaments saturated with TRITC–ph (or FITC–ph) exhibit substantial homo-FRET as expected from the Förster distance for the FITC–FITC pair (4.0 nm[27]; also see [1], [6], and [25] in this volume[28–30]). Therefore it is important to consider homo-FRET and/or local motion of the probe, as sources of the depolarization, when measuring FP values. Substoichiometric labeling will eliminate this artifact.

Calculating Fluorescence Polarization and Fluorescence Resonance Energy Transfer Efficiency from Image Pairs

Pixel-by-Pixel Calculation of Fluorescence Polarization. Images of the polarized components (parallel and perpendicular) of the fluorescence emission must be highly registered before calculating FP values on a pixel-by-pixel basis. We

[27] L. Erejman and G. Weber, *Biochemistry* **30,** 1595 (1991).
[28] D. M. Jameson, J. C. Croney, and P. D. J. Moens, *Methods Enzymol.* **360,** [1], 2003 (this volume).
[29] V. Subramaniam, Q. S. Hanley, A. H. A. Clayton, and T. M. Jovin, *Methods Enzymol.* **360,** [6], 2003 (this volume).
[30] M. Tramier, T. Piolot, I. Gautier, V. Mignotte, J. Coppey, K. Kemnitz, C. Durieux, and M. Coppey-Moisan, *Methods Enzymol.* **360,** [25], 2003 (this volume).

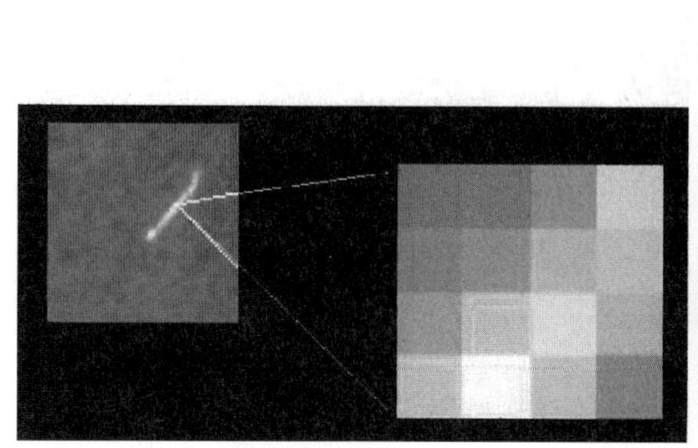

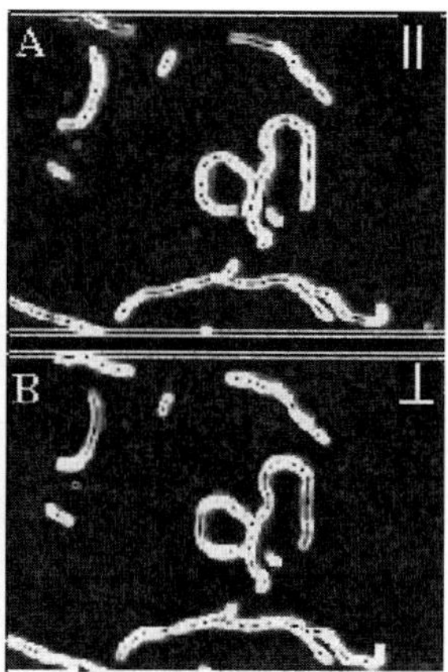

FIG. 7. Image registration of W-view images, based on contour profile images. *Left:* A typical filament with a width of 4 pixels across. *Right:* (A) Contour image of the parallel emission; (B) contour image of the perpendicular emission.

perform image registration by the cross-correlation technique, in which the two simultaneously recorded images are matched by shifting and rotating one of the images and comparing the overlap with the other image until a maximum correlation coefficient is obtained.[31] Because the intensity profiles of the two images are not identical, image registration for the pair of polarized images cannot be achieved by using the intensity profile as the basis for obtaining the maximum correlation. To solve this problem, we first conduct contour detection (Fig. 7, right) and then equalize the intensities within the detected contours. Clearly, when the two images are in perfect register the contours of the filaments should overlay with each other and this case will reflect a maximum correlation of the two contour profile images. In our microscope workstation, a typical actin filament has a width of 4 pixels (Fig. 7, left). Figure 8 shows an example of the FP images of actin filaments after image registration. Figure 8A shows the parallel component of the fluorescence emission, Figure 8B shows the perpendicular component of the emission, and Figure 8C is the FP image of actin filaments. Image registration is accurate to within 1 pixel. The polarization image (Fig. 8C) is obtained by calculating the FP value from parallel and perpendicular images on a pixel-by-pixel basis, using the Perrin equation. The polarization values range from +0.62 for filaments congruent to the parallel direction to −0.60 for filaments aligned in the perpendicular direction, whereas filaments aligned close to the magic angle have polarization values close to zero.

Relating Fluorescence Polarization Values to Orientations of Dipoles. We can relate the FP value (or anisotropy value r) to the angle the absorption/emission

[31] W. K. Pratt, "Digital Image Processing," 2nd Ed. Wiley-Interscience, New York, 1991.

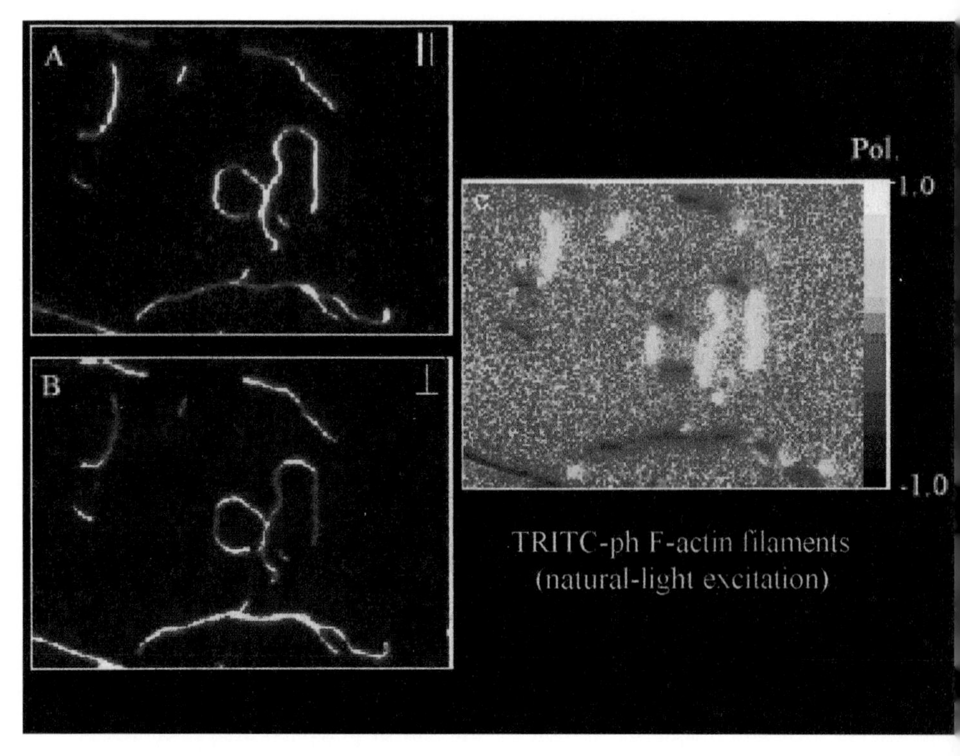

FIG. 8. FP images of F-actin filaments after image registration. (A) Parallel component; (B) perpendicular component; (C) FP image.

dipole moment makes with the long axis of the actin filament, using the following relationship[1]:

$$r = \frac{3\langle\cos^2\theta\rangle - 1}{2} \tag{12}$$

where θ is the angle between the dipole moment and the filament axis; and r is the anisotropy value, which is related to the FP value (P) by Eq. (13):

$$r = \frac{2P}{3 - P} \tag{13}$$

The angle for FITC–ph or TRITC–ph is 30° (see Kinosita *et al.*[6]).

Calculating κ^2 Value. The κ^2 value for FITC–ph and TRITC–ph on adjacent protomers in the filament is calculated as follows: each actin protomer in the filament is related to its neighbor by an axial translation of 2.75 nm and a rotation of 166°.[8] The radial coordinate of the phalloidin molecule is calculated to be 1.45 nm on the basis of the calculated distance between phalloidin molecules on

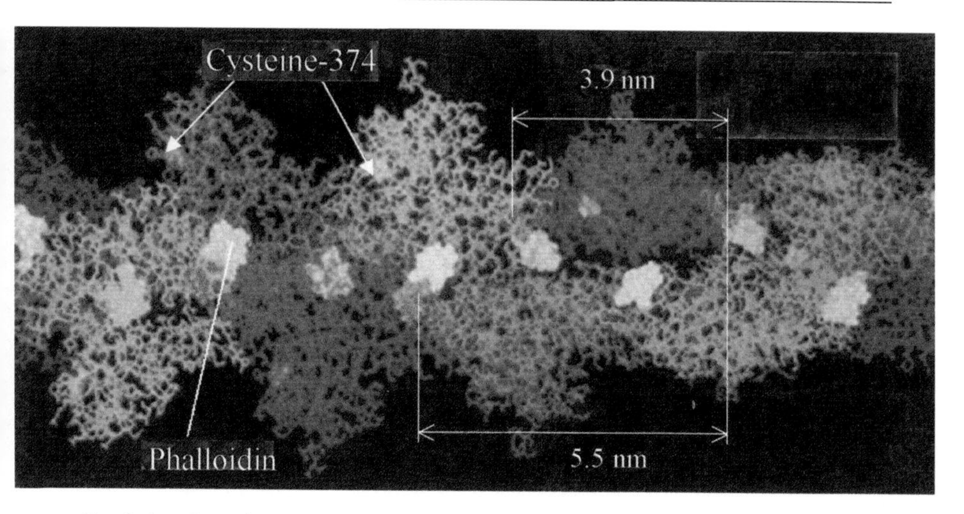

FIG. 9. Atomic model of a stretch of the F-actin filament in its complex with phalloidin (white) [M. Lorenz, D. Popp, and K. C. Holmes, *J. Mol. Biol.* **234,** 826 (1993)]; showing the likely positions of FITC (green) and TRITC (red). The distance between the centers of the conjugated ring system of probes attached to adjacent actin monomers on opposite helices is 3.9 nm. The corresponding distance between adjacent actin monomers on the same helix is 5.5 nm.

opposite actin helices, using the approach described by Taylor *et al.*,[8] because the donor or acceptor fluorophores are expected to have identical angles with respect to the helix axis ($\beta = \gamma = 30°$; Kinosita *et al.*[6]). Using these data we calculate a κ^2 value of 1.42 for FITC–ph and TRITC–ph, based on an assumption that each probe is immobile over the fluorescence lifetime. The difference between the FRET-based distance measurement for FITC–ph and TRITC–ph on actin, using a κ^2 value of 1.42 versus 0.67 (random orientation), is 13%.

Molecular Modeling of Fluorescence Resonance Energy Transfer Distances on Actin Filament

Molecular modeling of the FITC and TRITC probes in the Lorenz model of phalloidin-bound actin[32] shows that the distance between the center of the conjugated aromatic ring on each probe is 3.9 nm for probes on adjacent actin protomers on opposite helices ($n + 1$) and 5.5 nm for the same sites along the same helix ($n + 2$) (Fig. 9). The former value is in reasonable agreement with both solution- and microscope-based FRET distance determinations (3.7 vs. 3.6 nm). These distances are close to the R_0 value (distance at 50% transfer efficiency) between FITC–ph and TRITC–ph on F-actin (3.7 nm). On the other hand, the efficiency of FRET between probes along the same helix (5.5 nm) is only 0.08. Correspondingly,

[32] M. Lorenz, D. Popp, and K. C. Holmes, *J. Mol. Biol.* **234,** 826 (1993).

these probes primarily report on the distance between phalloidin-binding sites on opposite actin helices (Fig. 9). A unique feature of the FRETIM/FPIM approach as applied to the actomyosin motility assay is that it provides an opportunity to measure changes in proximity and orientation[6] of probes on single filaments while they slide at 5 μm/sec on a monolayer of HMM in the presence of ATP.[2] Analysis of these FRET and FP data recorded at video rate shows that neither the distance between adjacent phalloidin probes nor the orientation of these probes changes during the cross-bridge cycle.

Application of the FRETIM/FPIM approach to study protein structure and protein structural dynamics in macromolecular systems is not limited to single actin filaments; other systems that could be studied include Ca^{2+}-regulated thin filaments sliding on myosin, microtubules sliding on motor proteins, and the F_1-ATPase rotary motor (see Adachi *et al.*[33]). In principle, κ^2 values can also be measured by FPIM for other protein systems by covalently cross-linking the donor- and acceptor-labeled proteins to a unique site on actin, for example, through Cys-374. Finally, we believe that knowledge of the κ^2 value greatly improves the precision of FRET-based distance determinations and it may be used as a new parameter for studying protein structural dynamics.[16]

[33] K. Adachi, H. Noji, and K. Kinosita, *Methods Enzymol.* **361,** [11], 2003 (in press).

[25] Homo-FRET versus Hetero-FRET to Probe Homodimers in Living Cells

By Marc Tramier, Tristan Piolot, Isabelle Gautier, Vincent Mignotte, Jacques Coppey, Klaus Kemnitz, Christiane Durieux, and Maïté Coppey-Moisan

Introduction

Fluorescence resonance energy transfer (FRET) is a nonradiative phenomenon in which energy is transferred from a donor fluorophore to an acceptor chromophore with an efficiency that depends on the distance between the two chromophores, the extent of overlap between the donor emission and acceptor excitation spectra, the quantum yield of the donor, and the relative orientation of the donor and acceptor. FRET efficiency (E) decreases as the sixth power of the distance (R) between the donor and the acceptor as shown in Eq. (1):

$$E = 1/[1 + (R/R_0)^6] \quad (1)$$

0076-6879/03 $35.00

where R_0 is the Förster radius, the distance between the donor and acceptor at which 50% energy transfer takes place. It is given by

$$R_0 = [(8.79 \times 10^{-5}) J \kappa^2 Q_D n^{-4}]^{1/6} \text{ Å} \tag{2}$$

where Q_D is the donor quantum yield, n is the refractive index, and J is the spectral overlap integral, calculated from the donor fluorescence (f_D) and acceptor absorption (ε_A) spectra,

$$J = \int f_D(\lambda)\varepsilon_A(\lambda)\lambda^4 d\lambda \bigg/ \int f_D(\lambda)\, d\lambda \tag{3}$$

and κ is the orientation factor for dipole–dipole coupling. The mathematical expression for κ is given by

$$\kappa = \cos\theta_T - 3\cos\theta_A \cos\theta_D \tag{4}$$

where θ_T is the angle between the donor emission and acceptor excitation dipoles, and θ_A and θ_D are the angles between these dipoles and a vector joining the donor–acceptor pair.

The possible values for κ are $0 < \kappa^2 < 4$. A mean value of $\langle\kappa^2\rangle = 2/3$ is taken when random local motion of the chromophores occurs.

Owing to the short working range (i.e., 2–8 nm) of FRET, specific interactions between two fluorescently tagged proteins can be determined. Confocal fluorescence microscopy can colocalize two populations of differently fluorescent molecules in a subcellular volume two orders of magnitude larger than the volume determined by the interacting distance. Thus, FRET detection in microscopy has become a widely used sensor for protein–protein interactions in living cells.[1,2]

The Förster mechanism of electronic energy transfer[3] can occur between unlike (hetero-FRET) as well as between like (homo-FRET) chromophores. In the case of hetero-FRET, the fluorescence lifetime of the donor decreases, or falls off, owing to the depopulation of its excited state by nonradiative energy transfer to the acceptor, according to the relation

$$\tau_{DA} = \tau_D / [1 + (R_0/R)^6] \tag{5}$$

τ_D is the fluorescence lifetime of the donor in the absence of acceptor and τ_{DA} is the fluorescence lifetime of the donor in the presence of acceptor at a distance R. Because the lifetime is independent of fluorophore concentration and light path, parameters difficult to control in live cells, fluorescence lifetime imaging microscopy gives more reliable hetero-FRET determination than the intensity-based FRET detection methods in microscopy.[1]

[1] P. I. Bastiaens and A. Squire, *Trends Cell Biol.* **9,** 48 (1999).
[2] P. R. Selvin, *Nat. Struct. Biol.* **7,** 730 (2000).
[3] T. Förster, *Ann. Phys.* **2,** 55 (1948).

For homo-FRET, because the photophysical properties of the two donor molecules are the same, the excitation energy is reversibly transferred between the fluorescent tags. This transfer does not change fluorescence lifetime properties and can be monitored only by fluorescence anisotropy.[4] Time-resolved fluorescence anisotropy monitors any process that changes the polarization of the emitted fluorescence during the excited state. Consequently, the fluorescence anisotropy decay depends on (1) rotational movements of the fluorescent molecules and (2) energy transfer taking place within the fluorescence time scale. In the case of dimeric proteins composed of identical monomers, and assuming that (1) the dimers are equivalent to each other (symmetrical configuration); (2) rotation of the dimer during fluorescence lifetime is negligible; (3) there is no reorientation of the fluorescence dipoles during lifetime (static model); and (4) absorption and emission transition moments of the fluorophore are parallel (which is the case for GFP), the time-dependent anisotropy is given by

$$r(t) = 1/10[(3 - 3\cos^2\theta)e^{-2\omega t} + 3\cos^2\theta + 1] \tag{6}$$

where θ is the orientation between the two chromophores. The transfer rate, ω, is linked to the distance R between the two interacting chromophores by

$$\omega = 3/2\langle\kappa^2\rangle(R_0'/R)^6\tau^{-1} \tag{7}$$

where R_0' is the R_0 value for a random orientation between the donor and the acceptor and τ is the fluorescence lifetime.

As for hetero-FRET, if random local motion of the chromophores occurs, the mean value of $\langle\kappa^2\rangle$ is 2/3. In this case, the transfer rate is directly linked to the distance between the two chromophores by

$$\omega = (R_0'/R)^6\tau^{-1} \tag{8}$$

The availability of spectral mutants of the green fluorescent protein (GFP)[5] and of DsRed[6] offers a choice of suitable pairs of chromophores that can be covalently linked to proteins of interest. By tagging each subunit of a homodimeric complex of thymidine kinase of herpes simplex virus with either identical or different fluorescent proteins and by using the ultrasensitive time-correlated single-photon counting technique adapted to microscopy, we could compare, in living mammalian cells, homo-FRET (between GFP chromophores) and hetero-FRET [between cyan fluorescent protein (CFP) and yellow fluorescent protein (YFP)]. In addition, we show that the donor–acceptor pair GFP–DsRed allows us to determine the hetero-FRET efficiency and the fraction of the donor within the heterodimer (exemplified by

[4] G. Weber, *Trans. Faraday Soc.* **50,** 552 (1954).

[5] R. Y. Tsien, *Annu. Rev. Biochem.* **67,** 509 (1998).

[6] M. V. Matz, A. F. Fradkov, Y. A. Labas, A. P. Savitsky, A. G. Zaraisky, M. L. Markelov, and S. A. Lukyanov, *Nat. Biotechnol.* **17,** 969 (1999).

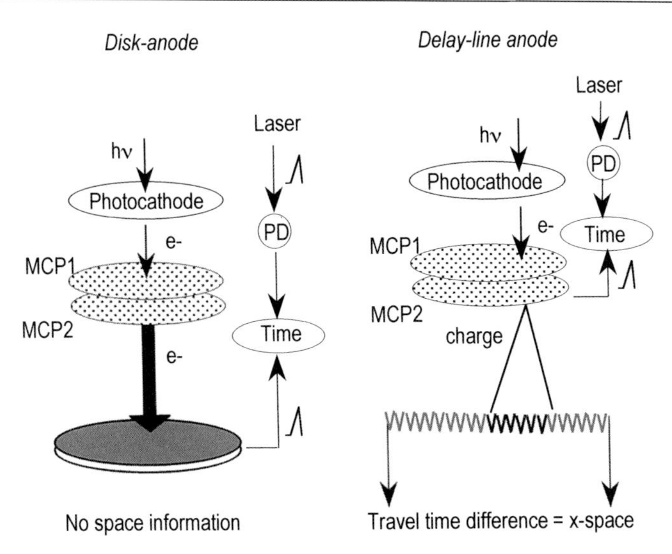

FIG. 1. Comparison between standard MCP-PMT disk-anode and delay-line anode detectors. MCP, Microchannel plate; PD, photodiode.

the heterodimeric transcriptional factor, NF-E2) in contrast to the results obtained with the donor–acceptor pair CFP–YFP.

Time-Domain Picosecond Fluorescence Lifetime Microscopy

Instrumentation

Acquisition of fluorescence dynamics on the picosecond time scale of weakly emitting sources from the subcellular area in living cells is carried out by using a time-correlated single photon-counting (TCSPC) method.[7] Two different detectors are adapted to the microscope: (1) a standard MCP–PMT (microchannel plate–photomultiplier tube) with no space information (Fig. 1, left), and (2) a DL (delay-line) detector, which gives the linear space location of the incoming photon (Fig. 1, right).[8] In this detector the incident photon produces a cone-shaped cloud of electrons at the output face of the second MCP that hits the DL at a spatial position corresponding to that of the photon. The electric pulse is split in two pulses amplified, discriminated, and used as the start and stop pulses in a time-to-amplitude converter. The travel–time difference indicates the x position.

[7] D. V. O'Connor and D. Phillips, "Time-Correlated Single Photon Counting." Academic Press, New York, 1984.

[8] K. Kemnitz, L. Pfeifer, R. Paul, and M. Coppey-Moisan, *J. Fluoresc.* **7,** 93 (1997).

A mode-locked titanium–sapphire laser (Millennia 5W/Tsunami 3960-M3BB-UPG kit; Spectra-Physics, Mountain View, CA) is tuned to 880 or 960 nm to obtain, after frequency doubling, a 440- or 480-nm excitation for CFP and GFP fluorescence decay measurement, respectively. The repetition rate is 4 MHz after pulse selection (Pulse Picker; Spectra-Physics). The laser beam is expanded to obtain an illumination field of 80 μm in the object plane with a $\times 100$ objective. The fluorescence corresponding to the different proteins is selected by using one of the following filter set combinations (Omega; Optophotonics, Eaubonne, France): 505DRLFP dichroic beam splitter/535AF45 emission filter (GFP) or 460DRLP dichroic beam splitter/480AF30 emission filter (CFP). A polarizer is placed after the emission filter and is oriented in a position corresponding to the "magic angle." A galvanometer system, oscillating at 60 Hz, is used to average the interference fringes generated by the passage of the coherent laser beam through the optics of the microscope. The laser power at the entrance of the microscope is adjusted to 20 nW by using neutral density filters. A rotating mirror is allowed to switch between Hg lamp and laser illumination. A rectangular area of the sample 60 μm long (x direction) and 5 μm wide is imaged on the sensitive area of the DL detector. The typical count rate is between 500 and 2000 counts per second (cps). The two-dimensional (2-D) multichannel analyzer combines time and space coordinates, which are displayed as a contour map in pseudo-colors with 256 and 2048 channels for space and time dimensions, respectively. The instrument response function (IRF) is measured by reflection of the laser beam on the surface of a microscope glass slide and by removing the cutoff filter.

Data Analysis

Fluorescence decay corresponding to different sample areas is obtained by gathering data over five contiguous channels. Fluorescence decays are deconvoluted with the IRF and fitted by a Marquardt nonlinear least-squares algorithm with Globals Unlimited (University of Illinois, Urbana-Champaign, IL) with the model function

$$i(t) = \mathrm{IRF}(t) * \sum a_i e^{-t/\tau_i} \tag{9}$$

where a_i corresponds to the relative contribution of the i fluorescent species characterized by its fluorescence lifetime τ_i, and $*$ is the convolution product. Static background, as measured by the mean number of counts before the rise of fluorescence, is treated as a fit parameter by the kinetic analysis software. Dynamic background is acquired from the control sample (e.g., medium) under identical conditions and incorporated into the analysis. In a first step, the decays are gathered and analyzed with one- and two-exponential models. The space-resolved decays are then fitted with a global analysis by linking the lifetime values determined in the first analysis. When the fluorescence decays are better fitted with a biexponential kinetic, the respective populations of the unbound and bound GFP-tagged protein are determined from the preexponential amplitude values. Fluorescence decays of CFP chromophore are always better fitted by a biexponential kinetic model than by a single-exponential model, even when the fluorescent protein is not involved in a hetero-FRET process. When CFP–YFP is used as donor–acceptor pair, hetero-FRET is determined by the decrease in the mean fluorescence lifetime value calculated from the biexponential fit of the fluorescence decay of the CFP chromophore in the presence of the acceptor.

Fluorescence Anisotropy Decay Microscopy

Experimental Setup and Practical Considerations

The fluorescence anisotropy decay is obtained from a subcellular volume by focusing the picosecond laser beam (Tsunami; Spectra-Physics) through a $\times 100$ objective (NA 0.8–1.3) of a Nikon epifluorescence inverted microscope. The fluorescence photons emitted from the illuminated volume of 1 μm^3 are collected by the objective and conducted through an optic fiber (400-μm diameter) to a standard MCP–PMT with no space information (Fig. 1, left) (Hamamatsu Photonics). Different subcellular localizations of the excited volume (cytoplasm and nucleus) can be chosen to measure fluorescence decays. A Fresnel rotator is placed in the excitation laser beam and a polarizer is placed before the optical fiber in the emission path. The optical design of the microscope results in four geometric components of the fluorescence polarization, where i_{vh} and i_{hv} pertain to the parallel direction and i_{vv} and i_{hh} to the perpendicular direction, relative to the direction of laser excitation. Parallel $[i_{vh}(t)]$ and perpendicular $[i_{hh}(t)]$ decays are acquired sequentially from the same sample spot.

Data Analysis

The two decays are normalized to correct for the different transmission efficiency of each geometric component of the excited and emitted polarized light, as well as for polarizing (or depolarizing) effects of the microscope optics.[9] The normalized experimental decays, $i_{vh}^{N}(t)$ and $i_{hh}^{N}(t)$, are distorted by the measurement apparatus, and are related to the real-time behavior, $i_{par}(t)$ and $i_{per}(t)$, by the convolution product of the instrument response function IRF(t),

$$i_{vh}^{N}(t) = \mathrm{IRF}(t) * i_{par}(t) \tag{10}$$

and

$$i_{hh}^{N}(t) = \mathrm{IRF}(t) * i_{per}(t) \tag{11}$$

[9] M. Tramier, K. Kemnitz, C. Durieux, J. Coppey, P. Denjean, R. B. Pansu, and M. Coppey-Moisan, *Biophys. J.* **78,** 2614 (2000).

The anisotropy function $r(t)$ is defined by

$$r(t) = D(t)/S(t) \tag{12}$$

where

$$D(t) = i_{\text{par}}(t) - i_{\text{per}}(t) \tag{13}$$

is the difference between the parallel and the perpendicular decay, and

$$S(t) = i_{\text{par}}(t) + 2i_{\text{per}}(t) \tag{14}$$

is the decay of total intensity.

By combining Eqs. (12), (13), and (14),

$$3i_{\text{par}}(t) = S(t)[1 + 2r(t)] \tag{15}$$

and

$$3i_{\text{per}}(t) = S(t)[1 - r(t)] \tag{16}$$

For GFP, the total intensity decay is monoexponential in solution[10] and in living cells (see page 587). $S(t)$ can be fitted with the model function,

$$S(t) = a/3e^{-t/\tau} \tag{17}$$

where τ is the fluorescence lifetime and a is a constant.

To fit the experimental data, two phenomenological functions are used:

$$r(t) = r_0 e^{-t/\Phi} \tag{18}$$

or

$$r(t) = r_0[(1-b)e^{-t/\Phi_1} + be^{-t/\Phi_2}] \tag{19}$$

where r_0 is the initial anisotropy, Φ_i are correlation relaxation times, and b is a constant.

The two normalized experimental decays are fitted by a nonlinear least square algorithm, using the expressions

$$i_{\text{vh}}^{\text{N}}(t) = \text{IRF}(t) * [ae^{-t/\tau}(1 + 2r_0 e^{-t/\Phi})] \tag{20}$$

and

$$i_{\text{hh}}^{\text{N}}(t) = \text{IRF}(t) * [ae^{-t/\tau}(1 - r_0 e^{-t/\Phi})] \tag{21}$$

or, if the experimental data are better fitted with a two-exponential anisotropy decay,

$$i_{\text{vh}}^{\text{N}}(t) = \text{IRF}(t) * (ae^{-t/\tau}\{1 + 2r_0[(1-b)e^{-t/\Phi_1} + be^{-t/\Phi_2}]\}) \tag{22}$$

[10] A. Volkmer, V. Subramaniam, D. J. S. Birch, and T. M. Jovin, *Biophys. J.* **78,** 1589 (2000).

and

$$i^{\mathrm{N}}_{\mathrm{hh}}(t) = \mathrm{IRF}(t) * (ae^{-t/\tau}\{1 - r_0[(1-b)e^{-t/\Phi_1} + be^{-t/\Phi_2}]\}) \quad (23)$$

Analysis is performed with Global analysis software (Globals Unlimited).

Hetero-FRET between GFP and DsRed Chromophores within Heterodimeric Protein NF-E2

NF-E2 is a heterodimeric member of the basic region-leucine zipper (bZip) class of transcription factors. Its large subunit, called p45, is expressed mainly in two hematopoietic lineages, the erythroid cell lineage (precursor of the red cells) and megakaryocytes (the precursors of blood platelets).[11] It associates with a more widely expressed subunit, MafG, which is a member of the small-Maf family of proteins.[12]

Here, p45 and MafG are used as a model of heterodimeric nuclear protein. The GFP-coding sequence is tethered to the amino terminus of the p45-coding sequence, to generate GFP–p45 fusion cDNA. Similarly, a DsRed–MafG expression vector is constructed. These vectors are transfected either separately or together, into HeLa cells, which do not express endogenous p45 or MafG.

Quantitative Determination of Hetero-FRET between GFP–p45 and DsRed–MafG

Space-resolved picosecond fluorescence decays of GFP–p45 are acquired in living mono- and cotransfected cells by using the DL detector. Individual emitted photons are counted according to (1) their coordinate along the detector axis (white rectangle in Fig. 2A and C) and (2) the time delay between the laser pulse and the arrival of the emitted photon (Fig. 2). The results of photon counting are shown in Fig. 2B and D for mono- and cotransfected cells, respectively. The corresponding kinetics of the fluorescence decays gathered over the whole space channels from each experiment are displayed in Fig. 2E. The green curve corresponds to the GFP–p45-monotransfected cells and can be fitted by a single-exponential decay with a lifetime of 2.6 ns. The red curve, which corresponds to the GFP–p45/DsRed–MafG-cotransfected cells, is better fitted with a biexponential decay with the two lifetimes values, $\tau_1 = 2.83$ ns and $\tau_2 = 0.74$ ns, than with a single exponential kinetic (Fig. 3). The former value corresponds to the typical fluorescence kinetic

[11] N. C. Andrews, H. Erdjument-Bromage, M. B. Davidson, P. Tempst, and S. H. Orkin, *Nature (London)* **362,** 722 (1993).

[12] K. Igarashi, K. Kataoka, K. Itoh, N. Hayashi, M. Nishizawa, and M. Yamamoto, *Nature (London)* **367,** 568 (1994).

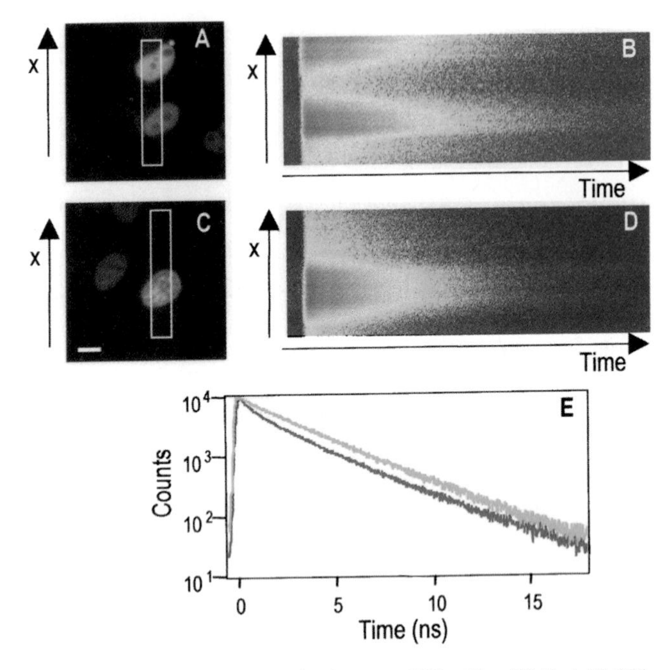

FIG. 2. FRET determination of the interaction between GFP–p45 and DsRed–MafG by picosecond fluorescence decay microscopy. GFP–p45 and DsRed–MafG were coexpressed in HeLa cells and the fluorescence lifetime of GFP was determined with a time- and space-correlated single photon-counting DL detector. (A) CCD fluorescence image of GFP–p45 after single transfection. (C) CCD fluorescence image after cotransfection of GFP–p45 and DsRed–MafG. Magnification bar (C): 10 μm. The excitation source was a 50-W high-pressure mercury lamp. The white rectangle in (A) and (C) represents the region of the sample imaged in the active area of the delay-line detector. (B and D) Two-dimensional histogram [horizontal, time after laser pulse; vertical, space (x direction) along the delay line] of single counted photons, from GFP fluorescence of GFP–p45 expressed in the absence and presence of DsRed–MafG, respectively. The excitation source was a titanium–sapphire (Ti:Sa) laser, tuned to 960 nm (480 nm after doubling), with wide-field illumination. (E) Fluorescence decays (green and red) of GFP–p45 (collected from the entire channels along the x direction) corresponding to the 2-D histograms [(B) and (D), respectively].

properties of the GFP chromophore[13–15] and is not significantly different from the values found in different monotransfected cells (e.g., $\tau = 2.6 \pm 0.2$ ns). The latter value is characteristic of the fluorescence emitted by GFP chromophore engaged in hetero-FRET process. This hetero-FRET is not due to direct interaction between

[13] I. Gautier, M. Tramier, C. Durieux, J. Coppey, R. B. Pansu, J.-C. Nicolas, K. Kemnitz, and M. Coppey-Moisan, *Biophys. J.* **80,** 3000 (2001).

[14] R. Pepperkok, A. Squire, S. Geley, and P. I. Bastiaens, *Curr. Biol.* **9,** 269 (1999).

[15] R. Swaminathan, S. Bicknese, N. Periasamy, and A. S. Verkman, *Biophys. J.* **71,** 1140 (1996).

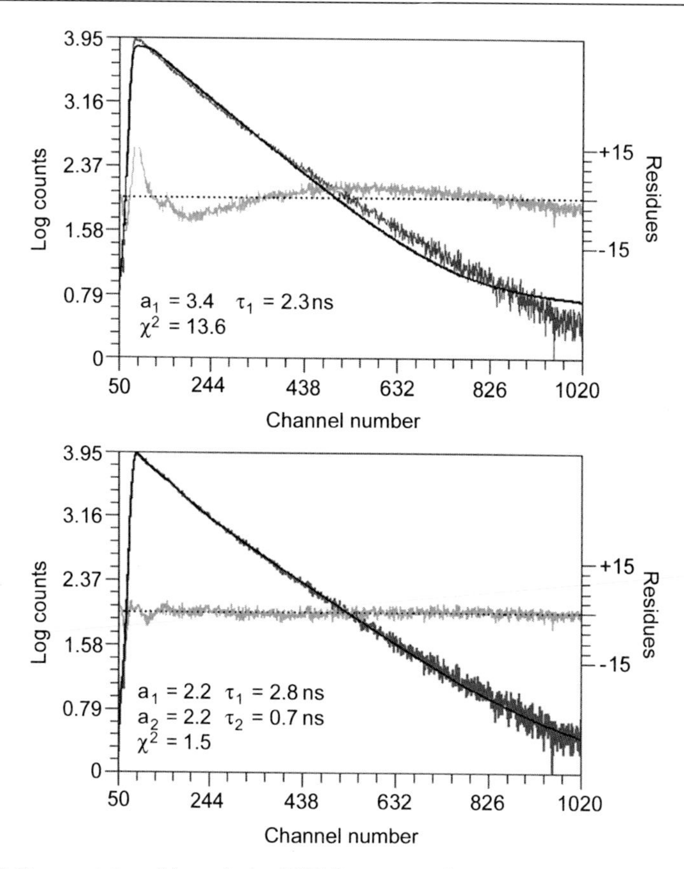

FIG. 3. Representation of the analysis of GFP fluorescence decay in the presence of acceptor, fitted with Eq. (9) by using Globals Unlimited software. *Top:* Single exponential model. *Bottom:* Biexponential model. The experimental data (dark gray) are plotted with their respective fit (plain curve) and their respective residues (light gray). Channel time, 20 ps; τ_i, fluorescence lifetime; a_i, preexponential factor. Occurrence of hetero-FRET between GFP and DsRed fused to p45 and MafG, respectively, is determined by the existence of the short fluorescence lifetime (0.74 ns). The long fluorescence (2.83 ns) lifetime component corresponds to GFP-p45 molecules unbound to DsRed–MafG. The fraction of GFP–p45 bound to DsRed–MafG is determined from the ratio of preexponential factors: $a_2/(a_1 + a_2)$.

GFP and DsRed because a short fluorescence lifetime is not detected when GFP is expressed with the protein DsRed.

Analysis of the picosecond fluorescence decay of GFP chromophore gives two pieces of information with a single acquisition under the microscope: (1) the hetero-FRET efficiency, which depends on the interchromophoric distance between donor and acceptor, and (2) the fraction of the GFP-tagged protein that is in the heterodimer. Indeed, in the analysis (results shown in Fig. 3) of the experiment presented in Fig. 2C and D, 50% of the GFP–p45 is bound to DsRed–MafG to form a heterodimer in which the interchromophoric distance is 41 Å (corresponding to an FRET efficiency of 0.69) [calculated from Eqs. (1) and (4) and by taking $R_0 = 52$ Å for the GFP–DsRed couple, with a random orientation between donor and acceptor].

A global analysis of the space-resolved decays shows that the fluorescence decay of GFP–p45 in the presence of DsRed–MafG is biexponential, with the same two lifetime values of 2.83 and 0.74 ns, and that the fraction of GFP–p45 molecules in interaction with DsRed–MafG is constant within a single nucleus, but differs from cell to cell (in the range of 30 to 60%; data not shown). This can be explained by the variable expression level ratio of GFP–p45 and DsRed–MafG from cell to cell. The fact that the fluorescence lifetime value of GFP–p45 involved in the interaction with DsRed–MafG is the same whatever the cell and its nuclear location means that the distance between the green and the red chromophores is constant, which allows us to expect that the NF-E2 complex would have a similar conformation within the nucleus.

Practical Considerations

Several artifacts can occur, inducing false-positive FRET determination. DsRed is known to emit weak green fluorescence before its maturation into a red-emitting species *in vitro*.[16] The 0.74-ns green fluorescence lifetime cannot be ascribed to the immature green-emitting DsRed or to cellular autofluorescence because the fluorescence contribution of the 0.74-ns lifetime species is higher than the green nuclear fluorescence of cells expressing DsRed–MafG without GFP–p45.

DsRed is known to form a tetramer *in vitro*.[17] Therefore it could be assumed that hetero-FRET between the immature green-emitting intermediate of DsRed protein and the mature red-emitting molecule within the tetrameric form of DsRed protein might be the main process involved in the occurrence of the short green fluorescence lifetime. This possibility is ruled out, however, because the value of 0.74 ns does not correspond to the expected value for hetero-FRET between green- and red-emitting chromophores within the tetramer. Indeed, fluorescence anisotropy decay in solution gave a transfer rate of 2.36 ns^{-1} for the energy transfer between the red-emitting chromophores within the mature DsRed tetramer[17] and the corresponding calculated lifetime value of the green-emitting chromophore within the tetramer would be less than 0.43 ns. In addition, such a short lifetime is not detected when DsRed is expressed with GFP, meaning that FRET within DsRed tetramer is not detected under these conditions. Thus, the lifetime

[16] G. S. Baird, D. A. Zacharias, and R. Y. Tsien, *Proc. Natl. Acad. Sci. U.S.A.* **97,** 11984 (2000).

[17] A. A. Heikal, S. T. Hess, G. S. Baird, R. Y. Tsien, and W. W. Webb, *Proc. Natl. Acad. Sci. U.S.A.* **97,** 11996 (2000).

of 0.74 ns, observed in GFP–p45 and DsRed–MafG-coexpressing cells, is not an artifact due to multimerization of DsRed moieties resulting from the existence of the immature green-emitting state but can be ascribed to GFP–p45/DsRed–MafG interaction.

Hetero-FRET between CFP and YFP Chromophores within Homodimeric Protein Thymidine Kinase of Herpes Simplex Virus

The thymidine kinase (TK) of herpes simplex virus type 1 phosphorylates a wide range of nucleoside analogs. Crystallized in the presence of substrate, TK forms homodimers[18] that seem to be the active form.[19] Thus, TK is used as a model of homodimeric interactions in living cells.

Homodimerization of TK_{366} Probe by Hetero-FRET between TK_{366}–CFP and TK_{366}–YFP

Two different chromophores are used as a donor–acceptor couple to measure hetero-FRET between TK monomers within the dimer by acquisition of picosecond fluorescence decays of the donor. CFP and YFP, a couple widely used for fluorescence intensity-based FRET detection, were fused to the carboxyl terminus of the TK shortened of the last 10 amino acids (TK_{366}–CFP and TK_{366}–YFP). The donor–acceptor pair GFP–DsRed cannot be used because the DsRed-tagged TK strongly aggregates and is not colocalized with the GFP-tagged TK (not shown). Vero cells are transfected either with the expression vector for TK_{366}–CFP or cotransfected with expression vectors for TK_{366}–CFP and TK_{366}–YFP and analyzed 72 hr after transfection. Figure 4A and C presents the fluorescence with TK_{366}–CFP expressed alone or with TK_{366}–YFP, respectively. CFP and YFP fluorescence are detected in the cytoplasm and in the nucleus. An additional punctate cyan (and yellow) fluorescence appears in a proportion of cells, which increases with time after transfection, as yet described for TK_{366}–GFP.[13] The space-resolved picosecond fluorescence decays of the CFP molecules in these cells are shown in Fig. 4B and D for mono- and cotransfected cells, respectively. Qualitatively, the CFP fluorescence decays are globally faster when TK_{366}–YFP is present as observed in the 2-D histograms (Fig. 4B and D) as well as in the kinetics of the fluorescence decays gathered over the whole space channels from each experiment (Fig. 4E). This is likely to be ascribed to hetero-FRET from CFP to YFP chromophores within the TK_{366}–CFP/TK_{366}–YFP dimer. Analysis of these fluorescence decays yields lifetimes of 1.19 ± 0.10 and 3.79 ± 0.52 ns for TK_{366}–CFP

[18] K. Wild, T. Bohner, G. Folkers, and G. E. Schulz, *Protein Sci.* **6,** 2097 (1997).

[19] J. Fetzer, M. Michael, T. Bohner, R. Hofbauer, and G. Folkers, *Protein Expr. Purif.* **5,** 432 (1994).

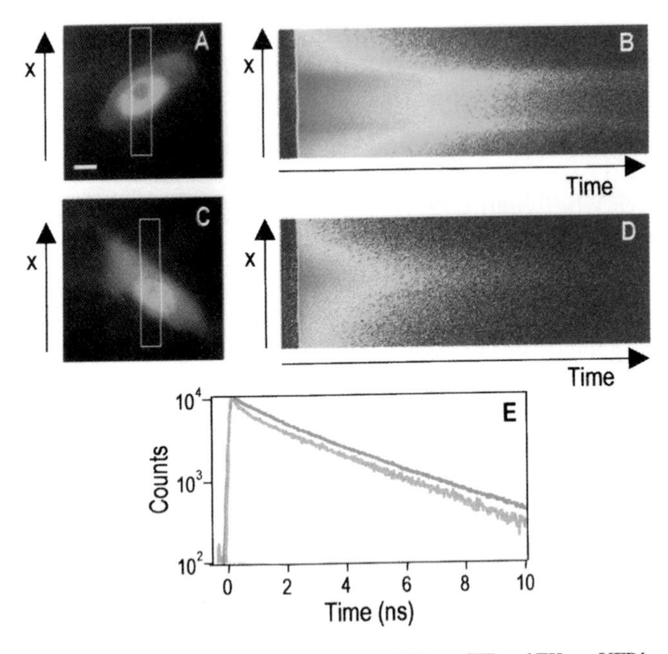

FIG. 4. FRET determination of the interaction between TK_{366}–CFP and TK_{366}–YFP by picosecond fluorescence decay microscopy. TK_{366}–CFP was expressed or coexpressed with TK_{366}–YFP in Vero cells and the fluorescence lifetime of CFP was determined with a time- and space-correlated single photon-counting DL detector. (A) CCD fluorescence image of TK_{366}–CFP after single transfection. Magnification bar (A): 10 μm. (C) CCD fluorescence image after cotransfection of TK_{366}–CFP and TK_{366}–YFP. The excitation source was a 50-W high-pressure mercury lamp. The white rectangles in (A) and (C) represent the region of the sample imaged in the active area of the delay-line detector. (B and D). Two-dimensional histograms [horizontal, time after laser pulse; vertical, space (x direction) along the delay line] of single counted photons from CFP fluorescence of TK_{366}–CFP expressed in the absence and presence of TK_{366}–YFP, respectively. The excitation source was a Ti:Sa laser, tuned to 880 nm (440 nm after doubling), with wide-field illumination. (E) Fluorescence decays (blue and orange) of TK_{366}–CFP (collected from the entire channels along the x direction) corresponding to the 2-D histograms (B and D, respectively).

expressed alone and 0.86 ± 0.18 and 3.73 ± 0.08 ns for TK_{366}–CFP expressed together with TK_{366}–YFP (Table I). The quantification of this hetero-FRET is, however, complicated by the fact that the CFP chromophore presents a biexponential decay as shown for TK_{366}–CFP proteins (Table I), in agreement with a previous study.[14] The molecular and/or photophysical bases for such fluorescence lifetimes have not yet been elucidated. By taking the mean fluorescence lifetime of CFP kinetics for TK_{366}–CFP in the absence ($\langle\tau\rangle = 2.56$ ns) and in the presence of TK_{366}–YFP ($\langle\tau\rangle = 2.18$ ns), an apparent FRET efficiency of 0.13 could be calculated, evidencing the dimerization of TK. Hetero-FRET is detected 3 days

TABLE I
CYAN FLUORESCENCE DECAY KINETIC ANALYSIS OF CFP-TAGGED TK_{366} AND FRET CHARACTERIZATION IN LIVING CELLS[a]

Parameter	Protein	α_1	τ_1 (ns)	α_2	τ_2 (ns)	$\langle\tau\rangle$ (ns)
J_1	TK_{366}–CFP	0.62 ± 0.03	1.50 ± 0.11	0.38 ± 0.03	4.25 ± 0.15	2.55 ± 0.08
	TK_{366}–CFP/TK_{366}–YFP	0.62 ± 0.03	1.54 ± 0.14	0.38 ± 0.03	4.45 ± 0.22	2.65 ± 0.09
J_3	TK_{366}–CFP	0.46 ± 0.07	1.19 ± 0.10	0.53 ± 0.07	3.79 ± 0.52	2.56 ± 0.12
	TK_{366}–CFP/TK_{366}–YFP	0.54 ± 0.06	0.86 ± 0.18	0.42 ± 0.03	3.73 ± 0.08	2.18 ± 0.09

[a] a_i is the normalized fluorescence relative contribution ($\sum a_i = 1$) of the i fluorescent species characterized by its fluorescence lifetime τ_i. The error corresponds to the standard deviation. $n = 3$. J_1 and J_3 correspond to experiments carried out 1 and 3 days after transfection, respectively.

after the transfection, which corresponds to the slow dimerization of TK_{366}–CFP (or GFP) in living cells as previously described.[13]

Limitations of Donor–Acceptor Pair CFP–YFP for Homodimer

The hetero-FRET efficiency determined here cannot be used to determine the distance between CFP and YFP chromophores in the dimer or the fraction of TK_{366}–CFP molecules involved in the FRET process. In addition, when TK_{366}–CFP associates as homodimer, the transfer taking place in the dual CFP-tagged dimers does not change fluorescence lifetime properties and cannot be detected in these experiments. These limitations of quantification do not arise from the technical approach, but from the complex fluorescence properties of the CFP chromophore and from homodimer formation. Similar limitations occur in fluorescence intensity-based FRET detection and in the fluorescence lifetime imaging microscopy approach.[1] Because of the monoexponential fluorescence decay property of GFP,[13–15] it seems that GFP–DsRed would be a better chromophore pair than CFP–YFP for picosecond FRET measurements. Unfortunately, TK_{366}–DsRed shows an abnormal cellular localization and aggregation, precluding its use. In conclusion, despite the fact that TK_{366}–CFP can dimerize with TK_{366}–CFP instead of TK_{366}–YFP, and despite the fact that the CFP fluorescence decay property appears to be biexponential, it is possible, under our conditions, to detect hetero-FRET.

Homo-FRET between GFP Chromophores within Homodimeric Protein Thymidine Kinase of Herpes Simplex Virus

The use of CFP chromophore as donor, owing to its complex fluorescence decay kinetics, cannot allow hetero-FRET efficiency to be determined. The situation is even more complex for quantifying the homodimerization of proteins, because of the formation of CFP–CFP dimers that participate in the fluorescence decay

TABLE II
FLUORESCENCE DYNAMICS AND HOMO-TRANSFER PARAMETERS OF TK_{366}–GFP EXPRESSED IN LIVING CELLS[a]

Protein	Lifetime (ns)	Initial anisotropy r_0	Relaxation time Φ_1 (ns)	Relaxation time Φ_2 (ns)	Anisotropy contribution of Φ_2	θ	ω (ns^{-1})	R (Å)
GFP	2.58 ± 0.02	0.26 ± 0.01	23.4 ± 2.3					
TK_{366}–GFP monomer	2.47 ± 0.20	0.26 ± 0.01	81.0 ± 15.3					
TK_{366}–GFP dimer	2.47 ± 0.20	0.23 ± 0.01	2.4 ± 0.3	≥ 200 (∞)	0.63 ± 0.02	$44.8° \pm 1°$	0.210 ± 0.028	$70.1 \pm 1.5^*$ $52.0 \pm 0.2^\dagger$

[a] Errors are standard deviations. The anisotropy contribution of Φ_2 corresponds to parameter b in Eq. (19). The r_0 values are similar to the value previously determined by microscopy with an NA = 1.3 objective and a chromophore for which the orientations of the excitation and emission dipoles are identical (see Ref. 9). θ, the mutual orientation between the two GFP chromophores, and ω, the energy transfer rate, are calculated from Eq. (6); R, the distance between the two GFP chromophores, is calculated from Eq. (7) with $\kappa^2 = 4^*$ or with $\kappa^2 = 2/3^\dagger$, and $R'_0 = 47$ Å.

kinetics but without changing their fluorescence lifetime values. Here, we show that homo-FRET is an alternative method to study homodimerization of protein.

Because of its monoexponential fluorescence decay kinetics, GFP was chosen to tag TK_{366}. One day after transfection into Vero cells, anisotropy decay analysis gives a single rotational time that corresponds to the monomer of TK_{366}–GFP (Table II). In contrast, 3 days after transfection fluorescence anisotropy decays obtained from subcellular volumes are biexponential (Fig. 5, right). The punctate fluorescence that appears in the cytoplasm corresponds to the aggregation of

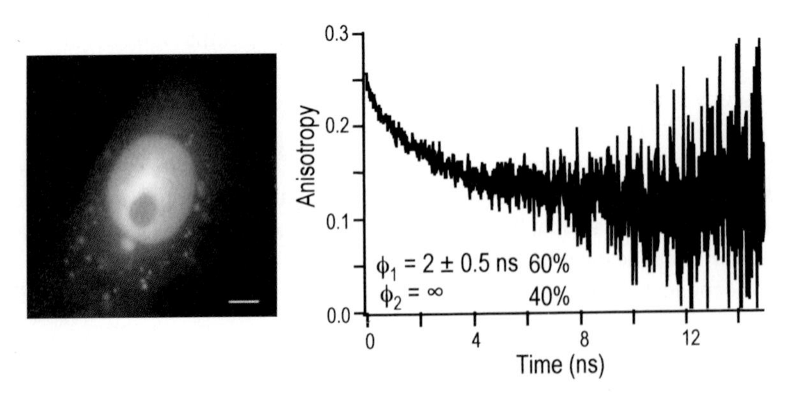

FIG. 5. Fluorescence anisotropy decay of TK_{366}–GFP to probe homodimerization by homo-FRET. *Left:* Vero cell transfected with plasmid expressing TK_{366}–GFP, 3 days before fluorescence microscopy observation. *Right:* Typical fluorescence anisotropy decay obtained from a subcellular volume (1 μm^3) either from nucleus, cytoplasm, or aggregates in this cell. Analysis of this anisotropy decay gives a fast depolarization component $\Phi_1 = 2 \pm 0.5$ ns for 60% and a residual anisotropy $\Phi_2 = \infty$ for 40%. The fast component corresponds to the occurrence of homotransfer. Magnification bar: 5 μm.

TK_{366}–GFP several days after transfection.[13] Localization and structure of these aggregates are different from those occurring with DsRed-tagged TK (not shown).

The experimental fluorescence anisotropy decay is theoretically analyzed in the frame of a symmetric dimer model [Eq. (6)]. The fit parameters of the biexponential relaxation equation [Eq. (19)] and the parameters of homotransfer obtained in the frame of the theoretical dimer model [Eq. (7)] are reported in Table II. The value of relaxation time Φ_2 is found to be $\geq$200 ns and, thus, is interpreted as the residual anisotropy, r_∞, in the context of one-step energy migration between two GFP chromophores within dimers. From the experimental value of r_∞, the angle θ, corresponding to the mutual orientation between the two GFP chromophores, is determined [constant term in Eq. (6)] and found to be equal to $44.6 \pm 1.6°$. The fast relaxation time Φ_1 allows us to calculate the transfer rate from one GFP chromophore to another within the dimer, i.e., $\omega = 0.21$ ns^{-1}. From this value, the distance between the two interacting chromophores, R, can be calculated if the orientational factor, $\langle \kappa^2 \rangle$, is known [Eq. (7)]. If the link between GFP and the protein is not rigid, allowing random motion, a value of $\langle \kappa^2 \rangle = 2/3$ (averaging over all positions) can be used. This leads to a computed value of 52 Å for the distance between the two interacting chromophores [Eq. (8)]. However, for TK_{366}–GFP, random local motion of GFP or its chromophore cannot be evidenced from the anisotropy decay of the monomer.[13] In this case no average calculation for κ^2 can be used. $\langle \kappa^2 \rangle$ depends on the mutual orientation of the transition dipoles of the two chromophores, θ, but also on their orientation with respect to the axis joining them [see Eq. (4)]. This last orientation is unknown in the TK_{366}–GFP dimer. Thus, the upper limit of R is calculated from the maximal value of $\langle \kappa^2 \rangle$ and is 70 Å (Table II), the minimal value of R being 24 Å, which corresponds to close contact between two GFP barrels.

Comparison between Donor–Acceptor Pairs GFP–DsRed and CFP–YFP for Picosecond Fluorescence Decay Microscopy

The best donor–acceptor pair of fluorescent tags is determined and the features of hetero-FRET- and homo-FRET-based interaction detection are compared in relation to the formation of heterodimer and homodimer.

GFP and DsRed provide a useful donor–acceptor pair to probe heterodimer in living cells when DsRed can be fused to one of the subunits of the heterodimer without perturbation of the dimerization, as exemplified by the GFP–p45 and DsRed–MafG interaction. The measurement of the fluorescence lifetime of the donor, as carried out here, allows detection of hetero-FRET despite excitation of the acceptor (DsRed) when exciting the donor. Furthermore, acquisition of the picosecond fluorescence decay of the GFP chromophore provides a simple method to determine (1) the FRET efficiency (expressed in terms of distance or of relative orientation between donor and acceptor) and (2) the fraction of donor molecules within the heterodimer. The simplicity of the method relies on the single-exponential decay of the GFP chromophore, a property maintained within the chimeric protein. In contrast, three lifetimes were obtained for GFP in solution from other experiments.[20] Such a discrepancy could arise from altered protein folding due to extraction and purification procedures as well as the occurrence of photoconversion processes, as previously evidenced.[10,21] A control measurement of the fluorescence decay of the GFP-tagged protein in the absence of acceptor must be carried out for each hetero-FRET experiment.

The donor–acceptor pair CFP–YFP is widely used in intensity-based FRET experiments. However, as shown here, the fluorescence decay of CFP is biexponential even in the absence of the acceptor. Therefore, a detailed study of the fluorescence properties of CFP is required before using this chromophore for detection of interactions based on quantitative FRET.

Comparison between Homo-FRET and Hetero-FRET for Homodimeric Protein

Fluorescence anisotropy decay measurement is difficult to carry out under the microscope[9,15] from several viewpoints. The method requires (1) correction for the different transmission efficiency of each geometric component of the excited and emitted polarized light and for the depolarizing (or polarizing) effects of the microscope optics, (2) correction for laser fluctuations, acquisition times, and potential photobleaching, and (3) a longer acquisition time than for fluorescence decay analysis.

However, the use of this approach for homo-FRET determination affords several advantages for studying oligomerization of protein. First, it would be possible to determine the degree of multimerization from the shape of the fluorescence anisotropy decay curve, by unraveling the number of energy step migrations and using models for fitting anisotropy decay curves. Second, structural information can be obtained: the distance between the two interacting chromophores and the spatial orientation between the chromophores within the dimer. In the model of a symmetric dimer, the distance between the chromophores is deduced from the determination of the transfer rate (ω) [Eq. (8) providing a mean value for κ^2] and the orientation (θ) between the two chromophores is deduced from the value of residual anisotropy. Third, a weaker FRET efficiency (large distance between chromophores and/or κ^2 close to zero) can be determined by homo-FRET rather than by hetero-FRET. Indeed, a low energy transfer rate is easier to detect from

[20] M. A. Hink, R. A. Griep, J. W. Borst, A. van Hoek, M. H. M. Eppink, A. Schots, and A. J. W. G. Visser, *J. Biol. Chem.* **275,** 17556 (2000).

[21] T. M. H. Creemers, A. J. Lock, V. Subramaniam, T. M. Jovin, and S. Völker, *Proc. Natl. Acad. Sci. U.S.A.* **97,** 2974 (2000).

fluorescence anisotropy decay than from the fluorescence decay of a GFP-tagged protein. The homotransfer is detected by a fast fluorescence depolarization, much faster than the fluorescence depolarization due to rotation of the dimeric protein, whereas a weak hetero-FRET efficiency is detected by a small decrease in fluorescence lifetime.

Here we show that, although FRET-based interaction determination is a widely used method, its application to living cell studies requires the performance of extensive quantitative analyses of fluorescence properties (kinetics of fluorescence decays and fluorescence anisotropy decays). In addition, according to the type of interaction, hetero- or homodimer, the methodology, hetero- or homo-FRET, must be judiciously chosen to obtain the best information about structural data within the macromolecular complex.

Acknowledgments

We are indebted to Dr. Robert B. Pansu and Patrick Denjean for technical advice, to Prof. Jean-Claude Nicolas for fruitful discussion, and to Dr. Fabien Gerbal for critical reading of the manuscript. This work was supported by grants from the European Union (BIO4 CT97 2177), the Association pour la Recherche sur le Cancer (Grants 9222 and 5632 to J.C. and Grants 9518 and 5936 to V.M.), the Groupement des Entreprises Françaises de Lutte contre le Cancer, and le Laboratoire Glaxo Wellcome. M. Tramier and T. Piolot were supported by European Union fellowships.

[26] Spinning Disk Confocal Microscope System for Rapid High-Resolution, Multimode, Fluorescence Speckle Microscopy and Green Fluorescent Protein Imaging in Living Cells

By PAUL S. MADDOX, BEN MOREE, JULIE C. CANMAN, and E. D. SALMON

Introduction

We describe here a spinning disk confocal fluorescence microscope system we initially assembled for fluorescent speckle microscopy (FSM) of the assembly dynamics and movements of individual microtubules and actin filament arrays within cells.[1–4] In this application, image contrast is generated by polymer assembly from

[1] C. M. Waterman-Storer and E. D. Salmon, *Biophys. J.* **75,** 2059 (1998).
[2] C. M. Waterman-Storer, A. Desai, and E. D. Salmon, *Curr. Biol.* **8,** 1227 (1998).
[3] C. M. Waterman-Storer and E. D. Salmon, *FASEB J.* **13,** S225 (1999).
[4] P. Maddox, K. Bloom, and E. D. Salmon, *Nat. Cell Biol.* **2,** 36 (2000).

0076-6879/03 $35.00

a cytoplasmic subunit pool containing a small fraction of fluorescently labeled subunits (typically 1% or less).[1,3] The random nature of subunit association creates a nonuniform fluorescent "speckle" pattern along the polymer lattice during polymerization. FSM can also be used to record the binding and release of microtubule-associated proteins (MAPs) on the microtubule lattice.[5] Optimum contrast is obtained for fluorescent speckles containing only a few fluorophores, often five or fewer, within the maximum resolution limits of the light microscope.[3] This requires a high-resolution imaging system designed for maximum sensitivity to detect few fluorophores without significant photobleaching problems. Previously, we have used wide-field epifluorescence microscopy and a cooled charge-coupled device (CCD) camera with high quantum efficiency and low noise.[6] Similar instrumentation has worked well for thin specimens such as the lamella of mammalian tissue culture cells,[7] neuronal growth cones,[8] flattened axons,[9] and newt cells in mitosis, which remain flat.[2] However, for thicker polymer arrays such as the mitotic spindle in most tissue cells and embryos, better depth discrimination is needed to permit resolution of specific fibers, such as bundles of kinetochore microtubules, and to prevent the loss of speckle contrast produced by out-of-focus fluorescence from speckles in polymers above and below the plane of focus.

Inoué and co-workers have described the advantages of the Yokogawa CSU-10 Nipkow spinning disk confocal scanning unit for obtaining high-quality fluorescent images with brief exposures and low fluorescence bleaching.[10,11] Stimulated by these studies, we have combined the CSU-10 unit with efficient microscope optics and a panchromatic CCD camera with the high quantum efficiency and speed to facilitate high spatial and temporal resolution fluorescent imaging at multiple wavelengths. In addition to applications in FSM, the resolution and sensitivity of this instrument have proved valuable for live cell imaging of green fluorescent protein (GFP) fusion proteins.[12] Also, the high signal-to-noise ratio of images obtained with this instrument has provided the opportunity to obtain three-dimensional (3-D) immunofluorescent images of extraordinary resolution and image quality by constrained iterative deconvolution.[13]

[5] J. C. Bulinski, D. J. Odde, B. J. Howell, E. D. Salmon, and C. Waterman-Storer, *J. Cell Sci.* **114,** 3885 (2001).

[6] E. D. Salmon, S. L. Shaw, J. Waters, C. M. Waterman-Storer, P. S. Maddox, E. Yeh, and K. Bloom, *Methods Cell Biol.* **56,** 185 (1998).

[7] C. M. Waterman-Storer and E. D. Salmon, *J. Cell Biol.* **139,** 417 (1997).

[8] N. Kabir, A. W. Schaefer, A. Nakhost, W. S. Sossin, and P. Forscher, *J. Cell Biol.* **152,** 1033 (2001).

[9] S. Chang, T. M. Svitkina, G. G. Borisy, and S. V. Popov, *Nat. Cell Biol.* **1,** 399 (1999).

[10] S. Inoué and T. Inoué, *Methods Cell Biol.* **38,** in press (2002).

[11] P. Maddox, A. Desai, E. D. Salmon, T. J. Mitchison, K. Oogema, T. Kapoor, B. Matsumoto, and S. Inoué, *Biol. Bull.* **197,** 263 (1999).

[12] C. P. Pearson, P. S. Maddox, E. D. Salmon, and K. Bloom, *J. Cell Biol.* **152,** 1255 (2001).

[13] D. A. Agard, Y. Hiraoka, P. Shaw, and J. W. Sedat, *Methods Cell Biol.* **30,** 353 (1989).

The major features of the Yokogawa CSU-10 spinning disk confocal unit have been described in detail elsewhere.[10,14–16] Here we describe our system integration and its important features for FSM and other applications. Waterman-Storer *et al.*[17–19] provide detailed methods for FSM analysis of microtubule and actin filament assembly dynamics and motility within cells, including biochemical methods for obtaining fluorescent tubulins and actins, procedures for their incorporation into living cells, and kymograph analysis of fluorescent speckle motility.

Overview of Instrument Components

Table I lists the major components, accessories, and sources for our spinning disk microscope and the deconvolution systems. Figure 1 shows the major components of the spinning disk microscope system. A Nikon TE-300 inverted microscope equipped with epifluorescence optics, phase-contrast optics, and a transmitted light shutter from Vincent Associates (Rochester, NY) is mounted on a vibration isolation table. A Nikon (Melville, NY) remote-focusing device uses a stepping motor to control microscope *z*-axis focus. The Yokogawa CSU-10 is fastened to a standard "C-mount" adaptor on the side port of the microscope as seen in more detail in Fig. 2. The CSU-10 obtained from PerkinElmer Wallac (Bethesda, MD) has a Sutter filter wheel mounted on the back for controlling wavelength selection. Excitation light is supplied from an optical fiber connected to a 100-W argon–krypton air-cooled laser. A Hamamatsu Orca ER cooled CCD camera (Hamamatsu Photonics, Bridgewater, NJ) is connected to the C-mount connector at the output of the CSU-10. Shutter control, wave-length selection, focus, time-lapse image acquisition, image storage, and routine image processing are controlled by MetaMorph digital imaging software (Universal Imaging, Downingtown, PA) in a PC computer system (Fig. 1, left). Image deconvolution from a stack of optical sections through a specimen is performed with an off-line DeltaVision deconvolution system and Softworx software (not shown; Applied Precision, Issaquah, WA).

[14] C. Genka, H. Ishida, K. Ichiomoi, Y. Hirota, T. Tanaami, and H. Nakazawa, *Cell Calcium* **25,** 199 (1999).

[15] A. Ichihara, T. Tanaami, K. Isozaki, Y. Sugiyama, Y. Kosugi, K. Mikuriya, M. Abe, and I. Uemure, *Bioimages* **4,** 57 (1996).

[16] A. Ichihara, T. Tanaami, H. Ishida, and M. Shimizu, *in* "Fluorescent and Luminescent Probes," 2nd Ed., p. 344. Academic Press, New York, 1999.

[17] C. M. Waterman-Storer, A. Desai, and E. D. Salmon, *Methods Cell Biol.* **61,** 155 (1999).

[18] C. M. Waterman-Storer, *in* "Current Protocols in Cell Biology" (J. L. Schwartz, ed.). John Wiley & Sons, New York, 2001.

[19] C. M. Waterman-Storer, E. D. Salmon, and W. M. Bement, *J. Cell Biol.* **150,** 361 (2000).

TABLE I

Instrument Components and Their Sources

Nikon TE-300 epifluorescence inverted microscope (Nikon USA, Melville, NY)
Lens options
×100/1.4-NA Planachromatic bright field, ×100/1.4-NA Planachromatic phase 3, ×60/1.4-NA Planachromatic bright field
Epi-fluorescence filter sets
DAPI, HiQ FITC, and HiQ Texas Red (Nikon USA, Melville, NY)
Air table
6 ft by 4 ft by 8 inches (Technical Manufacturing, Peabody, MA)
Transmitted light
Lamp house (HMX-2)
Unibliz shutter for transmitted light (Vincent Associates, Rochester, NY)
IR-cut filter
Phase optics in condenser
0.85-NA dry condenser lens
Nikon focus accessory
Spinning disk confocal unit
Yokogawa CSU-10 (PerkinElmer Wallac, Bethesda, MD)
488 single-line filter set
568 single-line filter set
Sutter Lambda 10.2 filter wheel (PerkinElmer Wallace)
Custom dual-dichroic and emission filter set (488- and 568-nm excitation; Chroma Technology, Brattleboro, VT)
Laser
Omnichrome 100-mW argon–krypton mixed gas air-cooled laser (488- and 568-nm laser lines; PerkinElmer Wallac)
Single-mode fiber optic cable with manipulator (PerkinElmer Wallac)
Power meter
Orion-PD (Ophir Optronics, Danvers, MA)
Cooled CCD camera
Hamamatsu Photonics CCD Orca ER (Hamamatsu Photonics, Bridgewater, NJ)
Digital image acquisition and processing system
IBM-compatible PC
750-MHz Pentium III processor
768-MB RAM
50-GB hard drive
8× CDR
21-inch Sony Trinitron monitor
MetaMorph imaging software, version 4.6 (Universal Imaging, Downingtown, PA)
Fluorescence overlay option
Motion analysis option
Hamamatsu acquisition option
MuTech mv-1500 digitizing board (CCD interface)
Deconvolution system
Silicon Graphics Octane 1 computer (Silicon Graphics, Mountain View, CA)
Dual processors (300-MHz MIPS r12000 ip30 processors)
1.2-GB RAM
Primary hard drive, 17 GB
Secondary external SCSI hard drive, 34 GB
8× CDR
IRIX64 release 6.5 operating system
Softworx software version 2.5 (Applied Precision, Issaquah, WA)

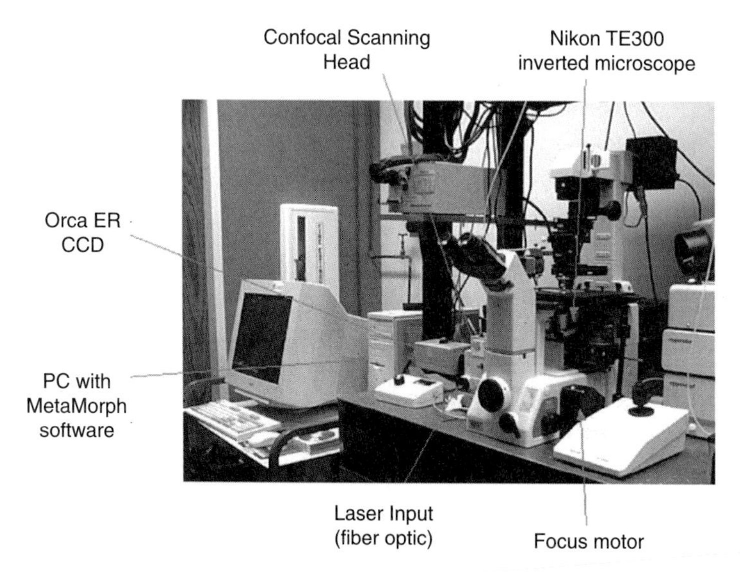

FIG. 1. Photograph of the spinning disk confocal microscope system in the Salmon laboratory. Key parts are labeled. See Table I for details.

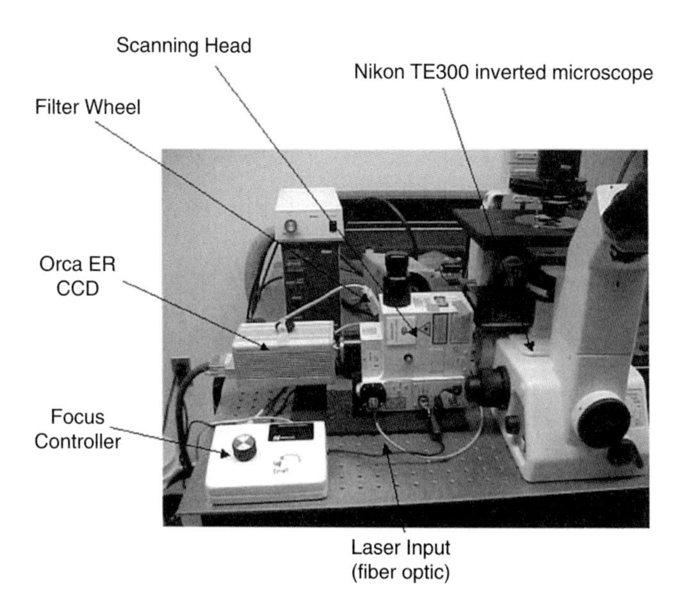

FIG. 2. Close-up view of the Yokogawa CSU-10 mounted at the output port of the Nikon TE-300 inverted microscope. The prototype Hamamatsu Orca ER camera is attached to the output port of the CSU-10. Note that the fiber optic cable from the laser attaches to the CSU-10 on the lower left-hand side. The Sutter excitation filter wheel is mounted to the CSU-10 on the back side (not in view).

Rationale for Component Selection

Inverted Epifluorescence Microscope

Our imaging system is based on a Nikon TE-300 inverted microscope stand. We chose an inverted stand for several reasons. First, the stage on the Nikon inverted microscope is stable and resists focal drift (less than 2 μm/hr). A stable stage is extremely important for time lapse imaging of living specimens and accurate z series. Second, manipulations of samples (for instance microinjection) are facilitated with the use of a long working distance condenser. Finally, the inverted stand allows for easy access to the optical paths of the microscope. Focus is controlled by using a Nikon stepper motor, capable of accurate steps of 100 nm or greater for optical sections. We use standard epifluorescence optics and filter cubes for locating specimens by eye and then switch to an open filter slot for confocal imaging.

We typically use a Nikon $\times$100/1.4-NA (numerical aperture) Plan Apochromat objective for maximum confocal effect. It is light efficient and the $\times$100 magnification is needed to make a 0.5-μm-diameter illumination spot on the specimen by light passing through the 50-μm-diameter pinholes in the spinning disk. Lower magnification objectives increase proportionally spot size and potential out-of-focus light.

The $\times$100 magnification is also needed to match the optical resolution of a 1.4-NA objective to the 50-μm diameter of the pinhole in the spinning disk.[15,16] Fluorescent point sources in the specimen are spread out at the image plane by the diffraction of light at the objective aperture (the point-spread function, PSF, of the objective[20]). The Airy pattern describes the intensity distribution in the x–y plane of focus as a function of the radial distance from the optical axis. Optical resolution is given by the radius, r, of the first minimum of the central Airy disk by Eq. (1)[21]:

$$r = 0.61\lambda/NA_{\text{obj}} \tag{1}$$

For GFP fluorescence at 510 nm, and $\text{NA}_{\text{obj}} = 1.4$, then $r = 0.22$ μm and the predicted diameter of the Airy disk is 0.44 μm. For a $\times$100 objective, this Airy disk diameter at the spinning disk will be 44 μm, slightly smaller than the 50-μm pinhole diameter. For 600-nm wavelength red fluorescence excited by the 563-nm laser line, $r = 0.27$ μm, and the diameter of the Airy disk central maximum at the spinning disk is 53 μm, for the $\times$100 objective, slightly larger than the pinhole diameter.

The $\times$100 magnification slightly oversamples the CCD we currently use (see Cooled Charge-Coupled Device Camera, below), ensuring that resolution is limited by the optics and not the pixel size of the CCD chip. The optimal magnification, M,

[20] E. H. Stelzer, *in* "Imaging Neurons: A Laboratory Manual" (R. Yuste, F. Lanni, and A. Konnerth, eds.), p. 12.1. Cold Spring Harbor Laboratory Press, Cold Spring Harbor, NY, 2000.

[21] S. Inoué and K. R. Spring, "Video Microscopy—The Fundamentals," 2nd Ed. Plenum Press, New York, 1997.

needed for sufficient contrast to resolve overlapping Airy disks separated by radius r depends both on the sampling frequency of the pixel detector elements in the detector and the signal noise.[20] For FSM we have used the following criteria[1,3]:

$$M = (3 \times \text{pixel width})/\text{optical resolution} \tag{2}$$

The optimal magnification from Eq. (2) for $r = 0.22\ \mu\text{m}$ is 85, for the 6.45-μm pixel size of the CCD chip in our Orca ER camera.

For samples requiring phase-contrast as well as fluorescence imaging, we use a Nikon ×100/1.4-NA Plan Apo Phase 3 objective. Using a phase lens is somewhat problematic for low light-level, high-resolution applications because the phase ring in the back aperture of the objective reduces the signal by 10–15% and slightly distorts the objective point-spread function (PSF; see Image Acquisition, Storage, and Processing, below). A Vincent shutter is placed in front of the 100-W quartz halogen lamp used for transmitted light illumination and this shutter is used to control illumination for phase-contrast recordings. A heat-reflecting filter is placed in the illumination light path to prevent specimen damage and to remove infrared light from the camera image.

For samples with extremely low fluorescence imaging, we use one of two options to increase the fluorescence signal without overilluminating (bleaching) the sample. The first is to bin the CCD chip 2 × 2 (binning increases the signal by 4-fold without increasing the readout noise). Using the ×100/1.4-NA objective and binning the CCD result in loss of lateral resolution in the image by a factor of about 2, because of undersampling by the camera detector, but does not decrease the confocal performance of the instrument. Another option that does not sacrifice as much lateral resolution is to use a ×60/1.4-NA objective and 1 × 1 binning. The ×60 lens passes more light than the ×100 lens to the image ($I_{\text{image}} \approx 1/M^2$) but only slightly undersamples the CCD chip. However, the ×60 lens reduces the confocal ability of the microscope by creating an illumination spot size larger than optimal and producing Airy disk images of points in the specimen smaller than the 50-μm size of the pinholes in the spinning disk. This effect is not as significant for the larger diameter Airy disks of red fluorescence compared with green.

Yokogawa CSU-10 Spinning Disk Confocal Scanner

The CSU-10 mounts between a standard output port of the microscope and the camera. As mentioned above, the CSU-10 is mounted to a C-mount adaptor on the side port of the TE-300 Nikon inverted microscope and the cooled CCD camera is mounted on a C-mount adaptor on the output port of CSU-10 (Fig. 2). A slight adjustment of the ocular focus makes viewing the specimen by eye using the standard epifluorescence illumination in the microscope parfocal with the camera. CSU-10 illumination and imaging is achieved by moving the microscope filter cube carrier to an open position to allow illumination and image formation at the microscope side port without interference from any filter cubes.

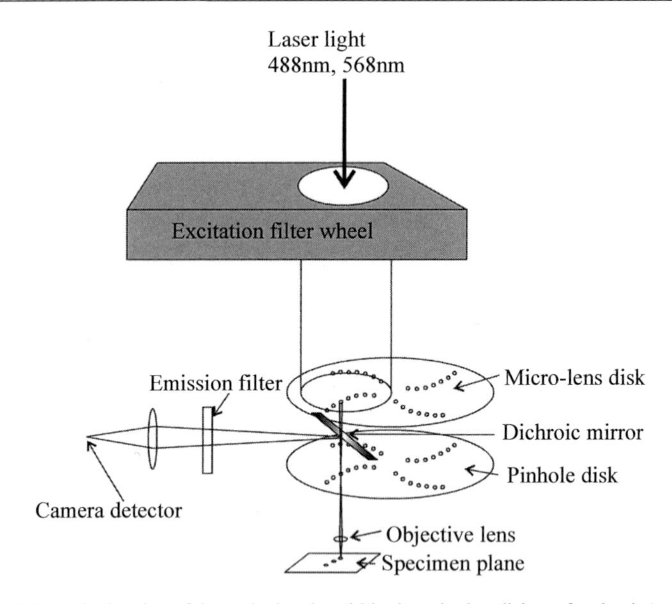

FIG. 3. Schematic drawing of the optical paths within the spinning disk confocal unit (see Refs. 10, 15, and 16). Laser light enters the unit via a fiber optic cable (not shown). The excitation wavelength is chosen by the filter wheel, and then the light passes through the microlens disk (see text). Excitation light from the microlenses passes through the dichromatic mirror and is focused on the pinholes in the second spinning disk. Emission light is collected by the objective lens and then descanned through the pinholes. The dichromatic mirror reflects the emission light through an emission filter and relay optics focus the light onto the camera detector.

The major features of the CSU-10 confocal scanner are diagrammed in Fig. 3. The scanner uses a Nipkow disk positioned at an intermediate image plane so that the holes in the disk are par-focal with the specimen plane of the microscope. Figure 4 shows an image of the Nipkow disk obtained with the Orca ER camera when the motor that spins the disk is turned off. As mentioned above, the pinholes are fixed in size, about 50 μm in diameter. They are spaced in a constant pitch helical array[10,15] with a spacing between the centers of the holes of about 250 μm. This spacing is needed to significantly prevent fluorescent light excited by illumination through one hole from entering an adjacent hole (Fig. 4). With these dimensions, there are about 1000 beams within a 7×10 mm imaging frame at the Nipkow disk. The illumination area on the specimen depends inversely on objective magnification, and for the $\times 100$ objective it is about 70×100 μm. In the smaller field of view of our CCD camera, there are about 800 pinholes at any one time (Fig. 4). The disk spins at a constant speed of 1800 rpm. The pinhole array on the Nipkow disk of the CSU-10 is designed to raster scan 12 image frames in one rotation. At 1800 rpm, the disk scans 360 frames/sec.[10,15,16]

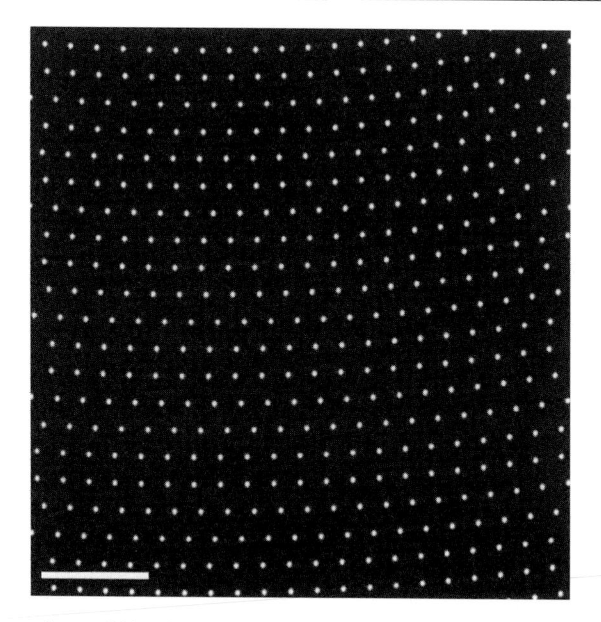

FIG. 4. Camera image (900 × 900 pixels) of the pinholes. The motor for the spinning disk was turned off and an image was recorded, using a 200-ms exposure with transmitted light illumination of an empty slide–coverslip preparation viewed with a ×100/1.4-NA bright-field objective. Note that the pinholes have about a 50-μm diameter and occupy a small percentage of the disk area, about 5%. Bar: 5 μm.

A novel feature of the CSU-10 is that it has two disks: one is the pinhole Nipkow disk described above and the other is an upper disk that contains microlenses at positions aligned with the pinholes on the lower Nipkow disk (Fig. 3). Both upper and lower disks are mechanically connected and spun at the same rate. Our measurements from images like those in Fig. 4 indicate that the holes in the Nipkow disk represent about 4% of the disk area. A novel feature of the CSU-10 is that the lenses on the upper disk pass about 40% of the light incident on the upper disk through the pinholes. This helps significantly to overcome a major defect of the Nipkow disk design: poor incident light illumination intensity.[10,15,16]

The excitation light path within the CSU-10 begins with the output from the fiber optic cable connected to the laser light source. Lenses expand the light from the end of the fiber into a wide beam capable of illuminating the 7 × 10 mm imaging area of the disks (Fig. 3). The illumination light passes an 8-position Sutter Lambda 10.2 filter wheel (provided by PerkinElmer Wallac; Table I) before becoming incident on the spinning disks. The Sutter wheel has variable speeds (adjustable by software) and is attached directly to the confocal unit (see Fig. 2). We typically use two excitation filters in the wheel: 488 and 568 nm. All other positions in the wheel contain aluminum disks to shutter the light. The electronic shutter within the CSU-10 unit requires about 0.5 sec to open or close. This is much too slow for our applications. Therefore, we use the filter wheel for both excitation wavelength selection as well as to shutter the excitation light between exposures. This filter wheel can switch between positions within 50 ms, but we typically use 100 ms to prevent vibrations from reducing image resolution. To eliminate this vibration problem completely, it would be better to place the device for wavelength selection between the laser and the optical fiber input to the CSU-10.

Another major optical feature of the CSU-10 design is that the dichromatic mirror is placed between the upper and lower disks (Fig. 3). Excitation light is collected by the microlenses on the upper disk and projected straight through the dichromatic mirror to the conjugate pinholes of the lower disk and focused by the objective lens onto the specimen. Fluorescent light from an image point in the specimen excited by light from an individual pinhole is collected by the objective and focused back through the same pinhole. It then is reflected from the dichromatic mirror, redirected by relay optics through an emission filter, and then projected to the output port to the camera (Fig. 3). Note that the emission light does not go through the microlens disk and, unlike conventional Nipkow disk scanners, the excitation light that reflects off the front side of the microlens disk is separated from the emission light pathway. This design helps the CSU-10 achieve the sensitivity and high signal-to-noise ratio (SNR) in images needed to resolve the pattern of fluorescent speckles containing few fluorophores in our FSM applications.

Individual filter sets were available from the manufacturer for maximizing fluorescence for single-wavelength excitations of green or red fluorescence, using 488- and 568-nm laser lines, respectively (Table I). These emission filters have broadband emission spectra. In our living cell FSM studies, we were particularly interested in dual-wavelength imaging using fluorophores excited sequentially by the 488- or 568-nm laser lines by simply changing excitation filters, using the excitation filter wheel as described for wide-field FSM microscopy.[6] For this application, Chroma Technology constructed multiple bandpass dichromatic mirror and emission filters (Table I) that have excellent transmission efficiency within their bandwidths and low levels of "bleedthrough" between the different fluorescent channels. Our measured bleedthrough between green and recording red is 1.4% and between red and green is 0.02%.

To maximize the light throughput of the system, it is crucial to ensure that the spinning disk is lined up on the optical axis of the side port of the microscope. To test this, the spinning disk is switched on and illuminated with either 488- or 568-nm light. An arc lamp focusing device is placed on the objective turret and the position of the illuminating light observed (as if you were focusing the wide-field epifluorescence arc lamp). The circular pattern should be centered on the focusing device. If it is not, the screws on the side port of the microscope are loosened and the confocal head shimmed until it is centered. Then the screws are retightened. We also use stiff foam placed under the camera to support the camera and to dampen the vibration produced by the excitation filter wheel.

Laser Light Source

We currently use a 100-mW argon–krypton air-cooled laser that has most of its energy in either the 488-nm line or the 568-nm line (Table I). The first adjustment is properly aligning the laser to the fiber optic cable. The laser light is coupled by a single-mode fiber optical cable between the laser and the CSU-10 (Table I). The laser can be aligned so that about 50–70 mW passes through the fiber optic cable into the confocal head. Alignment of the fiber should be done by using a high-sensitivity light power meter (Table I) with the laser on the lowest power setting, stand-by, to reduce risk of injury to persons and the tips of the fiber optic cable.

Cooled Charge-Coupled Device Camera

For FSM we needed diffraction-limited resolution and as close as possible to photon-limited sensitivity for resolving and detecting red or green fluorescent speckles containing only a few fluorophores so that excessive photobleaching would not be a major problem. In addition, we also had several applications requiring image acquisition at 3–5 frames/sec.

Fortunately, at the time we were assembling this imaging system, we were able to obtain a prototype camera from Hamamatsu that contained a newly developed CCD chip that meets our needs (Fig. 2). The current camera, an Orca ER, has about 60–70% quantum efficiency over a broad spectrum including the green and red fluorescent wavelengths (Fig. 5). There is no shutter in this camera so that shutter vibration and speed are not a problem. The CCD is an interline chip design, in which alternate columns of pixel elements are masked.[21] To achieve high efficiency

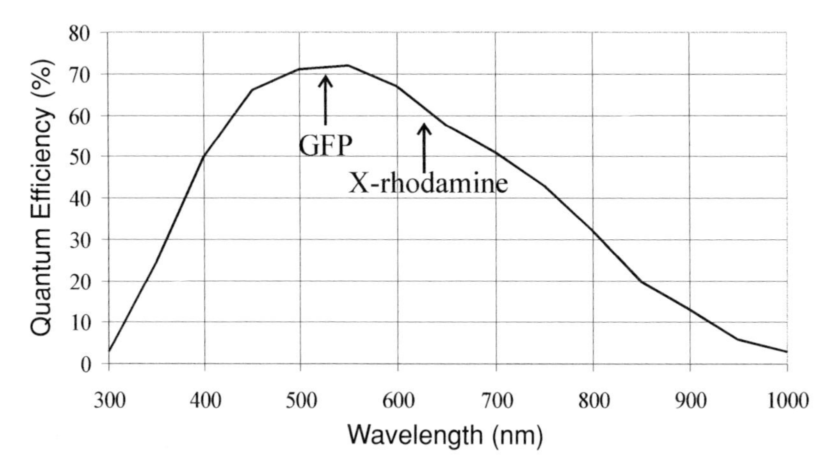

FIG. 5. Spectral response curve of the Orca ER CCD camera. The graph shows quantum efficiency as a function of wavelength. The CCD has a high (~60–70%) quantum efficiency over a broad visible spectrum (450–650 nm). [From www.hamamatsucameras.com\sys-biomedical\orca-er\default.htm]

of light collection to the pixels in the unmasked columns, microlenses are placed over each pixel to collect light from both the open and masked regions. A full frame for the imaging system reads 1280×1024 light-collecting pixel elements in the chip. These pixels have a 6.45-μm center-to-center spacing along their rows and between columns.

During an exposure, photoelectrons are collected by the unmasked pixels and at the end of the exposure, the collected electrons are rapidly transferred to adjacent pixels in the masked rows and readout to the computer through high-speed digital connections. Exposures can be selected from 1 ms to 10 sec. We typically use between 200- and 1000-ms exposures, depending on sample fluorescence intensity.

Each pixel has a maximum well capacity of about 20,000 photoelectrons. Counting these electrons uses 12-bit analog-to-digital conversion, where 2^{12} is equal to 4096 gray levels in an image. The conversion rate is about 4.9 electrons per digital count, so that 4096 gray levels corresponds approximately to the well capacity of 20,000 electrons.

The readout rate for the camera is 14.7 megapixels/sec, which corresponds to a full frame transfer time to the computer of about 120 ms. Faster rates of image transfer to the computer can be achieved by reading out subarrays within the chip. We often use this feature of the camera for imaging small cells such as budding yeast or regions within larger cells.

The other remarkable feature of this camera is that the readout noise at the 14.7 megapixels/sec readout rate is equivalent to about 8 electrons. This low noise is important for detecting weak fluorescence such as that from a few fluorophores. For a full well capacity of 20,000, the useful dynamic range of the camera is $20{,}000/8 = 2500$.

To increase sensitivity with the Orca ER camera, we often use 2×2 binning. This combines the electrons collected in four adjacent pixels on the chip and reads them out as 1 pixel. This gives a 4-fold increase in signal-to-readout noise by sacrificing a 2-fold decrease in lateral resolution and no effect on z axis resolution (see above). In many applications where fluorescence is weak and photobleaching is a problem, 2×2 binning gives a sufficient signal-to-noise ratio (SNR) and a proportional reduction in the rate of photobleaching by reducing the intensity of illumination. For FSM, 2×2 binning reduces the resolution of fluorescent speckles along microtubules or within bundles of microtubules, so we try to avoid binning if possible.[3]

Image Acquisition, Storage, and Processing

The filter wheel, focus motor, transmitted light shutter, camera settings, and image acquisition are controlled via a PC computer (Table I) running MetaMorph imaging software (Table I). The main features of the computer are 768 MB of RAM memory, a 50-GB hard disk, an 8× CD writer, and a 21-inch monitor with a large-enough viewing screen to display three channels of images as well as windows for controlling image acquisition. The Hamamatsu Orca ER camera is connected to the computer bus through a MuTech digitizing board that is controlled by the MetaMorph imaging software. Image acquisition, display, and storage are done with journal routines similar to those we have described previously for wide-field multimode, multiwavelength digital imaging microscopy.[6] The green fluorescent, red fluorescent, and phase-contrast images are each displayed and stored as an image stack, with each image marked with the time of initial acquisition.

An important feature of the MetaMorph software for live cell applications is that each image stack can be reviewed on the screen in between image acquisitions or by pausing image acquisition during time-lapse mode. This allows the investigator to judge if focus needs correction or to judge when another experimental manipulation is needed. All fluorescent images are stored as 12-bit images while transmitted light (e.g., phase-contrast) images are converted to 8-bit images before storage to save memory. Currently we transfer image data to CD-ROM for permanent storage.

The MetaMorph software also has extensive analysis tools for image intensity measurement, contrast enhancement, arithmetic motion analysis, and color encoding.[6,17,18] The software also has subroutines we often use for generating red, green, and blue 24-bit color image stacks from three 12-bit image stacks, aligning images within a stack for analysis by both translation and rotation, making montages of sequential images, and generating kymographs of fluorescent speckle movement.[17,18]

Delta Vision Off-Line Deconvolution System

We have been performing constrained iterative deconvolution and three-dimensional (3-D) image reconstruction of confocal immunofluorescent images, using a Delta Vision off-line system (Table I[13]). This software produces significant improvement in z-axis resolution, enhances lateral resolution and, in general, greatly improves the contrast of fine structural detail in the specimen as well as providing 3-D views. We purchased the Softworx software with the ability to directly import MetaMorph image stacks. These image stacks are sent from the PC computer to the Silicon Graphics (Mountain View, CA) workstation (Table I) by file transfer protocol (FTP) through the Internet. To deconvolve the confocal images, the software is calibrated to the microscope as described in detail by Agard *et al.*[13] Briefly, a PSF is experimentally obtained by acquiring a z series of 100 images at 100-nm steps through a 175-nm fluorescent bead (Molecular Probes, Eugene, OR) and stored as a MetaMorph image stack. A different z-series image stack through a fluorescent bead is acquired for each objective, and also for each wavelength. These z-series image stacks are then loaded into the Silicon Graphics computer and a subroutine in a pull-down menu is used to calculate the "optical transfer function" (OTF) for each objective at the different wavelengths used. These

OTFs are then applied to experimental data, producing the deconvolved data by a constrained iterative deconvolution method.[13] The deconvolved data can then be rendered into a 3-D projection for accurate views of complex structures. Deconvolution is also effective in removing the weak "cross-talk" (about 0.5%) that occurs between adjacent pinholes of the spinning disk. Cross-talk means that emission light from a single point is collected by multiple pinholes, causing some loss of contrast. This combination of software and hardware creates a robust platform for high-resolution, high-contrast digital imaging.

Listed below are the specific steps we use for deconvolution and 3-D image reconstruction, because they have not been described in detail elsewhere.

1. Image stacks (.stk) are acquired by MetaMorph. For immunofluorescent specimens we typically acquire 100 images separated by 100-nm steps along the z axis.
2. Files are then transferred by FTP to the Delta Vision computer (SGI Octane Silicon Graphics) in the .stk format.
3. In the SGI, files are converted from MetaMorph Stacks (.stk) to Delta Vision (_dv) files by manually entering x and y pixel spacing as well as z step size, from a pull-down menu.
4. The .stk_dv files are then deconvolved, using a wavelength- specific optical transfer function (OTF) file.
5. The deconvolution method is done by using 15 constrained iterations (the default settings). This "ratio" or "conservative" method will more reliably come to a solution. For finer resolution more iterations can be used.
6. On deconvolution, the files (now called .stk_dv_d3d) are then color combined by using the Image Fusion subroutine in Softworx.
7. The color-combined files are then processed to create a 3-D reconstruction by using the Volume Viewer in Softworx. The standard 3-D reconstruction can be done around either the x or y axis for 180 or 360° with 15 projections or 45° around the normal or z axis. The viewing parameters that are used include maximal intensity method, best quality for optimal z section interpolation, and best z resolution. Custom rotations can be done in which one varies the number of the projections (and thus the angle between each projection), the start angle on the rotation, or the axis of rotation.
8. Quick projections, that form an image by taking the maximal intensity for each point through the stack, can also by made by using the Quick Projection subroutine for the _d3d files.

Examples

Our major applications for the spinning disk confocal microscope are for imaging structural dynamics in living cells, cytoplasmic extracts, or reconstituted preparations. Imaging fluorescently labeled structures in live cells presents several practical problems. First, living cells do not react well with high-intensity light and photobleaching is a major problem, particularly for FSM, where speckles contain few fluorophores. Second, live cells are highly dynamic. Third, many studies involving live cells focus on subcellular structures and require high resolution. The spinning disk confocal microscope described here uses a sensitive, low-noise CCD camera for detection that allows the use of low-excitation light levels and provides longer observation periods with acceptable image SNR.

We have found our spinning disk instrument to be superior to laser scanning confocals for time-lapse imaging of live cells, including mammalian tissue cells, yeast cells, and *Drosophila* embryos. One advantage is that much less excitation light intensity is required for the same total exposure time.[10] For example, to obtain a 1000×1000 pixel image using a laser point scanning instrument requires that fluorescence for each pixel element be recorded within 1 μs. This will require about 50,000 times greater intensity than needed for a 1-sec exposure, using our spinning disk instrument (the spacing between the pinholes reduces the net illumination time during the 1-sec exposure). Lower excitation light intensities also avoid the problem of fluorescence saturation that can occur with high-intensity illumination.[22] The high quantum efficiency and low noise of the CCD camera also appear much better than that of the scanners, photomultipliers, and readout electronics commonly used with the laser scanning confocal instruments.[23] The CCD also has a rapid readout, making it possible to acquire data at 5 frames/sec and that offers insight into dynamic processes. Combining a low-noise, fast-readout CCD with the spinning disk confocal unit and a high-numerical aperture objective lens produces a high-contrast, high-resolution image without the use of image-processing tools. However, postprocessing by deconvolution can enhance image quality if needed. Below is a brief description of several of our applications for the spinning disk confocal microscope that highlight its performance features.

Our spinning disk confocal system has important advantages for *in vitro* assays of molecular dynamics. It can image clearly individual fluorescent speckle microtubules assembled *in vitro* in thick chambers, where out-of-focus fluorescence makes individual microtubules nearly invisible by normal wide-field microscopy.[24] Figure 6 compares images of fluorescent speckled microtubules bound to the inner surface of a coverslip that sits on top of a 70-μm-thick chamber of fluorescent microtubules at steady state assembly with a free tubulin subunit pool. About 2% of the tubulins are labeled with X-rhodamine. Individual microtubules are barely detectable from the background fluorescence by wide-field imaging with the CCD camera. In contrast, views of the same specimen (different region of the coverslip) with the spinning disk and CCD camera system and the same objective yield

[22] M. A. Weber, F. Stracke, and A. J. Meixner, *Cytometry* **36,** 217 (1999).
[23] J. R. Swedlow, K. Hu, P. D. Andrews, D. S. Roos, and J. M. Murray, *Proc. Natl. Acad. Sci. U.S.A.* **99,** 2014 (2002).
[24] S. V. Grego, V. Catillana, and E. D. Salmon, *Biophys. J.* **81,** 66 (2001).

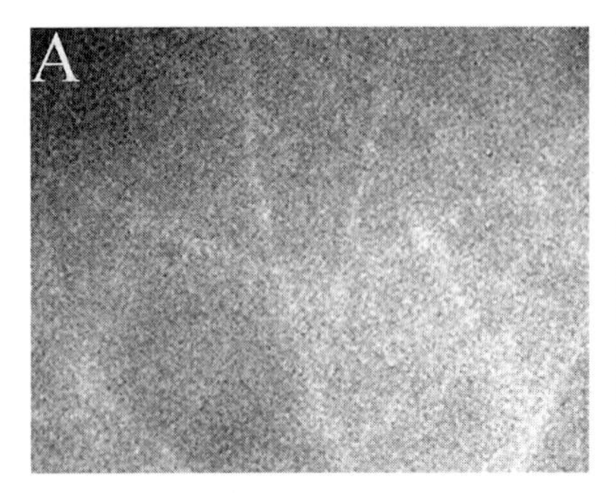

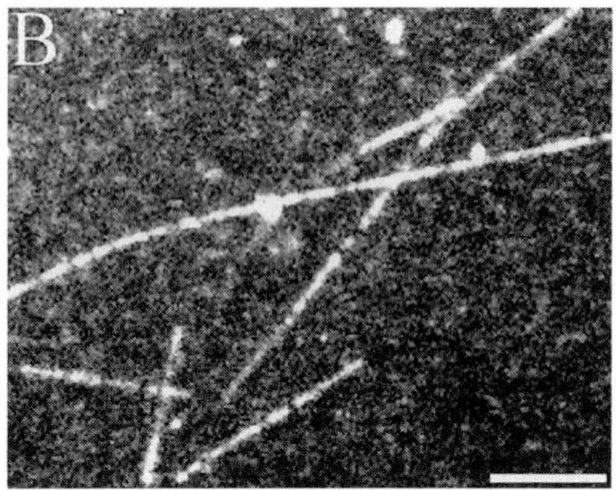

FIG. 6. Comparison of (A) wide-field and (B) spinning disk confocal images of fluorescent speckle microtubules attached to the inner surface of a coverslip by kinesin motor proteins in a 70-μm-thick chamber of self-assembled microtubules and unassembled tubulin dimer subunits. About 2% of the tubulins are labeled with X-rhodamine to produce fluorescent speckled microtubules as they polymerize. Images were obtained with the Nikon $\times$100/1.4-NA Plan Apochromat objective and no binning in the camera. See S. V. Grego, V. Catillana, and E. D. Salmon, *Biophys. J.* **81,** 66 (2001) for details (reprinted with permission of the Biophysical Society). Bar: 5 μm.

high-contrast images of both microtubules and the fluorescent speckles along their lattice. The ability of the spinning disk confocal system to reject out-of-focus fluorescence is a major feature of the instrument important for FSM of cells, embryos, and extracts.

For live cell imaging, we have used our spinning disk confocal system to rapidly image GFP–tubulin-labeled microtubules and other spindle proteins in small cells such as budding yeast,[12] in large embryonic cells such as those of *Caenorhabditis elegans* (Fig. 7), and in mammalian tissue cells for long periods of time ($\sim$100–1000 images collected) with minimal photobleaching. Figure 7 shows an image of spindle and astral microtubules in a *C. elegans* embryo in first division. This embryo is expressing low levels of GFP-α tubulin[25] and microtubules near the coverslip exhibit fluorescent speckles (data not shown). The image shown is at the center of the 40-μm-thick embryo. Light scattering by the yolk platelets makes imaging microtubule fluorescent speckles nearly impossible, but the rejection of out-of-focus fluorescence still provides images of excellent clarity for spindle and astral microtubule organization within the embryo. The spinning disk confocal allows fast, high-resolution imaging of the rapid embryonic mitoses of this genetic model system.

Drosophila melanogaster is another genetic model organism that can be difficult to image. The first 14 divisions are syncytial and the cell cycle time is short, about 15 min.[26] The spinning disk confocal has also provided a fast, high-contrast, high-resolution solution to imaging microtubule dynamics of the early embryonic

[25] K. Oegema, A. Desai, S. Rybina, M. Kirkham, and A. A. Hyman, *J. Cell Biol.* **153,** 1209 (2001).
[26] V. E. Foe, C. M. Field, and G. M. Odell, *Development* **127,** 1767 (2000).

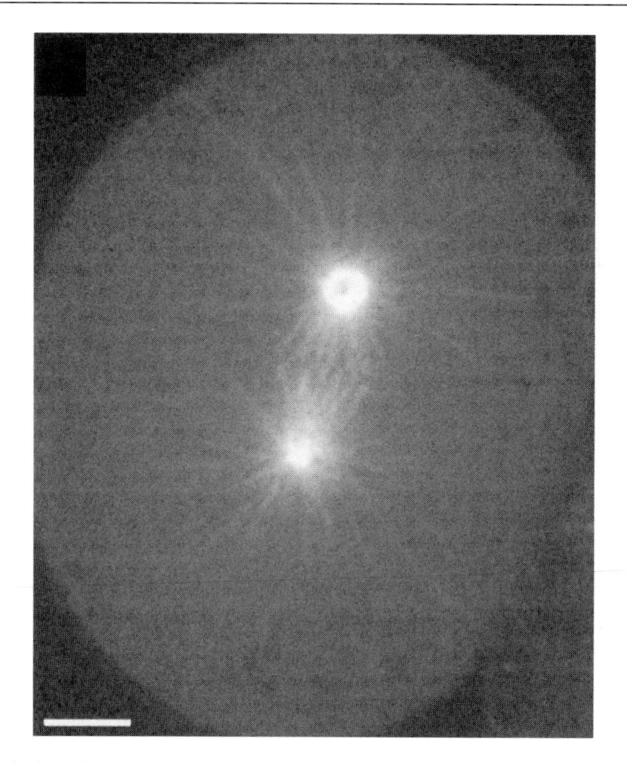

FIG. 7. Rejection of out-of-focus fluorescence by the spinning disk confocal system produces clear images of the assembly dynamics of fluorescent microtubules in the first division of *C. elegans* embryos. A frame from a time-lapse record is shown. The worms are constitutively expressing GFP–α-tubulin. Images were obtained with the Nikon ×100/1.4-NA oil immersion objective using 2 × 2 binning in the camera. Bar: 5 μm.

divisions. In *Drosophila* embryos, where the mitotic spindles draw close to the embryo surface, we have been able to achieve excellent FSM of microtubule assembly dynamics within the spindles (data not shown) with our spinning disk confocal system.

In budding yeast, which are about 5 μm in diameter, we have been able to image GFP–tubulin fluorescent speckles within individual astral microtubules.[4] Tran *et al.*[27] have shown that the spinning disk confocal is effective for FSM of interphase microtubule dynamics within fission yeast.

S. Inoué, in pioneering work with polarization microscopy (reviewed in Inoué and Salmon[28]), discovered by the early 1950s that mitotic spindle fibers were assembled in a dynamic equilibrium with a cellular pool of subunits (now known

[27] P. T. Tran, L. Marsh, V. Doye, S. Inoue, and F. Chang, *J. Cell Biol.* **153,** 397 (2001).

[28] S. Inoué and E. D. Salmon, *Mol. Biol. Cell* **6,** 1619 (1995).

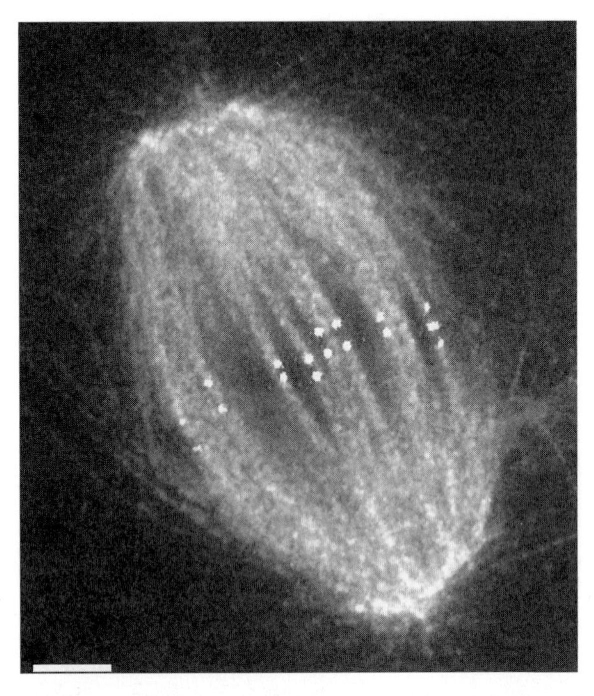

FIG. 8. High-resolution spinning disk confocal FSM of microtubules relative to kinetochores in a *Xenopus* extract spindle. Shown is an image from a time-lapse series recorded with the Nikon ×100/1.4-NA Plan Apochromat objective and no binning in the camera. Red fluorescent speckles on the microtubules were obtained by adding low amounts of X-rhodamine tubulin to the *Xenopus* egg extracts during spindle assembly. Kinetochores were selectively labeled with green fluorescent antibodies to the kinetochore-specific protein CENP-A. In this black-and-white image, the bright green kinetochores are superimposed on the red fluorescent speckle image of spindle microtubules. Bar: 5 μm.

to be the tubulin subunits of microtubules). Fluorescence speckle and confocal microscopy with our spinning disk system show with great temporal and spatial resolution the growth dynamics and poleward flux of individual microtubules within the spindle responsible for the spindle dynamic equilibrium. Figure 8 shows an FSM image taken with our spinning disk confocal system of a metaphase spindle assembled in isolated cytoplasmic extracts of *Xenopus* egg extracts.[2,29] With the spinning disk confocal system we are able to resolve both fluorescent speckles within microtubules and the fibrous structure of the spindle fibers. By wide-field imaging, the density of fluorescence needed to detect individual bundles of microtubules within the spindle, like kinetochore fiber microtubule bundles, makes detection of fluorescent speckle contrast nearly impossible. In Fig. 8, kinetochores are marked by the bright white dots of fluorescence in the black-and-white image

[29] A. Desai, P. S. Maddox, T. J. Mitchison, and E. D. Salmon, *J. Cell Biol.* **141,** 703 (1998).

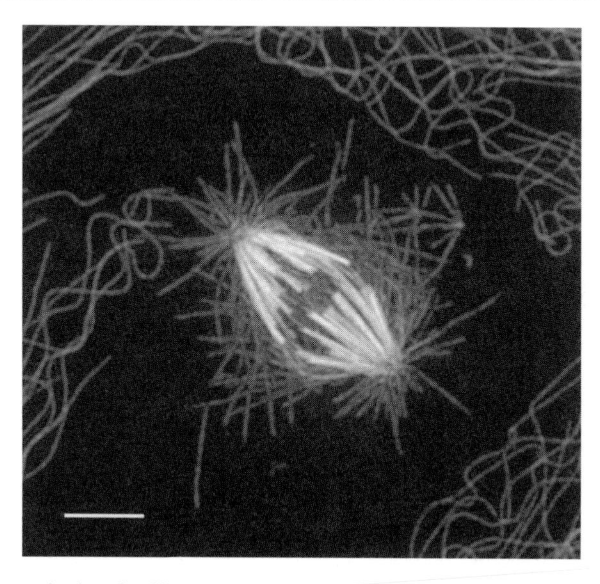

FIG. 9. Deconvolved confocal images of immunofluorescence specimens yield a view of extraordinary high resolution, clarity, and SNR of immunofluorescently labeled microtubules in a mitotic spindle and surrounding interphase PtK1 mammalian tissue cells. Red fluorescent confocal optical sections were obtained with the Nikon ×100/1.4-NA Plan Apochromat objective at 100-nm *z*-axis steps though the specimen with 20 planes included in the stack for both regions above and below the specimen. The total image stack was deconvolved in the DeltaVision system and rendered in 3-D. The image shown is a maximum-intensity projection along the *z* axis through the 3-D volume. Bar: 5 μm.

of the red fluorescent speckles within the spindle and astral microtubules. These dots were obtained from the green fluorescence of a kinetochore marker as described in the caption to Fig. 8. These dots mark the kinetochore ends of bundles of kinetochore microtubules. Time-lapse movies of images similar to that in Fig. 8 show that fluorescent speckles form at kinetochores and flux poleward at the same rate as the majority of microtubules in the spindle. These data were used to show that at metaphase, microtubule polymerization occurs at kinetochores at the rate of poleward microtubule flux (about 2 μm/min at 19°) and that microtubule depolymerization occurs near the spindle poles at the same rate because the spindle maintains a steady state constant length.[11]

In addition to our live cell and extract studies, we have been able to use our spinning disk confocal system in combination with deconvolution to obtain high-resolution, high-quality, multicolor 3-D images of the distribution of the molecular components of the mitotic spindle, chromosomes, and kinetochores within immunofluorescently stained PtK1 mammalian tissue cells.[30] Figure 9 shows the microtubule channel from a multicolor deconvolved image of microtubules (red)

[30] D. Cimini, B. Howell, P. Maddox, F. Degrassi, and E. D. Salmon, *J. Cell Biol.* **153,** 517 (2001).

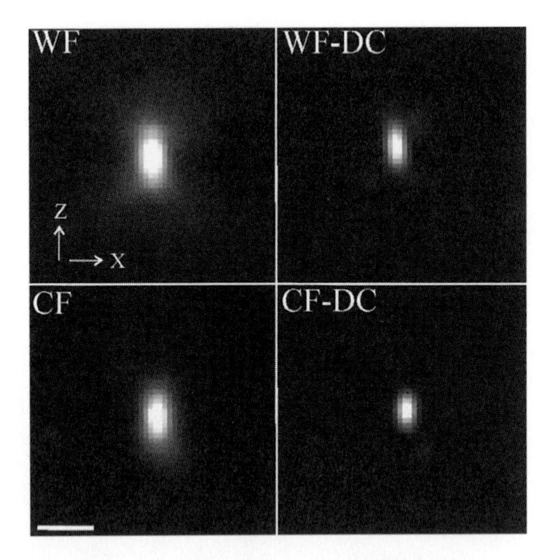

FIG. 10. Comparisons of objective point-spread functions (PSFs) in the x–z plane (side view) that were obtained for green fluorescent beads (175 ± 5 nm; Molecular Probes, Eugene, OR) by wide-field (WF) and spinning disk confocal (CF) before (WF, CF) and after deconvolution (WF-DC and CF-DC). Images were obtained with the Nikon ×100/1.4-NA Plan Apochromat objective and no binning in the camera. Bar: 1 μm.

and chromosomes (green, not shown). The remarkable SNR of the spinning disk images makes the exceptional clarity of the microtubules images possible.

Our last example is a comparison of the lateral and z-axis image resolution of a 175-nm green fluorescent bead obtained by wide-field, spinning disk confocal and by deconvolution of wide-field and spinning disk z-axis image stacks. A Nikon ×100/1.4-NA Plan Apochromat objective was used to obtain 100 images at 100-nm steps though the specimen for both the wide-field and spinning disk image stacks. These stacks were then sent to the DeltaVision system and 3-D views obtained. In Fig. 10, we show the x–z intensity pattern of the bead PSF along the optical axis through the center of the bead. The width of the pattern in the x direction at focus is the lateral width of the Airy disk image of the bead, while the width of the pattern along the z axis is the depth of the central maximum of the Airy pattern along the z axis.[20,31]

One measure commonly used in confocal microscopy for lateral and longitudinal resolution[31] is to take x-axis and z-axis line scans through the point of focus and measure values for the full width of the profile at half the intensity of the peak (FWHM). Our measured FWHM values from x-axis scans were 330 nm for

[31] T. Wilson, *in* "Handbook of Biological Confocal Microscopy" (J. B. Pawley, ed.), p. 167. Plenum Press, New York, 1996.

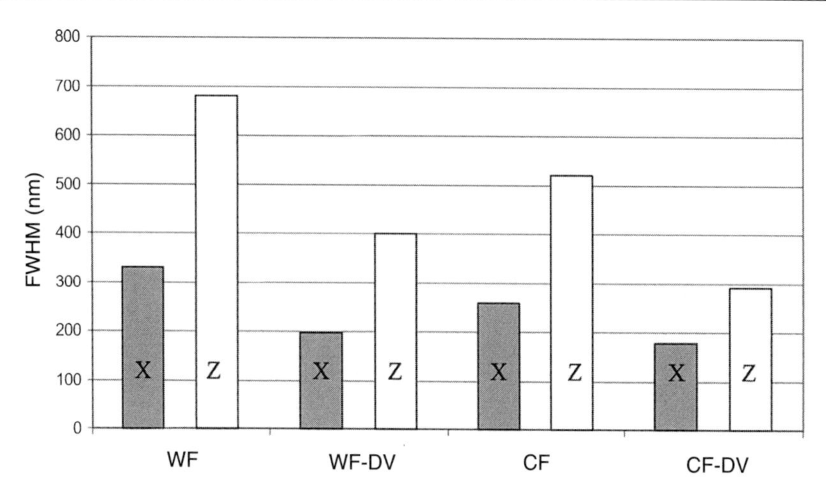

FIG. 11. Comparison of full-width half-maximal (FWHM) values obtained from line scans in the x-axis (X) and z-axis (Z) directions through the focal point of the PSFs shown in Fig. 10 for wide-field (WF), confocal (CF), and deconvolved wide-field (WF-DC) and confocal (CF-DC) images.

wide-field, 200 nm for wide-field with deconvolution, 260 nm for confocal, and 180 nm for confocal with deconvolution. FWHM values from z-axis scans were 680 nm for wide-field, 400 nm for wide-field with deconvolution, 520 nm for confocal, and 290 nm for confocal with deconvolution. These measurements are plotted in Fig. 11 for comparison. The confocal improves both lateral (by 21%) and z-axis resolution (by 24%) compared with wide-field imaging, but most significant is the improvement in z-axis resolution. For both lateral and z-axis resolution, wide-field plus deconvolution gives better performance than the spinning disk confocal without deconvolution, 40 and 41%, respectively. However, deconvolution of the spinning disk images gives the best resolution, by about 9% over wide-field plus deconvolution for lateral and 27% for z-axis resolution.

Acknowledgments

We thank many people for input in the development of the instrumentation described here for our FSM, live cell, and high-resolution imaging studies of microtubules and mitosis. They include Shinya Inoue, Clare Waterman-Storer, Wendy Salmon, Phong Tran, Arshad Desai, Karen Oegema, Tim Mitchison, Chris Field, Sonia Grego, Daniella Cimini, Jennifer Turneuer, Jean-Claude Labbe, Chad Pearson, Dale Beach, Kerry Bloom, Elaine Yeh, Tarun Kapoor, Lisa Cameron, and Bonnie Howell. We also thank Hamamatsu for support in providing the prototype Orca ER, Universal Imaging Corporation for needed modification to MetaMorph software, and the Cell Division Group for support during summers at the Marine Biological Laboratory where practical aspects of this instrumentation were tested and improved. Major support was provided by NIH GM60678.

[27] Single-Particle Tracking Image Microscopy

By KEN RITCHIE and AKIHIRO KUSUMI

Introduction

The classic model of Singer and Nicholson (1972),[1] known as the fluid mosaic model, suggested the importance of lateral diffusion of membrane proteins as a key process for their function and interactions with other membrane components. This inspired cell biologists to study their diffusivity. In this model, the lipids act as a two-dimensional (2-D) fluid solvent in which the proteins freely diffuse. However, Bulk techniques suggested that the membrane was not a simple fluid, but rather contained mobile and immobilized (possibly through cytoskeletal attachment) protein and perhaps even a compartmentalized structure. This led to investigations by single-molecule techniques. We now think that the plasma membrane of the cell contains assemblies of lipids that may act as stable or transient platforms for macromolecular signaling complexes[2] (see Fig. 1).

Although the ability to monitor single molecules is now available, the question arises as to what problems will require the viewing of single molecular behavior to obtain new information. Problems requiring single-molecule techniques include those in which (1) one is looking for an infrequent event, (2) one is looking for a process that averages out in bulk (i.e., an unsynchronized event over a large population where averaging removes the signal), and (3) the direct interaction between two or more molecules is of interest. The time required to gain sufficient data to determine average values makes single-molecule techniques unsuitable for finding parameters that are easily determined through bulk averaging methods such as fluorescence recovery after photobleaching (FRAP, also known as fluorescence photobleaching recovery, FPR) for determining the average diffusion coefficient of a membrane species and in determining the immobile fraction on a given time scale.[3] On the other hand, single-molecule tracking, as is described in detail, can specifically determine whether the immobile fraction is bound and stationary or freely moving but restricted from long-range diffusion.

This article explains the technique of single-particle tracking (SPT) and the associated technique of applying optical tweezers to the SPT probe. The technique of SPT involves the binding of a small colloidal particle, in our case a 40-nm-diameter gold particle, through an antibody IgG or its Fab fragment to a membrane-bound molecule, such as a protein or lipid. This small gold particle can then

[1] S. J. Singer and G. L. Nicolson, *Science* **175,** 720 (1972).
[2] K. Simons and E. Ikonen, *Nature (London)* **387,** 569 (1997).
[3] K. Jacobson, A. Ishihara, and R. Inman, *Annu. Rev. Physiol.* **49,** 163 (1987).

0076-6879/03 $35.00

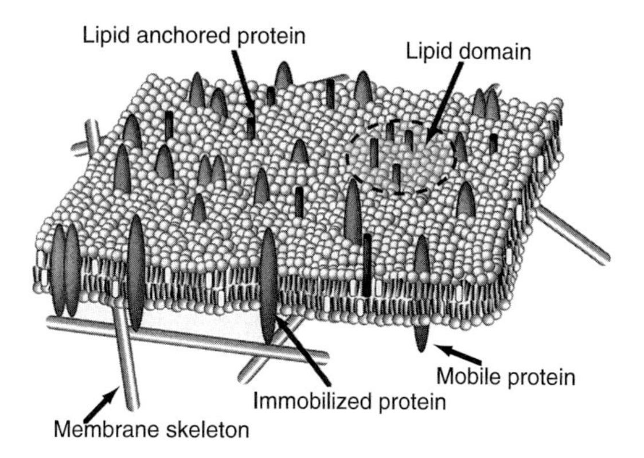

FIG. 1. Diagram of the plasma membrane of a cell, consisting of a fluid bilayer composed of lipids and cholesterol with embedded proteins. Proteins can be lipid anchored or span the membrane. Some membrane-spanning protein is attached to the underlying actin cytoskeleton and hence is immobile. Other proteins are free to diffuse along this crowded interface. Lipid and cholesterol-based domains may increase association among some membrane molecules. The exterior extracellular matrix of secreted polymers and proteins is not shown.

be directly visualized through bright-field, contrast-enhanced, video microscopy. After accurately tracking the position of the gold particle over time, one obtains the trajectory of the labeled molecule through the membrane. As such, SPT can monitor the motion of molecules in the plasma membrane of live cells.

A direct image of the trajectory of a molecule in the membrane allows one to see the diffusive characteristics of the molecule under scrutiny. If this is a receptor molecule, the addition of ligand may induce, say, a short-time temporary localization of the molecule during signaling that would be difficult to visualize by other techniques. Furthermore, the basic properties of the plasma membrane are probed by knowledge of the diffusion of protein and lipid, challenging the simple models of the fluid bilayer.

Background

When discussing SPT the strong complementary technique of FRAP[3,4] should also be mentioned. FRAP is a technique in which a fluorescently labeled species in the membrane is selectively photobleached by an intense laser pulse in a defined area. Recovery of fluorescence due to diffusion from surrounding regions indicates both the diffusion coefficient (or multiple diffusion coefficients if more than one exists, such as for activated and quiescent molecules) and the immobile fraction, that is, the amount of fluorescence that does not recover in the time scale set by

[4] D. Axelrod, D. E. Koppel, J. Schlessinger, E. Elson, and W. W. Webb, *Biophys. J.* **16,** 1055 (1976).

the experiment. The limitation of such a bulk technique is that it is not possible to determine the actual mode of diffusion uniquely or to isolate events in the diffusion of a given molecule (i.e., stopping during signaling).

In 1985, the technique of single-particle tracking was first used to directly image the diffusion of transmembrane proteins.[5] De Brabander *et al.* successfully applied the use of nanoscopic colloidal gold probes to measure the native movements of molecules.[6–9] The group of Sheetz in 1989 extended the technique to view the movement of single molecules at nanometer-level spatial resolution.[10,11] From this, they could begin to probe the modes of motion of single molecules in live cell membranes and from that infer the control mechanism in place that organizes the membrane molecules.[12] They also applied this technique to the nanometer-scale motion of motor proteins.[13]

Single-Particle Tracking

Introduction

The technique of single-particle tracking was developed to be able to visualize the movements of single (or small numbers of) surface molecules, primarily protein at first, on the plasma membrane of live cells by optical microscopy. Because direct optical imaging of a desired molecule is not possible, a microscopic probe particle with large refractive index, such as a colloidal gold particle, provides a marker that can be tracked. The probe particles, invisible to the eye under an optical microscope, can be seen with a bright-field or Nomarski microscope with electronic contrast enhancement. The results show a direct view of the random walk/directed motion of the constituents of the membrane. To date, many proteins and lipids have been studied in attempts to determine the control mechanisms in place in the plasma membrane with as little perturbation of the cell as possible. To limit the possibly undesirable effects of cross-linking many proteins under a larger probe, we focus on the preparation and use of a 40-nm colloidal gold probe.

[5] M. de Brabander, G. Geuens, R. Nuydens, M. Moeremans, and J. Demey, *Cytobios* **43,** 273 (1985).

[6] M. de Brabander, R. Nuydens, H. Geerts, and C. R. Hopkins, *Cell Motil. Cytoskel.* **9,** 30 (1988).

[7] M. de Brabander, R. Nuydens, G. Geuens, M. Moeremans, and J. Demey, *Cell Motil. Cytoskel.* **6,** 105 (1986).

[8] M. de Brabander, R. Nuydens, A. Ishihara, B. Holifield, K. Jacobson, and H. Geerts, *J. Cell Biol.* **112,** 111 (1991).

[9] H. Geerts, M. Debrabander, R. Nuydens, S. Geuens, M. Moeremans, J. Demey, and P. Hollenbeck, *Biophys. J.* **52,** 775 (1987).

[10] M. P. Sheetz, S. Turney, H. Qian, and E. L. Elson, *Nature (London)* **340,** 284 (1989).

[11] D. F. Kucik, E. L. Elson, and M. P. Sheetz, *Nature (London)* **340,** 315 (1989).

[12] M. P. Sheetz, N. L. Baumrind, D. B. Wayne, and A. L. Pearlman, *Cell* **61,** 231 (1990).

[13] J. Gelles, B. J. Schnapp, and M. P. Sheetz, *Nature (London)* **331,** 450 (1988).

Gold Preparation

The preparation of colloidal gold probes is a critical part of single-particle tracking. If one expects to obtain reproducible results on a day-to-day basis, the gold probe preparation must be carefully controlled. Our recipes are given for (1) determining the minimal protecting amount of protein required to stabilize the gold colloid, (2) gold probe production for experiment, and (3) cell preparation and gold attachment to cells for experiment, including the appropriate controls. In all cases, care must be taken to ensure that all tubes, pipette tips, and so on, are extremely clean and all solutions are free of contamination. Solutions should be filtered (0.22-μm pore size) to remove contaminants. The presence of any contaminants causes gold particle aggregation, perhaps by induction of nucleation. The cleanliness of storage vessels and solutions correlates well with the shelf life of protected and unprotected colloidal gold particles.

Minimal Protecting Amount Determination. The minimal protecting amount (MPA) is the smallest concentration of the protein/molecule needed to coat the gold that will stabilize the colloid against high salts (i.e., removal of the electrostatic repulsion that usually stabilizes the colloid). This is monitored at an absorbance of 580 nm. The stabilized colloid has a reddish hue. If the colloid aggregates, the solution shifts to a bluish color with an increase in absorbance at 580 nm. This procedure follows that of Leunissen and de Mey[14] (see also Refs. 15–18 for gold preparation and labeling).

MATERIALS

Gold particles (40-nm diameter), pH 7.5 (British Biocell, Cardiff, UK)	6 ml
Protein or ligand at 0.2 mg/ml in buffer A	275 μl
Buffer A (2 m*M* phosphate at pH 7.0 or 2 m*M* borate at pH 9.0)	225 μl
NaCl, 10% (w/v)	500 μl

METHODS

1. To each of 10 Eppendorf tubes add 0.6 ml of the gold colloid.
2. Spin down for 1 min at 2000*g* at 4°. The pellet contains aggregated gold from the colloid and will not be used.

[14] J. L. M. Leunissen and J. R. de Mey, *in* "Immuno-Gold Labeling in Cell Biology" (A. J. Verkleij and J. L. M. Leunissen, eds.), p. 3. CRC Press, Boca Raton, FL, 1989.
[15] J. R. de Mey, *in* "Practical Applications in Pathology and Biology" (J. M. Polak and S. van Noorden, eds.), p. 82. Wright PSG, Bristol, 1983.
[16] J. M. Lucoq and W. Baschong, *Eur. J. Cell Biol.* **42,** 332 (1986).
[17] J. W. Slot and H. J. Geuze, *J. Cell Biol.* **90,** 533 (1981).
[18] P. M. P. V. Henegouwen and J. L. M. Leunissen, *Histochemistry* **85,** 81 (1986).

3. To 10 separate Eppendorf tubes add the protein/ligand and buffer such that tube 1 has 5 μl of protein/ligand and 45 μl of buffer, tube 2 has 10 μl of protein/ligand and 40 μl of buffer and so on, up to tube 10, which has 50 μl of protein/ligand only.

4. To each tube of step 3 add 0.5 ml of the supernatant from the colloidal gold spun down as in step 2.

5. Incubate the 10 tubes from step 4 at room temperature with light shaking for 1 hr.

6. Add 10 μl of 10% (w/v) NaCl to each tube from step 5 and incubate for 5 min.

7. Measure the absorbance at 580 nm for each sample in step 6.

The results of determining the minimal protecting amount for rabbit anti-mouse IgG (ICN Pharmaceuticals, Costa Mesa, CA), using the 2 m*M* phosphate buffer, are shown in Fig. 2. The abscissa presents the concentration of ligand per milliliter of the colloidal gold solution. The ordinate presents the absorbance at 580 nm, normalized by the maximum absorbance found. Note the sharp increase in absorbance at 1.2 μl/ml of gold colloid, indicating that at concentrations of ligand below this value, the amount of ligand on the gold is incapable of stabilizing the gold colloid under high-salt (reduced electrostatic repulsion) conditions. If the value is not found in the range used, the concentration of stock protein should be increased or decreased as necessary and the MPA determined in the same way as described above.

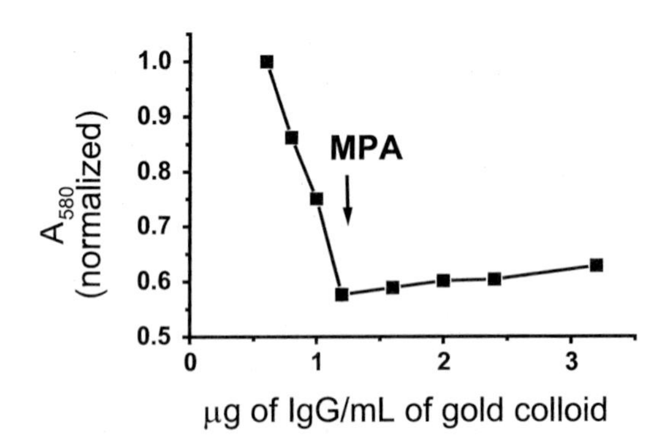

FIG. 2. Determination of the minimal protecting amount (MPA) for rabbit anti-mouse IgG on 40-nm-diameter colloidal gold. The absorbance of the gold–antibody solution at 580 nm increases sharply as the amount of protein decreases, signaling aggregation of the colloid due to insufficient coverage by the protein. The arrow shows the minimal amount of protein required to stabilize the colloid and is used as a reference point to determine the concentration of protein needed to coat the gold particles.

Preparation of Colloidal Gold Probes for Single-Particle Tracking. To perform SPT, at the single-molecule level, the amount of ligand/antibody on the gold particle may be reduced compared with the MPA. As such, a titration of 1×, 1/2×, 1/3×, 1/5×, 1/10×, and 1/30× the MPA of ligand/antibody on the gold should be prepared and tested on the cells to be examined. Analysis of the trajectories will show a convergence of the diffusion coefficient as the amount of ligand/antibody is reduced. To further promote one-to-one association of the gold to the membrane molecule, the Fab fragment of the antibody should be used. One might question that the divalency of the IgG compared with the Fab fragment would be an important factor when many antibodies may be present on the gold. It should be noted that although the gold may be coated with antibody/ligand, the molecules might not all be viable due to association with the gold (possible denaturation on the gold surface; tight binding, reducing freedom to find the correct orientation for binding; etc). As such, it is advantageous to make those molecules that are active, monovalent. We have found that in some cases there is a great reduction in the diffusion coefficient of the target molecule when IgGs are used (unpublished results, 2000).

To stabilize the colloid under isotonic conditions at the lower than MPA concentration, a 20-kDa polyethylene glycol (PEG; Sigma, St. Louis, MO) polymer is added before raising the salt concentration to fill the spaces on the gold surface.

MATERIALS

Gold particles (40-nm diameter), pH 7.5	0.6 ml
Protein or ligand at 0.2 mg/ml in buffer A	x μl
Buffer A (2 m*M* phosphate at pH 7.0 or 2 m*M* borate at pH 9.0)	y μl
PEG polymer (20 kDa), 10 mg/ml	30 μl
Buffer A with 20-kDa PEG at 0.5 mg/ml (buffer A + PEG)	1.5 ml
Hanks' balanced salt solution (HBSS) in 2 m*M* PIPES, pH 7.4, with 20-kDa PEG at 0.05 mg/ml (HBSS + PEG)	0.5 ml

METHODS

1. Put 0.6 ml of colloidal gold solution in an Eppendorf tube and spin down at 2000*g* at 4° for 1 min.

2. In a second Eppendorf tube mix x μl of protein/ligand at a concentration of 0.2 mg/ml with y μl of buffer A. *Note:* x and y are such that the final concentration of protein/ligand in the gold solution is the expected fraction of the MPA and that the volume added is 50 μl in total.

3. Add 0.5 ml of the supernatant from step 1 to the protein/ligand solution in step 2 and mix gently.

4. Incubate the gold and protein/ligand solution for 1 hr at room temperature with gentle mixing.

5. Add 30 μl of PEG, mix gently, and incubate for 15 min.

6. Wash the gold solution three times in buffer A + PEG, spinning down at 12,000g (10 min, 4°) each time, finally resuspending the gold in HBSS + PEG and storing on ice.

At this point, the gold is now conjugated with an antibody or ligand to a specific cell surface molecule. The gold should be used that day and made fresh again the next day if required. Before application to cells, the stock concentration of gold label should be diluted one-fifth by addition of HBSS + PEG.

Cell Preparation and Cell Labeling with Gold. To view the 40-nm colloidal gold probe in an optical microscope, a high-magnification, high-numerical aperture, oil immersion condenser and objective are required. As such, limitations are placed on the applicable cell types. To observe the colloidal particle over long times/distances a flat cell is necessary because of the thin focal plane. Also, the shallow working distance dictates that if the membrane under observation is the apical plasma membrane, where the gold most easily labels, it must be close to the coverslip, making thin adherent cells advantageous.

The following procedure is appropriate to label cell surface molecules with the colloidal gold made in the previous step. Cells are plated in growth medium, after splitting, onto clean coverslips. The restrictions imposed by the high-NA objective and matched condenser require that a thin chamber be constructed. As well, if during an experiment we wish to exchange media, replenish nutrients, or add a drug to the chamber during observation then the corners of the chamber must be left open. Generally, for SPT on an upright microscope, the chambers consist of inverting the cell-containing coverslip onto two L-shaped spacers (thickness, 0.15 mm) on a glass slide. The gold solution, diluted to one-fifth the stock concentration with HBSS + PEG, is added between the glass slide and coverslip. If one wants to label an abundant protein in the membrane with antibody, the possibility of cross-linking many proteins under a single gold particle is high. As such, free antibody may be added to the buffer to reduce the number of proteins to which the gold may attach.[19] The closed edges are sealed with melted paraffin wax. The chamber is then mounted onto the upright microscope. Every 15 min, fresh gold solution is perfused into the chamber through one of the open corners, while a small filter paper shard placed at the apposing corner removes the excess.

As a test for excessive cross-linking of molecules underneath the gold probe, the off-rate of gold particles from the cell should agree with known off-rates for the antibody/ligand used or be comparable to the off-rate of single fluorescently labeled antibody/ligand molecules under similar conditions. Tests for specificity of the probe are also required. After labeling cells, 100-fold excess free antibody can be perfused into the chamber to compete with the labels on the cell. After

[19] M. Tomishige, Y. Sako, and A. Kusumi, *J. Cell Biol.* **142,** 989 (1998).

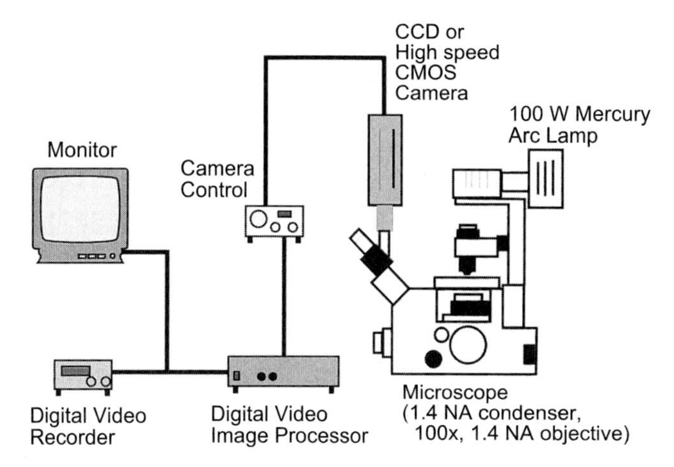

FIG. 3. Schematic of the instrumentation for SPT. The microscope, equipped with a 1.4-NA condenser and a 1.4-NA, ×100 objective, projects its image to a CCD or high-speed CMOS camera. Before recording by a digital video recorder, the signal is contrast enhanced through background subtraction, analogy gain and offset control, and digital enhancement by a digital video image processor.

some time, the particles should have naturally released from the cell and, due to the excess antibody/ligand, no new gold particles should bind. This loss of gold particles implies that the gold has specifically labeled the molecule of interest.

Single-fluorophore video imaging for live cells has become available.[20–25] We now always compare the diffusion data of gold-tagged protein with that of fluorescent probe-tagged protein (via its antibody Fab) to confirm that the gold probes are not cross-linking the target protein. Gold probes have advantages regarding observations for longer periods, higher time resolutions, and experiments using optical tweezers.

Image Acquisition. To image the 40-nm-diameter colloidal gold label, clean optics and well-aligned, Köhler illumination as well as a high-numerical aperture (NA), oil immersion condenser and ×100 objective (preferably an NA of 1.4) are required. A schematic of the SPT experimental setup is shown in Fig. 3. The microscope must also be placed on a floating vibration isolation table to reduce the effect of building vibrational noise on movement of the probe. The entire

[20] G. J. Schütz, G. Kada, V. P. Pastushenko, and H. Schindler, *EMBO J.* **19,** 892 (2000).

[21] G. J. Schütz, M. Sonnleitner, P. Hinterdorfer, and H. Schindler, *Mol. Membr. Biol.* **17,** 17 (2000).

[22] T. Schmidt, G. J. Schutz, W. Baumgartner, H. J. Gruber, and H. Schindler, *Proc. Natl. Acad. Sci. U.S.A.* **93,** 2926 (1996).

[23] G. S. Harms, *Biophys. J.* **81,** 2639 (2001).

[24] R. Iino, I. Koyama, and A. Kusumi, *Biophys. J.* **80,** 2667 (2001).

[25] R. Iino and A. Kusumi, *J. Fluoresc.* **11,** 187 (2001).

microscope is placed into a temperature-controlled chamber with access to the stage and focusing mechanisms to allow experiments at 37°.

Because single colloidal gold particles cannot be seen directly, the image must be enhanced after collection by a video camera. The image is captured either through a high-speed CMOS (complementary metal oxide semiconductor) camera or a charge-coupled device (CCD) camera at a pixel resolution of about 40 nm/pixel. Online image subtraction of a slightly out-of-focus empty background image is followed by analog amplification and digital contrast enhancement of the image. This is accomplished with a Hamamatsu Photonics (Bridgewater, NJ) DVS-3000, Hamamatsu Photonics Argus, or Olympus (Melville, NY) XL-20 in-line video correction unit. The image is recorded by a digital-video tape recorder [Sony (Tokyo, Japan) DSR-20].

High-Speed Single-Particle Tracking. A major advantage of the use of colloidal probes to locate the position of proteins and other membrane-bound molecules is the ability to view these particles by bright-field microscopy at framing rates that exceed the usual rates of the NTSC (National TV Standards Committee) (30 frames per second, fps) or PAL (phase-alternating line) (25 fps).[19] To obtain higher time resolution and faster framing rates, a camera capable of this is required; in addition, the light passing through the condenser must be increased.

The camera we employ is the Photron (San Diego, CA) Ultima-40 CMOS-based camera. It has a 256 × 256 CMOS array that can reach rates from 30 to 40,500 fps (with a limited area of 64 × 64 pixels). This camera has onboard memory for quick caching of images to be downloaded later. During downloading, in-line image processing is applied.

An increase in light along the illumination path is accomplished by removing diffusers and neutral density (ND) filters that may be in the path and, if necessary, changing to white light illumination. Although white light illumination will reduce the accuracy of position detection and will more quickly damage the cells under observation, the increased time resolution may be a worthwhile tradeoff in certain situations.

Analysis. Computer tracking of the positions of the colloidal gold marker is performed according to the method of Gelles *et al.*[13] Briefly, an image of the colloidal particle from, say, the first frame of a video segment is taken as the kernel. This kernel is then cross-correlated with subsequent frames. The correlation function is thresholded to isolate the peak and remove noise. The position of the particle, and hence the protein to which it is bound, is estimated from the geometric center of the correlation function above the threshold, local to the last position of the particle.

From the trajectory thus obtained, the mean squared displacement (MSD) of the probe versus time is calculated.[26] For simple Brownian motion, the MSD–t

[26] Q. A. Hong, M. P. Sheetz, and E. L. Elson, *Biophys. J.* **60,** 910 (1991).

plot should be linear, with slope dependent on the diffusion coefficient of the probe (twice the diffusion coefficient for one-dimensional diffusion, four times the diffusion coefficient for two-dimensional diffusion). Anomalous diffusion is shown by deviations in the MSD–t plot from linear. Of interest here is that when the MSD changes as the square of time the particle is undergoing directed motion (i.e., being specifically dragged in one direction) and when the MSD–t plot asymptotically approaches a finite value, $L^2/6$, the probe is trapped in a domain of linear size L. An intermediate case occurs when there is fast diffusion within a domain with infrequent transitions to an adjoining domain. In this case, the MSD–t plot displays a short time-linear region as well as a long time-linear region with a reduced slope. Statistical analysis of the MSD–t plot yields an estimate of both the domain size and the average residency time in each domain.[19,27]

Accuracy Concerns. When small colloidal particles are employed as reporters of the position of an underlying membrane-bound molecule, the position of the centroid of the particle can be determined with (usually) nanometer precision at video framing rates. One should not confuse the precision with which the center of a Gaussian pattern of colloidal gold can be detected with the resolving power of the microscope.

A perhaps more important question concerns how accurately colloidal probe diffusion represents the diffusion of the underlying molecule. That is, does the addition of this large colloidal marker (large when compared with the diameter of membrane-spanning proteins or lipids) affect the diffusion of the protein and/or is there a significant difference between the position of the molecule and the position reported by the marker.

The diffusion of a disk in a strictly two-dimensional plane suffers from the Stokes paradox, whereby the velocity fields around the diffusing disk do not decay at infinite distance (they have a logarithmic dependence on the radial distance from the disk). Of course, diffusion in a membrane is not strict 2-D diffusion. First, the membrane has a thickness, and thus we are talking of a cylinder diffusing through a thin shell. Second, the membrane is embedded in a three-dimensional, dissipative environment: its aqueous surroundings. Saffman and Delbrück[28] took such effects into account in their classic paper in 1975. They calculated that the diffusion coefficient for a cylinder of radius a, in a sheet of width h and viscosity μ, embedded in a medium of viscosity μ', is

$$D = \frac{k_B T}{4\pi \mu h} \log\left[\left(\frac{\mu h}{\mu' a}\right) - \gamma_E\right] \tag{1}$$

where $k_B T$ is the thermal energy and γ_E is the Euler constant.

[27] A. Kusumi, Y. Sako, and M. Yamamoto, *Biophys. J.* **65,** 2021 (1993).

[28] P. G. Saffman and M. Delbrück, *Proc. Natl. Acad. Sci. U.S.A.* **72,** 3111 (1975).

Thus for a 1-nm-diameter cylinder (to represent a transmembrane-spanning protein) in a 4-nm-thick sheet of viscosity 1 P embedded in a medium of viscosity 1 cP the diffusion coefficient is expected to be 4×10^{-8} cm^2/sec. The diffusion coefficient of a colloidal gold probe of radius $r = 20$ nm in the aqueous environment is (from $D = k_B T/6\pi\mu' r$) 1×10^{-7} cm^2/sec. Thus the diffusion of the gold should not seriously affect the diffusion of the protein under observation.

But does the position of the gold faithfully report the position of the underlying molecule under study? Consider a colloidal gold probe of radius R, anchored to a stationary point through a tether of length d. The fluctuation in position about the stationary point, δ is given by $\delta^2 = dR + d^2$. Thus, if the tether is of length $d = 2$ nm and the colloidal gold is of radius $R = 20$ nm, then the fluctuation would be $\delta = 6.6$ nm, which is comparable to the noise level found when tracking the position of a stationary gold probe [embedded in 10% (w/v) polyacrylamide on a poly-L-lysine-coated coverslip].[19]

Results

Essentially, any membrane molecule that presents an extracellular portion can be tagged by colloidal gold through a suitable linker, be that an antibody, ligand, or a tag such as a fluorescein group or Myc tag. The only restrictions are then operational.

Many transmembrane proteins have been tracked by using the colloidal gold system.[19,27,29–33] Shown in Fig. 4 is a representative trajectory of the transferrin receptor at video rate. Note the detail in the trajectory that cannot be observed by bulk averaging techniques. As can be seen in Fig. 4, the trajectories differ from that expected for free Brownian motion. The motion of the protein seems to be restricted to small domains at short times, followed be infrequent interdomain transitions. The slow long-time diffusion is the result of the slow transitions between domains, a mode of diffusion we term "hop" diffusion.[27]

Discussion

Single-particle tracking has shown itself to be a valuable tool in many applications. One of its strengths is the ability to image the trajectory of a molecule of interest at time resolutions 1000-fold greater than standard video framing rates. This window into short-time diffusive behavior allows the researcher to visualize

[29] E. D. Sheets, G. M. Lee, R. Simson, and K. Jacobson, *Biochemistry* **36,** 12449 (1997).
[30] R. Simson, E. D. Sheets, and K. Jacobson, *Biophys. J.* **69,** 989 (1995).
[31] M. Edidin, S. C. Kuo, and M. P. Sheetz, *Science* **254,** 1379 (1991).
[32] Y. Sako and A. Kusumi, *J. Cell Biol.* **125,** 1251 (1994).
[33] Y. Sako, A. Nagafuchi, S. Tsukita, M. Takeichi, and A. Kusumi, *J. Cell Biol.* **140,** 1227 (1998).

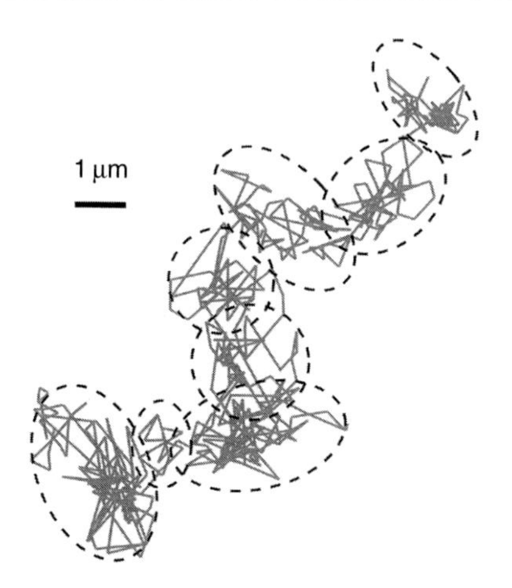

FIG. 4. Trajectory of the transferrin receptor in a live NRK cell as reported by a 40-nm-diameter colloidal gold label. The dashed circles represent areas where the protein resided for long times and are chosen to enhance the detection of the compartmentalization. Note that the protein tends to stay in well-defined compartments with infrequent intracompartmental hops. The data were collected at 30 Hz and 16.6 sec of the trajectory are displayed.

the fine structure of the membrane that is missed when using bulk or slower time resolution techniques.

Further, the use of a colloidal gold label significantly increases the observation time relative to similar fluorescence-based techniques. As such, not only can high time resolution be obtained, but also simultaneously long time observation of the same molecule/gold complex can be observed. Thus tracking to see an infrequent and slow-to-occur event is feasible.

The use of 40-nm-diameter colloidal gold has the added advantage of being able to label single (or few) molecules in the membrane with little disturbance to the diffusion of the labeled molecules. An extension of this technique is to use a colloidal gold particle as a cross-linker to see the motion of spatially clustered groups of molecules. In this way the effects of a cross-linking ligand can be easily simulated and monitored.

Optical Trapping and Optical Force Microscopy

Introduction

To further the ability to directly probe the fine structure of the plasma membrane of live cells, one can combine the above-described technique of SPT with an optical trap. Optical trapping (OT) was first demonstrated by Ashkin *et al.*

in 1986.[34] Here, we discuss the trapping of small, metallic spheres (of diameter less then the wavelength of the trapping beam), which occurs as each particle interacts with a highly focused light beam so that it is centered with a linear restoring force. This simple restoring force makes the trap act as a two-dimensional Hookean spring. Thus it acts both as a noninvasive nanotweezer as well as a force transducer, where the position of the particle reports directly the force the particle receives from its environment.

Combining OT and SPT allows the researcher to gently guide the motion of a membrane molecule. The gold acts dually as the reporter of the position of the molecule, as in SPT, but also as a handle by which to grab the molecule by OT. Early work on this combination has allowed direct and experimenter-controlled evidence of barriers to free diffusion in the membrane.[31,35,36]

The next advance was to use the SPT/OT combination in an imaging mode by which maps of the force found against a given probe molecule could be used to image the structure in the membrane. This technique has been termed scanning optical force microscopy (SOFM, or photonic force microscopy, PFM).[37–39] Briefly, a membrane molecule with a colloidal particle handle is trapped in a long-wavelength laser trap and scanned in a small area, say, 2×2 μm, along a flat section of a live cell plasma membrane.

Instrumentation

Optical Trap Setup. This section describes the essential parts of the optical trapping system. For a strong review of optical trapping methods and procedures, see Sheetz *et al.*[40] The following description refers to Fig. 5.

The trapping laser should be a stable long-wavelength laser, such as the neodymium-doped yttrium orthovanadate (Nd:YVO_4) 1064-nm wavelength laser that we use [or a neodymium-doped yttrium aluminum garnet (Nd:YAG) laser at 1064 nm]. A power of about 1 W would be sufficient in colloidal gold-trapping experiments. The trapping laser is first expanded to fill the entrance diameter of the objective to be used. In our case this takes the form of a 5× beam expander consisting of a 1.2 cm focal length objective, a 25-μm-diameter pinhole spatial filter at the focus, and a final 60-cm focal length lens. Next the beam is directed though two 10-cm focal length lenses that act as a 1× beam expander and can be used to focus the beam to

[34] A. Ashkin, J. M. Dziedzic, J. E. Bjorkholm, and S. Chu, *Optics Lett.* **11,** 288 (1986).

[35] L. P. Ghislain and W. W. Webb, *Optics Lett.* **18,** 1678 (1993).

[36] Y. Sako and A. Kusumi, *J. Cell Biol.* **129,** 1559 (1995).

[37] K. P. Ritchie and A. Kusumi, *J. Biol. Phys.,* in press (2002).

[38] A. Pralle, M. Prummer, E. L. Florin, E. H. K. Stelzer, and J. K. H. Horber, *Microsc. Res. Tech.* **44,** 378 (1999).

[39] E.-L. Florin, A. Pralle, J. K. H. Horber, and E. H. K. Stelzer, *J. Struct. Biol.* **119,** 202 (1997).

[40] M. P. Sheetz, L. Wilson, and P. Matsudaira, eds., "Lazer Tweezers in Cell Biology." Academic Press, San Diego, CA, 1998.

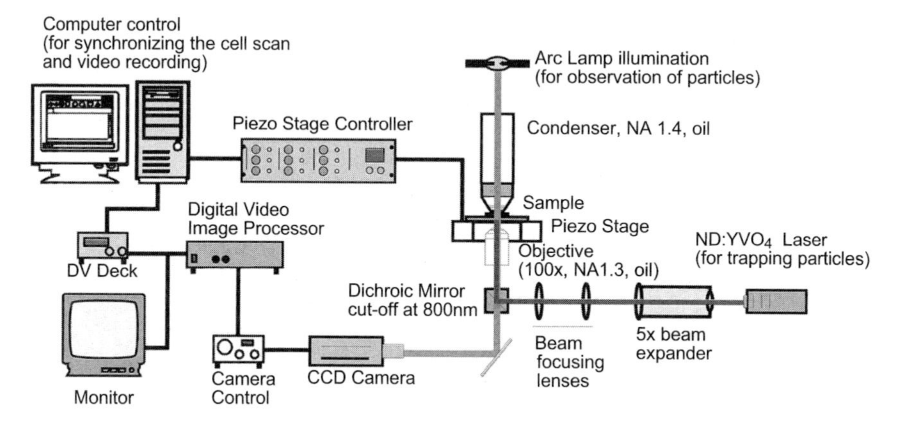

FIG. 5. Schematic of the instrumentation for scanning optical force microscopy. An arc lamp (using the 546.1-nm mercury line) is used for imaging through a 1.4-NA condenser and a 1.3-NA, ×100 objective, as in SPT. The trapping path consists of a Nd:YVO_4 1064-nm laser, a 5×beam expander, and a system of two lens (100-mm focal length) for focusing. The trapping beam is steered into the epiport, bottom port, or side of the microscope and directed to the sample by an 800-nm cutoff dichroic mirror. The imaging system consists of a CCD camera with postvideo enhancement as in SPT (see Fig. 3). A piezo scanning stage is computer controlled to scan the sample in an area under the trapping beam. Computer control also synchronizes video recording to the scanning process.

be parfocal with the image plane of the microscope. The beam is then directed into the microscope objective (×100 Plan Apochromat, 1.3 NA; Zeiss, Thornwood, NY) through an 800-nm cutoff dichroic mirror (Chroma, Brattleboro, VT). The beam can reach the dichroic mirror either through the epiillumination port, before a bottom camera port, or from the side directly. Directing the beam in from the side or bottom still allows use of the epiport for fluorescence measurements.

The power of the laser can be attenuated after expansion by addition of neutral density filters. The system of mirrors and lenses required to direct the beam into the microscope reduces the output power of the laser, in the image focal plane, to one-quarter of its original value (this will vary depending on the coatings on the optics, specifically those inside the objective used).

To drag molecules bound to gold probes around the cellular surface, a piezo-driven scanning stage is required. Computer control of the stage can be easily set up through a digital-to-analog board in the computer to set the control voltage on the piezo power supply. Further, computer control of the recording device to synchronize the scan to the position of the probe is also required.

Optical Trap Spring Constant Calibration. There are three simple methods to calibrate the force–distance relation for the colloidal gold in the laser trap used. The first relies on the thermal vibrations of a free colloidal particle in solution held in the trap. These vibrations will be random and depend directly on the spring constant, k_{sp}, of the trap. The mean squared displacement of the particle from the center of the trap, $\langle \Delta x^2 \rangle$, is set through the equipartition theorem ($k_B T$ is the thermal energy) to be

$$\frac{1}{2} k_{sp} \langle \Delta x^2 \rangle = \frac{1}{2} k_B T \tag{2}$$

With sufficiently large numbers of positions of the particle in the trap the spring constant can be estimated in two perpendicular directions. A caveat is that normal video time resolution is insufficient to accurately determine the position of the particle inside the trap. Consider that one is trying to determine the spring constant of a trap whose actual spring constant is 1 pN/μm. The diffusion coefficient, D_{free}, of a sphere of diameter, $R = 40$ nm, in water (viscosity, $\eta = 1$ cP), is $D_{free} = k_B T / 6\pi \eta R \approx 10^{-7}$ cm^2/sec. Thus in a 33-ms frame of normal video frame rates the particle can travel about 600 nm. The root mean squared width of the trap is only a mere 65 nm. Thus the time averaging of the image capture will completely blur the actual position of the particle. In water, a time resolution of 10 μs is required for 10-nm motions per frame. As an alternative, the viscosity of the solution can be increased with minimal effect on the spring constant through the use of glycerol (note that glycerol cannot be used for latex spheres).

The second method is akin to the first and requires that positions of the particle be obtained at high frequencies. Here, a fitting to the power spectrum, $S(f)$, of the motion of the gold as a function of frequency f gives an estimate of the spring constant. The power spectrum is Lorentzian and is given by

$$S(f) = \frac{k_B T}{\pi^2 \gamma (f^2 + f_c^2)} \tag{3}$$

where $\gamma = k_B T / D_{free}$ is the friction coefficient given by the Einstein equation. The parameter f_c sets the corner frequency. The corner frequency and friction coefficient can be found through fitting the experimental power spectrum. The corner frequency determines the spring constant of the trap through $k_{sp} = 2\pi \gamma f_c$.

A third method involves using the movable stage to apply a drag force on a trapped particle. The moving stage sets a fluid flow past the particle that is counterbalanced by the force of the trap. Changing the speed of chamber movement and/or the viscosity of the medium can allow a direct mapping of the force in the trap through $f = \gamma v$, where v is the velocity of the fluid relative to the gold particle and γ is the friction coefficient. The particle must be kept far away from the coverslip, at least four times its radius away,[41] to avoid additional effects.

Methods and Materials

Preparation of the gold probes and cells is the same as that reported under Single Particle Tracking (above), with the following exception. Because a laser

[41] J. Happel and H. Brenner, "Low Reynolds Number Hydrodynamics with Special Applications to Particulate Media." Noordhoff International Publishing, Leyden, The Netherlands, 1973.

trap is in use, excess gold probes cannot be allowed to float around in the medium above the cells. If such a condition exists, then each time the trap is engaged, many probes from solution will be attracted to the trap and will mask the view of the probe on the cell. As such, the colloidal gold solution should be incubated on the cells for 10 min at either 37°, room temperature, or on ice, depending on the experiment being performed. Excess gold can then be lightly washed off with HBSS and the chamber constructed with HBSS as medium. The incubation period allows the gold to attach to the surface before chamber construction.

Results

Application of laser tweezers to control the motion of transmembrane proteins has been performed to critically probe the barriers to free diffusion in the plasma membrane of live cells.[31,42] Sako and Kusumi[36] directly measured the force required to surpass the compartment barriers found through SPT. A transferrin receptor molecule was dragged laterally along the plasma membrane of NRK cells. Along its path, barriers were sensed through a retardation of the movement of the gold probe–membrane molecule relative to the trap. The lag behind the trap directly gives the force received by the probe as it encounters a barrier, until it either escapes from the trap or passes the barrier, at which time the particle returns to the center of the trap and the force returns to zero. They found that the barriers were elastic, with a spring constant of 1–10 pN/μm.

The extension of this technique is scanning optical force microscopy, where the stage is scanned over an area of the cell and the resistance to movement is used to force image the plasma membrane structure.[37–39] Figure 6 shows the force image produced by repeatedly scanning the same line across the membrane, using the transmembrane protein CD44 on NRK cells.[37] Forces reach at most 0.2–0.3 pN during the scan. Imaged as light areas, the barriers to free diffusion are apparent at a spacing that is consistent with that expected from SPT.

Discussion

There are many benefits to be had by imaging forces in the cell membrane with a colloidal gold handle manipulated deftly with soft optical tweezers. First, the small spring constant of the trap allows a range of forces around 1 pN to be measured. This is about the range expected in such systems as the kinesin motor, which pulls with piconewton levels of force.[43–45] The spring constant of the trap is also tunable, by attenuation of the laser power, allowing the range of applied force to be varied as required for a given experiment. As well, the optical trap

[42] M. Edidin, M. C. Zuniga, and M. P. Sheetz, *Proc. Natl. Acad. Sci. U.S.A.* **91,** 3378 (1994).
[43] C. M. Coppin, D. W. Pierce, L. Hsu, and R. D. Vale, *Proc. Natl. Acad. Sci. U.S.A* **94,** 8539 (1997).
[44] K. Svoboda and S. M. Block, *Cell* **77,** 773 (1994).
[45] H. Kojima, E. Muto, H. Higuchi, and T. Yanagida, *Biophys. J.* **73,** 2012 (1997).

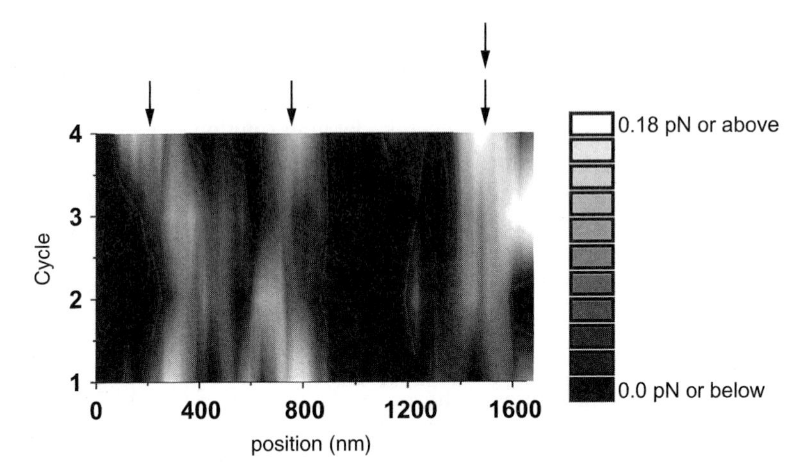

FIG. 6. Image of the forces felt by the transmembrane protein CD44 labeled by a 40-nm colloidal gold particle on repeated scanning of the same line. The array of high-force regions (shown as light regions and marked by arrows, maximum force level of 0.2–0.3 pN) implies the existence of barriers to free diffusion caused by the cytoskeletal meshwork. The tandem arrows show a position with high resistance, implying possibly that there are two barriers crossing at this position. Scan speed, 1.8 pN/sec; 4 cycles; forward scans only.

is essentially noninvasive (the problem of heating is minimized by moving the position of the laser relative to the cell during scanning).

Being a soft spring, the trap does not severely restrict the fluctuations of the probe and thus does not overpower the protein/lipid under observation. This subtle controllability allows many *in vivo* processes to be directly observed. As such, the OT/SPT system allows direct force imaging of molecular interactions, both entropic and enthalpic, with a minimum of disturbance to the system.

Summary

The techniques of single particle tracking (SPT) and optical force microscopy (OFM) as described above allow direct imaging of the motion of molecules in the membrane of live cells, and provide a means of controlling the movement by an almost noninvasive method. Combination of these techniques with other single-molecule methods, such as single-fluorophore imaging, allows direct comparison of motion at video rate (because faster than video rate imaging of fluorophore is still not generally feasible) to determine any effect due to the attached colloidal gold particle. Also, simultaneous use of the two techniques allows for monitoring two molecules, one at high time resolution. As such, the system can then be used in conjunction with green fluorescent protein (GFP) transfection to watch simultaneously the motion of an internal component of, say, a signaling pathway while seeing the motion of the transmembrane signaling receptor.

[28] Diffusion in Cells Measured by Fluorescence Recovery after Photobleaching

By ALAN S. VERKMAN

Introduction

Fluorescence recovery after photobleaching is a useful experimental method for measurement of the translational diffusion of fluorophores and fluorescently labeled macromolecules in cellular compartments. Fluorescent labels are targeted to cellular aqueous or membrane compartments by transfection methods (for the green fluorescent protein) or are incorporated by incubation or microinjection methods. Fluorophores in a defined sample volume are irreversibly bleached by a brief intense laser pulse. Using an attenuated probe beam, the diffusion of unbleached fluorophores into the bleached volume is then measured as a quantitative index of fluorophore translational diffusion. A variety of optical configurations, detection strategies, and analysis methods have been used to quantify diffusive phenomena in photobleaching measurements. This article describes the instrumentation, analytic methods, and fluorescent probes for measurement of diffusion in cells by photobleaching methods.

Instrumentation

The instrumentation for photobleaching consists of a laser/light source to bleach and probe a defined sample volume, a fluorescence microscope, and a detector. A spot photobleaching apparatus with microsecond response that was constructed in our laboratory is shown schematically in Fig. 1A.[1] The beam intensity of an argon ion laser is modulated by an acousto-optic modulator and directed onto a fluorescent sample, using an epifluorescence microscope with a dichroic mirror and objective lens. The excitation path also contains beam-focusing optics, a variable aperture, a variable neutral density filter, and a fast electronic shutter. The shutter blocks excitation light before bleaching and between successive measurements when following slow diffusive processes of seconds to minutes. For full-field illumination to visualize the sample, the zero-order beam is directed by a fiberoptic and lens/shutter system onto a mirror. The emission path contains an interference filter, pinhole (in the back focal plane), and gated photomultiplier detector. The photomultiplier is gated off during the intense bleach pulse, using a circuit that controls dynode voltage. The timing of acousto-optic modulator voltage, shutter

[1] H. P. Kao and A. S. Verkman, *Biophys. Chem.* **59,** 203 (1996).

0076-6879/03 $35.00

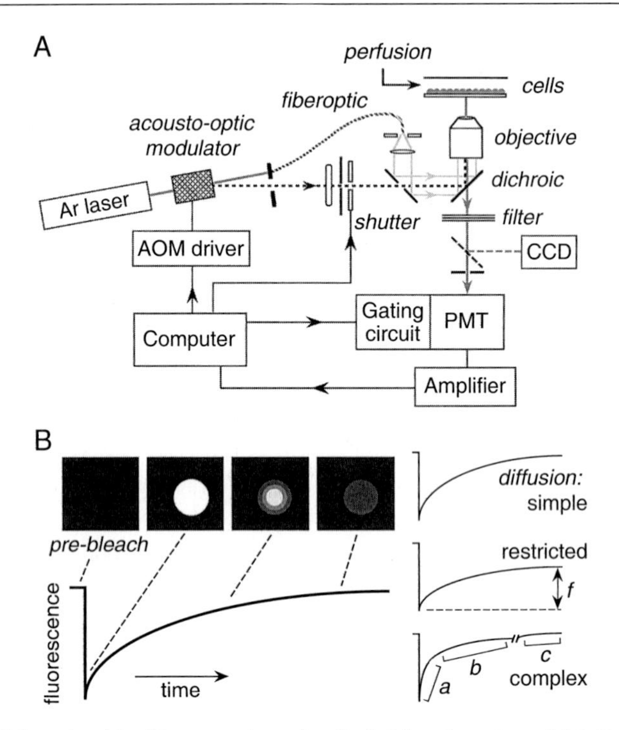

FIG. 1. (A) Spot photobleaching apparatus and method. A laser beam is modulated by an acousto-optic modulator and directed onto a sample, using a dichroic mirror and objective lens. Emitted fluorescence is collected, filtered, and detected by a gated photomultiplier tube (PMT). The undeflected zero-order beam is directed by a fiberoptic lens–mirror assembly for full-field illumination. (B) *Left:* Schematic of spot photobleaching showing fluorescence recovery into a circular bleached region by inward diffusion of unbleached fluorophores. *Right:* Fluorescence recovery curves for simple, restricted and complex diffusion. See text for explanations.

signal, photomultiplier protection circuitry, and data acquisition are centrally controlled. The probe-to-bleach beam attenuation ratio is typically 2000–10,000 : 1, the minimum bleach time is 1 μs, and processes down to ~20 μs can be followed. In cell studies it is important to record images of fluorescence recovery to complement the quantitative spot photobleaching measurements. We use a Nipkow-wheel confocal microscope and cooled charge-coupled device (CCD) camera detector to record full-field fluorescence images after bleaching a defined spot.[2] Similar semiquantitative imaging studies have been accomplished by scanning laser confocal microscopy to monitor relatively slow diffusive processes.[3]

[2] M. Dayel, E. Hom, and A. S. Verkman, *Biophys. J.* **76,** 2843 (1999).

[3] J. Lippincott-Schwartz, J. F. Presley, K. J. Zall, K. Hirschberg, C. D. Miller, and J. Ellenberg, *Methods Cell Biol.* **58,** 261 (1999).

Photobleaching measurements of cultured cells grown on transparent supports are done with a temperature-regulated perfusion chamber to immobilize the cell layer and allow changes in perfusate solution composition. An objective lens heater is used to prevent temperature differences between the sample and immersion objectives. Other cell and tissue preparations can be studied. For example, we have found that the viscosity of freshly secreted fluid from the submucosal glands of intact airways from cystic fibrosis patients is more viscous than that of normal subjects.[4] Fluorescent probes were microinjected into fluid droplets secreted onto the airway surface under oil, and spot photobleaching was carried out with an approximately parallel laser beam produced by a low-numerical aperture lens.

Alternative optical configurations have been used in photobleaching measurements. Total internal reflection–photobleaching has been accomplished by directing the laser beam onto a prism in optical contact with the fluorescent sample.[5] Fluorescence recovery in the evanescent field is recorded with an objective lens and detection optics as described above. A pair of modulated interfering laser beams has been used to generate a spatial fringe pattern at the sample to improve x–y resolution.[6] Spot photobleaching has also been done by two-photon excitation, using a focused Ti:sapphire femtosecond laser.[7] The restricted bleach volume in the z direction in two-photon excitation is a potential advantage in studying diffusion in three dimensions.

Quantitative Analysis of Spot Photobleaching Data

In spot photobleaching the recovery of fluorescence in a bleached spot is measured (Fig. 1B, left). For unrestricted diffusion of a single fluorescent species in a homogeneous environment, fluorescence recovers to the initial (prebleach) level and the recovery curve contains a single component whose shape depends on the geometry of the bleached region (Fig. 1B, right, simple diffusion). However, fluorescence recovery in cell systems is often incomplete (Fig. 1B, right, restricted diffusion) or multicomponent (Fig. 1B, right, complex diffusion). Incomplete recovery is often taken to indicate that a fraction of the fluorescent molecules are immobile because of binding to slowly moving cellular components or trapping in noncontiguous compartments. However, laser-induced photodamage, probe beam photobleaching, and other phenomena can produce apparent incomplete recovery. Moreover, because the extent of fluorescence recovery is measured at a finite time

[4] S. Jayaraman, N. S. Joo, B. Reitz, J. J. Wine, and A. S. Verkman, *Proc. Natl. Acad. Sci. U.S.A.* **98,** 8119 (2001).

[5] R. Swaminathan, N. Periasamy, S. Bicknese, and A. S. Verkman, *Biophys. J.* **71,** 1140 (1996).

[6] A. M. Dupont, F. Foucault, M. Vacher, P. F. Devaux, and S. Cribier, *Biophys. J.* **78,** 901 (2000).

[7] E. B. Brown, E. S. Wu, W. Zipfel, and W. W. Webb, *Biophys. J.* **77,** 2837 (1999).

after photobleaching, slow processes (process *c* in Fig. 1B, right) may be neglected and thus the presence of restricted diffusion misinterpreted.[8]

The deduction of diffusion mechanisms from fluorescence recovery curves, $F(t)$, requires quantitative analysis procedures. For spot photobleaching in two dimensions (as in membranes) in which bleach and probe regions are identical, the diffusion equation has been solved analytically for a Gaussian beam profile and short bleach time to compute $F(t)$ from solute diffusion coefficient and spot diameter.[9] The diffusion equation has been solved numerically for total internal reflection–FRAP (fluorescence recovery after photobleaching) measurements for arbitrary bleach time.[5] However, direct application of the diffusion equation has limited utility in most photobleaching measurements where beam profile is non-ideal, bleach time may not be negligible compared with recovery time, sample geometry is complex, and the assumption of simple diffusion of a single species is not valid. It is generally impractical to solve the diffusion equation for these nonideal situations and for diffusion in three dimensions.

For analysis of photobleaching data in three dimensions such as the aqueous phase of cell cytoplasm, we introduced a calibration procedure in which the half-time ($t_{1/2}$) for fluorescence recovery in cells is compared with $t_{1/2}$ measured in thin layers of fluorophores dissolved in artificial solutions of known viscosity.[10] Empirical equations have been given to fit $t_{1/2}$ values from recovery curves:

$$F(t) = [F_0 + F_1(t/t_{1/2})^\alpha]/[1 + (t/t_{1/2})^\alpha]$$

where F_0 is fluorescence before bleaching, F_1 is fluorescence at infinite time, and $\alpha(0 < \alpha < 1)$ is an empirical parameter related to solute restriction/anomalous diffusion.[11] The potential pitfalls in the assumption of simple diffusion and the significance of long-tail kinetics in diffusive phenomena have been recognized, and work has considered the interpretation of photobleaching data in systems with multicomponent or anomalous diffusion.[8,11] To analyze these more complex recovery kinetics, we introduced the idea that photobleaching recovery data, $F(t)$, can be resolved in terms of a continuous distribution of diffusion coefficients, $\alpha(D)$.[12] A regression method to recover $\alpha(D)$ from $F(t)$ was developed that utilizes a "basis" recovery curve for simple diffusion obtained with a reference sample such as a thin uniform film of fluorescein in saline. An independent method to identify anomalous diffusive processes from $F(t)$ data was developed in which an apparent time-dependent diffusion coefficient, $D(t)$, was computed directly

[8] J. F. Nagle, *Biophys. J.* **63,** 366 (1992).

[9] D. Alexrod, D. E. Koppel, J. Schlessinger, E. Elson, and W. W. Webb, *Biophys. J.* **16,** 1055 (1976).

[10] H. P. Kao, J. R. Abney, and A. S. Verkman, *J. Cell Biol.* **120,** 175 (1993).

[11] T. J. Feder, I. Brust-Mascher, J. P. Slattery, B. Baird, and W. W. Webb, *Biophys. J.* **70,** 2367 (1996).

[12] N. Periasamy and A. S. Verkman, *Biophys. J.* **75,** 557 (1998).

from $F(t)$ and the reference recovery curve. The approach utilizes a reference $F(t)$ curve for simple diffusion of a single species in order to deduce $D(t)$ by a convolution procedure. The determination of $\alpha(D)$ and $D(t)$ from photobleaching data provides a systematic approach to identify and quantify simple and anomalous diffusive phenomena, provided that high-quality recovery curves measured over long times are available.

Fluorescent Probes and Photophysics

Fluorescent probes for photobleaching measurements should ideally undergo irreversible photobleaching without reversible photophysical processes (e.g., triplet state relaxation, flicker) or photochemical reactions. The chromophore should be moderately sensitive to photobleaching in order to minimize bleach beam intensity and duration, but not so sensitive that significant bleaching occurs during the measurement of fluorescence recovery by the attenuated probe beam. Brightness, cellular toxicity, Stokes shift, and excitation by available laser lines are additional factors in chromophore selection. Commonly used fluorophores in photobleaching measurements include fluorescein and its derivatives, green fluorescent protein (GFP) and rhodamine; lipophilic dyes such as 1, 1′-dihexadecyl-3,3,3′, 3′-tetramethylindocarbocyanine perchlorate ($DiIC_{16}$) and octadecylrhodamine B chloride have been used to study diffusion in cell plasma membranes. Examples of spot photobleaching data are given in Fig. 2A, in which thin aqueous layers (5-μm thickness between coverslips) of fluorescein, GFP, fluorescein isothiocyanate (FITC)–dextran (500kDa), and an FITC-labeled double-stranded circular DNA plasmid (6 kilobases). The bleach pulse produced an immediate drop in fluorescence, followed by a monophasic recovery of fluorescence to its initial level. The different recovery rates reflect the size-dependent diffusion coefficients.

Reversible photobleaching, the spontaneous recovery of fluorescence without diffusion, is a concern in cellular photobleaching measurements. Neglect of reversible photobleaching has led to serious misinterpretations of data, in which fluorescence recovery was incorrectly assumed to indicate rapid fluorophore diffusion. All fluorescent molecules, including GFP, can undergo reversible photobleaching in which fluorescence is transiently bleached but reappears at a later time. There are several photophysical processes that can produce reversible recovery. The best characterized is triplet state relaxation, in which light exposure populates a triplet state from the excited singlet state. A nonradiative transition from the triplet state to the ground state produces apparent fluorescence recovery. In general, triplet relaxation processes are rapid (milliseconds or less), sensitive to oxygen and triplet state quenchers, and produced by relatively weak illumination. A very slow (>100 ms), oxygen-dependent reversible photobleaching process

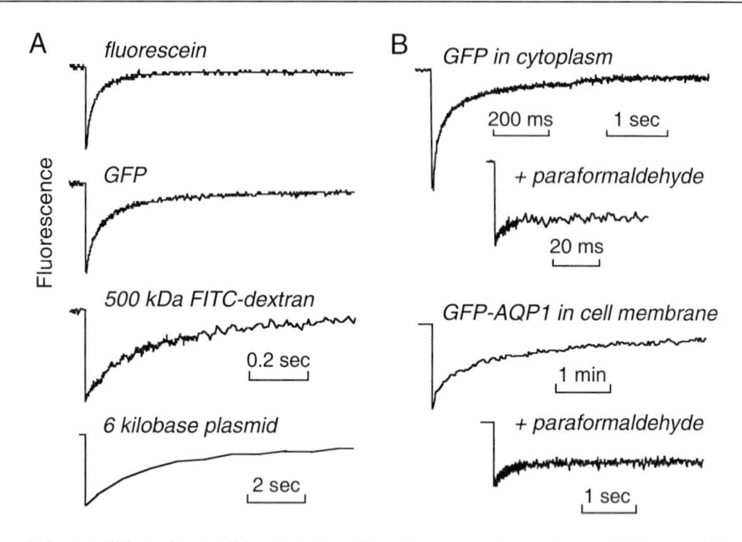

FIG. 2. Irreversible and reversible photobleaching of common fluorophores. (A) Irreversible photobleaching of aqueous solutions of fluorescein, purified recombinant GFP, 500-kDa FITC–dextrans, and a 6-kb FITC-labeled circular double-stranded plasmid. Thin solution layers (~5 μm) were bleached with a ×20 objective, producing a spot ~4 μm in diameter. Fluorescence was bleached by 20–30% by a brief, 488-nm laser pulse. Adapted from Refs. 2, 22, and 24. (B) *Top:* Reversible photobleaching of GFP expressed in the cytoplasm of CHO cells, and GFP–AQP1 expressed in the plasma membrane of LLC-PK1 cells. Cytoplasmic GFP was bleached by a brief, 488-nm pulse, using a ×40 objective. Where indicated, cells were fixed in 4% (w/v) paraformaldehyde to immobilize GFP. *Bottom:* Membrane-associated GFP–AQP1 fluorescence was bleached with a ×100 objective. Adapted from Refs. 2 and 14.

was reported for a surface-adsorbed FITC-labeled membrane protein,[13] although the physical mechanism for the reversible recovery was not established. We have observed slow reversible recovery processes of many milliseconds to seconds for GFP-labeled membrane proteins.[14] The physical origins of many reversible recovery processes remain unknown, emphasizing the need to rigorously distinguish between diffusion-related and reversible fluorescence recovery processes in FRAP measurements.

There are several practical maneuvers to distinguish irreversible from reversible photobleaching.[15,16] Irreversible bleaching produces a fluorescence recovery whose rate depends strongly on spot diameter, whereas recovery by reversible photobleaching is independent of spot size. Changing bleach beam

[13] A. L. Stout and D. Axelrod, *Photochem. Photobiol.* **62,** 239 (1995).
[14] F. Umenishi, J. M. Verbavatz, and A. S. Verkman, *Biophys. J.* **78,** 1024 (2000).
[15] N. Periasamy, S. Bicknese, and A. S. Verkman, *Photochem. Photobiol.* **63,** 265 (1996).
[16] R. Swaminathan, C. P. Hoang, and A. S. Verkman, *Biophys. J.* **72,** 1900 (1997).

intensity and time also gives predictable effects on reversible versus irreversible photobleaching efficiencies for reversible fluorescence recovery by triplet state relaxation. Subtle reversible photobleaching processes may be exposed by fluorophore immobilization to suppress diffusion-related recovery, as has been done for GFP in various cellular compartments using paraformaldehyde fixation. Figure 2B (top) shows GFP photobleaching in cytoplasm, using a $\times 40$ objective producing an ~ 2-μm diameter spot. The efficient recovery indicates that GFP is mobile and the recovery rate indicates that GFP diffusion in cytoplasm is slowed by 3- to 4-fold compared with that in water. Paraformaldehyde fixation abolished the slow recovery, but revealed a rapid (~ 5 ms) reversible recovery process whose recovery rate was independent of spot diameter. Figure 2B (bottom) shows the slow membrane diffusion of a GFP-tagged AQP1 water channel protein in epithelial cells. Paraformaldehyde fixation revealed a spot size-independent reversible recovery process of over ~ 100 ms. It is noted that the presence of reversible photobleaching does not preclude diffusion coefficient determination from analysis of the irreversible photobleaching component. The recovery from reversible photobleaching is often much faster than that from irreversible photobleaching, or the diffusive component can be slowed by increasing spot diameter. For some chromophores such as fluorescein, triplet state quenchers such as oxygen can be used to accelerate or eliminate the reversible photobleaching process, as was done in total internal reflection–FRAP measurements of solute diffusion in membrane-adjacent cytoplasm.[5] However, the reversible photobleaching of GFP is not sensitive to triplet state quenchers because of the inaccessibility of its triamino acid chromophore.

Diffusion in Cytoplasm

The diffusion of small and macromolecule-size solutes in cellular aqueous compartments is determined by solute properties, and the composition, organization, and geometry of the cellular compartment. For transport of small solutes such as metabolites, second messengers, and nucleotides, an important parameter describing cytoplasmic viscosity is the solute translational diffusion coefficient. We analyzed the determinants the translational mobility of the small fluorescent probe BCECF in cytoplasm, using spot photobleaching.[10] Diffusion of BCECF [2,′7′-bis-(2-carboxyethyl)-5-carboxyfluorescein] in cytoplasm was about four times slower than in water (diffusion in cytoplasm vs. water, $D_{cyto}/D_{water} \approx 0.25$). The slowed BCECF diffusion was analyzed in terms of three independently acting factors: (1) slowed diffusion in fluid-phase cytoplasm, (2) probe binding to intracellular components, and (3) probe collisions with intracellular components (molecular crowding). Slowed diffusion in fluid-phase cytoplasm, or "fluid-phase viscosity," is defined as the microviscosity sensed by a small solute in the absence of interactions with macromolecules and organelles. As measured by time-resolved anisotropy, the picosecond rotational correlation times of BCECF and other small solutes were only 10–30% slower in cytoplasm than in water, indicating that the fluid-phase viscosity of cytoplasm is not much greater than that of water.[17] Probe binding, as quantified independently by confocal microscopy in digitonin-permeabilized cells and by time-resolved anisotropy, accounted for ~20% of slowed diffusion. Probe collisions (molecular crowding), as assessed by measurements of cytoplasmic BCECF diffusion as a function of cell volume, were determined to be the principal diffusive barrier that accounted for the 4-fold slowed diffusion of BCECF in cytoplasm versus water. Similar slowing of BCECF diffusion was measured in membrane-adjacent cytoplasm, using total internal reflection–photobleaching.[5]

Several laboratories have studied the diffusion of larger molecules. Luby-Phelps and colleagues[18,19] used spot photobleaching to measure the translational diffusion of microinjected, fluorescently labeled dextrans and Ficolls. As dextran or Ficoll size was increased, diffusion in cytoplasm progressively decreased relative to that in water, suggesting a "sieving" mechanism that was proposed to involve the skeletal mesh. Qualitatively similar findings were reported for dextran diffusion in cytoplasm of developing nerve processes[20] and skeletal muscle cells.[21] Our laboratory carried out a series of quantitative comparisons of the diffusion of size-fractionated FITC–dextrans and FITC–Ficolls introduced into cytoplasm by microinjection[22] and green fluorescent protein (GFP) introduced by transfection.[16] Examples of photobleaching recovery curves measured in cytoplasm are shown in Fig. 3A. Figure 3B summarizes D_{cyto}/D_{water} (diffusion in cytoplasm vs. saline) for different solutes and macromolecules. Diffusion of GFP was 3- to 5-fold slowed in cytoplasm versus saline, as was the diffusion of FITC–dextrans and FITC–Ficolls of molecular size under ~500 kDa. Similar slowing of GFP diffusion in cytoplasm was reported in photobleaching measurements in amebas.[23] The diffusion of large molecules (e.g., 2000-kDa FITC–dextran; Fig. 3A) was remarkably impaired. Although the exact details of the D_{cyto}/D_{water} versus molecular size curve shape (Fig. 3B) probably depend on cell type and on analysis procedures (assumption of single component hindered diffusion vs. complex diffusion), the main message is that the diffusion of small macromolecules in cytoplasm is only mildly impaired whereas that for large macromolecules can be greatly impaired.

[17] K. Fushimi and A. S. Verkman, *J. Cell Biol.* **112,** 719 (1991).

[18] K. Luby-Phelps, D. L. Taylor, and F. Lanni, *J. Cell Biol.* **102,** 2015 (1986).

[19] K. Luby-Phelps, P. E. Castke, D. L. Taylor, and F. Lanni, *Proc. Natl. Acad. Sci. U.S.A.* **84,** 4910 (1987).

[20] S. Popov and M. M. Poo, *J. Neurosci.* **12,** 77 (1992).

[21] M. Arrio-Dupont, G. Foucault, M. Vacher, P. F. Devaux, and S. Cribier, *Biophys. J.* **78,** 901 (2000).

[22] O. Seksek, J. Biwersi, and A. S. Verkman, *J. Cell Biol.* **138,** 131 (1997).

[23] E. O. Potma, W. P. Boeij, L. Boxgraaf, J. Roelofs, P. J. van Haastert, and D. A. Wiersma, *Biophys. J.* **81,** 2010 (2001).

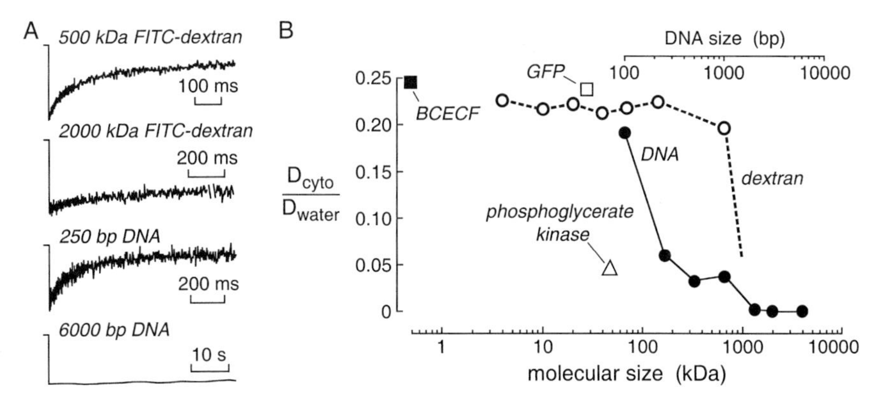

FIG. 3. Diffusion of macromolecules in cytoplasm. (A) Spot photobleaching (×60 objective, short bleaching time) of indicated fluorescein-labeled dextran and linear double-stranded DNA fragments in microinjected cells. (B) Ratio of diffusion coefficient in cytoplasm versus saline (D_{cyto}/D_{water}) for indicated solute/macromolecules. Data taken from Refs. 10, 16, 22, 24, and 26.

Although molecular crowding and sieving restrict the mobility of large solutes, binding can severely restrict the mobility of smaller solutes. The diffusional mobility of DNA fragments in cytoplasm is thought to be an important determinant of the efficacy of DNA delivery in gene therapy and antisense oligonucleotide therapy. We measured the translational diffusion of fluorescein-labeled double-stranded (naked) DNA fragments from oligonucleotide size (21 base pairs) to plasmid size (6000 base pairs) after microinjection into cytoplasm.[24] As shown in Fig. 3A and B, the diffusion of DNAs of size >250 base pairs was remarkably reduced whereas diffusion of comparably sized FITC–dextran was not impaired. DNA binding to cellular components and/or its nonspherical shape may account for its strongly size-dependent diffusion. Like DNA, the diffusion of proteins in cytoplasm can be impaired by binding interactions, as shown by measurements of labeled protein diffusion in nerve processes[20] and muscle cells.[21] Diffusion of several microinjected FITC-labeled glycolytic enzymes was impaired compared with that of comparably sized FITC–dextrans. D_{cyto}/D_{water} data for phosphoglycerate kinase are shown in Fig. 3B. There is evidence that the diffusion of some glycolytic enzymes might by regulated by cell metabolic state. Metabolic depletion by 2-deoxyglucose in 3T3 cells resulted in an ~20% increase in the fraction of mobile aldolase,[25] and we found that D_{cyto}/D_{water} for pyruvate kinase was reduced

[24] G. L. Lukacs, P. Haggie, O. Seksek, D. Lechardeur, N. Freedman, and A. S. Verkman, *J. Biol. Chem.* **275,** 1625 (2000).

[25] L. Pagliaro and D. L. Taylor, *J. Cell Biol.* **118,** 859 (1992).

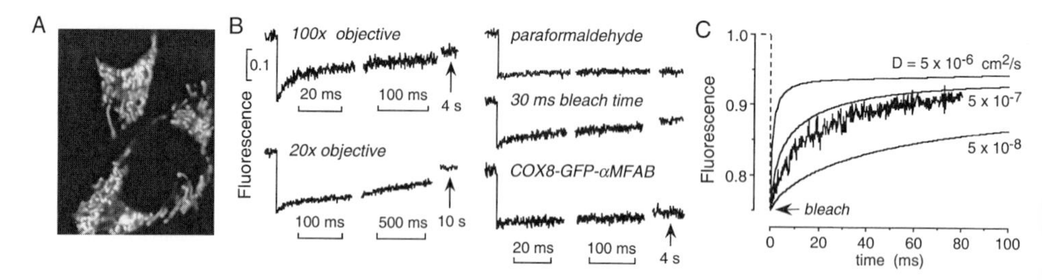

FIG. 4. Diffusion of green fluorescent protein in the mitochondrial matrix. (A) Fluorescence micrograph of GFP in the mitochondrial matrix of transfected CHO cells. (B) Spot photobleaching of unconjugated GFP (*left top*, ×100 lens; *left bottom*, ×20 lens), after paraformaldehyde fixation (*right top*) and with long (30-ms) bleach time (*right middle*). *Right bottom:* Photobleaching of a fusion protein of GFP with a matrix enzyme (α subunit of the mitochondrial β-fatty acid oxidation pathway). (C) Model predictions for fluorescence recovery for different diffusion coefficients (D) for bleaching with a ×100 objective and short bleach time. Adapted from Ref. 28.

from ~0.08 to 0.04 when cells were depleted of ATP.[26] Altered enzyme binding to slowly diffusing cytoplasmic components may be responsible for the dependence of mobility of some enzymes on metabolic state.

Diffusion in Intracellular Organelles

The mitochondrial matrix, the aqueous compartment enclosed by the inner mitochondrial membrane, is a major site of metabolic processes. Theoretical considerations suggest that the diffusion of metabolite- and enzyme-sized solutes might be severely restricted in the mitochondrial matrix because of its high density of proteins.[27] The ability to target GFP to the mitochondrial matrix provided a unique opportunity to test the hypothesis that solute diffusion is greatly slowed in the matrix.[28] Figure 4A shows mitochondrial-specific GFP targeting. Spot photobleaching of GFP with a ×100 objective (0.8-μm spot diameter) gave a half-time for fluorescence recovery of 15–19 ms with greater than 90% of the GFP mobile (Fig. 4B, left, top curve). As predicted for aqueous-phase diffusion in a confined compartment, fluorescence recovery was slowed or abolished by increased laser spot size (Fig. 4B, left, bottom curve), paraformaldehyde fixation (Fig. 4B, right, top curve), or increased bleach time (Fig. 4B, right, middle curve). The fluorescence recovery data were analyzed using a mathematical model of matrix diffusion. Fluorescence recoveries were computed from analytical solutions to the diffusion equation. Predicted recovery curves for different diffusion coefficients are shown

[26] P. Haggie, L. Vetrivel, and A. S. Verkman, *Biophys. J.* **80,** 280a (2001) [abstract].
[27] G. R. Welch and J. S. Easterby, *Trends Biochem. Sci.* **19,** 193 (1994).
[28] A. Partikian, B. Ölveczky, R. Swaminathan, Y. Li, and A. S. Verkman, *J. Cell Biol.* **140,** 821 (1998).

in Fig. 4C. The fitted value of D was $2\text{–}3 \times 10^{-7}$ cm^2/sec, only 3- to 4-fold less than that for GFP diffusion in water. Interestingly, little fluorescence recovery was found for bleaching of GFP in fusion with subunits of the fatty acid α-oxidation multienzyme complex (Fig. 4B, right, bottom curve) that are normally present in the matrix. The rapid and unrestricted diffusion of GFP in the mitochondrial matrix suggested that metabolite channeling may not be required. The clustering of matrix enzymes in membrane-associated complexes might serve to establish a relatively uncrowded aqueous space in which solutes can freely diffuse, thus reducing metabolite transit times and pool sizes.

Photobleaching was also used to measure solute diffusion in the endoplasmic reticulum (ER), the major compartment for the processing and quality control of newly synthesized proteins.[2] GFP was targeted to the aqueous ER lumen of CHO cells by transient transfection with cDNA encoding GFP with a C-terminal KDEL retention sequence and upstream preprolactin secretory sequence. Figure 5A shows a time series of fluorescence images of the GFP-labeled ER after bleaching a large circular spot. The darkened zone produced by the bleach pulse progressively filled in by GFP diffusion through the ER lumen. The fluorescence of an adjacent cell (Fig. 5A, white arrow) was not affected by the bleach. Repeated laser illumination at the same small spot resulted in complete bleaching of ER-associated GFP throughout the cell (Fig. 5B), indicating a continuous ER lumen. Interestingly, after 60 bleach pulses, the remaining fluorescence (<2% of original signal) had a Golgi-like pattern, suggesting imperfect KDEL retention. Quantitative spot

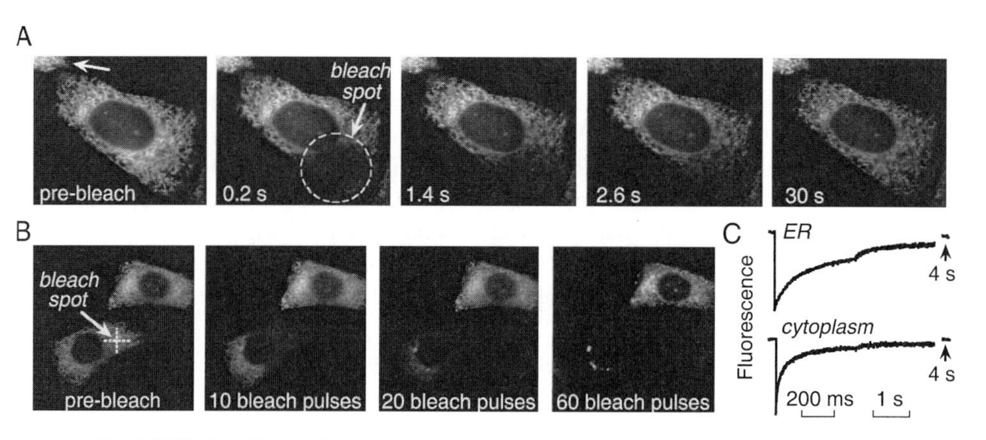

FIG. 5. Diffusion of green fluorescent protein in the lumen of the endoplasmic reticulum (ER) in transfected CHO cells. (A) Serial micrographs showing ER fluorescence before (prebleach) and at the indicated times after bleaching. Bleach spot is denoted by a dashed white circle. Arrow points to an adjacent unbleached cell. (B) Serial micrographs showing fluorescence depletion from the ER before and after the indicated number of bleach pulses, using a small bleach spot. (C) Spot photobleaching (×60 objective) of GFP in ER versus cytoplasm. Adapted from Ref. 2.

photobleaching with a single brief bleach laser pulse ($<$0.1 ms) indicated that GFP was fully mobile (Fig. 5C). Direct comparison of bleaching of GFP in ER versus cytoplasm in the same cell type indicated $\sim$3-fold apparent slowing in ER. The ER diffusion of GFP–nascent protein chimeras may be useful in evaluating protein folding and interactions with molecular chaperones.

Diffusion of Membrane Proteins

Diffusion in membranes is remarkably slower than in aqueous compartments and may be complex because of membrane crowding with proteins, the presence of distinct lipid domains, and membrane–cytoskeleton interactions. There is a large literature on the diffusion of membrane lipids and proteins.[3,29] Photobleaching measurements in membranes are generally technically easier than in aqueous compartments because of the slower recovery rates, giving improved detection of the signal-to-noise ratio. In addition, rapid reversible photobleaching processes can be less of a concern for measurement of slow lateral diffusion in membranes.

Photobleaching of GFP chimeras in the Golgi membrane was used to establish that membrane protein immobilization is not the mechanism of protein retention.[30] In a similar study, the mobility of cytochrome P450 was measured, a protein that is restricted to the ER membrane despite containing no known retention/retrieval signals.[31] A high diffusion coefficient again suggested that mechanisms other than immobilization were responsible for ER retention. We have measured the ER diffusion of AQP2-T126M, a mutant water channel that causes the hereditary disease nephrogenic diabetes insipidus.[32] Figure 6A (top) shows the diffusion of AQP2 in the ER (in brefeldin A-treated cells), which was remarkably slower than that of GFP in the ER lumen (see Fig. 5A). Spot photobleaching showed that the diffusion of GFP-labeled AQP2-T126M was similar to that of wild-type AQP2 (Fig. 6A, bottom), indicating that slowed ER diffusion is not responsible for its ER retention. Studies of the lamin B receptor–GFP fusion revealed a dynamic relationship between the nuclear envelope and the ER during the cell cycle.[33] Interphase cells contain a mobile fraction of lamin B in the ER membrane in addition to immobile lamin B in the nuclear envelope. Similarly, modulation of

[29] T. M. Jovin and W. L. Vaz, *Methods Enzymol.* **172,** 471 (1989).

[30] N. B. Cole, C. L. Smith, N. Sciaky, M. Terasaki, M. Edidin, and J. Lippincott-Schwartz, *Science* **273,** 797 (1996).

[31] E. Szczesna-Skorupa, C. D. Chen, S. Rodgers, and B. Kemper, *Proc. Natl. Acad. Sci. U.S.A.* **95,** 14793 (1998).

[32] M. H. Levin, P. M. Haggie, L. Vetrivel, and A. S. Verkman, *J. Biol. Chem.* **276,** 21331 (2001).

[33] J. Ellenberg, E. D. Siggia, J. E. Moreira, C. L. Smith, J. F. Presley, H. J. Worman, and J. Lippincott-Schwartz, *J. Cell Biol.* **138,** 1193 (1997).

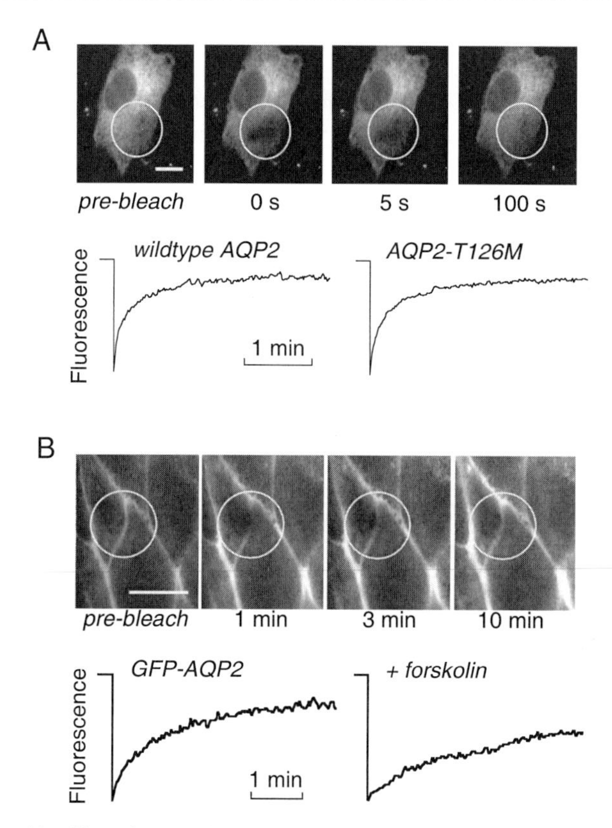

FIG. 6. Photobleaching of aquaporin–GFP chimeras in the ER and the cell plasma membrane. (A) GFP–AQP2 was trapped in the ER by brefeldin A treatment. *Top:* Serial micrographs showing cell ER fluorescence before (prebleach) and at the indicated times after bleaching. The bleach spot is denoted by a white circle. *Bottom:* Spot photobleaching (~0.8-μm-diameter spot size) of wild-type human AQP2 versus AQP2-T126M, an ER-retained mutant causing human nephrogenic diabetes insipidus. Adapted from Ref. 32. (B) *Top:* GFP–AQP1 was targeted to the plasma membrane of epithelial cells. A series of pre- and postbleach images is shown as in (A) (top). *Bottom:* Spot photobleaching of cells as in (A) (bottom), showing slowed GFP–AQP2 after forskolin treatment. Adapted from Ref. 14.

the mobility of cadherin has been followed by photobleaching of GFP fusion proteins during cell adhesion.[34] Cadherin–GFP was initially mobile but became immobilized and sequestered at sites of cell contact.

At the plasma membrane, photobleaching of GFP chimeras has been used to investigate the mobility of transport proteins such as the aquaporins,[14] and proteins

[34] C. L. Adams, Y. T. Chen, S. J. Smith, and W. J. Nelson, *J. Cell Biol.* **142,** 1105 (1998).

involved in signal transduction such as Ras[35] and nitric oxide synthase.[36] The rapid diffusion coefficient of Ras ($\sim 2 \times 10^{-9}$ cm^2/sec), only 1.6-fold slower than that of the membrane probe DiC_{16}, supported the contention that farnesyl moieties are involved in membrane attachment.[35] Further, the diffusion coefficient of Ras was increased by pharmacological agents that disrupt the interaction of Ras with the plasma membrane. Photobleaching experiments demonstrated that eNOS had more restricted mobility in the plasma membrane as compared with the Golgi.[36] Figure 6B (top) shows serial fluorescence images of epithelial cells expressing a GFP–AQP1 fusion protein. AQP1 is a small integral membrane protein that functions as a plasma membrane water channel. Fluorescence recovery into the large bleached zone occurred over several minutes. The fluorescence recovery was abolished by paraformaldehyde fixation and was strongly dependent on spot diameter, as expected for a diffusion-related process.[14] A GFP–AQP1 diffusion coefficient of $\sim 5 \times 10^{-11}$ cm^2/sec was determined. Similar fluorescence recovery was found for cAMP-regulated water channel GFP–AQP2. Figure 6B (bottom) shows a spot photobleaching measurement using a 0.8-μm diameter spot produced by a $\times 100$ objective. Fluorescence recovery was nearly complete, indicating that GFP–AQP2 was fully mobile. Stimulation of cells with the cAMP agonist forskolin resulted in remarkable slowing of GFP–AQP2 mobility, suggesting interactions of AQP2 with skeletal or other proteins. Photobleaching thus provides a unique biophysical tool to investigate protein–protein interactions in living cells that are difficult to study by classic biochemical methods.

Research Directions

Photobleaching and related dynamic fluorescence measurements have provided new insights into the dynamics of biologically important solutes and macromolecules in cells. The available instrumentation for photobleaching—lasers, beam modulators, and detectors—have adequate temporal and spatial resolution for most cellular applications. Rapid advances in fluorophore targeting methods make possible the selective labeling of cellular components with a wide variety of chromophores. There remain many interesting questions for which photobleaching and related methods are applicable. The mechanisms and regulation of membrane protein mobility remain poorly understood, as is the role of molecular crowding and supermolecular enzyme organization in cell metabolism. There are many potential implications for the understanding and therapy of human diseases, such as in drug and gene delivery into cells and in elucidating the mechanistic basis of inherited diseases of mitochondrial enzymes.

[35] H. Niv, O. Gutman, Y. I. Henis, and Y. Kloog, *J Biol. Chem.* **274,** 1606 (1999).

[36] G. Sowa, J. Liu, A. Papapetropoulos, M. Rex-Haffner, T. E. Hughes, and W. C. Sessa, *J. Biol. Chem.* **274,** 22524 (1999).

[29] Fluorophore-Assisted Light Inactivation for Multiplex Analysis of Protein Function in Cellular Processes

By BRENDA K. EUSTACE and DANIEL G. JAY

Introduction

The application of photons to the imaging and analysis of biological processes has made great strides. This includes the ability to observe single-molecule interactions and advances in light microscopy using green fluorescent protein (GFP) fusions and fluorescence energy transfer to show dynamic protein interactions in cells. Photons are also used to move or hold microscopic objects and to measure nanoscale forces using optical tweezers. One aspect of biophotonics that is less developed is the use of photons as a means to manipulate *in situ* function. Photodynamic therapy, the use of photosensitizing reagents in conjunction with light, has been applied to kill specific cells, for example, in cancer treatment.[1] We and others have extended this approach by developing chromophore-assisted laser inactivation (CALI). CALI is a method to disrupt protein function in cellular context[2] as a means to address the roles that specific proteins play in cells and model organisms.[3,4] Since its invention CALI has been applied to address *in situ* function for many proteins in a wide variety of cellular processes. CALI has thus far largely been applied in a hypothesis-driven manner, assessing one protein at a time. To take a more discovery-based approach, it is necessary to perform this functional inactivation in a high-throughput manner. For this purpose, we have developed fluorophore-assisted light inactivation (FALI).[5] FALI can be performed with diffuse light, such that many samples may be done in parallel using multiwell plates. This provides an unprecedented throughput in the ability to directly address the functional roles of specific proteins in cellular assays. This innovation allows for more global applications of light inactivation such that a functional proteomic approach to biological research is possible. In this article we describe CALI and FALI and their applications and methods. We include the set-up of light sources and labeling of reagents with isothiocyanate derivatives of malachite green and fluorescein. We close by describing methods to show how FALI may be applied

[1] L. Dalla Via and S. Marciani Magno, *Curr. Med. Chem.* **8,** 1405 (2001).

[2] D. G. Jay, *Proc. Natl. Acad. Sci. U.S.A.* **85,** 5454 (1988).

[3] A. E. Beermann and D. G. Jay, *Methods Cell Biol.* **44,** 715 (1994).

[4] E. V. Wong and D. G. Jay, *Methods Enzymol.* **325,** 482 (2000).

[5] S. Beck, T. Sakurai, B. Eustace, G. Beste, R. Schier, F. Rudert, and D. G. Jay, *Proteomics* **2,** 247 (2002).

0076-6879/03 $35.00

in a multiplex fashion in a particular cellular assay for cancer cell migration and invasion.

Advances in genomics and proteomics have provided a new challenge for biologists: assigning functional roles to the enormous number of proteins in the cell. It is estimated that there are 30,000 genes in the human genome. These genes translate into about 1 million different protein products generated by pre- and post-translational modifications that can each confer a unique function to the protein. One common approach to address *in situ* protein function is to use function-blocking antibodies, and these reagents have been used to identify important proteins in many cellular processes.[6–8] Function-blocking antibodies have a significant benefit because they may be used directly for therapeutics.[9] Unfortunately, only a small proportion (1–5%) of antibodies raised against proteins of interest is able to effectively and specifically block function.[10]

Because the vast majority of antibodies directed to particular proteins do not block function, we developed CALI to utilize these reagents to specifically inactivate proteins (Fig. 1A).[2] The non-function-blocking antibodies are multiply conjugated with a chromophore, malachite green (MG). When these antibodies are irradiated with high-powered pulsed laser light at a wavelength of 620 nm (absorbance of MG), short-lived hydroxyl radicals are formed. These short-lived radicals generated during CALI generate a half-maximal radius of damage of 15 Å,[11] causing oxidative damage to the bound antigen while neighboring proteins and subunits of the same protein are largely unaffected. This distance is compatible with the use of CALI in cells because the average intermolecular distance between proteins in cells is ~80 Å. This specificity was illustrated with the T cell receptor complex, in which one subunit could be inactivated in T cells while other associated subunits were functionally undamaged during CALI.[12] CALI has been used to study a diverse array of different proteins, including membrane proteins, signaling molecules, cytoskeletal elements, and transcription factors (Table I).

Micro-CALI, a related approach in which the laser is focused to micron diameters by microscope optics, has been used extensively to assess protein function in single cells.[13,14] Micro-CALI has been reviewed elsewhere, and is thus not

[6] D. M. O'Rourke and M. I. Greene, *Immunol. Res.* **17,** 179 (1998).

[7] J. Baselga, *Ann. Oncol.* **11,** 187 (2000).

[8] H. Hashida, A. Takabayashi, M. Adachi, T. Imai, K. Kondo, N. Kohno, Y. Yamaoka, and M. Miyake, *Int. J. Oncol.* **18,** 89 (2001).

[9] Z. Fan and J. Mendelsohn, *Curr. Opin. Oncol.* **10,** 67 (1998).

[10] B. K. Muller and F. Bonhoeffer, *Curr. Biol.* **5,** 1255 (1995).

[11] J. C. Liao, J. Roider, and D. G. Jay, *Proc. Natl. Acad. Sci. U.S.A.* **91,** 2659 (1994).

[12] J. C. Liao, L. J. Berg, and D. G. Jay, *Photochem. Photobiol.* **62,** 923 (1995).

[13] H. Y. Chang, K. Takei, A. M. Sydor, T. Born, F. Rusnak, and D. G. Jay, *Nature (London)* **376,** 686 (1995).

[14] P. Diamond, A. Mallavarapu, J. Schnipper, J. Booth, L. Park, T. P. O'Connor, and D. G. Jay, *Neuron* **11,** 409 (1993).

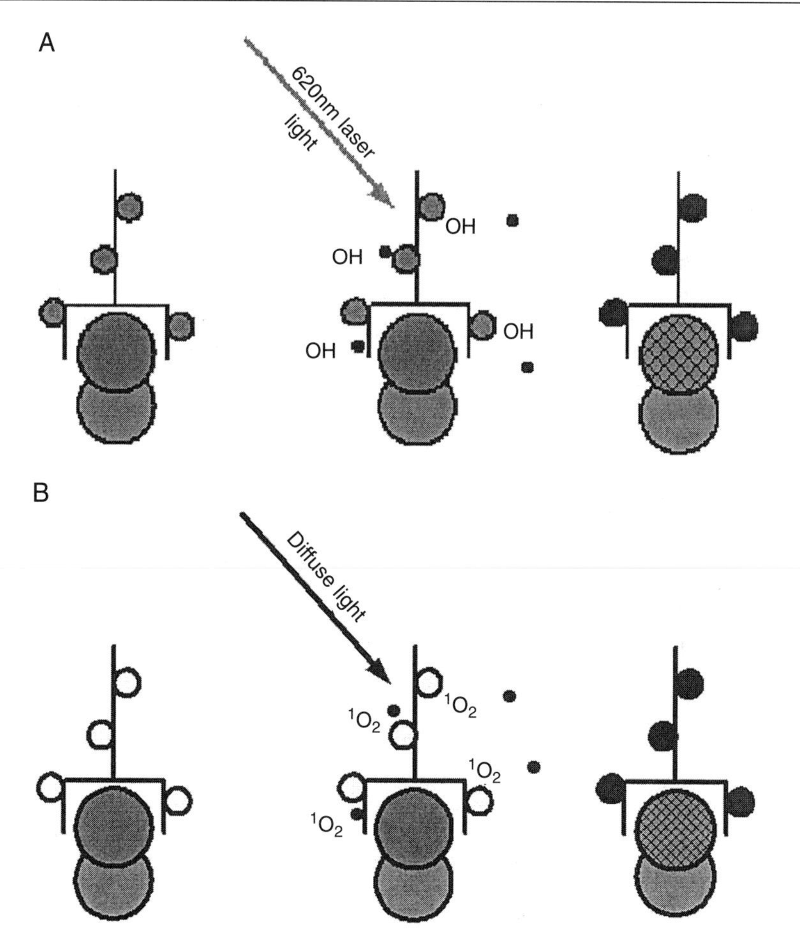

FIG. 1. Comparison of the mechanisms of CALI and FALI. (A) A non-function-blocking antibody is labeled with the chromophore malachite green (MG). This antibody binds to the protein and on excitation of MG by 620-nm pulsed laser light, hydroxyl radicals are generated. These free radicals specifically modify the protein bound to the antibody without affecting neighboring proteins. (B) A non-function-blocking antibody is labeled with the fluorophore fluorescein. This antibody binds to the protein and, on excitation of fluorescein by diffuse light (fluorescein excitation, 494 nm), singlet oxygen molecules are generated. These singlet oxygen molecules specifically modify the protein bound to the antibody without affecting neighboring proteins.

discussed here.[4,15] CALI is an effective technique to directly address protein function in cells and a high-throughput CALI approach would be beneficial for multiplex applications. To develop high-throughput applications, we have investigated an alternative approach to inactivate proteins in cells using diffuse light.

[15] A. Buchstaller and D. G. Jay, *Microsc. Res. Tech.* **48,** 97 (2000).

TABLE I
PROTEINS INACTIVATED BY CHROMOPHORE-ASSISTED LASER INACTIVATION

Class and protein	Ref.
Enzymes	
β-Galactosidase (*in vitro*)	2
Alkaline phosphatase (*in vitro*)	2
Acetylcholinesterase (*in vitro*)	2
Caspase 3 (*in vitro* and in cells)	17[a]
Signal transduction molecules	
Calcineurin (*in vitro* and in cells)	13
IP_3 receptor (*in vitro* and in cells)	29[a]
MAP kinase (*in vitro*)	17[a]
Surface proteins	
Grasshopper fasciclin I (*in vivo*)	31
Grasshopper fasciclin II (*in vivo*)	14
α Chain of T cell receptor (in cells)	12
β Chain of T cell receptor (in cells)	12
ε Chain of T cell receptor (in cells)	12
Drosophila patched protein (*in vivo,* mimicks hypomorphic mutation)	32
NCAM (*in vitro* and in cells)	33
L1 (*in vitro* and in cells)	33
RGM (*in vitro*)	10
FMRF amide receptor (*in vitro*)	34[a]
Fas receptor (in cells)	17[a]
β_1-Integrin (in cells)	5
Transcription factors	
Drosophila even skipped (*in vivo,* mimicks genetic loss of function)	35
Tribolium even skipped (*in vivo*)	36
Proteins not inactivated by CALI	
Hexokinase	2
Glyceraldehyde-3-phosphate dehydrogenase	2
Cytoskeletal proteins	
Myosin V (*in vitro* and in cells)	24
Talin (in cells)	25
Radixin (in cells)	26
Kinesin (*in vitro*)	16[a]
Hamartin (TSC I) (in cells)	27
Tau (*in vitro* and in cells)	28
Myosin 1β (in cells)	24
Vinculin (*in vitro* and in cells)	25
Ezrin (*in vitro* and in cells)	30

[a] Established by other groups independently.

Surrey *et al.*[16] showed that antibodies conjugated with fluorescein isothiocyanate (FITC) are much more efficient for CALI-like inactivation of protein *in vitro*. We have extended these studies to show that this technique is not only effective *in vitro,* but in cellular assays as well.[5] We have also shown that the spatial restriction of FALI is sufficient to allow for specific inactivation of proteins in a cellular context. We have named this approach fluorophore-assisted light inactivation (FALI) to distinguish it from CALI. Rubenwolf *et al.*[17] have used a continuous-wave argon laser for FALI such that each irradiation takes seconds and can be performed in an automated and sequential fashion. Alternatively, Beck *et al.*[5] have performed FALI with diffuse light such that many samples may be irradiated simultaneously in a 96-well plate. Thus far, several proteins have been inactivated in cells by FALI, including β_1-integrin, caspase 3, and the Fas receptor.[5,17]

FALI works by a slightly different mechanism than CALI (Fig. 1B). When FITC is excited, a type 2 photosensitization reaction occurs after excitation with 494-nm light. The transfer of energy from an excited photosensitizer, in this case fluorescein, to an adjacent oxygen molecule generates an excited singlet oxygen molecule. Singlet oxygen is known to have a longer lifetime than free radicals, visualized by extended reactions with neighboring substrates. The upper limit distance for inactivation for FALI has previously been estimated to be less than 300 Å.[16] We have estimated (using quenching data) that the half-maximal radius of damage is approximately 40 Å.[5] Although this value is more than twice as large as the half-maximal radius of damage for CALI, we have shown, using a variety of approaches, that it still shows sufficient specificity such that inactivating neighboring proteins is unlikely. In addition, although it has been shown that singlet oxygen is involved in FALI and that hydroxyl radicals are involved in CALI, there may be other mechanisms involved, as specific quenchers did not completely prevent inactivation.[5,11]

FALI provides significant advantages for high-throughput analysis. First, the efficiency of FITC-mediated damage is high enough that a laser is not needed to generate the appropriate light intensity for inactivation. Instead, we can use diffuse light sources such as a desk lamp or a slide projector, such that multiplex irradiation in 96-well plates can be done. In addition, FITC is a significantly less hydrophobic molecule than malachite green isothiocyanate (MGITC). Therefore aggregation during labeling is less problematic and labeling can easily be done in a high-throughput fashion. These advantages allow FALI to be used by a wider range of investigators and can be used to inactivate multiple samples simultaneously.

[16] T. Surrey, M. B. Elowitz, P. E. Wolf, F. Yang, F. Nedelec, K. Shokat, and S. Leibler, *Proc. Natl. Acad. Sci. U.S.A.* **95,** 4293 (1998).

[17] S. Rubenwolf, J. Niewohner, E. Meyer, C. Corinne Petit-Frere, F. Rudert, P. R. Hoffmann, and L. L. Ilag, *Proteomics* **2,** 241 (2002).

Methods

Antibody Preparation and Labeling with Malachite Green Isothiocyanate and Fluorescein Isothiocyanate

We generally employ non-function-blocking antibodies with high affinity for the target of interest for CALI and FALI. Traditional monoclonal or polyclonal antibodies can be used for inactivation, as well as recombinant single-chain variable fragments (scFvs). There is also evidence that ligands for receptors, such as inositol 1,4,5-trisphosphate and malachite green-labeled streptavidin bound to biotinylated enzymes, can also be used for inactivation with high specificity.[2,18] In addition, directed inactivation of RNA transcripts using aptamer sequences to bind MG has also been found to be effective.[19]

MGITC and FITC are used for labeling and react with amino groups on antibodies to form a stable thioester. Generally, lysine residues in the antibody provide these amino groups. Thus, one must be certain that the antibody preparation does not contain any contaminants with free amino groups, such as Tris- or glycine-based buffers, because the isothiocyanate group on FITC and MGITC readily reacts with such groups.

MGITC is hydrophobic when hydrolyzed, because of its capability for π-stacking interactions, and thus precipitation may be problematic. This problem is reduced somewhat by keeping the labeling reaction at a high pH (antibody in 0.5 *M* $NaHCO_3$, pH 9.5), resuspending the MGITC dye in dry dimethyl sulfoxide (DMSO), keeping the MGITC at high concentrations in the labeling mix, and adding bovine serum albumin (BSA) to prevent precipitation. MGITC-labeled BSA has not been found to cause damage to surrounding protein *in vitro* or *in vivo,* and thus can be used effectively to reduce hydrophobic MG interactions. A 10-mg/ml solution of MGITC (Molecular Probes, Eugene, OR) in dry DMSO is prepared fresh and added in a 1 : 5 (w/w) ratio with the lgG molecule. We generally label between 100 μg and 1 mg of protein in up to 1 ml of total reaction solution. The MGITC solution is added in three separate aliquots, with constant shaking, every 5 min. After the final addition of MGITC, the labeling mixture is allowed to incubate for 15 min. The free dye is separated from the MG-labeled antibodies by passage through a prepacked gel-filtration column (PD-10; Amersham Pharmacia Biotech, Uppsala, Sweden) in the buffer to be used for the experiment [Hanks' buffered saline (HBSS), phosphate-buffered saline (PBS), Dulbecco's modified Eagle's medium (DMEM), etc.].

We aim to reach a labeling ratio of four to eight dye molecules per lgG or about one or two dye molecules per scFv (a much smaller protein). This value is calculated by taking the optical density of the labeled solution, to determine dye

[18] T. Inoue, K. Kikuchi, K. Hirose, M. Iino, and T. Nagano, *Chem. Biol.* **8,** 9 (2001).

[19] D. Grate and C. Wilson, *Proc. Natl. Acad. Sci. U.S.A.* **96,** 6131 (1999).

concentration, at 620 nm (molar absorptivity of 150,000 M^{-1} cm^{-1}) and dividing that value by the antibody concentration. Because of the difficulty in measuring protein concentration after labeling, we assume complete protein recovery (although it is likely less than 100%) and use the optical density (OD) of the solution as a measure of the moles of label. Although this estimate is arbitrary, it is a standard of comparative scale used when determining the labeling versus efficacy for CALI and FALI. For some applications, a concentrated dye solution is required and the MG-labeled antibody solution is concentrated with a Centricon filter (Amicon, Danvers, MA). Storage is generally done in aliquots at −80°, and is stored for less than 6 months.

FITC is less prone to hydrophobic aggregation, and thus labeling is performed in a slightly different manner. Labeling is not often performed in the presence of BSA, because hydrophobic interactions between FITC molecules are less common. To stabilize the FITC-labeled antibody during storage, however, BSA is added to a concentration of 1 mg/ml after labeling. A freshly prepared 10-mg/ml solution of FITC (Molecular Probes) is made in dry DMSO and added to the antibody solution (in 0.5 *M* $NaHCO_3$, pH 9.5) in a 1 : 5 (w/w) ratio. As with MGITC labeling, we generally label between 100 μg and 1 mg of protein in up to 1 ml of total reaction solution. The FITC solution is added all at once to the antibody solution, and is incubated with constant rocking for 1 hr at room temperature. Free dye is separated from FITC-labeled antibody by gel filtration, using the same protocol as described above for MGITC. A hand-held UV lamp is used to detect fluorescence in the eluate to collect the fractions containing FITC-labeled protein.

For FALI, the elution buffer must be phenol red free. Phenol red is a pH indicator added to many culture media. It is an efficient quencher of singlet oxygen species and must be avoided for efficient FALI-generated inactivation. As for MG labeling, the goal is to have four to six FITC molecules per IgG molecule and this can be calculated by dividing the optical density at 494 nm (molar absorptivity of 68,000 M^{-1} cm^{-1}) by the antibody concentration. The antibody is concentrated with Centricon concentrators, if necessary, and storage is done at −80° in aliquots and stored for less than 6 months.

FITC labeling of many antibodies in parallel is possible as well. We routinely label 48 scFv molecules at one time in a 96-well plate, and full-chain antibodies could likely be done in the same manner. A solution containing between 50 and 100 μg of antibody in 200 μl is added to a standard 96-well plate along with 2 μl of 10-mg/ml FITC in DMSO and 25 μl of 1 *M* $NaHCO_3$, pH 9.5. This reaction is incubated for 2 hr at room temperature with rocking. The free dye is removed by passage through G-25 spin columns and associated multiplex-24 plate apparatus (Pharmacia, Piscataway, NJ). The labeling ratio is determined by taking the optical density of the FITC-labeled antibody solution at 494 nm in a 96-well Spectrafluor Plus fluorescence plate reader (Tecan, Durham, NC).

Chromophore-Assisted Laser Inactivation: Laser Setup and Irradiation Conditions

Nd:YAG (neodymium-doped yttrium aluminum garnet)-driven dye laser (GCR-11 with HG-2 doubling crystal, PDL-2; Spectra-Physics, Mountain View, CA)

DCM laser dye in methanol (Exciton, Dayton, OH): oscillator concentration, 175 mg/liter; amplifier concentration, 40 mg/liter

Right-angle prism, holder, and mounting rods (Newport, Fountain Valley, CA)

Convex lens and lens holder (Newport)

Laser parameters are as follows: peak power, 56 MW/cm^2; spot size, 2 mm; pulse width, 3 ns; frequency, 10 Hz; energy per pulse, 15 mJ.

An Nd:YAG-driven dye laser (GCR-11 with HG-2 doubling crystal, PDL-2; Spectra-Physics) along with DCM laser dye in methanol (Exciton; oscillator dye, 175 mg/liter; amplifier dye, 40 mg/liter) is used to generate the pulsed 620-nm laser light. The laser is mounted on a vibration-free table (Newport) and the laser beam is directed from the Nd:YAG-pumped dye laser to a right-angle prism mounted on a rod and prism holder approximately 50 cm from the exit port. The laser beam is directed down vertically onto the center of an interjected planar–convex lens (focal length, ~70 mm) attached to the same mounting rod as the prism. The laser spot size is determined by the distance between the prism and the lens, and is usually set to a size of 2 mm. The spot size can be determined by placing a black-and-white Polaroid photographic print on the table below the lens. Each pulse (3-ns pulse width) of the laser generates approximately 15 mJ of energy and forms a single, slightly oblong, uniformly bleached spot of 2 mm.

Protein or suspended cell samples to be used for CALI *in vitro* are placed in the wells of a Nunc transferrable solid-phase plate (Nunc International, Roskilde, Denmark). The wells of the plate are approximately 2 mm in diameter, and the laser beam should be centered as accurately as possible in the well. Pulsed irradiation is allowed to occur for 2 to 5 min, depending on the assay to be done. *In vitro* experiments are usually pulsed for 5 min, whereas assays involving cells are usually limited to 2 min. For cell-based assays and *in vitro* assays of temperature-sensitive enzymes, samples are incubated on ice during irradiation to reduce sample heating.

Whenever lasers are used, extreme caution is employed, with special concern for eye protection. The short pulse width of the Nd:YAG laser produces high peak power (megawatts) and a single stray beam could cause blindness. Protective goggles are always worn, unless specified, and beam blockers are placed to prevent stray reflections. Contact with skin should also be avoided. Investigators are advised to carefully follow laser safety protocols provided with the laser system.

Fluorophore-Assisted Light Inactivation: Slide Projector Setup and Irradiation Conditions

Ektographic III slide projector (Kodak, Rochester, NY)
Brilliant blue filter 69 (Roscolux, Stamford, CT)
Rectangular mirror, holder, and mounting rods (Newport)

Light parameters are as follows: power, 300 W.

Because FITC absorbs visible light (494 nm) and is more efficient than MGITC for inactivation, it is possible to use a variety of light sources for FALI. We have shown that continuous-wave laser light (argon ion) or diffuse light from an ordinary 60-W light bulb or from a 300-W slide projector are all effective light sources for FALI.[5] Routinely, we use a 300-W slide projector containing a blue filter in the slide slot, which limits the transmission of light so that greater than 50% of the transmitted light is between 420 and 500 nm (Brilliant blue filter 69; Roscolux).

Our setup is illustrated in Fig. 2. The slide projector is set up such that the light is directed onto a mirror that is oriented at a 45° angle approximately 19 cm from the projector and 10 cm from the sample. The light is thus projected downward onto the sample. Many samples can be illuminated at one time if a multiwell plate is used, because the projected light is not focused into a beam. However, the

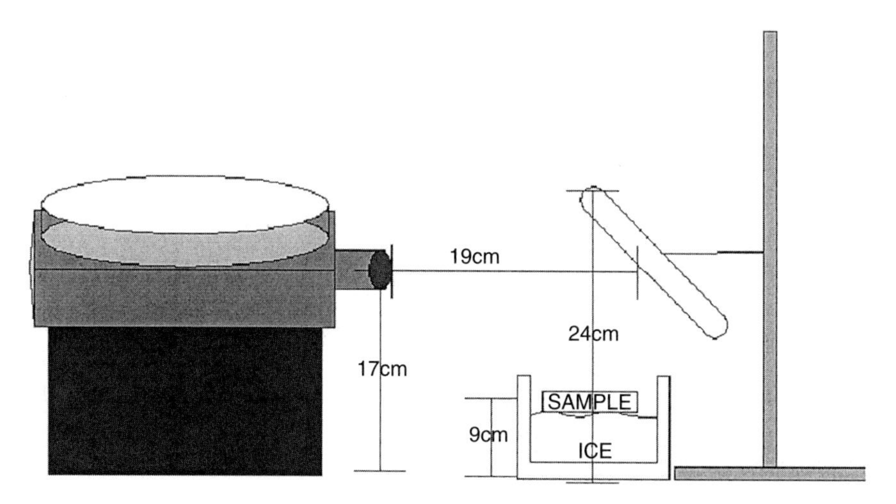

FIG. 2. The setup for FALI using a diffuse light source. A standard 300-W slide projector is raised approximately 10 cm from a base and focused onto a mirror attached to a stationary stand at an ~45° angle from the base. A blue filter (Brilliant Blue 69; Roscolux, Stamford, CT) is placed into the slide slot to restrict the transmission of light so that >50% of light is between 420 and 500 nm. The middle of the mirror is placed about 20 cm from the lens of the projector. The sample to be illuminated is placed under the mirror on ice, measured about 15 cm from the top of the mirror. Illumination is allowed to occur for 1 hr for cellular assays or for 30 min for *in vitro* assays.

irradiation time is much longer when using a diffuse light source than when using focused laser light. The $t_{1/2}$ for inactivation is ~10 min.[5] It is useful to perform a dose–response curve for FALI. To achieve maximal inactivation in cellular assays, samples are illuminated for 1 hr while incubated on ice to reduce sample heating. Generally, *in vitro* FALI assays are illuminated for 30 min for maximal inactivation.

Applying Fluorophore-Assisted Light Inactivation in Multiplex Assay for Cancer Cell Invasion

The multiplex capacity of FALI allows us to assess in parallel the role of many proteins in cellular processes using an array of antibody or scFv antibodies. We have coupled FALI to a cancer cell invasion assay for this purpose and tested it with antibodies against β_1-integrin, a well-validated protein target involved in invasiveness.[5,20] Cancer cell invasiveness is an area of high clinical importance as an early step in metastasis. A standard assay for cancer invasiveness is the *in vitro* transwell assay.[21] The transwell assay is based on the ability of cells to migrate through an 8-μm pore size filter coated with Matrigel, an active fraction of extracellular matrix that mimicks the basement membrane in cell culture.[22] Cells that migrate through the Matrigel are manually counted or measured colorimetrically after manual removal of the cells on the top layer (i.e., cells that fail to migrate).[23] The transwell assay discriminates well between invasive and noninvasive cells and there is an excellent correlation between cells that can cross the Matrigel-coated filter and metastasize in nude mice.[21] This assay has also been used to test the efficacy of anti-invasiveness drugs such as doxorubicin.

Our assay incorporates several existing technologies to generate a high-throughput version of the transwell assay compatible with FALI. Instead of the transwell, we employ a plate-sized filter (Neuroprobe, Gaithersburg, MD) with a 96-holed mask chamber. We use CellTracker Orange labeling (Molecular Probes) of cells (fluorescently labeled, but at a different absorption maximum than FITC) to quantitate the number of cells that cross the filter. After the invading cells cross the filter, the top side of the filter is scraped in single vertical and horizontal sweeps that are amenable to automation. The fluorescence of the underside of the filter is then quantitated in parallel using a fluorescence plate reader. This assay has several advantages over the conventional transwell assay. It is significantly less expensive per sample (approximately one-tenth the cost per single assay) and utilizes a fluorescent plate reader to read many samples on a filter simultaneously so

[20] R. Fassler, M. Pfaff, J. Murphy, A. A. Noegel, S. Johansson, R. Timpl, and R. Albrecht, *J. Cell Biol.* **128,** 979 (1995).

[21] A. Albini, Y. Iwamoto, H. K. Kleinman, G. R. Martin, S. A. Aaronson, J. M. Kozlowski, and R. N. McEwan, *Cancer Res.* **47,** 3239 (1987).

[22] V. P. Terranova, E. S. Hujanen, D. M. Loeb, G. R. Martin, L. Thornburg, and V. Glushko, *Proc. Natl. Acad. Sci. U.S.A.* **83,** 465 (1986).

[23] K. Saito, T. Oku, N. Ata, H. Miyashiro, M. Hattori, and I. Saiki, *Biol. Pharm. Bull.* **20,** 345 (1997).

that the results of an entire plate of assays can be read simultaneously. In addition, the volume is significantly lower than currently used for transwell assays, which allows the use of the assay for samples that are cell limited (e.g., clinically derived tumor samples). We currently assess 40,000 cells per assay and have evidence that fewer cells can be used with equal sensitivity.

Fluorophore-Assisted Light Inactivation: Coupled to Transwell Assay for Invasion

CellTracker Orange (Molecular Probes)
ChemoTx disposable chemotaxis system (Neuroprobe)
Matrigel (Becton Dickinson, Franklin Lakes, NJ)
Spectrafluor Plus fluorescence plate reader (Tecan)

Preparation of the membrane for assaying invasion is a critical step. Matrigel (13.3 μl of a 0.300-mg/ml solution in PBS) is coated on top of the membrane. The membrane is then placed in a desiccator (humidity, $<18\%$) overnight to dry. At least 2 hr before initiation of the assay, the Matrigel-coated wells are rehydrated with 0.1% (w/v) BSA–DMEM. Also before the assay, the cells are fluorescently labeled with CellTracker Orange (Molecular Probes). This process is done essentially as described by the manufacturer, with minor modifications. First, the cells are washed with 0.1% (w/v) BSA–DMEM. After the wash, a 3μM solution of CellTracker Orange (Molecular Probes) is added to the flask and incubated for 15 min. After this incubation, the cells are washed twice with 0.1% (w/v) BSA–DMEM and then allowed to recover in 0.1% (w/v) BSA–DMEM for an additional 15 min. To remove the cells from the flask, they are washed with HBSS (no Ca^{2+} or Mg^{2+}) once and then with 0.53 mM EDTA once. A small amount of 0.53 mM EDTA is then added back to the cells and allowed to incubate for 6 min at 37° and 7% CO_2. The cells are removed by gently washing the flask with 0.1% (w/v) BSA–DMEM. The cell suspension is subsequently pelleted at 1000 rpm, washed with 0.1% (w/v) BSA–HBSS (no phenol red), and repelleted. The pellet is resuspended in 0.1% (w/v) BSA–HBSS (no phenol red) and the cells are counted by trypan blue exclusion for viability. The cells are brought to a concentration of 8×10^6 cells/ml and an equal volume of 40-μg/ml antibody solution [in 0.1% (w/v) BSA–HBSS (no phenol red)] is added. The cells are aliquoted in 38-μl volumes to two non-tissue culture-treated 96-well plates and allowed to incubate for 1 hr. The plates are subsequently put on ice, and one plate is illuminated by Brilliant Blue filtered slide projector light for 1 hr. After illumination, 0.1% (w/v) BSA–DMEM is added to each well to a cell concentration of 8×10^5. Chemoattractant [5% (v/v) FCS in DMEM] is added to the bottom plate and the membrane is placed on top after removing hydration medium. Fifty microliters (40,000 cells) of the cell–antibody mixture is added to each well of the membrane and the membrane is incubated for 6 hr at 37° and 7% CO_2. After incubation, the top of the membrane is scraped with a cell lifter, rinsed with HBSS, swabbed with a cotton-tipped applicator, and dried with Whatman (Clifton, NJ) paper. The membrane is then read with a fluorescence plate reader.[23]

Concluding Remarks

FALI allows us to apply light-mediated protein inactivation to samples in a multiplex fashion to address protein function in cellular processes with potentially high throughput. Each step is compatible with robotic automation, and this should increase throughput and decrease variance. Although FALI is an effective technique for protein inactivation, many aspects of development of FALI are still in progress. For example, although the FITC dye is effective for FALI, it has low quantum efficiency for singlet oxygen generation (~2%). As such, more efficient light-induced singlet oxygen generators may allow for shorter irradiation times, more potent inactivation, and higher throughput. We expect that innovations in the selection of dyes and the possible use of expressed fluorophores (such as green fluorescent protein or its variants) will extend the variety of applications for high-throughput FALI. These types of tools will prove to be necessary when addressing the complex functional roles of the proteome.

Acknowledgments

The authors acknowledge support from NIH Grants CA81668, NEI 11992, and NS34699, a predoctoral training grant DK07542 to B.K.E., and the GRASP Center at the New England Medical Center for monoclonal library preparation. FALI was codeveloped and tested with Xerion Pharmaceuticals (Martinsried, Germany).

[24] F. S. Wang, J. S. Wolenski, R. E. Cheney, M. S. Mooseker, and D. G. Jay, *Science* **273,** 660 (1996).
[25] A. M. Sydor, A. L. Su, F. S. Wang. A. Xu, and D. G. Jay, *J. Cell Biol.* **134,** 1197 (1996).
[26] L. Castelo and D. G. Jay, *Mol. Biol. Cell* **10,** 1511 (1999).
[27] R. F. Lamb, C. Roy, T. J. Diefenbach, H. V. Vinters, M. W. Johnson, D. G. Jay, and A. Hall, *Nat. Cell Biol.* **2,** 281 (2000).
[28] C. W. Liu, G. Lee, and D. G. Jay, *Cell Motil. Cytoskel.* **43,** 232 (1999).
[29] K. Takei, R. M. Shin, T. Inoue, K. Kato, and K. Mikoshiba, *Science* **282,** 1705 (1998).
[30] R. F. Lamb, B. W. Ozanne, C. Roy, L. McGarry, C. Stipp, P. Mangeat, and D. G. Jay, *Curr. Biol.* **7,** 682 (1997).
[31] D. G. Jay and H. Keshishian, *Nature* (*London*) **348,** 548 (1990).
[32] D. Schmucker, A. L. Su, A. Beermann, H. Jackle, and D. G. Jay, *Proc. Natl. Acad. Sci. U.S.A.* **91,** 2664 (1994).
[33] K. Takei, T. A. Chan, F. S. Wang, H. Deng, U. Rutishauser, and D. G. Jay, *J. Neurosci.* **19,** 9469 (1999).
[34] J. J. Feigenbaum, M. D. Choubal, K. Payza, J. R. Kanofsky, and D. S. Crumrine, *Peptides* **17,** 991 (1996).
[35] M. Schroeder, S. Miller, V. Srivastava, E. Merriam-Crouch, S. Holt, V. Wilson, and D. Busbee, *Mutat. Res.* **316,** 237 (1996).
[36] R. Schroder, D. G. Jay, and D. Tautz, *Mech. Dev.* **80,** 191 (1999).

Author Index

Numbers in parentheses are footnote reference numbers and indicate that an author's work is referred to although the name is not cited in the text.

A

B

C

D

E

F

G

H

K

L

M

N

O

P

Q

R

S

T

U

V

W

X

Y

Z

Subject Index

A

H

I

K

L

M

O

T

V